2010 Proceedings of the European Solid-State Device Research Conference

(ESSDERC 2010)

Sevilla, Spain
14 – 16 September 2010

IEEE Catalog Number: CFP10543-PRT
ISBN: 978-1-4244-6658-0

Copyright © 2010 by the Institute of Electrical and Electronic Engineers, Inc
All Rights Reserved

Copyright and Reprint Permissions: Abstracting is permitted with credit to the source. Libraries are permitted to photocopy beyond the limit of U.S. copyright law for private use of patrons those articles in this volume that carry a code at the bottom of the first page, provided the per-copy fee indicated in the code is paid through Copyright Clearance Center, 222 Rosewood Drive, Danvers, MA 01923.

For other copying, reprint or republication permission, write to IEEE Copyrights Manager, IEEE Service Center, 445 Hoes Lane, Piscataway, NJ 08854. All rights reserved.

***This publication is a representation of what appears in the IEEE Digital Libraries. Some format issues inherent in the e-media version may also appear in this print version.**

IEEE Catalog Number: CFP10543-PRT
ISBN 13: 978-1-4244-6658-0
ISSN: 1930-8876

Additional Copies of This Publication Are Available From:

Curran Associates, Inc
57 Morehouse Lane
Red Hook, NY 12571 USA
Phone: (845) 758-0400
Fax: (845) 758-2633
E-mail: curran@proceedings.com
Web: www.proceedings.com

TABLE OF CONTENTS

ESSDERC PLENARY SESSIONS

GaN-on-Si Technology, A New Approach for Advanced Devices in Energy and Communications 1
Tomás Palacios, J. W. Chung, K. Ryu, B. Lu

High Power LEDs for Solid State Lighting 6
Berthold Hahn

Trends and Perspectives for Electrical Characterization and Reliability Assessment in Advanced
CMOS Technologies 13
Guido Groeseneken, Robin Degraeve, Ben Kaczer, Koen Martens

ASL-B PROCESS INTEGRATION: MORE THAN MOORE DEVICES ARCHITECTURES

Silicon Photodiodes for High-Efficiency Low-Energy Electron Detection 22
Agata Sakic, Lis Nanver, Tom Scholtes, Carel Heerkens, Gerard Van Veen, Kees Kooijman, Patrick Vogelsang

New Mechanism of Plasma Induced Damage on CMOS Image Sensor: Analysis and Process
Optimization 26
*Jean-Pierre Carrere, Jean-Pierre Oddou, Claire Richard, Maxime Gatefait, Sebastien Place, Christophe Aumont,
Arnaud Tournier, Francois Roy, Cecile Jenny*

Drain-Extended MOS Transistors Capable for Operation at 10V and at Radio Frequencies 30
Andreas Mai, Holger Rücker

Analysis of Silicon On-Chip Integrated Antennas for Intra- and Inter-Chip Wireless Interconnects 34
Takamaro Kikkawa, K. Kimoto, S. Kubota

Passive Components Integration in CMOS Technology 38
*Siamak Salimy, Fatiha Challali, Antoine Goullet, Marie-Paule Besland, Serge Toutain, Dominique Averty, Ahmed
Rhallabi, Jean-Pierre Landesman, Jean-Claude Saubat, Alain Charpentier*

A3L-C RELIABILITY, VARIABILITY AND MISMATCH

Drain Current Variability in 45nm Heavily Pocket-Implanted Bulk MOSFET 42
Cecilia Mezzomo, Aurelie Bajolet, Augustin Cathignol, Gérard Ghibaudo

Mismatch Sources in LDMOS Devices 46
Pietro Andricciola, Hans Tuinhout

Combining Process and Statistical Variability in the Evaluation of the Effectiveness of Corners in
Digital Circuit Parametric Yield Analysis 50
Plamen Asenov, Noor Ain Kamsani, David Reid, Campbell Millar, Scott Roy, Asen Asenov

Threshold Voltage Shift and Drain Current Degradation by NBT Stress in Si (110) pMOSFETs 54
Kensuke Ota, Masumi Saitoh, Yukio Nakabayashi, Takamitsu Ishihara, Toshinori Numata, Ken Uchida

Resistive Switching-Like Behaviour of the Dielectric Breakdown in Ultra-Thin HF Based Gate Stacks
in MOSFETs 58
*Albert Crespo-Yepes, Javier Martin-Martinez, Aude Rothschild, Rosana Rodriguez, Montse Nafria, Xavier
Aymerich*

A3L-D SRAM AND DRAM

Tri-Gate Bulk CMOS Technology for Improved SRAM Scalability 62
*Changhwan Shin, Chen Hua Tsai, Mei Hsuan Wu, Chung Fu Chang, You Ren Liu, Chih Yang Kao, Guan Shyan
Lin, Kai Ling Chiu, Chuan-Shian Fu, Cheng-Tzung Tsai, Chia Wen Liang, Borivoje Nikolic, Tsu-Jae King Liu*

Impact of Fast-Recovering NBTI Degradation on Stability of Large-Scale SRAM Arrays 66
Stefan Drapatz, Karl Hofmann, Georg Georgakos, Doris Schmitt-Landsiedel

Experimental Comparison of Programming Mechanisms in 1T-DRAM Cells with Variable Channel
Length 70
Alexandre Hubert, Maryline Bawedin, Georges Guegan, Sorin Cristoloveanu, Thomas Ernst, Olivier Faynot

Substrate Bias Dependency of Sense Margin and Retention in Bulk FinFET 1T-DRAM Cells 74
Nadine Collaert, Marc Aoulaiche, An De Keersgieter, Bart De Wachter, Malgorzata Jurczak, Laith Altimime

COLK Cell : A New Embedded DRAM Architecture for Advanced CMOS Nodes 78
Sébastien Cremer, Olivier Goducheau, Hervé Petiton, Sébastien Gaillard, Emek Yesilada, Marc Vernet, Cécile Jenny, Frédéric Lalanne

A4L-B ALTERNATIVE FETS

Sub-60nm Tunnel Field Effect Transistors with $I_{on} > 100\mu A/\mu m$ 82
Wei Yip Loh, Kanghoon Jeon, Chang Yong Kang, Jungwoo Oh, Pratik Patel, Casey Smith, Joel Barnett, Chanro Park, Tsu-Jae King Liu, Hsing-Huang Tseng, Prashant Majhi, Raj Jammy, Chenming Hu

High-Performance Enhancement-Mode $In_{0.53}Ga_{0.47}As$ Surface Channels n-MOSFET with Thin $In_{0.2}Ga_{0.8}As$ Capping and Laser Anneal Effect 86
Injo Ok, Pui-Yee Hung, Dmitry Veksler, Jungwoo Oh, Prashant Majhi, Raj Jammy

Optimization of Tunnel FETs: Impact of Gate Oxide Thickness, Implantation and Annealing Conditions 90
Daniele Leonelli, Anne Vandooren, Rita Rooyackers, Stefan De Gendt, Marc Heyns, Guido Groeseneken

A4L-C SPECIAL CHARACTERIZATION METHODS AND STRUCTURES

Test Structure and Method for the Experimental Investigation of Internal Voltage Amplification and Surface Potential of Ferroelectric MOSFETs 94
Alexandru Rusu, Giovanni Antonio Salvatore, Adrian Ionescu

Experimental Evidence of Unconventional Room-Temperature Quantum Hall Effect (RTQHE) in 65nm Si nMOSFETs at Very Low Magnetic Fields 98
Edmundo Gutierrez, Fernando Guarin

Thermal Broadening of Two-Dimensional Electron Gas Mobility Distribution in AlGaN/AlN/GaN Heterostructures 102
Gilberto Umana-Membreno, Tiziana Stomeo, Vittorianna Tasco, Adriana Passaseo, Massimo De Vittorio, Lorenzo Faraone

A4L-D ADVANCES IN ALGORITHMS AND SIMULATION METHODS

Hardware/Software Co-Simulation for the Rapid Prototyping of an Acceleration Sensor System with Force-Feedback Control 106
Ruslan Khalilyulin, Thomas Steinhuber, Gabriele Schrag, Gerhard Wachutka

On the Inclusion of Lorentz Force Effects in TCAD Simulations 110
Wim Schoenmaker, Peter Meuris, Jean Jimenez, Philippe Galy

Modeling Methodology of High-Voltage Substrate Minority and Majority Carrier Injections 114
Fabrizio Lo Conte, Jean-Michel Sallese, Maher Kayal

A6L-B SOI MOSFETS

Subthreshold FinFET SRAM Cell Optimization Considering Surface-Orientation Dependent Variability 118
Ming-Long Fan, Vita Pi-Ho Hu, Chien-Yu Hsieh, Pin Su, Ching-Te Chuang

Fin-Height Controlled PVD-TiN Gate FinFET SRAM for Enhancing Noise Margin 122
Yongxun Liu, K. Endo, Shin-Ichi O'Uchi, Junichi Tsukada, Hiromi Yamauchi, Yuki Ishikawa, Kunihiro Sakamoto, Takashi Matsukawa, Meishoku Masahara, T. Kamei, T. Hayashida, A. Ogura

Dual Channel and Strain for CMOS Co-Integration in FDSOI Device Architecture 126
Cyrille Le Royer, Mikaël Cassé François Andrieu, Olivier Weber, Laurent Brevard, Pierre Perreau, Jean-François Damlencourt, Sophie Baudot, Claude Tabone, Fabienne Allain, Pascal Scheiblin, Caroline Rauer, Louis Hutin, C. Figuet, C. Aulnette, N. Daval, Bich-Yen Nguyen, K. K. Bourdelle

UT2B-FDSOI Device Architecture Dedicated to Low Power Design Techniques 130
Jean-Philippe Noel, Olivier Thomas, Marie-Anne Jaud, Claire Fenouillet-Beranger, Pierrette Rivallin, Pascal Scheiblin, Thierry Poiroux, Frédéric Boeuf, François Andrieu, Olivier Weber, Olivier Faynot, Amara Amara

A6L-C FERROMAGNETIC AND POLYCRYSTALLINE DEVICES

Ultra-Low Volume Ferromagnetic Nanodots for Field-Coupled Computing Devices 134
Josef Kiermaier, Stephan Breitkreutz, Xueming Ju, Gyorgy Csaba, Doris Schmitt-Landsiedel, Markus Becherer

The Curie Temperature As a Key Design Parameter of Ferroelectric Field Effect Transistors 138
Giovanni Antonio Salvatore, Livio Lattanzio, Didier Bouvet, Adrian Ionescu

Performance Trade-Offs in Polysilicon Source-Gated Transistors 142
Radu Sporea, Mike Trainor, Nigel Young, John Shannon, Ravi Silva

Analysis and Modeling of Pseudo-Short-Channel Effects in ZnO-Nanoparticle Thin-Film Transistors 146
Karsten Wolff, Ulrich Hilleringmann

B3L-B SIMULATION OF ADVANCED SILICON DEVICES

Comparison of Strained SiGe Heterostructure-on-Insulator (001) and (110) PMOSFETs: C-V Characteristics, Mobility and ON Current 150
Anh-Tuan Pham, Christoph Jungemann, Bernd Meinerzhagen

A New Model for the Backscatter Coefficient in Nanoscale MOSFETs 154
Jan-Laurens Van Der Steen, Pierpaolo Palestri, David Esseni, Ray Hueting

Multi-Subband Monte Carlo Simulation of Bulk MOSFETs for the 32nm-Node and Beyond 158
Carlos Sampedro, Francisco Gámiz, Andrés Godoy, Raul Valín, Antonio Garcia-Loureiro, Noel Rodriguez

Modeling Study on Carrier Mobility in Ultra-Thin Body FinFETs with Circuit-Level Implications 162
Mirko Poljak, Vladimir Jovanovic, Tomislav Suligoj

TCAD Based Device Architecture Exploration Towards Half-Terahertz Silicon/Germanium Heterojunction Bipolar Technology 166
Arturo Sibaja-Hernandez, Shuzhen You, Stefaan Van Huylenbroeck, Rafael Venegas, Kristin De Meyer, Stefaan Decoutere

B3L-C PHOTODECTECTORS

Integrated Phototransistors in a CMOS Process for Optoelectronic Integrated Circuits 170
Plamen Kostov, Wolfgang Gaberl, Horst Zimmermann

CMOS Process Enhancement for High Precision Narrow Linewidth Applications 174
Frank Hochschulz, Uwe Paschen, Holger Vogt

A 2µm Diameter, 9Hz Dark Count, Single Photon Avalanche Diode in 130nm CMOS Technology 177
Justin Richardson, Lindsay Grant, Eric Webster, Robert Henderson

Buried Finger Concept for a Correlating Double Cathode Photodetector in BiCMOS 181
Alexander Nemecek, Horst Zimmermann

Understanding Dark Current in Pixels of Silicon Photomultipliers 185
Roberto Pagano, Salvatore Lombardo, Sebania Libertino, Giuseppina Valvo, Giovanni Condorelli, Beatrice Carbone, Delfo Nunzio Sanfilippo, Piero Giorgio Fallica

B3L-D SILICON AND GALLIUM NITRIDE POWER DEVICES

Hot-Carrier Stress Induced Degradation in Multi-STI-Finger LDMOS: an Experimental and Numerical Insight 189
Stefano Poli, Alberto Loi, Susanna Reggiani, Giorgio Baccarani, Elena Gnani, Antonio Gnudi, Marie Denison, Sameer Pendharkar, Rick Wise, Sridhar Seetharaman

Repetitive Avalanche Cycling of Low-Voltage Power Trench N-MOSFETs 193
Olayiwola Alatise, Ian Kennedy, George Petkos, Keith Heppenstall, Khalid Khan, Adrian Koh, Phil Rutter

Investigation of a Dual Channel N/P-LDMOS and Application to LDO Linear Voltage Regulation 197
Marie Denison, Yizhong Xie, Hannes Estl

High Transconductance AlGaN/GaN HEMT with Thin Barrier on Si(111) Substrate 201
François Lecourt, Yannick Douvry, Nicolas Defrance, Virginie Hoel, Jean-Claude De Jaeger, Samira Bouzid, Michel Renvoise, Derek Smith, Hassan Maher

Study of GaN HEMTs Electrical Degradation by Means of Numerical Simulations 205
Valerio Di Lecce, Michele Esposto, Matteo Bonaiuti, Fausto Fantini, Alessandro Chini

B4L-B MODELING OF TEMPERAURE AND STRESS IMPACTS

On the Influence of Flash Peak Temperature Variations on Schottky Contact Resistances of 6-T~SRAM~Cells 209
Christian Kampen, Alexander Burenkov, Jürgen Lorenz

Temperature Dependent Dielectric Absorption Characterization and Modeling for SiN, Al_2O_3 and Ta_2O_5 213
Hajro Muminovic, Philipp Riess, Peter Baumgartner, Peter Klein

Strained MOSFETs on Ordered SiGe Dots 217
Johann Cervenka, Hans Kosina, Siegfried Selberherr, Jianjun Zhang, Nina Hrauda, Julian Stangl, Guenther Bauer, Guglielmo Vastola, Anna Marzegalli, Leo Miglio

B4L-C ADVANCED FET CHARACTERIZATION

Optimized Oxygen Annealing Process for V_{th} Tuning of p-MOSFET with High-K/Metal Gate Stacks 221
Takamasa Kawanago, Yeonghun Lee, Kuniyuki Kakushima, Perhat Ahmet, Kazuo Tsutsui, Akira Nishiyama, Nobuyuki Sugii, Kenji Natori, Takeo Hattori, Hiroshi Iwai

Experimental Analysis of Surface Roughness Scattering in FinFET Devices 225
Jae Woo Lee, Doyoung Jang, Mireille Mouis, Gyu Tae Kim, Thomas Chiarella, Thomas Hoffmann, Gérard Ghibaudo

Parameter Extraction of Nano-Scale MOSFETs Using Modified Y Function Method 229
Subramanian Narasimhamoorthy, Gérard Ghibaudo, Mireille Mouis

B4L-D PHASE CHANGE MEMORIES

Carbon-Doped GeTe Phase-Change Memory Featuring Remarkable Reset Current Reduction 233
Giovanni Betti Beneventi, Luca Perniola, Andrea Fantini, Denis Blachier, Alain Toffoli, Emmanuel Gourvest, Sylvain Maitrejean, Veronique Sousa, Carine Jahan, Jean-François Nodin, Alain Persico, Sebastien Loubriat, Anne Roule, S. Lhostis, H. Feldis, Gilles Reimbold, T. Billon, B. De Salvo, L. Larcher, Paolo Pavan, D. Bensahel, P. Mazoyer, R. Annunziata, F. Boulanger

Current Distributions of BJT-Based Decoding Array for Phase Change Memory 237
Domenico Ventrice, Alessandro Calderoni, Paolo Fantini

SET Switching Effects on PCM Endurance 241
Vincenzo Della Marca, Francesca Carboni, Luca Larcher, Andrea Padovani, Paolo Pavan

B6L-B LEAKAGE CURRENT AND TRAPS

Study of N-Induced Traps Due to Nitrided Metal Gate in HK/MG nMOSFETs 245
Mikaël Cassé, Xavier Garros, Olivier Weber, François Andrieu, Gilles Reimbold, Fabien Boulanger

Carbon Junction Implant: Effect on Leakage Currents and Defect Distribution 249
Guntrade Roll, Stefan Jakschik, Matthias Goldbach, Thomas Mikolajick, Lothar Frey

Grain Boundary-Driven Leakage Path Formation in HfO_2 Dielectrics 253
Gennadi Bersuker, Jung Yum, Vanessa Iglesias, Marc Porti, Montse Nafria, Keith McKenna, Alex Shluger, Paul Kirsch, Raj Jammy

Extracting Accurate Position and Energy Level of Oxide Trap Generating Random Telegraph Noise(RTN) in Recessed Channel MOSFET's 257
Sunyoung Park, Sanghoon Lee, Yeonsung Kang, Byung-Gook Park, Jong-Ho Lee, Jooyoung Lee, Gyoyoung Jin, Hyungcheol Shin

B6L-C TUNNELING FET DEVICES

SOI TFETs: Suppression of Ambipolar Leakage and Low-Frequency Noise Behavior 261
Jing Wan, Cyrille Le Royer, Alexander Zaslavsky, Sorin Cristoloveanu

A Simulation-Based Study of Sensitivity to Parameter Fluctuations of Silicon Tunnel FETs 265
Kathy Boucart, Walter Riess, Adrian Ionescu

Impact of Electron Velocity on the I_{on} of n-TFETs 269
Hasanali Virani, David Esseni, Anil Kottantharayil

Abrupt Switch Based on Internally Combined Band-to-Band and Barrier Tunneling Mechanisms 273
Livio Lattanzio, Luca De Michielis, Arnab Biswas, Adrian Ionescu

C3L-B NANOWIRE TRANSISTORS

Junctionless Nanowire Transistor (JNT): Properties and Design Guidelines 277
Abhinav Kranti, Ran Yan, Chi-Woo Lee, Isabelle Ferain, Ran Yu, Nima Dehdashti Akhavan, Pedram Razavi, Jean-Pierre Colinge

Gate Semi-Around Si Nanowire FET Fabricated by Conventional CMOS Process with Very High Drivability 281
Soshi Sato, Yeonghun Lee, Kuniyuki Kakushima, Parhat Ahmet, Kenji Ohmori, Kenji Natori, Keisaku Yamada, Hiroshi Iwai

Dopant-Independent and Voltage-Selectable Silicon-Nanowire-CMOS Technology for Reconfigurable Logic Applications 285
Frank Wessely, Tillmann Krauss, Udo Schwalke

3D Source/Drain Doping Optimization in Multi-Channel MOSFET 288
Kiichi Tachi, Nathalie Vulliet, Sylvain Barraud, Bernard Guillaumot, Virginie Maffini-Alvaro, Christian Vizioz, Christian Arvet, Jean-Michel Hartmann, Thomas Skotnicki, Sorin Cristoloveanu, Hiroshi Iwai, Olivier Faynot, Thomas Ernst

Hole Mobilities and Electrical Characteristics of Ω-Gated Silicon Nanowire Array FETs with 110- and 100-Channel Orientation 292
Stefan Habicht, Sebastian Feste, Qing Tai Zhao, Siegfried Mantl

C3L-C DEVICE STEEP SLOPE AND LEAKAGE

Breaching the kT/q Limit with Dopant Segregated Schottky Barrier Resonant Tunneling MOSFETs: a Computationnal Study 296
Aryan Afzalian, Denis Flandre

Steep-Slope Nanowire FET with a Superlattice in the Source Extension 300
Elena Gnani, Susanna Reggiani, Antonio Gnudi, Giorgio Baccarani

Modeling Impact of Electric Field and Strain on the Leakage of Embedded SiGe Source/Drain Junctions 304
Abraham Luque Rodriguez, Juan Antonio Jimenez Tejada, Salvador Rodríguez Bolivar, Mireia Bargallo Gonzalez, Geert Eneman, Cor Claeys, Eddy Simoen

Modeling Temperature Dependency (6 - 400K) of the Leakage Current Through the SiO_2/High-K Stacks 308
Luca Vandelli, Andrea Padovani, Luca Larcher, Richard Southwick, Bill Knowlton, Gennadi Bersuker

C3L-D ADVANCED MEMORIES

Oxide-Based RRAM: Physical Based Retention Projection 312
Bin Gao, Jinfeng Kang, Haowei Zhang, Bing Sun, Bing Chen, Lifeng Liu, Xiaoyan Liu, Ruqi Han, Yangyuan Wang, Zheng Fang, Hongyu Yu, Bin Yu, Dim-Lee Kwong

A Stochastic Model of Bipolar Resistive Switching in Metal-Oxide-Based Memory 316
Alexander Makarov, Viktor Sverdlov, Siegfried Selberherr

Dependence of the Switching Characteristics of Resistance Random Access Memory on the Type of Transition Metal Oxide 320
Wan Gee Kim, Min Gyu Sung, Sook Joo Kim, Ja Yong Kim, Ji Won Moon, Sung Joon Yoon, Jung Nam Kim, Byung Gu Gyun, Taeh Wan Kim, Chi Ho Kim, Jun Young Byun, Won Kim, Te One Youn, Jong Hee Yoo, Jang Won Oh, Ho Joung Kim, Moon Sig Joo, Jae Sung Roh, Sung Ki Park

A 3D Stackable Carbon Nanotube-Based Nonvolatile Memory (NRAM) 324
Sohrab Kianian, Glen Rosendale, Monte Manning, Darlene Hamilton, Xue Ming Henry Huang, Karl Robinson, Young Weon Kim, Thomas Rueckes

Experimental and Simulation Study of the Program Efficiency of HfO_2 Based Charge Trapping Memories 328
Sabina Spiga, Gabriele Congedo, Ugo Russo, Alessio Lamperti, Olivier Salicio, Francesco Driussi, Elisa Vianello

C4L-B CHANNEL AND GATE STACK ENGINEERING

Carrier Transport Characteristics of Strained N-MOSFET Featuring Channel Proximate Silicon-Carbon Source/Drain Stressors for Performance Boost 332
Shao-Ming Koh, Peng Zhang, Shu-Feng Ren, Chee-Mang Ng, Ganesh S. Samudra, Yee-Chia Yeo

Fluorinated CMOS HfO₂ for High Performance (HP) and Low Stand-by Power (LSTP) Application by Pre- and Post-CF₄ Plasma Passivation 336

Woei-Cherng Wu, Chao-Sung Lai, Huai-Hsien Chiu, Jer-Chyi Wang, Pai-Chi Chou, Tien-Sheng Chao

Origins of Universal Mobility Violation in SOI MOSFETs 340

Noel Rodriguez, Sorin Cristoloveanu, Francisco Gámiz

C4L-C SIMULATION OF III/V DEVICES

Optimization of III-V FETs Architecture for High Frequency and Low Consumption Applications 344

Ming Shi, Jérôme Saint-Martin, Arnaud Bournel, Philippe Dollfus

Vertical Design of InN Field Effect Transistors 348

Ralf Granzner, Vladimir Polyakov, Mario Kittler, Frank Schwierz

A Continuous Physics-Based Electrothermal Compact Model for the Study of Non-Linearities in III-V HEMTs 352

Toufik Sadi, Frank Schwierz

C4L-D CHARGE TRAP NAND FLASH

Investigation of Rare-Earth Aluminates As Alternative Trapping Materials in Flash Memories 356

Antonio Cacciato, Amit Suhane, Olivier Richard, Antonio Arreghini, Christoph Adelmann, Johan Swerts, Aude Rothschild, Geert Van Den Bosch, Hugo Bender, Malgorzata Jurczak, Ingrid Debusschere, Jorge Kittl, Jan Van Houdt

Optimization of the Crystallization Phase of Rare-Earth Aluminates for Blocking Dielectric Application in TANOS Type Flash Memories 360

Laurent Breuil, Christoph Adelmann, Geert Van Den Bosch, Antonio Cacciato, Mohammed B. Zahid, Maria Toledano-Luque, Amit Suhane, Antonio Arreghini, Robin Degraeve, Ingrid Debusschere, Jorge Kittl, Margorzata Jurczak, Jan Van Houdt

Investigation of the ISPP Dynamics and of the Programming Efficiency of Charge-Trap Memories 364

Alessandro Maconi, Christian Monzio Compagnoni, Salvatore M. Amoroso, Evelyne Mascellino, Michele Ghidotti, Giorgio Padovini, Alessandro S. Spinelli, Andrea Lacaita, Aurelio Mauri, Gabriella Ghidini, Nadia Galbiati, Alessandro Sebastiani, Claudia Scozzari, Eugenio Greco, Elisa Camozzi, Paolo Tessariol

C6L-B ANALYTICAL/COMPACT MODELS

3D Analytical Modelling of Subthreshold Characteristics in Pi-Gate FinFET Transistors 368

Romain Ritzenthaler, Francois Lime, Olivier Faynot, Sorin Cristoloveanu, Benjamin Iniguez

A Compact Model for Double Gate Carbon Nanotube FET 372

Sebastien Fregonese, Cristell Maneux, Thomas Zimmer

Modeling of Partial-RESET Dynamics in Phase Change Memories 376

Stefania Braga, Alessandro Sanasi, Alessandro Cabrini, Guido Torelli

On the Modelling and Optimisation of a Novel Schottky Based Silicon Rectifier 380

Tom Van Hemert, Ray Hueting, Bijoy Rajasekharan, Cora Salm, Jurriaan Schmitz

C6L-C ELECTROMECHANICAL DEVICES

CMOS-MEMS Free-Free Beam Resonators 384

Nuria Barniol, Joan Lluis Lopez, Eloi Marigo, Joan Giner, Jose Luis Munoz-Gamarra, Gabriel Vidal, Francesc Torres, Arantxa Uranga

Electro-Thermal Analysis of RF MEM Capacitive Switches for High-Power Applications 388

Francesco Solazzi, Cristiano Palego, Subrata Halder, James C. M. Hwang, Alessandro Faes, Viviana Mulloni, Benno Margesin, Paola Farinelli, Roberto Sorrentino

Active NEM Filters for Communications Applications Based on Vibrating Body Transistors 392

Andrea Lovera, Sebastian Bartsch, Daniel Grogg, Suat Ayoz, Risto Kaunisto, Adrian Ionescu

Piezoresistivity and Electrical Properties of Poly-SiGe Deposited at CMOS-Compatible Temperatures 396

Pilar Gonzalez, Luc Haspeslagh, Simone Severi, Kristin De Meyer, Ann Witvrouw

Author Index

GaN-on-Si Technology, A New Approach for Advanced Devices in Energy and Communications

J. W. Chung, K. Ryu, B. Lu, and Tomás Palacios
Department of Electrical Engineering and Computer Science
Massachusetts Institute of Technology
Cambridge, USA
tpalacios@mit.edu

Abstract—The Si substrate of GaN-on-Si wafers offers new opportunities to increase the functionality and performance of nitride-based devices. This paper will review three examples of these new devices/systems. First, GaN-on-Si substrates allow the on-chip heterogeneous integration of GaN and Si electronics. Second, the easy removal of the Si substrate through dry or wet etching gives access to the N-face of the GaN layer, and all the new device structures that this orientation enables. Finally, the use of Si substrates for the growth of GaN high voltage switches makes the cost of these devices competitive with Si devices, and the total or partial etch of Si brings a new degree of freedom to increase the breakdown and performance of GaN transistors.

I. INTRODUCTION

Nitride semiconductors are quickly transforming our world by enabling new solid-state lighting, highly efficient amplifiers for wireless communications, advanced power electronics with unprecedentedly low losses, and a large array of new high performance devices. Out of the different substrates available for the growth of nitride-based semiconductors (i.e. bulk GaN, Si, sapphire and SiC), the Si substrate has traditionally been considered a low-cost, lower-performance option. In this paper we will show that the use of a Si substrate, far from hindering the performance of nitride semiconductors, is a powerful tool that gives unprecedented flexibility for the fabrication of advanced new nitride-based devices.

Three different examples will be discussed. First, the wafer bonding of a GaN-on-Si sample to a Si (100) wafer has been used to demonstrate the first on-wafer integration of GaN electronics and Si (100) CMOS circuits [1]. This integration allows the combination of the high complexity and flexibility of Si circuits with the vast array of new devices enabled by GaN: transistors, LEDs, energy harvesting devices, and filters. A second example involves the fabrication of N-face GaN transistors by removing the Si substrate of a Ga-face-grown sample [2]. These new devices offer new opportunities for improved heat dissipation and high frequency electronics. Finally, the large wafer diameters available for GaN-on-Si samples provide an important competitive advantage for the development of GaN-based high performance power electronics [3]. These devices, with an estimated potential market above $10 billion dollar, could be key to energy savings equivalent to 10% of the global electricity consumption.

II. ON-WAFER INTEGRATION OF GaN TRANSISTORS AND Si (100) CMOS ELECTRONICS

The unique properties of AlGaN/GaN High Electron Mobility Transistors (HEMTs) have made them the best option for many RF amplifiers. The unsurpassed high current levels possible in these devices, in combination with their very high breakdown voltage allow almost 10 times higher maximum power density than GaAs amplifiers [4]. In

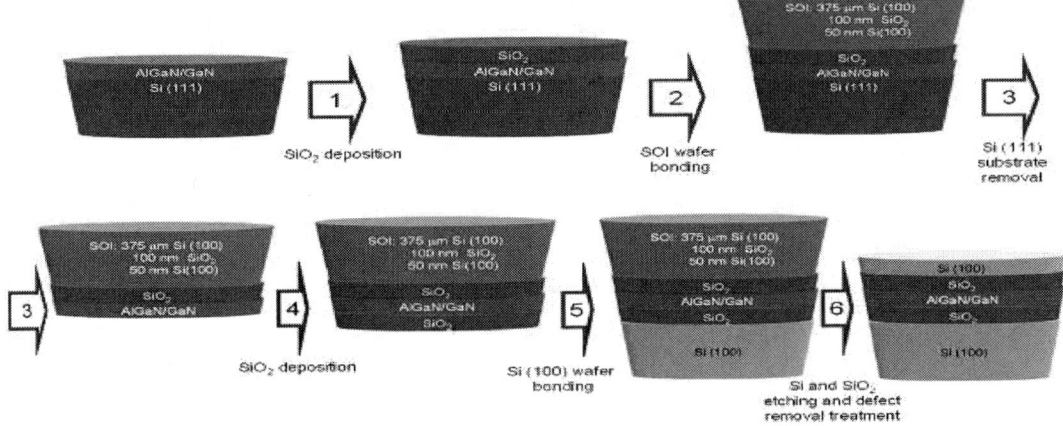

Figure 1. Main steps in the fabrication of a Si/GaN/Si hybrid wafer.

978-1-4244-6658-0/10 $26.00 © 2010 IEEE

addition, their high frequency performance enables extremely high power added efficiencies by maximizing the available power gain, and their high output resistance significantly simplifies the design of the matching networks in RF amplifiers. Finally, the recent demonstration of device lifetimes in excess of 10^6 hours at a channel temperature of 175°C make this technology one of the most reliable semiconductor technologies.

In spite of the excellent performance demonstrated by nitride transistors, these devices cannot compete with Si MOSFETs in terms of scalability and level of integration. On the other hand, traditional Si electronics is facing tremendous challenges to continue its scaling and performance improvement. The on-chip heterogeneous integration of GaN power devices and Si (100) CMOS electronics would enable unprecedented flexibility for advanced circuits, from highly linear power amplifiers, to new power distribution networks in Si microsystems.

The high thermal stability of GaN-based semiconductors makes them ideal for their integration with Si, as the GaN material can survive the very demanding thermal budget of Si processing. We have recently demonstrated a technology to integrate, for the first time, these two devices. This technology is based on the fabrication of a hybrid Si (001) / GaN / Si (001) substrate by wafer bonding (Figures 1 and 2). Due to the high thermal stability of GaN, Si CMOS electronics can then be processed in this new substrate without affecting the nitride layers underneath the surface (Figure 3). After the Si devices are fabricated, the Si material is removed from the regions where nitride devices are needed. Then, the nitride devices (transistors, LEDs, lasers or sensors) are processed and, finally, an interconnection layer forms the final hybrid circuits.

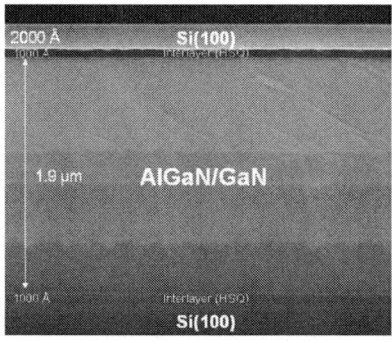

Figure 2. Scanning electron micrograph of the cross-section of a Si/nitride/Si wafer.

Figure 4 shows the scanning electron micrograph of a finished sample where AlGaN/GaN HEMTs were fabricated in very close proximity to Si (100) pMOSFETs following the process flow outlined in Figure 3. These devices show excellent DC performance and we are currently developing new circuits to take advantage of the flexibility enabled by this integration. Some of these circuits include:

- Integrated GaN power switches and Si control electronics
- High power digital-to-analog converters (DACs)
- On-wafer wireless transmitters
- Driver stages for on-wafer optoelectronics
- Power amplifiers coupled to Si linearizer circuits
- High speed (high power) differential amplifiers
- Normally-off power transistors based on the cascaded connection on a Si normally-off transistor and a normally-on GaN HEMT.

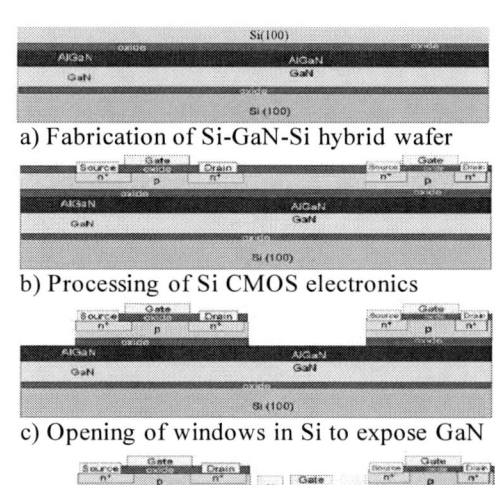

a) Fabrication of Si-GaN-Si hybrid wafer

b) Processing of Si CMOS electronics

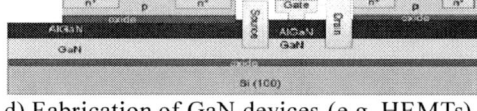

c) Opening of windows in Si to expose GaN

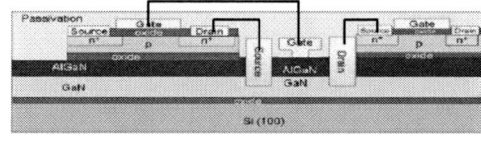

d) Fabrication of GaN devices (e.g. HEMTs)

e) Planarization and interconnection layers

Figure 3. Main processing steps for the fabrication of GaN-Si hybrid

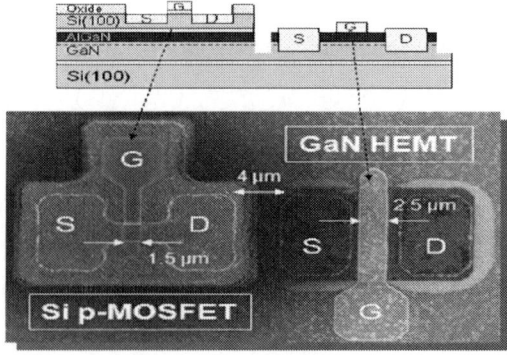

Figure 4. Scanning electron micrograph of the heterogeneous integration of GaN and Si electronics.

978-1-4244-6658-0/10 $26.00 © 2010 IEEE

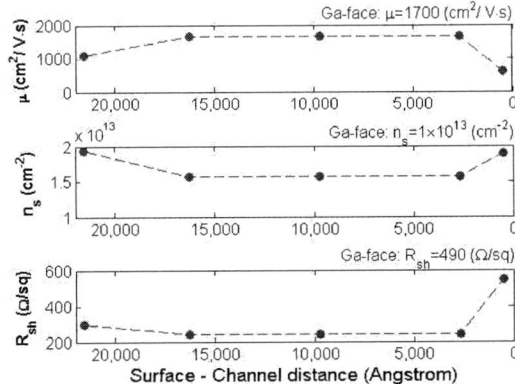

Figure 5. Electron mobility (μ), 2DEG density (n_s), and sheet resistance (R_{sh}) of N-face GaN as a function of the distance between N-face surface and 2DEG channel estimated by room temperature Hall measurement.

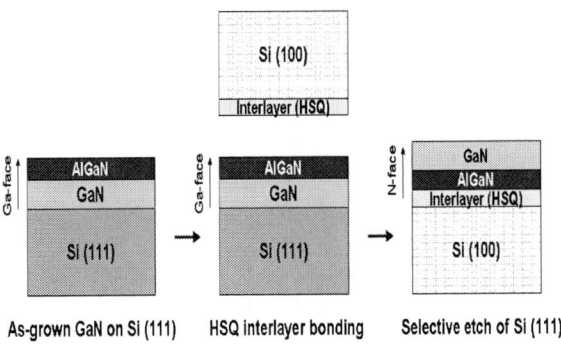

Figure 6. Main processing steps in the fabrication of N-face GaN on Si (100) substrate through substrate removal.

- New enhancement-mode power transistors
- Buffer stages for ultra-low-power electronics

III. N-FACE GaN ELECTRONICS FROM GA-FACE GAN-ON-SI WAFERS

Although most of the reported GaN devices have been fabricated on nitride structures grown along the c-direction (i.e. Ga-face), N-face GaN/AlGaN transistors have the potential of higher electron confinement and lower contact resistances [5]. However, in spite of this promise, the performance of N-face devices is still much lower than in Ga-face devices due to the inferior material quality. Although N-face devices have been grown by molecular beam epitaxy and, recently [6], by metal-organic chemical vapor deposition (MOCVD) [7], the growth of N-face nitrides is much more challenging than the growth of the more stable Ga-face structure. The use of Si substrates provides a new approach for the fabrication of N-face nitride wafers, not through growth but through the substrate removal of a Ga-face-grown GaN samples, as we will demonstrate in this section.

The Ga-face AlGaN/GaN transistor structures used in this work were grown on Si (111) substrates by MOCVD at Nitronex with a structure similar to the one described in Reference [8]. To have access to the N-face of these samples, we have developed the substrate transfer technology shown in Figure 6. First, the Ga-face surface was bonded to a Si (100) carrier wafer by using a hydrogen silsesquioxane (HSQ) interlayer. HSQ is a flowable oxide with excellent thermal stability, which stands the high thermal budget required during the processing of GaN devices. The HSQ film is spin-coated on Si (100) to a thickness of about 1000 Å and baked sequentially on hot plates at 150 °C and 200 °C for 1 minute each. Then, the HSQ-coated Si (100) substrate is attached to the as-grown AlGaN/GaN layer and thermally compressed at 400 °C for an hour. The elevated temperature hardens the HSQ layer and forms an extremely stable bond between the GaN wafer and the Si carrier wafer. After the wafer bonding, the original Si (111) substrate is completely removed by dry etching using an SF$_6$-based dry etch. The GaN buffer is an effective etch-stop layer for the SF$_6$ plasma etch and a smooth N-face GaN surface is obtained at the end of the etch. After the substrate transfer, the N-face GaN buffer is etched by electron cyclotron resonance reactive ion etching (ECR-RIE) with Cl$_2$/BCl$_3$ gas mixture until the desired distance between

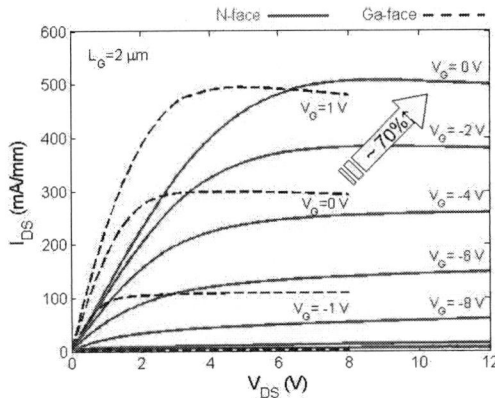

Figure 7. DC current-voltage characteristics of N-face (solid line) and Ga-face (dashed line) HEMTs with a gate length (L_G) of 2 µm. Almost 70% higher maximum current at V_G=0 V is achieved in N-face HEMTs. Higher on-resistance (R_{on}) in N-face HEMTs is due to non-optimized ohmic contacts.

the N-face GaN surface and the AlGaN/GaN interface is achieved.

The effect of the GaN buffer thickness on the electron transport was evaluated by four-point van der Pauw Hall measurements. The GaN buffer was etched as described in the previous paragraph and the remaining GaN thickness was measured with an interferometer (NanoSpec). The main transport properties are almost constant with buffer thickness (μ=1670 cm^2/V·s, n_s=1.6×10^3 cm^{-2}, R_{sh}=240 Ω/sq). The sheet resistance is 50% lower than the sheet resistance measured in the Ga-face of the sample. This important reduction in the sheet resistance is due to an increase in the mobility and electron density during the substrate removal process. The change in the Fermi level pinning when the N-face surface is

978-1-4244-6658-0/10 $26.00 © 2010 IEEE

exposed after the substrate removal is believed to be partly responsible for this improvement although other causes are still under investigation.

N-face GaN/AlGaN structures fabricated through the substrate removal process described previously have been used in the fabrication of N-face high electron mobility transistors (HEMTs). In this samples, the distance between the N-face surface and the 2DEG was reduced to 1000 Å by a Cl_2-based dry etching process. A Ti/Al/Ni/Au multilayer was deposited for the ohmic contacts and annealed at 870°C for 30 s in a N_2 atmosphere. Then Cl_2/BCl_3 plasma was used for the mesa isolation. Finally, a 2 μm-length gate is defined by photolithography and a Ni/Au/Ni metallization was deposited for the Schottky contact. The drain current versus drain voltage characteristic of the N-face device is shown in Figure 7 and it is compared to a Ga-face HEMT used as a reference. For a gate voltage of 0 V, the maximum current in the N-face device is almost 70% higher than in the Ga-face device. This difference is mainly due to the higher charge density in the N-face device. The transconductance in this initial N-face deviced is low due to the large gate-to-channel distance.

To improve the performance and yield of this technology, we are currently focused on increasing the wafer diameter and in developing a selective etch technique to reproducibly thin down these N-face wafers and improve, in that way, the device transconductance.

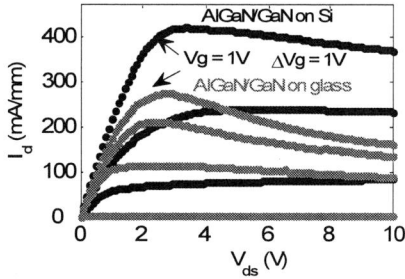

Figure 8. I-V characteristics of an AlGaN/GaN HEMT before and after transferring it to a glass substrate.

IV. GaN-on-Si Power Electronics

AlGaN/GaN HEMTs grown on Si substrates have also attracted a great interest for power electronics applications. However, despite the low cost of the Si substrate, the breakdown voltage (V_{bk}) of AlGaN/GaN HEMTs grown on Si (less than 600 V for 2 μm total nitride epilayer) is much lower than that grown on SiC (1.9 kV for 2 μm total epi-layer), which severely reduces device performance. In this section we propose the total or partial removal of the Si substrate to increase the breakdown voltage of GaN power switches grown on Si substrates [3].

For the device demonstration, we used the same standard epitaxial structure described in Section III. In this case, the devices have a gate width of 100 μm, a 2 μm gate length, and varying gate-to-drain distances. To eliminate the vertical breakdown of the AlGaN/GaN HEMTs on Si, we removed

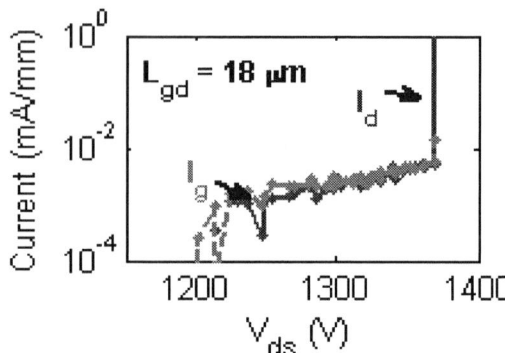

Figure 9. Three terminal breakdown of an AlGaN/GaN device on glass with L_{gd} = 18 μm and V_{gs} = - 8 V.

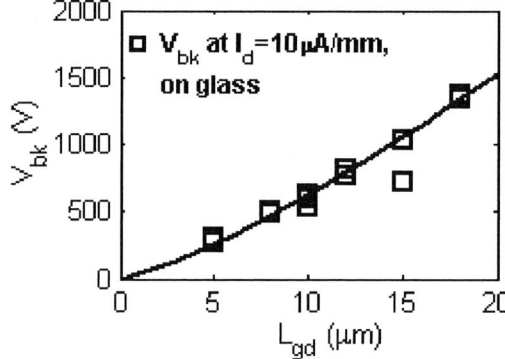

Figure 10. Three terminal breakdown of AlGaN/GaN HEMTs with different L_{gd} transferred to a glass substrate.

the Si substrate and transferred the AlGaN/GaN HEMTs to a glass wafer through wafer bonding.

The device maximum current (Figure 8) is lower after bonding to the glass substrate due to increased self-heating, however the breakdown voltage increases significantly. A device with L_{gd} = 18 μm shows breakdown of 1370 V and on-resistance of 4.3 mΩ•cm^2 with very low leakage current (< 10 μA/mm) (Figure 9). Three terminal breakdown voltage as a function of gate-to-drain spacing (L_{gd}) is shown in Figure 10. More than 1450 V breakdown and an on-resistance of 5.3 mΩ•cm^2 is achieved on devices with L_{gd} = 20 μm, which is beyond our power supply maximum output voltage. The breakdown voltage of these devices before the Si removal was below 600 V. These results show the tremendous potential of GaN heterostructures grown on Si for very high voltage applications, once that the Si substrate has been removed. It should be noted that the removal of the Si substrate can be integrated with the packaging of the device, which will reduce cost and increase performance.

V. Conclusions

In summary, the micromachining and processing of the Si substrate in GaN-on-Si wafers is a new tool that device engineers can use to improve the performance and increase

978-1-4244-6658-0/10 $26.00 © 2010 IEEE

the flexibility of GaN electronics. This paper has described three different devices where the partial or total removal of the Si substrate is beneficial: heterogeneous on-chip integration of GaN and Si transistors, N-face GaN devices, and ultra high voltage GaN power switches. The substrate should, therefore, no longer be considered a secondary part of modern devices GaN-on-Si devices, but an integral part of the device with which achieve higher device performance in applications such as power electronics and high speed communications.

REFERENCES

[1] J. Chung, J. Lee, E. Piner, and T. Palacios, "Seamless On-Wafer Integration of Si(100) MOSFETs and GaN HEMTs," *IEEE Electron Device Letters*, vol. 30, Oct. 2009, pp. 1015-1017.

[2] J. Chung, E. Piner, and T. Palacios, "N-Face GaN/AlGaN HEMTs Fabricated Through Layer Transfer Technology," *IEEE Electron Device Letters*, vol. 30, 2009, pp. 113-116.

[3] B. Lu and T. Palacios, "High Breakdown (>1500 V) AlGaN/GaN\line HEMTs by Substrate Transfer Technology," *IEEE Electron Device Letters* (in press).

[4] U.K. Mishra, P. Parikh, Y.F. Wu, and others, "AlGaN/GaN HEMTs-an overview of device operation and applications," *Proceedings of the IEEE*, vol. 90, 2002, pp. 1022–1031.

[5] M.H. Wong, Y. Pei, T. Palacios, L. Shen, A. Chakraborty, L.S. McCarthy, S. Keller, S.P. DenBaars, J.S. Speck, and U.K. Mishra, "Low nonalloyed Ohmic contact resistance to nitride high electron mobility transistors using N-face growth," *Applied Physics Letters*, vol. 91, 2007, p. 232103.

[6] S. Rajan, A. Chini, M.H. Wong, J.S. Speck, and U.K. Mishra, "N-polar GaN/ AlGaN/ GaN high electron mobility transistors," *Journal of Applied Physics*, vol. 102, 2007, p. 044501.

[7] S. Keller, N.A. Fichtenbaum, F. Wu, D. Brown, A. Rosales, S.P. DenBaars, J.S. Speck, and U.K. Mishra, "Influence of the substrate misorientation on the properties of N-polar GaN films grown by metal organic chemical vapor deposition," *Journal of Applied Physics*, vol. 102, 2007, p. 083546.

[8] J.W. Johnson, E.L. Piner, A. Vescan, R. Therrien, P. Rajagopal, J.C. Roberts, J.D. Brown, S. Singhal, K.J. Linthicum, and R. Therrien, "12 W/mm AlGaN-GaN HFETs on silicon substrates," *IEEE Electron Device Letters*, vol. 25, 2004, pp. 459–461.

978-1-4244-6658-0/10 $26.00 © 2010 IEEE

High Power LEDs for Solid State Lighting

Berthold Hahn
Osram Opto Semiconductors
93155 Regensburg
Berthold.hahn@osram-os.com

Abstract— For solid stated lighting high light output in combination with high conversion efficacy is essential. High efficiencies are relatively easy to realize at low current densities, but efficiency tends to decline as the current is cranked up. In order to overcome the barriers for high flux LEDs, both epitaxy and chip design have to be optimized.

In this paper we report on the improvement of ThinGaN®-PowerLED structures in epitaxy, chip design, phosphor efficiency and package design. A key for improving LED performance is understanding the carrier loss mechanisms in blue and green epitaxy structures. Assuming an indirect Auger effect as one of the major loss mechanisms in InGaN LEDs, a reduction of the carrier density per emitting well is enabling efficiency improvement for blue/green LEDs.

Along with improved epi designs the extraction efficiency had to be improved. A new chip design allows high current operation in combination with efficiencies beyond 100lm/W for white. New conversion schemes allow the fabrication of extremely efficient green light sources, which enable new generations of high flux projection applications.

I. INTRODUCTION

The lighting picture of today is still dominated by standard lighting devices. They are widely spread, well known, have a well established infra structure and still set standards on lm/€ values. But the transition phase has already started to modern LED lighting which is driven by the tremendous improvements of LED performance with respect to efficacy, lifetime robustness as well as their cost.

Figure 1. Street illumination with LED luminaires compared to conentional high power sodium. The LED illumination saves >50% energy combined with a better quality of light

In street lighting, LEDs give the possibility to focus light where it is needed thus avoiding light pollution on one hand, generating security for people due to improved recognition of objects and individuals on the street on the other hand (Figure 1.). The possibility to direct light where and when it is needed enables high energy savings and is driving the market.

In modern architecture, LEDs give the possibility to architects to illuminate facades differently at different day and night times as can be seen at the Malmö turntower, Sweden or the Lincoln memorial, Washington USA.

In most of these lighting applications energy efficiency, infinite color tuneability, light quality, design flexibility, possibility of dimming and safety are considered to be unique selling points for LEDs.

The application of LEDs in solid state lighting can be separated in different sectors (Figure 2.):

- Warm white applications with a CCT of 2700-3000K for interior lighting, which mimics the emission of conventional lighting with high quality of light (CRI>80)

- Daylight with a CCT of 5000K and a high quality of light for office lighting

- Ultrawhite with a CCT of 6500K and low CRI for highest efficacy for street and outdoor lighting,

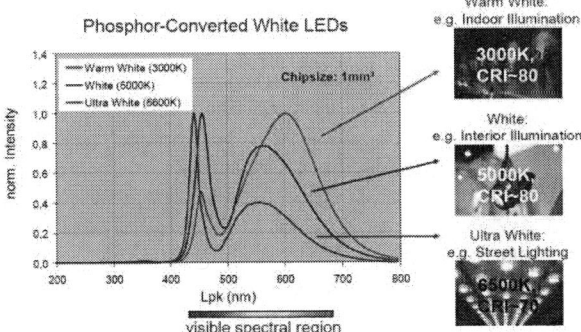

Figure 2.   Typical CCTs and spectra of LEDs used for solid state lighting

Beside these main sectors, which is today mainly achieved by phosphor converted LEDs, also special lighting with adaptive color control by RGB LEDs is widely used.

II.   DOES LED LIGHT REALLY SAVE ENERGY ?

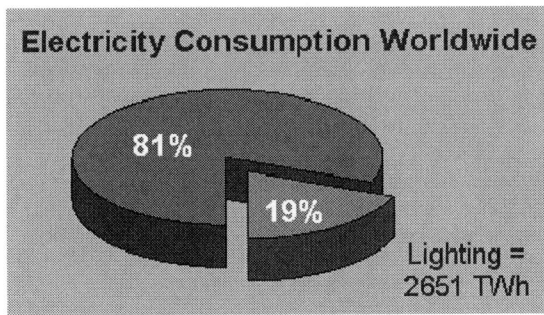

Figure 3.   Consumtion of electrical energy world wide per year

The main driver for the adoption of solid sate lighting is the potential of energy saving. About 20% of today's electrical energy consumption is linked with lighting. This is equivalent to 2651 TWh electricity per year (Figure 3.). By using the best technologies available today we could already save 30% of this amount. An additional 40% could be saved by using intelligent lighting systems by applying sensors and software in order to supply light only where and when it is needed. LEDs are the ideal light sources for this new kind of lighting systems as they offer the possibility of fast switching, instant on lighting and their easy interfacing to low voltage control, which simplifies the integration to digital control. Looking out into 2030 the "ITC and energy efficiency" of the European Union predicts savings of 1300TWh of electrical energy.

All of these estimations are based on the expected energy savings of LED systems during operation. As LEDs are rather complex light sources it is often questioned, how much energy LED lighting systems really consume during their complete lifecycle including their production and disposal at end of life[1].

Figure 4.   Life cycle of lighting systems

In order to clarify this issue a comparative study of already existing lighting systems as conventional incandescent lamps (GLS), compact fluorescent lamps (CFL) and LEDs has been performed and published [1].

The study is based on comparable light output and lifetime of the different devices as one 8W LED lamp (Parathom) with 25.000h lifetime, 2.5 CFL 8W each with 10.000 lifetime and 25 GLS with 40W and 1000h lifetime (Figure 5.)

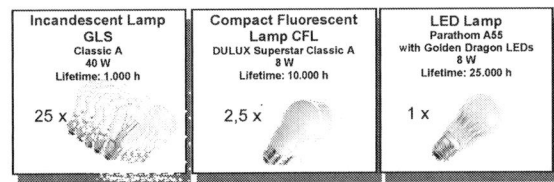

Figure 5.   Devices producing equivalent light output and lifetime

The result of the total lifetime assesment is shown in Figure 6.   The study shows that today's commercial LED systems are already competitive with CFL and far superior to conventional lighting systems on a power consumption level. In comparison to the energy consumption of 3.302kWh both LEDs and CFL consume less than 670kWh during their lifetime. For all considered systems the energy consumed during operation is the main contributor. The higher effort for producing solid state lighting is already more than matched by the long life time of these devices.

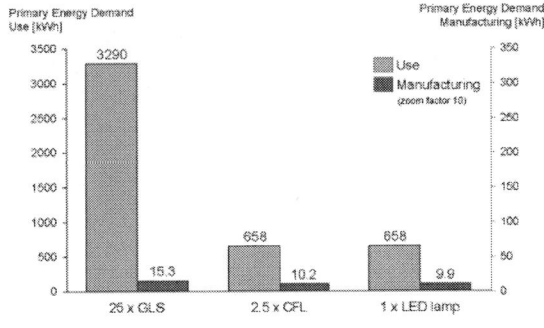

Figure 6.   Comparative total life time assesment of lighting systems

978-1-4244-6658-0/10 $26.00 © 2010 IEEE

An additional factor for the environmental impact of solid state lighting is the absence of hazardous materials like mercury in the bill off material and the low use of scarce materials as rare earths for conversion materials.

This study showed clearly, that the key success factors for solid state lighting is life time and energy efficiency. However for solid state lighting the complete system has to be optimized. The energy balance of a typical LED lighting system is shown in **Fehler! Verweisquelle konnte nicht gefunden werden.**

The main contributors are optical and thermal losses due to lens design and thermal management within the LED lamp and electrical driver losses. Each of these loss channels is contributing with 10%-20% to the overall loss. This sums up to 30%-40% system loss in today's design. This means, that for constructing a 70 lm/W lamp today at least 110 lm/W are needed on LED level.

The same considerations account for lifetime. Therefore high efficiency and long life designs on the optical and driver side are crucial for the success of solid sate lighting.

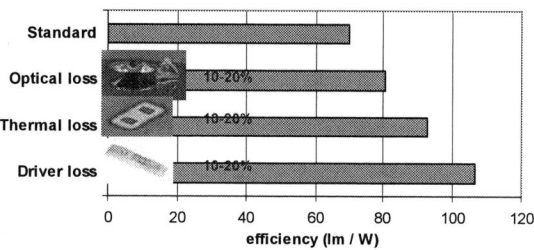

Figure 7.   Major loss factors for solid state lighting systems

III.   CHIP TECHNOLOGY

In order to optimize the LED output the main loss contributors have to be analyzed. These are

- Constraints in the internal quantum efficiency (IQE), e.g. losses within the active region of the LED structure resulting in a non optimum conversion of electrical carriers into photons

- A non optimum light extraction efficiency (LEE) resulting in a reabsorbtion of generated photons within the LED structure

- Electrical losses resulting in ohmic losses due to contact resistance and piezo barriers within the LED structure

- Conversion losses due to the Stokes loss and optical loss and scattered light within the applied converting materials used to generate white light out of the monochromatic blue pump light of the LED chip.

One of the most advanced LED chip architectures is the thinfilm principle, which was commercialized by OSRAM by the ThinGaN®-Technology[2].

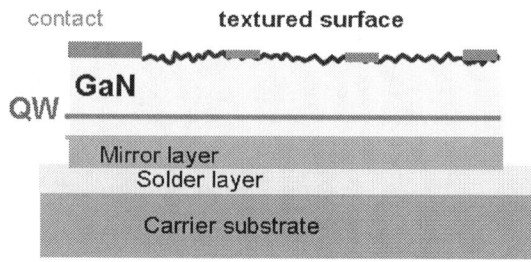

Figure 8.   Schematic of a ThinGaN® LED structure

The main constraint for extracting light out of the generating crystal is the difference of the indices of extraction of GaN and the packaging material e.g. silicone and the resulting Fresnel losses. In order to overcome this barrier, light is given multiple chances to be extracted by a random redirection at a textured surface and a highly reflective mirror at the back of the device (Figure 8.). In order to obtain high extraction efficiencies it is also mandatory to reduce absorption within the structure by using as thin GaN layers and avoid light guiding by removing the sapphire substrate which is used for epitaxial growth by a laser lift off process. With this process light extraction efficiencies of up to 80% have been demonstrated.

The device is a pure surface emitter, which allows easy scaling of chip size without efficiency loss.

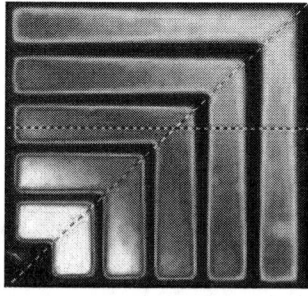

Luminance @ 1,4A

Figure 9.   Non uniform light distibution at high current operation (1mm chip size)

Major problems still come from the metallization layers needed on top of the device which are necessary to allow current distribution. In order to reduce absorption the dimensions of these have to be minimized, e.g. optimized to 350mA operation. However this leads to a highly non uniform luminance pattern and loss, if the device is driven at high currents (Figure 9.).

The newly designed UX:3-technology allows to have both, n- and p-contact buried to the back side of the device, which is possible due to a ground breaking via hole contact structure. Using this, even the built in contacts can be made reflective. Resulting in light extraction efficiencies of up to 90% and reduced losses due to reabsorbtion of scattered light generated by the conversion layers and an improved high current characteristics due to an optimum current distribution (Figure 10.).

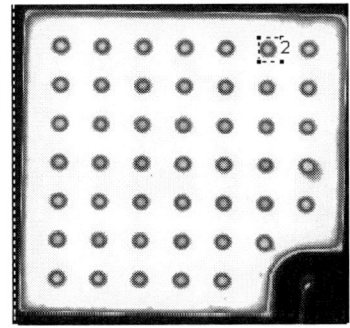

Luminance @ 2,8A

Figure 10. Highly uniform light extration of the newly developed UX:3 technology even in the high current regime of 2.8A (1mm) chip size

IV. PROPERTIES OF WHITE LEDS

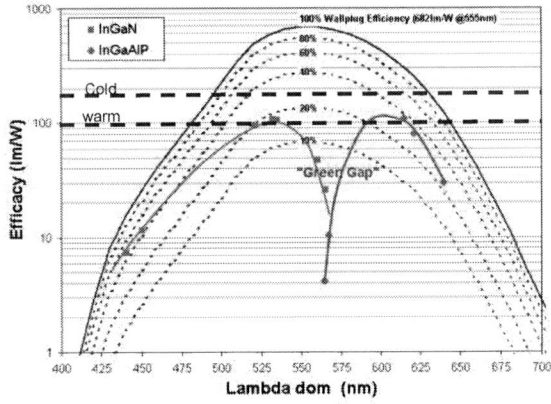

Figure 11. R&D Efficacies for AlInGalP and InGaN LEDs at operating conditions

For generating white light, there exist two main routes:

- Mixing of the emission of red, green and blue chips

- Downconversion of parts of the emitted blue LED light (Phosphor conversion (PC))

The comparison of these two systems shows a clear better efficiency of the phosphor conversion system in comparison to the direct conversion. This originates from the strongly decreasing efficiency of green LEDs with increasing wavelength (**Fehler! Verweisquelle konnte nicht gefunden werden.**). The maximum efficiency of today's InGaN LEDs is centered at 430 nm. As also the AlInGaP shows no efficient light generation at maximum of the eye sensitivity, which would yield high brightness devices, the so called "green gap" exists in today's LED technology, which is limiting RGB based systems as needed e.g. in projection systems.

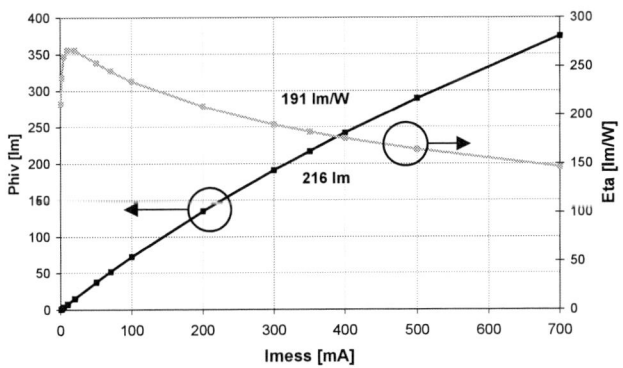

Figure 12. Brightness and efficacy of a 540nm green emitter based on 3M full conversion technology[17]

A way to overcome this shortcoming is the use of narrow band with converting systems as II-VI semiconductors which are pumped by a high efficiency blue LED. Corresponding results with conversion layers supplied by 3M in combination with UX:3 chips resulted in efficiencies of 190lm/W at 350mA and 540nm which is about twice the efficiency which can be achieved by direct emitted green (**Fehler! Verweisquelle konnte nicht gefunden werden.**)[17]. With this new technology, projection systems in the 1000 lm class have been demonstrated.

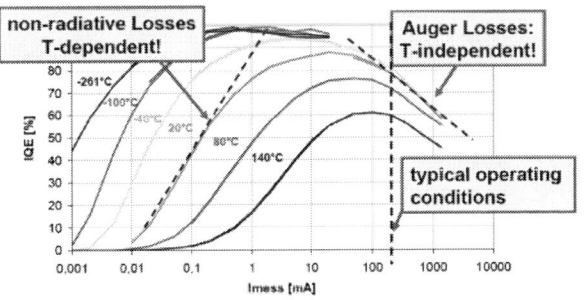

978-1-4244-6658-0/10 $26.00 © 2010 IEEE

Figure 13. Tyical temperature dependence of the IQE vs drive current characteristics of InGaN devices

The origin of the low efficiency of long wavelength InGaN devices is still a topic of world wide research.

A detailed analysis of the temperature dependent output characteristics reveals two main contributions (Figure 13.):

- A strongly temperature dependent behavior for low currents which is originating in classical non radiative losses
- The high current regime is temperature independent

In typical drive conditions (>350mA) for 1mm LED chips the device is mainly defined by the temperature independent high current characteristics, yielding the high thermal stability of InGaN based devices.

The dependence of the light emission density on the current density can be modeled by the simple rate equation of:

$$J \sim An + Bn^2 + Cn^3$$

where n is the carrier concentration, A,B and C are the non-, radiative and Auger-like recombination parameters.

The origin of the high current efficiency droop is still a topic of current discussion. Even so there are numerous onsets of explanation such as carrier loss and escape [3][4][5], defect and dislocation induced losses [6][7][8], piezoelectric field induced losses [9][10] as well as losses due to Auger recombination [11][12], this effect is not yet fully understood.

However, according to experimental and theoretical analyses performed at OSRAM, clear evidence is given that the observed droop effect is mainly dominated by Auger-like loss mechanisms. According to published data [15,16] the loss mechanism is depending on the QW-internal carrier density, strongly increasing towards higher carrier densities or operating current.

In more detail, experimental analysis of green light emitting quantum well structures were carried out at two different temperatures of 300 and 4K using resonant photoluminescence and electroluminescence (Figure 14.). The structures analyzed were single quantum well structures in order to eliminated interwell effects as well as uncertainties by diffusion generated recombination centers as a dominating loss mechanism, for example in the uppermost quantum well on the p-side. As shown in Figure 14. , decreasing efficiencies are observed on one hand at high optical or electrical excitation. On the other hand the efficiency increases as the excitation and therewith the carrier density decreases. After calibrating the electrical and optical excitation power, even the efficiency maximum of electroluminescence and resonant photoluminescence are located at almost identical excitation densities. More details about the design of the samples, the experimental setup and analyses can be found in Figure 16. [14].

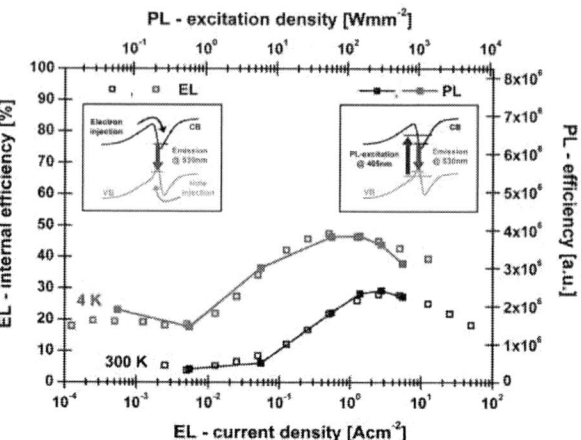

Figure 14. Internal efficiency study of a green light emitting single quantum well LED, by electroluminescence (open symbols) and resonant photoluminescence (full symbols). Both experiments were done at 4K and 300K. In the inset, carrier generation and recombnation is schematically shown for both excitation methods

As the typical drive condition is already dominated by the Auger-like regime, the main route to increase the efficiency is to reduce the carrier density in the active region of the LED. This can be done by thicker wells and the use of multiquantum wells. Since piezoelectric fields increase for thicker quantum wells and the material quality for indium rich quantum wells decreases significantly, multi quantum wells seem to be the answer to reduce the carrier density in technical designs.

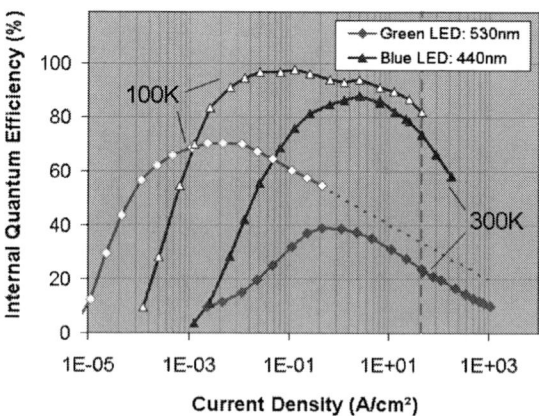

Figure 15. Internal quantum efficiency of blue and green LEDs as a function of drive current at 300 and 100K. The data are derived from 1mm² ThinGaN devices

The detailed analysis of green and blue MQW LEDs shows a basically similar behavior (Figure 15.). However the IQE of green LEDs peaks at much lower current densities showing IQE values of up to 70%. For green the piezoelectric field

978-1-4244-6658-0/10 $26.00 © 2010 IEEE   10

becomes dominant, thus hindering the multiquantum well operation which is effective for blue. However more research is necessary to completely clarify the low efficiency of green LEDs.

Another way to decrease the carrier density and thus increasing the efficiency is to increase the chip size at constant drive conditions. This is possible for fully scalable designs like the ThinGaN® platform. In order to clarify this effect a 1mm² ThinGaN® chip was mounted in a cold white Golden dragon device and the Wall plug efficiency as a function of drive current was measured (Figure 16. 1 mm² curve). As the design was proven in earlier experiments to be fully scalable, the big chips were simulated by simply renormalizing the drive currents. By taking the efficiencies at a drive current of 350mA the nominal efficiency of the same device can be easily tuned from originally 110lm/W to 85 lm/W for a 0.5 mm² device to up to 140lm/W for a 4 mm² device.

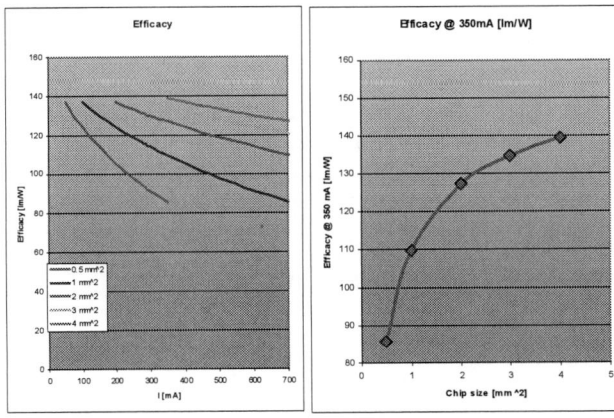

Figure 16. Scaling effects on the nominal efficacy of high power devices. The 1mm² curve is experimentally obained on 1mm² ultrawhite Golden Dragon devices. The curves for the other devices are obtained by renormizing the drive current asuming a full scalability

V. PHOSPOR CONVERTED WHITE LEDs

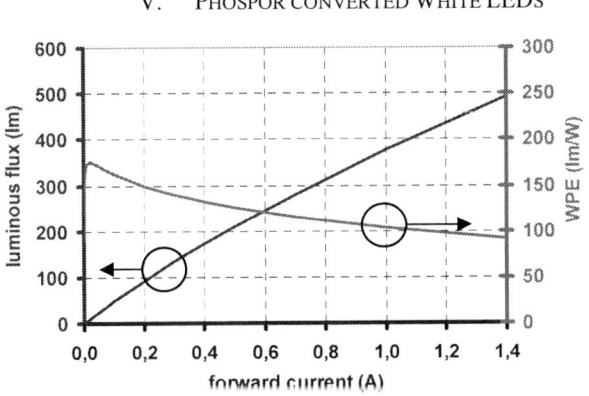

Figure 17. Brightness and wall plug efficiency as a function of drive current of a UX:3 device at a CCT of 5000K

By using 1mm2 UX:3 chips and ultrawhite converter efficiencies of up to 138lm/W could be achieved at a CCT of 5000K. The devices emitted up to 154lm at 350mA (Figure 17.).

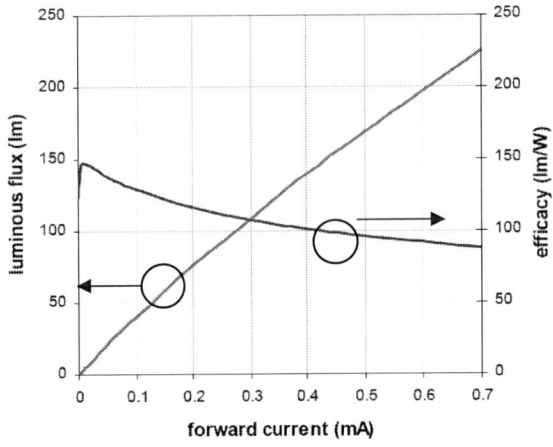

Figure 18. Brightness and wall plug efficiency as a function of drive current of a UX:3 device at a CCT of 3000K

Also for warm white devices the performance is significantly improved. When looking at 3000K and a CRI of >80, a luminous flux of 124lm or an efficacy of 104lm/W can be achieved at 350mA drive current. This is demonstrated in Figure 18.

For high quality light phosphors with a very wide spectrum have been developed. They offer a CRI > 95 at a CCT of 3000K. With a moderate brightness loss of 15% compared to a CRI 80 solution.

However in order to achieve a high CRI at CCT <3000K, which is equivalent to the emission of conventional halogen lamps a large fraction of light has to be converted to the red region of the spectrum (Figure 19.). This increases the Stokes losses significantly thereby reducing the efficiency of the system.

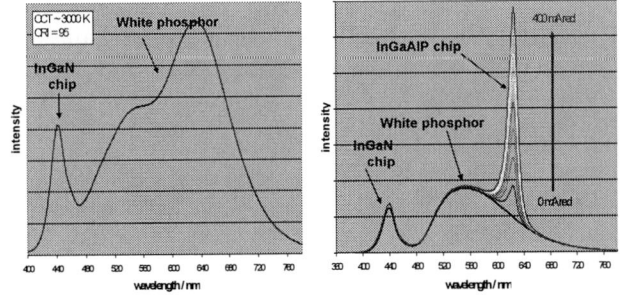

Figure 19. Spectra of a high CRI phosphor converted LED (left) and a hybrid phosphor converted/InGaAlP LED

An alternative is the combination of a highly efficient red InGaAlP emitter with a phosphor converted white LED (Figure 19.).

Besides a CRI of 90 this solution also provides the possibility to tune the CCT between 2700K and 5000K by changing the drive current of the InGaAlP LED.

In order to enable high efficiency devices on the technology stated above, the efficiency of the ultrared devices had to be optimized. By using an optimized epi design it was possible to increase the efficacy of the 1 mm² InGaAlP thinfilm devices to 135lm/W for amber at 350mA and peak wall plug efficiencies of up to 62% for hyperred (658nm) (Figure 20.).

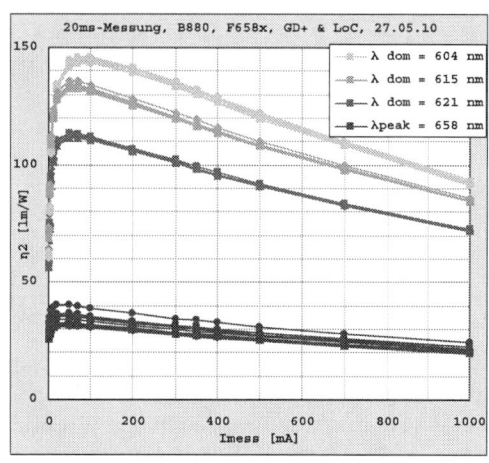

Figure 20. Wall plug efficiencies as a function of drive current of 1mm² high power InGaAlP devices set up in Golden Dragon

high quality white a combination of InGaAlP red with a phosphor converted LED was used. By improving the epi structures peak wall plug efficiencies > 60% for hyperred devices have been demonstrated.

ACKNOWLEDGMENT

Thanks to the entire R&D-Team at OSRAM Opto Semiconductors

Part of this work has been founded by BMBF and the European Commision

REFERENCES

[1]   www.osram-os.com/life-cycle-assessment
[2]   V.Härle et al, Proc.SPIE 4996, 133 (2003)
[3]   Kim et al, Appl. Phys. Lett. 91, 183507 (2007)
[4]   M.F.Schubert et al, Appl. Phys. Lett. 93, 041102 (2008)
[5]   I.A.Pope et al, Appl.Phys.Lett. 82, 17 (2003)
[6]   M.F.Schubert et al, Appl. Phys. Lett. 91, 23114 (2007)
[7]   A.Hangleiter et al, Phys. Rev. Lett. 95, 127402 (2005)
[8]   B.Monemar et al, Appl. Phys. Lett. 91 181103 (2007)
[9]   T.Schwarz, M.Kneissl, Phys.Stat.Sol. RRL 1, 3 A 44-A46 (2007)
[10] Y.C.Shen et al, Appl.Phys.Lett. 91, 141101 (2007)
[11] J. Hader et al, Appl. Phys.Lett. 92, 261103 (2008)
[12] M.Baeumler et al, Phys.Stat.Sol.(a) 204, 1018 (2007)
[13] M. Peter et al, Phys.Stat.Sol.C 5, 2050 (2008)
[14] A. Laubsch et al, to be published in IEEE
[15] A. Laubsch et al, Proceedings of IWN 2008
[16] David et al,Appl.Phys.Lett 92 053502 (2008)
[17] T. Miller et al, Proceedings of the SPIE, Volume 7617, pp. 76171A-76171A-10 (2010).

VI.   CONCLUSION

In this paper, it has been demonstrated that by addressing a total cost of ownership analysis LED illumination is already at least on par with best in-class conventional lighting. A key for further improvement will be optimized system efficiency. To do so decreasing active layer carrier density is central to reduce the droop effect and improve internal quantum efficiency at high operating current. Two path to do so are wide quantum wells and as well as multi quantum well structure. It has also been shown the UX:3 is significantly improving device performance compared to ThinGaN by reduced contact shading, yielding also in 9% reduced droop effects when moving from 350mA to 1A. Efficacies of these chip variants of up to 136lm/W for cold white and 104lm/W for warm white were demonstrated. For high quality warm white a special phosphor with CRI> 90 was developed. As a highly efficient alternative to wide spectrum phosphors for

Trends and perspectives for electrical characterization and reliability assessment in advanced CMOS technologies

Guido. Groeseneken[1], R. Degraeve, B. Kaczer, K. Martens

imec , Kapeldreef 75, B-3001 Leuven, Belgium
[1]also at K.U. Leuven, ESAT Department, Belgium
Email: Guido.groeseneken@imec.be

Abstract— **In this paper we give a brief historical review of the evolution of device reliability research over the past decades. Then we give some examples on how established characterization techniques that were developed for silicon based devices can be completely misinterpreted when applied to Ge or III-V based MOS-structures, and how a simple modification of the technique can ensure a correct interpretation. We also show how novel techniques, such as TSCIS (Trap spectroscopy by Charge Injection and Sensing), were developed recently to overcome the problem of dielectric material screening for logic and memory applications. With the scaling of the devices into the nanometer regime single traps are causing large variations in the device parameters, which leads to a time-dependent variability, which makes lifetime analysis difficult. Finally we show that when using the classical reliability assessment methodology based on accelerated testing, the available reliability margins are strongly reduced, in some cases even down to zero, especially for sub-1nm EOT (Effective Oxide Thickness) devices. As a result, we argue that the reliability community will have to look for alternative ways to ensure and guarantee the lifetime of future products.**

I. INTRODUCTION

With the continuous downscaling of CMOS technologies, reliability is more and more becoming a major bottleneck and this for several reasons. First of all the electric fields and current and power densities have increased continuously and are now reaching the maximum values that can be allowed for reliable operation. At the same time an impressive effort is taking place introducing new materials and novel device architectures to maintain the effective performance scaling. New materials like high k dielectrics and metal gates for both logic and memory technologies have already been introduced, while Ge or III-V materials for high mobility devices and novel device concepts such as Multiple gate FET's are under investigation. These new materials and devices often have unknown reliability behavior and/or introduce new failure mechanisms, whereas their speed of introduction exceeds the capabilities to explore their reliability performance in great detail. In order to understand or screen these new materials

and devices often novel or revised characterization techniques are required. Finally, the market is continuously demanding higher reliability levels, with single digit failure rates in FIT units (1 FIT= 1 failure per 10^9 operating hours) for present technologies. In the past, the technological reliability margins that were available to achieve the required failure rate levels where always sufficiently high, but in some of the technologies under development this becomes more and more cumbersome.

In this paper we discuss some of the implications these trends have on the characterization and reliability assessment of such novel technologies. We do this by first giving some examples of recent work done on novel materials and devices, followed by some considerations on paradigm shifts that are happening in guaranteeing circuit lifetime for future applications.

In the next section we give a brief historical review of the evolution of device reliability research. In section III we show how an established characterization technique, such as the conductance technique on MOS-structures, can be completely misinterpreted when applied to Ge or III-V based MOS-structures, and how a simple modification of the technique can ensure a correct interpretation. In section IV we discuss a recently introduced characterization technique, called TSCIS (Trap spectroscopy by Charge Injection and Sensing). This technique allows overcoming the problem of dielectric material screening for logic and memory applications. We also show how it can help to better understand the threshold voltage control in high k/metal gate devices. In section V we demonstrate that in very narrow and short devices even single traps can cause large threshold voltage shifts in nanometer devices, in this way increasing the time-dependent variability of the device parameters. In section VI we then show that when using the classical reliability assessment methodology based on accelerated testing, the available reliability margins are strongly reduced, in some cases even down to zero, especially for sub-1nm EOT (Effective Oxide Thickness)

978-1-4244-6658-0/10 $26.00 © 2010 IEEE

devices. As a result, the reliability community will have to look more in detail into what exactly determines these margins, and how the reliability assessment methodology can be changed in order to gain new margin for the most advanced technologies. More and more reliability engineers will have to account for the impact of failures on circuit functionality in order to guarantee sufficient and realistic lifetimes for the products, and as a consequence more interaction with the designers will become necessary in the future.

II. HISTORY OF RELIABILITY ANALYSIS

As an example of the trends in the electric fields existing in the transistors under operating conditions, Fig. 1 shows the evolution over the past 40 years of the oxide and silicon fields as a function of the gate length. Clearly 3 periods can be distinguished: a first constant voltage scaling period in the seventies and eighties, in which the power supply voltage was not reduced when scaling the geometries, and consequently the fields increased continuously with scaling. This was followed by a more or less constant field scaling period, in which the power supply voltages were reduced with every new technology node, so that the fields saturated at a certain plateau. Since the 65 nm node, however, the power supply voltages are saturating at a level around 1V, and can no further be reduced because of the non-scaling sub-threshold slopes of the MOSFET's. As a result we observe again a further increase in the electric fields[1], which starts to put new constraints on the reliability of the devices.

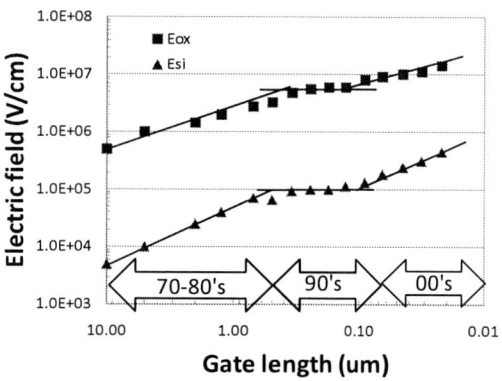

Figure 1.   Evolution of oxide and silicon electric fields showing 3 different scaling scenario periods.

Moreover, the power density has also continuously increased, as shown on Fig. 2 [1], which leads to higher chip temperatures, and consequently even a stronger acceleration of the degradation mechanisms. A particular trend here is that the static power density has been rising at a much faster rate than the dynamic power, which is again due to the aggressive power supply voltage scaling and accompanying Vt-reduction, which has led to an exponentially increasing off-current. Based on this trend the static power starts to dominate the

---

[1] For the oxide field we used EOT to calculate the effective field in the interfacial oxide layer, which is the maximum field in the dielectric stack

dynamic power from the 45 nm node on unless we can break through the sub-threshold slope barrier and develop devices that can operate at power supply voltages as low as 0.3V [2-4]. All of this leads to a strong reduction of the reliability margins for most failure mechanisms mentioned above.

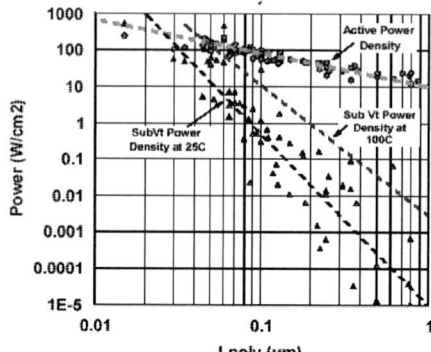

Figure 2.   Dynamic and static power density trends with scaling feature size [1].

Over the past decades reliability research can also be divided in three main periods. In a **first period** in the seventies and the eighties reliability research focused mainly on understanding the physics of the failure mechanisms, like oxide breakdown [5], bias-temperature-instability [6] and hot carrier degradation[7-8] at the device level, electromigration [9], stress voiding [10] and intermetal dielectric integrity at the interconnect level [11]. Based on the physical understanding of the failure mechanisms acceleration models for the wear-out testing of these mechanisms were developed, based upon which the lifetime of the various technology modules could be predicted. This was possible because the materials that were used in the devices and technologies were not changing very much with every new technology node, as the performance gain was mainly caused by pure geometrical scaling.

In a **second period** during the nineties the attention started to shift towards the impact of failures on the circuit operation [12-16]. With the decreasing reliability margins, the reliability community started to realize that a failure at the device level does not necessarily lead to a failure at the circuit level, which created room for new reliability margin, as will be further discussed in section VI [17].

In the **third period**, which started more or less after the turn of the century, the attention shifted again to the effect of introducing novel materials, like low k dielectrics, Cu and barrier layers in the backend, high k dielectrics and metal gates for logic devices, MIM capacitors and Flash memory interpoly dielecrics , and metal oxides for RRAM technology. Recently and even the replacement of Si by Ge or GaAs is under consideration. As a result of this doing a thorough analysis of all failure mechanisms for all possible material combinations is no longer possible. Of course a lot of learning from the 'classical materials can be reused, but dedicated techniques for screening of novel materials become necessary, whereas some of the 'classical' techniques need to be revised when applied to new material combinations, as will be illustrated with a few examples in the next sections.

III. CHARACTERIZATION FOR Ge AND III-V MOS STRUCTURES

In the period of the "classical geometrical scaling" of silicon technology the techniques that were applied to characterize the semiconductor-gate dielectric interface, such as low frequency and high frequency C-V analysis, conductance technique and charge pumping were well established and could be applied without too many problems caused by the scaling. With the scaling of the oxide thickness in the sub 5nm regime, however, the increased direct tunneling leakage currents were jeopardizing the use of these techniques [18, 19], and alternatives such as RF-CV [19, 20] or RF Charge pumping [21] have been successfully proposed to cope with this problem. In both cases the measurement frequency was increased to compensate for the high gate leakage currents, but fundamentally the interpretation of the techniques were not changed.

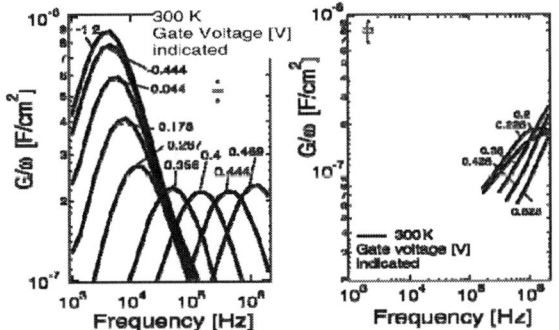

Figure 3. MOS-capacitor (left) conductance induced by weak inversion and inversion leads to overestimation of Dit ($\propto G/\omega$) and is eliminated by measuring full conductance (right), requiring shorting source, drain and bulk [31].

With the end of the classical Si-scaling in sight, alternative channel materials like germanium show promise for enhancing CMOS performance beyond silicon capabilities [22-28]. Like for Si, a crucial issue for the development of alternative channel material MOSFETs is the optimization of the electrical interface properties of non-Si/SiO₂ interfaces. Correctly determining the interface passivation of high-mobility semiconductors is of key importance for understanding and resolving interface passivation issues. The conductance method [29-30] is one of the most reliable and commonly adopted interface trap density (Dit) extraction techniques used to evaluate the passivation of interfaces. But blindly applying the conventional Si-based interface state density extraction methodology from the conductance technique on alternative substrates can lead to incorrect conclusions by which it is possible to both under- and overestimate the interface trap density by more than an order of magnitude [31].

A first issue of the conventional conductance technique is the possible confusion of a weak inversion with a depletion interface trap response, which occurs for MOS-devices with smaller bandgap materials then Si and at higher temperatures. Because of the smaller Ge bandgap, the interface traps in the minority half of the bandgap contribute to the C-V and G-V at room temperature in the 1kHz - 1MHz range [31-32]. The weak inversion (WI) response in Ge has similar features in C-V and G-V characteristics as a depletion bias response, and can therefore easily be confused. This can lead to erroneous extraction of D_{it} when using the conductance method, which implicitly assumes a depletion bias. Similar issues occur for inversion generation.

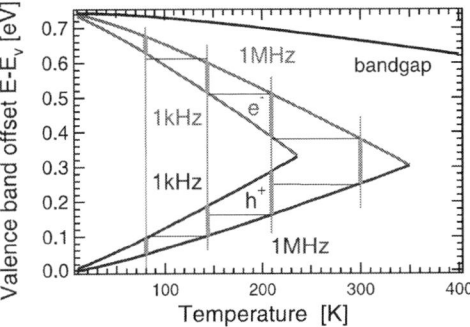

Figure 4. Calculated energy ranges in the bandgap in which traps are visible with the conductance method (1kHz-1MHz) as a function of temperature.

For Ge-based MOS structures the weak inversion response and inversion generation usually dominate the conductance at room temperature, and hence this jeopardizes the use of the conductance method. This can be easily avoided by applying the so-called full conductance measurement [31], in which the measurement is carried out on a MOSFET instead of a MOSCAP, and for which the source and drain are shorted to the substrate contact. This solves the two issues (see Figure 3).

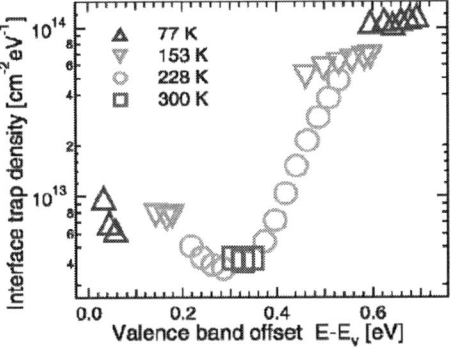

Figure 5. Interface trap distribution result of the full conductance technique applied on a Si-passivated HfO₂ Ge MOSFET [28].

At 300K traps near the band edges are not observable with the conductance technique at typical C-V frequencies of 1kHz - 1MHz (see Figure 4). This is less of a problem for Si/SiO₂, but is crucial for novel interfaces with substrates such as Ge, as often the majority of the traps are located at these energy levels. As a result germanium requires low temperature measurements (see figure 5). To prove the successful electrical passivation of novel dielectric-semiconductor interfaces with the conductance technique, the most reliable admittance based technique, full conductance measurements should be done at different temperatures to cover the entire bandgap.

This adaptation of the conductance technique, the full conductance technique, will avoid large errors in Dit-extractions.

IV. TRAP SPECTROSCOPY BY CHARGE INJECTION AND SENSING (TSCIS)

As already mentioned before, several novel dielectric materials are being introduced, both for logic and for memory devices. These novel dielectrics usually contain much more electronic defects than conventional high quality thermal oxide, which went through decades of process optimization. For these novel dielectrics, however, there can only be a few years between the first time processing and their possible introduction at product level. Since it is too time consuming to perform the classical reliability tests, such as TDDB for logic devices or memory retention tests for non-volatile memory devices, adequate rapid screening techniques are needed to get quantitative information on the quality of these materials.

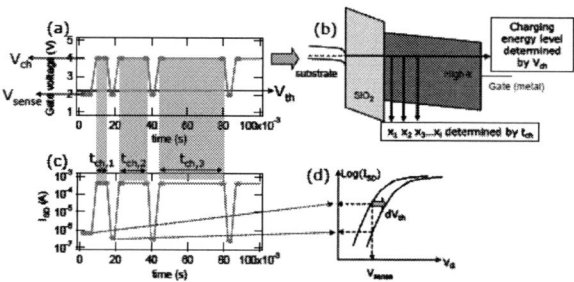

Figure 6. The operating principle of Trap Spectroscopy by Charge Injection and Sensing (TSCIS) for nMOS transistors.

The higher defectivity of the novel dielectrics was already discovered at the very beginning of the high k development. In the past years, several characterization techniques that are either based on controlled charge trapping, such as pulsed IV [33-35], bulk charge pumping techniques like VT²ACP ((Variable T_{charge}-$T_{discharge}$ Charge Pumping) [36-37]) or leakage current analysis such as SILC spectroscopy [38]) have been developed or modified. These techniques allowed to quantify the trap density levels in the novel dielectrics and to characterize their properties.

One of the important defect properties is their energy and spatial distribution, as they have an important impact on the performance of the devices and memory cells in which the novel materials are implemented. For example in the field of non-volatile memory devices, the TANOS (TaN or TiN/Al$_2$O$_3$/Si$_3$N$_4$/SiO$_2$/Si) gate stack has attracted considerable interest, but for optimizing charge storage in the nitride layer and, simultaneously, optimizing the Al$_2$O$_3$ blocking layer, detailed information on the energy and spatial distribution of the electronic defects in both layers is required.

A new technique that belongs to the category of charge trapping techniques is Trap Spectroscopy by Charge Injection and Sensing (TSCIS), which can directly provide quantitative data on trap energy and spatial position in dielectrics [39]. The principle of TSCIS is illustrated in Fig. 6 for an nMOS transistor, but it can also be adapted to capacitors. A charging gate voltage V_{ch} (>V_{th}) is applied during time intervals with

increasing length (~10 ms up to 1000s, Fig. 6.a), in which bulk dielectric traps are charged by direct tunneling of electrons from the inversion layer (Fig. 6b). In between the charging intervals, the gate voltage is switched for ~3ms to V_{sense} (with 0<V_{sense}<V_{th}) and the source-to-drain current I_{SD} (with V_D=0.1V) is measured (Fig. 6c). The drop of I_{SD} at Vsense is converted into a V_{th}-shift using an initially measured I_{SD}-V_G characteristic (Fig. 6d). This measurement methodology has been modified from and is similar to fast V_{th}-evaluation methods developed for minimizing the relaxation during NBTI tests, and uses the same equipment and software [40].

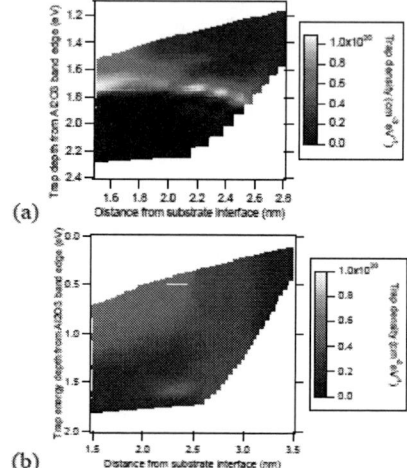

Figure 7. The trap density plotted vs energy and distance frmo the substrate interface as measured with TSCIS in (a) crystalline Al$_2$O$_3$ material and (b) amorphous Al$_2$O$_3$.

After discharging the sample, V_{ch} is incremented and the sample is again charged. The result of this measurement yields V_{th}-shift vs. t_{ch} measured for a range of V_{ch}-values, which can then be transformed into a trap density profile following a self-consistent algorithm involving WKB-approximation for determining the tunneling distance, a Poisson solver for finding the band bending in the presence of an arbitrary charge profile and the detailed balance between the electron injection level and trapped charge.

As a typical result we show in Fig. 7 the defect density vs. trap position and energy obtained on a 1/10 nm SiO$_2$/Al$_2$O$_3$ stack. We clearly observe a defect band between ~1.7 and 2.0 eV below the Al$_2$O$_3$ conduction band edge for a crystalline dielectric (Fig. 7.a), whereas for amorphous Al$_2$O$_3$ (with PDA @ 700C instead of 1000C) no distinct defect band signature is present (Fig. 7b) and instead, all trap energy levels are equally present.

In a similar way, different processing options of Si$_3$N$_4$ have been investigated with the aim of selecting the best suited charge storage layer for TANOS. For optimized Si$_3$N$_4$ charge traps at ~1.6 eV below the conduction band edge and with increasing density as function of distance to the tunnel oxide are found [41].

Independent retention modeling of TANOS required the presence of an Al$_2$O$_3$ defect band with identical energy level

978-1-4244-6658-0/10 $26.00 © 2010 IEEE

and trap density as found with TSCIS [42], and also yielded an Si₃N₄ charge profile that matched very well the TSCIS result [41], thus confirming the consistency of the technique. As a result the technique can be used to predict the retention performance of memory cells.

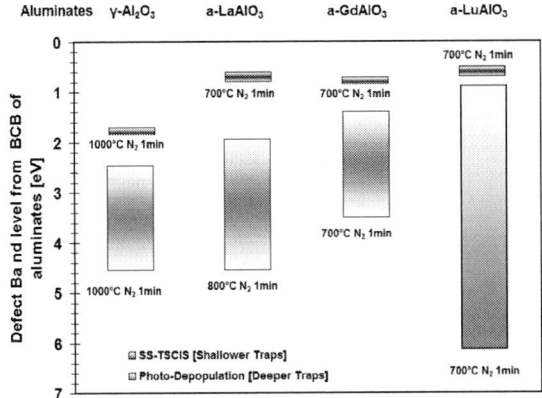

Figure 8. Representation of the energy distribution of charge defects in Al₂O₃ and three aluminates. In each case a sharply defined shallow defect band and a broader deep energy band is found. The figure combines results of Substrate-Side TSCIS and photo-depopulation.

By reversing the charging polarity, TSCIS can be expanded to detect traps close to the gate. This revealed the presence of an additional high density of charge defects at deeper energy level (~3.5 eV) in Al₂O₃, causing anomalous erase behavior and device instability [43]. With the aid of complementary techniques like 2-pulse CV [44] and photo-depopulation [45], the two-defect band structure of Al₂O₃ could be consolidated and a similar structure is found in several rare-earth aluminates as shown in Fig. 8.

TSCIS has recently also been extended to thin dielectric layers (1/3 nm SiO₂/HfSiO) used in transistors. This happened at the cost of losing the spatial resolution because defect discharging to the gate electrode cannot be neglected and requires complex models to account for. Yet, the spectroscopic capability in terms of energy remains intact and reveals how for instance Ar or As implantation modifies the defect structure in HfSiO by eliminating deep energy level traps and introducing shallow traps (Fig. 9) [46]. The density of deep traps could be directly related to the V_{th} of the transistor, suggesting that the mechanism of threshold voltage adjustments in HfSiO/TiN/metal gate transistors is in fact controlled by the bulk trap density in the dielectric.

V. STOCHASTIC NATURE OF RELIABILITY

Another trend that is observed in reliability characterization is the impact of increasing statistical variability of the degradation effects, comparable to the increasing variability of the initial parameters.

Until now, the large, micrometer-sized FET devices of the past CMOS technologies were considered identical in terms of electrical performance. Similarly, the application of a given stress resulted in an identical parameter shift in all devices.

With the gradual downscaling of the FET devices, the oxide dielectric was the first to reach nanometer dimensions, thus introducing the first stochastically distributed reliability mechanism—the time dependent dielectric breakdown [47]. With the shrinking of lateral device dimensions to atomic levels, variations between devices appear due to effects such as random dopant fluctuations and line edge roughness [48-49].

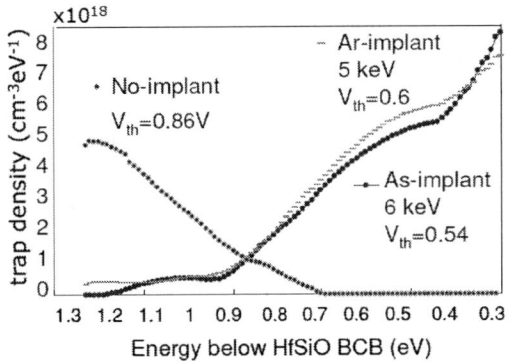

Figure 9. The energy distribution of the trap density in 3 nm HfSiO measured by TSCIS. As or Ar implantation removes all detectable deep energy defects, but introduces shallow defects.

Similarly, application of a fixed stress in such devices results in a distribution of the parameter shifts [50-51]. Understanding these distributions is crucial for correctly predicting the reliability of future deeply downscaled technologies [52].

Fig. 10a shows a typical result of a Measure-Stress-Measure (MSM) measurement of a relaxation transient following NBTI (Negative-Bias-Temperature Instability) stress. As already reported previously [50, 51], clear steps caused by single discharge events are visible in the NBTI relaxation transients. For larger device sizes these relaxation transients are continuous and spread over several decades in time. In this case, however, the average step height is significantly larger than reported earlier. The down-steps were detected in relaxation traces (Fig.10a) in all measured pFETs and a histogram of the step heights was constructed (Fig. 10b). It is important to note here that the steps corresponding to *a single discharging event in some devices exceed 30 mV*, the NBTI lifetime criterion presently used by some groups, which means that 1 single charge can cause threshold voltage shifts as high as the failure criterion.

An accurate reproduction of this process is typically done through computation-intensive physics-based device simulations with random dopant fluctuations, line edge roughness, and other realistic effects [53, 54]. The essence of this process can be captured, however, in a simplified channel percolation model without the need of a full device simulation with random dopant fluctuations [52]. The threshold voltage of such a FET corresponds to carrier energies sufficient to generate conduction (percolation) path(s) in the random dopant potential between source and drain. Depending on the position of the NBTI-stress-generated oxide charge, the

978-1-4244-6658-0/10 $26.00 © 2010 IEEE

conduction path(s) could be either unaffected or obstructed by the newly charged defect. In the latter case, the drop in the current has to be compensated by an increase of the gate voltage, resulting in the observed ΔVth.

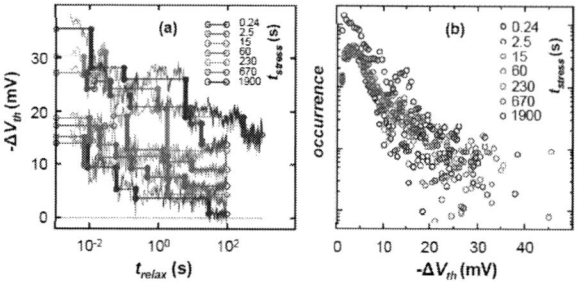

Figure 10. A typical result of 7 NBTI relaxation transients following the indicated stress times. Individual downsteps are marked with vertical abscissas. (b) Histogram of NBTI transient step heights for 72 devices shows a clear exponential distribution

The resulting time dependent variability of the degraded transistors complicates the lifetime extrapolation which will lead to a paradigm shift in the future reliability assessment and lifetime prediction, as will be further discussed in the next section.

VI.   NEW PARADIGM FOR RELIABILITY ASSURANCE

Until recently, reliability assessment was mainly carried out at the technology level, through accelerated testing for each major failure mechanism. Lifetime prediction is then based on an accelerated model for each failure mechanism, like hot carrier degradation [7, 8], time-dependent dielectric breakdown (TDDB) [5], negative-bias- temperature instability (NBTI) [6], electromigration [9], stress voiding [10], interconnect dielectric instability and breakdown [11]. For all these mechanisms, tests at increased stress levels (temperature, field, current density) are carried out on dedicated test structures. Such a test consists in measuring time-to-failure distributions on these test structures, followed by three extrapolations of the results:

- towards lower percentiles (typically 0.01%), based on the assumed statistical distribution (Weibull, Lognormal, etc.);

- towards larger chip areas, based on the series rule for failure distributions [55];

- towards operating conditions, based on the acceleration models for the specific failure mechanism.

In this classical methodology, a number of important assumptions are made:

- The failure criteria are often chosen arbitrarily, without a clear link with circuit (dis)functionality. E.g. the failure criterion for hot carrier degradation is 10% shift in g_m, for NBTI it is 30 mV shift in V_t.. These values are chosen arbitrarily, but have a strong impact on the predicted lifetime.

- It is assumed that the first failure anywhere in the chip will cause the full chip to fail. This is not necessarily true, *and it has been demonstrated [12-16]* that some circuits can tolerate more than one failure without stopping to function.

- The impact of the failure at the circuit level is not considered in the reliability assessment. Designers do not (have to/want to) care about reliability or take reliability into consideration in the design: reliability is qualified at technology level, and is mainly considered to be the job of the technology/process engineers.

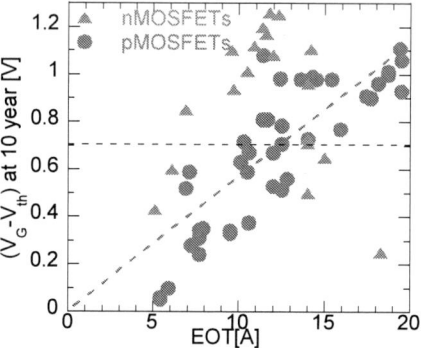

Figure 11. Trends of maximum overhead gate voltage for 10 years NBTI lifetime as a function of Effective Oxide Thickness (EOT) of the MOSFET's

Due to the trends discussed in the introduction and in sections II and V, however, reliability margins are reduced, in some cases even to zero. As an example, Fig. 11 shows the maximum gate voltage overdrive for nMOS and pMOS transistors that can be allowed for 10 years lifetime under Bias-Temperature-Instability conditions as a function of the effective oxide thickness (EOT). As one can see, the maximum gate overdrive drops below the level of 0.7V for sub-1nm EOT devices, and even to zero for 0.5 nm EOT. As a result, the reliability engineers have started to reconsider the classical approach of reliability assessment.

Various factors are present that allow us to gain new reliability margin. For example, it was found that for ultra-thin oxides the MOSFET device characteristics do not change dramatically after (soft) breakdown [56, 57]. The consequence of this is that circuits are relatively immune to oxide breakdown. Digital circuits, such as SRAM's or logic gates are very robust and a lot of leakage current is needed before it induces malfunction of the circuit [12-14]. That means that the attention in the reliability assessment is shifting from the pre-breakdown period (time-to-breakdown) towards the post-breakdown behavior. Failure of the circuit will not necessarily occur at the first transistor breakdown, but when the breakdown causes a sufficiently high leakage current. As an example, figure 12 shows the simulated static noise margin (SNM) reduction of an SRAM cell as a function of the current through the breakdown spot for breakdowns at the common p- and n-drain, at the n-source and at the p-source, respectively [14].

978-1-4244-6658-0/10 $26.00 © 2010 IEEE   18

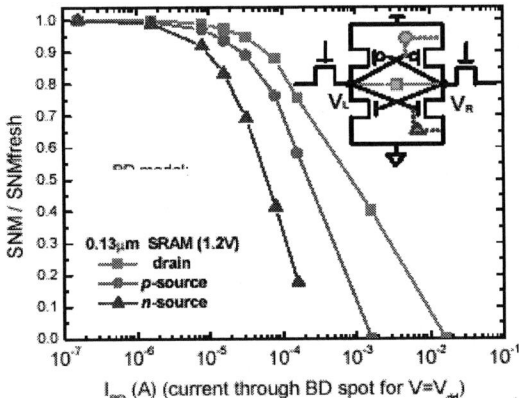

Figure 12. Simulated Static Noise Margin reduction of an SRAM cell as a function of leakage current through the BD path for three types of breakdown (drain, p-source, n-source [14].

It is clear that the SRAM remains functional, and the noise margin is reduced with 50% only after 100 µA current flows through the breakdown spot. This means that immediately after soft breakdown (when currents in the order of nA are flowing) the SRAM will be fully functional. As a result the reliability assessment will have to take into account a new circuit failure criterion for the post-breakdown behavior (e.g. maximum gate leakage current). Of course other types of circuits, like analog and/or RF-circuits [15] or dynamic logic circuits and DRAM's [16], are much more sensitive to breakdown induced leakage current. For such circuits the failure criteria will have to be more stringent then for digital or static circuits. The main conclusion here is that in order to determine the new reliability margins and set realistic reliability specs, input from and interaction with the designers is more and more necessary.

Another factor we can change in the reliability assessment is to omit the assumption that the first breakdown will lead to chip breakdown. By changing this criterion the statistical consequences are considerable and again a lot of new reliability margin can be gained [58-60].

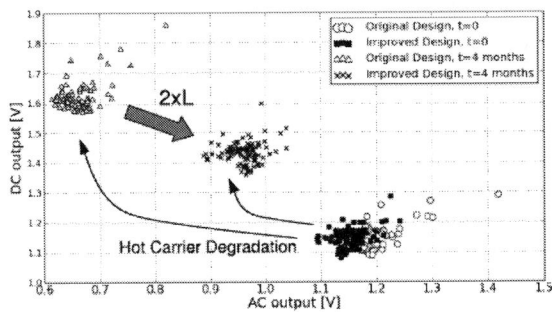

Figure 13. Simulation result for two versions of a one-stage amplifier over a life-time of 4 months [63]

All these factors will fundamentally change the reliability paradigm in the future. Ultimately, the future integrated system reliability will not be guaranteed anymore at the technology level, but at the design level. Designers have already started to take into account the effects of failures, degradation and time-dependent variability in a so-called reliability-aware design approach [61-65], and are thus developing design technologies to design reliable systems with more and more unreliable components or technologies. For this purpose similar techniques are used as the ones applied to cope with increasing variability, caused by e.g. random dopant fluctuations or line edge roughness effects. As was discussed in Section VI degradation mechanisms are indeed more and more appearing as time-dependent variability, rather than as abrupt catastrophic failures or uniform parametric shifts.

As a simple example Fig. 13 shows the simulated spread in a one stage amplifier in a 90 nm technology [63]. Two performance parameters were monitored: the AC output voltage and the DC output voltage. The circuit was simulated over a life-time of 4 months in which the circuit degrades due to hot carrier effects. The data show simulated results for two designs with different channel lengths to improve the hot carrier degradation. The variability was modeled with a special random fluctuation model for 50 circuits. As one can see, due to circuit ageing, both circuit performance parameters change over time.

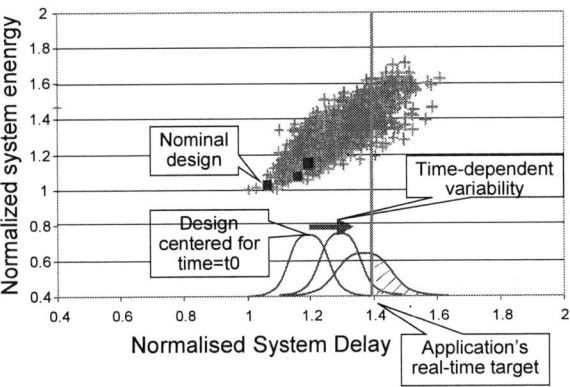

Figure 14. Effect of time-zero and time-dependent variability on energy vs delay trade off of an SRAM circuit.

However, the improved circuit degrades much less compared to the original circuit and is therefore much more reliable.

Another example is shown on Fig. 14, which shows the calculated spread in an SRAM circuit due to variability of threshold voltages in the circuit's FET's. The data are represented in a normalized energy vs delay plot. The spread in data points at time zero is represented by a Gaussian distribution. When degradation mechanisms, such as NBTI, are added the distribution shifts to the right, i.e. towards more delay for the same energy. When soft-breakdown occurs the distribution might even shift to higher values, beyond the application real-time target [65].

Such reliability-aware design techniques to cope with the effects during run-time will become extremely important and necessary in the future !

978-1-4244-6658-0/10 $26.00 © 2010 IEEE

VII. CONCLUSIONS

With the continuous scaling of CMOS technologies the electrical characterization and reliability assessment has been changing as well. With the introduction of novel materials to ensure effective rather than geometrical scaling novel techniques such as full conductance, VT^2-Charge pumping, pulsed I-V and TSCIS have been introduced. Moreover due to the scaling the effect of single trapped charges can become dominant in the degradation of devices and lead to threshold voltage shifts as high as 30mV. This leads to increasing time-dependent variability of the devices which has to be taken into account in the lifetime prediction. All this will ultimately lead to a paradigm shift in the reliability assessment of future, which will have to be guaranteed at the system design level rather then at the device and technology level. Research is underway to develop such reliability-aware design technologies which will change the operation conditions of the critical transistors during run-time.

ACKNOWLEDGMENT

The authors would like to acknowledge the support of the CMOST process technology development departments for logic and memory devices, and the members of the Device Reliability and Electrical characterization and Memory characterization groups.

This work was supported by imec's Core partner program on logic and memory devices.

REFERENCES

[1] D. Cox, IRPS Tutorial 2004.

[2] F. Maeyer, C. Le Royer, J.-F Damlencourt, K. Romanjek, F. Andrieu, C. Tabone, B. Previtali and S. Deleonibus, IEDM Tech Dig., p. 163, 2008.C. Hu, D. Chou, P. Patel, A. Bowonder, IEEE Int. Symp. VLSI Technology, Systems and Applications, p. 14, 2008.

[3] K. Boucart, W. Riess and A.M. Ionescu, IEEE Electron Device Letters, vol. 30, p. 656, 2009.

[4] A. Verhulst, W. Vandenberghe, K. Maex, G. Groeseneken, Appl. Phys. Lett., vol. 91, p. 053102-1-053 102-3, 2007.

[5] J. S. Suehle, IEEE Trans. El. Dev., vol. 49, p. 958, 2002.

[6] S. Ogawa, M. Shimaya, and N. Shiono, J. Appl. Phys. 77, 1137, 1995.

[7] C. Hu, S.C. Tam, F.C. Hsu, P.K. Ko, K.W. Terrill, IEEE Trans. El. Dev., vol. 32, p. 375, 1985

[8] E. Takeda and N. Suzuki,, IEEE El. Dev. Lett., vol. 4, p. 111, 1983

[9] J.R. Black, IEEE Trans. El. Dev., vol. 16, p. 338, 1969.

[10] J. McPherson and C.F. Dunn, J. Vac. Sci and Techn., vol. B.5, p. 1321, 1987.

[11] R. Tsu, J.W. McPherson, W.R. McKee, Proceedings IRPS, p. 348, 2000.

[12] B. Kaczer, R. Degraeve, M. Rasras, K. Van de Mieroop, P. J. Roussel, and G. Groeseneken, IEEE Transactions on Electron Devices, vol. 49, p. 500, 2002.

[13] B. Kaczer, R. Degraeve, E. Augendre, M. Jurczak and G. Groeseneken, Proceedings of ESSDERC-conference, Ed. J. Franca & P. Freitas, p. 75, 2003.

[14] R. Rodríguez, J. H. Stathis, B. P. Linder, S. Kowalczyk, C. T. Chuang, R. V. Joshi, G. Northrop, K. Bernstein, A. J. Bhavnagarwala and S. Lombardo, IEEE El. Dev. Lett., vol. 23, p. 559, 2002

[15] L. Pantisano, D. Schreurs, B. Kaczer, W. Jeamsaksiri, R. Venegas, R. Degraeve, K.P. Cheung, G. Groeseneken, IEDM Tech. Digest, p. 181, 2003

[16] B. Kaczer and G. Groeseneken, IEEE El. Dev. Lett., vol. 24, p. 742, 2003.

[17] G. Groeseneken, R. Degraeve, B. Kaczer and Ph. Roussel, Proceedings of the 14th IEEE International Symposium on the Physical and Failure Analysis of Integrated Circuits (IPFA), p. 1-9, 2007

[18] P. Masson, J.L. Autran and J. Brini, IEEE Electron Dev. Lett., p. 92, 1999

[19] J. Schmitz, F. Cubaynes, R.J. Havens, R. de Kort, A.J. Scholten and , L.F. Tiemeijer, IEEE Electron Dev. Lett., p. 37, 2003

[20] E. San Andres Serrano, L. Pantisano, J. Ramos, S. Severi, L. Trojman, S. De Gendt and G. Groeseneken,, IEEE Electron Device Letters, vol. 27, p. 772-774, 2006.

[21] G. Sasse and J. Schmitz, IEEE Trans. El. Dev., p. 881, 2008.

[22] K. Saraswat, C.O. Chui, K. Donghyun, T. Krishnamohan, A. Pethe,.IEDM Tech. Dig., p. 659, 2006

[23] C.O. Chui, F. Ito, K.C. Saraswat, IEEE Electron Dev.Lett., p. 1501, 2006

[24] T. Krishnamohan, , N.A., K. Saraswat, IEDM Tech Dig., p. 937, 2006

[25] C. Le Loyer, X. Garros, C. Tabone, L. Clavelier, Y. Morand, J.-M. Hartmann, Y. Campidelli, O. Kermarrec, V. Loup, E. Martinez, O. Renault, B. Guigues, V. Cosnier, S. Deleonibus, Proceedings of ESSDERC, p. 97-100, 2005.

[26] T. Maeda, M. Nishizawa, Y. Morita, Appl. Phys. Lett. vol. 90, 072911-1-3, 2007

[27] G. Mavrou, S. Galata, P. Tsipas, A. Sotiropoulos, Y. Panayiotatos, A. Dimoulas, E.K. Evangelou, J.W. Seo, Ch. Dieker, J. of Appl. Phys., Vol. 103, p. 014506-014506-9, 2008.

[28] P. Zimmerman, G. Nicholas, B. De Jaeger, B. Kaczer, A. Stesmans,. L.-Å. Ragnarsson, D. P. Brunco, F. E. Leys, M. Caymax, G. Winderickx, K. Opsomer, M. Meuris, and M. M. Heyns, IEDM Tech. Dig., 2006.

[29] E.H. Nicollian and A. Goetzberger, Bell Syst. Tech. J., vol. 46, p. 1055, December 1967.

[30] E.H. Nicollian, Solid-State Electronics, p. 937., 1969.

[31] K. Martens, C.O. Chui, G. Brammertz, B. De Jaeger, D. Kuzum, M. Meuris, M. Heyns, T. Krishnamohan, K. Saraswat, H.E. Maes, G. Groeseneken, "On the Correct Extraction of Interface Trap Density of MOS Devices With High-Mobility Semiconductor Substrates," IEEE Transactions on Electron Devices, Vol. 55, no. 2, 2008.

[32] P. Batude, X. Garros, L. Clavelier, C. Le Royer, J.M. Hartmann, V. Loup, P. Besson, L. Vandroux, Y. Campidelli, S. Deleonibus, F. Boulanger, "Insights on fundamental mechanisms impacting Ge metal oxide semiconductor capacitors with high-k metal gate stacks," J. Appl. Phys., Vol. 102, 034514, 2007.

[33] A. Kerber, E. Cartier, L. Pantisano, R. Degraeve, T. Kauerauf, Y. Kim, A. Hou, G. Groeseneken, H.E. Maes, U. Schwalke, IEEE Electron Device Letters, vol. 24, p. 87-89, 2003.

[34] A. Kerber, E. Cartier, R. Degraeve, P. Roussel, L. Pantisano, T. Kauerauf, G. Groeseneken, S. De Gendt and M. Heyns, Microelectronic Engineering, vol. , p. 45-52, 2003.

[35] A. Shanware, M. R. Visokay, J. J. Chambers, A. L. P. Rotondaro, J. McPherson, L. Colombo, G. A. Brown, C. H. Lee, Y. Kim, M. Gardner, and R. W. Murto, IEDM Tech.Dig., pp. 939–942, 2003.

[36] M. B. Zahid, R. Degraeve, J. F. Zhang, G. Groeseneken, Proceedings of the IEEE IRPS-conference (International Reliability Physics Symposium), p. 55-60, 2007

[37] M. B. Zahid, R. Degraeve, J. F. Zhang and G. Groeseneken, Microelectronics Engineering, vol. 84, p. 1951-1955, 2007.

[38] R. O'Connor, L. Pantisano, R. Degraeve, T. Kauerauf, B. Kaczer, P. Roussel, G. Groeseneken, Proceedings IRPS, p. 324, 2008.

[39] R. Degraeve, M. Cho, B. Govoreanu, B. Kaczer, M.B. Zahid, J. Van Houdt, M. Jurcak and G. Groeseneken, IEDM Tech. Dig., p. 775, 2008.

[40] B. Kaczer, T. Grasser, P.J. Roussel, J. Martin-Martinez, R. O'Connor, B.J. O'Sullivan and G. Groeseneken, Proceedings of the IEEE IRPS-conference, p. 20, 2008.

978-1-4244-6658-0/10 $26.00 © 2010 IEEE

[41] A. Suhane, A. Arreghini, R. Degraeve, G. Van den bosch, L. Breuil, M. Zahid, M. Jurczak, K. De Meyer, J. Van Houdt, IEEE El. Dev. Lett., vol. 31, p. 77, 2010.

[42] B. Govoreanu, R. Degraeve, J. Van Houdt and M. Jurczac., IEDM Tech. Dig., p. 353, 2008.

[43] R. Degraeve, M. Zahid, G. Van den bosch, P. Blomme, L. Breuil, B. Kaczer, M. Mercuri, A. Rothschild, A. Cacciato, M. Jurczak ,G. Groeseneken, J. Van Houdt, Proc. SSDM, p 428, 2009.

[44] D. Ruiz Aguado, B. Govoreanu, W. Zhang, M. Jurczak, K. De Meyer and J. Van Houdt, accepted for publication in IEEE Trans. El. Dev.

[45] W.C.Wang, M. Badylevich, V.V Afanas'ev, A.Stestmans, C. Adelmann, S. Van Elshocht, J. A. Kittl, M. Lukosius, Ch. Walczyk and Ch. Wenger, Applied Physics Letters , pp.95, 2009.

[46] S. Sahhaf, R. Degraeve, V. Srividya, B. Kaczer, D. Gealy, N. Horiguchi, M. Togo, T.Y. Hoffmann and G. Groeseneken, IEEE El. Dev. Lett., vol. 31, p. 272, 2010.

[47] R. Degraeve, G. Groeseneken, R. Bellens, J.-L. Ogier, M. Depas and H.E. Maes, IEEE Trans. El. Dev., p. 904, 1998.

[48] A. Asenov, IEEE Trans. El. Dev., vol. 45, p. 2505, 1998

[49] A. Asenov, S. Roy, R.A. Brown, G. Roy, C. Alexander, C. Riddet, C. Millar, B. Cheng, A. Martinez, N. Seoane, D. Reid, M.F. Bukhori, X. Wang, U. Kovac, IEDM Tech Dig. P. 421, 2008.

[50] S. E. Rauch, IEEE Trans. Dev. Mat. Rel., p. 524, 2007.

[51] V. Huard C. Parthasarathy, C. Guerin, T. Valentin, E. Pion, M. Mammasse, N. Planes, L. Camus, Proc. IRPS., p. 289, 2008.

[52] B. Kaczer, T. Grasser, Ph. J. Roussel, J. Franco, R. Degraeve, L.-A. Ragnarsson, E. Simoen, G. Groeseneken, H. Reisinger, Proc. IRPS, p. 26, 2010.

[53] A. Ghetti, C. M. Compagnoni, A. S. Spinelli, and A. Visconti, IEEE T. Electron Dev., p. 1746, 2009.

[54] M. F. Bukhori, S. Roy, A. Asenov, Int. Integ. Rel. Workshop, 2009.

[55] M. Ohring, "Reliability and failure of elecronic materials and devices", Academic Press, Ch. 4, p. 203, 1998.

[56] S.H. Lee, B.J. Cho, J.C. Kim and S.H. Choi, IEDM Tec. Dig., p.605, 1994.

[57] T. Hosoi, P. Lo Re, Y. Kamakura, K. Taniguchi, IEDM Tec. Dig., p. 155, 2002.

[58] M. A. Alam, R. K. Smith, B. E. Weir, and P. J. Silverman, IEDM Tech. Dig., p. 151, 2002

[59] M. A. Alam and R. K. Smith, Proc. IRPS, p. 406, 2003.

[60] J. Suñé, E. Wu and W.L. Lai, IEEE Trans. El. Dev. Vol. 51, p. 1584, 2004

[61] M. Alam, Microelectronics Reliability, vol. 48, p. 1114, 2008

[62] G. Gielen et al, Proceedings IEEE Design Automation Conferfence, p. 1164, 2008

[63] E. Maricau, G. Gielen, Proceedings IEEE Design Automation Conference, 2010.

[64] S.V. Umar, S.H. Im, S. Sapatnekar, Proceedings IEEE Design Automation Conferfence, p. 370, 2007.

[65] H. Wang, M. Miranda Corbalan, F. Catthoor, W. Dehaene, Proc. Memory Technology, Design and Testing Conference, Taipei, Taiwan, p. 71-76, 2006.

Silicon Photodiodes for High-Efficiency Low-Energy Electron Detection

Agata Šakić, Lis K. Nanver, T. L. M. Scholtes, and Carel Th. H. Heerkens
DIMES, Delft University of Technology
Feldmannweg 17, 2628 CT, Delft, The Netherlands
A.Sakic@tudelft.nl

Gerard van Veen, Kees Kooijman, and Patrick Vogelsang
FEI Company
Achtseweg Noord 5 5651GG Eindhoven, The Netherlands
Gerard.van.Veen@fei.com

Abstract—Solid-state electron detectors have been fabricated using a p⁺n silicon photodiode where the p⁺ region is created by a chemical-vapor deposition (CVD) surface doping from diborane B₂H₆. The as-obtained nm-deep p-type layer is resistant to conventional metal etchants, which allows elimination of both entrance contacts and protection layers from the photosensitive surface. This approach lowers the dead layer energy loss, while keeping near theoretical efficiency at high electron energies. The photodiodes have outstanding performance in terms of electron signal gain at low energies achieving 60% and 74% of the theoretical gain value at 500 eV and 1 keV, respectively. The ideal I-V characteristics and the small over-the-wafer spread of the dark current indicate a defect-free p⁺n junction, as well as a reliable and reproducible process.

I. INTRODUCTION

Recent developments in solid-state charged-particle detectors have made it possible to detect electrons with energies as low as 1 keV and with near 90% efficiency at energies > 3 keV [1]. Such improvements in low-energy charged-particle detection are of great interest for a large variety of applications such as medical diagnostics [2], space missions [3], scanning electron microscopy [4], and electron beam lithography [5]. In the past, photodiode detectors could only measure at electron energies greater than 5 keV. At lower energies, the impinging electrons could not reach the depletion region and generated carriers were more likely to recombine than to be collected. Thus, applications such as plasma physics measurements in space have required the addition of expensive equipment such as retarding field analyzers and Faraday cups for detection down to a few keV and below [3].

In this work, we present for the first time the use of the pure-B deposited silicon photodiodes to achieve near theoretical efficiency at high electron energies and outstanding signal gain in ≤ 1 keV region. They are fabricated in an ultrashallow junction technology using pure boron depositions [6] that in previous work have been successfully applied for the detection of deep/extreme ultra-violet light [7]. In contrast to these applications, the detection of very low-energy electrons does not allow the capping of the silicon diode with extra layers of any significant thickness. Both the distance in the Si to the active photodiode drift region and the thickness of any capping layers must be limited to a few nanometers because of the pulse height defect described in detail in [8] . In the past, nm-thin oxide layers have been used to passivate the Si surface but they have inherent instabilities related to the charge content of the oxide and interface state density that can vary from process to process and during use.

The present pure B-deposited diodes are reliably passivated by a pure B-layer that can be deposited uniformly and controllably down to about 2 nm. The depth of the p⁺n junction formed by boron doping of the Si during the deposition is also in that range. The responsivity of these photodiodes to low-energy electrons is investigated for different deposition conditions and capping layers. The results, particularly in the 0.5 - 2 keV range, are compared to detectors currently used in commercial scanning electron microscopes.

II. DEVICE FABRICATION

A. The p⁺n-junction Formation

The starting substrates are n-type 2-5 Ωcm Si(100) wafers on which 300 nm of silicon-oxide is thermally grown. Openings in the oxide are wet-etched and a B-layer is deposited by atmospheric pressure chemical vapor deposition (APCVD) in an ASM Epsilon One reactor using diborane (B₂H₆) as a gas source and hydrogen as both carrier gas and doping source dilutant [6]. The boron atoms deposit selectively on a clean Si rather than SiO₂ surface, so an oxide-free Si surface is prepared by dipping the wafers in HF (0.55%), Marangoni drying, and, finally, in-situ pre-baking at 900 °C in a H₂ atmosphere. At 700 °C, diborane decomposes forming an electrically active boron layer, part of which segregates on the silicon surface in an amorphous phase (α-B), while at the interface a boron-silicide layer (BₓSi₁₋ₓ) is formed. The surface silicon is doped to a level determined by high B gradient at the surface and the diffusivity and solid-solubility of B at 700 °C. The junction depth is in the order of nanometers, increasing slowly with the deposition time, which in some cases is followed by an in-situ anneal.

B. Anode Contact Definition

The B-deposited junctions are contacted with a layer of pure aluminium. This material is chosen for three reasons: it gives good ohmic contact to the B-layer, does not react with the B-layer [9], and can also be removed selectively to this layer. The processing sequence is illustrated in Fig. 1. First a 675-nm-thick layer of pure Al is sputtered at 350 °C (Fig. 1a). This layer is patterned in two steps: the first one to define the tracks and metal contacting of the diode and the next one to define the active area of the diode where the Al must be completely removed. In the first step the metal is plasma etched down to the SiO₂, leaving the whole diode area and a ring around the perimeter covered with Al (Fig. 1b). In the second step the Al in the active area is first etched down to about 100 nm by plasma etching, which is not selective to B (Fig. 1c). Therefore, the actual opening of the photosensitive surface is performed by a wet etching step in diluted HF to which the B-layer is highly resistant (Fig. 1d). In this manner the front entrance window is only covered with the B-layer, which does not oxidize significantly in air and is also resistant to oxidation in low-power oxygen plasmas that are often used to clean the detector surface. Additional capping layers can be added and patterned as shown in Fig. 1e. In the present work a 15 nm AlN dielectric layer was deposited by PVD and patterned by plasma-etching windows to the Al pads.

C. Post-Processing Variations

The junction depth is controlled by the thermal processing steps. At 700 °C the solid solubility is 2×10^{19} cm^{-3} and for a deposition time up to 10 min a sheet resistance of ~10 kΩ/sq is obtained. This value is dominated by the contribution from the doped c-Si layer since the α-B layer has a resistivity of $\sim 10^4$ Ωcm. A lower series resistance can be achieved by an in-situ thermal annealing step that provides higher dopant activation without considerable boron diffusion: for example for 1 min at 850 °C [6]. In addition, an extra B-doped epitaxial Si layer can be selectively grown which can improve drivability of the active surface layer. Coating layers as AlN can be added as well to protect the surface, or serve as radiation filters with accustomed pass-bands. However, it has already been proven that a drawback to adding an additional layer is a significant reduction in the detector sensitivity. Fabricated samples are designed to test the previously given options, combining the thin 1.8 nm boron layer with 15 nm of AlN, with and without the thermal annealing step and 50 nm of epi-Si. The samples discussed in the following sections are presented in Table I.

III. EXPERIMENTAL RESULTS

Fig. 2 shows the measurement set-up used to determine the electron signal gain. The beam of the scanning electron microscope (SEM) is used as an electron source and the photodiode is mounted on the multi-stub of the SEM next to the Faraday cup. In this way, calibration measurements on the Faraday cup and the gain measurements on the photodiode can be done in a fast sequence ensuring invariable conditions. The electron beam with the energy E_{beam} and a constant spot-size is first focused on a Faraday cup where the incident beam current I_{beam} is determined. It is expected that the

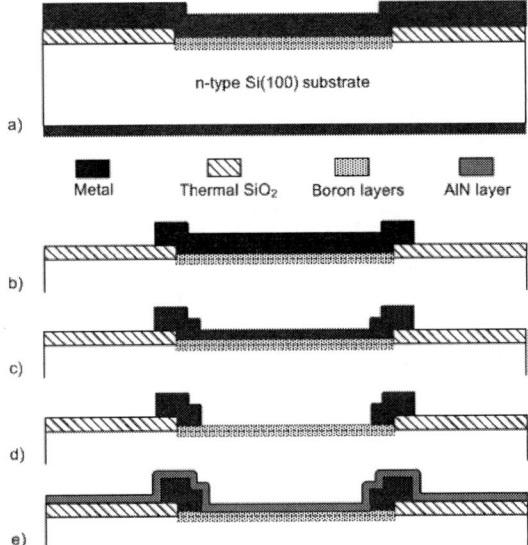

Figure 1. Schematic of process flow of the anode contact formation showing a) pure Al deposition, b) anode contact definition, c) dry etching step, d) dilute HF etch stop on the boron layer, and e) AlN depsition.

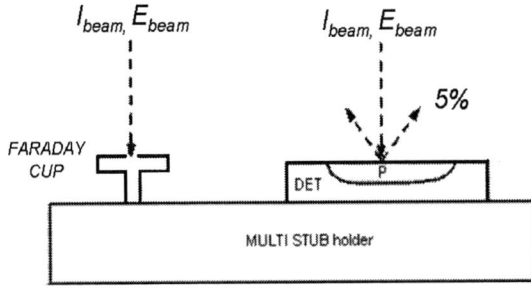

Figure 2. Faraday cup and the photodiode mounted on the SEM stub and exposed to the electron beam.

current is stable over the whole range of acceleration voltages and, accordingly, in all performed measurements it was ~ 5.3 nA. Immediately after, the beam was directed towards the photodiode with 0 V bias and the photodiode current I_{ph} was measured. With the acquired data, the relative gain of the photodiode can be calculated as

$$G_R = \frac{G_{PH}}{G_{TH}} \qquad (1)$$

where G_{PH} is the gain of the photodiode I_{ph} / I_{beam}, and G_{TH} can be related to the maximum theoretical number of generated carriers using the expression

$$G_{TH}(E_{beam}) = \frac{E_{beam}(1-\eta)}{e_0}, \qquad (2)$$

978-1-4244-6658-0/10 $26.00 © 2010 IEEE

TABLE 1 DESCRIPTION OF SAMPLE PROCESS VARIATIONS AND ELECTRON SIGNAL GAIN OF MEASURED PHOTODIODES AT 500 eV AND 1 keV

SAMPLE	Sample description				Electron signal gain at 500eV			Electron signal gain at 1 keV		
	B-layer [nm]	Anneal [min, °C]	epi-Si [nm]	AlN [nm]	G_{TH}	G_{PH}	$G_R[\%]$	G_{TH}	G_{PH}	$G_R[\%]$
1	1.8	-	-	-	131.58	78	59.3	263.16	196	74.5
2	5	-	-	-		30	22.8		147	55.9
3	1.8	-	-	15		-	-		34	12.9
4	1.8	1 min, 850 °C	50	15		-	-		-	-
BSE	commercially available detectors					18	13.7		46	17.5
vCD						52	39.5		156	59.3

where e_o is the mean energy ~ 3.61 eV required to produce an electron-hole pair in silicon [10] and η is the backscattered loss assumed to be 0.05. In general, an ideal number of generated carriers is E_{beam}/e_o. However, the corrections are added due to backscattering losses to assure that the created electron-hole pairs are defined per absorbed and not per incident electron energy, in analogy to internal and external quantum efficiency. Therefore, we can define the relative gain as

$$G_R(E_{beam}) = \frac{I_{ph}}{I_{beam}(E_{beam}/e_0)(1-\eta)},\qquad (3)$$

which corresponds to the photodiode responsivity, a common measure of the number of collected electron-hole pairs generated in the active region of a photodiode with respect to the total energy deposited by the particles to be detected in the detector [3].

Sample 1 with a 1.8 nm boron layer thickness shows the superior performance at low-energy electrons compared to currently used detectors in scanning electron microscopy (Fig. 3). The results are expressed as the ratio between the measured and the theoretical gain: for a value of 0 nothing is detected, and 1 is the theoretical limit at which all absorbed energy is used for electron-hole creation. While state-of-art photodiodes approach the theoretical limits for detection of electrons with energies > 3 keV, low energy electron detection remains a technological challenge.

The described silicon-based boron-doped photodiode achieves 60% of the theoretical gain value at 500 eV electron beam energy as compared to 14% of a commercially available backscattered-electron detector (BSE), and 40% for a standard "low Voltage high Contrast Detector" (vCD) detector [4]. Similarly, at 1 keV gain it reaches 74% of the theoretical value that corresponds to 4.1 and 1.5 times improvement over the BSE and vCD detectors, respectively. A significant performance advantage is maintained up to 10 keV after which they all tend towards the 97% level. The ultrashallow junction and the fully exposed detector area minimize the dead layer loss which is the main loss mechanism at low-energy operation, as reported in [11]. As described, measured responsivity can be defined as

$$R_M = R_{TH}(1 - \Delta_{DL} - \Delta_B - \Delta_R) \qquad (4)$$

where Δ_{DL} is the above mentioned dead layer loss, Δ_B is the backscatter loss, and Δ_R are the residual energy losses, e.g.

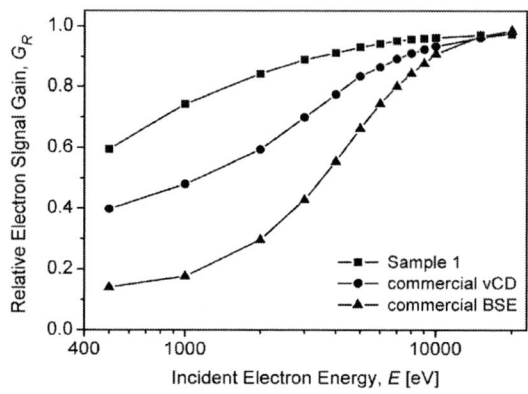

Figure 3. Measured relative electron signal gain for Sample 1 with a 1.8 nm boron layer, a commercially available backscattered-electron detector and a low Voltage high Contrast Detector.

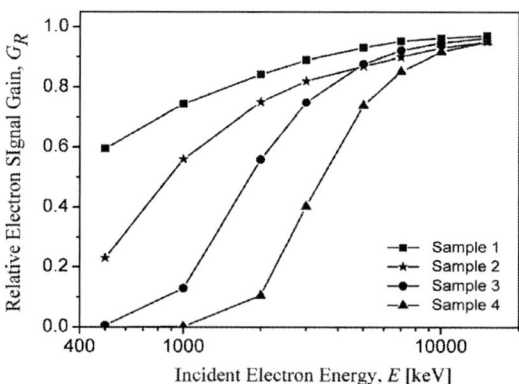

Figure 4. Measured relative electron signal gain for Sample 1 with a 1.8 nm boron layer, Sample 2 with a 5 nm boron layer, Sample 3 with a 1.8 nm boron layer and 15 nm AlN, and Sample 4 with a 1.8 nm boron layer, annealing 1 min at 850 °C, 50 nm epi-Si, and finally 15 nm

electron-hole recombination within or at the boundaries of the active area of the detector. It is stated that Δ_B plays the most significant role for energies > 1.5 keV. Looking at the results in Fig. 4, we can confirm that it is indeed the dead layer loss, and not the backscattering that governs the behavior at low energies: a slight increase in the boron layer thickness, up to 5 nm, considerably affects low-energy electron detection lowering it from 60% to 23% for 500 eV, and from 74% to

978-1-4244-6658-0/10 $26.00 © 2010 IEEE

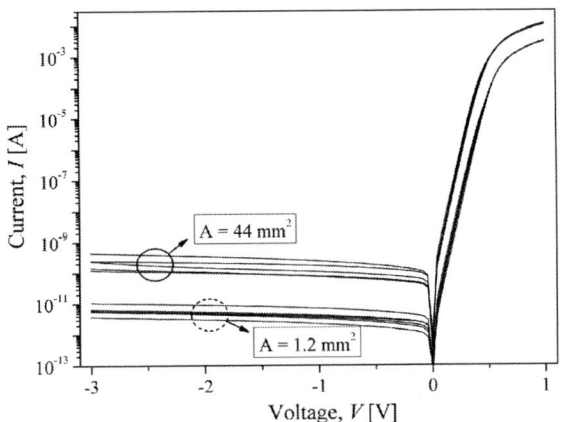

Figure 5. Measured I-V characteristics on six different positions on the wafer for photodiodes with areas of 44 mm^2 and 1.2 mm^2.

56% for 1 keV. Adding 15 nm of AlN layer on top of the 1.8 nm boron layer in Sample 3 suppresses 500 eV detection, while an extra annealing step with 50 nm epi-Si and 15 nm AlN for Sample 4 completely blocks the photodiode operations for energies $\leq$ 1 keV (Table 1).

In Fig. 5, I-V characteristics measured over the wafer are shown for Sample 1 for two tested areas, 1.2 mm^2 and 44 mm^2. The small spread of values substantiates that the boron layer is reliably deposited over the wafer, even for a thickness of only 1.8 nm. Moreover, the near-ideal electrical behavior confirms that an ideal pn-junction is formed. Here the diodes profit from the fact that the B-layer reduces the carrier injection from the substrate, and low saturation current levels comparable to deep-junction photodiodes are obtained.

IV. CONCLUSIONS

Using a pure boron chemical vapor deposition, we have fabricated ultrashallow p$^+$n photodiodes that have excellent performance as solid state electron detectors. Particularly in the low energy regime, the performance is well above that of state-of-the-art commercial photodiode detectors. Having the advantage of boron layer robustness to the conventional metal etchant, i.e. HF (0.55%), commonly used additional layers such as an etch-stop layer to form the metal contacts or a surface protection layer have been bypassed. Hence, the main energy loss mechanism in the below-keV region – dead layer loss – is substantially eliminated. Moreover, 1.8 nm boron deposited photodiodes show not only low dark current and excellent detection properties, but also a uniform current spread over the wafer which demonstrates that this is a high yield process that can be valuable for the future commercial applications.

V. ACKNOWLEDGMENT

The authors would like to thank the staff of the DIMES ICP cleanrooms and measurement room for their support in the fabrication and measurement of the experimental material. The support of FEI is gratefully acknowledged, in particular,

Ivanka Spee for electron signal gain measurements. Furthermore, we would like to thank to Sebastiaan Maas for the technical support. Finally, very special thanks to Dr. Francesco Sarubbi for his valuable research contribution.

REFERENCES

[1] C. S. Tindall, N. P. Palaio, B. A. Ludewigt, S. E. Holland, D. E. Larson, D. W. Curtis, S. E. McBride, T. Moreau R. P. Lin, and V. Angelopoulos, "Silicon detectors for low energy particle detection," IEEE Trans. On Nuclear Science, vol. 55, no.2 pp. 797–801, April 2008.

[2] J.Kataoka, T. Saito, Y. Kuramoto, T. Ikagawa, Y. Yatsu, J. Kotoku, M. Arimoto, N. Kawai, Y. Ishikawa and N. Kawabata, "Recent progress of avalanche photodiodes in high-resolution X-rays and γ-rays detection," Nuclear instruments and Methods in Physics Research Section A, vol. 541, Issues 1-2, pp. 398-404, April 2005.

[3] S.M.Ritzau, H.O. Funsten and J.E. Borovsky, "Solid State Detection of Low Energy Ions and Electrons for Constellation Missions," Science Closure and Enabling Technologies for Constellation Class Missions, pp. 131-135, UC Berkeley, California, December 1998.

[4] Laurent Y. Roussel, Debbie J. Stokes, Ingo Gestmann, Mark Darus and Richard J. Young, "Extreme high resolution scanning electron microscopy(XHR SEM) and beyond," Proc. of SPIE, Vol. 7378 73780W-1, 2009.

[5] C. S. Silver, J. P. Spallas, and L. P. Muray, "Silicon photodiodes for low-voltage electron detection in scanning electron microscopy and electron beam lithography," J. Vac. Sci. Technol. B, vol.24, Issue 6, pp. 2951-2955, 2006.

[6] F. Sarubbi, T.L.M. Scholtes, and L. K. Nanver in Journal of Electron. Mater., vol. 39, no. 2, pp. 162-137, February 2010.

[7] F. Sarubbi, L. K. Nanver, T. L. M. Scholtes, S. N. Nihtianov, and F. Scholze, "Pure boron-doped photodiodes: a solution for radiation detection in EUV lithography," Proceedings of IEEE 38th European Solid-State Device Research Conference (ESSDERC 2008), Edinburgh, Scotland, UK, pp. 278–281, September 2008.

[8] H. O. Funsten, S. M. Ritzau, and R. W. Harper, "Fundamental limits to detection of low-energy ions using siliconsolid-state detectors," Applied Physics Letter, Vol. 84, no. 18, pp. 3552-3554, 2004.

[9] A. Šakić, G. Lorito, F. Sarubbi, T.L.M. Scholtes, J. van der Cingel and L.K. Nanver, "Application of amorphous boron layer as diffusion barrier for pure aluminium," Proc. of SAFE 2009, pp. 112-115.

[10] F. Scholze, H. Rabus, and G. Ulm, "Mean energy required to produce an electron-hole pair in silicon for photons of energies between 50 and 1500 eV," Journal of Applied Physics, vol. 84, nr. 5, pp. 2926-2939, 1998.

[11] H. O. Funsten, D. M. Suszcynsky, S. M. Ritzau, and R. Korde, "Response of 100% Internal Quantum Efficiency Silicon Photodiodes to 200 eV–40 keV Electrons," in IEEE Transaction on Nuclear Sciences, vol. 44, no. 6, 1997.

New Mechanism of Plasma Induced Damage on CMOS Image Sensor: Analysis and Process Optimization

JP Carrère, JP Oddou, C. Richard, C. Jenny, M. Gatefait, S. Place, C. Aumont, A. Tournier, F. Roy

STMicroelectronics, R&D Technology dpt., 820, rue Jean Monnet, 38926 Crolles Cedex, France

Email address : jean-pierre.carrere@st.com

Abstract—A new plasma induced damage mechanism on CMOS image sensor is analyzed. An increase of the mean pixel dark current is observed after the plasma etch of a cavity on the pixel area. The degradation increases non-linearly when the dielectric layers between the photodiode and the plasma become thinner. This can be explained by a photo generation phenomenon in the dielectric nitride layer, induced by the plasma UV, and assisted by the wafer surface charge. This mechanism leads to a positive fix charge creation on the pixel surface, which can next deplete the top P layer of the pinched photodiode. Process and pixel architecture optimization ways are finally proposed.

I. INTRODUCTION

CMOS image sensor technologies are now scaling down to micron-size pixel, and with such smaller pixels, noise due to dark current is a major issue. It has been previously reported [1,2,3] that such optical devices are very sensitive to energetic photons damage, showing dark current increase after irradiation. Such irradiation can also occur during the sensor manufacturing, when plasma processes are used. It has already been reported [4] that plasma damage leads to an increase of some white pixel defects. The origin of this damage has been established as a pure electrical charging effect.

We present in this paper a new type of plasma damage, which combines some electrical and irradiation stresses. This can strongly impact the CMOS active image sensor quality. First, the process origin of the damage will be shown. Next, based on simple experiments, the degradation mechanisms will be analyzed and discussed. Finally, process optimization will be presented.

II. EXPERIMENTAL

The device used in this paper is a CMOS imager called Active Pixel Sensor (APS). A pinched N-diode (called PD in Fig.1a) is used to collect photo electrons, next the signal is treated by four NMOS transistors, embedded at pixel level: transfer gate (TG), source follower (SF), Read (RD) and Reset (RST) devices. Here a 1.4µm pixel platform is used, based on a 90nm CMOS technology made on 300mm wafers. The device is covered by a nitride capping layer, which prevents the silicon salicidation on the photodiode area, and allows a

good optical index adaptation between the silicon and the dielectric layers isolating the metal back-end lines. Three copper metal levels are used to connect the devices.

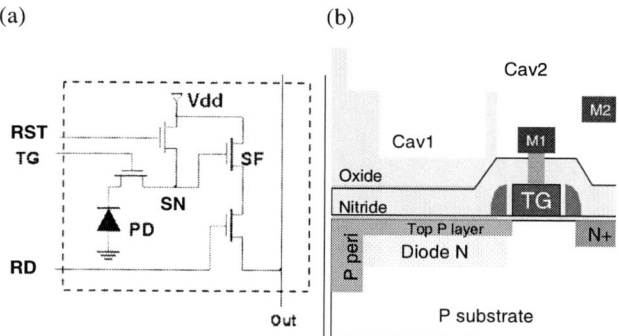

Figure 1. (a) CMOS Active Pixel Sensor, with a pinned photodiode (b) Cross-section schematic of the pixel during the Back-end cavity etch

The paper will be next focused on the process plasma steps used to etch two cavity levels in the back-end dielectrics above the pixel diode: they are called "Cav1" and "Cav2" in Fig. 1b. The goal of these cavities is to improve the total optical window and thus to maximize the pixel quantum efficiency. This is done by minimizing the dielectrics thickness on top of the photodiode. Cavity etch is performed with commercial Magnetic Enhanced Reactive Ion Etching (MERIE) tools. Finally, color Bayer filters and micro-lens (with the focus length adapted to the total back-end stack thickness) are patterned on top of the sensor. Note that a N_2/H_2 passivation anneal is performed, but before the cavity etch. After the dielectric layers opening, this anneal becomes no more efficient to cure defects because the encapsulation of the hydrogen species inside the back-end dielectrics is not performed during the anneal, as explained in ref [5].

III. RESULTS

A. Dark current process origin decorrelation

Large dark current excursions have been observed on image sensors with cavities fully processed. Fig. 2 clearly shows that the main origin of this dark current degradation is the cavity etch processes. The degradation seems to be cumulative during Cav1 and Cav 2 etch steps.

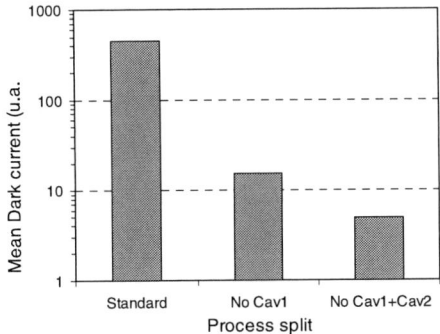

Figure 2.   Mean dark current measured on 1M pixel matrix (arbitrary unit), as a function of the Back-end cavity splits.

B.   Interaction with the dielectric thickness under cavity

The dark current distribution shows a strong wafer dispersion, with good dies located at the wafer edge, and the most degraded pixel sensors being at the center of the wafer, as shown on Fig. 3.

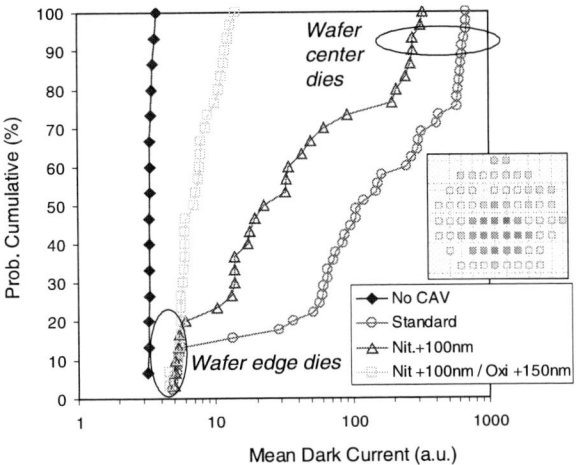

Figure 3.   Mean dark current measured on 1M pixel matrix (arbitrary unit), as a function of the dielectric thickness under the cavity. Inserted a typical wafer mapping of the worst dark current distribution.

The dark current degradation level is next very sensitive to the dielectric thickness between the cavity bottom and the photodiode surface: the thinner the dielectric layer, the stronger the degradation. An increase of about 100nm of both the nitride and the oxide layers under the Cav1 cavity is almost sufficient to prevent the dark current degradation.

IV. DISCUSSION

The origin of such a plasma damage effect can be attributed to the main interactions existing between the plasma

and the wafer, as shown in Fig. 6a. First, the wafer surface receives a positive ions flux, which is balanced by an electrons flux. But different factors can create local non-equilibrium between the two fluxes, leading to a charge formation on the wafer surfaces. Global plasma non-uniformity [6], and electron shadowing effect [7] are the most usual electrical charging origins. Next, the wafer is also exposed to a strong deep UV irradiation during the plasma process. As reported in [8], UV rays with an energy higher than 10eV have already been measured.

A.   Wafer charge damage mechanism:

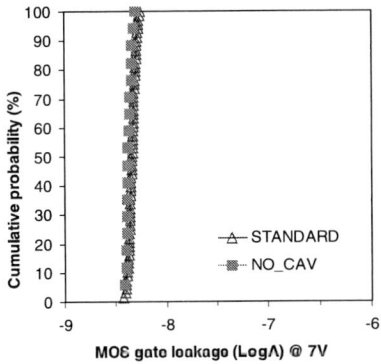

Figure 4.   Gate oxide integrity checked on pixel MOS device structure with large metal antenna ratio. No damage is observed even with the cavity etch.

In the study reported in [4], some MOS pixel gate oxide degradation due to plasma damage was identified as the root-cause of a white-pixel defectiveness. However, in the present study, no gate oxide damage was measured on the MOS pixel after the cavity etch process, as shown in Fig. 4. This can be explained by the absence of any metal area connected to the MOS gate, and exposed to the plasma process. So there is no electrical charging stress on the pixel MOS gate during the plasma process, this prevents the gate oxide damage.

B.   Deep UV irradiation impact on pixel:

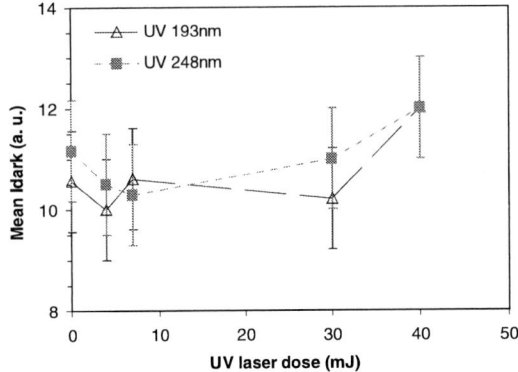

Figure 5.   Mean dark current on 1M pixels as a function of laser UV irradiation dose, for two laser sources (193nm and 248nm).

Many papers [8] show that UV photons have an energy high enough to create electron-hole pairs into oxide or nitride dielectrics layers. In order to simulate the impact of such plasma *VUV* irradiation on the image sensor, we have partially exposed a wafer after the etch of Cav1 and Cav2 to a supplementary laser UV irradiation. Two laser sources from typical photo-lithography tools have been used. The impact on dark current is shown on Fig. 5: no significant degradation is observed whatever the UV dose. This result looks surprising, as the laser UV photons energy is higher than 5eV. This should be energetic enough to create electron-hole pairs into the nitride layer which is located on top of the diode, and has a band-gap about 4.5eV. We believe that most of these e-h pairs probably recombine, as no electric field is applied through the nitride layer in this experiment.

C. Most probable damaging mechanism:

Based on the previous results, we propose the following pixel dark current origin, which is also illustrated on Fig. 6b. The cavity etch plasma creates a wafer charge, which is probably more positive in the bottom of the cavity, where the plasma electron flux is partially shielded [7].

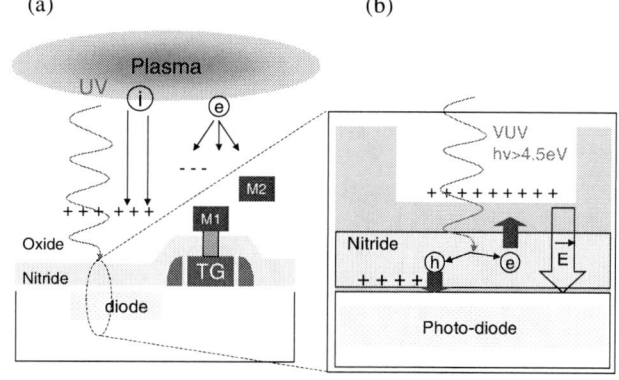

Figure 6. (a) Schematic of the typical interaction between a plasma process and the wafer surface: UV irradation, electron and ions flux. (b) Positive fix charge creation into the nitride layer on top of the photodiode.

Next, the plasma VUV irradiation creates e-h pairs into the dielectrics layers under the cavity, most probably in the nitride layer which has a lower band-gap than the oxide. The presence of the electric field generated by the positive plasma charge at the cavity bottom results in the separation of the radiation generated e-h pairs, and this prevents the carriers to recombine. This is also why the phenomenon is so sensitive to the dielectrics thickness, as it straightly modulates the internal electric field under the cavity. The strong dispersion of the degradation over the wafer can also be explained by the electrical wafer surface potential mapping, where the strongest potential is often concentrated at the center. Finally, because of their higher mobility, the generated electrons are easily evacuated, whereas the slower holes stay trapped near the bottom nitride interface. Similar mechanisms have been observed in [9] during a TG reliability light ageing test, where both illumination and electric field (in this case a TG gate bias) were needed to create a nitride fix charge on top of the silicon.

Fig. 7 explains how this fix positive charge in the top diode nitride layer can activate a dark current generation mechanism. In the absence of plasma damage, i.e. no nitride charge, the top diode interface is correctly filled with a P doping layer (Fig. 7a). On the other hand, when a fixed positive charge has been plasma induced in this nitride layer, it changes the internal field over the silicon, and the diode depletion area can extend up to the silicon/oxide top interface. In this case, the interface states are no longer passivated by the P doping and the dark current generation is strongly enhanced.

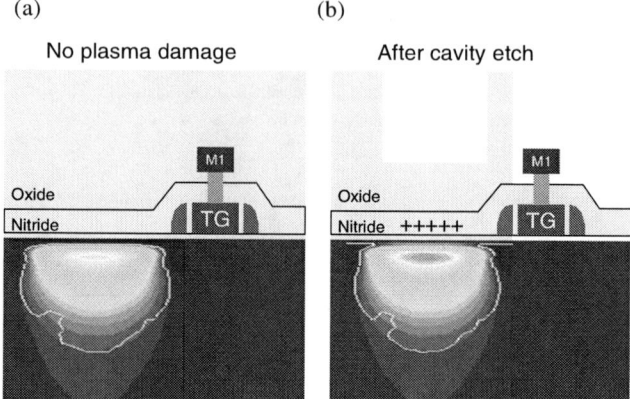

Figure 7. Shematic of the pinched photodiode potential from TCAD simulation: (a) initial (b) after nitride positive fix charge generation during cavity etch. The white line delimits the depletion areas.

V. PROCESS & PIXEL OPTIMIZATION

This plasma damage issue can be solved in many ways. First, the most obvious solution is to increase the dielectric layer thickness under the cavity. This action should decrease the electric constraint during the plasma exposition. The data shown in Fig. 3 shows that an increase of about 300nm is sufficient to prevent the damage. But this solution is not the best one because it shows the inconvenience to degrade the optical performance of the sensor... Another process optimization is to reduce the cause of the damage, i.e. work on the cavity plasma etch process.

A. Cavity etch plasma process optimization:

As most of the damage occurs at the end of the cav1 etch, when the dielectrics thickness is the thinnest, process optimization works have been focused on the plasma cavity overetch (OE) step. New plasma recipes have been developed by for example decreasing the OE time, or by making a low power RF bias process... Results shown in the Fig. 8a prove that it is the most efficient way to prevent the damage. Indeed, with the new cavity etch process, the dark current level is now as good as the reference sample where cavity has not been realized. Moreover, as shown in Fig. 8b, the dielectrics thickness under the cavity does not modulate anymore the dark current level value, as previously observed in Fig. 3. This will allow to strongly reduce the image sensors dark current die-to-die dispersion.

978-1-4244-6658-0/10 $26.00 © 2010 IEEE

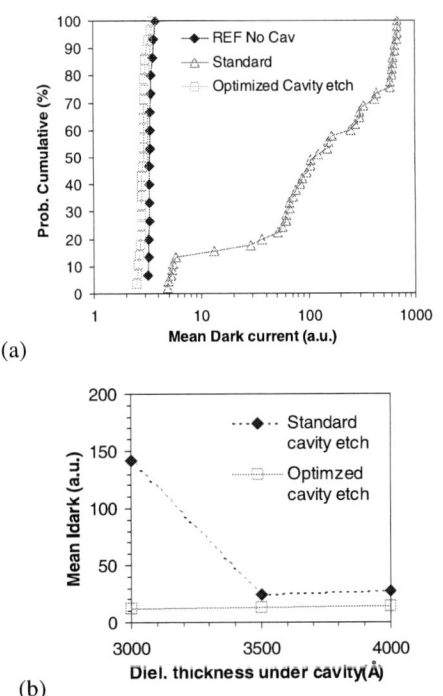

(a)

(b)

Figure 8. (a) Dark current reduction due to optimized cavity etch plasma. (b) The degradation dependance with the dielectric thickness under cavity. No more sensitivity is now observed with new cavity etch process.

B. Pixel architecture robustness increase:

Another way to secure the image sensor production against such plasma damage issue is to work on the pixel architecture in order to make it more robust against such issues. As the plasma induced positive fix charge tends to deplete the top diode P layer, an increase of its P-doping level should be sufficient to prevent the phenomenon, and so avoid the extension of the depletion zone towards the interface. This has also been realized, and the results are shown on Fig. 9. The dark current level has been clearly reduced.

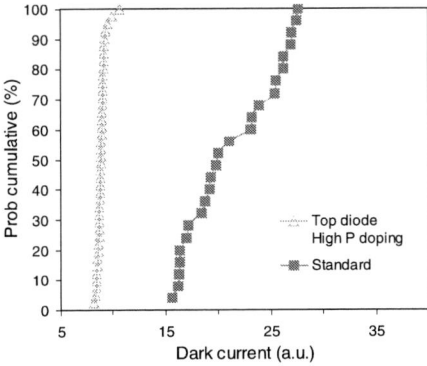

Figure 9. Dark current reduction can also be achieved by increasing the top diode P doping level. Sample where cavities have been etched.

VI. CONCLUSION

A new plasma induced damage mechanism has been put into evidence concerning the production of CMOS active pixel image sensor. This damage occurs during the final plasma etch process of cavities into the dielectrics of the metallization layers. The consequence is a strong increase of the sensor mean dark current level. The degradation level increases non-linearly when the dielectric thickness under the cavity decreases. We have shown that two factors are needed to create the plasma damage: first, deep UV irradiation during the plasma exposition generates electrons-holes pairs into the nitride dielectric layer above the photodiode. Next, these carriers are separated by the electric field created by the electrical plasma charging on the wafer surface. Afterwards, holes, which have a lower mobility than electrons, stay trapped in the nitride layer, and create a fix positive charge over the silicon. This fix positive charge can be sufficient to deplete the top diode interface, causing a strong increase of the dark current generation.

Finally, process optimization solution have been proposed to prevent the plasma damage. We have shown that both the plasma process optimization and a pixel architecture with an higher P doping level above the diode allow the CMOS image sensor production to be more robust against such damage.

ACKNOWLEDGMENT

The authors want to sincerely thank for their help and fruitful discussion J Prima, D. Benoit, C. De Buttet, P. Boulenc, S. Ricq, D. Lopez, D. Herault, S. Hulot, C. Baron, F. Lalanne.

REFERENCES

[1] M. Cohen and JP. David, "Radiation-Induced Dark Current in CMOS Active Pixel Sensors", IEEE Trans. Nucl. Sci., vol. 47, pp. 2485–2491, Dec. 2000.

[2] Flora M. Li, Nixon O, and Arokia Nathan, "Degradation Behavior and Damage Mechanisms of CCD Image Sensor With Deep-UV Laser Radiation", IEEE Trans. on Elec Dev., vol. 51, n° 12, pp. 2229-2236, 2004.

[3] V. Goiffon, E. Magali, P. Magnan, « Overview of Ionizing Radiation Effects in Image Sensors Fabricated in a Deep-Submicrometer", IEEE trans. on electron devices A. 2009, vol. 56, n° 11, pp. 2594-2601

[4] Ken Tokashiki, KeunHee Bai, KyeHyun Baek, Yongjin Kim, Gyungjin Min, Changjin Kang, Hanku Cho, Jootae Moon, "Study of plasma charging-induced white pixel defect increase in CMOS active pixel sensor", Thin Solid Films 515, pp. 4864–4868, 2007.

[5] D. Benoit, J. Regolini, P. Morin, "Hydrogen desorption and diffusion in PECVD silicon nitride. Application to passivation of CMOS active pixel sensors", Microelectronic engineering., vol. 84, n° 9-10, pp. 2169-2172, 2007.

[6] K. P. Cheung and C. P. Chang, "Plasma-Charging Damage : A Physical Model", J. Appl. Phys. Vol 75 , pp. 4415, 1994.

[7] K.P. Giapis, G.S. Hwang, "Pattern-dependant and the role of electron tunneling", Jpn. J. Appl. Phys., vol. 37, n°4B, pp. 2281-2290, 1998.

[8] T. Yunogami, T. Mizutani, K. Tsujimoto and K. Suzuki, "Mechanism of Radaition Damage in SiO₂/Si Induced by vuv Photons" Jpn. J. Appl. Phys., vol. 29, n°10, pp. 2269-2272, October 1990.

[9] Diana Lopez , Cédric Leyris , Stéphane Ricq , Francis Balestra, "Noise as a characterization tool for reliability under illumination of transfer gate transistor for image sensors applications", Proc. of ESSDERC conf, pp. 395-398, 2009.

Drain-Extended MOS Transistors Capable for Operation at 10 V and at Radio Frequencies

Andreas Mai and Holger Rücker

IHP, Im Technolgiepark 25, 15236 Frankfurt (Oder), Germany

Email: mai@ihp-microeelctronics.com

Abstract—This work reports on the integration of *n*-type lateral-drain-extended MOS transistors (LDMOS) in a 0.13 μm SiGe BiCMOS technology. The transistors are realized with no additional process steps using the core dual-gate-oxide CMOS flow only. LDMOS drift regions are formed by compensating lightly-doped drain (LDD) implantations of NMOS and PMOS transistors of the baseline process. Stable operation with less than 10% parameter variations in 10 years is achieved up to operating voltages $V_{DD,max}$ of 10 V for devices with breakdown voltages $BV_{DSS} = 30$ V and on-resistances $R_{ON} = 7.3$ Ωmm. Devices for different operating voltages $V_{DD,max}$ are realized by layout variations. Devices with $V_{DD,max} = 6$ V demonstrate breakdown voltages $BV_{DSS} = 25$ V, on-resistances $R_{ON} = 4.9$ Ωmm, and peak transit frequencies $f_T = 32$ GHz.

I. INTRODUCTION

High-voltage transistors are required in system-on-chip (SoC) technologies for a variety of applications ranging from power management circuits and driver circuits for non-volatile memories or light-emitting diodes to RF power amplifiers. These demands have been addressed in several generations of CMOS and BiCMOS technology platforms by using drain-extended MOS devices [1]. A major challenge encountered when integrating these devices is the restriction of the number of added process steps to only a few or none in order to restrict additional costs. Several concepts utilize dedicated ion implantation steps for the formation of drift and well regions of lateral-drain-extended MOS transistors (LDMOS) [2]–[6]. High-voltage devices with no added process steps have been realized by using the CMOS well implants for the formation of the extended drain regions [7]–[9]. An alternative concept for the integration of LDMOS transistors without any additional processing effort was proposed in [10]. There, the lightly-doped drain (LDD) implantations of the 3.3 V CMOS transistors were used for the formation of the drift regions.

Here, we present *n*-type LDMOS transistors fabricated in a dual-gate-oxide 0.13 μm CMOS flow without additional process steps. The drift regions of the LDMOS transistors are formed by a combination of the LDD implantations of the standard CMOS devices. In difference to [10], the drift region is divided in two parts with different doping concentrations resulting in reduced on-resistance for devices with the same breakdown voltage. The devices are capable for stable operation up to 10 V with less then 10% degradation of any device parameter in 10 years.

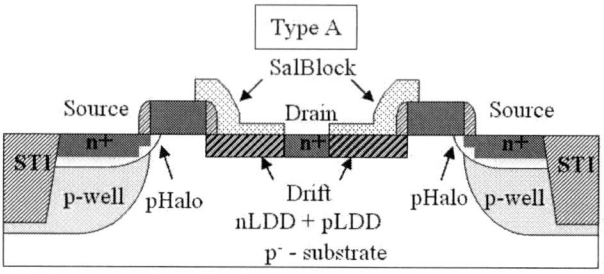

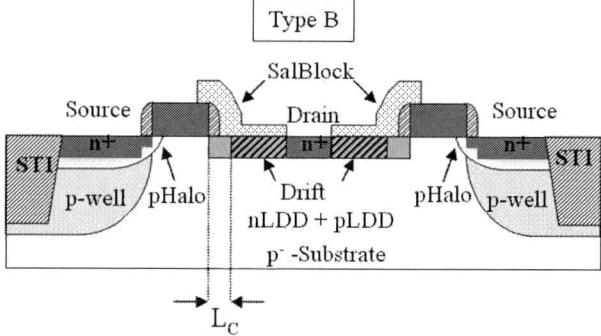

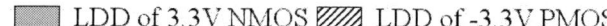

LDD of 3.3 V NMOS LDD of -3.3 V PMOS

Fig. 1. Cross sections of the two types of LDMOS transistors. For type A, the entire drift region is formed by a superposition of the *n*-LDD and *p*-LDD implantations of the 3.3 V NMOS and PMOS devices. For type B, only part of the drift region is compensated. The drift region near the gate edge (width L_c) is formed by the *n*-LDD implantation only.

II. DESIGN DESCRIPTION

The LDMOS devices were integrated in the 0.13 μm SiGe BiCMOS process of IHP [11]. This process addresses RF and mm-wave applications with a high level of integration. The technology offers high-speed SiGe heterojunction bipolar transistors (f_T =240 GHz, f_{max}=330 GHz) together with 1.2 V and 3.3 V CMOS transistors, a set of passive RF components, and seven layers of aluminum interconnects. The integration of LDMOS devices presented here is independent of the HBT process and uses only process steps of the core CMOS flow. The CMOS process features two gate oxide thicknesses of 2 nm and 7 nm for 1.2 V and 3.3 V CMOS transistors, respectively.

978-1-4244-6658-0/10 $26.00 © 2010 IEEE

Schematic cross sections of the LDMOS devices are shown in Figure 1 for two layout variants. The LDMOS devices share the 7 nm gate oxide of the 3.3 V MOS transistors. The doping profile at the source side is defined by the standard p-well, the heavily-doped drain (HDD) and the halo of the 1.2V-NMOS transistors (p-halo). These implants form a strongly asymmetric channel profile which is beneficial for DC and RF performance. The asymmetric channel profile supports high transconductance g_m and high transit frequencies f_T. The p-well implantation is restricted to the source regions in order to reduces the lateral electrical field at the drain side of the gate. This improves the breakdown voltage BV_{DSS} as well as the device stability against hot carrier injection [12]. The threshold voltage V_T of the device is mainly controlled by the halo implant since it creates a much higher doping concentration near the surface than the p-well. This fact effectively suppresses the sensitivity of V_T to alignment variations of the p-well mask edge [10].

The lightly-doped drift regions of the LDMOS are realized by a superposition of the n-LDD and p-LDD implantations of the 3.3 V NMOS and PMOS devices. For type A, the entire drift region is formed by a superposition of the two compensating implants. The net dose of this region is $3 \cdot 10^{12}\,\mathrm{cm}^{-2}$. For type B, only part of the drift region is compensated (Figure 1). The p-LDD implantation is separated from the gate edge by a distance L_C=0.1 μm resulting in a higher net doping of the drift region near the gate edge. The salicide formation is blocked in the drift region by a nitride layer which is used in the baseline process for the formation of unsalicided poly-silicon resistors.

III. RESULTS AND DISCUSSION

A. Impact of the drift region partitioning

Figure 2 show the output (a) and breakdown (b) characteristics for the two types of LDMOS layouts. The devices have drift lengths L_D of 0.4 μm and gate lengths L_G of 0.2 μm. While both types of devices show the same breakdown voltage BV_{DSS} of 17 V they differ strongly in on-resistance R_{ON}. The type A device has a R_{ON} of 6.8 Ωmm. This value decreases to 3 Ωmm for the type B device due to the increased doping concentration of the drift region near the gate edge. The higher doping concentration at the gate edge results also in a significantly improved saturation behavior for the type B device. These improvements of the DC characteristics of type B devices are most pronounced for short gate lengths. This is due to the closeness of the p-region at the source side to the lightly-doped drift region. The depletion of the drift region near the gate edge is stronger for type A devices which have a lower drift-region doping there.

Despite of the different doping of the drift regions near the gate edge, both types of devices exhibit almost the same breakdown voltage. This is illustrated in Figure 3 by simulated potential distributions at a drain voltage of 16 V. At high drain voltages, the drift regions are depleted for both device types. The main potential drop occurs across the compensated part of the drift region. High breakdown voltages are maintained for

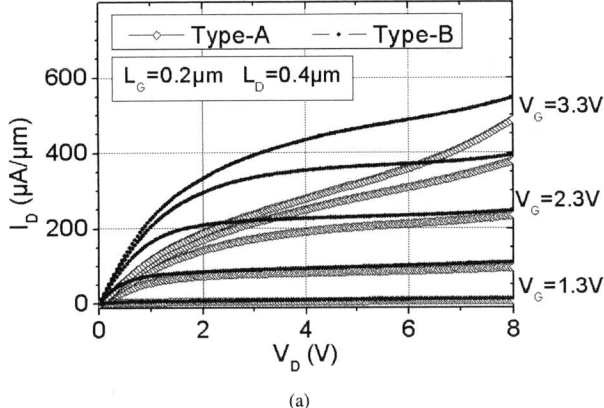

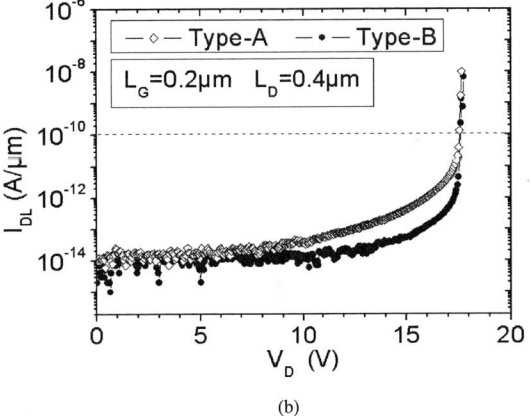

Fig. 2. Output (a) and breakdown characteristics (b) for the two layout variations of the LDMOS transistors.

the type B devices due to the low doping of this compensated part of the drift region. In consequence, one gets an improved trade-off between R_{ON} and BV_{DSS} for type B devices with the partitioned drift region. The remaining part of this paper is restricted to devices of type B.

B. High-voltage capability

Next, we address the impact of the design parameters L_D and L_G on the high-voltage capability of the LDMOS devices. Increasing the drift lengths and gate lengths result in higher breakdown voltages but also in higher on-resistances (see Table I). BV_{DSS} increases from 17 V for L_D=0.4 μm to 25 V for $L_D \geq 0.8\,\mu$m at a constant gate length L_G of 0.2 μm. A further increase of BV_{DSS} to 30 V is obtained for larger L_G of 0.5 μm at L_D=1.2 μm.

Voltage-stress measurements were performed to evaluate the long-term stability of the transistors. The devices were stressed at different drain voltages for time intervals up to 10^4 s. The gates were biased to $\approx$1.8 V during the stress corresponding to the maximum substrate current and the worst

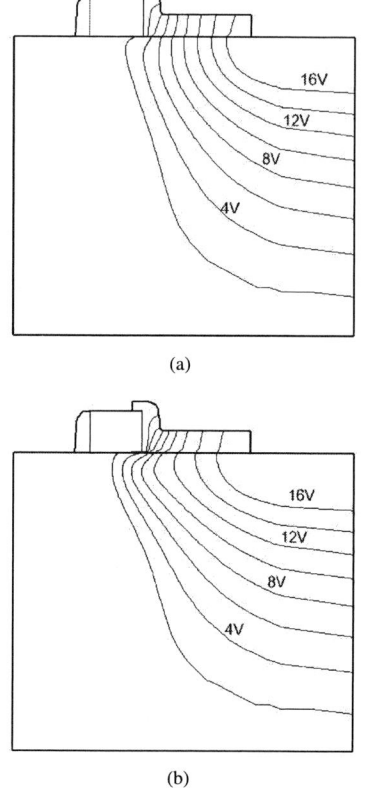

(a)

(b)

Fig. 3. Simulated potential distribution for LDMOS devices of type A (a) and type B (b) at V_G=0 V. Despite to the higher net doping near the gate for type-B, the potential drop in the drift region is similar for the two designs.

condition for high-voltage stress degradation of the devices. The on-resistance turned out to be the most sensitive parameter against high-voltage stress. Figure 4 shows the relative change of R_{ON} for different stress voltages and device geometries. The minimum size device with L_D=0.4 μm and L_G=0.2 μm is most vulnerable to high-voltage stress. For this device, a maximum operation voltage $V_{DD,max}$=4 V is extrapolated for the condition of less than 10% R_{ON}-degradation in 10 years. However, the devices degradation due to hot carrier injection can be reduced significantly for larger drift lengths. This is due to the reduced electrical field strength at the drain-sided gate edge. $V_{DD,max}$ increases to 8 V for devices with L_D=1.2 μm and L_G=0.2 μm. A further increase of the maximum operating voltage $V_{DD,max}$ to 10 V is obtained for devices with a larger gate length L_G=0.5 μm (Figure 5).

C. RF performance

S-parameter measurements were used to study RF characteristics of devices with 0.2 μm gate length and 200 μm gate width. Transit frequencies f_T and maximum oscillation frequencies f_{max} were extrapolated from the current gain $|h_{21}|$ at 20 GHz and the maximum available gain MAG at 40 GHz, respectively. The extracted values of f_T and f_{max} are plotted

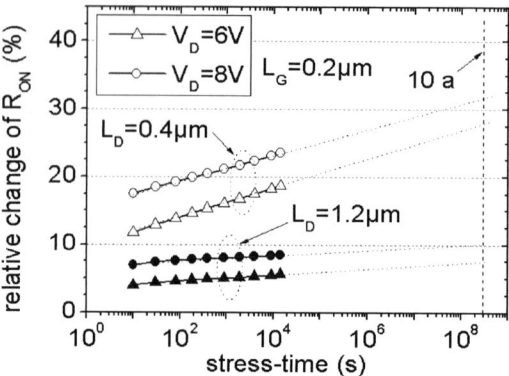

Fig. 4. R_{ON} degradation vs. stress time for devices with two drift lengths.

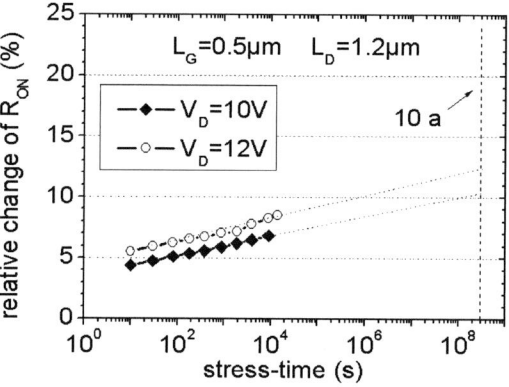

Fig. 5. R_{ON} degradation vs. stress time for a LDMOS with L_G=0.5 μm and L_D=1.2 μm. A maximum operating voltage of 10 V is obtained based on a R_{ON} degradation below 10 % in 10 years.

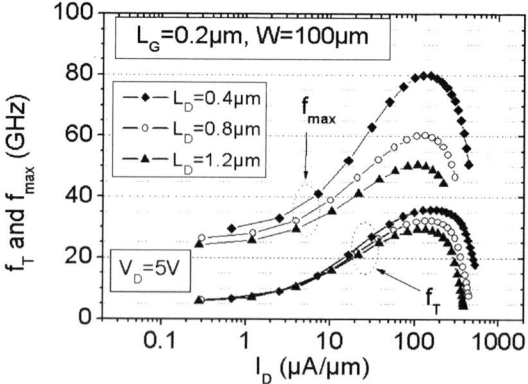

Fig. 6. Transit frequencies f_T and maximum oscillation frequencies f_{max} versus drain current for devices whit different drift lengths.

978-1-4244-6658-0/10 $26.00 © 2010 IEEE

TABLE I
DC AND RF PARAMETERS OF LDMOS TRANSISTORS WITH DIFFERENT DRIFT LENGTHS AND GATE LENGTHS. $V_{DD,max}$ RELATES TO AN ON-RESISTANCE DEGRADATION OF 10% IN 10 YEARS UNDER WORST BIAS CONDITIONS.

L_G [μm]	L_D [μm]	I_{Off} [$pA/\mu m$]	I_{ON} [$\mu A/\mu m$]	R_{ON} [Ωmm]	BV_{DSS} [V]	$V_{DD,max}$ [V]	$f_{T,max}$ [GHz]
		$V_D=10V$ $V_G=0V$	$V_{DD,max}$ $V_G=3.3V$	$V_D=0.1V$ $V_G=3.3V$	$V_G=0V$		$V_D=5V$
0.2	0.4	<1	450	3	17	4	36
0.2	0.8	<1	430	4.9	25	6	32
0.2	1.2	<1	400	6.8	25	8	29
0.5	1.2	<1	430	7.3	30	10	

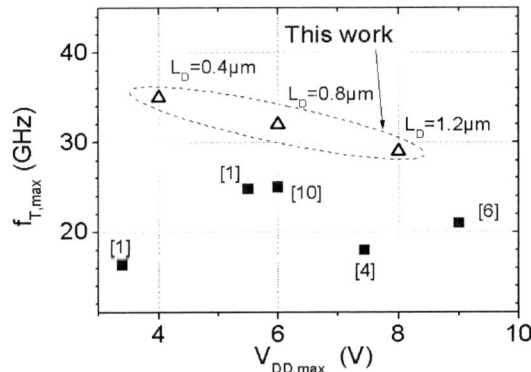

Fig. 7. Comparison of peak-f_T values vs. maximum operating voltage for various published RF-LDMOS devices.

in figure 6 as a function of the drain current for devices with various drift lengths. The values of f_T and f_{max} increase for devices with decreasing drift lengths due to an increased transconductance g_m and a decreased resistance for the drift region.

Figure 7 shows a comparison of the peak-f_T and $V_{DD,max}$ data of this work and previously reported data for integrated RF-LDMOS devices. Despite to the fact that the transistors presented here were fabricated with no additional process steps they exhibit highest peak-f_T values for the corresponding maximum operating voltages $V_{DD,max}$. The high speed of these devices is mainly due to their short gate lengths of $0.2\,\mu m$. The use of such short gate lengths for devices with breakdown voltages of up to 25 V is facilitated by the device construction. The implantation of the drift regions and as well as the major channel implant are self-aligned to the gate. Moreover, the inhomogeneous channel profile supports high transconductances g_m.

IV. CONCLUSION

Drain-extended MOS transistors for operating voltages up to 10 V were realized in a $0.13\,\mu m$ SiGe BiCMOS technology without additional process effort. The 10 V devices exhibit breakdown voltages of 30 V and on-resistances of 7.3 Ωmm.

Devices for different application areas can be realized by layout variations. The proposed LDMOS design facilitated the use of gate lengths down to $0.2\,\mu m$. These devices are suitable for application in the radio-frequency range. LDMOS transistors with a maximum operating voltage of 6 V and a breakdown voltage of 25 V demonstrate record peak-f_T values of 32 GHz.

ACKNOWLEDGMENT

The authors would like to thank the IHP clean room staff for the excellent support and D. Schmidt and C. Wipf for measurements.

REFERENCES

[1] R.A. Bianchi, C. Raynaud, F. Blanchet, F. Monsieur, O. Noblanc, "High voltage devices in advanced CMOS technologies," *Proc. of IEEE CICC*, pp. 363–366, 2009.

[2] D. Muller et al., "High performance 15-V novel LDMOS Transistor architecture in a0.25μm BiCMOS Process for RF-Power Applications," *IEEE Trans. on Electr. Dev. Let.*, vol. Vol. 54 No. 4, pp. 861–868, 2007.

[3] B. Szelag et al., "Integration and Optimization of a high performance RF lateral DMOS in an advanced BiCMOS Technology," *Proc. of IEEE ESDERC*, pp. 39–42, 2003.

[4] Z. Lee et. al, "A modular 0.18μm Analog RFCMOS Technology Comprising 32GHz F_T RF-LDMOS and 40V Complementary MOSFET devices," *Proc. of IEEE BCTM*, pp. 126–129, 2006.

[5] K.E. Ehwald et al. , "A two mask complementary LDMOS module integrated in a 0.25μm SiGe:C-BiCMOS Platform," *Proceedings of ESSDERC*, pp. 121–124, 2004.

[6] N.R. Mohapatra et al., "A Complementary RF-LDMOS Architecture compatible with a 0.13μm CMOS Technology," *Proc. of IEEE ISPSD*, pp. 37–40, 2006.

[7] G. Baldwin et al., "90nm CMOS RF Technology with 9.0V I/O Capability for Single-Chip Radio," *Digest of Symp. VLSI Technology*, pp. 62–63, 2003.

[8] J.C. Mitros et. al, "High-Voltage Drain Extended MOS Transistor for 0.18-μm Logic CMOS Process," *IEEE Trans. on Electron Devices*, vol. 48, no. 8, pp. 1751–1755, August 2001.

[9] T. Yan, H. Liao, Y. Z. Xiong, J. Shi, R. Huang, "Cost effective integrated RF power transistor in 0.18-μm CMOS Technology," *IEEE Electron Device Letters*, vol. 27, no. 10, pp. 856–858, October 2004.

[10] A. Mai, H. Rücker, R. Sorge, D. Schmidt, C. Wipf, "Cost-Effective Integration of RF-LDMOS Transistors in 0.13μm CMOS Technology," *Proc. of IEEE SiRF*, pp. 124–128, 2009.

[11] H. Rücker et al., "A 0.13 μm SiGe BiCMOS Technology Featuring fT / fMAX of 240 / 330 GHz and Gate Delays Below 3 ps," *Proc. of IEEE BCTM*, pp. 166–169, 2009.

[12] N.R. Mohapatra et al., "The impact of Channel engineering on the performance and reliability of LDMOS transistors," *Proceedings of IEEE-ESSDERC*, pp. 481–484, 2005.

Analysis of Silicon On-Chip Integrated Antennas for Intra- and Inter-Chip Wireless Interconncts

T. Kikkawa, K. Kimoto and S. Kubota

Research Institute for Nanodevice and Bio Systems,
Hiroshima University
1-4-2 Kagamiyama, Higashi-hiroshima, Hiroshima, Japan
e-mail:kikkawat@hiroshima-u.ac.jp

Abstract—Silicon on-chip integrated antennas have been developed for intra- and inter-chip signal transmission. However, the radiation efficiency and transmission gain of the antenna were degraded due to conductive loss of Si substrates. In order to solve these problems, a bypass planar wave guide which was composed of a high dielectric constant substrate was developed to reduce the propagation loss through Si chips. In this paper the effects of dielectric constant and thickness of the dielectric wave guide on radiation efficiency and transmission gain of integrated antenna are investigated.

I. INTRODUCTION

Complementary-metal-oxide-transistor (CMOS) cutoff frequency increases up to 400 GHz for 32-nm CMOS technology node. However, global clock frequency and data-rates for the transmission between CMOS large-scale integrated circuits (LSI) are limited below 4 GHz due to parasitic resistance-capacitance (RC) delay of long metal interconnects. In order to get rid of parasitic capacitances and resistances of metal interconnects, wireless interconnects have been developed [1-8]. The inductive couplings by use of spiral inductors have been developed for inter-chip short distance communication less than 100 μm for the inductor diameters less than 100 μm. Transverse electromagnetic (TEM) wave propagation by use of silicon on-chip integrated antennas has been developed for the distance longer than 1 mm. However, conventional on-chip integrated antennas had drawbacks in radiation efficiency and transmission gain due to conductive loss of Si substrate. The radiation efficiency and transmission gain are approximately 3% and -40 dB, respectively, at 5 mm distance when standard 10 Ω•cm resistivity silicon substrate is used. In order to solve these problems, reduction of Si substrate loss and improvement of transmission gain are necessary. In this paper analysis and optimization of radiation efficiency and transmission gain are conducted by increasing the resistivity of Si substrate, by thinning the Si substrate thickness and by introducing dielectric waveguides under Si chips.

II. STRUCTURES OF ON-CHIP ANTENNAS

Transmitter and reciever antennas were fabricated on Si substrates. The antenna length (L) and the separation distance between transmitting and receiving antennas (d) were changed for investigating intra-chip and inter-chip transmission characteristics as shown in Fig. 1. A dielectric substrate waveguide was formed under Si chips. The thickness (t_{int}) and relative dielectric constant (ε_r) of the wave guide were changed to investigate the influence on the transmission gain and radiation efficiency.

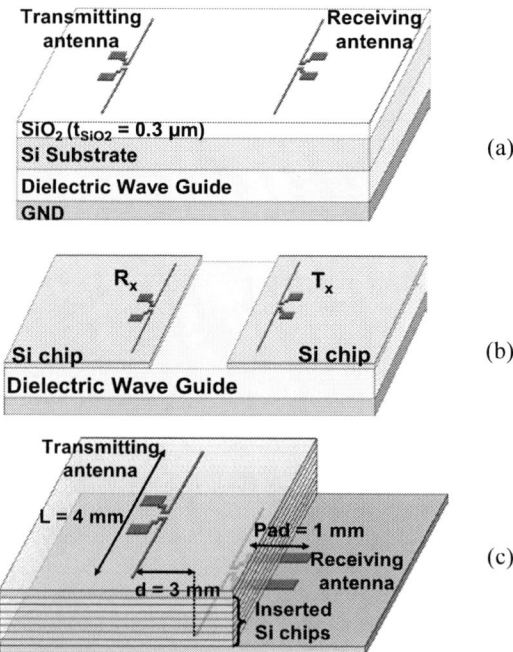

Fig. 1. Structures of Si on-chip integrated antennas. (a) Intra-chip antenna structure. (b) Inter-chip antenna structure. (c) Three-dimensional stacked chip structure.

where c is speed of light. The effective dielectric constant (ε_{eff}) is described in eq. (2).

$$\left|\varepsilon_{eff}\left(\omega\right)\right| = \sqrt{\varepsilon_{Si}^{2} + \left(\frac{\sigma}{\omega}\right)^{2}} \qquad (2)$$

When the thickness of Si substrate increased, the transmission loss increased due to the low resistivity of the Si substrate (ρ=10 Ω•cm). Therefore, S_{21} increased with decreasing Si substrate thickness at higher frequency above 15 GHz. On the other hand, S_{21} decreased with decreasing Si substrate thickness in lower frequency range. This is because the radiated wave in the near-field region propagates by capacitance coupling between transmitter and receiver antennas through the Si substrate and the effective capacitance decreased with decreasing the thickness.

Figures 3(a) and 3(b) show e_{rad} and S_{21} as functions of Si substrate thickness and substrate resistivity, respectively. The e_{rad} is defined as a ratio of radiated (P_{rad}) and input (P_{in}) powers. The e_{rad} and S_{21} were improved from 2.8% to 22.5% and from -30.3 dB to -18.4 dB by decreasing Si substrate thickness from 260 to 10 µm. It is attributed to the reduction of the loss of Si substrate. The e_{rad} and S_{21} were improved up to 31% and −15.7 dB by using high resistivity Si substrate

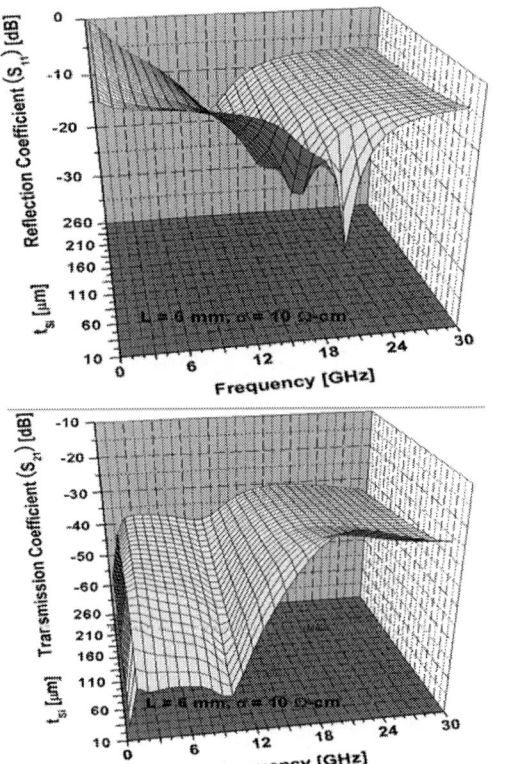

Fig. 2. Reflection coefficient (S_{11}) and transmission coefficient (S_{21}) versus frequency and silicon substrate thickness. (a) S_{11} versus frequency and thickness. (b) S_{21} versus frequency and thickness.

The thicknesses of Si substrate (t_{Si}) were also changed from 260 to 10 µm so that the transmission characteristics and radiation efficiency (e_{rad}) of the integrated antenna were investigated.

III. RESULTS AND DISCUSSION

Figures 2(a) and 2(b) show influences of Si substrate thickness and frequency on reflection coefficient (S_{11}) and transmission coefficient (S_{21}), respectively. The antenna length was L=6 mm, the antenna distance was d=5 mm and Si substrate resistivity was ρ=10 Ω•cm. The bandwidth of S_{11} at -10 dB return loss decreased and the resonant frequency (f_{res}) increased with decreasing Si substrate thickness. This is because the dielectric constant (ε_{Si}) of Si substrate was 12 so that EM wave propagated through the Si substrate and the effective wavelength decreased due to the higher dielectric constant of the substrate than that of air as eq. (1),

$$f_{res} = \frac{c}{2L\sqrt{\left|\varepsilon_{eff}\right|}} \qquad (1)$$

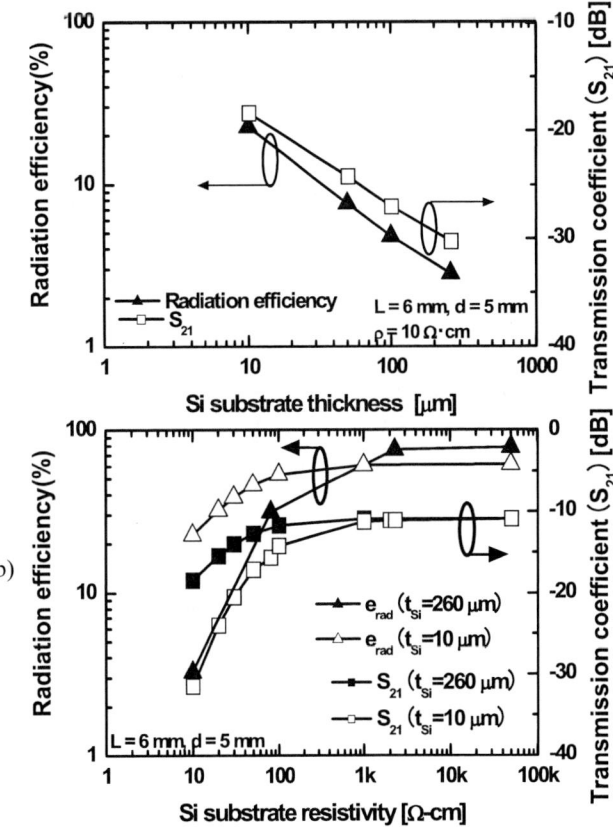

Fig. 3. (a) Dependence of radiation efficiency of a dipole antenna and transmission coefficient between two antennas on Si substrate thickness. (b) Dependence of radiation efficiency of a dipole antenna and transmission coefficient between two antennas on Si substrate resistivity.

978-1-4244-6658-0/10 $26.00 © 2010 IEEE

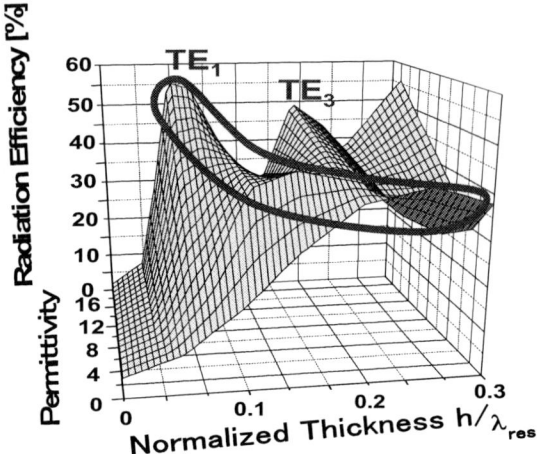

Fig. 4. Radiation efficiency as functions of permittivity and normalized thickness of the wave guide inserted between Si substrate and ground metal.

(ρ=79.6 Ω•cm). The maximum e_{rad} and S_{21} were approximately 60% and -10 dB for the Si substrate resistivity of ρ=100kΩ•cm. Figure 4 shows radiation efficiency versus permittivity and normalized thickness of wave guide which is divided by resonance wavelength (h/λ_{res}).

The thickness (t_{int}) of wave guide were changed from 1 to 10 mm and the relative dielectric constant (ε_r) were changed from 2.15 to 20. The effects of t_{int} and ε_r on the e_{rad} were investigated. The cutoff frequency (f_c) of surface wave mode is defined as eq. (3),

$$f_c = \frac{nc}{4h\sqrt{\varepsilon_r - 1}} \qquad (3)$$

The f_c is cutoff frequency of transverse magnetic (TM_n) mode for n=0, 2, 4,…, and transverse electric (TE_n) mode for n=1, 3, 5,…. The TM_0 mode is a lossy mode and it has no cutoff frequency. The e_{rad} increased with increasing the dielectric constant of the wave guide. The peak values of e_{rad} shifted with t_{int} for each ε_r, which correspond to the cutoff frequencies of TE_1 modes. Since the EM wave propagates in higher dielectric materials, it propagates in the dielectric wave guide which has higher dielectric constants than Si. The e_{rad} increased with increasing ε_r. The e_{rad} was 50% at the cutoff frequency of TE_1 mode for ε_r=20 (h/λ_{res}=0.06).

Figures 5(a) and 5(b) show S_{21} versus distance for intra- and inter-chip transmissions, respectively. Fig. 5(c) shows S_{21} versus the number of chips for a 3D stacked chip stucture. S_{21} was lower than -60 dB at the distance of 20 mm when the resistivity and the thickness of the Si substrate were 10 Ω·cm and 260 µm. The transmission gain was improved by approximately +20 dB at the distance of 20 mm by increasing the resistivity of Si substrates up to 79.6 Ω·cm. The same improvement could be obtained by thinning the thicknesss of Si substrate down to 10µm.

The attenuation rates of S_{21} with respect to distance were -0.62 dB/mm for intra-chip signal transmission, -0.64 dB/mm for inter-chip signal transmission, and -0.53 dB/mm for through stacked Si chips.

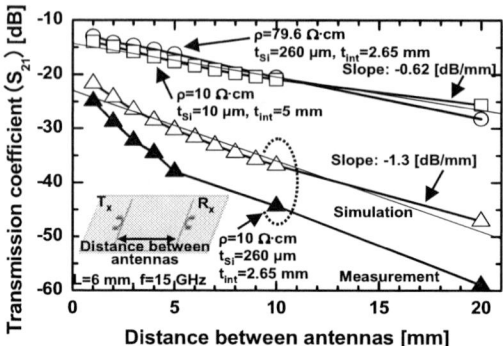

(b)

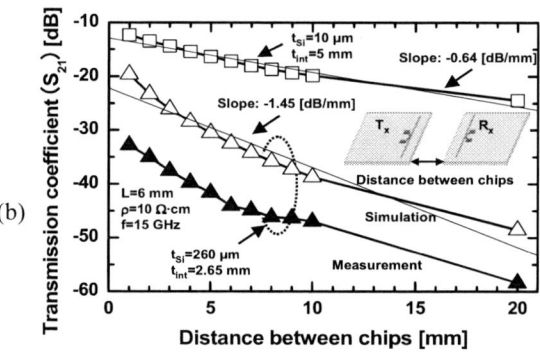

(c)

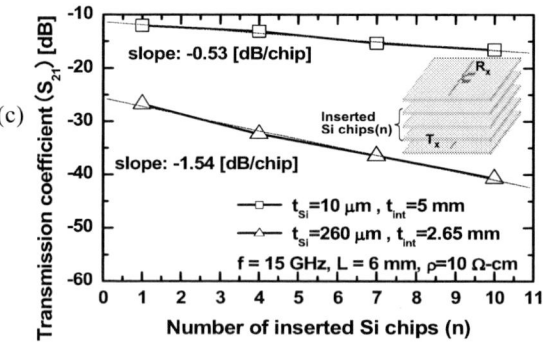

Fig. 5. Transmission coefficient (S_{21}) versus distance or line of sight for intra- and inter-chip and through Si chip transmissions. (a) S_{21} versus distance for intra-chip transmission. (b) S_{21} versus distance for inter-chip transmission. (c) S_{21} versus the number of chips for a 3D stacked chip stucture.

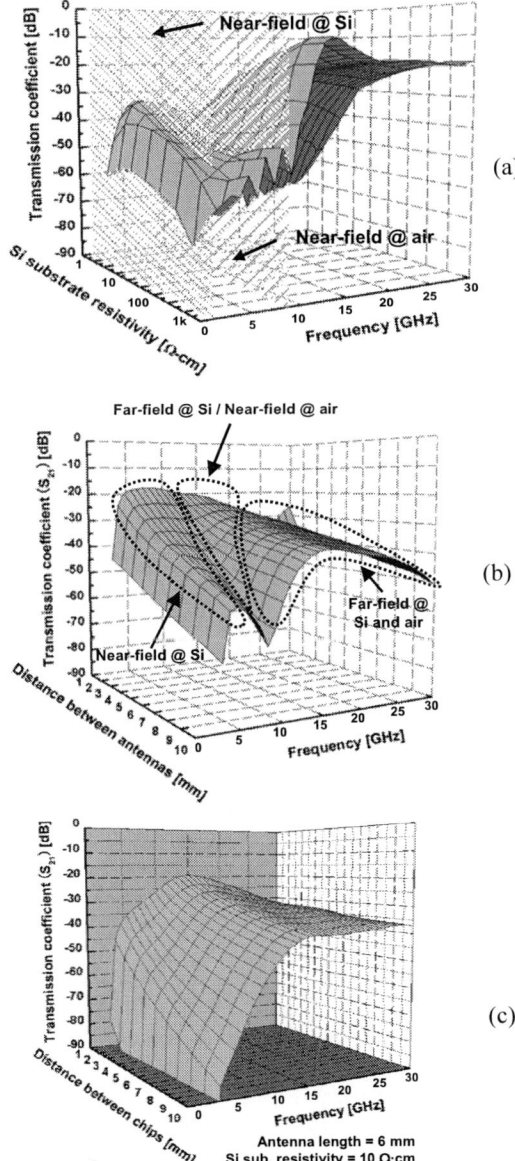

silicon and air. S_{21} decreased with increasing the resistivity along with the ridge of near-field propagation in silicon. Figure 6(b) shows S_{21} versus distance and frequency for intra-chip transmission. S_{21} decreases with increasing distance along with the steeper slope in the near-field. In the far field propagation in silicon and air, the direct wave in air and the surface wave in Silicon propagated along with the gradual slope. The valley between the two ridges is the interface of far-field in Si and near-field in air. Figure 6(c) shows S_{21} versus distance and frequency for inter-chip transmission. The direct wave in air and the surface wave in Silicon propagated along the ridge with the gradual slope in the far field as intra-chip propagation.

IV. CONCLUSION

Transmission gain of Si on-chip antenna was improved by use of a dielectric substrate wave guide having higher dielectric constant, on which Si chips are mounted. By optimizing the thickness and the dielectric constant of the wave guide, the antenna radiation efficiency and antenna transmission gain at the transmission distance of 20 mm were improved approximately up to 50% and -25 dB, respectively. The higher dielectric constant of the wave guide than that of Si leads the EM wave into the dielectric wave guide so that the gain was also improved by thinning the Si substrate thickness.

REFERENCES

[1] B. A. Floyd, C. Hung and Kenneth K. O, "Intra-chip wireless interconnects for clock distribution implemented with integrated antennas, receivers, and transmitters," IEEE J. Solid-State Circuits, vol. 37, no. 5, pp. 543-552, May 2002.

[2] A.B.M. H. Rashid, S. Watanabe and T. Kikkawa, "High Transmission Gain Integrated Antenna on Extremely High Resistivity Si for ULSI Wireless Interconnect," IEEE Electron Device Letters, Vol. 23, No.12, pp.731-733, December 2002.

[3] K. K. O, K. Kim, B. A. Floyd, J. L. Mehta, H.Yoon, C-M Hung, D. Bravo, T. O. Dickson, X. Guo, R. Li, N. Trichy, J. Caserta, W. R. Bomstad, II, J. Branch, D-J Yang, J. Bohorquez, E. Seok, L. Gao, A. Sugavanam, J.-J. Lin, J. Chen, and J. E. Brewer, "On-Chip Antennas in Silicon ICs and Their Application," IEEE Trans. Electron Devices Vol. 52, No. 7, pp.1312-1322, July 2005.

[4] T. Kikkawa, P. K. Saha, N. Sasaki, and K. Kimoto, "Gaussian Monocycle Pulse Transmitter Using 0.18 μm CMOS Technology With On-Chip Integrated Antennas for Inter-Chip UWB Communication," IEEE Journal of Solid-State Circuits, Vol. 43, No. 5, pp.1303-1312, May 2008.

[5] K. Kimoto, N. Sasaki, P. K. Saha, M. Nitta, T. Kikkawa and M. Sasaki, "Analysis of Transmission Characteristics of Gaussian Monocycle Pulse for Silicon Integrated Antennas," Japanese Journal of Applied Physics, Vol. 45, No. 4B, pp.3272-3278, 2006.

[6] N. Sasaki, K. Kimoto, W. Moriyama, and T. Kikkawa, "A Single-Chip Ultra-Wideband Receiver with Silicon Integrated Antennas for Inter-Chip Wireless Interconnection," IEEE Journal of Solid-State Circuits, Vol. 44, No. 2, pp.382-393, February 2009.

[7] P. K. Saha, N. Sasaki, K. Kimoto, and T. Kikkawa, "A 2.4 GHz Differential Wavelet Generator in 0.18 μm Complementary Metal-Oxide-Semiconductor for 1.4 Gbps Ualtra-Wideband Impulse Radio in Wireless Inter/Intra-Chip Data Communication," Japanese Journal of Appl. Phys., Vol. 45, No. 4B, pp.3279-3285, 2006.

[8] K. Kimoto and T. Kikkawa, "Signal Transmission Characteristics between Si Chips with Air Gap using Si Integrated Dipole Antennas," Japanese Journal of Appl. Phys., Vol. 45, No. 6A, pp.4968-4976, 2006.

Fig. 6. 3D diagrams of intra-chip and inter-chip transmission characteristics (S_{21}). (a) S_{21} versus Si substrate resistivity and frequency for intra-chip transmission. (b) S_{21} versus distance and frequency for intra-chip transmission. (c) S_{21} versus distance and frequency for inter-chip transmission.

Figure 6(a) shows S_{21} versus Si substrate resistivity and frequency for intra-chip transmission. Two ridges were observed. One is far-field propagation in Silicon substrate, the other is near-field propagation. S_{21} increased with increasing the resistivity along with the ridge of far field propagation in

Passive Components Integration in CMOS Technology

S. Salimy, S. Toutain, D. Averty

Institut de Recherche en Electrotechnique
et Electronique de Nantes Atlantique
(IREENA), Nantes University, 2, rue de
la Houssinière, 44322 Nantes Cedex 3,
France
siamak.salimy@univ-nantes.fr

F. Challali, A. Goullet, M-P Besland,
A. Rhallabi, J-P. Landesman
Institut des Matériaux Jean Rouxel
(IMN), Nantes University, UMR CNRS
6502, 2 rue de la Houssinière, BP 32229,
44322 Nantes Cedex 3, France

J-C Saubat, A. Charpentier,
MHS Electronics France,
92, route de Gachet 44306 Nantes
Cedex 3, France

Abstract — **The present paper aims at integrating thin films as passive components in the Back End of Line of an industrial Si-based CMOS technology while keeping limited additional technological steps. TiN_xO_y and Ti_xTa_yO thin films deposited by magnetron sputtering were respectively investigated as resistive and high-k materials dedicated to highly integrated resistors and capacitors. We report here on electrical characterizations of thin films of both materials regarding the performances criteria of the component versus thin film characteristics.**

I. INTRODUCTION

HIGH density integrated passive in standard CMOS technology is a key for the development of miniaturized Silicon Radio Frequency Integrated Circuits (Si-RFIC) such as amplifiers, oscillators and mixers. Increasing the density of the passive components and enhancing their electrical characteristics becomes today a challenge to optimize the Si-RFIC circuits. Several approaches have been investigated to follow this way. The use of Multi-Chip Module (MCM) technologies has been first investigated by stacking several chips on a motherboard, including one with processes and materials dedicated for passive components [1]. However, the MCM increases the assembly costs, and the interconnection between assembled chips needs to be well controlled and monitored. Another way to increase the density of passive components in CMOS technology is to build 3D architectures such as 3D MIM capacitors [2]. This approach, developed for Cu-Damascene process, provides large capacitance densities and allows the passive integration directly on the same chip as the active devices. However, for industrial standard Al/Cu CMOS technology, such modification of the manufacturing process requires a large investment. Moreover, opening the required trench to build the 3D capacitors cannot be done with a standard equipment of Al/Cu CMOS technology. After all, increasing the density of integrated passive components by involving solutions on the design of the component, from stack on different layers, is also limited in the considered CMOS technology, with three metal layers, contrary to technologies involving five metal layers.

Submicron CMOS industrial manufacturing processes including three metal layers have been initially developed and optimized to realize digital circuit. However, considering the increasing development of RF and mixed signal systems, the optimization of passive components performances by increasing their density to values higher than $10k\Omega/sq$ for resistances and $10fF/\mu m^2$ for capacitances represents a low cost solution to improve the capability of such technology. We propose here to integrate high density resistors and capacitors in a standard Al/Cu CMOS technology, including three metal levels. The surface resistance and capacitance constraints of the passive component are applied on the thin film material properties. Such strategy allows the use of common planar passive architectures, including manufactured components with available equipments of standard CMOS technology. In addition, integration of both active and passive devices on the same chip with the same manufacturing process would be obviously promising for further development of fully integrated RF and mixed signal devices. This study presents electrical characterizations of passive components integrated in a $0.5\mu m$ CMOS technology based on resistive (TiN_xO_y) and dielectric ($Ti_xTa_{1-x}O$) thin films dedicated to high density resistors and capacitors respectively. Thin films are deposited by physical vapor deposition technique which is already widely used for metal electrodes deposition in CMOS integrated circuits manufacturing.

A. Thin Film Resistors

Manufacturing of thin film resistors exhibiting very large range values and occupying the smallest surface area is required for the realization of CMOS multi-functional System on Chip. Low (few $100\Omega/sq$) and very high surface resistance value (up to $1M\Omega/sq$) are needed respectively for radiofrequency devices and digital systems. The potential of Titanium Oxynitride (TiN_xO_y) material to achieve weak dependence versus temperature has been reported in literature [3]-[4]. However, very few studies are dedicated to the resistance sensitivity of TiN_xO_y thin films based resistors to the applied voltage and typical density reported for TiN_xO_y

978-1-4244-6658-0/10 $26.00 © 2010 IEEE

Si-based resistors is limited to 2.5kΩ/sq [4]. In addition, to our knowledge, the integration of TiN_xO_y as resistive thin films in an industrial standard CMOS process has never been reported. In that study, TiN_xO_y resistive thin films were integrated in the Back End Of Line (BEOL). We investigated the effect of oxygen content in the film on the square resistance value. Voltage linearity versus material chemical composition was also considered.

B. Thin Film Capacitors

The development of high-κ materials in thin film is one alternative to increase capacitance density while maintaining acceptable electrical performances [5]. $Ti_xTa_{1-x}O$ has been recently identified as one promising high-κ dielectric to reach the required performances [6]-[8]. In addition, titanium and tantalum oxides are already integrated in CMOS technology and will make easier the integration of this material in the CMOS manufacturing process. We present here the extraction of intrinsic dielectric constant of $Ti_xTa_{1-x}O$ thin films from electrical characterization on MOS capacitors. The proposed extraction method allows overcoming the parasite contribution of silicon oxide interfacial layer formed under the $Ti_xTa_{1-x}O$ thin film. Electrical characteristics like capacitance density and leakage current are compared with the targeted specifications announced by the ITRS roadmap for 2015 [9].

II. Experimental

TiN_xO_y thin films were deposited at 25°C and 350°C by DC magnetron sputtering of a titanium target in reactive $Ar/N_2/O_2$ plasma, with a power density ranging from 2.5 to 4.5 W.cm^{-2} and a deposition pressure in the 1×10^{-3} to 2.7×10^{-3} mbar range. A 30mn post deposition annealing at 450°C was performed on selected TiN_xO_y films. The integration of a 50 nm thick TiNxOy resistive layer was performed in a 0.5μm-CMOS process by introducing the oxide Inter Metal Dielectric (IMD) between two BEOL metal layers, including W plugs for connection to the Al/Cu metal.

$Ti_xTa_{1-x}O$ thin films were deposited at 350°C on n-type silicon substrates by RF reactive sputtering of a $Ti_{0.6}Ta_{0.4}O$ target in reactive Ar/O_2 mixture, with a power density ranging from 2.5 to 4.5 W.cm^{-2} and pressure deposition in the 9.6×10^{-4} to 4.5×10^{-3} mbar range. Electrical measurements were performed on MOS capacitors, where square electrodes

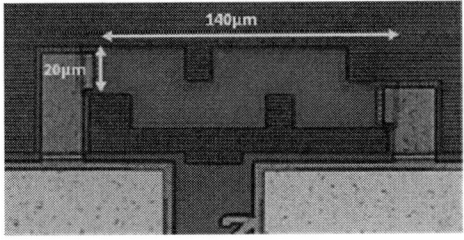

Fig.1 Optical image of a 10-square resistor integrated in the BEOL of 0.5μm CMOS technology.

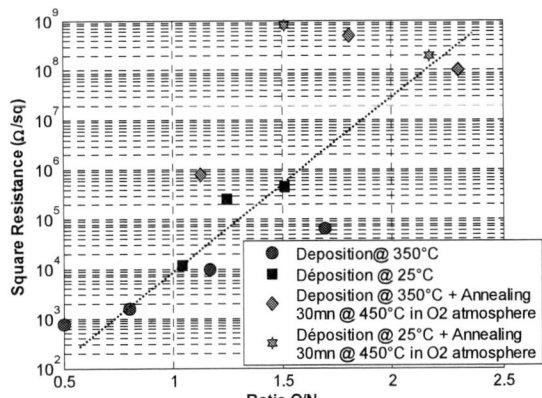

Fig.2 Variation of square resistance versus the O/N ratio in TiON thin films integrated in CMOS technology.

were elaborated by evaporation of 500nm thick and 500μm wide aluminum contacts (i.e. surface of 2.5×10^{-4} cm²). $Ti_xTa_{1-x}O$ thin films with thickness ranging from 25 to 320nm have been considered.

The bulk chemical compositions of TiN_xO_y and $Ti_xTa_{1-x}O$ thin films were determined by Energy Dispersive Spectroscopy (EDS) with a JEOL 5800-LV at 7kV. Electrical I-V characterizations on Al/ $Ti_xTa_{1-x}O$ /Si MOS capacitors and TiN_xO_y thin film resistors were performed using an Agilent 4155C parameters analyzer for leakage measurements. C-V characteristics of MOS capacitors were investigated thanks to an HP4194 impedance analyzer.

III. Results and Discussion

A. TiN_xO_y Thin Films

Fig.1 shows an optical image of a 10-square resistive TiN_xO_y thin film integrated in the Back End Of Line (BEOL) of a 0.5μm CMOS technology. Square resistance values were deduced from the ratio of resistance value extracted from I-V characterization and number of squares. Deposited TiN_xO_y thin films of various chemical compositions have been integrated in CMOS. As displayed in Fig.2, the square resistance of TiN_xO_y films, deposited at 25 and 350°C, is varying versus the O/N atomic ratio checked by EDS analysis. A 30 min ex-situ annealing at 450°C under flowing oxygen leads to a further increase of the square resistance value. As the O/N ratio is increasing from 0.4 to 2.4, the square resistance is varying over five decades from 5kΩ/sq to 500MΩ/sq. The square resistance increase appears thus related to the larger oxygen content which contributes to higher bulk resistivity. The variation of the Voltage Coefficient of Resistor (VCR) as a function of the O/N ratio is shown in Fig. 3. It appears that VCR is also correlated to the oxygen content in the film. Moreover, VCR can be negative for an O/N ratio lower than 1.5 and a corresponding square resistance larger than 1MΩ/sq. It is worth noting that such negative VCR value allows obtaining resistor with high value and a VCR coefficient almost equal to zero thanks to an

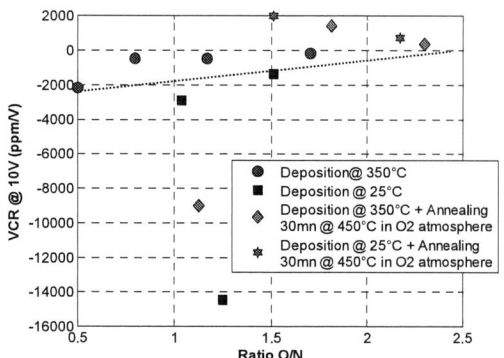

Fig.3 Variation of the VCR Coefficient versus the O/N ratio for TiN_xO_y thin films integrated as resistors in CMOS technology.

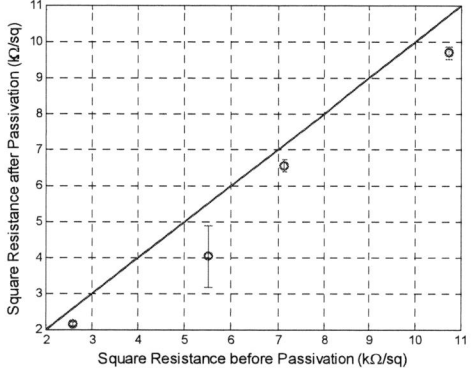

Fig.4 Comparison of square resistance value measured before and after the passivation process involving a two-layer dielectric (240 nm SiO_2/540 nm SI_3N_4) deposition and post-annealing at 390°C.

accurate control of the O/N ratio or by using two resistive layers exhibiting close chemical composition but opposite VCR values.

Further investigations of the CMOS back end process influence on the TiN_xO_y resistors were performed by depositing on selected samples a bi-layer dielectric (240nm SiO_2/540nm Si_3N_4) followed by a 45min annealing at 390°C. Square resistances were measured before and after such passivation procedure. As displayed in Fig.4, in all cases, i.e. independently of the O/N ratio value, the square resistance is lowered by the passivation process. Such variation can be attributed to a thermal effect during the annealing involved in the passivation process.

B. $Ti_xTa_{1-x}O$ Thin Films

C-V and *I-V* measurements were performed on MOS (Al/$Ti_xTa_{1-x}O$/Si) capacitors for $Ti_xTa_{1-x}O$ films with thickness in the 25-320nm range deposited on n-type silicon substrate. Similarly to previous published results [10], all MOS structures based on $Ti_xTa_{1-x}O$ films deposited in pure Ar were in short-circuit. We assume that due to the absence of oxygen in the gas phase during deposition, oxygen vacancies can be generated in the layer bulk leading to such behavior [11]. In the case of $Ti_xTa_{1-x}O$ films deposited in

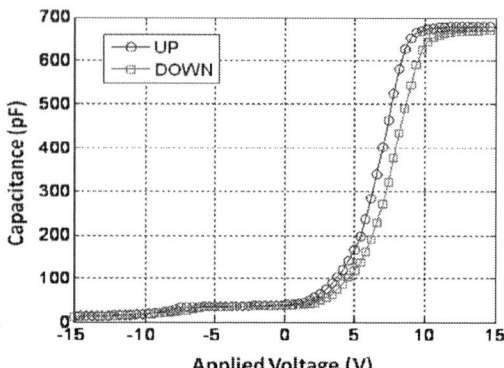

Fig.5 Forward and reverse C-V measurements for n-MOS capacitor with a 100nm thick $Ti_{0.3}Ta_{0.2}O$ film. *Deposition parameters: RF power density 4.5 W.cm^{-2}, deposition pressure 1.2x10^{-3}mbar, 20% oxygen content.*

reactive Ar/O_2 atmosphere, the permittivity tends to decrease while increasing the oxygen content from 5% to 30 % in the gas phase. Fig.5 shows an example of *C-V* curves measured at 1MHz for a MOS capacitor including a 100nm thick $Ti_xTa_{1-x}O$ film, obtained at a RF power density of 4.5 W.cm^{-2}, deposition pressure of 1.2x10^{-3}mbar and 20% oxygen content. The three typical regimes of a MOS capacitor are recorded with a small hysteresis between the forward and reverse measurements. The extracted flat band voltage V_{FB} is equal to 5.8 V, significant for a trap density located in the bulk dielectric or at the interfaces Si/dielectric or dielectric/Al. The accumulation regime allows deducing a dielectric permittivity for voltage higher than 10V, i.e. with silicon substrate acting as bottom electrode. Dielectric constant was extracted for $Ti_{0.3}Ta_{0.2}O$ films with thicknesses in the 25 to 320nm range as shown in Fig.5. The permittivity variation versus the physical oxide thickness (as measured from SEM observations) is shown in Fig.6. The intrinsic permittivity of $Ti_xTa_{1-x}O$ material was determined while considering the Capacitance Equivalent Thickness (CET). The slope of plot in Fig.7 allows extracting an intrinsic

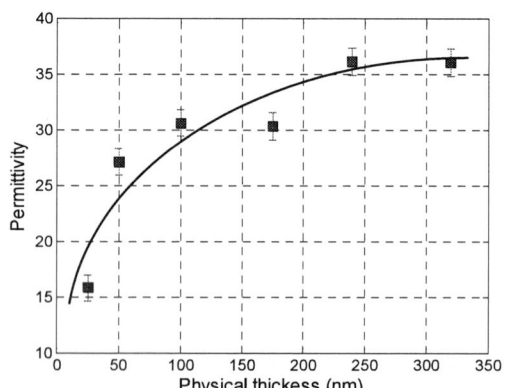

Fig.6 Variation of permittivity versus physical thickness of $Ti_xTa_{1-x}O$ thin films extracted from the C-V curves at 1MHz on MOS (Al/ $Ti_xTa_{1-x}O$ /Si) structures in accumulation regime.

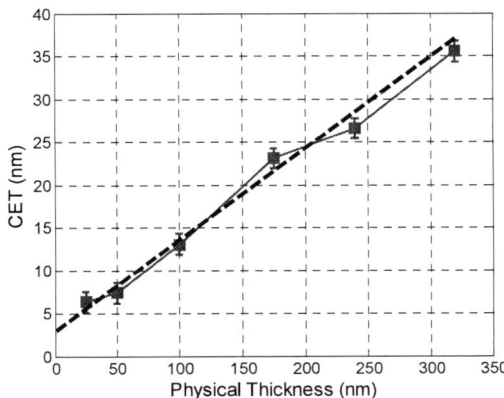

Fig. 7 Correlation between capacitance equivalent thickness (CET) versus physical thickness of $Ti_xTa_{1-x}O$ thin films

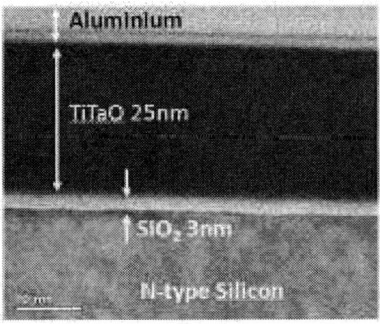

Fig.8 HRTEM image of a (Si/ $Ti_xTa_{1-x}O$ (25 nm)/Al) structure starting from the bottom interface on n-Si substrate with evidence for a 3 nm thick interfacial SiO_2 layer.

relative permittivity of 40 for $Ti_xTa_{1-x}O$ films with thickness ranging from 25 to 320nm. A capacitance density of $15fF/\mu m^2$ can thus be achieved with $Ti_{0.3}Ta_{0.2}O$ thin films. The intersection of a plot linear fit (dashed line) with the Y axis leads to an interfacial layer thickness of 3nm, which thus modify the stacked equivalent permittivity of the structure shown in Fig.5. HRTEM observations of a MOS (Al/ $Ti_xTa_{1-x}O$ (25 nm)/Si) structure confirm the existence of a 3 nm thick interfacial dioxide layer at the bottom TiTaO/Si interface (Fig. 8). For 100nm thick $Ti_{0.3}Ta_{0.2}O$ films, a leakage current density of $10^{-6}A/cm^2$ has been measured at 5V and 25°C.

IV. CONCLUSION

CMOS high density resistive and dielectric thin layers have been developed using PVD process compatible with CMOS technology. On one hand, TiN_xO_y resistive thin films, deposited by magnetron sputtering in $Ar/O_2/N_2$ mixture, have been integrated in the BEOL of a 0.5μm-CMOS technology. An accurate control of the O/N atomic ratio allows varying the square resistance value over a large range from 500Ω/sq to 500MΩ/sq. Such large scale of surface resistance values achieved with TiN_xO_y material leads to a significant

improvement of the actual state of art. Our results make thus TiN_xO_y as a very promising material for both RF and digital applications, since such applications are requiring low or high resistance values respectively. In addition, resistors with very small value of VCR coefficient can be manufactured. On the other hand, $Ti_xTa_{1-x}O$ thin films, deposited by magnetron sputtering in Ar/O_2 plasma, were also developed for integrated capacitors. Intrinsic dielectric constant of 40 has been obtained for $Ti_{0.3}Ta_{0.2}O$ thin films with thickness ranging from 25 to 320nm, allowing achieving a capacitance density of $15fF/\mu m^2$ for 25nm thick films. The concept of capacitance equivalent thickness was involved to identify a 3 nm thick interfacial dioxide layer, which has been validated by HRTEM observations of a MOS (Al/ $Ti_xTa_{1-x}O$ (25 nm)/Si) structure. Performances of passive components achieved in this study are in good agreement with ITRS roadmap requirements for 2015.

REFERENCES

[1] B.K.Gilbert and G.W. Pan, "MCM Packaging for Present and Next Generation High Clock-Rate Digital and Mixed-Signal Electronic System: Areas for Development", *IEEE Trans. on Microwave Theory and Techniques*, Vol.45, No10, pp 1819-1835, Oct.1997.

[2] M. Thomas , A. Farcy, N. Gaillard, C. Perrot, M. Gros-Jean, I. Matko, M. Cordeau, W. Saikaly, M. Proust, P. Caubet, E. Deloffre, S. Cremer, S. Bruyère, B. Chenevier, J. Torres, "Integration of a high density Ta2O5 MIM capacitor following 3D damascene architecture compatible with copper interconnects", *Microelectronic Engineering*, Vol. 83, pp. 2163–2168, Oct. 2006.

[3] N. D. Cuong, D-J. Kim, B.-D. Kang, S.-G. Yoon, " Structural and Electrical Properties of TiNxOy Thin-Film Resistors for 30 dB Applications of Pi-type Attenuator", *Journal of The Electrochemical Society*, No.153, Vol.9, pp.856-859, 2006

[4] A. Shibuya, K. Matsui, K. Takahashi, and A. Kawatani, "Embedded TiNxOy Thin-Film Resistors in a Build-Up CSP for 10 Gbps Optical Transmitter and Receiver Modules", *IEEE Transaction on Advanced Packaging*, Vol.25, No.3, pp 448-453, Aug. 2002.

[5] C. H. Ng, C.-S. Ho, S.-F. S. Chu, S.-C. Sun, "MIM Capacitor Integration for Mixed-Signal/RF Applications", *IEEE Trans. On Electron Devices*, Vol. 52, No. 7, July 2005.

[6] K.C. Chiang, C.H. Lai, A. Chin, H.L. Kao, S.P. McAlister, C.C. Chi, "Very high density RF MIM capacitor compatible with VLSI", *Microwave Symposium Digest, IEEE MTT-S International*, pp4, 2005

[7] K.C. Chiang, A. Chin, C.H. Lai, W.J. Chen, C.R. Cheng, B.R. Hung, C.C. Liao, "Very high K and high density TiTaO MIM capacitors for analog and RF applications", *VLSI Technology, 2005. Digest of Technical Papers. Symposium*, pp62 – 63, 2005.

[8] K. C. Chiang, C. C. Huang, A. Chin, W. J. Chen, S. P. McAlister, H. F. Chiu, J.-R. Chen, and C.C. Chi, "High-_ Ir/TiTaO/TaN Capacitors Suitable for Analog IC Applications", *IEEE Electron Device Letters*, IEEE Vol.26, No.7, pp504-506, July 2005.

[9] The International Technology Roadmap for Semiconductors, 2003. Semicond. Ind. Assoc.

[10] L. Martinu, D. Poitras, "Plasma deposition of optical films and coatings: A review", *J. Vac. Sci. Technol.*, Vol. A, No. 18, pp.2619-2644, 2000.

[11] F. Challali, M.P. Besland, D. Benzeggouta, C. Borderon, M.C. Hugon, S. Salimy, J.C. Saubat, A. Charpentier, D. Averty, A. Goullet, J.P. Landesman, Thin Solid Films (2009), in press, doi:10.1016/j.tsf.2009.12.045.

Drain Current Variability in 45nm Heavily Pocket-implanted Bulk MOSFET

Cecilia M. Mezzomo[1,2], Aurélie Bajolet[1], Augustin Cathignol[3] and Gérard Ghibaudo[2]

[1] STMicroelectronics, 850 rue Jean Monnet, 38926 Crolles, France
[2] IMEP-LAHC, Minatec, INPG, 3 Parvis Louis Néel, BP 257, 38016 Grenoble, France
[3] IBM France, 870 rue Jean Monnet, 38926 Crolles, France

Author's e-mail: cecilia.mezzomo@st.com

Abstract — **Pocket architecture is a useful technique to eliminate short channel effects to provide smaller transistors sizes. However, it has been shown that it has an important drawback on mismatch. In this paper, the drain-current mismatch $\sigma(\Delta Id/Id)$ is characterized for transistors without pockets and for heavily pocket-implanted transistors. These characterizations are performed from linear to saturation regime. A drain-current mismatch model as a function of drain voltage valid from weak to strong inversion region is also presented. For the first time, the drain current mismatch parameter is analyzed from linear to saturation regime for pocket devices. Thus, a comparison between transistors without pocket and transistors with pocket is performed and an important drain-current mismatch enhancement in the latter case is reported and discussed.**

I. INTRODUCTION

Transistors are scaled down to deep-submicrometer feature size to provide more performance and smaller circuits. To provide smaller transistor sizes, pocket implant technology has been introduced, reducing the short channel effect. By scaling down, the variability between two supposed identical transistors increases considerably and it becomes a major difficulty for process development. This local variation is commonly known under the terms of mismatch, variability or statistical fluctuations [1]. Among these fluctuations, the random dopant fluctuations, the line edge roughness and the poly gate granularity have been shown to be dominant ones in modern MOSFET technologies [2]. To quantify these fluctuations, pioneer models [1][3] have demonstrated a $(WL)^{-1/2}$ geometry scaling law in current mismatch, where W is the gate width and L the gate length, based on the fluctuation of SPICE model parameters, Vt (threshold voltage) and β (gain factor).

The increased Vt mismatch for short device has been widely observed and was explained by the global increase of impurity concentration in the channel [4]. For long lengths, the mismatch was shown not following the scaling law anymore, and this was attributed to the pocket implants [5][6][7]. Reference [8] pointed out that pockets dopants shrink the area which controls Vt, while [9] observed the impact of the high potential barriers on long transistors and

the gate bias dependence. Therefore, all these studies show the importance of the pocket-implanted transistors impact on the mismatch.

A mismatch model valid for all regions of operation has been published [10]. It describes mismatch in the drain current ($\Delta Id/Id$) as a function of mismatch in the threshold voltage (ΔVt) and a mismatch in the current factor ($\Delta\beta/\beta$). It analyzes the mismatch as a function of Vg and of the transistor geometries. This model gives a good estimation of mismatch in saturation for not too high inversion levels. In, [11][12][13] new approaches were presented based on an accurate physics-based model, which allows the assessment of mismatch from process parameter and is valid for any operating region. They analyzed the $\sigma(\Delta Id/Id)$ mismatch for drain-to-source voltage. However, the devices analyzed do not have pocket implants.

As we could observe, the new transistors generation incorporates pocket-implanted regions and these have very significant role on variability. Also, the threshold voltage, the gain factor and the drain current mismatch are usually analyzed as a function of the transistor geometries, the gate bias or the drain current. In this context, this paper presents a quasi-analytical matching model valid for all operation regimes from linear to saturation regime and from weak to strong inversions. The characterization of the local drain-current variability in pocket-implanted 45nm bulk MOSFET technologies is also presented. For the first time, the drain current mismatch is characterized for drain bias conditions for heavily pocket-implanted devices.

II. DRAIN-CURRENT MISMATCH MODEL

A general drain current mismatch model has been developed by considering the impact of a local shift, δV_t, in a portion of the channel, of area δa, as in the RTS noise approach [14]. In this approach, the relative drain current change due to a weak local conductivity $\delta\sigma$ variation reads (1),

$$\frac{\Delta Id}{Id} = \frac{\delta a}{WL} \cdot \frac{\delta\sigma}{\sigma} = \frac{1}{\sigma}\frac{\partial\sigma}{\partial Vt} \cdot \frac{\delta a}{WL} \cdot \delta Vt \qquad (1)$$

978-1-4244-6658-0/10 $26.00 © 2010 IEEE

By integrating these fluctuations over the channel surface yields for the drain current variance (2):

$$\sigma_{\frac{\Delta Id}{Id}}^{2} = \int \left(\frac{\partial \ln(\mu_{eff} Q_i)}{\partial Vt} \right)^2 \cdot \frac{A_{Vt}^2}{(WL)^2} \cdot dx\, dy \qquad (2)$$

where A_{Vt} is Vt mismatch parameter, μ_{eff} the mobility and Q_i the inversion charge.

Within the gradual channel approximation, accounting for current conservation along the channel enables the drain current variance to be equated to (3),

$$\sigma_{\frac{\Delta Id}{Id}}^{2} = \frac{\int_0^{Vd} \left(\frac{\partial \ln(\mu_{eff} Q_i)}{\partial Vt} \right)^2 \cdot \frac{A_{Vt}^2}{W L} \cdot \mu_{eff} \cdot Q_i \cdot dUc}{\int_0^{Vd} \mu_{eff} \cdot Q_i \cdot dUc} \qquad (3)$$

where Uc is the quasi Fermi level shift along the channel.

The drain current variance has been converted to calculate the equivalent gate voltage fluctuations variance, thanks to (4) [9], versus drain voltage for various cases: i) constant mobility (δVt & μ_{eff}=const.), ii) E_{eff}(Vg) dependent mobility (δVt & μ_{eff}(F_{eff})) and iii) correlated doping mobility fluctuations (δVt & $\delta\mu_{eff}$). In the latter case, the term (5) in Eq. 3 is multiplied by the factor (6), Na being the doping level. In all cases A_{Vt} has been evaluated from random doping fluctuations [15]. Fig. 1 represents these three cases numerically. In addition to the shift between the curves it is important to notice the discrepancy between them in the shape of the curve in the linear region.

$$\sigma_{\Delta Vg} \sqrt{WL} = \frac{\sigma_{\Delta Id/Id}}{(g_m/Id)} \sqrt{WL} \qquad (4)$$

$$\frac{\delta \ln(\mu_{eff} Q_i)}{\delta Vt} \qquad (5)$$

$$1 + \frac{\mu_{eff} / \delta Na}{\delta Vt / \delta Na} \qquad (6)$$

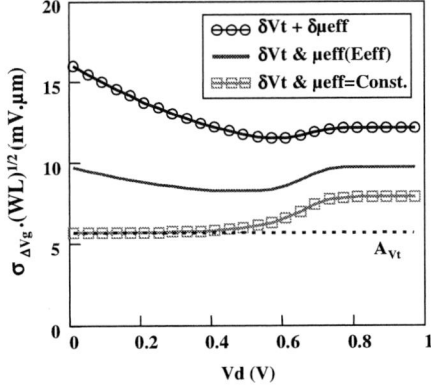

Figure 1.   Normalized drain-current variance versus Vd (Vg=1.2V, Na=10^{17}/cm^3, toxeff=2nm).

III.   EXPERIMENTAL SETUP

A.   Test Structures

Dedicated test structures are used to provide characterization of local fluctuations. These structures are composed of two supposedly identical MOSFET transistors (transistor pair), at a minimum design rule, in an identical environment. They have common bulk and separate gate (G), drain (D), source (S) and symmetrical connections.

The devices under test are 45nm bulk N-MOSFET. Some of them have heavily pocket-implanted regions, called here as "pocket (P)" devices. The gate width varies from 0.12μm to 5μm and the gate length from 0.04μm to 5μm. The other do not have pocket-implanted regions, called here as "no-pocket (NP)" devices. The gate width of these devices varies from 0.12μm to 5μm and the gate length from 0.11μm to 10μm.

Measurements are performed at ambient temperature (25°C).

B.   Test Setup

To ensure an acceptable statistical accuracy of matching measurements, at least 70 pairs of transistors are measured. Considering δP (δP = ΔP or ΔP/P) the difference of P parameter measured between the two paired devices, matching characterization studies result in evaluating both the mean δP and the standard deviation $\sigma(\delta$P) from δP Gaussian distribution.

In this paper, two mismatch parameters are used. The first one is the ΔId/Id, which is extracted directly from Id-Vd transistors characteristics. The drain voltage studied varies from linear to saturation regime. Then, the standard deviation $\sigma(\Delta$Id/Id) is estimated. This drain-current mismatch parameter is analyzed for different gate voltage. The other mismatch parameter is the threshold voltage Vt. In linear region, it is extracted by the extrapolation method, at maximum of the transistor transconductance. In saturation region, it is defined as the intersection of $\sqrt{Id} = 0$ line and the tangent line of $Vg - \sqrt{Id}$ curve at the point where the slope of the curve is maximum [16]. Then, the $\sigma(\Delta$Vt) is estimated.

IV.   CHARACTERIZATION OF THE DRAIN CURRENT VARIABILITY

Commonly, the threshold voltage, the gain factor and the drain current mismatch are analyzed as a function of the transistor geometries, the drain current or the gate bias. For the first time, the drain-current variability as a function of drain bias (linear to saturation regime) is characterized for heavily pocket-implanted devices in addition to the no-pocket devices. To perform these characterizations, the drain-current variability behavior is first analyzed for no-pocket devices, followed by the analysis of the pocket devices.

A. Drain-current variability on no-pocket transistors

The drain-current fluctuations σ(ΔId/Id) as a function of the drain-to-source voltage for no-pocket transistors is presented in Fig. 2. The Vd range is from 25mV (linear region) to 1.1V (saturation region), for different gate bias (Vg from 0.6V to 1.1V). The Id-Vd is also plotted to observe the transistor regimes. In this case, the current standard deviation increases until the transistor get into the saturation region. After this point, the standard deviation is constant. These mismatch behavior is in agreement with the one presented by [14]. In Fig. 2, only one transistor geometry is shown as the other no-pocket transistors have this same behavior.

B. Drain-current variability on pocket transistors

The drain-current fluctuations σ(ΔId/Id) as a function of Vd is now analyzed for pocket transistors. Fig. 3 shows these fluctuations and the correspondent Id(Vd) for a transistor with long gate length (L=5μm). This transistor has the same geometry as the no-pocket device presented in Fig. 2. In this case, the standard deviation increases much more than for the same geometry in a no-pocket device and it starts to decrease when the transistor passes from linear to saturation region, forming a hump. After this point, the standard deviation tends to a constant value.

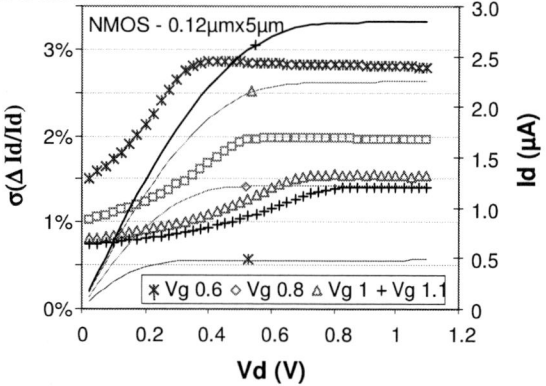

Figure 2. The σ(ΔId/Id) fluctuations as a function of drain bias (left axis, dashed lines) for a no-pocket device for different gate voltages, with W/L = 0.12μm/5μm. The Id(Vd) corresponding to each Vg is also plotted (right axis, solid lines).

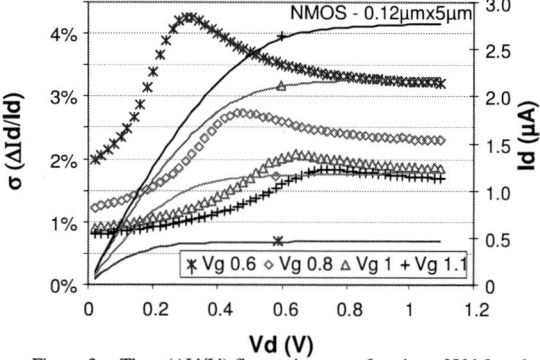

Figure 3. The σ(ΔId/Id) fluctuations as a function of Vd for a long transistor (W/L = 0.12μm/5μm) is shown for a pocket device for different gate voltages (left axis, dashed lines) and the correspondent Id(Vd) (right axis, solid lines).

In Fig. 4, the same plot is shown but for small transistor gate length (L=0.05μm). In this case, as the gate length is small, the source pocket region is close to the drain pocket region, even superimposed. Thus, the channel here is homogeneous as a no-pocket device, but heavily doped. Therefore, it presents the same behavior than a no-pocket device: the current standard deviation increases until the transistor get into the saturation region. After this point, the standard deviation is constant. This shows that the hump appears only for devices with long channel. It is attributed to the non-homogeneous channel, due the presence of heavily pocket-implanted regions in a weakly doped channel. Reference [9] reported the impact on Vt mismatch of long pocket devices as a function of Vg on linear region. The Vt fluctuations are extracted thanks to Id fluctuations. They have noticed unusual fluctuations for long devices, which are attributed to the presence of additional pocket induced potential barriers at both source and drain and these potential barriers are responsible for the high electrical fluctuations level in relatively long transistors.

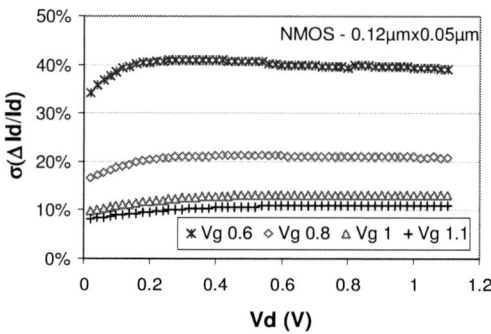

Figure 4. σ(ΔId/Id) fluctuations as a function of drain bias for a short pocket device for different gate voltages, with W/L = 0.12μm/0.05μm.

C. Gate voltage fluctuations

To analyze long transistors, [9] introduced a gate-bias-dependent threshold voltage (7), which provides the Vg fluctuations, extracted thanks to Id fluctuations and g_m/Id extraction.

$$\sigma_{\Delta Vg} = \frac{\sigma_{\Delta Id/Id}}{g_m/Id} \qquad (7)$$

Therefore, the σ(ΔId/Id) is now normalized by g_m/Id to perform the Vg fluctuations analysis (Fig. 5). In this figure, the σ(ΔVt) in the linear and in the saturation region for pocket devices is also plotted. In linear region (Vd=50mV), the difference between the σ(ΔId/Id)/(g_m/Id) for pocket and no-pocket devices decreases while Vg increases, as explained in [9]. For both pocket and no-pocket devices, $\sigma_{\Delta Vg}$ decreases from linear to saturation region. The hump noticed before for pocket devices is also observed. Analyzing the g_m/Id behavior (Fig. 6), no difference among pocket and no-pocket transistors for any Vg values is observed. Therefore, describing the drain-current mismatch simply as $(g_m/Id)\sigma(\Delta Vt)$ is not sufficient to model the mismatch behavior versus drain bias. Note from Fig. 1 that $\sigma_{\Delta Vg}$

mismatch with Vd cannot also be modeled by considering a constant mobility neither an E_{eff} dependent one. It appears from the model results of Fig. 1 that only the Vt and correlated mobility fluctuations due to random doping is appropriate for interpreting the experimental data (Fig. 5).

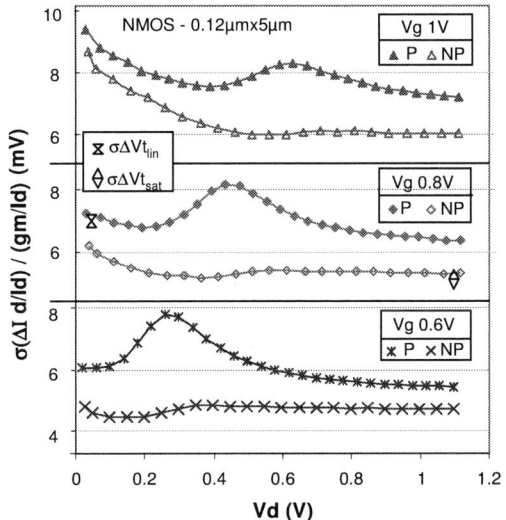

Figure 5. Gate voltage fluctuations $\sigma_{\Delta Vg}$ are shown for pocket (P) and no-pocket (NP) devices, with W=0.12µm and L=5µm. The $\sigma(\Delta Vt)$ extracted in the linear and in the saturation regime is also represented.

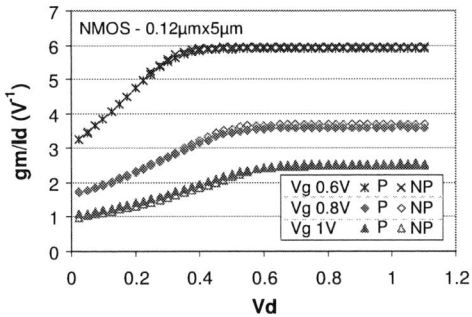

Figure 6. The g_m/Id is shown from linear to saturation region for different gate bias for pocket (P) and no-pocket (NP) devices (W/L =0.12/5µm).

V. CONCLUSION

A general drain current mismatch model has been proposed valid from linear to saturation regime and from weak to strong inversions. It has been developed by considering the impact of a local shift, δV_t, in a portion of the channel, as in the RTS noise approach. Only the case considering Vt and correlated mobility fluctuations due to random doping is appropriate for interpreting the

experimental data. In addition, the drain-current variability as a function of the drain voltage has been characterized for no-pocket and heavily pocket-implanted devices, in 45nm bulk MOSFET technology. For the first time a significantly different behavior has been shown for long pocket transistors, exhibiting more drain current fluctuations when the transistor changes from linear to saturation regime. This pocket induced extra drain current fluctuations behavior should be carefully taken into account from a design point of view.

ACKNOWLEDGMENT

This work has been partially supported by JU/ENIAC project MODERN.

REFERENCES

[1] Lakshmikumar K et al, "Characterisation and modeling of mismatch in MOS transistors for precision analog design". IEEE Journal of Solid-State Circuits, vol. 21, n° 6, pp. 1057-1066, 1986.

[2] A. Cathignol et al. "Quantitative evaluation of statistical variability sources in a 45nm technological node LP N-MOSFET", IEEE Electron Devices Letter, v. 29, n° 6, pp. 609-11, 2008.

[3] M. Pelgrom et al, "Matching properties of MOS transistors". IEEE Journal of Solid State Circuits, vol.24, n° 5, pp. 1433-40, 1989.

[4] R. Difrenza et al, "Dependence of channel width and length on MOSFET matching for 0.18 µm CMOS technology". ESSDERC, pp. 584-87, 2000.

[5] J. Croon et al,. "A simple and accurate deep submicron mismatch model". ESSDERC, pp. 356-59, 2000.

[6] K. Rochereau et al. "Impact of pocket implant on MOSFET mismatch for advanced CMOS technology". ICMTS, pp. 123-26, 2004.

[7] Stolk P et al, "Modeling Statistical Dopant Fluctuations in MOS Transistors". IEEE Transactions on Electron Devices, v. 45, n° 9, pp. 1960-71, 1998.

[8] T. Tanaka et al, "Vth fluctuation induced by statistical variation of pocket dopant profile", IEDM, p. 271 – 274, 2000.

[9] A. Cathignol et al, "Abnormally high local electrical fluctuatios in heavily pocket-implanted bulk long MOSFET". Journal of Solid-States Electronics, v. 53, n° 2, pp. 127-33, 2009.

[10] J. Croon et al, "Influence of doping profile and halo implantation on the threshold voltage mismatch of a 0.13 µm CMOS technology". ESSDERC, pp. 579-82, 2002.

[11] H. Klimach et al, "Consistent model for drain current mismatch in MOSFETs using the carrier number fluctuation theory", Proc. 2004 IEEE Int. Symposium on Circuits and Systems, 2004.

[12] C Galup-Montoro et al, "A compact model of MOSFET mismatch for circuit design". IEEE journal of solid-state circuits, vol. 40, n°8, pp. 1649-1657, 2005.

[13] H. Klimach et al, "MOSFET Mismatch Modeling: A New Approach", IEEE Design & Test, v.23 n.1, p.20-29, 2006.

[14] O. Roux et al, Model for drain current RTS amplitude in small area MOS transistors, Solid State Electronics, 35, 1273, 1992.

[15] A. Asenov et al "Suppression of random dopant-induced threshold voltage fluctuationsin sub-0.1-µm MOSFET's with epitaxial and δ-doped channels", IEEE Transactions on Electron Devices, v. 46, n° 8, pp. 1718-24, 1999.

[16] Y. Shimizu et al, "Test structure for precise statistical characteristics measurement of MOSFETs", ICMTS, v. 15, pp. 49-54, 2002.

Mismatch Sources in LDMOS Devices

Pietro Andricciola and Hans Tuinhout
Device Modeling and Characterization
NXP Semiconductors, Eindhoven, The Netherlands
Email: pietro.andricciola@nxp.com

Abstract—This paper discusses the influence of different sources of DC parametric mismatch in an LDMOS. By comparing measurements and statistical simulations the impact on mismatch of the most important fluctuation causes is qualitatively evaluated. We demonstrate that, whereas the shape of the doping profile in the channel has little effect, both interface states and series resistances play a major role in the mismatch. This work forms a crucial first step towards a better understanding of the random fluctuation mechanisms present in LDMOS devices used in MMICs.

I. INTRODUCTION

The lateral diffused MOS transistor (LDMOS) is widely used in base stations, radar and broadcast applications because of its capabilities to sustain high voltages, delivering substantial power and having good RF performance [1]. However, in monolithic microwave integrated circuits, these devices are employed in analog circuit block implementations under relatively "low power" conditions. The DC parametric mismatch performance becomes important for the functionality of such blocks. Yet the information in the literature about matching of LDMOS [2] is scarce and the fluctuation sources are not analyzed. In this paper we show, for the first time, simulations and measurements on the mismatch fluctuation behavior of LDMOS transistors. In particular, we address the influence of channel doping profiles, random doping fluctuations, interface states fluctuations and series resistance on the different operating regions of the device. The impact of these sources of fluctuation is verified using statistical device simulations and analyzed through parameter extraction and mismatch signatures applied both on these simulations and measurements.

II. DEVICE DESCRIPTION

The device investigated in this work is representative of a class of RF-LDMOS transistors fabricated by NXP Semiconductors. A slightly simplified version of the actual LDMOS is reproduced with Synopsys' Structure Editor and simulated with the 2-D Sentaurus Device simulator [3]. A schematic cross section of such a structure is shown in Fig. 1. The channel region is defined by the lateral diffusion of a p-type implantation from the source side of the transistor. This means that the doping in the channel area has a strong non-uniformity along the lateral direction. It is typically about 0.3-μm long. A lightly n-type doped region (drift region) defines the end

This work has been supported by the European Commission under the Marie Curie Action MOICCO and the ENIAC MODERN project.

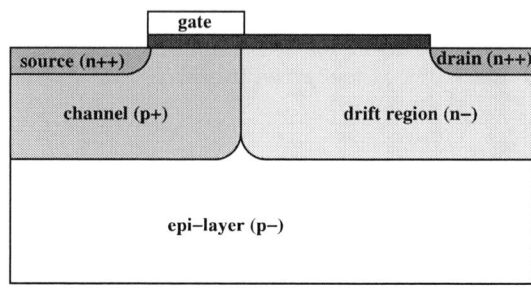

Fig. 1.   Schematic cross section of the simulated LDMOS transistor.

of the channel area and extends for about 3 μm in these simulations. The channel and the drift region are overlaid with silicon oxide (several tens of nm thick). In this study, the gate electrode covers only the actual channel area leaving the drift region gate-bias independent as is generally the case in LDMOS transistors for RF applications.

III. MEASUREMENTS

A. Experimental methodology

Parametric mismatch is evaluated by sequentially measuring two identical transistors (width of approximately 1300 μm) on two adjacent reticle placements of the same transistor. In this case implies a distance of about 1.1 mm between the two transistors of the pair. A population of 84 of these so-created "pairs" is spread out evenly over a 200-mm wafer. Measurement results reported in this paper are for two populations positioned on the same reticle field (in the rest of the paper these will be called population 'A' and 'B'). Although devices in a pair are not at the small distance normally applied for matched pairs (< 100 μm), the contribution of the deterministic gradient across the wafer was verified to be negligible compared to the random fluctuations.

The drain current of the LDMOS under test is measured at two different drain biases (V_{ds} = 50 mV and 1.05 V) while sweeping the gate voltage, V_{gs} = 0 to 3.0 V with 25-mV steps. The current of a typical device is plotted and compared with the results of TCAD simulations in Fig. 2. The simulated LDMOS reproduces the measurements very well, with the exception of low gate biases where the measurements and the simulations are dominated by junction leakage which is not tuned for this study. In the remainder of the paper the mismatch analysis will be done on measured (or simulated)

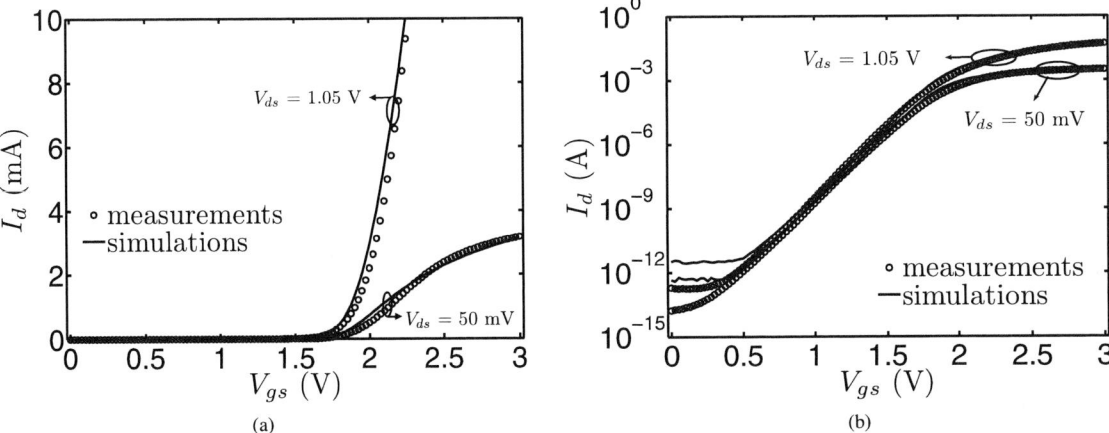

Fig. 2.   Median drain currents of measurements (symbols) and simulations (lines) for both drain-source biases plotted on linear scale (a) and on logarithmic scale (b). The scale of the graph (a) is limited to 10 mA for a better visualization of the curve at V_{ds} = 50 mV.

curves at V_{ds} = 50 mV as these are most suitable to analyze the principal device properties.

B. Results

Measurements of the full gate voltage sweeps as described in the previous section allow evaluation of the so-called mismatch signature. This signature consists of the mismatch fluctuation sweep ($\sigma_{\Delta I_d/I_d}$ vs. V_{gs} [4]), and a well-chosen autocorrelation coefficient curve (correlation between $\Delta_{I_d/I_d}(V_{gs})$ and $\Delta_{I_d/I_d}(V_{gs} = V_T)$ plotted vs. V_{gs}) [5]. Furthermore, the threshold voltage, V_T, and the current factor, β, are extracted (from measurements with V_{ds} = 50 mV) employing three-point extraction with fixed gate overdrive, as for instance described in [6].

The mismatch signatures of the two measured populations are depicted in Fig. 3. As the mismatch below 0.75 V is dominated by junction leakage fluctuations this region will not be considered hereafter. The behavior in weak and moderate inversion of both populations is similar to the one generally observed for conventional CMOS transistors. However, the signature of population 'B' indicates that an additional fluctuating component dominates the mismatch fluctuations in strong inversion (V_{gs} > 2.2 V). Also, the strong de-correlation between the mismatch at threshold voltage and at biases above threshold is more pronounced for population 'B'. This large fluctuation is ascribed to a series resistance variation attributed to a non-ideal probe-to-pad contact. This assumption will be verified with simulations in the next section.

In an attempt to avoid that these resistance fluctuations hamper the extraction of V_T and β, the gate overdrives were limited up to 0.7 V (V_{gs} < 2.5 V). This attempt led to approximately the same standard deviations of V_T and β mismatch for both populations but we had to conclude that the result for β is a grossly exaggerated value. In fact, we calculate a standard deviation of the threshold voltage mismatch of $\sigma_{\Delta V_T}$ = 2.0 mV and a standard deviation of the relative current factor mismatch of $\sigma_{\Delta\beta/\beta}$ = 1.2 %, yielding area factors of

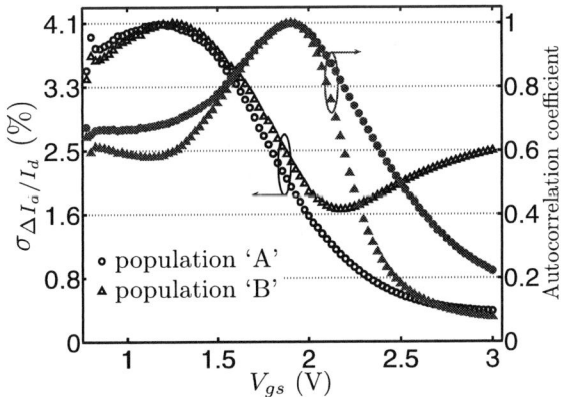

Fig. 3.   Mismatch signature of the two measured populations (V_{ds} = 50 mV).

$A_{\Delta V_T} \approx 35$ mVμm and $A_{\Delta\beta/\beta} \approx 21$ %μm when the σ's are scaled with the estimated channel area [7]. This area factor of beta mismatch is extremely large when compared to what is expected in standard CMOS technologies (less than 2 %μm [8]). Also, the relative drain current mismatch for population 'A' goes down to about 0.5 % for V_{gs} > 2.5 V (in Fig. 3), while, in principle, the drain current mismatch in the strong inversion regime should be very close to the standard deviation of beta fluctuations [7].

For these devices a combined use of three-point extraction (for threshold voltage evaluation) and mismatch signatures (for drain current mismatch evaluation) is the best analysis method. This method is therefore also used to study how intrinsic device characteristic variations such as different channel doping profiles, dopants and interface states fluctuations as well as external factors, e.g. probe-pad contact resistance fluctuations, affect LDMOS transistors mismatch fluctuations.

IV. SIMULATIONS

The study of the possible effect of well defined perturbations to the LDMOS device DC parametric mismatch is performed

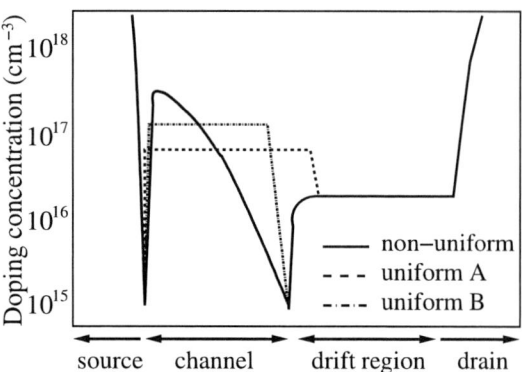

Fig. 4. Representation of the three doping profiles used in RDF-only simulations.

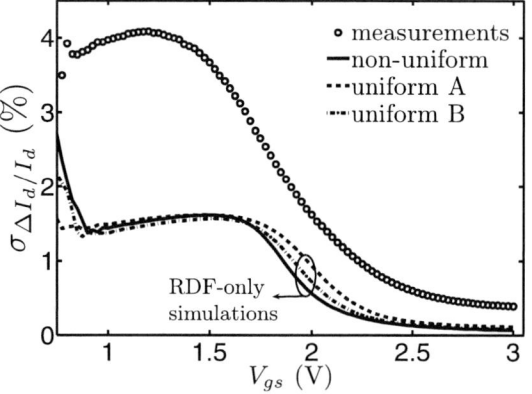

Fig. 5. Comparison between the fluctuation sweep of measured devices (population 'A') and RDF-only simulations with three different doping profiles.

using random device simulations. Three different mismatch contributors are applied to a tuned device through our doping and interface states randomizer [5] and then simulated using Sentaurus Device. The area, over which the doping and the interface states randomizations are applied, covers the channel as well as the drift region. For a good trade-off between computing time and statistical uncertainty the chosen population size is 51.

A. Influence of channel doping profile

An original concern was that the observed mismatch enhancements mentioned in section III-B could arise from the lateral non-uniformity of the channel doping. This has been reported, for instance, in long MOSFETs with pocket implantations [9]. To investigate this, random doping fluctuation (RDF) perturbations have been applied not only on a representation of a realistic laterally-diffused channel doping profile but also on two other devices with artificial constant doping levels. A schematic representation of the three lateral doping profiles is sketched in Fig. 4. The two constant doping profiles have different levels and channel lengths. The doping levels and the channel lengths have been chosen to reproduce approximately the measured electrical performance (for the

considered biases).

The fluctuation sweeps of the three configurations, simulated with RDF-only, are compared with the measurements of population 'A' in Fig. 5. For a fair comparison, the results from the device with a longer channel have been scaled with $\sqrt{L}$. All three simulated curves show similar behavior but they all deviate strongly from the measured levels. Thus, when only RDF is taken into account, shape and level of the channel doping apparently have little influence on the overall mismatch performance. Furthermore, the standard deviations of the threshold voltage mismatch and the relative beta mismatch extracted for the three configurations are approximately the same and much lower than the measured values. We calculate for example for the 'uniform A' device (easily comparable with Stolk's theory [10]), $\sigma_{\Delta V_T} = 1.1$ mV and $\sigma_{\Delta\beta/\beta} = 0.1$ %. So, calculating the area factors, we obtain: $A_{\Delta V_T} \approx 20$ mVμm ($A_{\Delta V_{T\text{theory}}} \approx 19$ mVμm) and $A_{\Delta\beta/\beta} \approx 1.8$ %μm. Given these observations, the introduction of other sources of fluctuation becomes absolutely unavoidable for a better representation of the relative drain current matching behavior.

B. Other sources of mismatch

The strength of random-fluctuation simulations lies in the possibility of combining and precisely controlling alternative fluctuation causes on a level that is unachievable by technological experiments and measurements. We introduce two additional sources of fluctuations: random interface states (RIF) and random series resistance fluctuations (RSR). It is worth pointing out that the aim of these simulations is to obtain a qualitative description of the impact that a certain fluctuating source has on the mismatch behavior, rather than give a quantitative analysis. This study is primarily focused at identifying the mechanisms that can be held responsible for the observed matching degeneration. Thus, it should not be interpreted as an alternative method for extracting interface state densities or series resistance fluctuations.

As explained in [5], interface states with random energy, concentration and position are assigned to the interface between the gate oxide and the silicon. The energy is randomly selected in the bandgap of the silicon and the nominal concentration follows a parabolic shape that ranges from 1×10 cm^{-2} at midgap to 5×10 cm^{-2} at the extremes. RIF should affect primarily the mismatch below threshold.

Above threshold, however, series resistance fluctuations may dominate the fluctuations. The currents delivered by these 1300-μm wide test devices are of the order of milliAmps (Fig. 2). This means that a significant potential can drop over a contact resistance of few tenths of an Ohm. To investigate this impact on the mismatch signature, two series resistances are randomly varied and assigned to the source and drain electrodes respectively. In order to reach the two levels of fluctuation in strong inversion (for the two measured populations), as shown in Fig. 3, two different ranges of resistances, representing a 'bad' and a 'good' probe-pad contact, are simulated. The variations, around a median value of 1 Ω, are

978-1-4244-6658-0/10 $26.00 © 2010 IEEE

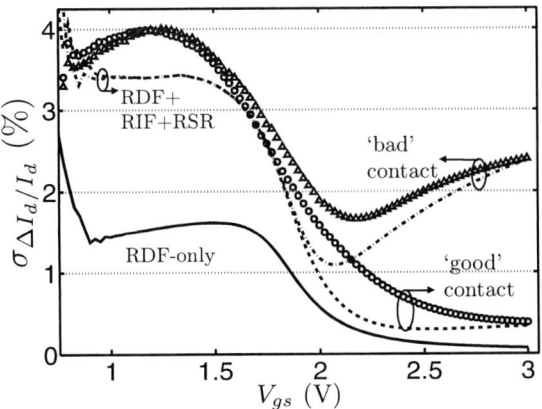

Fig. 6. Fluctuation sweeps for simulations of different sources of mismatch (lines) and measurements (symbols).

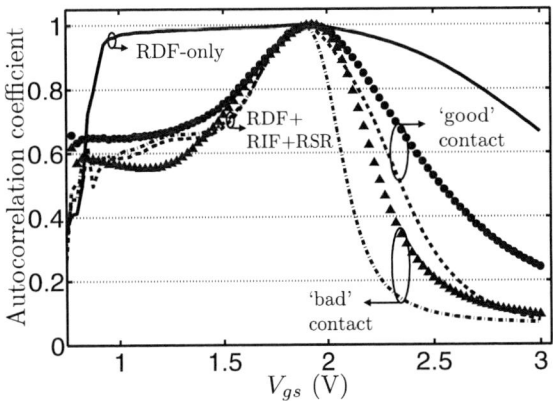

Fig. 7. Autocorrelation plot for simulations of different sources of mismatch (lines) and measurements (symbols).

chosen from a uniform distribution at ± 0.3 Ω and ± 0.05 Ω for the 'bad' and the 'good' contact respectively.

C. Discussion

Figures 6 and 7 show the mismatch signatures when RIF and RSR are added to the original RDF randomized device population with the non-uniform doping.

The level of mismatch in weak and moderate inversion is much better described by the combination of RIF and RDF (the current is too low in this region for the RSR to contribute). Also, the standard deviations of V_T and β mismatch now reach values comparable to the measurements: 2.3 mV and 0.8 % for $\sigma_{\Delta V_T}$ and $\sigma_{\Delta \beta / \beta}$, respectively. It is worth noticing that, unlike what was found for standard CMOS technologies, RIF strongly affects beta mismatch in LDMOS device populations when characterized near the peak transconductance point.

In the strong inversion region, the simulations with the two levels of RSR match very well with the two measured populations. The autocorrelation plot confirms the need of these additional independent fluctuation sources for a good description of the overall mismatch behavior. However, it cannot be denied that some discrepancies still remain in the moderate inversion region of the fluctuation sweeps (between 1.8 V and 2.5 V in Fig. 6), and between the autocorrelation plot of the simulated 'good' contact and the population 'A' ($V_{gs} > 2$ V in Fig. 7). These differences can most likely be minimized by optimizing the interface state density and energy distributions. This is however beyond the scope of this paper.

This analysis represents an important step towards performance improvements for MMICs through a better understanding of the mismatch dynamics in LDMOS devices.

V. Conclusion

This paper reports for the first time a study on parametric mismatch fluctuation causes in LDMOS devices. Measurements on transistor pairs show relatively large drain current mismatch fluctuations in all regions of operation. Three

sources of fluctuations have been analyzed by statistical simulations. We found that, if only random dopant fluctuations are taken into account, the shape and level of the channel doping cannot explain the observed mismatch behavior of this category of MOS devices. On the other hand, random interface states significantly affect the behavior of the device in subthreshold as well as in moderate inversion, also increasing the fluctuations of beta. Finally, we showed that particular care must be taken during the characterization of these devices in terms of probe-pad contact resistance, as series resistance variation can easily dominate the fluctuations at high gate biases.

Acknowledgment

The authors would like to thank their colleagues Jan de Boet and Henk Jan Peuscher for bringing this interesting subject to their attention.

References

[1] [Online]. Available: http://www.nxp.com/rfpower
[2] W. Posch, C. Murhammer, and E. Seebacher, "Test structure for high-voltage LD-MOSFET mismatch characterization in 0.35μm HV-CMOS," in *Proc. ICMTS*, 2009, pp. 96–101.
[3] *Sentaurus Device User Guide (A-9.2008)*, Synopsys, 2008.
[4] N. A. H. Wils, H. P. Tuinhout, and M. Meijer, "Influence of STI stress on drain current matching in advanced CMOS," in *Proc. ICMTS*, 2008, pp. 238–243.
[5] P. Andricciola, H. P. Tuinhout, B. de Vries, N. A. H. Wils, A. J. Scholten, and D. B. M. Klaassen, "Impact of interface states on MOS transistor mismatch," in *IEDM Tech. Dig. Papers*, 2009, pp. 711–714.
[6] J. A. Croon, *et al.*, "A comparison of extraction techniques for threshold voltage mismatch," in *Proc. ICMTS*, 2002, pp. 235–239.
[7] M. J. M. Pelgrom, A. C. J. Duinmaijer, and A. P. G. Webers, "Matching properties of MOS transistors," *IEEE J. Solid-State Circuits*, vol. 24, pp. 1433–1440, Oct. 1989.
[8] H. P. Tuinhout, "Electrical characterisation of matched pairs for evaluation of integrated circuit technologies," Ph.D. dissertation, Delft University of Technology, 2005.
[9] A. Cathignol, S. Bordez, A. Cros, K. Rochereau, and G. Ghibaudo, "Abnormally high local electrical fluctuations in heavily pocket-implanted bulk long MOSFET," *Solid-State Electronics*, vol. 53, pp. 127–133, Feb. 2009.
[10] P. Stolk, F. P. Widdershoven, and D. B. M. Klaassen, "Modeling statistical dopant fluctuations in MOS transistors," *IEEE Trans. Electron Devices*, vol. 45, no. 9, pp. 1960–1971, Sept. 1998.

Combining Process and Statistical Variability in the Evaluation of the Effectiveness of Corners in Digital Circuit Parametric Yield Analysis

P. Asenov, N. A. Kamsani, D. Reid, C. Millar, S. Roy, A. Asenov
Device Modeling Group, Department of Electronics and Electrical Engineering
Email : p.asenov@elec.gla.ac.uk

Abstract— **This paper focuses on two main types of MOSFET variability – systematic (process) and statistical (random) variability and discusses the use of process corners as a measure of yield and circuit performance. We provide a methodology for performing large-scale statistical SPICE simulations as a means of evaluating the accuracy of corners in a system dominated by statistical variability and then expand the methodology to include both systematic and statistical variability within the same large-scale SPICE simulations. This large-scale statistical/systematic approach is compared to the "global + local" statistical corner approach, which consists of statistical simulations around the process corners. Finally 2D kernel density estimates are used to extract yield data from the statistical simulations to allow energy/delay/yield optimization to be performed. This in turn highlights the deficiencies of the statistical corner approach.**

I. INTRODUCTION

Two main types of variability persist in modern silicon chips; process variability (part of which is the systematic layout and strain induced variability) and statistical (random) variability[1]. Process variability is introduced through process variations and as such is complex, but usually well characterized and in many cases predictable [2][3][4]. There are also methods available by which the impact of process variability can be ameliorated, e.g. adaptive body biasing[5][6] and gate length biasing[7]. Due to its nature, process variability presents as a gradual change in parameters across chip, across wafer and wafer to wafer (see figure 1, left plot). Statistical variability, which is exacerbated by aggressive scaling, is caused by atomic scale charge and device structure variations including random discrete dopants (RDD)[8], line edge roughness (LER)[9] and polysilicon/metal gate granularity[10]. Unlike systematic variability, statistical variability is truly stochastic and varies on a transistor to transistor level (see figure 1 right plot). It has been recently shown that in modern devices, unlike in previous technology generations, the level of statistical variability is approaching that of process variability[11].

Circuit design success in the presence of variability is evaluated through the calculation and evaluation of parametric yield. Historically a close to 100% yield guarantee has been possible by evaluating system performance at process corners, and guaranteeing these corners are within design specifications. These process corners are usually set at measured 3 sigma (σ) points of the distribution of process variability. This paper investigates the reliability of parametric yield estimates provided by process corners in the presence of statistical variability, through large-scale statistical SPICE simulations with statistical BSIM[12] compact models.

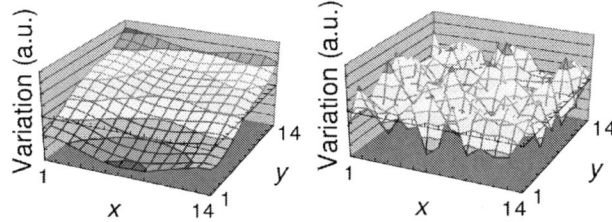

Figure 1 : Process variability (left) statistical variability (right) after M. Muracata (STARC)

II. SIMULATION METHODOLOGY

A. Process Corners

Process corners are defined in the design kits provided by foundries. They are generally based on physical measurements of large numbers of simple circuits like ring oscillators. Using these measurements it is possible to estimate process corners at set standard deviations from the mean values of the measured distribution of the desired device parameters. The usual methodology employed is to use these process corners within STA timing and power analysis calculations to determine the limits of circuit performance which guarantee a required yield. For the sake of simplicity and ease of comparison, our corners are defined as one, two and three standard deviations (σ) of normally distributed values of the threshold voltage V_t of devices, represented with the BSIM compact model parameter vth0, where the mean of the distribution is defined by the threshold voltage of a nominal compact model from the design kit. In order to calculate energy and delay data for these corners, all transistors in the spice simulation ensemble have the compact model parameter vth0 set to the desired corner value.

B. Statistical Variability

In order to obtain an accurate distribution of the timing and energy for a circuit, large ensembles of SPICE simulations are performed using Randomspice, a tool developed at the University of Glasgow which is capable of generating random

978-1-4244-6658-0/10 $26.00 © 2010 IEEE

instances of a seed circuit, allowing the accurate simulation of the impact of device variability on circuits. Simulations are performed in parallel on multiple cores of a large high performance compute (HPC) cluster. The first stage of the simulation process is to introduce statistical variability into the circuit simulation. Randomspice modifies the value of vth0 in each transistor in the circuit under test. In this case using a Gaussian distribution with a mean value taken from the simulated technology ($\mu = V_t$ of the typical model) and a σ calculated as a percentage of the baseline threshold voltage for the device technology.

In the following simulation results the assumptions made are as follows:

a) 1st order statistical variability is captured by the distribution of the device threshold voltage (vth0) - this distribution is assumed to be Gaussian

b) Corners are defined by 1 and 2 σ of the threshold voltage (vth0) parameter distribution.

c) σ is chosen as 15%, 20% and 30% of the mean vth0 (V_t) of the typical device performance.

Large ensembles of randomized circuits are simulated from which timing and energy data is extracted. In order to ensure accurate average timing and energy data, circuits are simulated over all possible input transitions. The dynamic current is measured over all transitions and average energy per transition is calculated. Delay for each transition is measured and maximum delay for each circuit instance is found. Maximum delay is selected, as this is the significant metric bounding circuit performance. In order to evaluate the effectiveness of corner based analysis, all devices in the circuit under test have their threshold voltage set to μ V_t +/-1 and 2 σ of the statistical distribution. These corners are then evaluated through single SPICE simulations. The final stage of the simulation methodology is to introduce process variability. The amount of process variability introduced was equal to that of the statistical. Due to the fact that process variability presents itself as a gradual change across chip/wafer/die-to-die the methodology assumes every transistor in the circuit is affected in the same way, so all vth0 parameters are shifted in the same direction. The correlation between the process variability related values of the n- and p-MOSFETs can vary between 0 and 100%. For process variability, a Gaussian distribution is defined as σ/μ, with typical model $V_t = \mu$, from which Randomspice it selects a new mean V_t for all devices in the circuit form this distribution. After this, random variability is again injected, in the form of statistical uncorrelated vth0 variability around this new mean.

III. TEST CIRCUITS

A. Inverter and 1-bit Adder

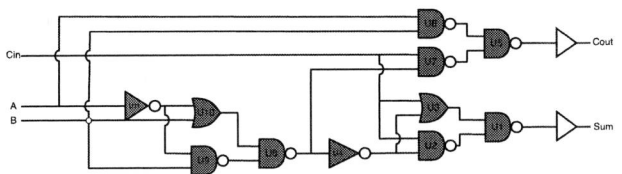

Figure 2: 1 bit adder schematic

The chosen test circuits are a simple CMOS inverter and a one bit adder shown in Fig. 2. The inverter is simulated as part of an inverter chain. The only randomized inverter in the chain, is the one being measured, this is to avoid introducing large circuit randomization effects in the inverter but maintain a realistic test signal.

The adder circuit was designed, fully laid out and extracted with RC interconnect parasitics post place and route. It is important to include parasitic connections as they have a large impact on energy and delay. The adder is treated as a "large" test circuit; it has 13 logic gates, and 52 transistors.

The simulation of a large and small circuit was chosen to illustrate the pessimism of corners in the presence of statistical variability only, and its exponential increase with the increase of circuit size.

IV. RESULT & DISCUSSIONS

A. Power-Delay Relationship

Figures 3 and 4 show average energy per transition against delay for the small and large test circuits. The plots of the adder and inverter consist of point clouds representing large-scale statistical simulations and corners. The results for the inverter, show that one-σ corners capture the edges of the inverter performance within ~95%; this trend continues to be true for increasing amounts of variability.

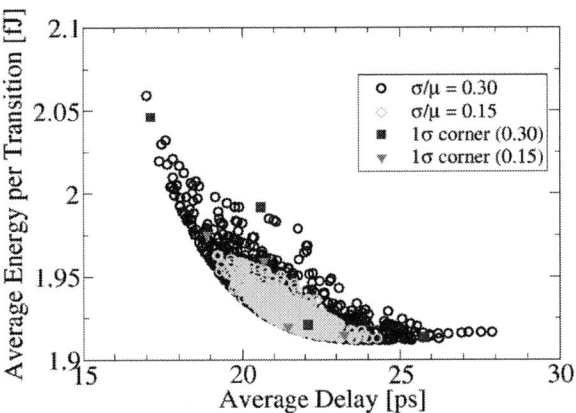

Figure 3 : Average energy per transition with maximum delay of an inverter at multiple levels of variability with corners

In the larger system, the adder, corners based on statistical variability only are overly pessimistic. This is due to the fact that uncorrelated random probability is statistically additive. In a system with 52 devices the probability that one device is at 2 σ is $P_1 = 0.05$, and that all devices are at 2 σ is represented by:

$$P_1 = 0.05$$
$$P_{52} = P_1^{52} = 2.22045 \times 10^{-68}$$

This shows that is it is practically impossible to get an adder at the 2 σ corner due to statistical variability. For the one-σ corners (all devices at 1 σ vth0 value, probability 1.85267e⁻²⁶) we get ~ 95% power enclosure [FnFp corner], we get 100% timing enclosure [SnSp corner] with a considerable amount of slack. 2 σ corners guarantee 100% power and timing enclosure, however we have a large amount of power and timing slack. This is an altogether expected result due to the uncorrelated nature of statistical variability.

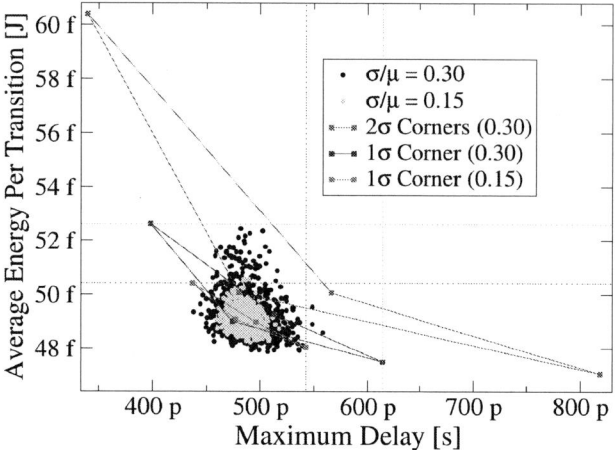

Figure 4 : Average energy per transition with maximum delay of an adder with corners

B. Combining Process and Statistical Variability

In order to introduce slowly changing global variability in the Randomspice simulation methodology, when each circuit is generated, Randomspice randomly generates a mean threshold voltage for all p- and all n-MOSFETs, thus assuming all devices in the circuit are affected in the same way due to global variability. Each device in the circuit is then randomized with statistical local variability. The level of variability injected for process was equal to that of the statistical σ = 15%. The results of the adder simulations are shown in Fig. 5.

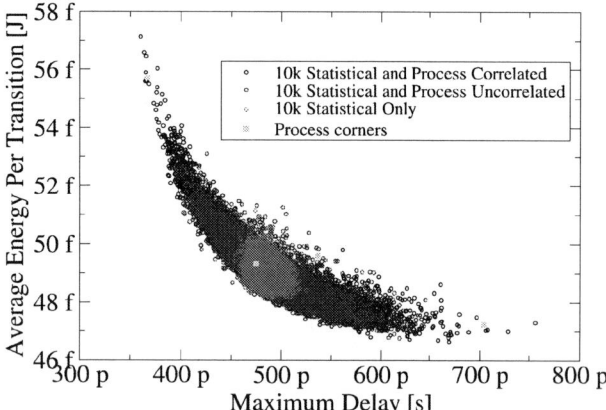

Figure 5 : Average energy per transition with maximum delay of an adder with global and local variability.

The simulations were performed assuming both 0% and 100% correlation between the threshold voltages of the p- and the n-MOSFETs related to the process variability. The combination of correlated process variability in the presence of statistical variability gives the widest power/delay distribution. Due to the nature of process variability, which shifts the means of all transistors in the circuit, in the presence of the same amount of statistical variability, process variability is still dominant. Statistical variability does however push some of the circuit instances outside the 3-σ process corner, which lowers yield to below 100%, which could be an undesirable effect. In the case of uncorrelated p- and the n-MOSFET process variability the distribution shrinks inside the process corners, which now guarantee close to 100 % yield.

Another approach to deal with the statistical variability is to replace the traditional "total corners" with "global corners" that excludes statistical variability. Then in order to evaluate the yield, perform monte-carlo analysis around the global corners. This approach was also evaluated with Randomspice:

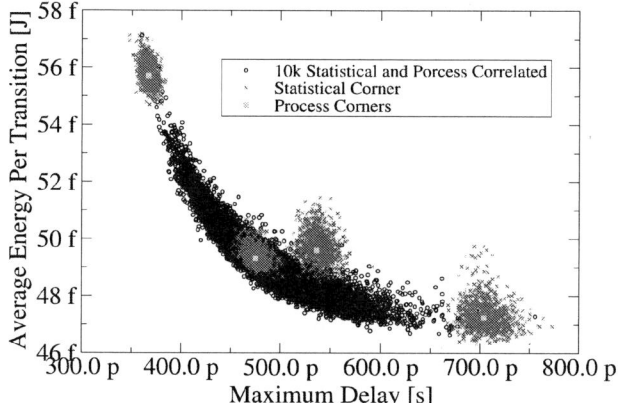

Figure 6 : Corner simulations with statistical distributions

In the following simulations the previous 3-σ corners were assumed to be "global corners". The results guarantee close to 100% yield but are significantly more pessimistic compared to the full statistical process with random variability simulations. Further to this, these "global corners" with random variability simulations do not allow for the energy/delay/yield design tradeoff process for optimization, as the complete circuit distribution is not present. In order to perform energy/delay/yield optimization it is required to extrapolate the cumulative distribution function (CDF) derived from the simulated data points in two correlated dimensions. This is achieved through a 2D kernel density estimate (KDE)[13]. This process replaces each data point with a 2D normal Gaussian distribution to give a continuous distribution function. The equi-potential lines along the CDF produced represent the equi-yield contours for those data points.

Figure 7 shows that global corners with random variability simulations provide an erroneous CDF distribution compared to the statistical process with random variability simulations. Therefore they will not allow an energy/delay/yield design tradeoff. Also the 95% yield curves show quantitatively the extra margins that the global corners with random variability analysis introduces.

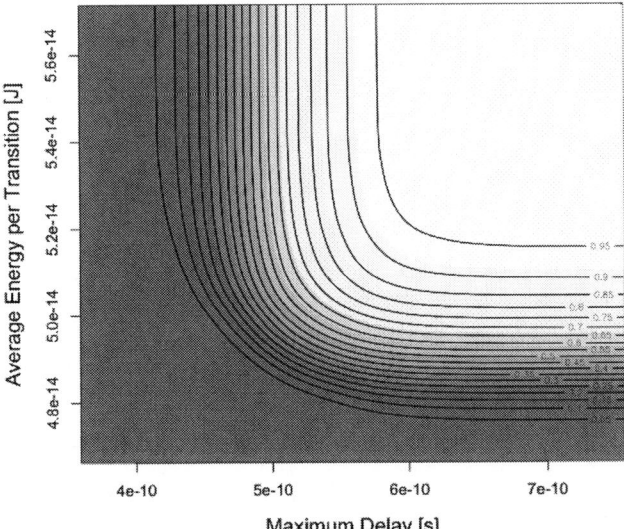

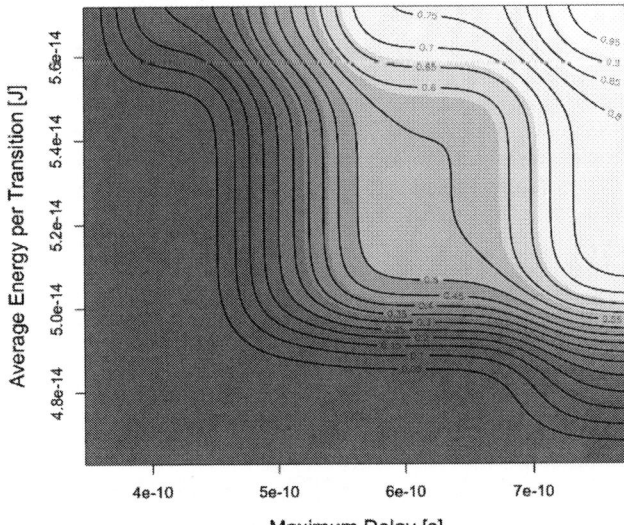

Figure 7 : CDF yield plots for continuous distribution of simulations of statistical and process variability (up) and statistical corner simulations (down)

V. CONCLUSIONS

This paper highlights problems of corner based design verification in the presence of significant statistical variability only. The corners, while fairly accurate for small cells, become increasingly pessimistic for larger circuits due to the additive nature or random uncorrelated statistics. If corner analysis alone is used as the defining measure of a technology dominated by statistical variability, the design will be extremely pessimistic with a large amount of slack in both

timing and energy. Statistical circuit simulations that combine equal amounts of process and statistical variability indicate that 3-σ process corners still can guarantee close to 100% yield in the presence of fully correlated process p- and n-MOSFET process variability but become pessimistic in the case of uncorrelated p- and n-MOSFET process variability. The "global corners" with random variability simulations, advocated by some of the foundries guarantee close to 100% yield but introduce approximately 20% extra margins. They also do not allow for the energy/delay/yield design tradeoff, which can differentiate high performance premium price chips from mass consumer chips. Future research includes the simulation of a system with more statistical variability that process to evaluate the effectiveness of corners with statistics in this eventuality.

REFERENCES

[1] Y Cao, P Gupta, AB Kahng, D Sylvester, "Design Sensitivities to Variability, Extrapolations and Assessments in Nanometer VLSI" Proc. ASiC/SOC, 2002

[2] K. Okada, H. Onodera, "Statistical Modeling of Device Characteristics with Systematic Variability" IEICE Trans. Fundementals, vol. E84–A, no.2, February 2001

[3] C. Proglera, A. Bornab, D. Blaauwb, P. Sixta, "Impact of lithography variability on statistical timing behavior" Proc. SPIE, Vol. 5379, 101 (2004)

[4] Borivoje Nikolić, Liang-Teck Pang, "Measurements and Analysis of Process Variability in 90nm CMOS" ICSICT '06. 8th International Conference on Solid-State and Integrated Circuit Technology, 2006.

[5] J. Tschanz, J. Kao, S. Narendra, R. Nair, D. Antoniadis, A. Chandrakasan, V. De. "Adaptive body bias for reducing impacts of die-to-die and within-die parameter variations on microprocessor frequency and leakage" in IEEE J. Solid-State Circuits, p 1396-1402, Nov. 2002

[6] J. Tschanz, S. Narendra, R. Nair, V. De, "Effectiveness of adaptive supply voltage and body bias for reducing the impact of parameter fluctuations in low power and high performance microprocessors" in IEEE J. Solid-State Circuits, p 826-829, May. 2003

[7] P. Gupta, A. B. Kahng, P. Sharma, D. Sylvester, "Selective Gate-Length Biasing for Cost-Effective Runtime Leakage Control" DAC 2004, June 7–11, 2004

[8] A. Asenov, G. Slavcheva, A. R. Brown, J. H. Davies and S. Saini, "Increase in the random dopant induced threshold fluctuations and lowering in sub-100nm MOSFETs due to quantum effects: A 3-D density-gradient simulation study," IEEE Trans. Electron Devices, vol. 48, no.4, 2001, pp. 271–350.

[9] A. Asenov, S. Kaya, A. R. Brown, "Intrinsic parameter fluctuations in decananometer MOSFETs introduced by gate line edge roughness," IEEE Trans. Electron Devices, vol. 50, no.5, 2003, pp. 1254–1260.

[10] A. R. Brown, G. Roy and A. Asenov, "Impact of Fermi level pinning at polysilicon gate grain boundaries on nano-MOSFET variability: A 3D simulation study," in Proc. 36th ESSDERC, 2006, pp.451-454.

[11] H. Aikawa, T. Sanuki, A. Sakata, E. Morifuji, H. Yoshimura, T. Asami, H. Otani and H. Oyamatsu "Compact Model for Layout Dependent Variability" IEDM09 P699-702

[12] B.Cheng, D.Dideban, N.Moezi, C.Millar, G.Roy, X.Wang, S.Roy, A.Asenov, "Benchmarking Statistical Compact Modeling Strategies for Capturing Device Intrinsic Parameter Fluctuations in BSIM4 and PSP", submitted to the IEEE journal of "Test and Design of Computers".

[13] B. W. Silverman, "Density estimation for statistics and data analysis," Chapman & Hall/CRC, 1986, p. 76

Threshold voltage shift and drain current degradation by NBT stress in Si (110) pMOSFETs

Kensuke Ota, Masumi Saitoh, Yukio Nakabayashi, Takamitsu Ishihara, Toshinori Numata, and Ken Uchida[1]

Advanced LSI Technology Laboratory, Corporate Research and Development Center, Toshiba Corporation, 8, Shinsugita-cho, Isogo-ku, Yokohama 235-8522, Japan, [1]Tokyo Institute of Technology, Phone: +81-45-776-4163 E-mail: kensuke.ota@toshiba.co.jp

Abstract—Threshold voltage shift and drain current degradation by NBT stress in Si (100) and (110) pMOSFETs are systematically studied. Threshold voltage shift in (110) pFET is larger than that in (100) pFET. However, time and temperature dependence of NBTI suggest that the mechanisms of the NBTI degradation are independent of the surface orientations. It is newly found that the drain current degradation in (110) pFET is severer than that in (100) pFET even when the same amount of charges at the interface is generated. This can be explained by larger mobility degradation in (110) pFETs due to the generated interface traps.

I. Introduction

The surface orientation engineering becomes a key technique to enhance the CMOS performance and to optimize the multi-gate structures [1,2]. Especially, (110) plane is usually used for the side-surface channel of FinFET or nanowire FET. While the performance of scaled (110) CMOS has been intensively studied, the reliabilities of (110) FETs have not been sufficiently understood. Recently, negative bias temperature instability (NBTI), one of the most crucial issues in scaled pFETs, have been reported for (110)-plane pFETs [3,4]. However, the difference of NBTI between (100) and (110) pFETs is still controversial. In addition, the performance degradation, that is, the drain current reduction in (110) pFETs by NBT stress has not been examined yet.

In this work, we present a systematic study of the threshold voltage (V_{th}) shift and the drain current reduction in (100) and (110) pFETs under the NBT stress and elucidate the difference of the degradation mechanisms between (100) and (110) pFETs. Charge pumping current of the stressed devices is also measured to confirm the generation of the interface traps.

II. V_{th} shift by NBT stress in (110) pFETs with poly Si/SiON stacks

NBTI measurements are performed in (100) and (110) bulk pFETs. Gate stacks are poly-Si gate and SiON dielectric with equivalent oxide thickness (EOT) of ~1 nm. The channel and halo dopants are As. Figures 1(a) and 1(b) show the typical V_g dependence of the channel current (I_{ch}) and that of the gate leakage current (I_g), respectively. In order to remove the I_g component from I_d, the channel current I_{ch} (= (I_d - I_s) / 2) is evaluated instead of I_d itself in this study. Because of the high mobility in (110) pFET, I_{ch} of (110) pFET is larger than that of (100) pFET at high V_g. On the other hand, the threshold voltage (V_{th}) of (110) pFET is nearly equal to that of (100) pFET due to almost the same channel impurity concentration. Moreover, I_g of (100) and (110) pFETs are almost the same indicating that the quality of the gate oxide is independent of the surface orientations. Similar V_{th}, I_g, and EOT between (100) and (110) pFETs enables us to fairly compare the NBTI degradation of both devices by applying the same amount of stress.

The time dependences of the threshold voltage shifts (ΔV_{th}) at different temperatures are shown in Fig. 2. Here, V_{th} was determined as V_g with the drain current of 1×10^{-9} A. ΔV_{th} of (110) pFET is larger than that of (100) pFET in all the temperature range. Although the absolute values of ΔV_{th} vary with the surface orientations, the slopes against time are almost the same.

Figures 3(a) and 3(b) show the temperature and stress field dependence of ΔV_{th} after the same stress time. ΔV_{th} in (110) pFET is larger than that in (100) pFET in all the temperature and the stress field range, whereas its dependence on temperature or the stress field is nearly equal. The activation energy extracted from the temperature dependence is about 0.07 eV which is almost the same between (100) and (110) pFETs. The same dependence of ΔV_{th} on time, temperature, and stress field suggests that the generation mechanisms of the interface traps or of the fixed charges, which cause ΔV_{th}, are independent of the surface orientations, although the amount of generated traps/charges is larger in (110) pFETs than in (100) pFETs. The amount of the generated interface traps is measured and discussed in the next section.

978-1-4244-6658-0/10 $26.00 © 2010 IEEE

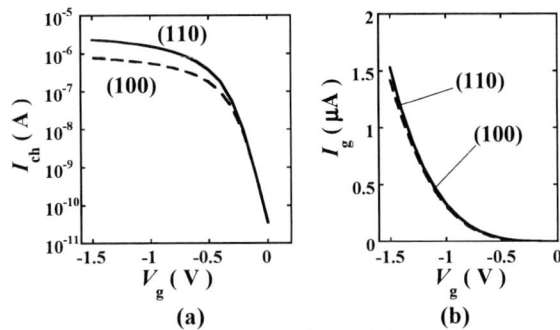

Fig.1 Typical (a) I_{ch}-V_g and (b) I_g-V_g characteristics at room temperature. L/W are 1 μm/1 μm.

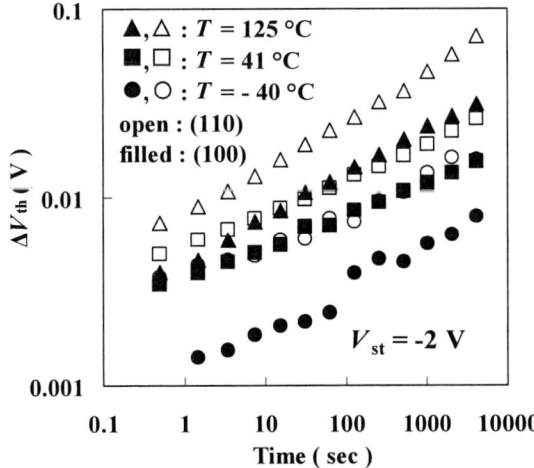

Fig.2 Time dependence of ΔV_{th} at different temperatures in (100) and (110) pFETs. Stress voltage of -2 V was applied.

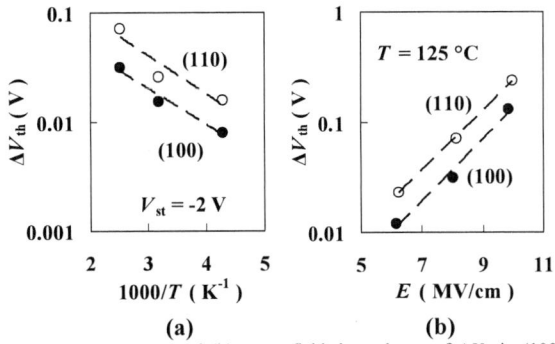

Fig.3 (a) Temperature and (b) stress field dependence of ΔV_{th} in (100) and (110) pFETs after the stress time of 4096 sec.

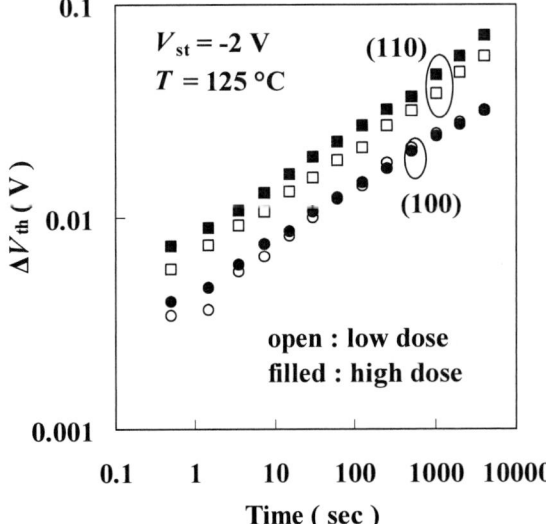

Fig.4 Time dependence of ΔV_{th} in (100) and (110) pFETs with different channel dose. Stress voltage of -2 V was applied.

Moreover, the channel dose dependences of ΔV_{th} in (100) and (110) pFETs are examined. Effective channel impurity concentration (N_{sub}) was set to be 1.6×10^{18} cm^{-3} (high dose pFETs) in the above ΔV_{th} measurement. Besides, pFETs with N_{sub} of 1.3×10^{18} cm^{-3} (low dose pFETs) are measured. Figure 4 shows the time dependence of ΔV_{th} in pFET with the different channel dose. ΔV_{th} of (100) pFET is independent of the channel dose, while ΔV_{th} of (110) pFET with high dose is slightly larger than that with low dose. In general, As tends to be segregated at the Si/SiO$_2$ interface [5]. Since (110) surface has more dangling bonds than (100) surface, the amount of segregated As at the interface is possibly larger in (110) pFETs. Consequently, higher As doping increases the segregated As, which degrades the interface quality and enhances the generation of interface traps by NBT stress in (110) pFETs.

III. Charge pumping measurements in (110) pFETs with poly Si/SiO$_2$ stacks

In order to investigate the correlation between ΔV_{th} and the interface trapped charge (Q_{it}), the charge pumping measurements were performed in (100) and (110) pFETs with poly-Si gate and SiO$_2$. The frequency (f) of the applied pulsed gate voltage was set to be 1 MHz. Figure 5 shows the time dependence of the increase in the charge pumping current (ΔI_{cp}). As well as ΔV_{th}, ΔI_{cp} of (110) pFET is larger than that of (100) pFET after the same stress conditions. This directly indicates that the number of the generated interface traps in (110) pFET is larger than that in (100) pFET.

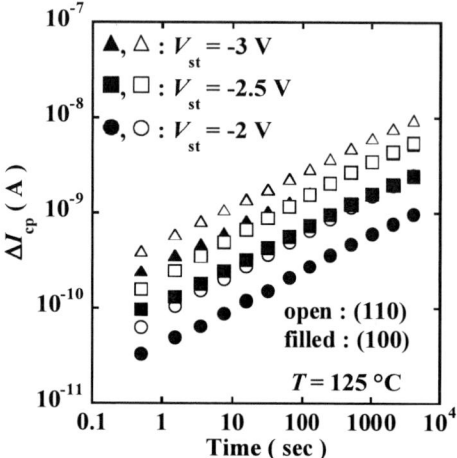

Fig.5 Time dependence of ΔI_{cp} in (100) and (110) pFETs under the different stress voltages. L/W and T_{ox} are 1 µm/5 µm and 2 nm, respectively.

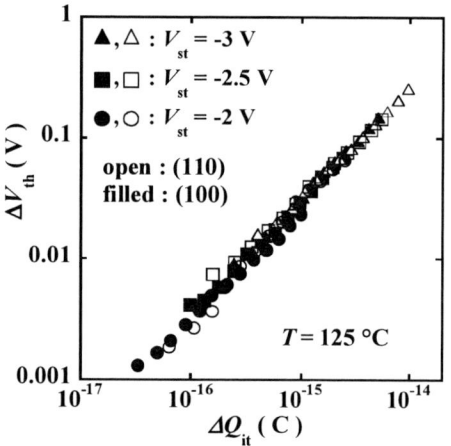

Fig.6 ΔQ_{it} vs ΔV_{th} in (100) and (110) pFETs under the various stress conditions.

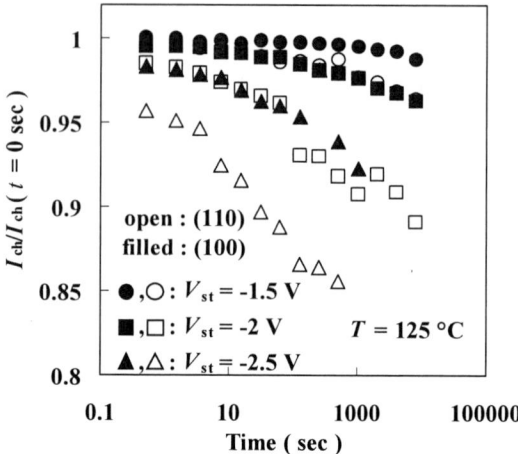

Fig.7 Stress time dependence of I_{ch} under the various stress conditions. I_{ch} is normalized by I_{ch} before the NBT stress was applied.

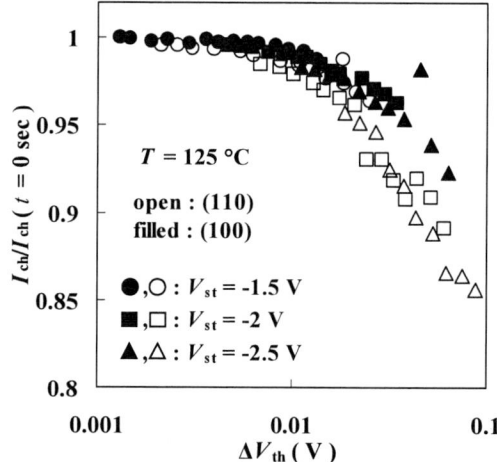

Fig.8 ΔV_{th} dependence of the normalized I_{ch} under the various stress conditions

Correlation between ΔV_{th} and ΔQ_{it} is shown in Fig. 6, where Q_{it} was extracted by the relation, $Q_{it} = I_{cp}/f$. It is notable that the correlation between ΔV_{th} and ΔQ_{it} is independent of the stress voltage and the surface orientations. This result confirms that the origin of ΔV_{th} is related to the interface traps and is independent of the surface orientations. Larger generation of interface traps in (110) pFETs, in spite of the same mechanisms of NBTI, can be explained by the higher Si bonds availability at the (110) interface. The density of Si bonds in (110) plane is more than that in (100) plane [6]. Thereby, the higher density of Si bonds increases the possibility of the de-passivated Si bonds at the interface which leads to larger number of the generated interface traps or the larger NBTI [7].

IV. Impact of NBT stress on drain current degradation

We, furthermore, examined the impact of the NBT stress on the drain current degradation with the pFETs whose ΔV_{th} was measured in Section II. The channel current at the gate overdrive voltage ($V_g - V_{th}$) of 1 V was measured after the NBT stress. By fixing the gate overdrive voltage, the effect of V_{th} increase, or the reduction of inversion carrier density, can be removed from the drain current degradation. Figure 7 shows the time dependence of the channel current degradation, which is normalized by the channel current before the NBT stress was applied. The channel current reduction of (110) pFET is larger than that of (100) pFET under the same stress conditions.

978-1-4244-6658-0/10 $26.00 © 2010 IEEE

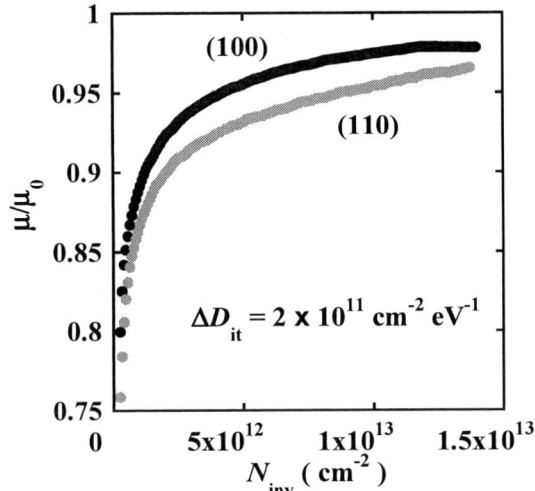

Fig.9 Normalized mobility vs inversion carrier density after the generation of interface traps by FN tunneling injection.

However, it is notable that the channel current degradation of (110) pFET is still larger than that of (100) pFET with the same ΔV_{th}, which is shown in Fig. 8. Since ΔV_{th} corresponds to the charges which is generated at the interface, the same amount of generated charges leads to more channel current degradation in (110) pFET than that in (100) pFET. In general, the channel current degradation at the fixed gate overdrive is dominated by the mobility degradation by the interface charges. Therefore, the mobility degradation was measured after the interface trap generation of 2×10^{11} cm^{-2} eV^{-1} by the FN tunneling injection in pMOSFETs with T_{ox} = 5 nm. Figure 9 shows the inversion carrier density dependence of the normalized mobility degradation (the ratio of the degraded mobility to the initial mobility). Apparently, the mobility degradation of (110) pFET is larger than that of (100) pFET. This experimental result indicates that the Coulomb scattering due to the same amount of interface trapped charges degrades the total mobility of (110) pFET more severely than that of (100) pFET, because the total mobility of (110) pFETs at fresh conditions is much higher than that of (100) pFETs.

V. Conclusion

NBTI measurements of (100) and (110) pFETs were performed. ΔV_{th} of (110) pFET was larger than that of (100) pFET as a result of larger number of the generated interface charges in (110) pFETs. Whereas, the mechanisms of NBTI degradation is the same between (100) and (110) pFETs. Drain current degradation under the NBT stress was found to be more serious in (110) pFET than that in (100) pFET, because the mobility reduction in (110) pFET was much larger than that in (100) pFET even when the same amount of the interface trap is generated. In order to realize highly-reliable (110) devices, the improvement of the interface property is crucial.

Acknowledgment

This work was partly supported by NEDO's Development of Nanoelectronic Device Technology. The authors thank Drs. N. Yasutake and K. Tatsumura for the sample provision.

References

[1] K. Uchida, A. Kinoshita, and M. Saitoh, "Carrier Transport in (110) nMOSFETs: Subband Structures, Non-Parabolicity, Mobility Characteristics, and Uniaxial Stress Engineering," *IEDM Tech. Dig*, pp. 1019-1021, 2006.

[2] M. Saitoh, S. Kobayashi, and K. Uchida, "Physical Understanding of Fundamental Properties of Si (110) pMOSFETs ~ Inversion-Layer Capacitance, Mobility Universality, and Uniaxial Stress Effects ~," *IEDM Tech. Dig*, pp. 711-714, 2007.

[3] M. Sato, Y. Sugita, T. Aoyama, Y. Nara, and Y. Ohji, "Impact of the Different Nature of Interface Defect States on the NBTI and 1/f noise of High-k / Metal Gate pMOSFETs between (100) and (110) Crystal Orientations," *Dig. Symp. VLSI Tech.*, pp. 64-65, 2008.

[4] S. Zafar, M. Yang, E. Gusev, A. Callegari, J. Stathis, T. Ning, R. Jammy, and M. Ieong, "A Comparative Study of NBTI as a function of Si Substrate Orientation and Gate Dielectrics (SiON and SiON/HfO$_2$)," *Dig. Symp. VLSI-TSA Tech.*, pp. 128-129, 2005.

[5] K. Inoue, F. Yano, A. Nishida, H. Takamizawa, T. Tsunomura, "Dopant distribution in gate electrode of n- and p- type metal-oxide-semiconductor field effect transistor by laser-assisted atom probe," Appl. Phys. Lett., vol. 95, p. 043502, 2009.

[6] S. Sze, editor. *VLSI Technology*. 2nd ed. NY: McGraw-Hill, 1988, p. 110.

[7] S. Maeda, J-A. Choi, J-H. Yang, Y-S. Jin, S-K.Bae, Y-W. Kim, and K-P. Suh , "Negative Bias Temperature Instability in Triple Gate Transistors," *Pros. IRPS*, pp. 8-12, 2004.

978-1-4244-6658-0/10 $26.00 © 2010 IEEE

RESISTIVE SWITCHING-LIKE BEHAVIOUR OF THE DIELECTRIC BREAKDOWN IN ULTRA-THIN Hf BASED GATE STACKS IN MOSFETs.

A. Crespo-Yepes, J. Martin-Martinez, R. Rodriguez,
M.Nafria and X. Aymerich
Departament d'Enginyeria Electrònica
Universitat Autònoma de Barcelona (UAB)
08193, Bellaterra (Spain)
albert.crespo@uab.es

A. Rothschild
IMEC
Kapledref 75
30001 Leuven (Belgium)

Abstract—**The gate dielectric breakdown (BD) reversibility in MOSFETs with ultra-thin hafnium based high-k dielectric is studied. The phenomenology is analyzed in detail and the similarities with the resistive switching phenomenon emphasized. The results suggest that the conductive path in the dielectric after BD can be 'opened' and 'closed' many times and that the BD recovery partially restores not only the current through the gate, but also the MOSFET channel related electrical characteristics.**

I. INTRODUCTION

The resistive switching (RS) phenomenon has recently acquired an increasing importance for non volatile storage applications since the resulting devices combine fast operation, compatibility with the CMOS process and huge scaling potential [1]. The phenomenon has been usually studied in MIM structures with a thick insulator (several tenths of nm), based in non standard CMOS chemical elements [2], in which the conductivity of the dielectric can be switched between a high and a low conductivity states. In this regard, the dielectric breakdown (BD), one of the most relevant failure mechanisms in CMOS technologies [3], is also characterized by a change of the insulator conductivity state from a low (pre-BD state) to a higher one (post-BD condition) whose currents can differ in several orders of magnitude. Traditionally, and contrarily to the RS phenomenon, the conductivity change due to the BD has been considered to be irreversible. However, several years ago it was shown that in SiO_2, in some occasions, BD could be reversible [4], i.e., a low conductivity state could be reached after BD. More recently, we have reported the presence of two interchangeable conductivity states after BD in ultrathin Hf based gate stacks [5] when the current during de BD transient is limited, so that the insulator properties of the dielectric can be at least partially recovered [2, 6].

In this work, we discuss the similarities of BD and RS phenomena and analyze in detail the change of dielectric conductivity between two states, in MOSFETs with ultra-thin Hf-based high-k dielectric. The procedure to switch between the two states, their local nature and reproducibility are discussed from the gate current characterization. Finally, the

This work has been partially supported by the Spanish MICINN (TEC2007-61294/MIC) and the Generalitat de Catalunya (2009SGR-783).

transistors performance is analyzed (I_D-V_D and I_D-V_G characteristics) for both dielectric conduction states.

II. SAMPLES AND EXPERIMENTAL

The samples used in this work were pMOSFETs with FUSI gate electrode and a dielectric stack formed of a (2.9nm) HfSiON film on top of a 1.2 nm SiO_2 interfacial layer (EOT=1.9 nm). Different combinations of channel width and length have been studied, ranging between W/L=0.25µm/0.15µm and W/L=1µm/0.5µm. The samples were subjected to a sequence of a current limited ramped voltage stress (CL-RVS) to induce the BD (high conductivity state), plus a ramped voltage stress (RVS) without current limitation, to switch to the low conductivity state, following a measurement-stress-measurement scheme shown in Figure 1. Though this work is focused in the characterization of the phenomenon by means of ramp voltage tests, the BD recovery can be also observed if the gate voltage is kept constant during

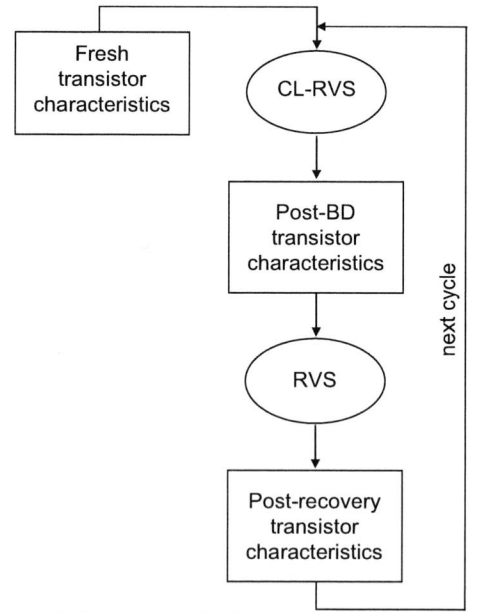

Figure 1. Stress sequence for the analysis of the dielectric BD reversibility. The compliance of the measuring equipment (Keithley 4200-SCS) was used to limit the current during the CL-RVS.

the stresses [6]. The stresses were applied to the gate with the rest of the transistor terminals grounded and I_G, I_D, and I_S were simultaneously registered. To observe the changes in the device performance after the different stresses, the I_D-V_D and I_D-V_G characteristics of the fresh sample and those after the successive CL-RVS and RVS steps of the sequence were registered.

III. BD REVERSIBILITY PHENOMENOLOGY

Figure 2 shows typical gate currents measured in the same sample during the CL-RVS and RVS for the initial cycles of the measurement sequence illustrated in Figure 1. Curve I_F (thick line) corresponds to the gate current registered during the first CL-RVS (fresh device). When dielectric BD takes place, at V_{BD}, a fast current increase is observed until reaching the current limit (500µA). At low voltages, the post-BD gate current (I_{BD}) obtained during the next RVS is, as expected, much larger than I_F. However, if the gate voltage continues increasing, at a given voltage (V_R), the I_{BD} current suddenly decreases several orders of magnitude. During the CL-RVS in the next cycle, at low voltages, the gate current (I_R) is larger than the fresh current but lower than I_{BD}. This indicates a partial recovery of the insulator properties of the gate dielectric, suggesting that, in some conditions, BD in ultra-thin Hf-based high-k oxides is a reversible phenomenon, i.e., a BD path that was 'opened' can be 'closed'. If V_G continues increasing during the CL-RVS and V_{BD} is reached, BD is observed, so that a high current level is measured again. However, in the next RVS, after V_R, the current decreases

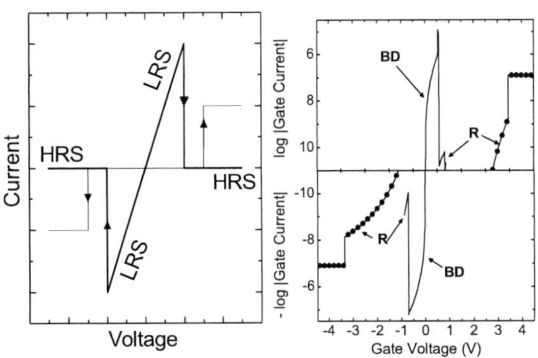

Figure 3. Left: schematic picture of the typical I-V characteristic of the RS phenomenon obtained in MIM structures [1]. Right: I-V characteristics obtained in MOS capacitors for the BD reversibility [5].

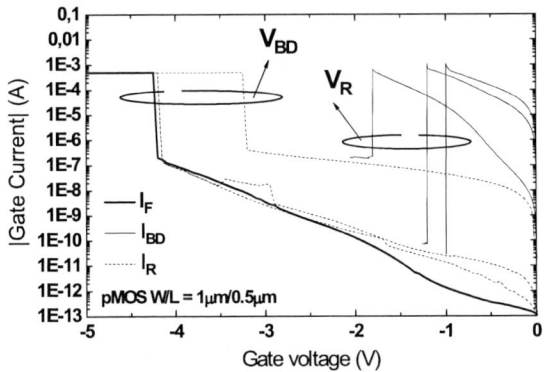

Figure 2. I_G-V_G curves measured in a pMOSFET with W/L=1µm/0.5µm after successive CLR-RVS + RVS iterations. A high current is registered after the current limited BD (I_{BD}), which suddenly drops after V_R. During the CL-RVS in the next cycle, at low voltages the gate current (I_R) is larger than the fresh current, but lower than I_{BD}, which indicates a partial recovery of the dielectric properties. The switching between both conductivity states can be provoked in successive cycles of CL-RVS + RVS.

once more. This behavior can be observed for many iterations of the stress sequence (more than 250 cycles in this work). Moreover, the phenomenon is qualitatively repetitive from sample to sample. It must be emphasized that if a BD appeared during the RVS or during the measurement of the transistor characteristics, the BD would become irreversible (the recovery is no longer observed), and the cycling would

not be possible anymore. For this reason, the maximum voltage value of the RVS applied to return to the low conductivity state should be kept below the typical values of V_{BD}. These results indicate that, after a current limited BD, two conduction states are allowed in the dielectric, a high conductivity one (BD state, with gate current I_{BD}) and a low one (R state, with gate current I_R). Switching between both states occurs when the two threshold voltages are reached (V_{BD} and V_R) and only if BD is produced under current limited conditions.

Therefore, a clear similarity, phenomenologically speaking, between the RS and BD mechanisms exists. This can be made evident by comparing the plots in Figure 3, where typical I-V characteristic for RS in MIM structures is schematically reproduced [1] (left) and an experimental I-V obtained in a MOS capacitor from the same wafer, extracted from the data presented at [5] are shown. In both graphs the existence of the two conductivity states (LRS and HRS in RS terminology, or BD and R respectively, in the terminology adopted for dielectric breakdown) are evident.

IV. VOLTAGES AND CURRENTS DISTRIBUTIONS

Figure 4 (top) shows the V_{BD} and V_R values obtained on a single device subjected to more than 250 cycles of the stress sequence. The values of V_{BD} and V_R rapidly decrease (in absolute value) in the first cycles and their mean values remain constant in the following ones. The dashed line in Fig.4 (top) indicates the maximum voltage value imposed (|1.3V|) to the RVS after the transient in order to avoid non desired BD events. Fig 4 (bottom) shows the cumulative distribution of V_{BD} and V_R (in this representation, a normal distribution corresponds to a straight line). As can be observed, the mean values of the distributions are separated around 1.2V and the spread is larger for V_{BD}.

978-1-4244-6658-0/10 $26.00 © 2010 IEEE   59

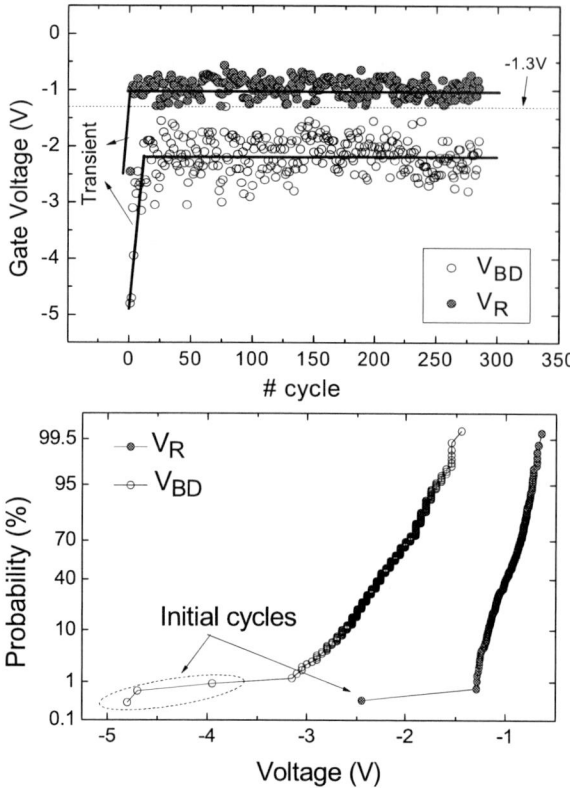

Figure 4: Top: BD voltage (V_{BD}) and R voltage (V_R) evolution during cycling. A transient behaviour is observed at the initial cycles, after the transient the mean values of V_{BD} and V_R remain stable.
Bottom: Cumulative probability of V_{BD} and V_R registered during the cycling on the same device.

The cumulative probability distributions of I_{BD} and I_R measured at -0.5V are shown in Figure 5. In this case the transient behavior during the first cycles is not so evident, and the tail observed in I_R corresponds to values randomly between the 1st and the 120th cycles. Concerning to the I_{BD} distributions, two modes are distinguished, that have been classified here as SBD and HBD attending to the current measured. Note that the current measured at the HBD mode is limited by the external current limitation imposed in the CL-RVS [5].

V. AREA DEPENDENCE AND CONDUCTION PATH LOCATION

Transistors with different areas have been analyzed to obtain more information about the nature of the R state. Figure 6 shows I_{BD} and I_R as a function of the transistor area. No area dependence of I_{BD} and I_R is observed, which indicates a localized gate current. The question to be answered is whether the BD and R states are controlled by the same conduction path or not. To check this point the location of the conductive path in the dielectric along the channel [7] has been studied. To do so, the magnitude α, defined as $I_D/(I_D+I_S)$, has been calculated for all the cycles at the BD and R states. α values close to 1 or 0 indicate that the conductive path is located close to the drain or source, respectively. Figure 7 shows the α evolution with during cycling for the R

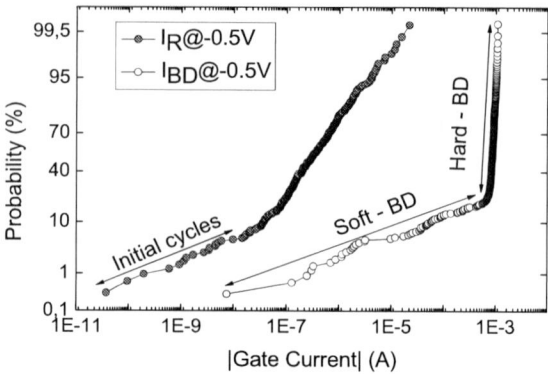

Figure 5: The cumulative probability function of I_{BD} (solid circles) and I_R (open circles) registered at $V_G = -0.5V$ during more than 250 measuring cycles on the same device.

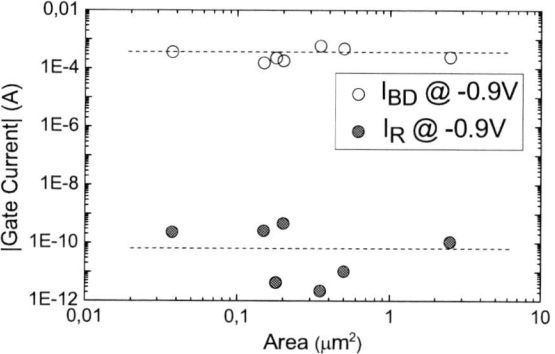

Figure 6: I_{BD} and I_R measured at $V_G=-0.9V$ as a function of the transistor area. No area dependence is observed in any of the two conductivity states.

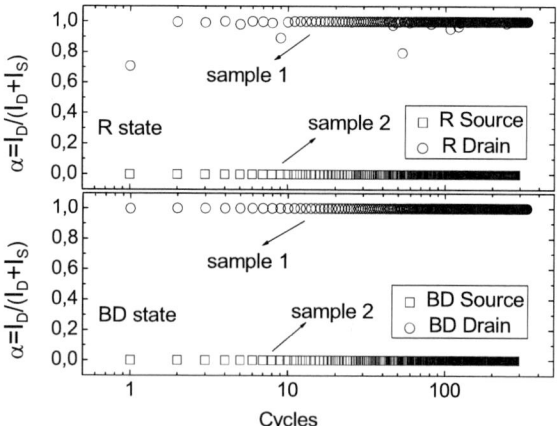

Figure 7: The quotient $\alpha=I_D/(I_D+I_S)$ allows to locate the conduction path in the dielectric along the channel [7]. Two samples, with BD located at the extremes of the channel, are considered. For each sample, the location of the conduction path in the BD and R states is the same and does not change during successive cycles.

(top) and BD (bottom) states for two MOSFETs. In each sample, the location of the conductive path through the oxide during the BD and R states is the same and does not change during the successive measuring cycles. This confirms the local nature of the R state and indicates that the conduction in both states should be attributed to the same path, i.e. the conductive path is 'opened' during the BD state and 'closed' during the R state. So that, the partial recovery of the high-k properties in the R state should be attributed to changes in the atomic structure of the conductive path [8].

VI. IMPACT ON THE TRANSISTOR CHARACTERISTICS

The transistor characteristics related to the channel conduction have been analyzed. Continuous lines in Figure 8a and 8b show the typical I_D-V_D and I_D-V_G characteristics, respectively, obtained in fresh transistors. Figures 8c and 8d show the curves when the conductive path has been formed closed to the drain and the device is at the BD state. In this case, the transistor characteristics are completely distorted, and I_D-V_G changes its sign being impossible to evaluate basic parameters such as the threshold voltage and the saturation current [9]. If the conduction path, at the BD state is located close to the source the drain current is of the order of nA (fig. 8e and 8f), because most of the current flows between source and gate. However, when the gate dielectric is switched to the R state, the I_D-V_D and I_D-V_G curves are partially recovered (solid circles and open triangles in figure 8a and 8b, respectively), with a large increase of the threshold voltage and decrease of the saturation current. In summary, a catastrophic change in the channel conduction is observed when the dielectric is at the BD state. However the channel

electric properties can be partially restored when the conductive path is switched to the R state, with a larger threshold voltage and smaller saturation current.

VII. CONCLUSIONS

In MOSFETs with ultra-thin high-k Hf based dielectrics, after the BD path has been created, two conductivity states are allowed in the insulator, BD and R, being the current during the BD state (I_{BD}) larger than during the R state (I_R). Switching between the two states is possible when two threshold voltages are applied, V_{BD} and V_R, respectively, and only if the current was limited during the BD transient. The effect has strong similarities with the resistive switching phenomenon observed in MIM structures. The value of V_{BD} and V_R decrease quickly in the first cycles of switching between the BD and R states, showing a transient behavior, not so clearly observed in the distributions of I_{BD} and I_R. In the I_{BD} distributions, two BD modes are distinguished, that have been classified here as SBD and HBD, attending to the measured current. The area dependence of the I_{BD} and I_R currents and their location along the transistor channel suggest that the conduction in both states is local and controlled by the same BD path. Finally, the electrical characteristics of MOSFETs show that, when the oxide conductivity returns to the R state, not only the gate current (I_G-V_G characteristics) is recovered but also the channel current (I_D-V_D curves). The restoration of the MOSFET performance could have an impact in the circuit functionality, which will have to be analyzed to make accurate reliability predictions.

REFERENCES

[1] R. Waser, M. Aono, "Nanoionics-based resistive switching memories," *Nature Materials*, vol. 6, pp. 833-840, 2007.

[2] W. H. Liu, K. L. Pey, X. Li, M. Bosman, "Observations of Switching behaviors in post-breakdown conduction in NiSi-gat stacks," *International Electron Devices Meeting*, pp. 1-4, 2009.

[3] E. Y. Wu, J. Suñé, "Power low voltage acceleration: A key element for ultr-thin gate oxide reliability," *Microelectronics and Reliability*, vol. 45, pp. 1809-1834, 2005.

[4] M. Nafria, J. Suñé, X. Aymerich, "Exploratory observations of post-breakdown conduction in polycrystalline-silicon and metal-gate thin-oxide metal-oxide-semiconductor capacitors," *Journal of Applied Physics*, vol. 73, pp. 205-215, 1993.

[5] A. Crespo-Yepes, J. Martin-Martinez, R. Rodriguez, M. Nafria, X. Aymerich, "Reversible dielectric breakdown in ultrathin Hf based high-k stacks under current limited stresses," *Microelectronics Reliability*, vol. 49, pp. 1024-1028, 2009.

[6] A. Crespo-Yepes, J. Martin-Martinez, A. Rothschild, R. Rodriguez, M. Nafria, X. Aymerich, "Recovery of the MOSFET and circuit functionality after the dielectric breakdown of ultr-thin high-k gate stacks," *Electron device Letters*, (accepted for publication), 2010.

[7] R. Degraeve, B. Kaczer, A. De Keersgieter, G. Groeseneken, "Relation between breakdown mode and breakdown location in short channel MOSFETs and its impact on reliability specifications," International Reliability Physics Symposium, pp. 360, 2001.

[8] J. Suñe, E. miranda, M. Nafria and X. Aymerich, "Modeling the breakdown spots in silicon dioxide films as point contacts", Applied Physics Letters, vol. 75, pp. 959-961, 1999.

[9] R. Fernández, R. Rodríguez, M. Nafría, X. Aymerich, "MOSFET output characteristics after oxide breakdown," Microelectronics Engineering, vol. 84, pp. 31-36, 2007.

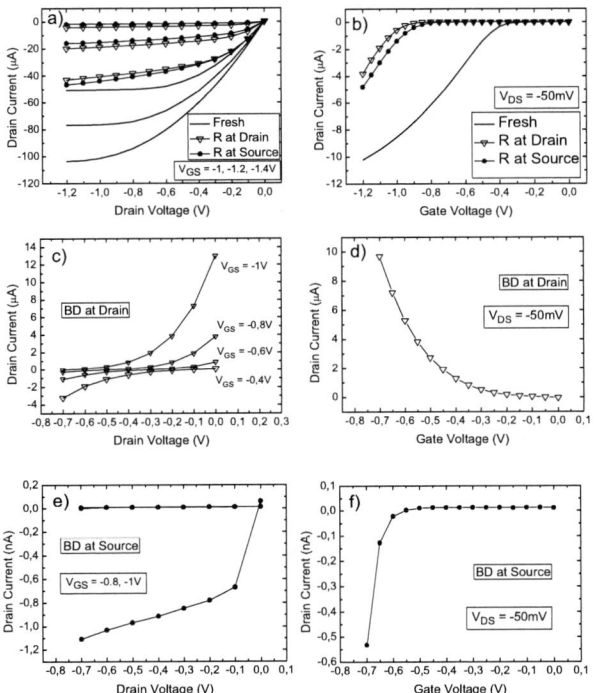

Figure 8: (a) I_D-V_D characteristics for fresh (solid lines) and at the R state (dots) transistor. (b) I_D-V_G characteristics for fresh (solid lines) and recovered (dots) transistor. (c) I_D-V_D characteristics with BD located at drain, and (e) with BD located at source. (d) I_D-V_G characteristics with BD located at drain, and (f) with BD located at source.

Tri-Gate Bulk CMOS Technology for Improved SRAM Scalability

Changhwan Shin, Borivoje Nikolić, Tsu-Jae King Liu
Department of Electrical Engineering and Computer Sciences
University of California, Berkeley
Berkeley, CA 94720-1770 USA
{shinch, bora, tking}@eecs.berkeley.edu

Chen Hua Tsai, Mei Hsuan Wu, Chung Fu Chang, You
Ren Liu, Chih Yang Kao, Guan Shyan Lin, Kai Ling Chiu,
Chuan-Shian Fu, Cheng-tzung Tsai, Chia Wen Liang
United Microelectronics Corporation
Hsinchu, Taiwan R.O.C.

Abstract — **A simple approach for manufacturing quasi-planar tri-gate bulk MOSFET structures is demonstrated and shown to be effective for reducing variation in 6T-SRAM read and write margins, in an early 28nm CMOS technology. With optimization of the pocket implant doses, quasi-planar bulk CMOS technology can facilitate voltage scaling. It also provides a means to achieve high yield with a notch-less 6T-SRAM cell layout, to facilitate area scaling.**

I. INTRODUCTION

A challenge for continued SRAM scaling is threshold voltage (V_T) mismatch due to process-induced variations [1], which degrades the minimum operating voltage (V_{min}) of an SRAM array [2]. To address this challenge, an improved transistor design that provides for reduced short-channel effects (*i.e.*, improved gate control over the channel potential) is required. The quasi-planar tri-gate bulk MOSFET [3] is an example of such a design; it utilizes a gate electrode that is physically wrapped around the top portion of the channel region to provide for greater capacitive coupling between the gate and the channel region [4]. In this work, a timed etch in dilute hydrofluoric (HF) acid solution is used to recess the isolation oxide prior to gate-stack formation, to form quasi-planar bulk MOSFETs using an otherwise conventional fabrication process flow. Improvements in transistor performance and SRAM yield are demonstrated in an early 28nm CMOS technology. The benefits of quasi-planar bulk MOSFET technology for voltage and area scaling are then assessed using three-dimensional (3-D) device simulations with atomistic doping profiles and analytical modeling to estimate 6T-SRAM cell yield for 22nm CMOS technology.

STI Formation
Well & V_T Implantation
STI Oxide Recess
Gox/Poly-Si Deposition
Gate (2P2E) Patterning
LDD & Pocket Implant
Spacer & S/D Formation
Activation Process
Salicidation

Figure 1. Front-end-of-line fabrication process steps used in this work.

II. DEVICE FABRICATION AND RESULTS

Individual logic transistors and 6T-SRAM arrays (~2500 cells per DUT) were fabricated using a standard test-chip mask set with an early 28nm low power CMOS technology which incorporates only dual stress liners for performance enhancement.

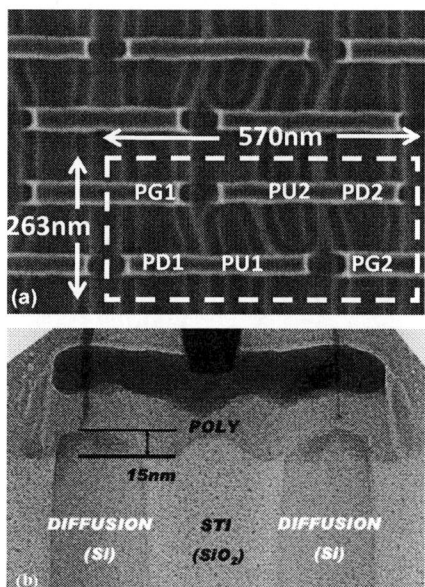

Figure 2. (a) 0.149um² SRAM cell plan-view CDSEM image after gate patterning. (b) XTEM taken along a poly-Si gate electrode in an SRAM array, for 15nm nominal STI recess depth.

The sequence of front-end-of-line fabrication process steps is illustrated in **Fig. 1**. After conventional shallow trench isolation (STI), well formation and V_T-adjust ion implantation, the STI oxide was recessed by a small amount (15nm) just prior to gate-stack formation. **Fig. 2** shows plan-view scanning electron microscopy and cross-sectional transmission electron microscopy images of a fabricated SRAM cell. Due to improved gate control, the quasi-planar structure achieves higher drive current (I_{ON}) for comparable off-state leakage current (I_{OFF}), as shown in **Fig. 3**. On average, I_{ON} is improved by 82%, 50%, and 79% for the pass-gate (PG), pull-

978-1-4244-6658-0/10 $26.00 © 2010 IEEE

down (PD), and pull-up (PU) devices, respectively. The standard compact model can be well fit to quasi-planar bulk MOSFET characteristics, including the body effect (**Fig. 4**).

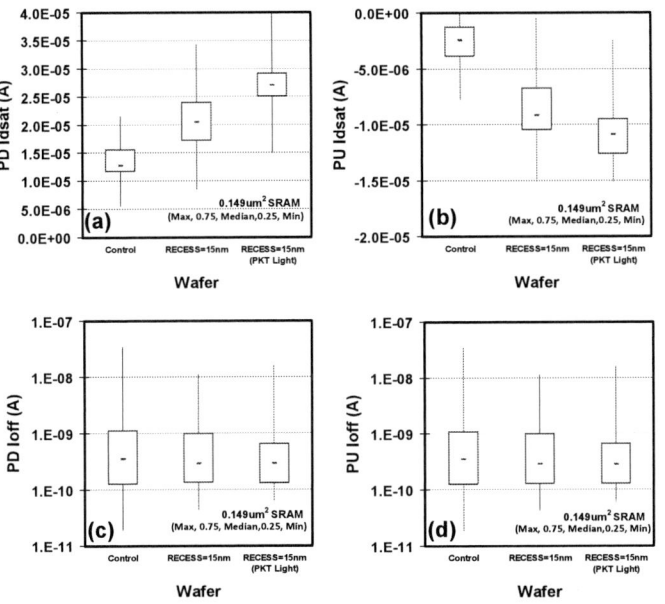

Figure 3. Comparison of on/off current statistics for planar (Control) *vs.* quasi-planar (RECESS=15nm) bulk MOSFETs in SRAM cells. (a) pull-down NMOS I_{ON} (b) pull-up PMOS I_{ON} (c) pull-down NMOS I_{OFF} (d) pull-up PMOS I_{OFF}.

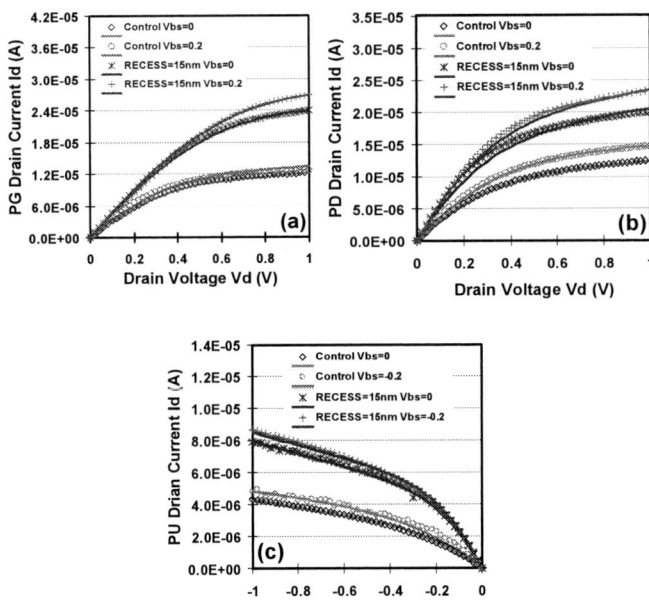

Figure 4. Comparison of measured output characteristics for planar (Control) vs. quasi-planar (RECESS=15nm) bulk MOSFETs in SRAM cells, for $|V_{GS}| = 1.0V$. The effect of forward body biasing is also shown. **(a)** pass-gate NMOS, **(b)** pull-down NMOS, **(c)** pull-up PMOS. The symbols are measured data; the lines show the fitted compact model.

V_T statistics are shown in **Fig. 5** for the devices. Due to improved gate control (resulting in steeper subthreshold swing) V_T is lower for the quasi-planar devices. For the NMOS devices, variation in V_T increases slightly due to relatively heavy pocket doping which results in more significant impact of random dopant fluctuations (RDF) for the gated sidewalls. This undesirable effect is eliminated by using a lighter pocket implant dose, as shown in **Figs. 5a and 5b** (PKT Light). This further lowers V_T and increases I_{ON} (Figs. 3a and 3b) without increasing I_{OFF} (Figs. 3c and 3d).

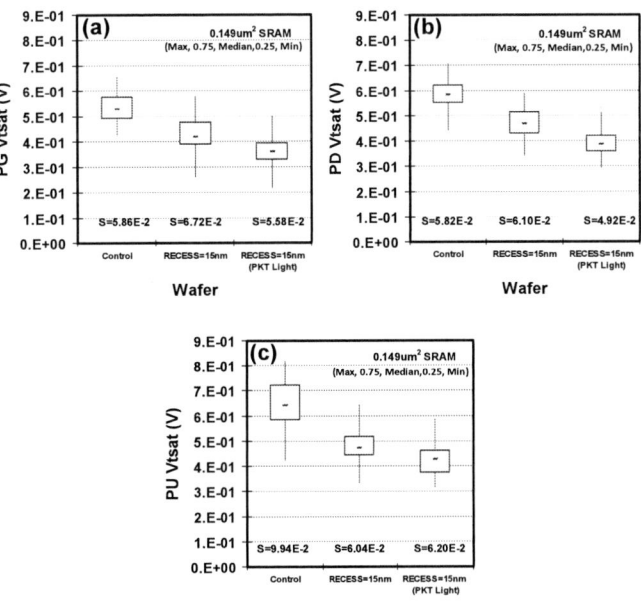

Figure 5. Comparison of saturation V_T statistics for planar (Control) vs. quasi-planar (RECESS=15nm) bulk MOSFETs in SRAM cells: (a) pass-gate NMOS, (b) pull-down NMOS, (c) pull-up PMOS.

Good short-channel control is maintained with the quasi-planar structure despite the lower pocket doping (**Fig. 6a**). In this early 28nm technology, the PMOS devices have lighter pocket doping than the NMOS devices, so that variation in V_T decreases when the STI oxide is recessed, due to the improved electrostatic integrity of the quasi-planar structure. However, if an even lighter pocket implant dose is used, then V_T variation increases slightly, as shown in **Fig. 5c**, due to slightly degraded short-channel control (**Fig. 6b**).

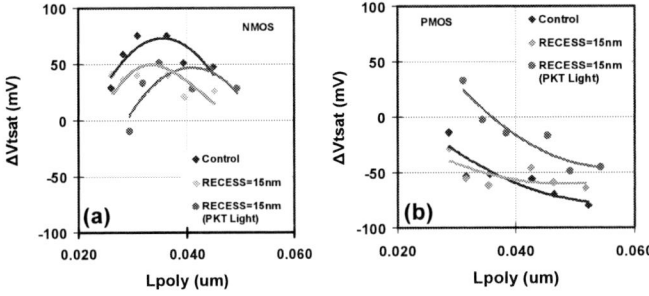

Figure 6. Change in saturation threshold voltage with decreasing gate length, for logic devices with $0.25\mu m$ drawn width. **(a)** NMOS **(b)** PMOS.

III. MEASURED 28NM SRAM RESULTS

With lighter pocket doping to reduce the impact of RDF, the quasi-planar structure results in reduced variability in SRAM read margin (SNM) and write margin (WRM) when the STI oxide is recessed, as shown in **Fig. 7**.

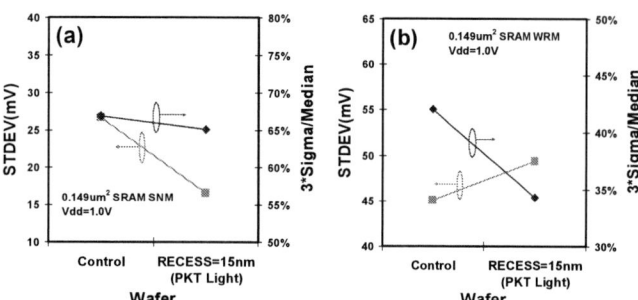

Figure 7. Sigma and 3sigma/median values for **(a)** read margin (SNM) and **(b)** write margin (WRM). V_{dd} = 1.0V.

Supply-voltage (V_{dd}) reduction is desirable to reduce power density and/or facilitate increased transistor density. Generally, however, relative variability increases as the gate overdrive (V_{dd}-V_T) decreases, so that yield (gauged by 3-sigma/median) is degraded. **Fig. 8** shows that the degradation in SNM yield with V_{dd} scaling can be dramatically reduced for quasi-planar bulk CMOS technology. With optimized pocket implant doses for both NMOS and PMOS devices (not achieved here), similar improvement is expected for WRM yield.

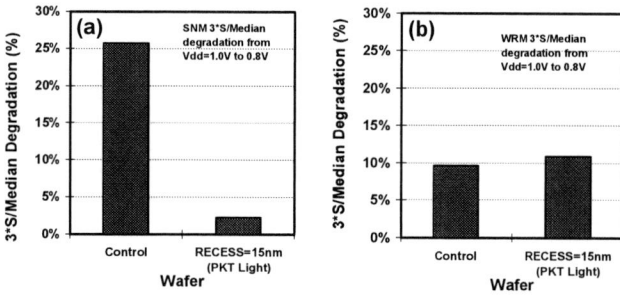

Figure 8. Degradation in 3-sigma/median for (a) SNM and (b) WRM as V_{dd} is reduced from 1.0V to 0.8V.

The reverse narrow width effect, *i.e.*, V_T reduction with decreasing channel width (W), stems from increased gate control for narrower channel width (due to fringing electric fields between the gate electrode and channel sidewalls). This effect is slightly worsened when the STI oxide is recessed, *i.e.*, quasi-planar devices show slightly increased sensitivity of V_T to variations in W, as shown in **Fig. 9**. Overall, however, variability is reduced for quasi-planar devices due to improved short channel control, which provides for the improved SRAM yield seen in Figs. 7 and 8.

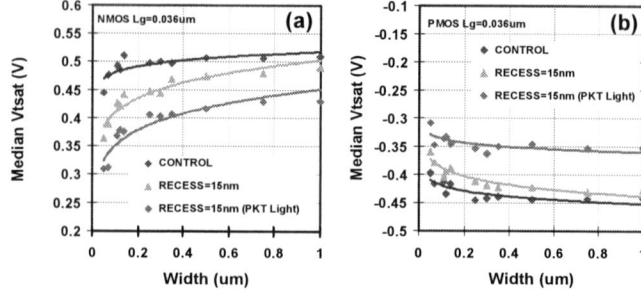

Figure 9. Measured reverse narrow width effect for devices with 36nm gate length: (a) NMOS (b) PMOS. Median V_T is lower when the STI oxide is recessed, due to improved gate control over the channel potential.

IV. 22NM BULK SRAM CELL DESIGN STUDY

Layout dimensions for 22nm (25nm drawn gate length) 6-T SRAM cells were selected based on recent publications [6-10] and are summarized in **Table I** for a conventional notched cell layout.

Table I. 6T- SRAM cell layout dimensions. Schematic half-cell plan views are shown for a conventional notched layout (upper left) and a smaller notchless layout (lower left) with W_{PD} = W_{PG} = W_{PU} = 35nm (area = 0.0684 μm^2).

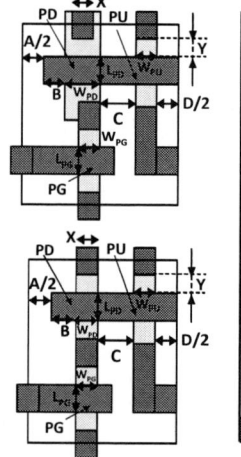

	Design rules	Symbol	Size [nm]
Cell Height	PG CH length	L_{PG}	25
	PD CH length	L_{PD}	25
	CONT size	X	30
	Gate-to-CONT	Y	20
	Total		**190**
Cell Width	POLY-to-POLY	A	30
	POLY-to-DIF ext	B	20
	PD Width	W_{PD}	55
	N/P isolation	C	50
	PU width	W_{PU}	30
	DIF-DIF (min)	D	50
	PG width	W_{PG}	30
	H_{si} for quasi-planar	H_{si}	10
	Total		**390**
	A SRAM cell area		**0.0741 μm^2**

The quasi-planar bulk MOSFET design was optimized via 3-D device simulations to achieve the highest I_{ON} for I_{OFF} = 3nA/um, at V_{dd} = 1V: electrical channel length (distance between the points where the source/drain doping profiles fall to $2 \times 10^{19} cm^{-3}$) L_{eff} = 27nm; effective oxide thickness EOT = 9Å; source/drain extension junction depth $X_{J,ext}$ = 10nm. Near-band-edge gate work functions (4.2eV for NMOS, 5.1eV for PMOS) are assumed. The STI recess depth (H_{si}) is 10 nm, to maintain a low aspect ratio for ease of fabrication. The retrograde channel doping profile is assumed to have a gradient of 4nm/dec and peak doping concentration = $10^{19} cm^{-3}$ at a depth T_{si} below the top channel surface. The planar bulk MOSFET design (for comparison) was optimized in the same manner.

The benefit of quasi-planar bulk CMOS technology for improved SRAM cell yield is assessed using the concept of cell sigma, defined as the minimum amount of variation for read/write failure [11-12]. Random V_T variation due to gate

978-1-4244-6658-0/10 $26.00 © 2010 IEEE

line-edge-roughness (LER) and RDF was estimated from 3-D Monte Carlo device simulations with realistic gate profiles and atomistic doping profiles, as described in [12]. Random V_T variation due to gate work function variations was estimated from [13]. Variations in transistor performance due to systematic variations in L_{eff}, W, EOT, and H_{si} (each assumed to have Gaussian distributions, with $\pm10\%$ corresponding to 3-sigma variation) as well as random V_T variations are considered in estimating SRAM cell yield.

As shown in **Fig. 10**, quasi-planar SRAM cells are projected to provide for >1 sigma improvement in read and write yields as compared with the planar SRAM cell, across a wide range of V_{dd}, primarily due to reduced V_T variation. As described in [14], the quasi-planar bulk MOSFET offers a new method of V_T adjustment, via tuning of the retrograde channel doping depth, to mitigate the tradeoff between reduced RDF-induced random V_T variation and improved short-channel control. This feature is leveraged in the notch-less quasi-planar SRAM cell design, in which the PD and PU devices have deeper retrograde channel doping profiles (such that T_{Si} > H_{Si}) and hence lower V_T [14], to achieve cell beta ratio > 1. The optimal T_{si} values to satisfy the 6-sigma (read and write) yield requirement were determined to be 14nm/10nm/14nm for the PD/PG/PU devices. For V_{dd} down to ~0.8V, a notch-less quasi-planar cell design with $W_{PD} = W_{PG} = W_{PU} = 35nm$ can meet the 6-sigma yield requirement while providing for significant (~10%) cell area savings as compared to the conventional notched cell design. (Here W_{PU} is constrained to be equal to W_{PD} and W_{PG} so as to be compatible with a regularly corrugated starting substrate [5] for improved W control.) Interestingly, the nominal SNM for the notch-less quasi-planar cell design is less sensitive to V_{dd}: it decreases by only 52 mV (from 174 mV to 122 mV) as V_{dd} is decreased from 0.9V to 0.5V, whereas the nominal SNM for the notched planar cell design decreases by 92 mV (from 180mV to 88mV) over the same V_{dd} range. This is because the benefit of the quasi-planar structure (improved sub-threshold swing) is greater for the narrower PD devices used in the notch-less cell design, so that they operate in the linear regime down to lower V_{dd}. Since variability decreases with decreasing V_{dd} (due to reduced short-channel effect and drain-induced barrier lowering), the SNM cell sigma stays relatively constant with decreasing V_{dd}, down to 0.6V, for the notch-less quasi-planar cell design. It should be noted that the notch-less cell design is expected to provide for reduced transistor mismatch arising from systematic variations in W, but that this benefit is not presumed for the SRAM cell yield analysis herein.

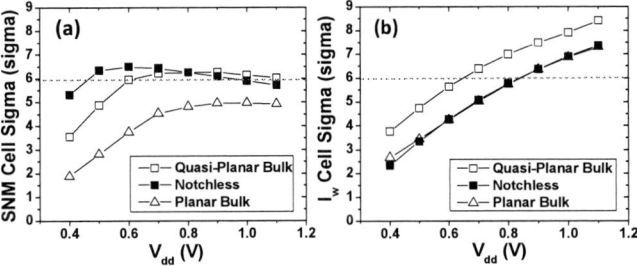

Figure 10. Comparison of cell sigma *vs.* V_{dd} for (a) read static noise margin (SNM) and (b) writeability current (I_W).

V. CONCLUSION

A simple process for achieving quasi-planar tri-gate bulk MOSFET structures is demonstrated in an early 28nm CMOS technology. With appropriate adjustments in the pocket implant doses, quasi-planar bulk CMOS technology can provide for improved performance and reduced variability, and thus can facilitate the scaling of SRAM operating voltage. Since the tradeoff between reduced RDF-induced random V_T variation and improved short-channel control is mitigated for quasi-planar CMOS technology, V_T (rather than W) adjustment can be used to tune the cell beta ratio and meet the 6-sigma yield requirements with a smaller cell.

ACKNOWLEDGMENTS

This work was supported in part by the Center for Circuit & System Solutions (C2S2) Focus Center, one of six research centers funded under the Focus Center Research Program, a Semiconductor Research Corporation program. C. Shin appreciates the support of the Korea Foundation for Advanced Studies (KFAS), and would like to thank Dr. Yasumasa Tsukamoto (Renesas Electronics, Japan) for his helpful discussion.

REFERENCES

[1] E. Josse *et al.*, "A cost-effective low power platform for the 45-nm technology node," *IEDM Tech. Dig.*, pp. 693-696, 2006.

[2] K. Nii *et al.*, "A 45-nm single-port and dual-port SRAM family with robust read/write stabilizing circuitry under DVFS environment," *Symp. VLSI Circuit Dig.*, pp. 212-213, 2008.

[3] X. Sun *et al.*, "Tri-gate bulk MOSFET design for CMOS scaling to the end of the roadmap," *IEEE Electron Device Letters*, Vol. 29, No. 5, pp. 491-493, 2008.

[4] M. Kito *et al.*, "Vertex channel array transistor (VCAT) featuring sub-60nm high performance and highly manufacturable trench capacitor DRAM," *Symp. VLSI Tech. Dig.*, pp. 32-33, 2005.

[5] U.S. Patent 7,190,050.

[6] H.S. Yang *et al.*, "Scaling of 32nm low power SRAM with high-K metal gate," *IEDM Tech. Dig.*, pp. 233-236, 2008.

[7] H. Kawasaki *et al.*, "Demonstration of highly scaled FinFET SRAM cells with high-K/metal gate and investigation of characteristic variability for the 32nm node and beyond," *IEDM Tech. Dig.*, pp. 237-240, 2008.

[8] B.S. Haran *et al.*, "22nm technology compatible fully functional $0.1\mu m^2$ 6T-SRAM cell," *IEDM Tech. Dig.*, pp. 625-628, 2008.

[9] C.H. Diaz *et al.*, "32nm gate-first high-k/metal-gate technology for high performance low power applications," *IEDM Tech. Dig.*, pp. 629-632, 2008.

[10] F. Arnaud *et al.*, "32nm general purpose bulk CMOS technology for high performance applications at low voltage," *IEDM Tech. Dig.*, pp. 633-636, 2008.

[11] A.E. Carlson, "Device and Circuit Techniques for Reducing Variation in Nanoscale SRAM," Ph.D. dissertation, Univ. California Berkeley, 2008.

[12] C. Shin, *et al.*, "Performance and area scaling benefits of FD-SOI technology for 6-T SRAM cells at the 22nm node," *IEEE Trans. Electron Devices*, Vol. 57, No. 6, pp. 1301-1309, 2010.

[13] H. Dadgour, K. Endo, V. De, and K. Banerjee, "Modeling and analysis of grain-orientation effects in emerging metal-gate devices and implications for SRAM reliability," *IEDM Tech. Dig.*, pp. 705-708, 2008.

[14] C. Shin, X. Sun, and T.-J. King Liu, "Study of random-dopant-fluctuation (RDF) effects for the tri-gate bulk MOSFET," *IEEE Trans. Electron Devices*, Vol. 56, No. 7, pp. 1538-1542, 2009.

Impact of fast-recovering NBTI degradation on stability of large-scale SRAM arrays

Stefan Drapatz*, Karl Hofmann‡, Georg Georgakos‡, and Doris Schmitt-Landsiedel*

*Institute for Technical Electronics, Technische Universität München, 80290 Munich, Germany, Email: stefan.drapatz@tum.de

‡Infineon Technologies AG, 85579 Neubiberg, Germany

Abstract—**This paper presents stability analysis of large-scale SRAM arrays directly after terminating NBTI stress. While the impact of static NBTI is well examined for cells and arrays, the fast-recovering component was not yet measured on SRAM arrays. The novel method presented here analyzes the flipping of cells directly after the supply voltage was lowered to a specific value where the structure is most sensitive for NBTI induced cell flips. Thus, read margin criterion is used to characterize the decreasing cell stability due to NBTI degradation with a resolution down to 1 ms. Applying this method, the impact of static and dynamic NBTI is measured in a 65 nm low power CMOS technology. Between 1 ms and 10.000 s after stress, the NBTI induced number of cell flips decreases by almost one half.**

I. INTRODUCTION

On 6-T SRAM core cells, the degrading effect of Negative Bias Temperature Instability (NBTI) predominantely affects cell stability. Simulations show that -100 mV V_{th} shift of one pMOS pullup due to NBTI result in approx. -7% hold stability and -10% read stability, while writeability and speed are not negatively influenced. In the past, several approaches were developed to characterize the stability of SRAM cells [1] and SRAM arrays [2] [3], which could be and partly were used to measure the impact of static NBTI.

NBTI is a degradation effect that occurs on pMOS transistors in inversion at high gate voltage. The threshold voltage V_{th} is shifted to more negative values [3], which makes the device weaker. High temperature increases this voltage shift. NBTI is partly a static, but even more a dynamic process [7]. Examinations on single transistors have shown that directly after termination of NBTI stress conditions, V_{th} shift starts to drop (Fig. 1). Reisinger et al. [4] have shown on large single transistors (W/L=10 µm/0.12 µm) in a 90 nm technology that after 100 s of stress with 2.8 V, V_{th} shift drops from 32 mV at 1 µs after end of stress down to 7 mV at 10.000 s after end of stress, which is a ratio of 80%.

The annealing time and ratio depends on stress time and stress voltage. Generally, with high stress voltages this annealing process is very fast, so that after a couple of seconds a major part of the voltage shift has disappeared. NBTI measurements are done under accelerating stress conditions (high T, high V_{DD}). Cooling down after stress takes some time, and V_{th} shift meanwhile decreases to a quasi-static value which does not change considerably within the next hours or days, compare Fig. 1. This value therefore is called the constant or static NBTI, although it will decrease further on a logarithmic time

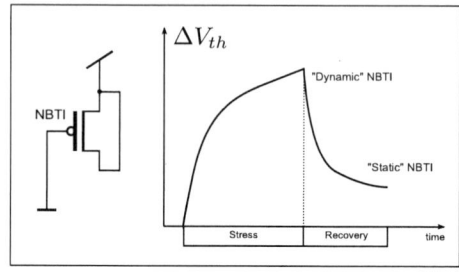

Fig. 1. Stress condition of NBTI and qualitative diagram of static and dynamic NBTI on single pMOS transistor: During stress, V_{th} shift grows quasi-logarithmically. Directly after end of stress, V_{th} rapidly decreases to a static value, which then only changes in very long timescales.

scale, with very long time constants (weeks to months). Altogether, this means that with nominal measurement techniques at room temperature, it is impossible to detect the dynamic part of NBTI degradation effect. This is why it has not been measured on SRAM arrays.

It is important to know this worst effect of NBTI on SRAM stability, i.e. what is the impact of V_{th} shift directly after end of stress. Especially in SRAM, storing one value for a long time represents the DC worst case, while in logic circuits, the dominating AC case allows continuous recovering. Yet all investigations on SRAM arrays only measure the static NBTI effects.

Static Noise Margin (SNM) [1] as a stability metrics is suitable for simulation. Other techniques, e.g. Read Margin (RM = difference between reduced core voltage where the cell flips and nominal voltage) determined by current measurement [2], are suitable for single cell analysis. The flipping cell analysis [3] is particularly suitable for fast array measurement. Therefore in this work, we have extended it for the task of dynamic NBTI characterization.

Test chip and the measurement setup incl. specific challenges are described in the next chapter. Results are presented in chapter III, followed by a conclusion.

II. EXPERIMENTAL SETUP

A. Universal SRAM test array

All measurements for this paper are performed on a 1 MBit SRAM array test structure in a 65 nm low power technology [5] with 2.816 cells per bitline and 384 cells per wordline

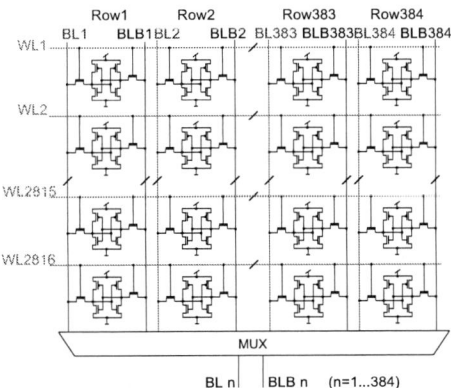

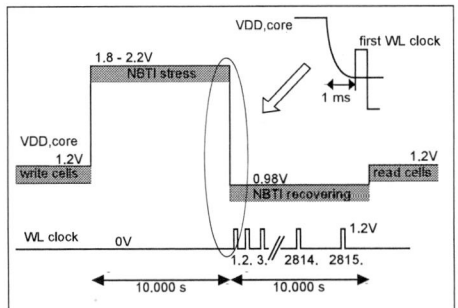

Fig. 2. 1 MBit 65 nm SRAM array with bitline- and wordline- shift registers and multiplexed bitline pairs for read current measurement in each cell.

Fig. 4. Schematic overview of measurement approach

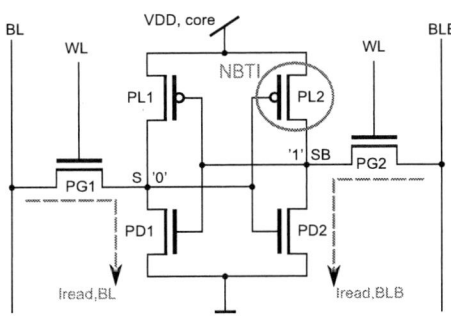

Fig. 3. Single 6T SRAM cell used in the testchip array. The read currents $I_{read,BL}$ and $I_{read,BLB}$ can be measured individually in each cell. This cell is in state '0', and pullup transistor PL2 is degraded during NBTI stress.

$delay = 1.0038^{\#WL}$		$delay = 1.007^{\#WL}$	
#WL	t betw. 2 WL	t after stress	t betw. 2 WL	#WL
1	1.0038 ms	1 ms	1.007 ms	1
10	1.038 ms	10 ms	1.07 ms	10
85	1.38 ms	100 ms	1.7 ms	76
413	4.78 ms	1 s	8 ms	298
965	39 ms	10 s	70 ms	610
1566	0.38 s	100 s	0.7 s	938
2173	3.8 s	1 ks	7 s	1268
2800	41 s	10 ks	70 s	1600

TABLE I
RELATION BETWEEN WL NUMBER AND TIME AFTER STRESS. FOR TIMES >0.1 S, THE WL RESULTS IN AN ALMOST EXPONENTIAL X AXIS. ONLY VERY SHORT TIMES ARE COMPRESSED.

(Fig. 2). Shift registers on bitlines and wordlines allow to select single cells in the array without a large number of address pins, and bitline-pairs are multiplexed to 2 external pins for direct read current measurement (Fig. 3) [6]. Selecting one bitline-pair causes all other bitlines being clamped to 1.2 V. Thus, all cells along the WL are in read condition, which is the so-called half-select state (WL=BL=BLB=1). All other cells in the array are in hold condition.

B. New Measurement Methodology

Stability analysis is based on the flipping behavior of the cells. The goal is to detect cell stability over time after stress with an ordinary SRAM array. Therefore, only a part of the cells should be sensitive to NBTI degradation in each time slot, while the rest is not affected at all. We start from the fact that read state is less stable than hold state, and that all 384 cells along one wordline (WL) can be set to read state, while the rest is kept in hold state. By sequentially activating only one WL at a time, the complete array serves as a time-dependent NBTI sensor: the cells in read state flip according to the actual NBTI degradation, while the cells in hold state are too stable to flip. This technique is illustrated in Fig. 4. After writing all cells in the array to one state (here '0'), $V_{DD,core}$ is ramped up to the NBTI acceleration voltage (in our case between 1.8 V and 2.2 V) for 10.000 s to stress pullup transistor PL2 (Fig. 3). To get sufficient NBTI acceleration and recovery conditions, the complete measurement is performed at 125 °C. The periphery voltage, i.e. all WL and BL drivers, is always kept at $V_{DD,nom}$=1.2 V. After stressing, $V_{DD,core}$ is lowered to a critical voltage of 0.98 V, and 1 ms later, the first WL clock activates the first WL, setting all 384 cells to read condition. Now these cells flip according to the actual NBTI-caused V_{th} drift.

Another 1 ms later, a shift register creates the next WL clock, switching off the first WL and activating the second WL. Then, the first WL is set back to hold condition and the second WL now is set to read condition. This is repeated until the WL clock was activated 2815 times and the whole array was sequentially set to read condition for a short fraction.

Switching the WLs is done on an exponential time scale. This means that after waiting $1.0038^{\#WL}$ ms at each WL, the shift register switches to the next WL, which results in an exponential-like x axis time scale, only the very short times are compressed (left half of Table I). Time between WL ♯1 and WL ♯2 therefore is 1.0038 ms, time between WL♯2799 and ♯2800 is approx. 41 s. Complete recovery time is therefore 10.000 s, which is as long as the stress time.

The cells are not read out during this procedure, this would take much too long. They are only set to read conditions all in parallel, which is a kind of 'simulated read'. After WL deactivation, the flip pattern is kept in the stable hold state, after all it is a memory array! Reading of the flipped cells is done after the relaxation process with nominal $V_{DD,core}$, when reading does not influence the cell states anymore. In

978-1-4244-6658-0/10 $26.00 © 2010 IEEE

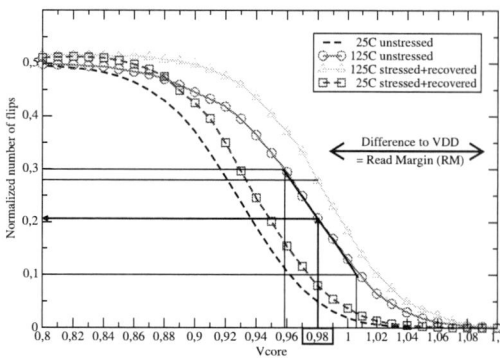

Fig. 5. Flip curve of WL ♯1500 at 25 °C and 125 °C in unstressed and stressed+recovered state. The highest NBTI sensitivity at 125 °C together with a low number of flips in unstressed condition could be found at 0.98 V.

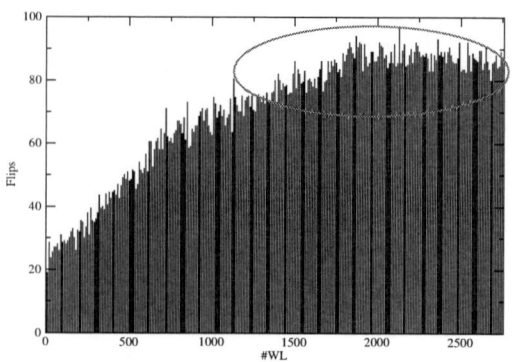

Fig. 6. Flips along 2800 WLs for **unstressed array** show non-constant behavior due to a design imperfection. The quasi-constant WLs ♯1150 to ♯2750 in the ellipse are used for measurements within this paper.

our approach, we only want to detect the flipping of a cell, so the measured BL current was used only to detect the cell state. This could be performed even faster in an ordinary SRAM array, as the bitline current measurement is no prerequisite of this approach. The only prerequisites are separate V_{DD} rails for the core array and for the periphery. Reduced core voltage is necessary for the flip-cell analysis (section II-B), enlarged core voltage is necessary for accelerated NBTI stress.

Read margin analysis (stepwise decreasing V_{core} and counting the number of flips [3]) of the unstressed and of the recovered array has given the flip curve in Fig. 5. From the 125 °C curves, the most NBTI sensitive core voltage was determined to 0.98 V: there the NBTI sensitivity was maximum ($\partial Flips/\partial Vth, NBTI \approx 1.5 Flips/mV$), together with only 20% of flips in unstressed condition, leaving enough headroom for the NBTI-induced V_{th}-shift.

C. Measurement challenges

Temperature is kept stable at 125 °C ± 0.05 °C during write, stress, recover and read. Otherwise, the flip count would vary due to temperature fluctuations. Also writing at 25 °C and heating up quickly can cause random cell flips. On Fig. 5 the shift of the flip curve between 25 °C and 125 °C can be seen. Simulations confirm 20% less stability at 125 °C compared to 25 °C. A Peltier Element together with a PID regulator enables the circuit to keep the temperature.

A fast falling voltage slope must be generated at switching from stress to recovery voltage to reach a high resolution time in the ms range. This is done with an active filter with a time constant of 100 μs. This guarantees a slope time of about 0.5 ms independent of the impedance of the array (which changes with temperature alteration). Using a switch was avoided to guarantee a well defined timing behavior. Furthermore unintended flips could occur if the slope was too fast. This does not happen with the used slope, which was verified via experiments. A stable voltage is required especially during the recovery period. The sensitivity to $V_{DD,core}$ is comparable to the sensitivity to V_{th} drift: ($\partial Flips/\partial VDD \approx \partial Flips/\partial Vth, NBTI \approx 1.5 Flips/mV$). So if a drift of some

tenths of Volts is to be measured, $V_{DD,core}$ must be stable to some mV.

D. Testchip challenges

In Fig. 6 the flip curve of the unstressed array is plotted (low-pass filtered over 10 WLs using a median filter). The unstressed 20% flips at 0.98 V refer to 384 cells, which equals approx. 80 flips that are visible at WL ♯1500. Ideally, this should show constant behavior without NBTI stress, but a gradient is visible. Examinations have shown that this is due to a design imperfection of the test chip. This was reported also at read current measurements with the same test chip in [6]. The reason is IR drop along the wordlines resulting in a gradient in access-transistor gate-voltages. This could be avoided by dividing the 1 MBit array into smaller subarrays. Here, the simple countermeasure was only to use the 1600 WLs ♯1150 to ♯2750 for the measurements. The x axis is then sampled with the law $1.007^{\sharp WL}$ ms to cover 10.000 s of recovery (right half of Table I).

III. RESULTS

The flip curves between WL ♯1150 and ♯2750 after stressing the array for 10.000 s with various stress voltages at 125 °C are presented. All graphs are low-pass filtered over 10 WLs using a median filter. Additionally, the trend of these filtered data obtained by another moving-average low pass filter is drawn in each diagram.

The first stress experiment was done with $V_{DD,core}$=1.8 V. Fig. 7 shows the flips directly after stress minus the flips before stress. The almost constant gradient towards less flips, i.e. more stability due to recovering NBTI-caused V_{th} shift, can be seen clearly. The same recovery curve is plotted in Fig. 8 for a stress voltage of 2.0 V and in Fig. 9(a) for a stress voltage of 2.2 V. After measurement with 2.2 V was finished, this experiment was repeated the next day without stressing the array again. This was done to check for the static NBTI component long after stress and to prove the validity of this technique, i.e. that the gradient has no other reason than NBTI recovery. Fig. 9(b) confirms a constant flip curve

978-1-4244-6658-0/10 $26.00 © 2010 IEEE

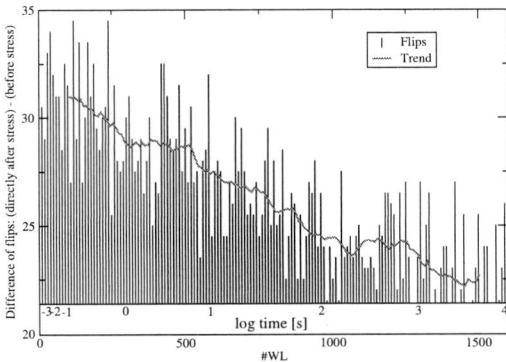

Fig. 7. Flips directly after stress minus flips before stress with **1.8 V stress voltage** show recovery behavior. The difference between 1 ms and 10.000 s after stress is approx. 10 flips.

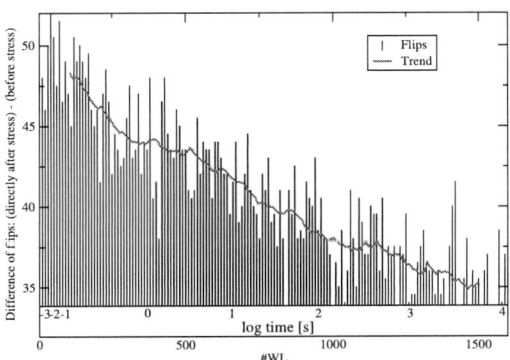

Fig. 8. Flips directly after stress minus flips before stress with **2.0 V stress voltage** show recovery behavior. The difference between 1 ms and 10.000 s after stress is approx. 15 flips.

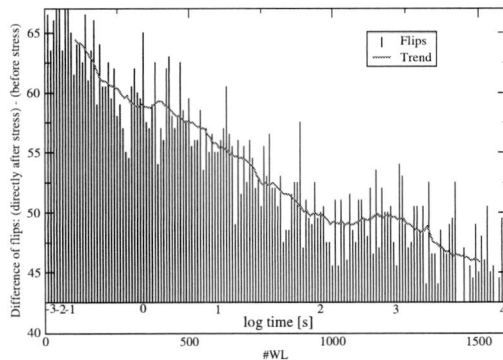

(a) Flips directly after stress minus flips before stress show recovery behavior. The difference between 1 ms and 10.000 s after stress is approx. 20 flips.

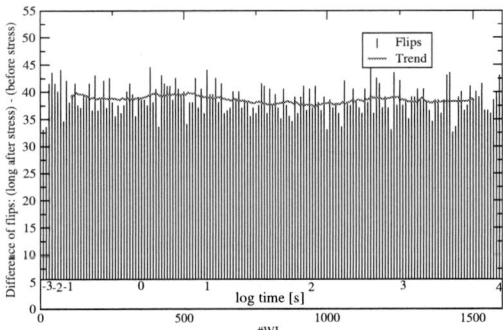

(b) Flips ≫10.000 s after stress minus flips before stress show constant behavior. Approx. 40 flips more in each WL due to static NBTI long after stress.

Fig. 9. Flip curves directly (a) and very long (b) after stress with **2.2 V stress voltage** show recovery and static behavior, respectively.

of approx. 40 flips difference between long after stress and before stress. This shows that during 10.000 s recovery time, the flips decrease from 65 to 45, then saturate at about 40. This experiment was also done for the 1.8V curve, again resulting in a constant flip difference of approx. 17 flips.

IV. CONCLUSION

A new concept to measure fast-recovering NBTI directly on large SRAM arrays was presented. With a resolution of 1 ms, the stability of a 1 MBit 65 nm low power SRAM array was analyzed directly after end of stress. The recovering part of NBTI could be identified, which was not possible with former stability measurements. The stability decreasing effect of NBTI was measured 1 ms after end of stress until 10.000 s later.

The least cell stability, caused from worst NBTI degradation, appears directly after end of stress and starts to recover. This results in more stable SRAM cells over time, which was identified by an approx. 40% decrease of cell flips. This confirms and quantifies that the worst effect of NBTI on SRAM is reading a cell directly after keeping one value for a long time.

ACKNOWLEDGMENT

The authors thank E. Amirante, P. Huber and M. Ostermayr for designing the universal SRAM test structure. This work has been supported by the German Ministry of Education and Research (BMBF) within the project 'Honey' (Project ID 01M3184A). The contents is the sole responsibility of the authors.

REFERENCES

[1] E. Seevinck, F. List, and J. Lohstroh, *Static Noise Margin Analysis of MOS SRAM Cells*, IEEE Journal of Solid State Circuits, Vol. 22, 1987
[2] Z. Guo et al., *Large-Scale Read/Write Margin Measurement in 45nm CMOS SRAM Arrays*, IEEE Symposium on VLSI Ciruits, June 2008
[3] S. Drapatz et al., *Fast stability analysis of large-scale SRAM arrays and the impact of NBTI degradation*, ESSDERC, 2009
[4] Reisinger, H. et al., *Analysis of NBTI Degradation- and Recovery-Behavior Based on Ultra Fast VT-Measurements*, Reliability Physics Symposium Proceedings, 2006
[5] Z. Luo et al., *High Performance and Low Power Transistors Integrated in 65nm Bulk CMOS Technology*, Electron Devices Meeting, 2004
[6] T. Fischer et al., *Analysis of Read Current and Write Trip Voltage Variability From a 1MB SRAM test structure*, IEEE Transactions on Semiconductor Manufacturing, pp. 534-541, 2008
[7] Chen, G. et al.: *Dynamic NBTI of PMOS transistors and its impact on device lifetime*, Reliability Physics Symposium Proceedings, 2003

Experimental comparison of programming mechanisms in 1T-DRAM cells with variable channel length

Alexandre Hubert[1], Maryline Bawedin[2], Georges Guegan[1], Sorin Cristoloveanu[2], Thomas Ernst[1], Olivier Faynot[1]

[1]CEA-LETI, MINATEC, 17 rue des Martyrs 38054 Grenoble Cedex 9, France
[2]IMEP, INPG-MINATEC, 3 Parvis Louis Néel, 38016 Grenoble Cedex 1, France
Email: alexandre.hubert@cea.fr or thomas.ernst@cea.fr

Abstract—**The bulk DRAM scaling requirements have lead to many different concepts of capacitor-less single-transistor (1T) DRAM. Amongst the various effects used to program the cell, this study is focused on the Impact Ionization (II), the most common mechanism to store charges in the body of the cell, and the Meta-Stable Dip (MSD) effect. Dynamic measurements are presented showing the impact of the gate length reduction on both the II and the MSD programming mechanisms. It is found that MSD is less impacted by the scaling of standard SOI MOSFETs without specific optimization. Those attractive performances result from the dynamic coupling between the front and back gates in Fully Depleted SOI (FDSOI) transistors.**

I. INTRODUCTION

The DRAM industry is now facing an important crisis due to the increasing complexity to further scale down the storage capacitor. An attractive solution is simply to suppress the capacitor using the floating body of a single SOI transistor as the storage node (Fig.1). The 'ON' state (bit '1') reflects the presence of the majority carrier excess in the body which increases its potential and, as a consequence, the drain current. The 'OFF' state (bit '0') features a lower current obtained by the charges removal from the body. Historically, the Partially Depleted SOI (PDSOI) was the first SOI cell to be used for 1T-DRAM applications [1].

Recently, Fully Depleted SOI (FDSOI) and double-gate transistor have also been considered as interesting candidates for 1T-DRAM cells because of their high scalability compared to PDSOI. The undoped channel of FDSOI transistors is also a strong asset compared to PDSOI for variability issues [2]. Various types of 1T-DRAM cells have been proposed over the last ten years. The cell families can be classified by their 'ON' state programming method: the excess of majority carriers are either generated by impact ionization (II) [1], bipolar junction transistor effect (BJT) [3] or band-to-band (BTB) tunneling [4]. The impact ionization effect operates efficiently in the channel pinch-off region when the transistor is in strong inversion and the drain voltage is higher than the gate voltage. In those conditions, the electrons are strongly energized and are able to create electron-hole pairs at the drain side. The hole excess can be maintained in the FDSOI transistor body thanks to a negative back-gate bias (Fig. 1 (a)). To remove the charges, the body-drain junction is usually forward biased by setting the drain voltage (V_D) to negative values.

A novel one-transistor memory, named MSDRAM, has been proposed recently [5-7]. It is based on the MSD (Meta-Stable Dip) effect in FDSOI and double-gate transistors. MSD consists in a drain current time-dependent hysteresis resulting from the combination of floating-body transients and interface dynamic coupling effects. The particularity of this concept is that the holes are generated and maintained at the front interface while the back channel is set to inversion mode. In those conditions, the front interface charge is read by capacitive coupling with the inverted back channel. To write bit '1', majority carriers (holes) are generated by band-to-band (BTB) tunneling in the front-gate-to-drain overlap region (Fig. 1 (b)) by applying a negative voltage at the front-gate. The cell erasing is performed in two steps; the front-gate bias is set (i) to 0V to evacuate the holes outside the body and next (ii) to a negative value to force the deep depletion in the whole body, thus suppressing the back inversion channel. A hysteresis loop can be observed in the static mode by doing a drain current double-sweep versus the front-gate voltage (V_{FG}) with V_{FG} between negative values and 0V (Fig. 2).

In this paper, we focused on both impact ionization and MSD programming mechanisms from the scalability point of view. In particular, we investigate the impact of the channel length on the programming window and the retention time at 25°C and 85°C.

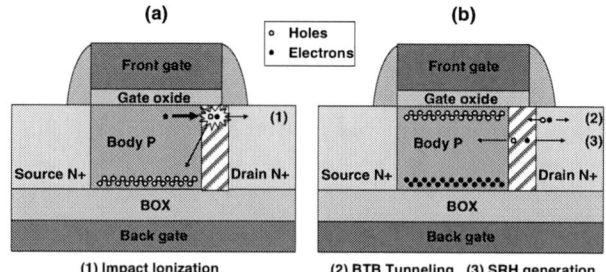

Figure 1. 'ON' state programming mechanisms for (a) impact ionization and (b) MSD.

978-1-4244-6658-0/10 $26.00 © 2010 IEEE

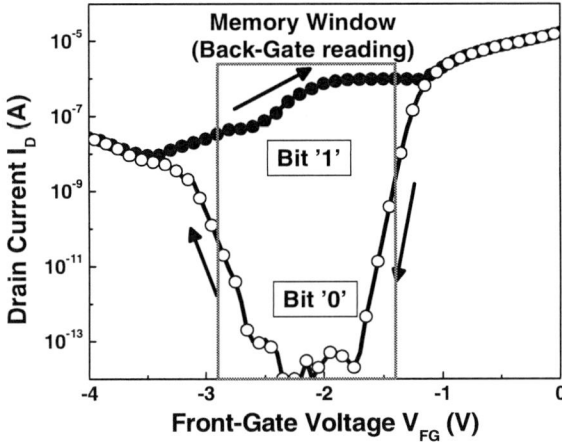

Figure 2. Double-sweep drain current characteristic versus the front-gate voltage obtained for FDSOI transistors.

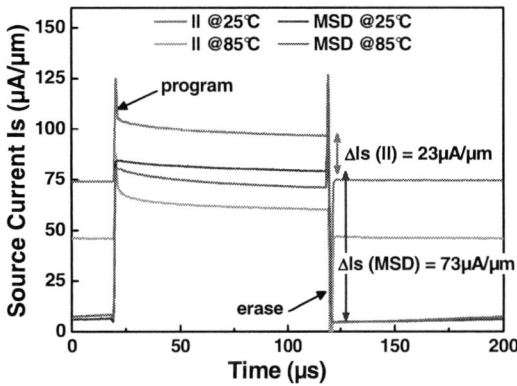

Figure 3. A FDSOI memory cell response to a cycle of writing (1µs) / reading (100µs) / erasing (1µs) / reading (100µs) for both impact ionization (II) and MSD effect at 25°C and 85°C on (W=5µm, L$_G$=0.35µm, T$_{Si}$=55nm). The programming window at 25°C obtained by MSD is more than three times larger than the one obtained by II.

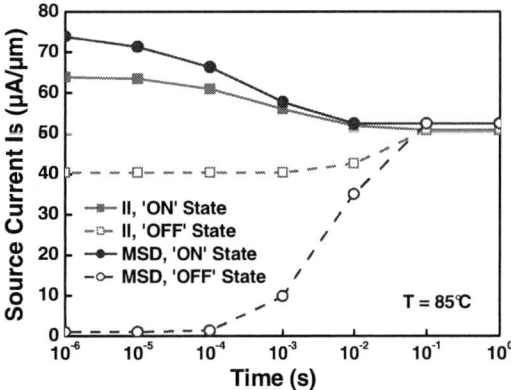

Figure 4. Data retention at 85°C for both impact ionization (II) and MSD effect on a FDSOI cell (W=5µm, L$_G$=0.35µm, T$_{Si}$=55nm). The retention at 85°C is almost the same for II and MSD induced programming mechanisms.

II. EXPERIMENT

A. Device design and experiment setup

MSD measurements were performed on standard FDSOI n-MOSFETs with undoped channel, 5-nm-thick gate oxide and 145-nm-thick buried oxide (BOX). The silicon film thickness T$_{Si}$ varied from 25 nm to 90 nm. Various channel lengths L$_G$ from 0.2µm to 2.5µm were investigated to understand their influence on both the impact ionization and the MSD effect. The measurements have been carried out using a specific setup to recover the dynamic response of the transistors. The drain and the front-gate were driven by pulse generators and the back-gate was set to an adequate value thanks to a constant voltage generator.

B. Measurements and signal optimisation

Two types of experiments have been carried out for each mechanism at both 25°C and 85°C. First, a cycle of writing (1µs) / reading (100µs) /erasing (1µs) / reading (100µs) pulses was applied to measure the amplitude of the memory window ΔIs (= Is 'ON state' – Is 'OFF state') (Fig. 3). The memory cell retention i.e. the source current difference ΔIs versus time was also measured (Fig. 4). The retention time T$_R$ is defined as the time when ΔIs is half of its initial value. The bias conditions have been set by searching the highest amplitude of the memory current window ΔIs for each transistor. The operation biases used at Fig. 3 and Fig. 4 are given in Table I.

III. IMPACT OF THE GATE LENGTH

A. Impact Ionization effect

1) Programming window ΔIs

The maximum ΔIs at the end of the reading pulse was obtained both at 25°C and 85°C for a 0.5µm-gate length (Fig. 5). For L$_G$ down to 0.5µm, the memory window increases when the gate length is reduced. This behavior is induced by the body volume reduction as L$_G$ decreases; which amplifies the impact of the amount of holes stored on the 'ON' state. On the other hand, when the gate length becomes too small (L$_G$ < 0.5µm), the programming window is drastically reduced: the current increase at the 'ON' state does not follow the amplification. This phenomenon seems to be induced by short channel effects (SCE) and drain induced barrier lowering (DIBL). These two phenomena reduce the efficiency of the barrier (built-in potential) at the body-source diode during the programming and enhance the hole leakage through the junctions as the gate length shrinks.

TABLE I. TYPICAL BIASES DURING MEMORY OPERATIONS FOR IMPACT IONIZATION AND MSD

		Drain Bias V$_D$ (V)	Front-Gate Bias V$_{FG}$ (V)	Back-Gate Bias V$_{BG}$ (V)
Impact Ionization Conditions	Program	1.65		
	Erase	-2.5	0.9	-17
	Read	0.5		
Meta-Stable Dip Conditions	Program	0.65	-4.25	
	Erase		0	18
	Read	0.5	-2.65	

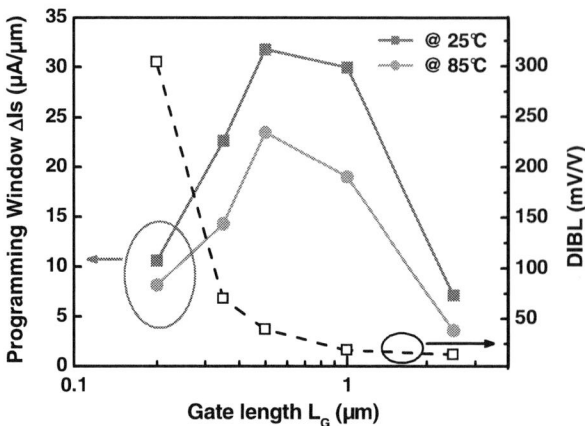

Figure 5. Impact of the gate length on the current margin with the impact ionization programming mechanism ($W = 5\mu m$ and $T_{Si} = 55nm$). The dashed curve shows the large DIBL obtained when $L_G < 0.5\mu m$ corresponding to the programming window drop.

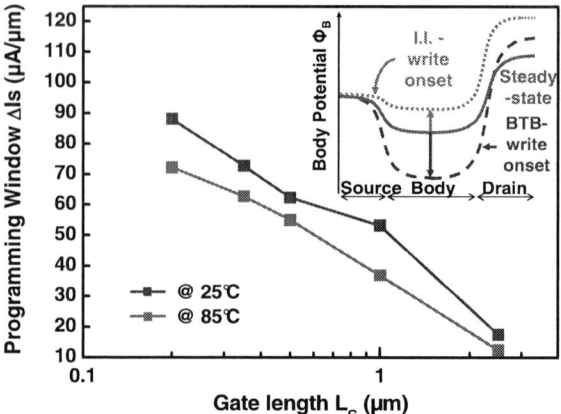

Figure 7. Impact of the gate length on the programming window of the MSD effect ($W = 5\mu m$ and $T_{Si} = 55nm$). Contrary to II induced programming, in the MSD mode ΔIs is not reduced by gate length scaling (inset).

2) Retention time T_R

The influence of the gate length on the retention time shows a sudden drop of T_R when L_G becomes lower than 1μm (Fig. 6). This drop is essentially due to a quicker diminution of the 'ON' state current when L_G is reduced. This effect is attributed to the increase of the junction leakage impact on the stored charges when the gate length is reduced. At 85°C, the retention time drops as soon as the gate length is reduced, demonstrating that the temperature degrades strongly the cell performances when the impact ionization mechanism is used.

B. Meta-stable dip effect

1) Programming window ΔIs

At 25°C, the influence of the gate length on the programming window in the MSD mode shows a quasi-linear increase of ΔIs when L_G is reduced. For the same reason as the impact ionization effect, when the gate length is reduced the memory window is amplified.

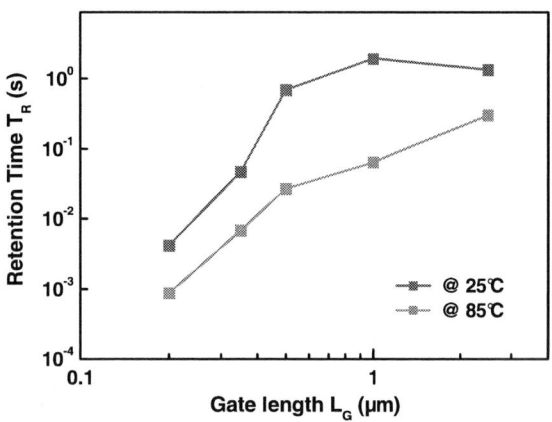

Figure 6. Influence of the gate length on the retention time at 50% of the impact ionisation effect ($W = 5\mu m$ and $T_{Si} = 55nm$).

However, regarding the impact ionization programming method, the DIBL effect is less efficient when the MSD is used. Indeed, the body potential becomes negative by dynamic capacitive coupling at the programming onset since the front accumulation layer cannot be formed instantly. Hence, the body-source junction is strongly reverse biased lessening the source hole leakage and keeping a high ON state current value for shorter channel lengths (i.e. <0.5V) (Fig. 7 inset).

2) Retention time T_R

The influence of the gate length on the retention time for the MSD operation mode is particular. Indeed, at 25°C the retention time is almost constant around 50ms for $L_G > 0.2\mu m$, then it suddenly drops for $L_G=0.2\mu m$. Compared to the impact ionization programming method, the MSD mechanism seems to be more resilient to the reduction of the gate length. On the other hand, the retention time resulting from the MSD programming is not only exposed to the loss of the holes stored in the body during the read/hold operation after writing the 'ON' state. Indeed, the parasitic BTB tunneling generation of holes at the front gate-to-drain overlap during the reading/holding of the 'OFF' state also degrades the current which tends to return to equilibrium faster. Depending on the cell gate length, the retention time is not driven by the same effect. When L_G is above 0.35μm, as the BTB tunneling is not impacted by the gate length, a slight T_R decrease can be observed. For smaller L_G, the lateral depletion at the drain and source-body junctions becomes significant and deteriorates the vertical dynamic gate coupling efficiency and consequently the potentials drop leading to the strong drop of the retention time. The 85°C curve shows the same tendency than for 25°C but all retention times are degraded by around one decade.

978-1-4244-6658-0/10 $26.00 © 2010 IEEE

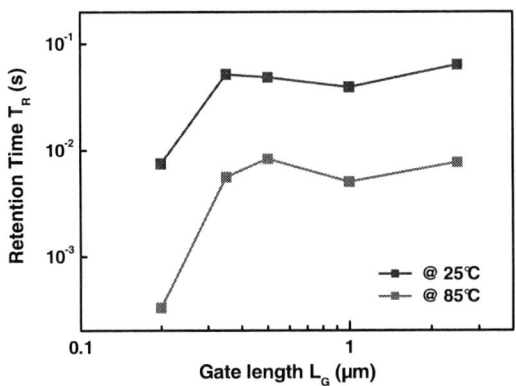

Figure 8. Influence of the gate length on the retention time at 50% of the MSD effect (W = 5µm and T_{Si} = 55nm). Until L_G=0.2µm, the retention of the MSD induced programming is almost not affected by the reduction of gate length contrary to the impact ionization programming method.

IV. CONCLUSION

Our experimental investigation sheds light on the influence of the gate length reduction on both the impact ionization and the meta-stable dip programming methods in FDSOI transistors. It provides direct information about the scalability of these effects. For the impact ionization programming method, our results confirms the previous ones [8,9] showing that the shrinking of the gate length, in a first time, increases the programming window. But when L_G becomes lower than 0.5µm, short-channel effects and more specifically DIBL strongly enhance the hole leakage through the body-source junction. On the other hand, since the MSD programming method uses the BTB mechanism and dynamic gate coupling, the programming window can be improved for shorter gate length. Thanks to this phenomenon, the programming window induced by MSD reaches 88µA/µm when L_G=0.2µm (almost 9 times the programming window obtained by impact ionization). The comparison of the impact of the gate length on the retention for the two floating body mechanisms shows once again that the MSD programming method seems more resilient to the gate length reduction than the impact ionization one. Indeed, for both 25°C and 85°C, the retention time for impact ionization is continuously decreasing with L_G, whereas for the MSD method, the gate length has to be reduced to 0.2µm to start to observe a significant drop of the retention time.

We have experimentally confirmed the MSD memory effect using dynamic measurement for the first time. A large usable memory window has been observed with a retention time of almost 10ms at room temperature. The transistors under test were not optimized for memory applications.

Optimization (via junction engineering) is expected to further improve the current ratio, the memory window, the retention time and the programming speed. Since the transistors were operated at low drain voltage, the power consumption and device reliability are very competitive.

A comparison with the impact ionization programming method has also been presented. This study demonstrates the strong potential of the MSD programming mechanism as a viable solution for low-power single-transistor DRAM memories.

ACKNOWLEDGMENT

The authors thank Dr F. Andrieu for device fabrication and pertinent scientific remarks. Nanosil, Eurosoi and WCU programs are acknowledged for their support.

REFERENCES

[1] S. Okhonin, M. Nagoga, J.M. Sallese and P. Fazan, "A SOI Capacitorless IT-DRAM Concept", IEEE Int. SOI Conf., 153-154 (2001)

[2] H. Furuhashi, T. Shino, T. Ohsawa, F. Matsuoka, T. Higashi, Y. Minami, H. Nakajima, K. Fujita, R. Fukuda, T. Hamamoto, and A. Nitayama, "Scaling Scenario of Floating Body Cell (FBC) Suppressing Vth Variation Due to Random Dopant Fluctuation", IEEE Int. SOI Conf., 33-34 (2008)

[3] S. Okhonin, M. Nagoga, E. Carman, R. Beffa, E. Faraoni, "New Generation of Z-RAM", IEDM Technical Digest. pp.925-928 (2007)

[4] C. Kuo, T.-J. King and C. Hu, "A capacitorless double gate DRAM technology for sub-100-nm embedded and stand-alone memory applications", TED, Vol. 50, n°12, 2408-2416 (2003)

[5] M. Bawedin, S. Cristoloveanu and D. Flandre, "A capacitor-less 1T-DRAM on SOI based on dynamic coupling and on double gate operation", IEEE Electron Device Letters, 29, 795-798 (2008)

[6] M. Bawedin, S. Cristoloveanu, J. G. Yun and D. Flandre, "A new memory effect (MSD) in fully depleted SOI MOSFETs", Solid-Sate Electronics, 49, 1547-1555 (2005)

[7] A. Hubert, M. Bawedin, S. Cristoloveanu, T. Ernst, "Dimensional effects and scalability of Meta-Stable Dip (MSD) memory effect for 1T-DRAM SOI MOSFETs", Solid-Sate Electronics, 53, 1280-1286 (2009)

[8] N. Collaert, M. Rosmeulen, M. Rakowskia, R. Rooyackers, L. Witters, A. Veloso, J. Van Houdt and M. Jurczak, "Comparison of scaled floating body RAM architectures", IEEE International SOI Conf., 35-36 (2008).

[9] T. Shino, N. Kusunoki, T. Higashi, T. Ohsawa, K. Fujita, K. Hatsuda, N. Ikumi, F. Matsuoka, Y. Kajitani, R. Fukuda, Y. Watanabe, Y. Minami, A. Sakamoto, J. Nishimura, H. Nakajima, M. Morikado, K. Inoh, T. Hamamoto, A. Nitayama, "Floating body RAM technology and its scalability to 32 nm node and beyond", IEDM Tech. Dig., pp. 569–572 (2006).

Substrate bias dependency of sense margin and retention in bulk FinFET 1T-DRAM cells

N. Collaert, M. Aoulaiche, A. De Keersgieter, B. De Wachter, M. Jurczak and L. Altimime

Imec, Kapeldreef 75, 3001 Heverlee, Belgium
E-mail: collaert@imec.be; Tel: +32 16 28 16 41

Abstract—The substrate bias in bulk FinFET devices can be used to increase both the sense margin and retention time in 1T memory cells. For given biasing conditions, a substrate bias can be found where sense margin and retention time are optimal. This substrate bias results from a trade-off between the storage of electrons and holes and the impact of the READ conditions.

I. INTRODUCTION

The use of floating body effects in single transistor (1T) memory cell applications goes back as far as the late 1970s [1]. With the increasing difficulties of scaling down the capacitor in DRAM, the use of capacitor-less 1T memory cells and especially floating body RAM (FB-RAM) has gained a lot of interest over the last decade. A wide variety of biasing schemes [2, 3] and device architectures have been investigated [3-5]. Apart from SOI-based cells, bulk devices have been studied in [5, 6] using either triple wells or the doping gradient to create a quasi floating body. In this work, we will focus on the use of bulk FinFET devices where the ground plane (GP) doping is used to create a potential well for electrons in the top part of the fin [6]. In this case, the substrate bias V_{BS} is an additional parameter which can be used to optimize the sense margin and retention. After a short description of the device fabrication in section II, we will first present the impact of the substrate bias through TCAD simulations. In section IV, we will show the measurement results. The V_{BS} dependence of sense margin and retention will be discussed and linked to the simulation results. Finally, a summary will be given in section V.

II. DEVICE FABRICATION

The starting material was a standard p-type bulk Si (100) wafer. Fin widths down to 10nm were fabricated using 193nm optical lithography and a hard mask (HM) based dry etch. After the formation of the Shallow Trench Isolation (STI), the field was recessed defining the fin height to be 60nm. The well and Ground Plane (GP) implantations were done and the gate stack consists of a 5nm thick SiO_2 gate dielectric and 5nm TiN as gate electrode capped with 100nm poly. After gate patterning, the extensions were implanted and the nitride spacer was formed. A NiPt-based salicide process was used after the deep S/D implants and spike

anneal. Finally, a standard Cu Back-End-Of-Line (BEOL) process was used to finish the devices. Figure 1 shows a tilted SEM picture of the bulk FinFET device after silicidation.

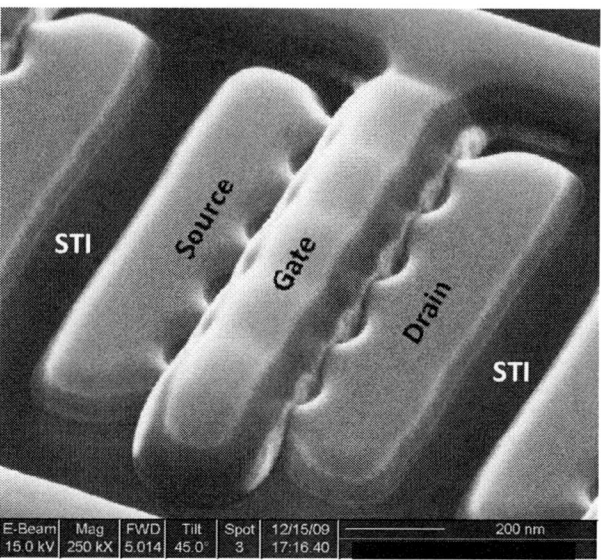

Figure 1. Tilted SEM of a bulk FinFET device after NiPt-silicidation; in this case the device dimensions are: W_{FIN}=30nm and L_G=130nm.

III. SIMULATION RESULTS

The biasing conditions used for the simulations and the electrical measurements are shown in Figure 2. Combined process and device simulations were used to investigate the transient behavior of the bulk FinFET devices using different substrate biasing conditions. In the next part of the section, mainly cuts along the gate will be used to explain the impact of the substrate bias. Figure 3 shows the electrical potential along the fin after WRITE "1" & HOLD for 1µs and WRITE "0" & HOLD for 1µs. A long channel device has been assumed. In this case, a substrate bias of 0V was applied. There is a clear difference in potential, extending over the entire fin height and thus affecting both the surface potential and the "body" potential in the fin, controlled by the gate.

978-1-4244-6658-0/10 $26.00 © 2010 IEEE

Simulations show that no excess holes are present in the fin after WRITE "1"& HOLD. As such, the electrical potential for state "1" is identical to the steady state. In Figure 4 the potential behavior is compared to the case when a negative back bias of -0.5V is applied.

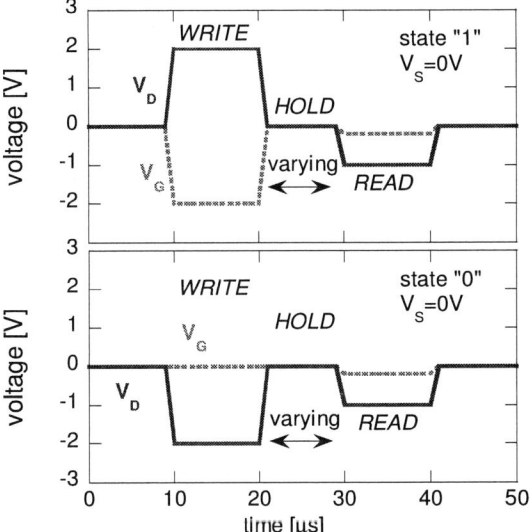

Figure 2. Biasing scheme for the bulk FinFET 1T-DRAM; the voltages are indicative and depending on the device dimensions it can be required to change the conditions slightly.

Again state "0" is compared to state "1" and the steady state. Comparing the two V_{BS} cases, one can see that firstly the difference between "1" and "0" is increased significantly when the substrate bias is negative. Secondly, both the storage of electrons and holes contribute to this potential difference. As the steady state potential in the fin is slightly negative for V_{BS}=-0.5V, holes can be temporarily stored.

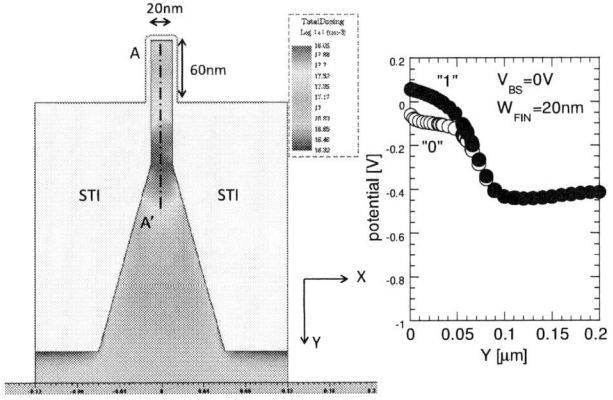

Figure 3. Schematic presentation of the bulk FinFET indicating the cut line AA' on the left side; on the right side the simulated electrical potential (along AA') after WRITE & HOLD; t_{HOLD}=1µs and W_{FIN}=20nm.

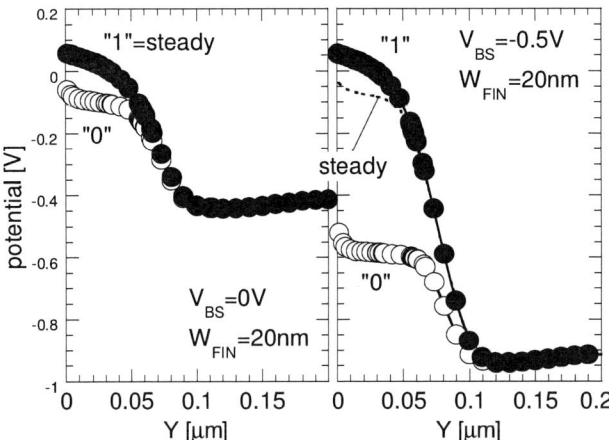

Figure 4. On the left side the simulated electrical potential after WRITE "1" & HOLD and WRITE "0" & HOLD when V_{BS}=0V and on the right hand side when V_{BS}=-0.5V.

However, as the potential gradient towards the substrate is not in favor of keeping the holes there (substrate is more negative) for a long time, the contribution of "1" to the overall retention behavior will still be limited. In order to assess the carrier loss to the substrate, the electron and hole concentrations were monitored as function of the hold time. The results are shown in Figure 5. After 10µs the hole concentration has dropped to the steady state level and there is no excess of holes that can contribute to the potential difference. Next to that, the negative drain bias during READ will easily remove the holes to the drain junction while the electron concentration remains stable over 100µs and longer (not shown here). Finally leakage towards the junctions and diffusion towards the substrate will lead to a loss of electrons and degradation of state "0".

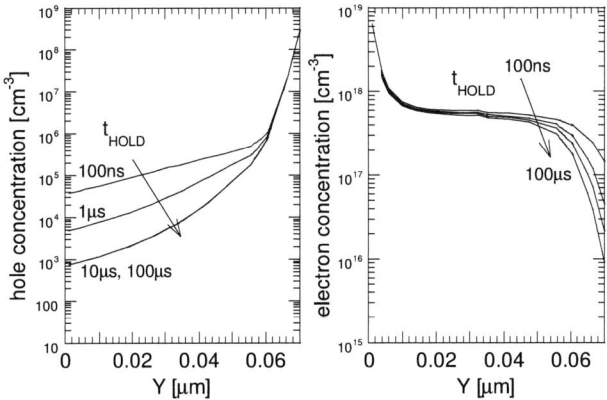

Figure 5. Hole and electron concentration along the fin height as function of the hold time; V_{BS}=-0.5V.

The surface potential as function of the substrate bias is shown in Figure 6. Only the surface potential of the top channel is shown, but the behavior of the sidewall and body potential is similar. In Figure 6, three different areas can be distinguished.

978-1-4244-6658-0/10 $26.00 © 2010 IEEE   75

At positive back bias, the storage of both holes and electrons is problematic, even at very short hold times, and no sense margin can be observed. In this case, the electrons are swept away towards the substrate.

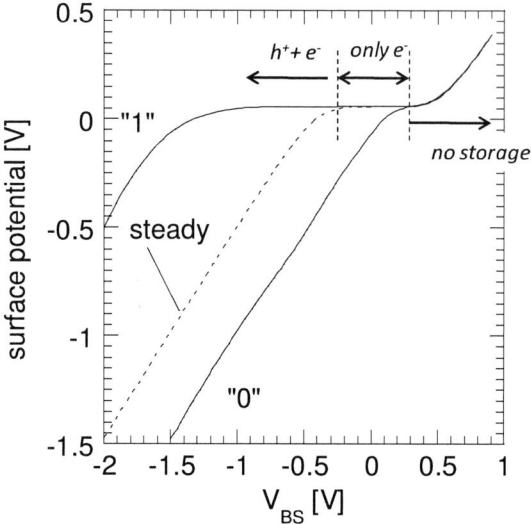

Figure 6. Surface potential (Y=0 μm) as function of the back bias; state "1" and state "0" are compared to the steady state potential at $t_{HOLD}=1\mu s$.

At substrate biases between -0.3V and 0.2V, only electrons will contribute to the sense margin. The state "1" is identical to the steady state potential and no excess holes are available. For more negative substrate biases, both holes and electrons contribute to the initial potential difference. Figure 6 indicates that the sense margin increases with more negative substrate bias conditions. Next to that, the impact of state "1" becomes larger. However, the figure only represents the potential difference at short hold times. By decreasing the substrate bias, the state "0" potential and also the steady state will become more negative and this will attract more holes towards the fin surface. These holes will recombine with the electrons leading to a faster degradation of state "0". This is shown in Figure 7, where the potential behavior is plotted as function of the hold time for two different V_{BS}. On the other hand, the stable state "1" implies that a change of READ conditions (positive versus negative) is needed in order to store the holes for a longer time. The impact of the READ disturb is also shown in Figure 7. The negative V_D during read-out (Figure 2) reduces the hole concentration dramatically. Both effects (recombination and READ) will finally lead to a reduced retention time and sense margin. It indicates that there is an optimal V_{BS} where both sense margin and retention are the highest and this optimal bias condition is very much dependent on the READ conditions. For more negative V_{BS}, where the impact of state "1" becomes larger, it is more beneficial to use READ conditions with positive V_D.

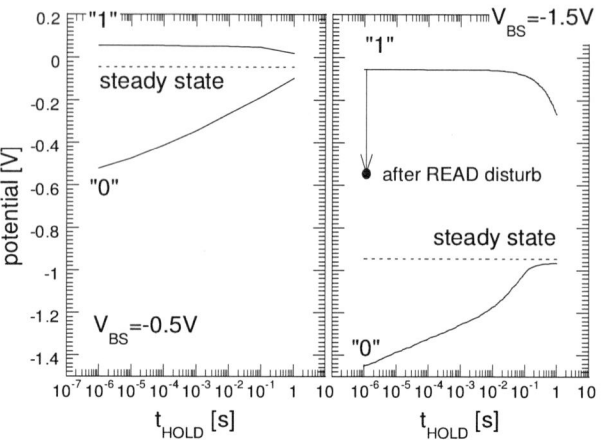

Figure 7. Simulated potential for state "1", state "0" and the steady state as function of the hold time for $V_{BS}=-0.5V$ (left) and -1.5V (right).

IV. 1T-DRAM RESULTS

The sense margin ΔI_S is determined as the current difference at 85°C between state "1" and "0" after a hold time of 100μs. The static retention time is defined as the time when $\Delta I_S=0$. The retention behavior at different back bias conditions is shown in Figure 8. For WRITE and READ, bias conditions as described in Figure 2 were used. As V_{BS} moves from 0.5V to -0.5V, the sense margin is increased which is in agreement with the simulation results. At the same time, there is a beneficial effect on the retention time.

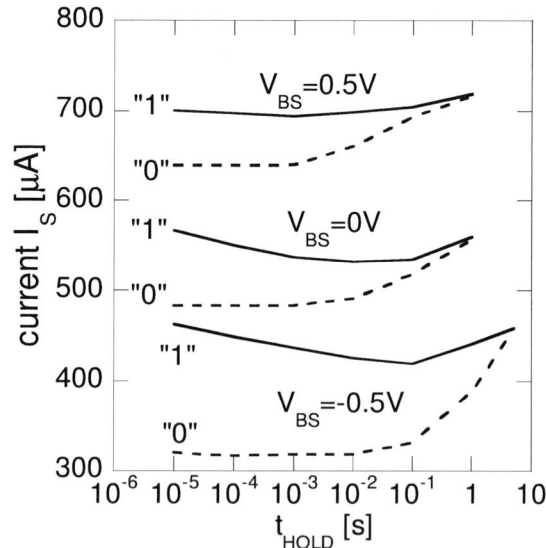

Figure 8. Retention characteristics at 85°C for a bulk FINFET device with $W_{FIN}=20nm$, $H_{FIN}=60nm$ and $L_G=90nm$ are compared for $V_{BS}=-0.5V$, 0V and 0.5V.

Figure 9 shows the V_{BS} dependence of the sense margin and retention time for devices with two different fin widths. A clear optimum is observed for both devices. For W_{FIN}=20nm, the highest sense margin can be obtained at V_{BS}=-0.5V. More positive V_{BS} reduces the amount of electrons available for state "0". Moreover, the increased junction leakage is detrimental for both sense margin and retention. For more negative V_{BS}, the sense margin is largely defined by state "1" and the READ disturbs will reduce the sense margin significantly. The optimal retention time is also achieved for substrate biases around -0.5 to -1V. There is a slight shift towards more positive substrate bias conditions for wider fins.

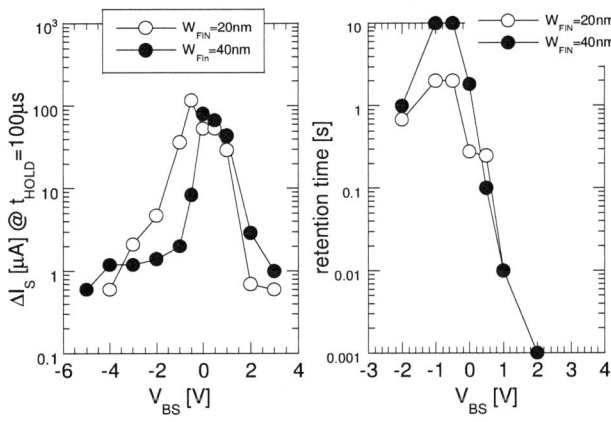

Figure 9. Sense margin (left) and retention time (right) at 85°C as function of back bias; devices with W_{FIN}=20nm and 40nm are compared; H_{FIN}=60nm and L_G=90nm.

The comparison of the potential in a device with W_{FIN}=20nm and 40nm is shown in Figure 10. Whereas the potential is almost constant over the entire fin height in the 20nm case, there is a large potential gradient for the 40nm devices.

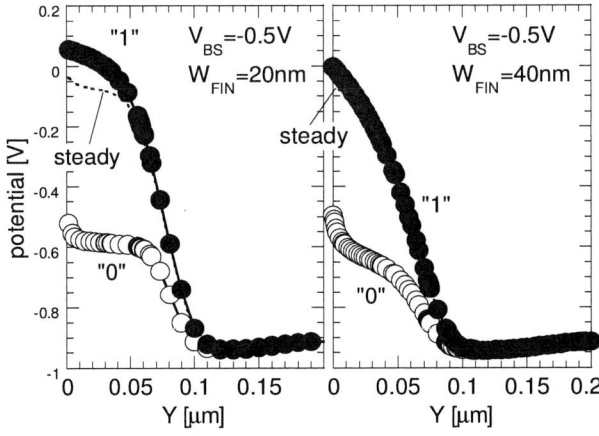

Figure 10. On the left side the simulated electrical potential after WRITE "1" & HOLD and WRITE "0" & HOLD for a device with W_{FIN}=20nm and on the right hand side for W_{FIN}=40nm; V_{BS}=-0.5V.

As such both the potential difference at the surface and in the fin body is reduced for the device with the wider fin. This results in a smaller sense margin. Next to that, for the same bias conditions, the contribution of holes is negligible in the case of 40nm fin width, indicating that the sense margin and retention time is only determined by state "0". This stable condition is achieved at slightly more positive substrate biases than in the case of 20nm fins. Simulations show that as the substrate bias becomes more negative the steady state potential is decreasing more rapidly for devices with W_{FIN}=40nm than 20nm, since the former devices behave more as partially depleted devices. As a consequence, the stored electrons in the top part of the fin will recombine more easily with the holes flowing towards the surface. A fast drop in sense margin and retention time can be seen.

V. CONCLUSIONS

In this work we have presented the impact of the substrate bias on the sense margin and retention characteristics of bulk FinFET 1T memory cells. Careful optimization of the back bias is needed to obtain the best trade-off between sense margin and retention time. Overall, negative substrate biases will increase the sense margin and retention time but the increasing current flow of holes towards the surface as the substrate bias becomes too negative will lead to a degradation of the state "0". Next to that, a change of READ conditions will be needed as the storage of holes will become more significant.

ACKNOWLEDGMENT

This work is supported by IMEC's partners and core partners on the emerging devices research program.

REFERENCES

[1] N. Sasaki, M. Nakano, T. Iwai and R. Togei, "Charge pumping SOS-MOS transistor memory", IEDM Tech. Dig., pp. 356–359, 1978

[2] S. Okhonin, M. Nagoga, J.M. Sallese and P. Fazan, "A SOI Capacitor-less 1T-DRAM Concept", 2001 IEEE International SOI Conference Proceedings, p. 153-154

[3] S. Okhonin, M. Nagoga, C.-W. Lee, J.-P. Colinge, A. Afzalian, R. Yan, N. Dehdashti Akhavan, W. Xiong , V. Sverdlov, S. Selberherr, C. Mazure, "Ultra-scaled Z-RAM cell", 2008 IEEE International SOI Conference Proceedings, p. 157-158

[4] U. E. Avci, I. Ban, D. L. Kencke and P.L.D. Chang, "Floating Body Cell (FBC) Memory for 16-nm technology with Low Variation on Thin Silicon and 10-nm BOX", in SOI Conf. Proc, 2008, pp. 29-30

[5] R. Ranica, A. Villaret, P. Malinge, G. Gasiot, P. Mazoyer, P. Roche, P. Candelier, F. Jacquet, P. Masson, R. Bouchakour, R. Fournel, J. P. Schoellkopf, and T. Skotnicki," Scaled 1T-Bulk devices built with CMOS 90nm technology for low-cost eDRAM applications", in VLSI Symp. Tech. Dig., 2005, pp. 38–39

[6] N. Collaert, M. Aoulaiche, M. Rakowski, A. Redolfi, B. De Wachter, J. Van Houdt, and M. Jurczak," Optimizing the Readout Bias for the Capacitorless 1T Bulk FinFET RAM Cell", Elec. Dev. Let., 2009, Vol. 30, No. 12, pp. 1377-1379

978-1-4244-6658-0/10 $26.00 © 2010 IEEE

COLK cell : A new embedded DRAM architecture for advanced CMOS nodes

S. Crémer, O. Goducheau, H. Petiton, S. Gaillard, E. Yesilada, M. Vernet, C. Jenny, F. Lalanne
STMicroelectronics, Crolles, France
Email : sebastien.cremer@st.com

Abstract— **This paper deals with a new and low cost embedded DRAM (eDRAM) architecture. COLK (Capacitor Over Low K) cell with capacitor placed in the first and thick SiO_2 dielectric has been successfully integrated. 4Mb eDRAM testchip using this new architecture is functional in 45nm node and presents good yield. Moreover we succeed to demonstrate the capability to continue downscaling of eDRAM for nodes down to 32nm and 22nm.**

Introduction

Major challenge for new eDRAM technology (using 1T/1C solution) is to maintain a high capacitance value while cell area is divided by two in comparison with previous node. The cell area reduction is crucial in order to stay competitive with SRAM solution.

While high-K material (HfO_2 or ZrO_2) has given some margin for 65nm and 45nm nodes, solution for next nodes is not yet demonstrated. Indeed capacitance value risks to be too low due to reduced cell area. Today three architectures are typically used in eDRAM world. Two of them are using stacked capacitor with either CUB cell [1] or COB cell [2] while the third one uses trench capacitor [3]. Major advantages and drawbacks including cost aspect of these 3 types of cells are summarized in Table 1. We propose here a new eDRAM architecture for advanced CMOS nodes.

Bitcell type	Advantages	Drawbacks
CUB	Medium cost	Capacitance value High aspect ratio contact
COB	Capacitance value	High cost High aspect ratio contact
Trench	High Capacitance value	Process complexity High cost

Table 1 : Comparison of existing eDRAM bitcell types

I. NEW EDRAM ARCHITECTURE

Schematic cross-section of the COLK cell is presented on Fig. 1. In order to keep high capacitance value and to suppress high aspect ratio (HAR) contact needed for COB and CUB cells, capacitor is placed in the first and thick SiO_2 dielectric of the back-end stack (for example above Metal 5 in 45nm technology having 7 metal layers : 5 thin metal layers and 2 thick). Capacitor process is quite similar to the one described in [1] for CUB cell : TiN/ZrO2/TiN is used for the Metal/Insulator/Metal (MIM) stack.

The HAR contact suppression leads to both a significant cost reduction and to process simplification: many yield detractors of logical and eDRAM circuits are then suppressed. Development of eDRAM capacitor and core process can be totally separated using this new concept. Moreover performance of core logic is no more degraded due to serial resistance increase when HAR contact is required.

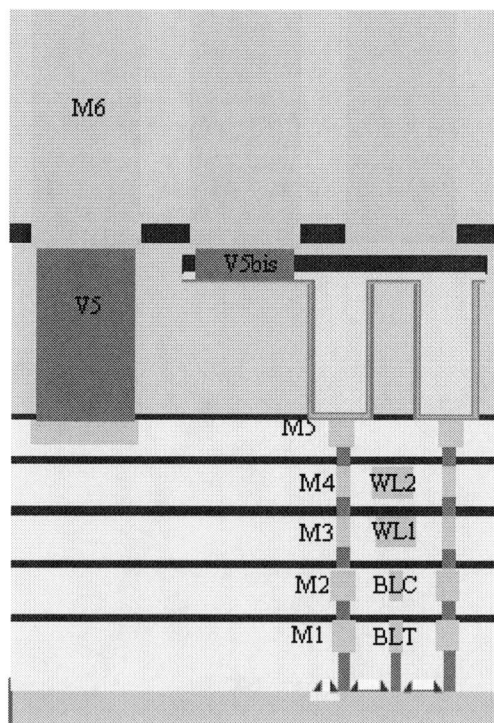

Figure 1. : COLK cell with MIM capacitor placed above M5

The COLK cell uses Metal 1 and Metal 2 of the core process for bitline routing leading to no extra cost in comparison with the specific level used in classical COB cell. Major advantages of this new cell are summarized in Table 2.

Bitcell type	Advantages
COLK (This work)	High capacitance value No impact on core process (yield, performance) Low cost

Table 2 : Advantage of the COLK cell

Placing the capacitor above bitline, like for COB cell, imposes to have an active shape oriented in non vertical direction (assuming poly of the eDRAM access transistor is horizontally oriented). Active layout has been optimized thanks to OPC simulation as shown on Fig. 2 where only active, poly and contact are represented.

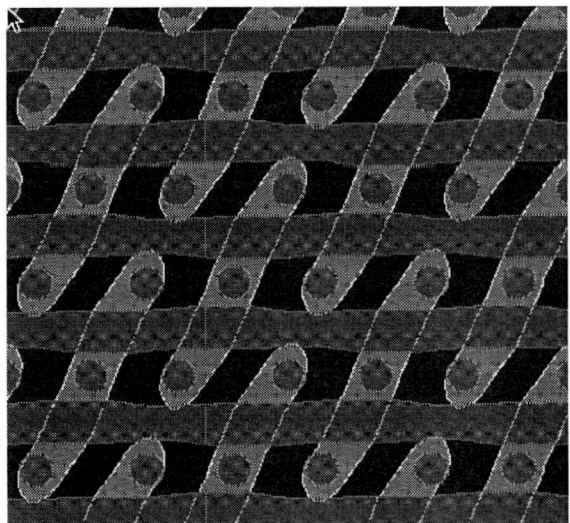

Figure 2. : OPC simulation for active, poly and contact

We can see on top view SEM photo (Fig. 3) an excellent agreement for active shape between OPC simulation and silicon results. Poly OPC has not been optimized on this experimental lot. Nevertheless this doesn't induce specific problem.

Back-end layout is shown on Fig. 4: only metal 1 (M1) and metal 3 (M3) levels are represented since metal 2 (M2) is superimposed to M1 and metal 4 (M4) is superimposed to M3. M1 and M2 are used for bitline routing (bitline true : BLT and BLC) while M3 and M4 allow wordline routing on two metal levels.

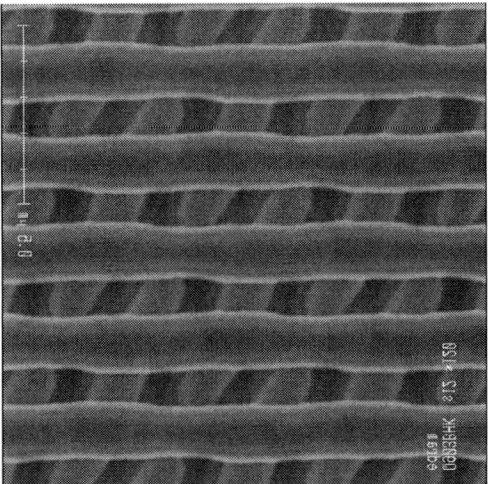

Figure 3. : SEM observation after poly gate patterning

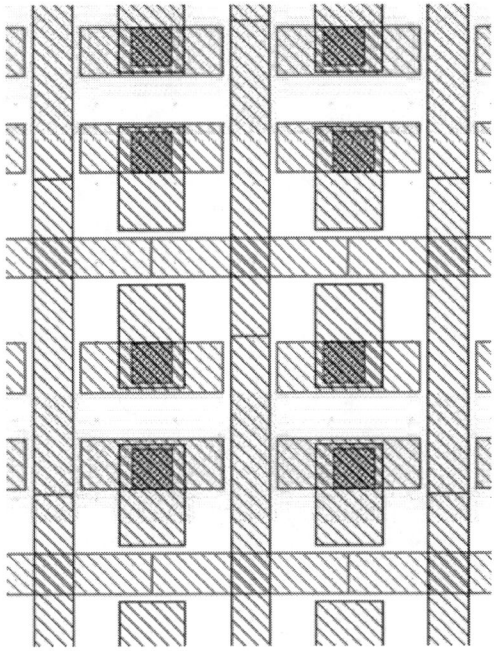

Figure 4. : Back-end layout (vertical M1, Via2 and horizontal M3)

Finally Fig.5 and Fig.6 are showing SEM top view and SEM cross-section of MIM capacitor (integrated on 45nm wafers) placed above M5 in the Via5 height. Only 1 critical mask is needed (for trench definition) for capacitor when at least 3 critical masks were needed for classical COB and CUB cells. Electrical results for the capacitor and for a testchip based on this new architecture are presented below.

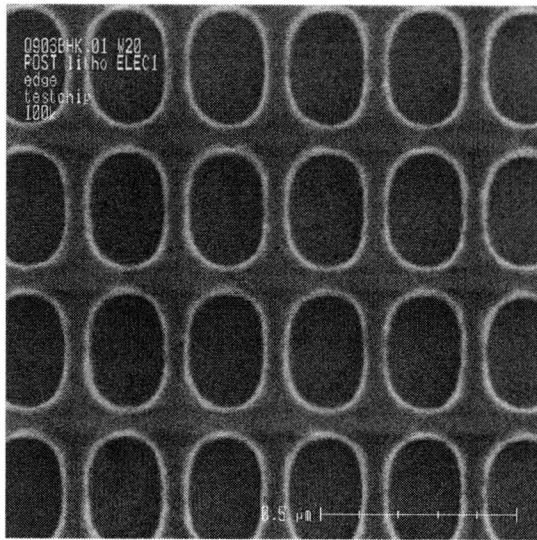

Figure 5. : SEM observation after capacitor patterning

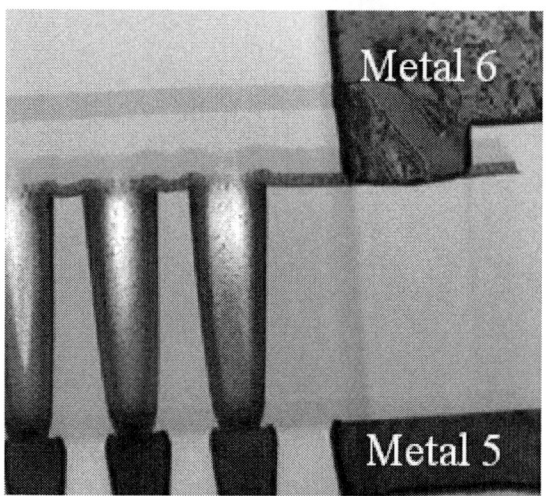

Figure 6. : SEM cross-section of MIM capacitor

II. ELECTRICAL RESULTS

Capacitance value versus cell area is shown on Fig. 7. 10fF/cell is reached in 0.05µm². This leads to capacitance density up to 200 fF/µm². Based on these data, simple geometric and electrical model has been built (Fig. 8) to evaluate capacitance value versus cell area and space between two capacitors. This model is taken into account geometric parameters (trench height, trench slope, trench dimensions in X and Y direction, high-K dielectric thickness) and electrical parameter (high-K dielectric constant)

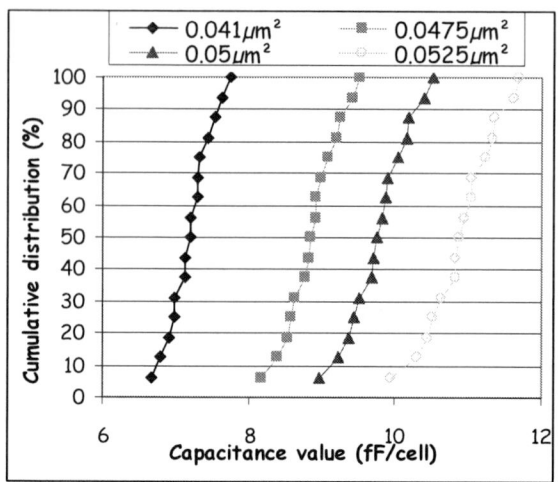

Figure 7. : Capacitance value as a function of cell area

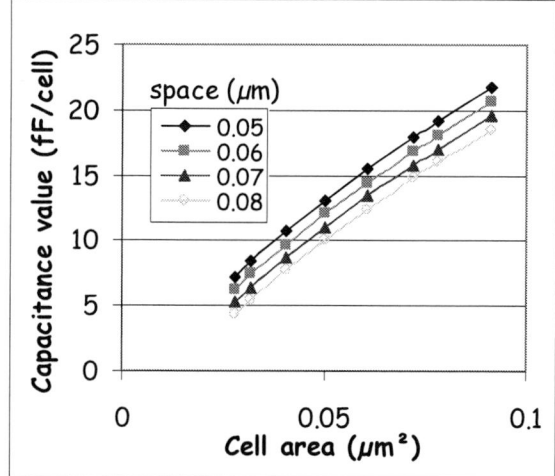

Figure 8. : Capacitance value model

Based on this model, we can conclude that capacitance value between 6-7 fF/cell is achievable for 0.025µm² cell. This cell area could be a competitive offer for eDRAM in 22nm node where the smallest SRAM bitcell will be probably around 0.075µm², i.e. 3 times bigger.

I(V) characteristics on 600k capacitors (Fig.9) demonstrated leakage current below 10 fA/cell at Vdd/2. This leakage level is sufficient to warranty good retention time : retention of eDRAM testchip will not be limited by leakage of the MIM capacitor in this case.

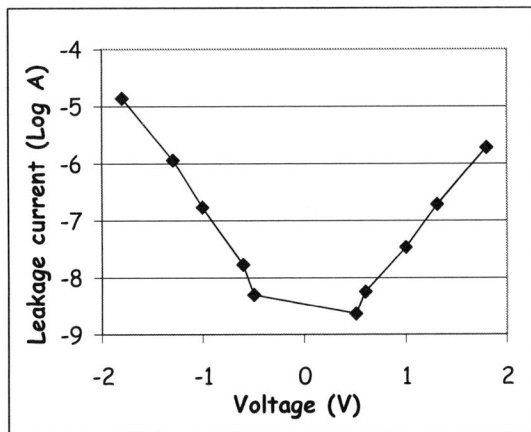

Figure 9. : I(V) on 600k capacitors for 0.05μm² cell

Transistor characteristics for 45nm and 32nm nodes are summarized in Table 3. Low Ioff current is maintained for 32nm node even if Vt has been significantly reduced in comparison with 45nm node. This Vt reduction will be helpful to increase Ion current and then to improve speed performance of eDRAM

Node (nm)	45	32
Transistor length (nm)	80	80
Vtsat (V)	1	0.7
Subthreshold slope (mV/dec)	90	80
Ion (μA/μm)	400	700
Ioff (log A/cell)	-14.6	-14
Sigma Vt (mV)	35	30

Table 3 : Transistor characteristics at 25°C

4Mb eDRAM testchip using this new architecture and 0.09μm² bitcell has been realized using 45nm process. Vdd/Vpp shmoo at 85°C using Marsh SS pattern on one 4Mb die (Fig. 10) indicates an excellent voltage window for functionality (Vdd min = 0.75V / Vpp min = 1.8V).

Figure 10. : Vdd/Vpp shmoo @85°C on 4Mb using MARSH SS pattern

Bit fail number histogram using full DRAM test flow (complex patterns but not retention patterns) with no hard redundancy and no ECC usage is shown on Fig. 11. 10% of

die have no fails and 65% of die have less than 4 bit fails i.e. less than 1 SBF / Mb. Retention curve at 85°C is shown on Fig. 12. Usage of redundancy and ECC would lead to 80% of repairable die on 4Mb. All these results validate this new architecture.

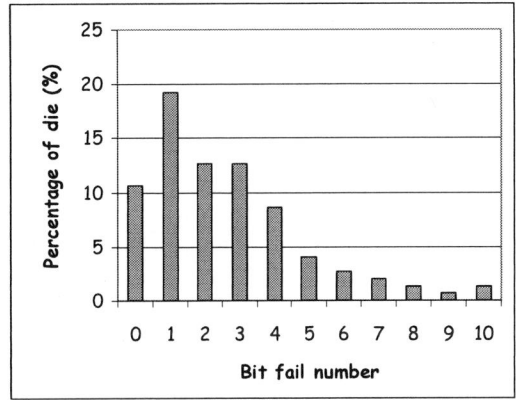

Figure 11. : histogram of bit fail number for 4Mb on 1 wafer

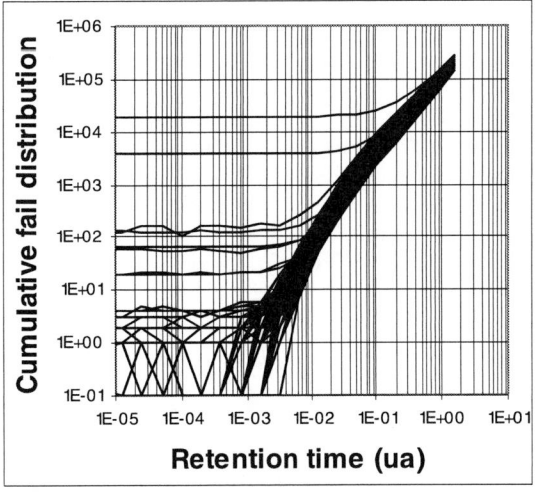

Figure 12. : retention curve at 85°C

CONCLUSIONS AND PERSPECTIVES

New architecture for eDRAM has been proposed and validated on 45nm node using 4Mb testchip. The COLK cell is very promising for 32nm and 22nm nodes.

REFERENCES

[1] A. Berthelot et al., "Highly Reliable TiN/ZrO2/TiN 3D Stacked Capacitors for 45nm Embedded DRAM Technologies, Proc. of ESSDERC 2006, p. 343-346

[2] Y. Yamagata et al., "Device Technology for embedded DRAM utilizing stacked MIM (Metal-Insulator-Metal) Capacitor, Proc. of CICC 2006, p.421-424

[3] G. Wang et al., "Scaling Deep Trench Based eDRAM on SOI to 32nm and Beyond", Proc. of IEDM 2009, p. 259-262

978-1-4244-6658-0/10 $26.00 © 2010 IEEE   81

Sub-60nm Si Tunnel Field Effect Transistors with $I_{on} > 100 \mu A/\mu m$

Wei-Yip Loh*[+], Kanghoon Jeon*[,2], Chang Yong Kang*, Jungwoo Oh*, Pratik Patel[2], Casey Smith*, Joel Barnett*,
Chanro Park*, Tsu-Jae King Liu[2], Hsing-Huang Tseng[3], Prashant Majhi*[,1], Raj Jammy* and Chenming Hu[2]

*SEMATECH, 2706 Montopolis Drive, Austin, TX 78741, USA, [1]Intel assignee
[2]EECS Dept. University of California, Berkeley, [3]Texas State University San Marcos, TX,
[+]Tel: (512) 356-3229, Fax: (512) 356-7640, email: wei-yip.loh@sematech.org

Abstract:

Si-tunneling field effect transistors (TFETs) with a record $I_{on} > 100 \mu A/\mu m$ and high I_{on}/I_{off} ratio ($> 10^5$) at $V_{ds}=1V$ are reported. Using an optimal spike and milli-sec flash anneal coupled with an engineered source-gate overlap through a gate-last process, Si TFETs have been demonstrated with 10 to 1000 times greater current than previously reported. The devices exhibit negative differential resistance and temperature dependencies consistent with band-to-band tunneling and current characteristics in excellent agreement with 2D TCAD simulations.

1. Introduction

MOSFET voltage scaling is limited due to the 60 mV/dec minimum subthreshold swing at room temperature. This inevitably limits threshold voltage scaling and hence supply voltage scaling, resulting in increased power density . Tunneling FETs (TFETs) [1-6] use band-to-band tunneling as the carrier-injection mechanism and do not have off-state leakage arising from the high-energy tail of the Fermi distribution of carriers; hence, they are not subject to the 60 mV/dec limit. However, previously reported Si and SiGe TFETs all have low on-state drive current (up to ~10 μA/μm). Recent experimental studies show that the drive currents of TFETs which employ low bandgap material like InGaAs [1] and Ge [2] are also relatively low, 10 and 20 μA/μm at $V_{ds}=1V$. This raises the question whether TFETs can ever provide drive currents comparable to MOSFETs, although simulations have predicted that TFETs, through innovative design, can achieve hundreds of μA/μm current at 1 V using low-bandgap materials or Si-Ge heterojunctions [7,8]. Such an energy-saving green transistor could eventually reduce IC power consumption by an order of magnitude, enabling decades more of IC growth without overburdening the world supply of electricity. By optimizing the source doping and using a deliberately overlapped source-gate structure, a 100 μA/μm TFET on silicon is achieved in this work, which represents a critical step toward achieving green transistors [7].

2. Experimental Procedure

TFETs were fabricated on 80 nm SOI substrates with a high-K/metal gate stack with a 1.1 nm EOT. The n+ source region was formed by a high-dose arsenic implant into half of the gate/channel region prior to the gate stack formation. The p+ drain was then formed by a boron

implant with an inverse mask after gate-stack formation. Spike annealing of the source and drain implants was performed at 1070°C; some wafers received additional flash annealing at 1200°C. **Fig. 1** shows a high-resolution TEM image of a fabricated device with 56 nm gate length and source/drain regions delineated. Due to the intentional ~15 nm overlap between the n+ source and gate, the effective channel length, L_{eff}, is approximately 40 nm.

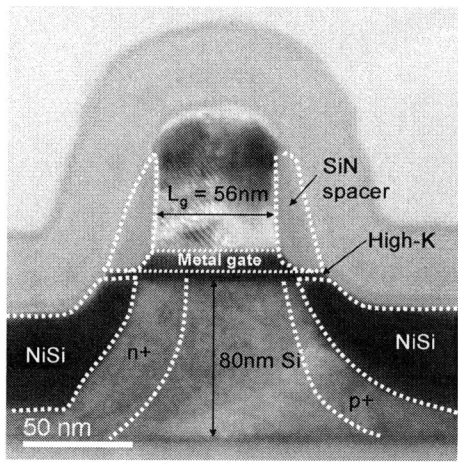

Fig. 1 High resolution XTEM micrograph of a TFET. The n+ source is implanted prior to gate stack formation while the p+ drain is implanted after gate stack formation.

3. Results and Discussion

3.1 TFET Characterization and Performance

Fabricated p-channel (n+ source) TFETs were characterized in a standard configuration with the n+ source grounded and p+ drain negatively biased. The measurements were taken at room temperature (300K) as well as lower temperature using a thermal chuck. **Fig. 2** shows the measured transfer characteristics for TFETs that underwent a spike anneal or spike+flash (thereafter referred to as "c-flash" for combination flash) anneal. Gate leakage in the fabricated TFETs is reasonably low, more than three orders of magnitude lower than the drive current, and hence does not affect the drive current. Compared to spike-annealed devices, c-flash devices show almost 10× higher drive current for comparable gate leakage current. The spike-annealed devices also show significant improvement in drive current compared to devices previously fabricated using a spike anneal but without a deliberate source-gate overlap structure (not shown). Band-to-band tunneling current in the p-channel

978-1-4244-6658-0/10 $26.00 © 2010 IEEE

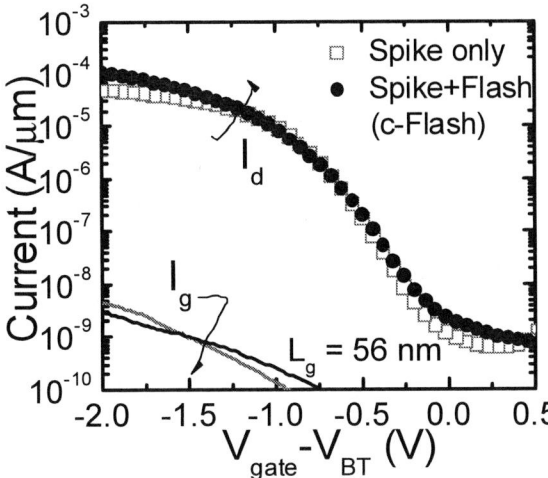

Fig. 2 Measured I_d-V_g characteristics for L_g=56nm TFETs which underwent either a spike anneal or spike+flash anneal. V_{ds} = -1.0V. V_{BT} is defined as the gate voltage corresponding to the onset of band-to-band tunneling.

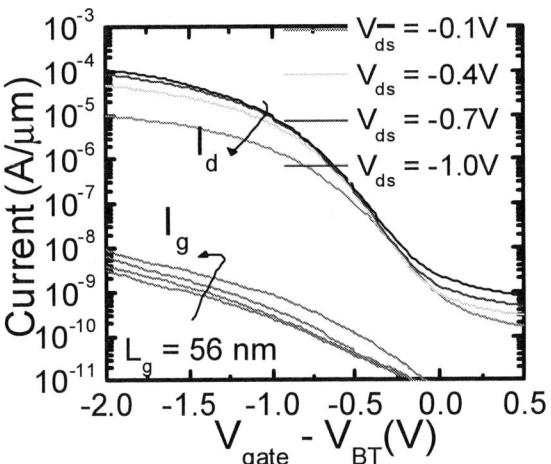

Fig. 3. Measured I_d-V_g characteristics of a TFET with L_g=56nm fabricated with a c-flash anneal, for different drain biases.

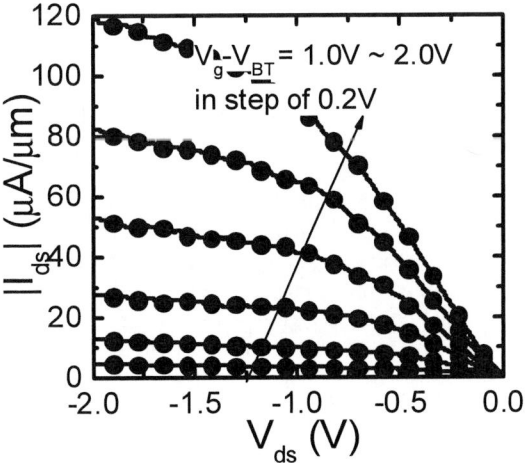

Fig. 4 Measured I_d-V_d characteristics for a TFET with L_g = 56nm fabricated with a c-flash anneal.

to the first order effect, is dependent on the transmission probability between the source and channel. Fermi function of the source is critical to the tunnelling current and hence the maximum level of dopant activation at the n+ source region is generally desired for TFETs. The advanced anneal technique using a spike anneal coupled with a millisecond flash anneal results in significantly higher tunneling current, indicating a higher level of dopant activation in the n+ source region. The highest I_{on} of 109 µA/µm is achieved at V_{ds} = -1V and gate overdrive of 2 V, as can be seen in Fig. 2.

The transfer characteristic of a c-flash TFET is shown in **Fig. 3**. Since the TFET has intrinsically both a n+source/channel and p+drain/channel junction, it can acts as both a n-type or p-type TFET depending on the gate bias. With positive gate bias, tunnelling is preferential in the p+drain/channel junction. In contrast, with negative gate bias, the TFET is operating as a p-type device with tunnelling occurring at the n+source/channel region. By using a gate last process, the n+source is deliberately overlapped with the gate while the p+drain is intentionally underlapped. This allows us to enhance the tunnelling current while suppressing the leakage current which can be attributed to the parasitic tunnelling across the p+drain/channel region. The c-flash samples demonstrate a non-linear subthreshold region, characteristics of TFET since it is not governed by the Fermi-tail thermonic emission in conventional MOSFETs. Subthreshold slope for the TFET ranges from 80 to 200 mV/dec. While sub-60 mV/dec swing is expected in a TFET, it was not achieved here and in most other studies, likely due to lateral implant straggle resulting in a non-abrupt junction [9]. This was observed in all our TFETs fabricated with conventional ion implantation.. Using silicide dopant segregation to pile up dopants close to the silicide/Si interface, TFETs with subthreshold swing of 46 mV/dec have been demonstrated in another study [6]. Further sharpening of

the lateral doping profile is expected to provide for further enhancement in the on-current [7,9]. Measured TFET output characteristics are shown in **Fig. 4**. Unlike conventional MOSFETs, drive current in TFETs is not expected to be strongly dependent on drain bias simply because the tunneling barrier is not strongly dependent on drain bias. Nevertheless, non-saturating behaviour is seen in Fig. 4. The drive currents, especially for c-flash devices, are exceptionally high compared to previously reported Si TFETs [2-6]. This demonstrates that high on-current in TFETs can be achieved with optimal source doping and source-gate overlap.

C-V measurements of spike and c-flash devices shown in **Fig. 5** show little difference between the two annealing schemes, yielding almost the same EOT of ~1.1 nm. (The C-V measurements were taken at 100kHz with both source and drain grounded.) Both samples show low interface trap density of 10^{11}/cm^2. This result indicates that the enhanced drive current of the c-flash device is due to optimized source design rather than improved gate electrostatic control.

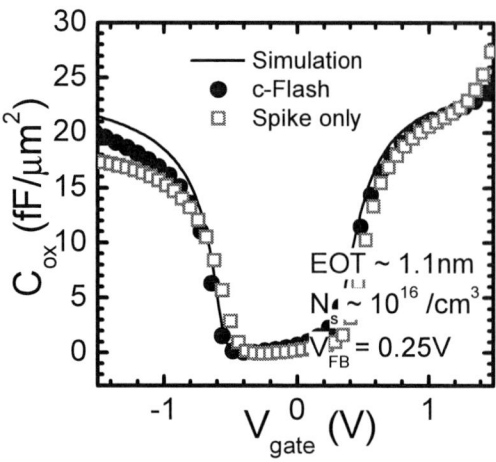

Fig. 5 C-V measurement of spike and c-flash TFETs taken at frequency = 100kHz. EOT of 1.1 nm is observed for both devices.

3.2 Device Simulation for Tunneling FETs

Since tunnel FETs (or TFETs) uses band-to-band tunnelling (BTBT) as source carrier injection, the tunnelling model in Si can be used to predict the Si TFET behaviour. **Fig. 6,** shows the device simulation obtained using the two-dimensional MEDICI device simulator with default tunnelling model parameters and with gate-source overlap and dopant concentration as key variable. Fig. 6(a) shows the tunnelling current as a function of varying gate-source overlap with overlap from zero to 10nm. With increasing overlap, tunnelling current increases because of enhanced electrostatic coupling of the gate to the n+source/channel junction although excessive overlap is also detrimental due to depletion at the n+source region. 2-D simulation shows that sufficient gate-source overlap is critical for high on-current. The 10 nm overlap required to explain the high drive current of 109 μA/μm is close to our designed overlap (see Fig. 1). With a lateral structure, exact carrier concentration at the n+source/p-channel is difficult to measure. Instead, we have performed 2-D simulation to attempt to match the theoretical BTBT tunnelling to our experimental results. The simulation shown in **Fig. 6(b)** show that dopant activation within the n+ source affects the on-current dramatically and that measured c-flash TFET results match the simulation for a carrier concentration of 5×10^{19}/cm^3. More importantly, Figs. 6(a) and (b) show that the drive current of our experimental Si TFET fit perfectly with the simulated tunnelling current, demonstrating the possibility of an optimized Si TFET with high output current.

3.3 Temperature Dependence of TFET Current

At high gate bias, TFET drive current is dominated by band-to-band tunneling which is relatively temperature insensitive.

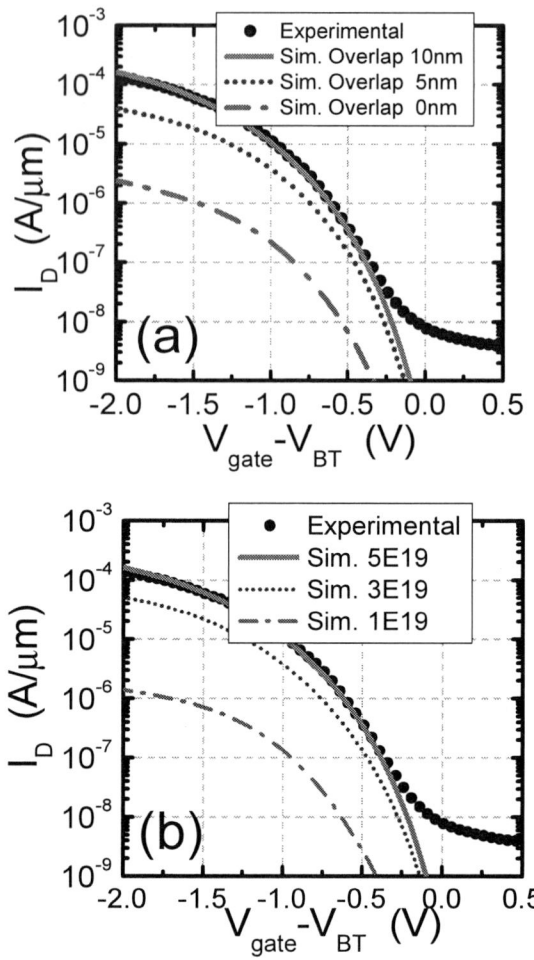

Fig. 6. Comparison of measured and simulated (ideal) Si TFET I_d-V_g characteristics. (a) Impact of gate-source overlap. (b) Impact of n+ source carrier concentration assuming 10nm overlap. The measured c-flash TFET current can be matched well by simulation.

Temperature dependence measurements for the c-flash TFETs show an exponentially increasing leakage floor (not shown) which can be explained by a SRH generation recombination process. In contrast, there is little temperature dependence of I_{on} at V_g-V_{BT} = -2V as shown in **Fig. 7**, indicating that the current-limiting mechanism is neither drift (mobility) nor thermal emission, but tunneling as designed. The subthreshold swing for c-flash TFETs shows a non-linear temperature dependence, distinctly different from MOSFETs as recently predicted by atomistic simulation [10]. In contrast, a conventional MOSFET will follow the linear kT/q slope as indicated in Fig. 7. The temperature dependence of the subthreshold slope in our TFETs demonstrates that the current-limiting mechanism in the c-flash TFET is not a thermal injection process.

978-1-4244-6658-0/10 $26.00 © 2010 IEEE

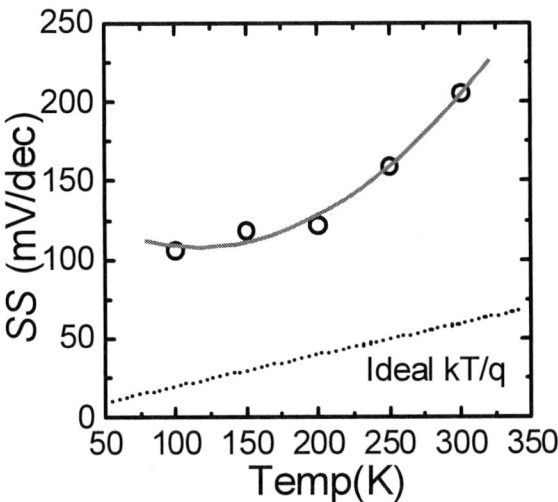

Fig. 7. Measured sub-threshold swing of a c-flash TFET as a function of temperature. Non-linear temperature dependence of the swing is characteristic of tunneling.

Fig. 8 shows measured negative differential resistance (NDR) behavior in c-flash TFET diodes due to inter-band tunneling. Esaki diode behaviour confirms a high active dopant concentration favorable to the tunneling process. A peak-to-valley ratio of 1.36 is observed at 213K under zero gate bias.

Table 1 compares reported TFET performances. This work shows the highest drive current reported for $V_{dd} = 1V$, with a competitive I_{on}/I_{off} ratio. Together with dopant-segregated Si gFETs [6] which show a subthreshold swing of 46 mV/dec, this suggests that Si is a compelling gFET material choice.

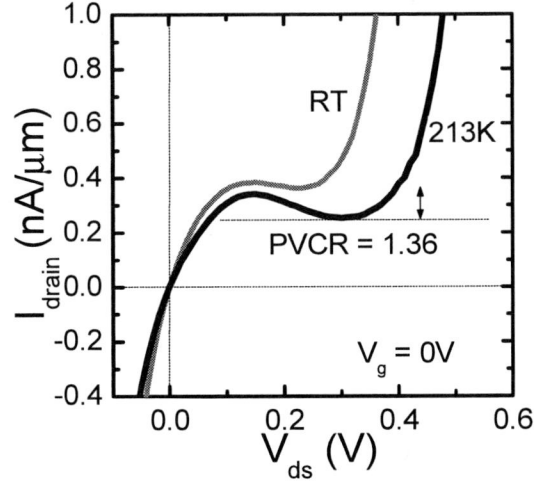

Fig. 8. Measured c-flash TFET diode behavior, showing negative differential resistance characteristics at 213K and 313K.

References	Channel Material	SS (mV/dec) @ RT	Ion[1] (μA/μm)	V_{ds} (V)	Ion/Ioff[1]
S. Mookerjea [1]	InGaAs	150~290	20	0.75	> 10^3
T. Krishnamohan [2]	Ge	50 ~ 60	10	1.00	10^6
T. Krishnamohan [2]	Si	460	10^{-4}	1.00	> 10^2
F. Mayer [3]	Ge	>400	4	0.80	> 10^2
F. Mayer [3]	Si	42 ~ 200	0.04	0.80	10^5
K. K. Bhuwalka [4]	Si	285	0.1	1.50	10^4
W. Y. Choi [5]	Si	52.8	12*	1.00	10^4
This work	Si	120 ~ 250	84	0.70	>10^5
			109	1.00	>10^4

[1]Ion is taken at overdrive of V_g-V_{BT} = 2.0V except for *. Ioff taken at onset of BT-BT, V_{BT}

Table 1. Comparison of reported TFET performances. This work shows the best I_{on} with good I_{on}/I_{off} ratio.

4. Conclusion

A well-designed Si TFET with the highest reported drive current of ~100 μA/μm at V_{ds}=1V and I_{on}/I_{off} of 10^5 has been demonstrated. This was achieved by optimal source doping and engineered gate-source overlap using a source-before-gate process. Tunneling conduction is verified by TCAD simulation, temperature dependence, and NDR behavior. This work shows the possibility of a practical tunneling transistor with performance approaching that of a MOSFET but consuming much less power for future green electronics.

Acknowledgement

This paper is based upon work supported by the Defense Advanced Research Project Agency under a SPAWAR Systems Center, San Diego contract, #N66001-08-C-2022.

5. References

[1] S. Mookerjea, D. Mohata, R. Krishnan, J. Singh, A. Vallett, A. Ali, T. Mayer, V. Narayanan, D. Schlom, A. Liu, and S. Datta, *IEDM Tech. Dig.* 2009, p.949.

[2] T. Krishnamohan., Donghyun Kim, Shyam Raghunathan and Krishna Saraswat, *IEDM Tech. Dig.* 2008, p. 947.

[3] F. Mayer, C. Le Royer, J.-F. Damlencourt, K. Romanjek, F. Andrieu, C. Tabone, B. Previtali, and S. Deleonibus, *IEDM Tech. Dig.* 2008, p. 163.

[4] K. K. Bhuwalka, S. Sedkmaier, A. K. Ludsteck, C. Tolksdorf, J. Schulze, and I. Eisele., *IEEE Trans, ED.*, vol 51(2), p. 279, 2004.

[5] W. Y. Choi, B-G. Park, J. D. Lee, and T.-J. King Liu., *IEEE EDL*, vol. 28(8), p. 743, 2007.

[6] K. Jeon, W.-Y. Loh, P. Patel, C. Y. Kang, J. Oh, A. Bowonder, C. Park, C.S. Park, C. Smith, P. Majhi, H.-H. Tseng, R. Jammy, T.-J. King Liu, and C. Hu, accepted for *VLSI Symp Tech.* 2010.

[7] C. Hu, D. Chou, P. Patel, A. Bowonder, *Intl Symp on VLSI Tech, Syst, Appls*, p. 14, 2008.

[8] A. Bownder, P. Patel, K. Jeon, J. Oh, P. Majhi, H.-H. Tseng, C. Hu, 8th *Int'l Workshop Junction Technology*, p.93, 2008

[9] P. Patel, K. Jeon, A. Bowonder, C. Hu, *SISPAD*,. p.1, 2009.

[10] Y. Yoon and S. Salahuddin, *Appl. Phys. Lett.* 96. 013510 (2010)

978-1-4244-6658-0/10 $26.00 © 2010 IEEE

High-performance Enhancement-Mode $In_{0.53}Ga_{0.47}As$ Surface Channels n-MOSFET with Thin $In_{0.2}Ga_{0.8}As$ capping and Laser anneal effect

I. Ok, P. Y. Hung, D. Veksler, J. Oh, P. Majhi R. Jammy

SEMATECH (Phone: 518-649-1045, FAX: 518-649-1322, e-mail: injo.ok@sematech.org), 257 Fuller Load, NY 12203

Abstract — In this work we have investigated the effect of thin $In_{0.2}Ga_{0.8}As$ capping layer and pulsed laser annealing (PLA) on self-aligned enhancement mode n-channel $In_{0.53}Ga_{0.47}As$ metal-oxide-semiconductor field effect transistor (MOSFET). We present the electrical and material characteristics of $TaN/ZrO_2/In_{0.2}Ga_{0.8}As/In_{0.53}Ga_{0.47}As$ n-MOSFET with atomic layer deposition (ALD) ZrO_2. Electrical characteristics with thin capacitance equivalent thickness (CET) (~0.8 nm), and improved drain current with thin capping layer were obtained. N-channel high-k InGaAs-MOSFETs with good transistor behavior, i.e. improved drain current after additional pulsed laser anneal with RTA anneal, on $In_{0.53}Ga_{0.47}As$ substrate have also been demonstrated.

I. INTRODUCTION

High mobility, narrow band gap group IV and III-V materials are considered strong contenders to replace strained-Si channels for logic applications beyond the 22 nm node due to their high electron mobility [1-3]. There have been a many efforts to fabricate III-V MOS capacitors and transistors [4-7]. While these materials typically provide very high bulk mobility, other features such as high channel mobility, high inversion charge, low series resistance, and low off-state leakage are also required to enable superior MOSFET performance [8]. fabrication of high performance self-aligned III-V MOSFET still remains challenging due to high temperature post metal anneal (PMA) for dopant activation in source and drain (S/D) regions after ion implantation in the fulfillment of an InGAs MOSFET, especially for n-channel devices. Subjecting the entire substrate to high temperature PMA can have deleterious effects on the InGaAs interface for the purpose of fabrication integrated circuits. This problem can be obviated by selective contact annealing with a laser beam. Pulsed laser annealing (PLA) may also be important in obtaining enhanced activation of implanted dopants [9-13]. In this work we have investigated the effect of thin $In_{0.2}Ga_{0.8}As$ capping layer with pulsed laser annealing (PLA) on self-aligned enhancement mode n-channel $In_{0.53}Ga_{0.47}As$ MOSFET. We present the electrical and material

characteristics of $TaN/ZrO_2/$ $In_{0.2}Ga_{0.8}As/In_{0.53}Ga_{0.47}As$ substrate self-aligned n-MOSFET with ALD ZrO_2. N-channel high-k $In_{0.53}Ga_{0.47}As$-MOSFETs with good transistor behavior, i.e. electrical characteristics with thin capacitance equivalent thickness (CET~ 0.8 nm), improved drain current after additional pulsed laser anneal with RTA anneal and with thin (2nm) $In_{0.2}Ga_{0.8}As$ capping layer on $In_{0.53}Ga_{0.47}As$ substrate have also been demonstrated. Excellent transistor characteristics with 23% of $G_{m,max}$ improvement with thin $In_{0.2}Ga_{0.8}As$ capping layer and with 5.4 % mobility improvement after PLA anneal have been obtained.

II. RESULTS AND DISCUSSION

A. Experiment

MOS capacitors were fabricated on molecular beam epitaxy (MBE) grown n-type (1×10^{18} /cm^3) $In_{0.53}Ga_{0.47}As$ wafers doped with Si on n-type InP substrate. ALD ZrO_2 (~5 nm) films were deposited followed by post-deposition anneal (PDA). ALD TiN (20nm)/ Physical vapor deposition (PVD) TaN (220 nm) were used as gate electrode. After gate patterning using reactive ion etching (RIE) based on CF_4 gas, low-resistance ohmic contact was formed by using AuGe/Ni/Au alloy on the source and drain side [14]. The samples were then annealed at 450°C for 30sec in nitrogen. Electrical characterization was performed on MOS capacitors and transistors. CET values were extracted from C-V.

B. Improving $In_{0.53}Ga_{0.47}As$ MOSFET property by $In_{0.2}Ga_{0.8}As$ capping

Based on the concept of Si capping for Si-Ge channels [15], we evaluated the impact of capping InGaAs with 20% of InGaAs. Note, pure GaAs was not selected as the capping layer due to the difficulty in un-pinning the Fermi level. Also, similar gate stacks on InGaAs 20% with surface modulation was demonstrated earlier [16]. The schematic cross section of n-MOSFET with a thin 2nm of 20% $In_{0.2}Ga_{0.8}As$ capping layer is shown in Fig. 12. A series of transistors with varying dopant activation anneals were processed.

978-1-4244-6658-0/10 $26.00 © 2010 IEEE

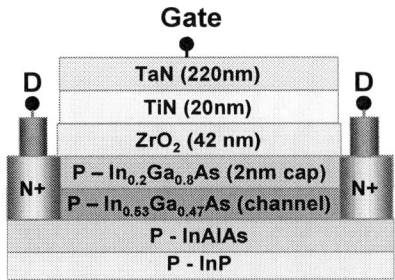

Fig. 1. Cross section of 20% $In_{0.2}Ga_{0.8}As$ capping transistor

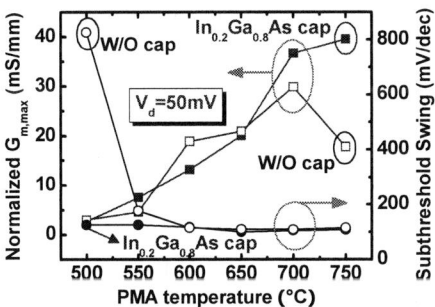

Fig. 2. $G_{m,max}$ and SS comparison for W/ and W/O $In_{0.2}Ga_{0.8}As$ capping layer as function of PMA

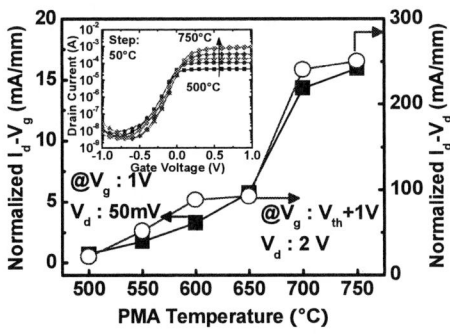

Fig. 3. Normalized I_d for I_d-V_g at V_g=1V and V_d=50mV and I_d-V_d at V_g=V_{th}+1V and V_d = 2V with $In_{0.2}Ga_{0.8}As$ capping layer

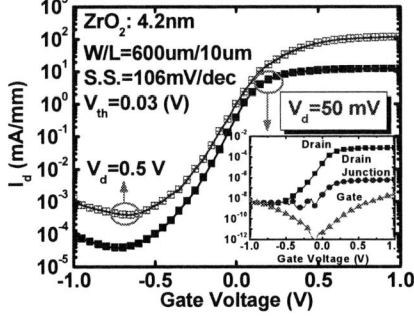

Fig. 4. I_d-V_g of 20% $In_{0.2}Ga_{0.8}As$ capping transistor in linear and saturation region (inset: I_d-V_g, drain leakage, and gate leakage at V_d = 50mV)

Figures 2, 3 and 4 show that the thermal stability of these MOSFETs can be increased to 750°C, while still demonstrating excellent $G_{m,max}$, SS, I_d-V_g, and I_d-V_d (SS = 94mV/dec [17], I_{on}/I_{off} = 1×10^6, G_m = 520mS/mm [18]). The drain current increases with a subthreshold swing of ~106 mV/decade providing Ion/Ioff ~10^5 at V_g = 0.5V. This number show high I_{on}/I_{off} ratio demonstrated for InGaAs MOSFET with I/I S/D since I/I damage has far stronger effects in compound semiconductors like III-V than in Si or SiGe.

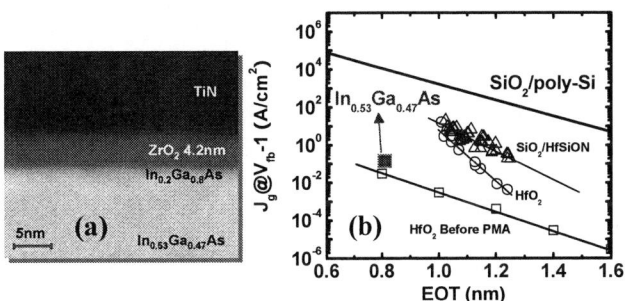

Fig. 5. (a) HRTEM on gate region with PMA of 700°C 10sec and (b) J_g Vs. EOT for 20% $In_{0.2}Ga_{0.8}As$ capping transistor

Fig. 5. (a) illustrates the TEM of the gate stack showing no distinct IL but also no indication of interfacial reaction between the ZrO_2 and $In_{0.2}Ga_{0.8}As$. This stack appears to scale to CET < 1nm and excellent leakage control (Fig. 5b).

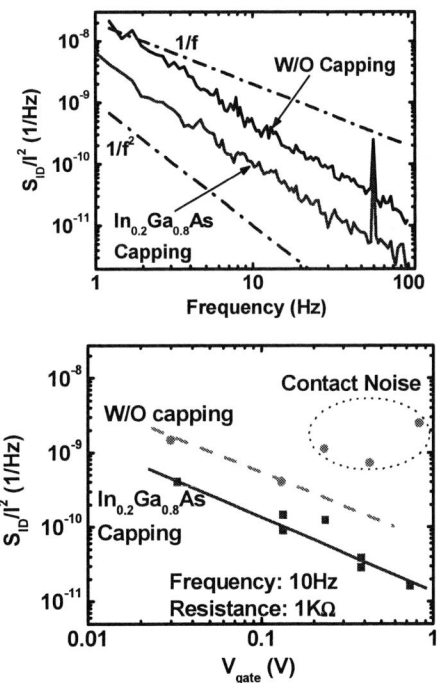

Fig. 6 (a) Drain noise current spectral density (b) Drain noise current spectral density as function of gate bias for w/ and w/o $In_{0.2}Ga_{0.8}As$ capping layer

978-1-4244-6658-0/10 $26.00 © 2010 IEEE

While the D_{it} numbers are still quite high (Fig 12, no laser anneal case) the noise measurements for the sample (Fig 18-20) clearly show that the capping layer improved bulk dielectric and interface quality, correlating with the transport data.

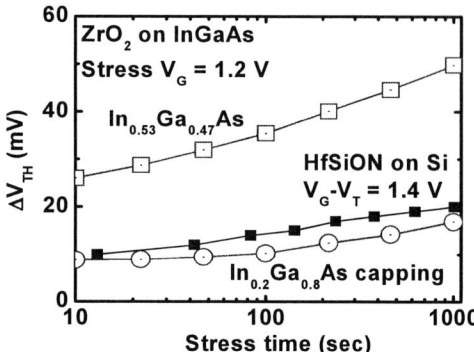

Fig. 7. PBTI (ΔV_T) of $In_{0.53}Ga_{0.47}As$ n-MOSFET with (ΔV_T ~10mV) and without $In_{0.2}Ga_{0.8}As$ capping

Reliability tests such as PBTI was conducted on these III-V channel MOSFETs showing that the ΔV_t, was comparable to good quality high-k on Si, indicating no catastrophic failure mechanism in III-V MOSFETs.

C. Combining Laser Anneal and InGaAs capping

To evaluate potential for further improvement in performance of the MOSFETs, $In_{0.2}Ga_{0.8}As$ capping + laser anneal was attempted.

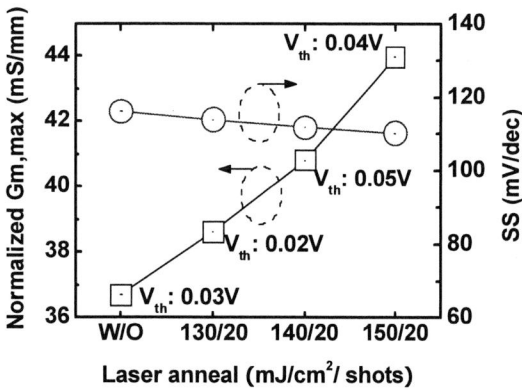

Fig. 8. Normalized $G_{m,max}$ and SS with different laser anneal condition on RTA 700°C 10sec

With increasing laser energy, $G_{m,max}$ and SS were improved (Fig. 8) along with I_d-V_g, I_d-V_d, and mobility (Fig. 9, 10, and 11). $In_{0.2}Ga_{0.8}As$ capping MOSFET resulted in lower Dit value compare to $In_{0.53}Ga_{0.47}As$ MOSFET measured by charge-pumping (CP) as function of frequency and temperature (Fig. 12). Additionally, a four-point wafer bending apparatus was used (figure 13) to evaluate the

additivity of uniaxial strain to the drive current and Gm enhancement of $In_{0.53}Ga_{0.47}As$ n-MOSFETs with stress.

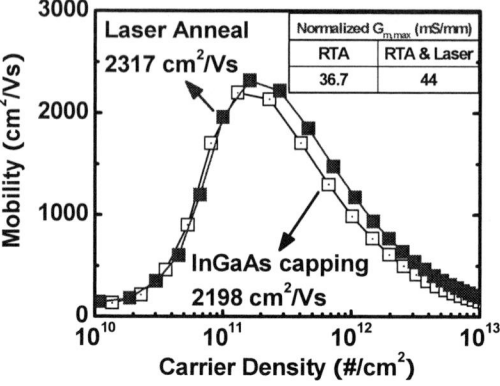

Fig. 9 Mobility with and without laser anneal for $In_{0.2}Ga_{0.8}As$ capped devices

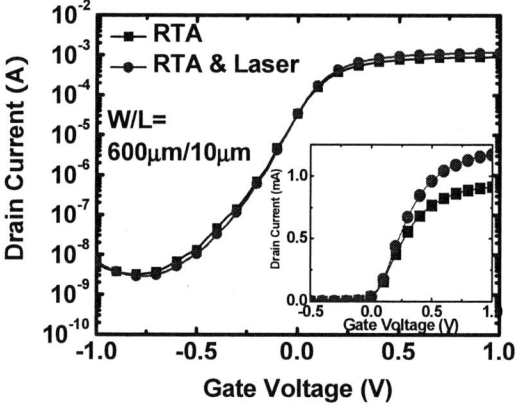

Fig. 10 I_d-V_g with and without laser anneal (150mJ/cm^2) for 20% $In_{0.2}Ga_{0.8}As$ capping transistor

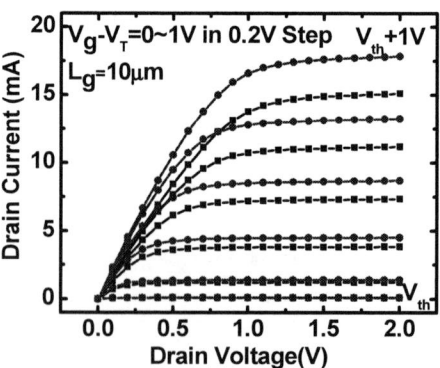

Fig. 11 D_{it} with and without laser anneal of $In_{0.2}Ga_{0.8}As$ capping transistor as function of frequency

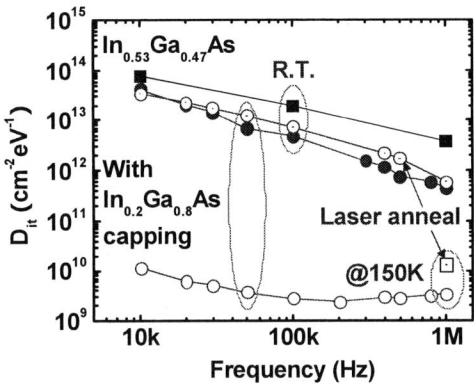

Fig. 12 I_d-V_d with and without laser anneal for 20% $In_{0.2}Ga_{0.8}As$ capping transistor

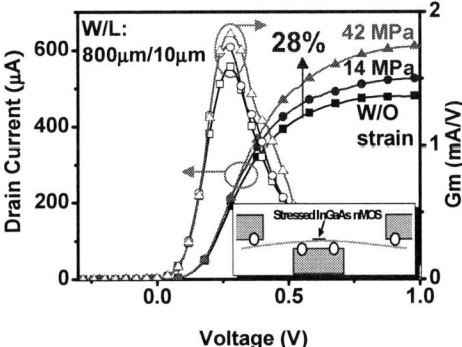

Fig. 13 Strain effect on I_d-V_g (inset: four-point wafer bending schematic)

Interestingly, a small 42 MPa uniaxial strain seems to improve the drain current by ~25 %. While this significant enhancement is surprising and needs support of simulation results, it provides another valuable knob to further enhance the performance of III-V MOSFETs along with improvement in interface quality and reduction in external resistance by advanced dopant activation technology.

CONCLUSION

Through a systematic evaluation of the thermal budget dependence of the structure and hence property of III-V MOSFETs, we have been able to demonstrate the potential of performance enehancement with $In_{0.2}Ga_{0.8}As$ capping, laser anneals, and uniaxial strain additivity. MOSFET devices with surface channel mobility ~ 3000 cm^2/V•sec, CET ~0.8 nm, Jg < 1A/cm^2, I_{on}/I_{off} ~ 1×10^5, uniaxial strain additivity and BTI < 40mV were demonstrated showing promise for insertion in future MOSFETs.

REFERENCES

[1] R. Chau, S. Datta, M. Doczy, B. Doyle, B. Jin, J. Kavalieros, "Benchmarking nanotechnology for high-performance and low-power logic transistor applications", IEEE Transactions on Nanotechnology, Volume 4, Issue 2, March 2005 Page 153 - 158

[2] M. K. Hudait, G. Dewey, S. Datta, J. M. Fastenau, J. Kavalieros, W. K. Liu, D. Lubyshev, "Heterogeneous Integration of Enhancement Mode $In_{0.7}Ga_{0.3}As$ Quantum Well Transistor on Silicon Substrate using Thin ($\leq$ 2µm) Composite Buffer Architecture for High-Speed and Low-voltage (0.5V) Logic Applications", in IEEE IEDM 2007.

[3] Suthram, P. Majhi, G. Sun, P. Kalra, H. R. Harris, K. J. Choi, "High Performance pMOSFETs Using Si/Si1-xGex/Si Quantum Wells with Highk/Metal Gate Stacks and Additive Uniaxial Strain for 22 nm Technology Node", in IEEE IEDM, pp. 727-730, 2007

[4] M. Passlack, J. Abrokwah, R. Droopad, Z. Yu, C. Overgaard, S. Yi, M. Hale, J. Sexton, and A. Kummel., "Self-aligned GaAs p-channel enhancement mode MOS heterostructure field-effect transistor", Electron Device Letters, IEEE, Volume 23, Issue 9, Sept. 2002 Page(s):508 - 510

[5] G.G. Fountain, S. V. Hattangady, D.J. Vitkavage, R. A. Rudder, and R. J. Markunas, "GaAs MIS structures with SiO2 using a thin silicon interlayer", Electron. Lettters 24, 1134 (1988)

[6] S. Tiwari, S. L. Wright, and J. Batey, "Unpinned GaAs MOS Capacitors and Transistors", IEEE Electron Device Letters. 9, 488 (1988).

[7] Meng Tao, Andrei E. Botchkarev, Daegyu Park, John Reed, S. Jay Chey, Joseph E. Van Nostrand, "Improved Si_3N_4/Si/GaAs metal-insulator-semiconductor interfaces by in situ anneal of the as-deposited Si", Journal of Applied Physics, 77 (8), 15 April 1995

[8] Lee Smith, Makoto Fujiwara, Krishna Saraswat, Yoshio Nishi, "Design Guidelines for High Mobility Channel Bulk n-MOSFETs", MRS Proc. V.995 0995-G01-05

[9] W. T. Anderson, A. Christou, J. F. Ciuliani, and H. B. Dietrich, "Laser Annealed and Thermal Annealed Refractory Ohmic Contacts to GaAs", IEEE Transactions on Industrial Electronics, Vol. IE-29, No. 2, May 1982

[10] Z. Jan, M, Petr, M. Vladimir, "Comparison of laser technology and RTA on Pt/Sn/Pd ohmic contacts to GaAs", in Proceedings of Society of Photographic Instrumentation Engineers (SPIE) Vol. 4016, 2000

[11] E. V. Monakhov, B. G. Svensson, M. K. Linnarsson, A. L. Magna, M. Italia, V. Privitera, G. Fortunato, M. Cuscuna, and L. Mariucci, "The effect of excimer laser pretreatment on diffusion an dactivation of boron implanted in silicon", Applied Physics Letters, 87, 192109, 2005

[12] Q. Zhang, J. Huang, N Wu, G. Chen, M. Hong, L. K. Bera, and C. Zhu, "Drive-Current Enhancement in Ge n-Channel MOSFET Using Laser Annealing for Source/Drain Activation", IEEE Electron Device Letters, Vol. 27, No. 9, September 2006

[13] A. Koh, R. lee, F. Liu, T. Liow, K. Tan, X. Wang, G. Samudra, N. Balasubramanian, D. Chi, and Y. Yeo, "Pulsed Laser Annealing of Silicon-Carbon Source/Drain in MuGFETs for Enhanced Dopant Activation and High Substitutional Carbon Concentration", IEEE Electron Device Letters, Vol. 29, No. 5, May 2008

[14] Jun-Kyu Yang, Min-Gu Kang, and Hyung-Ho Park, "Chemical and electrical characterization of Gd_2O_3/GaAs interface improved by sulfur passivation," J. Appl. Phys., Vol 96, No. 9, pp. 4811-4816, November 2004.

[15] S. H. Lee, J. Huang, P. Majhi, P.D. Kirsch, B-G. Min, C-S. Park, J. Oh, "Vth Variation and Strain Control of High Ge% Thin SiGe Channels by Millisecond Anneal Realizing High Performance pMOSFET beyond 16nm node" VLSI 2009

[16] I. Ok, H. Kim, M. Zhang, F. Zhu, S. Park, J Yum, and Jack C. Lee, "Metal Gate HfO2 MOS Structures on InGaAs Substrate with Varying Si Interface Passivation Layer and PDA Condition" , JVST B, 25 (4), Jul/Aug 2007

[17] N. Goel, D. Hehb, S. Koveshnikova, I. Ok, S. Oktyabrsky, V. Tokranov, "Addressing The Gate Stack Challenge For High Mobility InxGa1-xAs Channels For NFETs" IEDM 2008

[18] Y. Xuan, T. Shen, M. Xu, Y.Q. Wu, and P. D. Ye "High-performance Surface Channel In-rich $In_{0.75}Ga_{0.25}As$ MOSFETs with ALD High-k as Gate Dielectric", IEDM 2008

Optimization of Tunnel FETs: Impact of Gate Oxide Thickness, Implantation and Annealing Conditions

D. Leonelli[1,2], A. Vandooren[1], R. Rooyackers[1], S. De Gendt[1,3], M. M. Heyns[1,4], G. Groeseneken[1,2]

[1] Imec, Kapeldreef 75, B-3001 Leuven, Belgium
[2] Katholieke Universiteit Leuven, ESAT, Department of Electric Engineering, B-3000 Leuven, Belgium
[3] Katholieke Universiteit Leuven, Chemistry Department, B-3000 Leuven, Belgium
[4] Katholieke Universiteit Leuven, Department of Metallurgy and Materials Engineering, B-3000 Leuven, Belgium
Email: leonel@imec.be

Abstract—We show the impact of process parameters on the electrical performance of complementary Multiple-Gate Tunneling Field Effect Transistors (MuGTFETs), implemented in a MuGFET technology compatible with standard CMOS processing. Firstly, the impact of the gate oxide thickness and implant doping conditions on the tunneling performance is analyzed and compared with TCAD simulations. Secondly, three different annealing conditions are compared: spike anneal, sub-ms laser anneal and low temperature anneal for Solid Phase Epitaxy Regrowth (SPER). Surprisingly, the SPER anneal shows a strong enhanced tunneling current with a record drive current of $46\mu A/\mu m$ at V_{DD} of -1.2V and I_{OFF} of 5pA/μm for Si pTFETs.

I. INTRODUCTION

As CMOS scaling is reaching fundamental limits, alternative approaches are being pursued. One of the most promising options consists of changing the operation principle of the device, through the use of band-to-band-tunneling (BTBT) controlled by the gate. This device has the potential of achieving sub-60 mV/dec subthreshold swing (SS) and extremely low I_{off} because its carrier injection is no longer limited by carrier diffusion [1-2]. For this reason the Tunneling Field Effect Transistor (TFET) represents a good candidate for low standby power technology (LSPT).

In the last few years, an increasing number of publications regarding the simulation and the modeling of TFETs have been reported [3-4]. The outcome of these simulations depicts a series of challenges for the integration of such devices. Parameters such as the steepness of the tunneling junction, an extremely high doping concentration at the tunneling source and a very thin gate oxide, are essential to control and enhance the tunneling current. Furthermore, to become competitive with CMOS technology and further boost the tunneling current, the reduction of the bandgap at the tunneling junction is mandatory. This can be achieved by replacing silicon with germanium or III/V compounds to improve the performance of nTFETs and pTFETs, respectively [5]. Recently, strain engineering was also proposed as an additional booster [6].

Due to the limitations in the models used by TCAD simulators and the lack of model calibration in the case of 3 terminal TFETs, there is an increasing need for experimental results to assess the impact of processing parameters on the tunneling behavior. Because all-Si TFETs are directly compatible with standard CMOS processing, it is still high interest to investigate how far we can boost its performance while at the same time gain insights on the tunneling mechanisms and its optimization.

In this paper, we assess the impact of the gate oxide thickness, implantation and annealing conditions for complementary MuGTFETs. TCAD simulations are used to correlate the experimental data with theory. Regarding the dependence of the tunneling current with the annealing conditions, we observe that the solid phase epitaxy regrowth (SPER) anneal brings additional on-current enhancement. In particular, the value of $46\mu A/\mu m$ at V_{DD} of -1.2V for Si-pTFETs is measured. This is the first time such a high value of drive current is reported in an all-Si pTFET for this type of devices.

II. DEVICE FABRICATION

The tunneling devices were fabricated on a (100) SOI substrate with 65 nm thick Si film on top of a 145 nm of Buried Oxide (BOX). Fin widths down to 10 nm were patterned using 193nm optical lithography and aggressive resist and hardmask (HM) trimming. The channel of the device was left undoped. The gate stack consists of a 100 nm polycrystalline silicon (poly-Si) layer on top of a 5 nm TiN layer and 2 nm HfO_2 on a 1 nm interfacial SiO_2 layer (Fig. 1a). Atomic Layer Deposition (ALD) was used for the high-k deposition while Plasma Enhanced Atomic Layer Deposition (PEALD) was used for the deposition of the metal gate.

After gate patterning, complementary source/drain doping was obtained by an extension implantation of arsenic (As) and boron at 45 degrees tilt parallel to the gate using a modified mask design (Fig. 1b). The implantations were done with energy of 10 keV and a dose of 1×10^{15} atoms/cm^2 for As while an energy of 1 keV and a dose of 1.5×10^{15} atoms/cm^2

978-1-4244-6658-0/10 $26.00 © 2010 IEEE

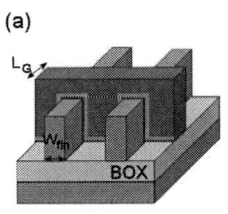 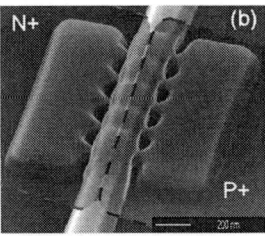

Figure 1. Schematic view of the tunneling MuGFET structure (a) and SEM image after the silicidation process of 10 nm wide fins (b). The overlay layers show how the doping is implemented.

for B. Then, 50 nm wide nitride spacers were formed on a 5 nm oxide liner. Next, the source/drain regions were highly doped by As/B implantation. After a 1050°C spike anneal for dopant activation and nickel silicidation, standard Cu backend processing was used.

This is the process of reference (POR). The impact of processing parameters is studied by modifying one or more steps of the POR as explained in the section III.

III. DEVICE PERFORMANCE

In this section we analyze the impact of the gate oxide thickness, dopants implantation and annealing conditions on the tunneling performance.

The POR conditions, described in section II, are used as a reference. Figure 2 shows the typical I-V curves for TFETs processed with standard POR conditions. The measurements are performed as described in the inset of Fig.2, keeping the tunneling junction of the device, called source, grounded. Therefore, the curves at different V_{DS} are superimposed. The drive current is rather low and an average subthreshold swing is extracted.

A dependence with the fin width is observed and was already reported in [7]. For this reason we analyze the behavior of wide fins and narrow fins separately.

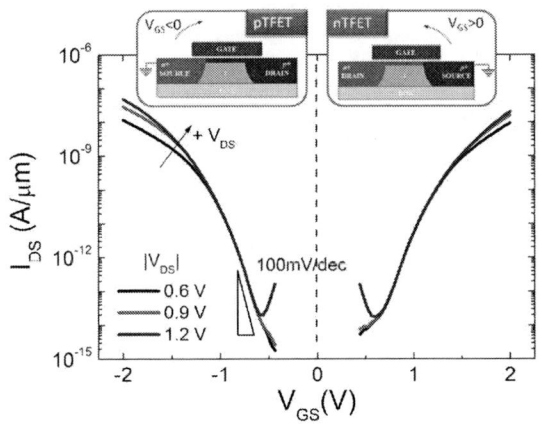

Figure 2. Transfer characteristics for pTFET and nTFET measured on the same device (W_{fin} 20nm, L_G 150nm and N_{fin} 15). nTFET and PTFET are measured keeping the side, where the tunneling occurs, grounded. The simmetry of their behavior is characteristic of the ambipolarity of tunneling devices.

A. Impact of the Gate Oxide Thickness

From models and simulations [8], the gate oxide thickness is expected to have has a strong impact on the device performance. Mainly, a decrease of the tunneling onset voltage is predicted along with a slight improvement of the subthreshold swing. In our experiment, we kept constant the interfacial oxide layer and modified the thickness of the high-K oxide. Two thicknesses of HfO_2 were implemented: 2nm and 4nm. An equivalent oxide thickness was extracted from CV measurements on test structures at $V_G = -1.5V$ (Fig.3). A T_{inv} of 1.7 nm was extracted for the reference sample of 2nm HfO_2 while a value of 2.6 nm was extracted for the thicker gate oxide.

Figure 4 shows the I_{DS}-V_{GS} characteristic of the measured devices. The impact of the oxide thickness is verified on narrow fins of 20nm and wide fins of 250nm. A decrease of onset voltage for thinner high k layers is observed in line with simulations for the pTFET with 2nm HfO_2 [3-4] . In addition, this is in line with the extraction of the threshold voltage for standard PMOS.

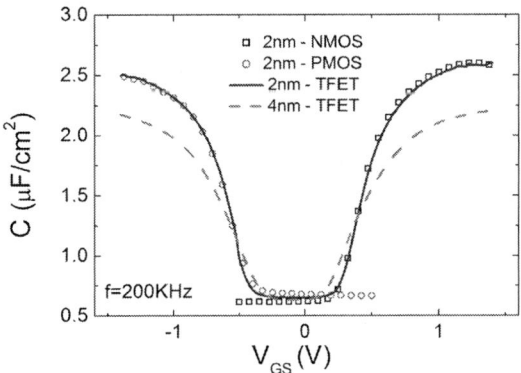

Figure 4. High Frequency CV measurements on PMOS and NMOS compared to TFET devices.

B. Impact of the doping conditions

The strategy used to tune the doping profile is based on engineering of the implantation angle and implantation dose.

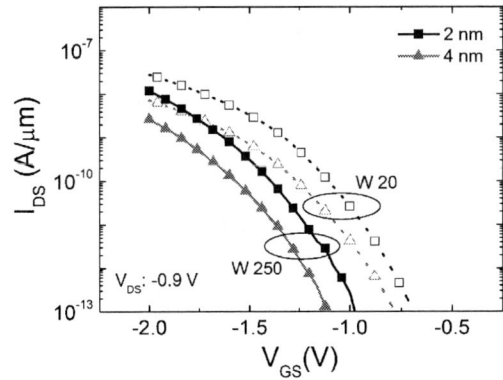

Figure 3. Average Transfer characteristic for pTFETs with 2nm and 4nm HfO_2. The narrow fins (dotted) have W_{fin} 20nm, L_G 150nm and N_{fin} 5 while the wide fins (solid) have W_{fin} 250nm, L_G 150nm and N_{fin} 5.

978-1-4244-6658-0/10 $26.00 © 2010 IEEE

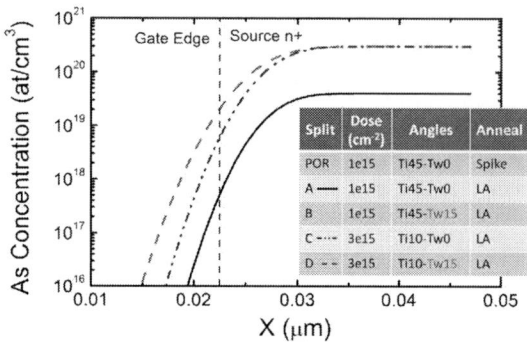

Figure 5. Simulated doping profiles after the extension implantation with As. The energy used is 10keV for all the splits. Cut is performed along the fin direction, in the middle of the fin and 5 nm from the top of the fin. The doping splits are described in the inset.

For all splits the implantation is always done in two quadrants and symmetric to the channel direction to assure a symmetrical doping profile along the fin. In addition, a sub-ms laser anneal activation step was used to increase the activation rate and the steepness of the junction as reported in [9]. The focus of this study is to optimize the tunneling at the n^+ side while the tunneling at the p+ side is not considered in this study because of the high diffusion of boron.

Figure 5 shows the simulated doping profiles for the conditions A, C and D (see inset of Fig.5). The doping profile is extracted at 5nm from the top of the fin along the channel direction in the middle of the fin. The POR condition described in section II is used as a reference. Two different doses are used to evaluate the impact of the total implantation dose. In addition, two different twist conditions are reported. The goal of this split is to increase the effective doping under the gate. The twist is the angle between the projection of the implantation vector on the wafer plane measured from the axis parallel to the gate direction. A positive value means that the dopants are implanted under the gate.

The electrical results obtained for the different splits are summarized in Fig. 6 for wide and narrow fins. Error bars are added to each value to show the uniformity of the results. First, a clear improvement is observed for the sub-ms laser anneal over the spike (POR) likely due to a steeper junction or to the better dopant activation as explained at the beginning of this section. Secondly, comparing splits A and C an additional improvement of current for wide fins occurs due to the increased doping concentration, as also shown in Fig.5. Thirdly, an additional increase of current is present in case of a larger twisted implantation. The twisted implantation increases the amount of the dopants under the gate as shown in figure 5. As a consequence, an increase of tunneling current is expected. On the other hand, for narrow fins, the impact of the implantation conditions is much less outspoken. The reason is that the variability for narrow fins is larger than for the wide fins. This might be due to the full amorphization of the fins during the implantation and affecting the stability of the performance of these devices.

In conclusion, the impact of the doping conditions studied here is relatively marginal (less than one decade). The same

behavior is observed in TCAD simulation (Fig. 7). Here, only a marginal increase of the current with the dose is visible in case of a graded junction.

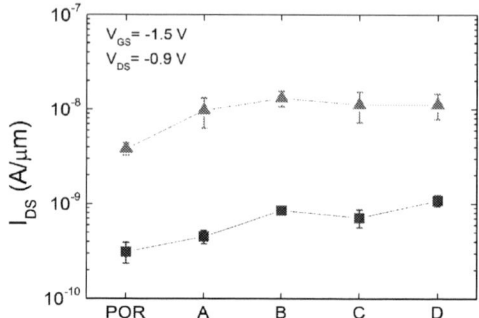

Figure 6. Extraction of the tunneling currents for the different annealing conditions. Trends for narrow fins (red) - W_{fin} 20nm - and wide fins (blue) – W_{fin} 250nm - are plotted. The values are extracet at V_{GS} of -1.5V.

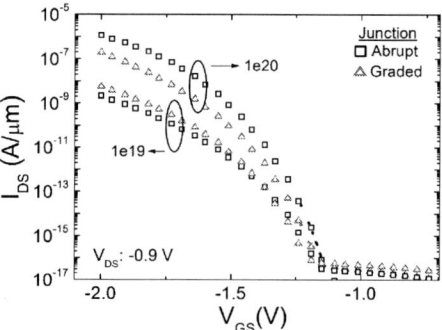

Figure 7. Simulated input characteristics for abrupt box profile and 1D graded profile along the channel. The graded profile is a gaussian profile with a gradient of about 6nm/decade.

C. Impact of the annealing conditions

Different anneals were performed after the highly doped drain (HDD) implantation. The processing prior to the annealing corresponds to the POR described in section II. The annealing conditions used are:

- conventional spike anneal (POR),

- sub-ms laser anneal (LA),

- low temperature anneal for Solid Phase Epitaxy Regrowth (SPER), 1' at 650°C.

Measurements on planar test structures, i.e. MOSFETs with large perimeter, were performed to extract the overlap capacitance and, consequently, the overlap length. The inset of figure 8 summarizes the results for the three different anneals at the n-doped region and the p-doped region. The outcome is not surprising: for the n+ side, the overlap length (L_{ov}) is comparable in all the splits due to the limited diffusion of the As atoms. In contrast, at the p+ side B atoms diffuse almost twice as fast in the case of spike anneal as in the case of LA or SPER anneal.

978-1-4244-6658-0/10 $26.00 © 2010 IEEE

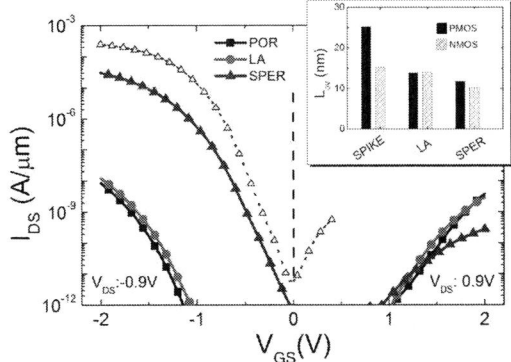

Figure 8. Tunneling current extracted at $|V_{DS}|$ =0.9V and for three different annealing processes. The device has W_{fin} 250nm (solid), L_G 150nm and N 5. The device with the best performance (dashed) has W_{fin} 10nm. In the inset, the extraction of the overlap lenghts from overlap capacitance measurements for NMOS and PMOS. The measured structures are shared on the same mask of the TFETs.

Figure 8 shows the corresponding tunneling performance for wide nTFETs and pTFETs. For nTFETs, the larger overlap of the p+ side does not to affect the device performance. The reason for this was already explained in section III.B through graded profile and low dose.

The pTFETs exhibit an unexpected behavior for the case of the SPER condition. The on current is much higher than the POR or LA conditions. This increase is observed for all fin widths down to 10nm with a good reproducibility.

TEM inspections were performed to understand the origin of this phenomenon. From the TEM analysis (Fig. 09) it is clear that silicide encroachment is present at the n+ side. This effect is related to the presence of defects introduced during the implantation. In the case of the SPER anneal, the defects are not completely cured. As a consequence, nickel piping is observed during the silicidation process [10]. The presence of silicide encroachment and the large defect density might explain the large on current. The presence of silicide encroachment leads to higher electric field at the tunneling junction because of dopants pile-up and/or reduced distance between the silicide and the gate. The presence of defects may reduce the effective bandgap of Si at the n+ side increasing the tunneling current.

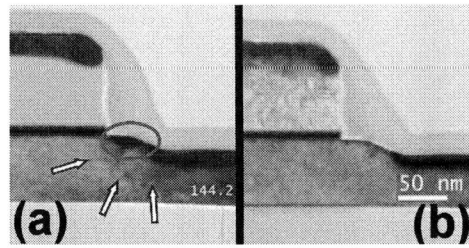

Figure 9. TEM images of the devices after SPER (a) and after the sub-ms Laser anneal (b). Silicide encroachment is present in (a) and defects are visible at the interface between the n+ region and i-channel region.

IV. CONCLUSION

In this paper complementary TFETs were fabricated and optimized through different process parameters.

Firstly, the reduction of the gate oxide thickness is beneficial for the TFETs but it mainly affects the onset of the tunneling. Further scaling of the oxide thickness will be a challenge since the direct tunneling through oxide will affect the tunneling characteristic.

Secondly, an improvement by 50% of the on current is achieved by engineering the implantation condition. However, this is not clearly visible for narrow fins due to the poor uniformity especially in case of high dose implantation.

At last, the impact of the anneal conditions reveals that the tunneling current can be greatly increased thanks to silicide encroachment. This effect is reproducible with low temperature anneal (SPER) but could also be implemented in different ways.

In conclusion, Si TFETs with record on current of $46\mu A/\mu m$ at V_{DD} of -1.2V and I_{OFF} of 5pA/μm have been shown for narrow fins. The optimization presented in this work has a marginal impact on the subthreshold swing of TFETs. Si TFETs are less complex to integrate and can quickly bring better understanding of the tunneling mechanisms involved. This learning can then readily be exported to the TFET optimization using smaller bandgap materials for increased current drive.

REFERENCES

[1] W. M. Reddick and G. A. J. Amaratunga, "Silicon surface tunnel transistor," *Appl. Phys. Lett.*, vol. 67, pp. 494-496, Jul 1995.

[2] J. Appenzeller, Y. M. Lin, J. Knoch, Z. H. Chen and P. Avouris, "Comparing carbon nanotube transistors - The ideal choice: A novel tunneling device design," *IEEE Trans. on Electron Devices*, vol. 52, pp. 2568-2576, 2005.

[3] K. Boucart and A. M. Ionescu, "Double-gate tunnel FET with high-kappa gate dielectric," *IEEE Trans. on Electron Devices*, vol. 54, pp. 1725-1733, Jul 2007.

[4] E.-H. Toh, G. H. Wang, G. Samudra and Y.-C. Yeo, "Device physics and design of germanium tunneling field-effect transistor with source and drain engineering for low power and high performance applications," *J. Appl. Phys.*, vol. 103, p. 104504, 2008.

[5] A. S. Verhulst, W. G. Vandenberghe, K. Maex and G. Groeseneken, "Boosting the on-current of a n-channel nanowire tunnel field-effect transistor by source material optimization," *J. Appl. Phys.*, vol. 104, 2008.

[6] K. Boucart, W. Riess and A. M. Ionescu, "Lateral Strain Profile as Key Technology Booster for All-Silicon Tunnel FETs," *IEEE Electron Device Letters*, vol. 30, pp. 656-658, Jun 2009.

[7] D. Leonelli, A. Vandooren, R. Rooyackers, A. S. Verhulst, S. De Gendt, M. M. Heyns and G. Groeseneken, "Performance Enhancement for Multi Gate Tunneling Field Effect Transistors by Scaling the Fin-Width," *Jpn. J. Appl. Phys.*, in press.

[8] K. Boucart and A. M. Ionescu, "Double Gate Tunnel FET with ultrathin silicon body and high-K gate dielectric," *ESSDERC Proceedings*, 2006, pp. 383-386.

[9] T. Noda, W. Vandervorst, S. Felch, V. Parihar, C. Vrancken, S. Severi, *et al.*, "Analysis of dopant diffusion and defect evolution during sub-millisecond non-melt laser annealing based on an atomistic kinetic Monte Carlo approach," *IEDM Tech. Dig.*, 2006, pp. 4.

[10] A. Lauwers, J. A. Kittl, M. J. H. Van Dal, O. Chamirian, M. A. Pawlak, M. de Potter, *et al.*, "Ni based silicides for 45 nm CMOS and beyond," *Mat. Sci. Eng. B-Solid.*, vol. 114, pp. 29-41, Dec 2004.

978-1-4244-6658-0/10 $26.00 © 2010 IEEE

Test Structure and Method for the Experimental Investigation of Internal Voltage Amplification and Surface Potential of Ferroelectric MOSFETs

Alexandru Rusu, Giovanni A. Salvatore, Adrian M. Ionescu
Nanolab, Ecole Polytechnique Fédérale de Lausanne, Switzerland
{alexandru.rusu},{adrian.ionescu}@epfl.ch

Abstract—**In this paper we report the fabrication and detailed electrical characterization of a novel test structure based on Metal-Ferroelectric-Oxide-Semiconductor transistor with internal metal contact, aiming at extracting the surface potential and the investigation of internal voltage amplification expected due to negative capacitance effect. The proposed test structure is p-Fe-FET with a thin Al contact in-between the PVDF ferroelectric and a pedestal oxide, enabling access to the internal voltage potential in all the regimes of operations, from weak to strong inversion. Moreover, the capacitances of reference MOS transistor and of Fe-FET can be independently probed. The test structure was fabricated on low doped silicon with STI isolation, in n-implanted well, with a gate stack including 6.5nm of SiO_2, 50nm of Al, 100nm of P(VDF-TrFE) and Au as top contact. The fabricated p-type Fe-FET has an excellent subthreshold slope of 75mV/decade, Ion/Ioff > 10^7 and Ioff in the pA range. Based on voltage and capacitive measurements, the Fe-FET surface potential is extracted for the first time. We demonstrate that the internal node voltage amplitude can be controlled by the sweeping conditions of the polarization loops. The test structure appears highly suited for the future investigation of the negative capacitances and of more complex ferroelectric gate stacks.**

I. Introduction

Ferroelectric FETs (Fe-FET) are considered promising devices for applications as one transistor memory cells [1] and/or or abrupt switching with an inverse subthreshold swing better than 60mV/decade at room temperature [2]. The steep slope application is intensely studied. However the concept of a negative capacitance is still controversial and experimental evidence is required [3]. Due of the complexity of the behavior of ferroelectric material, accurate analytical models for Fe-FETs has not yet been developed [4]. Therefore, the purpose of our test structure is to contribute to the characterization of the surface potential dependence on the gate voltage Fe-FETs and determine the exact value of the internal voltage drops on the capacitance divider in such a transistor. This is expected to enable the probing of the variation of the internal potential between the dielectric and the pedestal oxide, dV_{int}/dV_g, which is expected to be >1 in case of a negative capacitance effect. The ferroelectric material investigated here is a thin layer (100 nm) of PVDF that can be integrated with a low temperature process in the gate stack of a MOS transistor. Our test structure enables capacitive, drain current and internal voltage measurements, therefore offering all the conditions for the experimental extraction of Fe-FET surface potential, which, to the best of our knowledge has never been reported before.

II. Fabrication

We have integrated 100nm of P(VDF-TrFE) 70%-30% ferroelectric polymer in the gate stack of a regular MOS, adding Al contact in between the pedestal thermal SiO_2 and the ferroelectric. A low doped p-type (N_A=5 x 10^{15} cm^{-3}) wafer was processed. The fabrication process starts with Shallow Trench Isolation (STI) achieved by silicon dry-etching, LTO growth **(Fig. 1a)** and chemical mechanical polishing (CMP). A phosphorus (P) n-well is implanted using high energy and annealed at 1150°C for 60 minutes **(Fig. 1b)**. Source and drain are implanted using BF2 **(Fig 1c)**. The wafer is again annealed for 5 minutes at 900°C. The impurity doping is presented in Fig. 2, where one can see a well depth of 1.8 μm and a drain junction depth of 0.3μm.

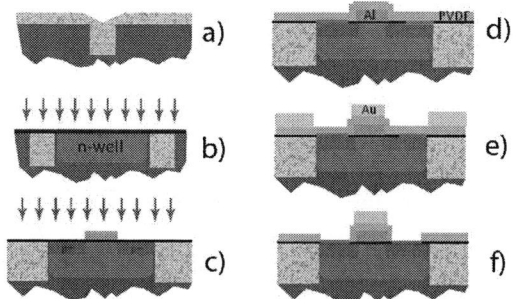

Figure 1: Process flow used for the fabrication of the Fe-MOSFET test structure: (a) STI definition followed by CMP, (b) creation of the n-well followed by (c) source and drain contact implant, (d) creation of the test transistor gate stack: thermal oxide growth, deposition and etching of Al intermediate contact for probing the internal gate voltage, V_{int}, and PVDF deposition, (e) evaporation, lithography and etching of Au contacts, (f) etching of PVDF using Au as a mask.

978-1-4244-6658-0/10 $26.00 © 2010 IEEE

Then, 6.5nm of thermal SiO_2 is grown as the gate oxide. 50nm of Al is sputtered in order to contact the source and drain, and to form the intermediate contact **(Fig 1d)**. A 1% solution of P(VDF-TrFE) using Methyl-Ethyl-Ketone as solvent was spincoated and baked for 5 minutes at 137°C, resulting in a 100nm thick layer [5]. Gold was evaporated and used as mask for the PVDF etching (Oxygen Plasma Etching) above the source, drain and intermediate contact **(Fig. 1e)**. After the PVDF etching, the gold is wet etched for the gate formation **(Fig. 1f)**. Fig. 2 presents the final cross section of the test device as well as its top-view with four contacts: gate, internal contact, source and drain. The sizes of the designed devices are large, with typical length and width of the Fe-FET of the order of 100 x 100µm².

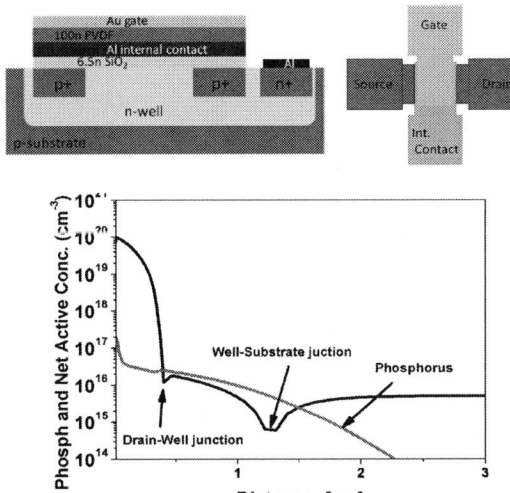

Figure 2: *Top:* cross section and top view of the proposed four-contact test Fe-FET device, *Bottom:* simulation of the implant profiles at the drain junction (black curve) and under the gate (red curve).

III. F_E-FET TEST STRUCTURE CHARACTERIZATION

In this section we report the capacitance and the static I_d-V_g and I_d-V_d characterization of the four contact Fe-FET test structures.

A first experiment was performed using the internal contact as gate; source, drain and substrate were grounded and a quasi-static capacitance measurement was made using the Agilent 4156C semiconductor parameter analyzer. We observe typical p-MOSFET gate capacitance characteristics without any hysteresis, as expected, **Fig. 3a**. Based on this measurement and using Tsividis MOS capacitance model [6], we have precisely extracted the flat-band voltage, V_{fb}=0.3V, the oxide thickness, t_{ox}=6.5nm and the doping achieved in the fabrication of the n-well, N_d=5 x 10^{17}cm^{-3}.

The second measurement consists in using the top gate as the Fe-FET gate and measuring the potential at the intermediate contact. **Fig. 3b** shows the clear effect of the in series 100nm ferroelectric layer in this case: the overall capacitance value is

reduced and a typical hysteresis is recorded. Based on our structure we have extracted the relative permittivity of the non-polarized ferroelectric, having the value of k=16.

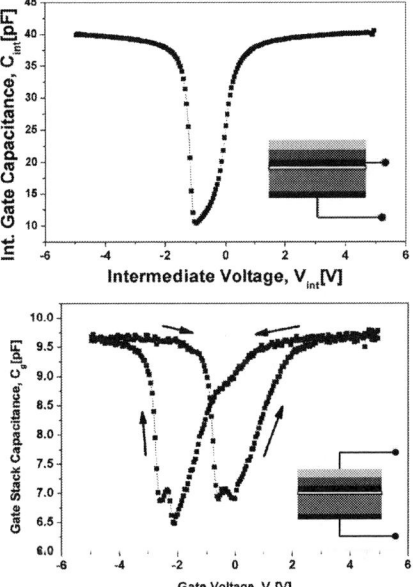

Figure 3: a) MOS capacitance versus internal gate contact, Vgint, measured by QSCV method, b) Complete gate stack capacitance, including ferrolectric layer and thermal (pedestal) oxide, for the 50µm x 150µm p-MOS transistor.

Drain current versus gate voltage measurements for voltage sweep of the *internal* and *external* gate voltage from -5V to +5V, maintaining the drain potential at three different values (V_d=100mV, 200mV and 300mV), are reported in **Figs. 4a and b**, respectively. **Fig. 4a** shows the characteristics of a 'normal' long-channel MOS transistor with a gate oxide of 6.5nm; a very low I_{off} current (smaller than 1pA) and an I_{on}/I_{off} ~10^7 are obtained, which enables a highly accurate investigation of the device in all the regions of operation: from depletion to strong inversion. **Fig. 4b** reports the hysteretic drain current characteristics when the top gate is biased and drain is set in same conditions as for **Fig. 4a**; on the same plot we report the measured value (using high precision voltmeter) of the internal gate voltage node, which shows a clear hysteresis and a non-linear variation. This potential is affected by two factors, first the polarization of the ferroelectric and, second, the capacitor voltage divider. Therefore, the internal voltage, V_{int}, is dictated by the field dependent ferroelectric capacitance and the subsequent MOS capacitance that is also varying with the V_{int}.

The inverse subthreshold slope, $S=[dlogI_d/dV_g]^{-1}$, was experimentally recorded for both ascending and descending I_d-V_g branches of the Fe-FET transistor and is presented in Fig. 5; we report excellent values of 80mV/decade for S in both cases, which reflect an excellent interface and small amount of interface traps.

978-1-4244-6658-0/10 $26.00 © 2010 IEEE

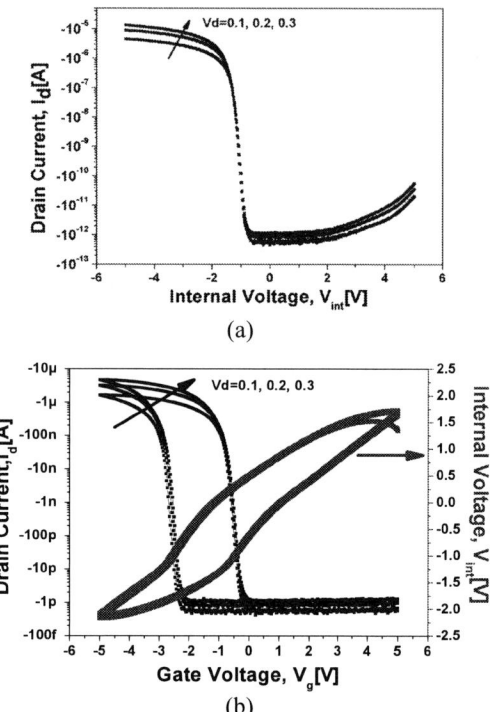

(a)

(b)

Figure 4: (a) I_d-V_g characteristics of the transistor using the intermediate contact as gate b) I_d-V_g characteristics of the full stack FeFET on the right axis and the measured intermediate contact voltage on the left axis.

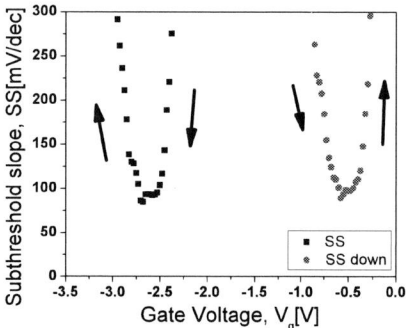

Figure 5: Experimental inverse subthreshold slope, $S=[dlogI_d/dV_{gs}]^{-1}$, of the Fe-FET; the left (black) curve is calculated when sweeping V_g down and the right curve (red) when sweeping it back up.

The last measurement using the intermediate contact as the gate was to sweep the V_d from -0.1V to -5V, using three different V_{int} voltages as gate voltage (-1V,-2V and -3V). A similar measurement has been carried out with the gate voltage applied on the full gate stack. **Fig. 6** presents the I_d-V_d characteristics resulting from these two experiments. A hysteresis is obtained in the second case only, mirroring the effect of V_{gd} on the ferroelectric polarization.

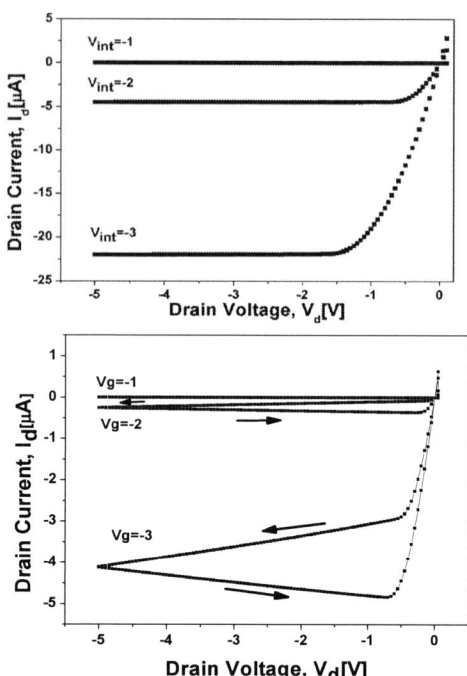

Figure 6: a) I_d-V_d output characteristics of the transistor using the intermediate contact, V_{int}, as gate b) I_d-V_d characteristic of the full Fe-FET transistor with hysteresis induced by V_{gd}.

IV. EXTRACTION OF INTERNAL VOLTAGES AND DISCUSSIONS

As mentioned earlier, the internal voltage is influenced by two factors: (i) the capacitive voltage divider between the ferroelectric capacitance and the oxide in series with the bulk capacitance and (ii) the key effect of polarization of the ferroelectric capacitance. The capacitance of the semiconductor is varying mostly in the depletion region so we can observe a steepening of the V_{int} dependence on Vg in that region. One objective was to study whether the influence of the polarization of the ferroelectric is strong enough to have a faster variation of the intermediate potential then the variation of the gate voltage, to record how high the value of dV_{int}/dV_g is and how it depends on the experimental conditions. The resulting slope would then be highly dependent on the polarization of the ferroelectric and negative capacitance could eventually be achieved under some particular conditions. Measurements were performed at different symmetrical sweeps of V_g, (±3V, ±5, ±7V) and they are presented in **Fig. 7**. The plot reported for a swing of 5V corresponds to the measurements on the Fe-FET transistor test structure (see Fig. 4). As one can observe, the slope of the V_{int} with respect to V_g greatly improves in the subthreshold region. The derivative of the V_{int} with respect to the V_g is presented in **Fig. 8**; this result is presented for the first time in a Fe-FET, being uniquely enabled by our test structure. An increasing V_{int} variation (peaks in Fig.8), potentially leading to amplification, is observed in the subthreshold region.

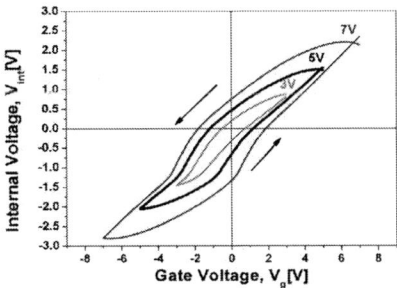

Figure 7: Internal voltage, V_{int}, measurement at different symmetric V_g sweeps in our test p-Fe-FET structure.

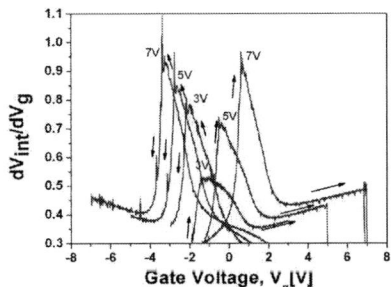

Figure 8: Experimental derivative dV_{int}/dV_g showing sub-unit tuneable gain in the various regions of operations of the Fe-FET. A value close to one is reported for the largest sweep.

In the case of the previous measurement of the I_d-V_g, at a sweep of ±5V the highest dV_{int}/dV_g values are found at -2.5V and at -0.5V. We can see from the graph in **Fig. 7** that the slope is dependent on the polarization, growing as the polarization grows. The maximum value obtained for the ±7V sweep, is 0.95, meaning that the potential on the intermediate point varies almost the same as the V_g. Larger dV_{int}/dV_g than unity would mean internal amplification due to negative capacitance and conditions to record sub-60mV/dec subthreshold slope. Hence, we experimentally demonstrate significantly higher internal node potential variations in the subthreshold region. Based on the internal voltage values and the semiconductor parameters extracted from the capacitance, the surface potential of the Fe-FET, Ψ_s, has been analytically calculated by numerically resolving numerically the following exact equation [7]:

$$V_g - V_{FB} - \psi_S = \gamma *$$
$$\sqrt{\psi_S + \phi_t \left[exp\left(-\frac{\psi_S}{\phi_t}\right) - 1 \right] + + \phi_t exp\left(\frac{-V - 2\phi_F}{\phi_t}\right) * \left[exp\left(\frac{\psi_S}{\phi_t}\right) - 1 \right]}$$

By knowing the relative dielectric constant k and the thickness of the PVDF, the equivalent oxide thickness was calculated (EOT=50nm) and the surface potential of equivalent MOS transistor was simulated and compared to the surface potential extracted for the measured V_{int} values. The extractions are presented in **Fig. 9**. **Fig. 10** clearly shows that the variation of the surface potential in a Fe-FET in weak inversion with the gate voltage is significantly faster that in a MOSFET with "passive" gate stack having the same EOT.

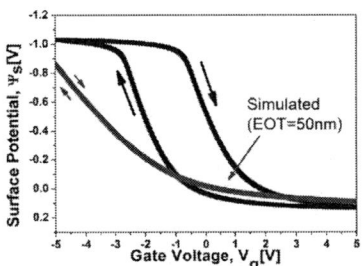

Figure 9: Absolute value of the surface potential with respect to the gate voltage for the ±5 sweep.

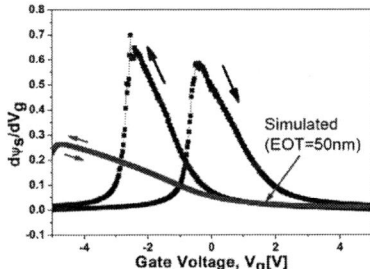

Figure 10: Slope of the surface potential variation with V_g versus gate voltage: measured versus simulated values.

Comparing the two transistors, the ferroelectric and the EOT one, we observe two phenomena; one is the shifting of the threshold voltage and the other is the faster variation of the surface potential in the subthreshold region. Both phenomena are due to the nonlinearity of the polarization in the ferroelectric. The surface potential is essential for developing a complete compact model. Going further with the Tsividis model, all the currents could be determined, leading to an accurate model for the transistor.

V. CONCLUSIONS

In this paper we have reported a novel test structure in order to characterize the internal voltage nodes in ferroelectric MOS transistor. For the first time we have been able to report the experimental hysteretic dependence of the internal node and of the surface potential of the external gate voltage in Fe-FETs with organic PVDF ferroelectric and showed that in weak inversion there is a peak of the internal potential variation. This opens new experimental for future experimental methods dedicated to the accurate characterization of Fe-FETs and to the investigation of the effect of negative capacitance.

REFERENCES :

[1] A. Gerber et al., J. Appl. Phys. v.100 024110 (2006).
[2] E. Supriyanto and H. Goebel, MIEL 2004, 2004, P571-P574
[3] S. Salahuddin, S. Datta, Nano Letters, Vol. 8, pp. 405-410, 2008.
[4] Hang-Ting Lue, Chien-Jang Wu, Tseung-Yuen Tseng, IEEE TRANSACTIONS ON ELECTRON DEVICES, VOL. 49, NO. 10, OCTOBER 2002
[5] S. Fujisaki and H Ishiwara, Appl. Phys. Letts. 90, (2007).
[6] Y. Tsividis, 2nd Edition, McGraw Hill, 1999.
[7] R. Langevelde, F. Klaassen, S.S. Electronics, Volume 44, Issue 3, 1 March 2000, Pages 409-418

Experimental evidence of unconventional Room-Temperature Quantum Hall effect (RTQHE) in 65nm Si nMOSFETs at very low magnetic fields

Edmundo A. Gutiérrez-D.
Department of Electronics
INAOE
Puebla, Mexico
e-mail:edmundo@inaoep.mx

Fernando Guarin
Semiconductor Research and Development Center
IBM Microelectronics
Hopewell Junction NY 12533, USA
e-mail:guarinf@us.ibm.com

Abstract— For the first time we introduce experimental evidence of an anomalous or unconventional Room-Temperature QHE at B fields at and below 50 mT (milli-Teslas). The observed Ultra-High-Conductivity-State (UHCS), and negative channel current are explained in terms of a fractional-dimensional charge transmission, quasi-particle-phonon resonance, and marginal phase (semiconductor-insulator) transition. We believe this work is fundamental to understand the physics of low- and fractional-dimensional devices, which may open a wider and deeper road for Si-based devices applications.

I. INTRODUCTION

Since the discovery of the integer and fractional Quantum Hall Effect (QHE) in the 80's [1,2], some of the research has been done at cryogenic temperatures and at low, medium and high magnetic B fields [3,4] for GaAs devices, and recently Novoselov reported QHE in Graphene layers [5] at room temperature and B=19 T. Following this trend of the research that focuses on understanding the physics behind 2DEG electron transport, which may be applied in developing ultra small devices (smaller than 10 nm), Gutiérrez-D. showed the possibility of using electro-magnetic techniques to study the Si-SiO$_2$ interface of 65nm nMOSFETs [6]. In such a work, the 2DEG layer was space and time modulated by a time-dependent B field, which gave a first clue of the influence of the level of "disorder" -atomic granularity or fractional-dimensionality- of the interface on the charge transport of devices with less than half a million of Si atoms. Therefore, in the quest for understanding the charge transport of low- and fractional-dimensional devices, where the electron may behave as a Boson, Anyon, or Skyrmion quasi-particle rather than a simple Fermion [7], we introduce experimental results where the action of an external B field on a 65nm nMOSFET, makes the transistor behave as if it had a fractional dimensionality between 2 and 1. In our experimental results this is manifested as an Ultra-High-Conductivity-State (UHCS), where the channel current shows up current spikes with an incremental value of up to one order of magnitude,

with a periodic voltage separation that suggests the existence of an unconventional quantum Hall effect. Our latest result includes "full channel backscattering" measured as negative channel current, where the electrons change the source-to-drain to a drain-to-source movement.

II. EXPERIMENTAL OBSERVATIONS AND HYPOTHESIS

Steady-state electro-magnetic measurements were done on nMOSFETs fabricated in a 65nm MOS technology with a gate nitride oxide thickness Tox of 2 nm. Devices with a constant

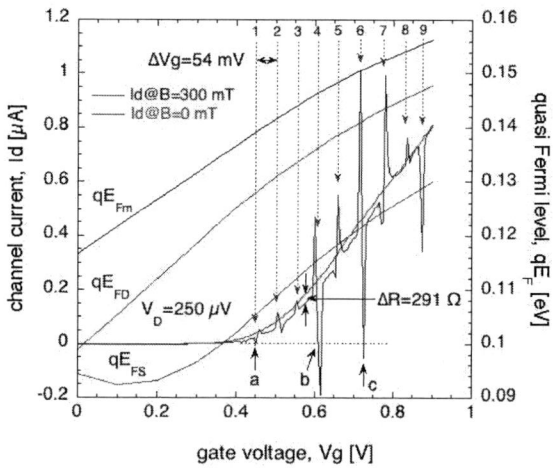

Figure 1.- Measured Id-Vg curve for Vd=250 µV at B=0 mT (red) and 300 mT (blue) of a (2µm/65nm) nMOSFET.

gate width W of 2 µm, and channel lengths L of 60, 65, and 70 nm were measured with an Agilent Semiconductor Device Analyzer B1500A, while assembled in between the poles of an electromagnet that produced a DC B field from about ten micro-Teslas (µT) up to 300 milli-Teslas (mT). Special tri-axial em-shielding and de-embedding was used to avoid any spurious signals. Figure 1 shows the Id-Vg characteristics of an (2µm/65nm) nMOSFET biased at Vd=250 µV, with and without a constant B field of 300 mT.

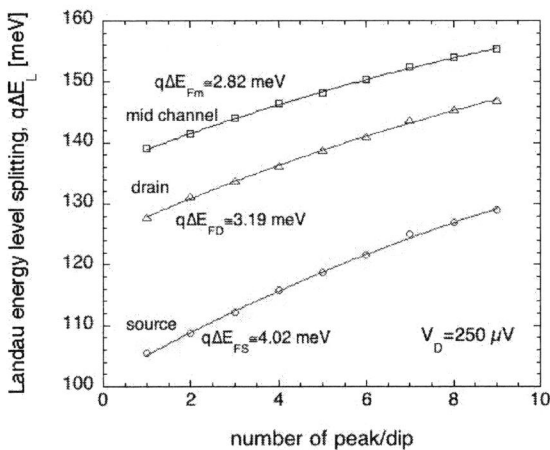

Figure 2.- Calculated Landau energy splitting for the 9 current peaks/dips of Figure 1.

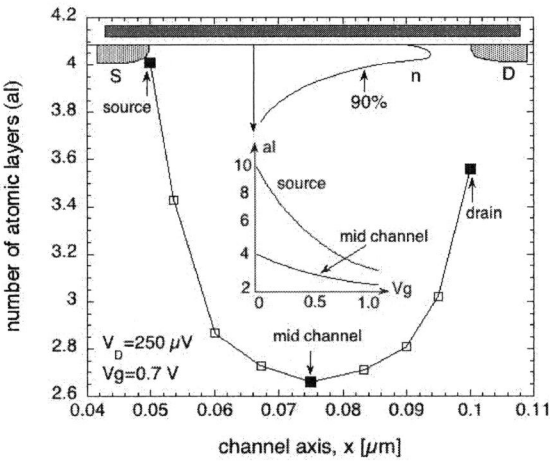

Figure 3.- Simulated number of atomic layers (al) along the channel axis for Vd=250 µV and Vg=0.7 V.

The B field is applied perpendicular to the surface, so channel electrons are swept and crowded to the left or right side along the width W. Accordingly to the conventional Hall effect the longitudinal channel resistance increases by an amount ΔR=291 Ω. However, 9 Id current peaks/dips are unexpectedly observed between Vg= 0.46 V and 0.88 V. Three of these dips, marked a, b, and c, are even negative. The peaks/dips are voltage equidistant by ΔVg=54 mV. The voltage periodicity of the peaks/dips suggests the presence of

an unconventional Landau energy level splitting [9], which manifests as a 2D to 1D quasi-transition of the density of states (DOS). One should expect that when the Fermi energy

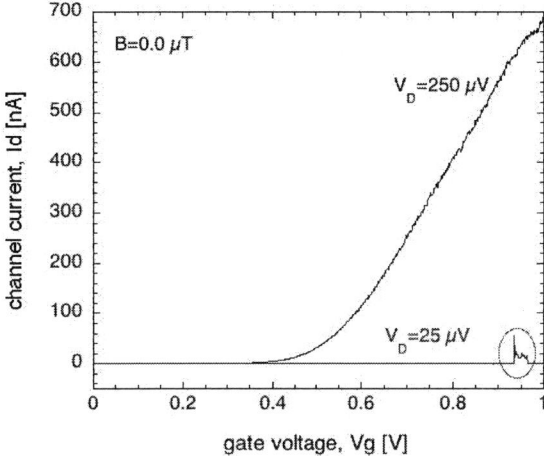

Figure 4.- Measured Id-Vg curve for B=0 µT for Vd=250 µV and 25 µV (blue line).

of the electrons channel crosses over the peaks of the DOS, the longitudinal conductance should show a sudden increase. Therefore we simulated [6] and calculated the quasi-Fermi energy level qE_F, for three different positions along the

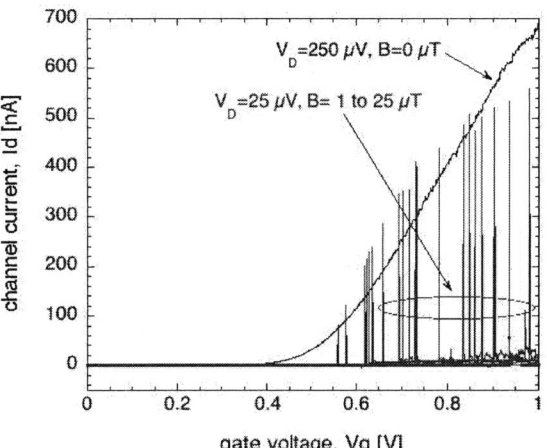

Figure 5.- Measured Id-Vg curve for Vd=250 µV (B=0 µT), and Id-Vg for 25 µV for B varying from 1 to 25 µT.

channel (right axis of Figure 1). These three positions are at source side (qE_{FS}), the mid channel (qE_{Fm}), and the drain side (qE_{FD}) as shown in Figure 3. The value of qE_F at the current peaks/dips is then translated into a Landau energy level splitting $q\Delta E_L$ versus the peak/dip number as shown in Figure 2. The variation of $q\Delta E_L$ with the gate voltage Vg is almost

978-1-4244-6658-0/10 $26.00 © 2010 IEEE   99

linear, which gives a qΔE_L of 4.02, 2.82, and 3.19 meV for the source, mid channel, and drain side, respectively. The calculations also indicate that 90% of the charge transport occurs through about 2.7 atomic layers in the mid channel and widens at the channel ends (source and drain) as shown in Figure 3. The number of atomic layers, where 90% of the current flows, reduces as the gate voltage increases as shown in the inset of Figure 3. For this calculation we assume the maximum of the inversion channel peaks under the Si-SiO$_2$ interface (see also inset of Figure 3).

The inversion channel thickness approaches that of a mono-atomic layer in the middle of the channel, which indicates the low-and fractional-dimensional effect is strengthen at the mid channel. On the other hand, because the cyclotron energy ω_c=361.6 μeV is smaller than the thermal voltage, the condition to have QHE is not fulfilled. Therefore, in order to change the energy of the electrons, and see if that modified the oscillatory behavior of figure 1, we observed that for Vd>375 μV the current peaks/dips fade away, and that for Vd<150 μV, the transistor started to shut off. There is then a low Vd range where the energy of electrons seems to resonate with the energy states of the fractional-dimensional channel that gives rise to the oscillatory behavior.

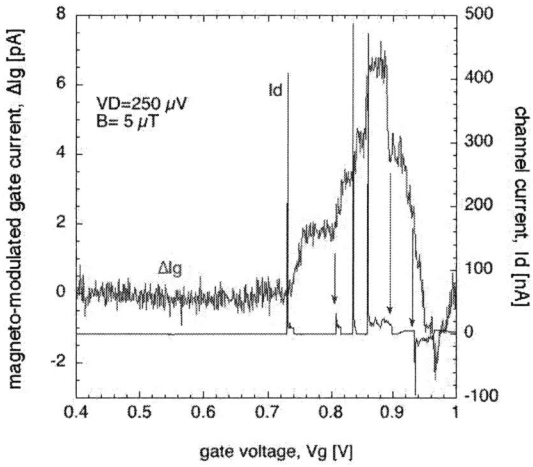

Figure 6.- Measured magneto-modulated gate ΔIg and Id currents versus Vg for Vd=250 μV and B=5 μT.

An approach to explain the RT unconventional QHE is by assuming the electrons as a quasi-particle with a boson-like behavior with a renormalized effective mass m* [7]. A renormalized effective mass m* implies very long lifetimes for low-energy quasi-particles, which is the case of MOSFET biased at low Vd voltages. Under these conditions, the 2DEG, now composed of quasi-particles and bare particles (electrons), can have a collective response under the action of an external B field. This collective mode response may result in zero sound quasi-particle-phonon resonance [9]. A resonant state, where the energy of the particle is above the Fermi level, can have more than one particles and one hole as calculated in

[9]. This change of the spectrum of the 2DEG energy system indicates a different quantity of quasi-particles and particles, which are the fundamentals for the Fermi liquid concept [10]. This is an important assumption because it implies a finite overlap between the wave function of a quasi-particle and the wave function of a fermion, so QHE can still be observed at low energies and small B fields. For a low energy case (low Vd), when the quasi-particle-particle interactions are weak, the scattering process of low-energy particles can be very small, which implies a quasi-particle with a reduced renormalized mass m* that may approach the "mass-less electron" concept [11]. A "mass-less " quasi-particle needs almost no energy to change its momentum, which may lead to a long straight free path, larger than the cyclotron radius, with no collisions. This situation may also lead to the formation of superconducting energy gaps, which may also explain the observed large current peaks. In order to explore the action of a very low B field in a system with very low energy above the thermal energy, we measured the Id-Vg characteristics with no B field at both Vd=250 μV and 25 μV. For Vd=25 μV the transistor is off except for a very small current spike at Vg=0.95 V (see

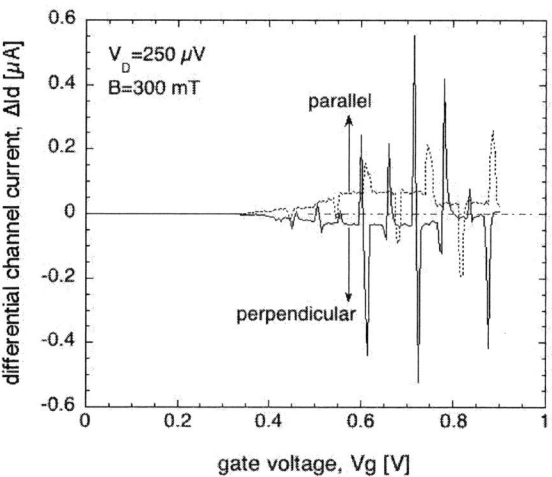

Figure 7.- Measured differential channel current ΔId versus Vg at Vd=250 μV for perpendicular and parallel B field of 300 mT.

Figure 4). Now we proceed to measure the Id-Vg curve under the action of a constant B field at different values from 1 μT up to 25 μT in steps of 2 μT. The results in Figure 5 show the occurrence of 19 different current deltas. For some values of B there is only one current delta, but for some other values, like B=5 μT, there are three positive deltas and one negative delta (see Figure 6). This looks as if the Vd voltage had been effectively increased from 25 μV to 250 μV, where the current deltas were trying to fill up the area below the Id-Vg (for B=0 T) curve. A possible explanation is that the existence of more than two quasi-particles, with Fermi energy above the seed Landau level, induces a further sub-splitting of the Landau level, leading to the occurrence of the current spikes of figure 5. We tried to fill up the whole area below the Id-Vg curve by changing the B field range and the magnitude of the B step,

978-1-4244-6658-0/10 $26.00 © 2010 IEEE

but we only got to fill some spaces that were empty before,

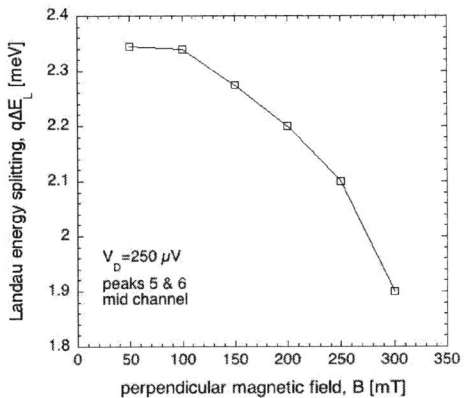

Figure 8.- Extracted energy level splitting for the peaks 5 and 6 at the mid channel.

but opened some others. Furthermore, to discard any possibility of electromagnetic random or spurious noise induction, we also measure the gate current I_g, and observed that any time an I_d current delta shows up, the differential gate current ΔI_g, which is the difference of the gate current measured under B field minus the I_g current with no B field, shows a transition (see Figure 6). The results shown in Figure 6 are for B=5 µT, but the effect is also seen for other values of B, not shown here for the sake of simplicity. These results indicate that the quantized energy transitions at the quasi-2D x-y plane of the MOSFET channel are also having a vertical quantum effect on the gate current (z direction). The charge source for the I_g current is the surface charge now composed of quasi-particles and fermions, which explains the magneto-modulation of the gate current. In the case of a B field applied parallel to the surface, the same effect is observed as shown in Figure 7. The differential channel current ΔI_d, which is the difference of the current with B field minus the current with no B field, is plotted for parallel and perpendicular B fields of 300 mT. In both cases the current peaks/dips are shown separated by equidistant ΔV_g voltages. However, for the perpendicular case the channel conductance decreases as predicted by the conventional QHE, while for the parallel case the channel conductance increases (dotted line of Figure 7). The B field applied parallel to the Si-SiO$_2$ surface pushes electrons up towards the interface, which explains the increase of the channel conductance. Finally, we extracted what we called the Landau energy level splitting $q\Delta E_L$ between the peaks 5 and 6, and plotted them as a function of the perpendicular B field (see Figure 8). We observe that for the mid channel the energy splitting of the current peaks decreases as the magnitude of the B field increases. This probably points out a large influence of the collective interaction of the quasi-particle system with the lattice perturbations.

III. PRELIMINARY CONCLUSIONS

For the first time we report the occurrence of magneto current oscillations in a Si device operated at room temperature at a magnetic field of 50 mT and below. As the condition ($\hbar\omega_c \gg kT$) for a conventional Quantum Hall Effect is not fulfilled for room temperature and low magnetic fields, we explain the occurrence of the periodic channel current peak/dips as a result of a quasi-particle-phonon resonance and phase transition, which transforms the conventional 2DEG into a fractional-dimensional channel with semiconducting-to-superconducting-to-insulating phase transitions. The magneto modulation of the channel current is correlated to the gate current as seen from Figure 6, which suggests for the development of an electro-magnetic technique for the characterization of the Si-insulator interface. Furthermore, the observance of the RTQHE in 65nm MOSFETs shed light on the potential of keeping Si as the material for the future development of nano-metric technologies. This work also serves as an empirical reference for a future understanding of charge transport at the atomic level.

ACKNOWLEDGMENTS

The first author wants to acknowledge IBM for providing the test samples, CONACyT for funding the research through grant 100028, and Intel for partial funding.

REFERENCES

[1] Klitzing, K. von; Dorda, G.; Pepper, M., "New Method for High-Accuracy Determination of the Fine-Structure Constant Based on Quantized Hall Resistance". Phys. Rev. Lett. 45 (6): 494–497, 1980.

[2] D. C. Tsui, H. L. Stormer, and A. C. Gossard, "Two-dimensional magneto transport in the extreme quantum limit, Phys. Rev. Lett., Vol. 48, pp. 1559-1562, 1982.

[3] Ch. Yang, "New quantum oscillations in magneto transport of a high-mobility two-dimensional electron system", PhD. Thesis, University of Utah, August 2004.

[4] A. A. Bykov, Jing-qiao Zhang, , and Sergey Vitkalov, "Zero-differential resistance state of two-dimensional electron systems in strong magnetic fields", Phys. Rev. Lett., Vol. 99, p. 116801-166804, September 2007.

[5] K. S. Novoselov, Z. Jiang, Y. Zhang, S. V. Morozov, H. L. Stormer, U. Zeitler, J. C. Maan, G. S. Boebinger, P. Kim, and A. K. Geim, "Room-temperature quantum Hall effect in Graphene", Science, Vol. 315, No. 5817, p. 1379, March 2007.

[6] Edmundo A. Gutierrez-D., J. Molina-R., P. J. Garcia-R., J. Martinez-C., and F. Guarin, "Magnetic field induced gate leakage current in 65nm nMOS transistors", Proc. ESSDERC 2009, pp. 185-189, Athens, Greece, September 2009.

[7] Zyun F. Ezawa, "Quantum Hall Effects", World Scientific, 2008.

[8] T. Kawarabayashi, Y. Ono, T. Ohtsuki, S. Kettemann, A. Struck, B. Kramer, "Unconventional conductance plateau transitions in quantum Hall wires with spatially correlated disorder", Phys. Rev. B75, 235317, 2007.

[9] Yu. A. Bychkov and A. V. Kolesnikov, "Slightly nonideal 2D Fermi gas in a magnetic field", JETP Letters Vol. 58, No. 5, 10 Sep. 1993, pp. 352-356.

[10] Y. M. Vilkk and A.-M. S. Tremblay, "destruction of Fermi-liquid quasi particles in two dimensions by critical fluctuations", Europhysics Letters, Vol. 32, Nr. 2, pp. 159-164, 1996.

[11] Ch.-H. Park and S. G. Louie, "Making massless Dirac fermions from a patterned two-dimensional electron gas", Nano Letters, Vol. 9, Nr. 5, pp. 1793-1797, April 2009..

Thermal broadening of two-dimensional electron gas mobility distribution in AlGaN/AlN/GaN heterostructures

G. A. Umana-Membreno[a], T. Stomeo[b], V. Tasco[b], A. Passaseo[b], M. de Vittorio[b], L. Faraone[a]

[a] School of Electrical, Electronic and Computer Engineering, The University of Western Australia,
35 Stirling Hwy, Crawley 6009, Australia
[b] National Nanotechnology Laboratory of CNR-INFM, Distretto Tecnologico-ISUFI,
Università del Salento, Via Arnesano, 73100 Lecce, Italy.

Abstract— **Two-dimensional electron gas (2DEG) transport in $Al_{0.3}Ga_{0.7}N$/AlN/GaN heterostructures has been studied using magnetic-field dependent Hall-effect measurements and advanced mobility spectrum analysis techniques over the temperature range from 95 K to 300 K. It is shown that electronic transport is due to a single well-defined 2DEG species, with room-temperature sheet concentration and average mobility of 9.3×10^{12} cm^{-2} and 1,880 cm^2/Vs, respectively. No parasitic conduction through the bulk GaN layer was detected. Importantly, it is shown that the 2DEG exhibits an approximately Gaussian mobility distribution, the linewidth of which broadens with increasing temperature. This is the first reported observation of thermal broadening effects in the 2DEG mobility distribution.**

Keywords- charged carrier mobility, quantitative mobility spectrum analysis, QMSA, high-resolution QMSA, 2DEG, HEMT, Hall effect, mobility distribution, AlGaN, GaN

I. INTRODUCTION

The group III-nitride compound semiconductors have become indispensable for the fabrication of short-wavelength laser and light-emitting diodes, and are becoming increasingly important for the fabrication of high-efficiency AlGaN/GaN-based high electron mobility transistors (HEMTs) aimed at applications demanding operation at high-temperatures, high-power levels, and at high frequencies [1]. Since the mobility and the carrier density in the two-dimensional electron gas (2DEG) formed at the AlGaN/GaN heterojunction are key parameters that are directly related to device performance, better understanding of the physical mechanisms influencing these parameters is necessary to enable enhanced device performance and/or to overcome present-day limitations. Precise evaluation of the electronic transport parameters is vital for material growth and device optimization.

Though conventional Hall-effect measurements at a single magnetic field are routinely used to probe the electronic transport parameters of 2DEG channels, such measurements assume the presence of a single carrier – with a well-defined and discrete mobility – and can only provide weighted averages of the electron mobility and carrier density in the case of multiple-channel transport. In the case of AlGaN/GaN heterostructures, conventional Hall-effect measurements are unable to separate the effect of parasitic conduction through the bulk GaN from the 2DEG channel conductivity [2, 3]. While this complexity can be somewhat circumvented using multicarrier fitting procedures [4], such approach has arbitrary elements that are inadequate for modern semiconductor structures. A more general approach is to employ quantitative mobility spectrum analysis (QMSA) of measured magnetic-field dependent Hall-effect and resistivity data to separate and resolve the conductivity-mobility spectra of all carriers present in a sample[5]. Furthermore, as envisioned by Beck and Anderson in their theoretical formulation of the mobility spectrum analysis, the carrier distribution's lineshape and its thermal broadening may yield important insight into carrier spatial fluctuations, the Hall-scattering factor, of relaxation-time distributions, and cyclotron frequency distributions (related to band non-parabolicity) [5]. Though over the last two decades, the mobility spectrum analysis technique has been enhanced and developed into a practical and powerful tool by numerous researchers [6–11], accurate extraction of mobility distributions and the observation of broadening effects has been curtailed by limited resolution capabilities in available QMSA algorithms.

In this work, 2DEG transport in $Al_{0.3}Ga_{0.7}N$/AlN/GaN heterostructures have been characterized employing a recently developed high-resolution quantitative mobility spectrum analysis algorithm (HR-QMSA) [12]. It is shown that transport in the channel is due to a single well-defined 2D electron species that exhibits an approximately Gaussian mobility distribution and thermal broadening with increasing temperature. To the best of our knowledge, this is the first reported observation of thermal broadening effects in 2DEG mobility spectra.

II. SAMPLE AND EXPERIMENTAL DETAILS

The samples studied were grown by metal organic chemical vapor deposition under Ga-rich conditions on c-plane sapphire substrates at the National Nanotechnology

---

This work was supported by the Australian Academy Science under the Australia-Italy Award scheme, and by the FIRB national project: "Tecnologie abilitanti e caratterizzazione per componenti elettronici integrati riconfigurabili a banda larga per alta frequenza" Prot. N° RBIP068LNE.

Laboratory, University of Salento. The epitaxial structure of the sample consisted of 100nm-tick AlN nucleation layer, followed by 1.8 μm GaN layer, 0.75 nm AlN spacer layer and 18 nm $Al_{0.3}Ga_{0.7}N$ as the topmost layer. No intentional doping impurities were introduced during growth.

The samples studied were patterned into Greek-cross geometry Van der Pauw test-structures using Ar/Cl_2 plasma-based dry-etching (arm length and width of 400μm and 200μm, respectively), with ohmic contacts formed by thermal evaporation of Al/Cr/Au followed by rapid thermal annealing at 800°C in nitrogen ambient for 60s. Magnetic-field-dependent sheet resistance and Hall-effect measurements were carried out at field intensities up to 2T, using a narrow gap electromagnet (1cm gap), at sample temperatures from 95K to 300K in a liquid-nitrogen cooled continuous flow cryostat.

A. Experimental Data Analysis

The longitudinal and transverse conductivity tensor components σ_{xx} and σ_{xy}, respectively, were obtained from the measured magnetic-field dependent sheet resistance and Hall coefficient using the following expressions [5]:

$$\sigma_{xx}(B) = \frac{1}{R_s(B)\left[1 + \left(\frac{R_H(B)\,B}{R_s(B)}\right)^2\right]}$$

$$\sigma_{xy}(B) = \frac{R_H(B)\,B}{R_s^2(B)\left[1 + \left(\frac{R_H(B)\,B}{R_s(B)}\right)^2\right]}$$

In the general case of a distribution of carriers, the contribution of all carriers in the sample to the conductivity tensors is described by the *mobility transforms* [10]:

$$\sigma_{xx}(B) = \int_0^\infty \frac{s_p(\mu) + s_n(\mu)}{1 + \mu^2 B^2}\,d\mu$$

$$\sigma_{xy}(B) = \int_0^\infty \frac{[s_p(\mu) - s_n(\mu)]\,\mu B}{1 + \mu^2 B^2}\,d\mu$$

where s_p and s_n are hole and electron mobility-conductivity density functions, respectively (assumed to be magnetic-field independent). In practice, the transforms are calculated from their discretized form [10]:

$$\sigma_{xx}(B) = \sum_{i=1}^N \frac{S_p(\mu_i) + S_n(\mu_i)}{1 + \mu_i^2 B^2}$$

$$\sigma_{xy}(B) = \sum_{i=1}^N \frac{[S_p(\mu_i) - S_n(\mu_i)]\,\mu B}{1 + \mu_i^2 B^2}$$

with S_n and S_p being the hole and electron mobility spectra, respectively.

In this work, the longitudinal and transverse conductivity tensor components were analyzed using a recently developed high-resolution quantitative mobility spectrum analysis algorithm (HR-QMSA). In contrast to commercially available algorithms (such as LakeShore Cryotronics' iQMSA), HR-QMSA does not utilize the derivatives of the measured conductivity tensor components, thus it is less sensitive to

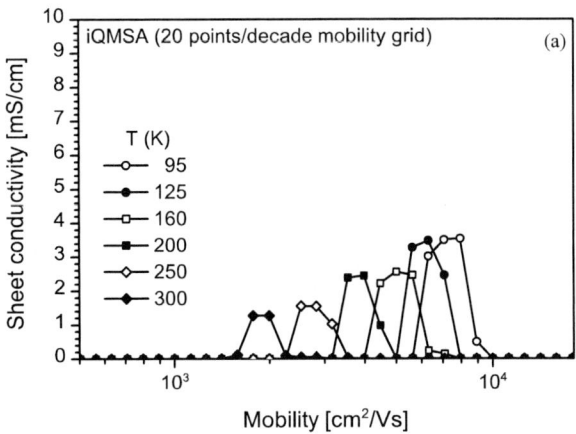

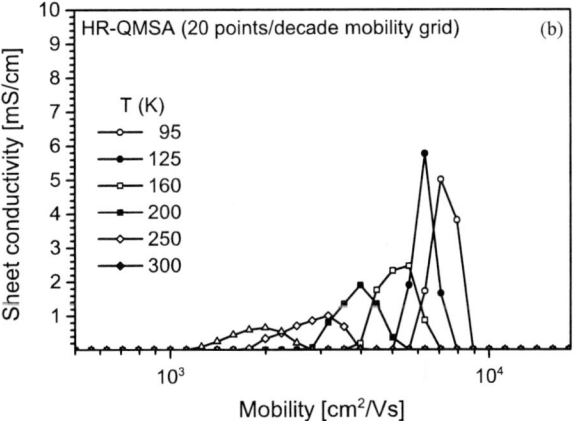

Fig. 1   Electron mobility spectra extracted using (a) iQMSA and (b) HR-QMSA employing a standard 20 points/decade mobility analysis grid. Note that in both cases the spectra are well-behaved, indicating the presence of a single electron associated with the 2DEG. The HR-QMSA spectra show a degree of linewidth broadening that is absent in the iQMSA spectra.

noise in the measured data and has shown greater resolution. HR-QMSA operates through modification of the whole electron and hole spectra at every iteration, and yields smooth spectra without the need for point-swapping procedures. However, HR-QMSA is very slow computationally; for a typical analysis requiring ~10,000 iterations, the algorithm runtime approaches 5 hrs in a computer with a 2GHz Pentium IV processor. Although further refinements and optimizations may be required to significantly reduce computation time, the HR-QMSA algorithm has undergone extensive testing employing synthetic datasets with varying degrees of complexity and noise-levels. These tests have demonstrated that HR-QMSA is able to better resolve closely-spaced mobility peaks arising from separate carriers as well as providing information on the spread and distribution of the mobility for each individual carrier species.

III.   RESULTS AND DISCUSSION

Fig. 1 shows the electron mobility spectra obtained using the iQMSA and HR-QMSA algorithms, using the iQMSA

978-1-4244-6658-0/10 $26.00 © 2010 IEEE

standard mobility analysis grid of 20 points/decade. In general, the mobility spectra obtained from both analysis procedures are similar, indicating the presence of a well-defined single electron species associated with the 2DEG. At room temperature the average mobility and sheet density were found to be 1,900±20 cm²/Vs and (9.2±0.1)×10¹² cm⁻², respectively (the error-bars indicating the small differences in the values extracted using the two different analysis algorithms). No parasitic conduction through the bulk GaN layer was detected, in agreement with the semi-insulating character of the underlying GaN layer as determined from separate measurements on 1μm-thick GaN films grown under identical conditions.

In order to refine the linewidth resolution of the 2DEG conductivity peaks in Fig. 1(b), the measured data was further analyzed using HR-QMSA a denser mobility grid of 50 points/decade. The results of this extended analysis are shown in Figs. 2 and 3. As shown in Fig. 3(a), the average mobility extracted from the spectra decreased with increasing temperature exhibiting an approximately $T^{-3/2}$ temperature dependence characteristic of non-polar phonon scattering limited mobility (e.g., acoustic deformation potential) rather than by LO polar optical phonon scattering. The electron density remained relatively constant over the whole temperature range as evidenced by the constant area under the conductivity peaks in Fig. 2 and the negligible thermal activation energy of 0.2 meV obtained from the linear fit shown in Fig. 3(b). From the room-temperature spectrum, the average mobility and the electron density were found to be 1,880 cm²/Vs and 9.3×10¹² cm⁻², respectively.

Particularly significant is the clear thermal broadening of the 2DEG mobility distribution that is evident in the spectra shown in Fig 2. Analysis of the linewidth of the 2DEG conductivity peaks indicate that the conductivity peaks are approximately Gaussian in shape. Thus, the spread of the mobility distribution is well described by a dimensionless broadening parameter $\Delta\mu/\mu$, with $\Delta\mu$ representing the full-width at half-maximum of the mobility distribution and μ the mean mobility. As shown in Figs. 2 and 4, the 2DEG mobility distribution becomes broader with increasing temperature exhibiting a thermal activation energy of 12±1 meV. It is important to note that this activation energy is associated solely with processes affecting the 2DEG mobility distribution and not the sheet electron density which is virtually constant over the whole temperature range.

Given the relatively high sample temperatures, the high 2DEG density and its temperature-independence, it is unlikely that ionized-impurity scattering would significantly influence the 2DEG mobility distribution for T > 95 K. Though the activation energy is too low for the broadening to be associated with high energy optical phonon (68–92 meV [13]). the broadening activation energy (12 meV) appears to be consistent with low-energy acoustic phonon modes (17–39 meV range) [13,14]. It should be noted that Brazis and Raguotis found it necessary to include such low-energy phonon modes in their Monte Carlo study of electron transport in GaN [14], and that their results have been found to be much

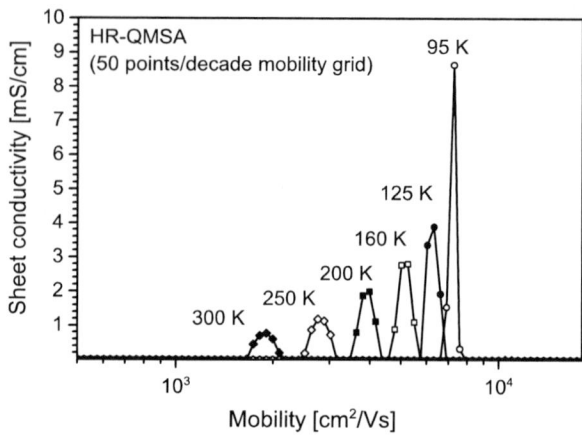

Fig. 2   Electron mobility spectra extracted using HR-QMSA with a 50 points/decade mobility grid. Note the significant improvement in resolution and the broadening effect of temperature on the linewidth of the conductivity peaks.

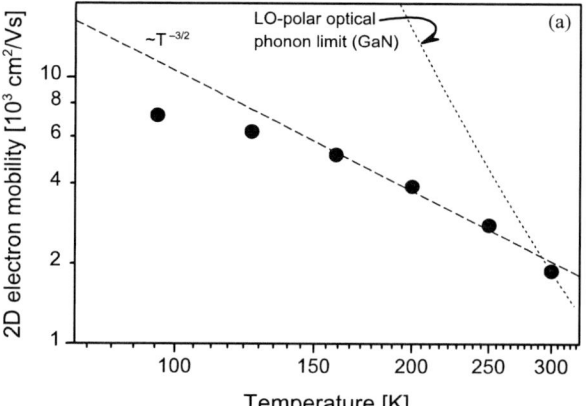

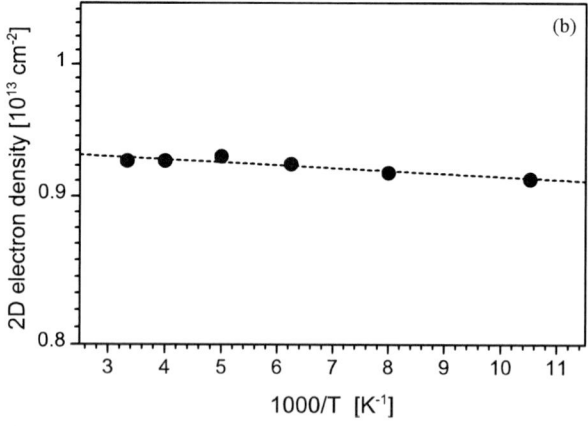

Fig. 3. (a) average 2DEG mobility and (b) electron density, both extracted from the mobility spectra in Fig. 2. The 2DEG mobility (b) exhibits a temperature dependence that is unlike that expected from LO-polar phonon scattering limited mobility. The electron density remained virtually constant over the whole temperature range (exhibiting a negligible thermal activation energy of 0.2meV, as indicated by the linear fit shown).

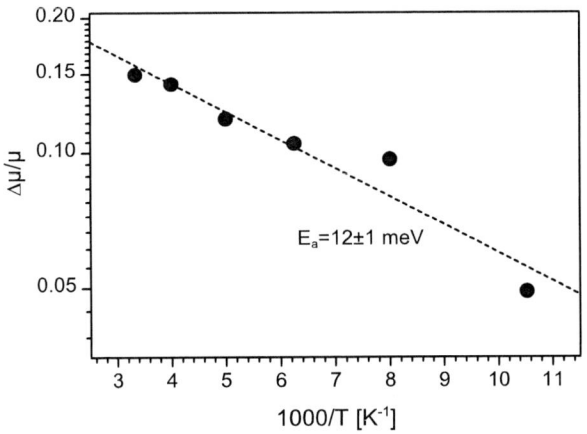

Fig. 4 Temperature dependence of the broadening parameter $\Delta\mu/\mu$. The linear fit shown indicates a thermal activation energy of 12 meV.

closer to reported experimental results than those found in previous studies [14, 15].

While at present, it is not possible to unequivocally ascertain whether the observed thermal broadening of the 2DEG mobility distribution is indeed due to scattering by low energy acoustic phonons, this first report of thermal broadening effects in the 2DEG mobility distribution prompts for further studies on scattering mechanisms and their effects on carrier mobility-spectra. Currently there are no adequate physical models to describe carrier transport in terms of mobility distributions (rather than as single discrete carriers). To a great extent, mobility distributions and broadening effects have not been studied due to a lack of robust mobility spectrum algorithms with sufficiently high resolution to accurately determine distribution linewidth. In this regard, the significantly better resolution, finer details, and better consistency attainable with the HR-QMSA technique represent a significant advance in this area.

IV. CONCLUSION

Two-dimensional electron gas transport in AlGaN/AlN/GaN heterostructures has been studied employing magnetic-field dependent Hall-effect and resistivity measurements, together with a recently developed high-resolution mobility spectrum analysis technique. A single well-defined 2DEG species was detected – no parasitic carrier conduction through the bulk GaN was detected – with a room-temperature mobility of 1,880 cm^2/Vs and a sheet electron density of 9.3×10^{12} cm^{-2}. The 2DEG mobility spectra were found to be approximately Gaussian in shape, with a mobility distribution that exhibited thermal linewidth broadening. The linewidth broadening was found to be thermally activated with an energy of 12 meV. This is the first reported observation of thermal broadening effects in 2DEG mobility spectra.

This study provides evidence of the need for magnetic field dependent magnetotransport characterization and QMSA analysis to study fundamental electronic transport, even when a single well-defined carrier is present. This study also prompts for further studies on the effects of scattering mechanisms on mobility-spectrum distributions.

ACKNOWLEDGMENT

The authors wish to thank Dr. Nicola Sasanelli and the Australian Academy of Science for financial support of this project via the Australia-Italy Award for early postdoctoral researchers. T. Stomeo thanks Prof. Faraone and the Microelectronics Research Group at The University of Western Australia for their support during her visit to Australia.

REFERENCES

[1] U. K. Mishra, P. Parikh, Y.-F. Wu, "AlGaN/GaN HEMTs—An Overview of Device Operation and Applications", Proc. IEEE, vol. 90, p.1022, 2002

[2] Z. Dziuba, J. Antoszewski, J. M. Dell, and L. Faraone, P. Kozodoy, S. Keller, B. Keller, S. P. DenBaars, and U. K. Mishra, "Magnetic field dependent Hall data analysis of electron transport in modulation-doped AlGaN/GaN heterostructures", J. Appl. Phys., vol. 82, p.2996, 1997

[3] S. Contreras, W. Knap, and E. Frayssinet, M. L. Sadowski, M. Goiran, M. Shur, "High magnetic field studies of two-dimensional electron gas in a GaN/GaAlN heterostructure: Mechanisms of parallel conduction", J. Appl. Phys., vol. 89, p.1251, 2001

[4] M.C. Gold, D. A. Nelson, "Variable magnetic field Hall effect measurements and analyses of high purity, Hg vacancy (p-type) HgCdTe", J. Vac. Sci. Technol. A, vol. 4, p. 2040, 1986.

[5] W A Beck, J R Anderson, "Determination of electrical transport properties using a novel magnetic filed dependent Hall technique", J.Appl.Phys., vol. 62, p.541, 1987.

[6] Z. Dziuba, M. Gorska, "Analysis of the electrical conduction using an iterative method", J. Phys. III France vol. 2, p.99, 1992.

[7] J. Antoszewski, D.J. Seymour, L.Faraone, J.R.Meyer, C.A. Hoffman, "Magneto-transport characterization using quantitative mobility-spectrum analysis", J. Electron. Mater., vol. 24, p.1255, 1995

[8] I. Vurgaftman, J.R. Meyer, C.A. Hoffman, D. Redfern, J. Antoszewski, L. Faraone, J.R. Lindemuth, "Improved quantitative mobility spectrum analysis for hall characterization", J. Appl. Phys., vol. 84, p.4966, 1998.

[9] S. Kiatgamolchai, M. Myronov, O.A. Mironov, V.G. Kantser, E.H.C. Parker, and T.E. Whall, "Mobility spectrum computational analysis using a maximum entropy approach", Phys. Rev. E, vol. 66, 036705, 2002.

[10] J. Antoszewski, L. Faraone, I. Vurgaftman, J.R. Meyer, and C.A. Hoffman, "Application of quantitative mobility-spectrum analysis to multilayer HgCdTe structures", J. Electron. Mat., vol. 33, p.673, 2004.

[11] J. Rothman, J. Meilhan, G. Perrais, J.-P. Belle, and O. Gravrand, "Maximum Entropy Mobility Spectrum Analysis of HgCdTe Heterostructures" J. Electron. Mat., vol. 35, p. 1174, 2006

[12] G. A. Umana-Membreno, J. Antoszewski, L. Faraone, E. P. G. Smith, G. M. Venzor, S. M. Johnson and V. Phillips, "Investigation of Multicarrier Transport in LPE-Grown HgCdTe Layers", J. Electron. Mat., in press (DOI: 10.1007/s11664-010-1086-7)

[13] V. Yu. Davydov, V. E. Kitaev, I. N. Goncharuk, A. N. Smirnov, J. Graul, O. Semchinova, D. Uffmann, M. B. Smirnov and A. P. Mirgorodsky, R. A. Evarestov, "Phonon dispersion and Raman scattering in hexagonal GaN and AlN", Phys. Rev. B., vol.58, p.12899, 1998

[14] R. Brazis, R. Raguotis, "Additional phonon modes and close satellite valleys crucial for electron transport in hexagonal gallium nitride", J. Appl. Phys., vol. 85, p. 609, 2004.

[15] S.K. O'Leary, B.E. Foutz, M.S. Shur, L. F. Eastman "Steady-State and Transient Electron Transport Within the III–V Nitride Semiconductors, GaN, AlN, and InN: A Review", J. Mater. Sci.: Mater. Electron., vol. 17, p. 87, 2006.

978-1-4244-6658-0/10 $26.00 © 2010 IEEE

Hardware/Software Co-Simulation for the Rapid Prototyping of an Acceleration Sensor System with Force-Feedback Control

Ruslan Khalilyulin, Thomas Steinhuber, Gabriele Schrag, Gerhard Wachutka
Institute for Physics of Electrotechnology, Munich University of Technology
Arcisstrasse 21, 80290 Munich, Germany
e-mail: {ruslan, schrag, wachutka}@tep.ei.tum.de

Abstract — **We present hardware-in-the-loop simulations of a MEMS acceleration sensor system with view to reduce the development time and complexity especially regarding design and test of dedicated control algorithms. To this end, the physically-based high-level Simulink model of the mechanical sensor is connected to a FPGA hardware board, on which the corresponding control algorithms (in our case PID and fuzzy logic) are implemented. The different control algorithms are described on the level of behavioral modeling in Simulink using the Xilinx System Generator tool, which then automatically generates a VHDL code of the models and transfers it to the hardware board. The advantage of these hardware/software co-simulations compared to "pure" simulations of the entire system in Simulink is a realistic estimation of the overall performance (run times, power consumption, efficiency, etc.) of the control circuitry even before the transducer is realized in hardware itself. This, on the other side, allows also modifications and optimizations not only on the control circuit but also on the transducer design in a very early stage of the design process.**

I. Introduction and Problem Description

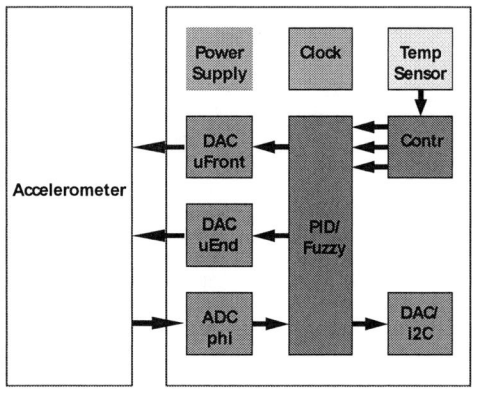

Figure 1. Schematic of the full accelaration sensor system.

Cost and duration aspects are key issues in modern electronic design. Today, sensors are complex systems equipped with "smart functionalities" and complex control electronics, which require many steps and iterations to be designed properly for the given application. Thus, virtual prototyping of entire systems comprising also realistic models of the surrounding control circuitry is more and more gaining in importance. Hardware/software co-simulations, where high-level transducer models are directly connected to FPGA hardware boards, on which the corresponding control algorithms are implemented, are an appropriate measure to obtain important information on the system performance already in a very early stage of the design phase, even before the system is transferred to the "real world".

We applied this approach for the micromechanical accelerometer sensor system depicted in Figure 1. It consists of the mechanical transducer (Figure 2) and an electronic part, which converts and transfers signals to the output of the sensor and controls the position of the transducer, viz. the membrane deflection, within a force feedback loop. However, proper and adequate concepts for reliable sensor control and signal conditioning can only be derived on the basis of an accurate transducer model, which is based on the profound understanding of the effects disturbing the sensor output signal.

To this end, we first derived a high-level model of the mechanical transducer, which comprises an accurate physical description of the significant impact factors (external and internal disturbances) on the sensor output signal.

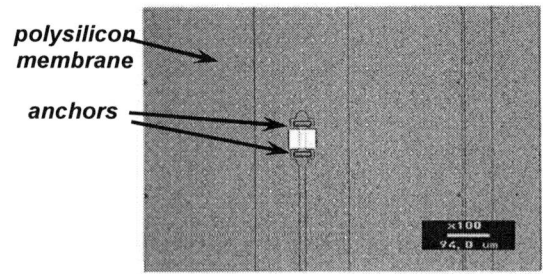

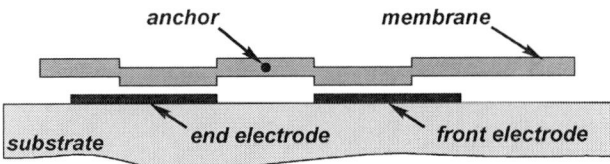

Figure 2. Top: Micrograph of the micromechanical accelerometer.
Bottom: schematic side view of the accelerometer.

The mechanical transducer consists of an asymmetrically anchored perforated polysilicon plate with a small gap and several read-out electrodes underneath. The transducer model of the accelerometer is based on a system of generic equations of motion:

$$J\ddot{\varphi} + D\dot{\varphi} + K_t\varphi = M_e \qquad (1)$$

Here, φ denotes the torsional angle, J the moment of inertia, D the damping factor due to viscous damping, K_t the torsional stiffness, and $M_e = \frac{U^2}{2} \cdot \frac{\partial C}{d\varphi}$ the moment acting on the device during electrostatic actuation. All mechanical model parameters like the moment of inertia J, the stiffness K_t, and the mechanical resonance frequency f_{sim} have been calculated from detailed FEM. The damping factor D can be determined by applying the mixed-level modeling approach proposed in [1]. This approach reduces the degree of complexity by replacing the non-linear and highly complicated Navier-Stokes equation by the well-known Reynolds' equation [1], which is discretized to form a fluidic Kirchhoffian network and then solved within a standard circuit simulator. Additional pressure drops at the edges and the perforations in the structure are taken into account by introducing physically-based compact models at the respective locations. However, tolerances due to manufacturing processes can drastically affect the mechanical and the electrostatic operation of the transducer as well as the fluidic damping forces. These impacts are difficult to predict by high-fidelity simulations alone. Hence, we calibrated the model parameters of the transducer applying a dedicated parameter extraction and calibration procedure using experimental data obtained by electrical and white light interferometer measurements under various ambient pressure conditions. This is described in detail in [2].

The core of the electronic part comprises the pressure compensated digital algorithm to control the position of the membrane deflection in case of occurring external disturbances. Two algorithms, namely a PID and fuzzy logic algorithm, are implemented in the hardware board and their performance is investigated with view to efficiency and applicability to the considered acceleration sensor system.

II. CONTROL BY PID ALGROTIHM

A well-established algorithm for the application in position control is the PID algorithm, which is based on an exact mathematical model [3]. For the investigated acceleration sensor the PID controller calculates an "error" value $e(t)$ from the difference between the measured and the desired capacity

at quiescent state of the membrane. The controller attempts to minimize this error value by adjusting the voltage at the front and end electrodes of the acceleration sensor. The PID controller algorithm involves three separate parameters K_p, K_i, K_d. The proportional value K_p determines the control action due to the current error, the integral value K_i determines the control action based on the sum of recent errors, and the derivative value K_d determines the control action based on the rate at which the error is changing. The output of the PID controller is defined as [3]:

$$u(t) = K_p e(t) + K_i \int_0^t e(\tau)d\tau + K_d \frac{d}{dt}e(t) \qquad (2)$$

Both, the transducer model and the controller model are implemented in Simulink through the Xilinx System Generator tool as behavioral models.

For the implementation in digital logic (time-discrete system) the continuous variable "time" t will be replaced by a system internal clock. In this case the integrator (block 3 in Fig. 3) and the differentiator (block 4) are realized through appropriate discrete operators. Finally, in block 5 the voltages at the front and end electrodes are calculated.

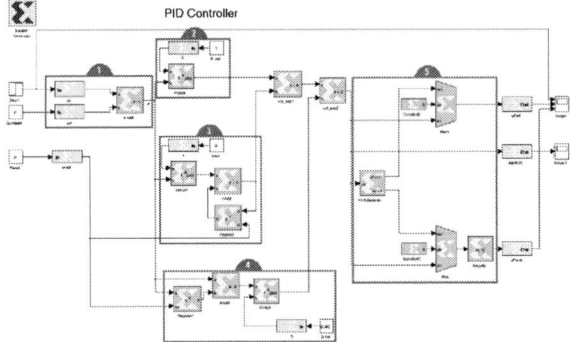

Figure 3. Model of the PID-controller in Simulink.

After having derived the behavioral control model, the transfer into hardware is accomplished automatically. The blocks that run directly on the FPGA board are depicted in blue color (dark grey) in Fig.3 and the yellow (light grey) coloured items represent the blocks providing the communication between the Simulink model and the control algorithm implemented on the hardware board.

III. CONTOL BY FUZZY LOGIC

As alternative control concept a fuzzy controller has been implemented to investigate its performance in combination with the sensor system. In contrary to PID controllers, which realize an exact mathematical scheme, fuzzy controllers are based on empirical rules and are, in some cases, well suited to realize low-cost implementations. The big advantage of such systems is that they can be upgraded easily by adding new rules to improve their performance or to add new features.

978-1-4244-6658-0/10 $26.00 © 2010 IEEE

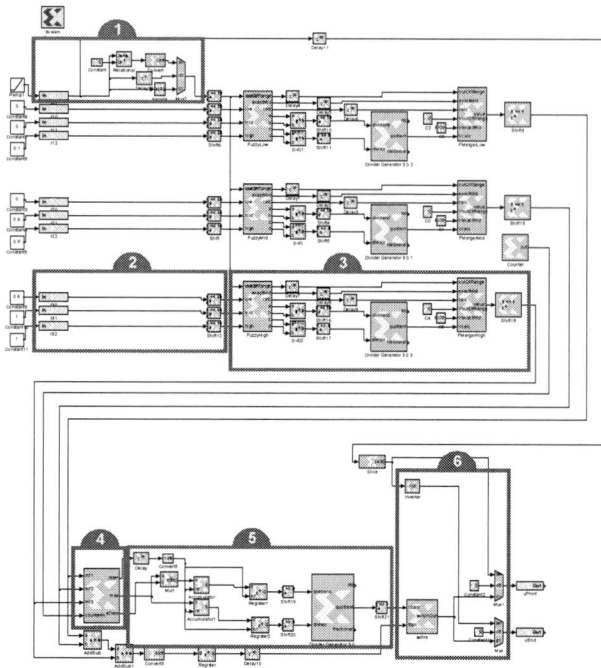

Figure 4. Model of the fuzzy controller in Simulink.

The concept of fuzzy controllers is very simple. It consists of an input stage (fuzzifier), a processing stage (fuzzy expert system), and an output stage (defuzzifier) [3]. The fuzzifier (block 3 in Figure 4) maps the absolute value of the membrane tilt angle φ (φ is zero, φ is small or φ is large) and the output voltage U (U = zero, U = low, U = high) to appropriate truth values and membership functions, for example triangular functions (see eq.3), which are implemented as lookup tables:

$$\mu_{inp}(x) = \begin{cases} 0, \text{ for } x < a_1 \\ (x-a_1)/(a_2-a_1), \text{ for } a_1 \leq x < a_2 \\ (a_3-x)/(a_3-a_2), \text{ for } a_2 \leq x \leq a_2 \\ 0, \text{ for } x > a_3 \end{cases} \quad (3)$$

Here a_1, a_2, a_3 denote x-coordinates of the triangular function and $\mu(x)$ the output of the fuzzifier. In Table 1 the problem-adapted parameter values a_1, a_2, a_3 for the used membership functions are given.

TABLE I. PARAMETERS OF THE MEMBERSHIP FUNCTIONS OPTIMIZED AT A PRESSURE OF 100 HPA

Membership function	a_1	a_2	a_3
φ = zero	0	0.3	0.6
φ = small	0	0.6	0.8
φ = large	0.6	1	1
U = zero	0	0	0.001
U = low	0	2.7	4
U = high	2.3	4	4

The fuzzy expert system then invokes any of the given problem-related, implemented rules, for example

rule 1: "if φ is *zero* then U is *zero*";

rule 2: "if φ is *small* then U is *low*";

rule 3: "if φ is *large* then U is *high*",

generates a result for each rule and combines these results appropriately for further processing in the defuzzifier.

Finally, the defuzzifier (block 5) converts the combined result back into specific voltage values to be applied to the front and end electrodes of the acceleration sensor using the so-called "center of area" method. This defuzzification method returns the center of area underneath the curve [3].

$$\int_0^{Uout} \mu_{out}(U)dU = \int_{Uout}^{U\max} \mu_{out}(U)dU \quad (4)$$

Here U_{out} denotes the output value of the fuzzy controller. Depending on the sign of the tilt angle φ this voltage is applied either to the front or to the end electrode underneath the membrane to regulate it to a quiescent state in case of external disturbances.

IV. COMPARISON OF CONTROL ALGORITHMS

In order to compare the performance of both algorithms, the parameters of both controllers (PID and fuzzy) have been optimized at a pressure of 100 hPa. Then they have been adjusted automatically at various pressure values ranging between 1hPa and 1000 hPa to ensure optimum operation of each control scheme. Subsequently, the sensor membrane is exposed to an external acceleration of 5g and the performance in readjusting the step response of the membrane displacement by the two control concepts is compared. The results are displayed in Fig. 5 (top), where the position of the sensor membrane is controlled by PID and Fig. 5 (bottom), where it is controlled by applying fuzzy logic.

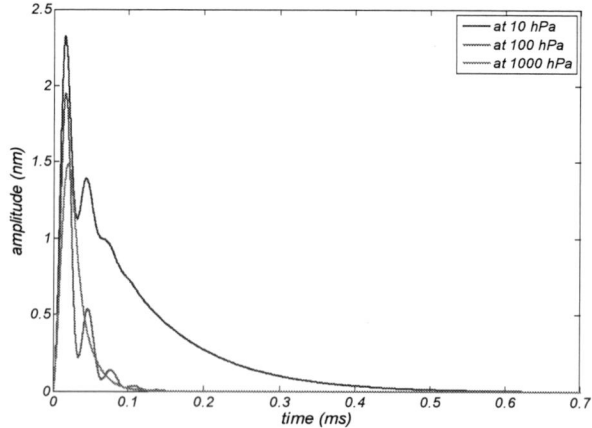

978-1-4244-6658-0/10 $26.00 © 2010 IEEE

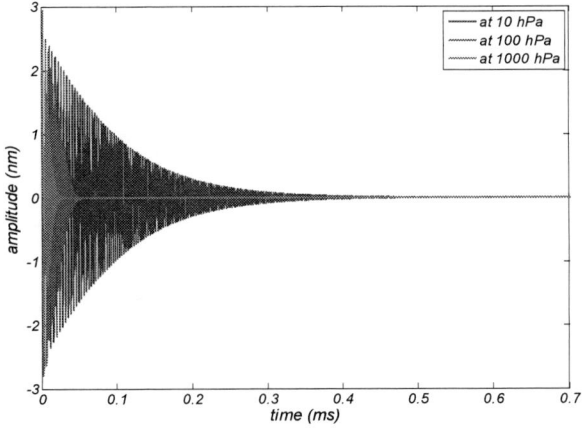

Figure 5. Simulated transients of the dynamic membrane displacement. Top: The position of the membrane is controlled by the PID algorithm. Bottom: The position of the membrane is controlled by applying fuzzy logic.

Although continuous, fast oscillations of the membrane are occurring in the signal controlled by fuzzy logic, the fuzzy controller exhibits a proper dynamics and, as a consequence, allows for measuring the acceleration faster than in the case of the PID controlled system.

The disadvantages of fuzzy controllers, however, are that more logic elements are utilized (see Table II) and, as a consequence, more power is consumed. The power consumption is determined approximately using the power analysis tool of the system generator at 1MHz clock rate and for a 1.2V power supply of the logic. These calculations reveal a difference in power consumption of about 30% between the PID controller ($P \approx 5mW$) and the fuzzy controller ($P \approx 7mW$).

TABLE II.   REQUIRED LOGIC UTILIZATION FOR PID AND FUZZY CONTROLLERS

Logic Utilization	PID	Fuzzy	Available
Number of Slice Flip Flops	2980	9297	33280
Number of 4 input LUTs	3332	5122	33280
Number of occupied Slices	1749	6073	16640
Number of Slices containing only related logic	1749	6073	6073
Number of Slices containing unrelated logic	0	0	1749
Total Number of 4 input LUTs	3332	5262	33280
Number of bonded IOBs	227	366	519
Number of BUFGMUXs	1	2	24
Number of DSP48As	0	11	84

In Table II "LUT" stands for the lookup table, "IOB" for the input-output block, "BUFGMUX" for the multiplexed global clock buffer and "DSP48A" for the block that is used for the multiplication operations.

V.   CONCLUSIONS

We presented a modeling approach for the rapid prototyping of entire microsystems comprising the transducer as well as the surrounding control and read-out electronics. The basis for the predictive design of such systems are physically transparent and calibrated transducer models that reflect all external and internal disturbing effects on a physical basis and, hence, allow for the proper derivation of suitable control algorithms using system simulation. The derived control concepts can then be implemented on the level of behavioral modeling into a hardware board (FPGA) and co-simulated with the Simulink model of the transducer offering a quick access for testing various control and circuit concepts in order to identify the most promising solution for the given application. Besides, the implementation of the control algorithm on a hardware board provides additional estimations of real run-times, power consumption and the degree of logic utilization.

We applied this methodology for a micromechanical acceleration sensor system, for which we derived a physically-based Simulink-model and carried out hardware co-simulations for two different control algorithms implemented on a FPGA board. The obtained simulations allowed a good insight and comparison of the performance of both solutions and showed satisfying results.

ACKNOWLEDGMENT

We would like to acknowledge Dr. Johannes Heidenhain-Stiftung GmbH (Germany) for financial support.

REFERENCES

[1] G. Schrag, G. Wachutka, Sensors and Actuators A, 111, 2004, p. 222-228.

[2] R. Khalilyulin, G. Schrag, G. Wachutka, in Proc. of the EUROSENSORS XXIII conference, Lausanne, Schweiz, September 06.-09 2009: Procedia Chemistry 1, 2009, pp 128-131.

[3] H. Lutz, W. Wendt, "Taschenbuch der Regelungstechnik", 4th ed., Frankfurt am Main: Harri Deutsch, 2002.

[4] R. Sattler, G. Wachutka, in Proc. of NSTI Nanotechnology Conference and Trade Show, Boston, MA, USA, Mar. 07-11, 2004, p. 243-246.

[5] J. Cheng, J. Zhe, X. Wu: Journal of Mechanical and Microengineering, 14, 2004, p. 57-68.

On the Inclusion of Lorentz Force Effects in TCAD Simulations

Wim Schoenmaker
MAGWEL NV
Martelarenplein 13
B-3000 Leuven, Belgium
Email: wim@magwel.com

Peter Meuris
MAGWEL NV
Martelarenplein 13
B-3000 Leuven, Belgium
Email: peter.meuris@magwel.com

Jean Jimenez
ST Microelectronics
850, Rue Jean Monnet
F-38926 Crolles, France
Email: jean-cro.jimenez@st.com

Philippe Galy
ST Microelectronics
850, Rue Jean Monnet
F-38926 Crolles, France
Email: philippe.galy@st.com

Abstract—A new implementation of the Lorentz force is presented in TCAD field solving. The method is applicable to metallic and semiconducting materials. Apart from external magnetic fields, the self-induced magnetic fields are also taken into account. The soundness of the implementation is demonstrated by comparing the numerical results with the outcomes of analytic solutions for a simple wire structure.

I. INTRODUCTION

Full-wave electromagnetic field solvers are commonly based on solving the electric field $\mathbf{E}$ and the magnetic field $\mathbf{B}$ using linear constitutive relations in the metallic regions. Even in the static regime it is of interest to be able to compute both the electric as well as the magnetic field and to study their induced effects in devices. For example, the Hall sensor's operation depends essentially on the fact that the magnetic field distorts the flow of currents due to the Lorentz force [1]. Steady-state self-induced Lorentz force effects are equally realistic but are in general considered to be of less importance due to the fact that they are a few orders of magnitude less pronounced. However, the steady-state solvers may be very useful for estimating effects in the transient regime by extending their use to the quasi-static regime. In the latter case one may assume large currents being present for a short time. In order to better understand their impacts it is desired to compute the corrections originating from the Lorentz force instead of performing a full transient computation. In particular, switching currents need to stay below design rules thresholds and since they are large, the Lorentz force corrections are not negligible. Furthermore, even if one aims at a complete large-signal or transient simulation it is needed as a first step to include the Lorentz force in a steady-state description. Encouraged by these considerations we have addressed the inclusion of the Lorentz force into the static sector of a combined device simulator for semiconductors, metals and insulators. The implementation not only considers the usual Hall effect originating from external magnetic fields but is also capable of dealing with 'self-induced' Lorentz forces. This requires a complete self-consistent approach. The self-induced Lorentz force represents a non-linear correction to the current density constitutive relation and the artillery of computational techniques for semiconductor simulations are required to arrive at a solution *even* for metallic domains. In this paper we

will present a new discretization of the Lorentz force term which exploits the geometrical discretization techniques of the underlying finite-integration schemes for the vector potential [2].

II. STEADY-STATE EQUATIONS

In this section we will summarize the equations that will serve as a starting point for the implementation of the Hall effect and the self-induced Lorentz forces. For semiconductors we refer to the usual classical drift-diffusion model at constant temperature. This model is sufficiently detailed to explain our approach. The current densities for holes and electrons in the drift-diffusion model *with* inclusion of the Lorentz force are

$$\mathbf{J}_n = q\mu_n n \left(\mathbf{E} + \mathbf{v}_n \times \mathbf{B}\right) + kT\mu_n \nabla n$$
$$\mathbf{J}_p = q\mu_p p \left(\mathbf{E} + \mathbf{v}_p \times \mathbf{B}\right) - kT\mu_p \nabla p \quad (1)$$

where $\mathbf{J}_n = -qn\mathbf{v}_n$ and $\mathbf{J}_p = qp\mathbf{v}_p$ are implicitly defining the carrier velocities. The latter can be eliminated in favor of the current densities and both equations can be expressed as

$$\mathbf{J} + s\mu\mathbf{B} \times \mathbf{J} = \mathbf{K}_{\mathrm{DD}} \quad (2)$$
$$\mathbf{K}_{\mathrm{DD}} = q\mu c\mathbf{E} - skT\mu\nabla c \quad (3)$$

In here, $c = p$ and $s = +1$ (holes) or $c = n$ and $s = -1$ (electrons). Were it not for the second term at the left-hand side, equation (2) would be the usual drift-diffusion expression for the current density. Although it is easily possible to perform a full inversion of the current density in terms of the drift-diffusion current density $\mathbf{K}_{\mathrm{DD}}$, the parameter $|\mu B| << 1$ allows to write

$$\mathbf{J} = \mathbf{K}_{\mathrm{DD}} + s\mu\mathbf{K}_{\mathrm{DD}} \times \mathbf{B} \quad (4)$$

The current-continuity equations and the Poisson equations complete the description of the drift-diffusion model, i.e.

$$\nabla \cdot \mathbf{J} + s(R - G) = 0 \quad (5)$$
$$\nabla(-\epsilon\nabla V) = \rho(V, \phi_p, \phi_n) \quad (6)$$

Inside metallic regions, we use the following expression for the current density with μ_{H} the Hall mobility

$$\mathbf{J} = \sigma\mathbf{E} + \mu_{\mathrm{H}}\mathbf{J} \times \mathbf{B} \quad (7)$$

which leads after using $|\mu_{\rm H} B| << 1$ to

$$\mathbf{J} = \sigma\mathbf{E} + \mu_{\rm H}\mathbf{E} \times \mathbf{B} \qquad (8)$$

The current density satisfies the current continuity equation $\nabla \cdot \mathbf{J} = 0$. The magnetic field consists of two terms of which one is representing the self-induced field and the other the external field $\mathbf{B}_{\rm ext}$

$$\mathbf{B} = \nabla \times \mathbf{A} + \mathbf{B}_{\rm ext} \qquad (9)$$

$$\nabla \times \left(\frac{1}{\mu_{\rm perm}} \nabla \times \mathbf{A} \right) = \mathbf{J} \qquad (10)$$

As can be seen, we now end up with a system of equations that should be solved self-consistently, i.e. $\mathbf{J} = \mathbf{J}(V, \mathbf{A})$ since $\mathbf{B} = \nabla \times \mathbf{A}$ and $\mathbf{A} = \mathbf{A}(\mathbf{J})$.

III. Discretization of the Lorentz current densities

The equations (4, 8) show that the total current density consists of two contributions. The usual finite-integration method will lead to the nodal current-balance equations

$$\sum_j \frac{d_{ij}}{h_{ij}} J_{ij}^{\rm standard} + \sum_j \frac{d_{ij}}{h_{ij}} J_{ij}^{\rm Lorentz} + s(R-G)_i \Delta v_i = 0 \quad (11)$$

where d_{ij} is the dual area and h_{ij} is the length of the link and $J_{ij}^{\rm standard}$ represents the current density expression as obtained without inclusion of the Lorentz force, e.g. for semiconductors it reads

$$J_{ij}^{\rm standard} = K_{ij} = \frac{\mu_{ij}}{d_{ij}} \left(c_i B(sX) - c_j B(-sX) \right) \qquad (12)$$

$$B(x) = \frac{x}{\exp(x) - 1} , \quad X = \frac{q}{kT}(V_i - V_j) \qquad (13)$$

and $J_{ij}^{\rm Lorentz}$ represent the correction due to the Lorentz force, e.g. for semiconductors it is

$$J_{ij}^{\rm Lorentz} = s\mu \left(\mathbf{K} \times \mathbf{B} \right) \cdot \mathbf{n}_{ij} \qquad (14)$$

In here, $\mathbf{n}_{ij}$ is the unit vector along the link $< ij >$. We will consider the discretization of (14). The current balance in each node is achieved by summing all contributions from each mesh element and its associated set of links that are attached to the node under consideration. In particular, a contributing link in some mesh element is an edge of two adjacent faces in the element. Figure 1 illustrates the situation for the link $< ij >$ and the faces $F1$ and $F2$.

Since we want the contribution of $\mathbf{K} \times \mathbf{B}$ in the direction $< ij >$, we merely need the components of $\mathbf{K}$ and $\mathbf{B}$ in the plane of the dual area d_{ij}. To compute $(\mathbf{K} \times \mathbf{B}) \cdot \mathbf{n}_{ij}$ we exploit two local coordinate frames in this plane: one coordinate frame $\{\mathbf{s}, \mathbf{t}\}$ is used to perform a decomposition of $\mathbf{B}$ and another coordinate frame $\{\mathbf{u}, \mathbf{v}\}$, whose base vectors are perpendicular to the first ones, is used to compute $\mathbf{K}$ in the volume segment spanned by the link $< ij >$ and its dual area d_{ij} in the element. Using these two frames, we obtain that

$$\begin{aligned} \mathbf{K} \times \mathbf{B} &= (K_u \mathbf{u} + K_v \mathbf{v}) \times (B_s \mathbf{s} + B_t \mathbf{t}) \\ &= K_u B_s (\mathbf{u} \times \mathbf{s}) + K_v B_t (\mathbf{v} \times \mathbf{t}) \\ &+ K_u B_t (\mathbf{u} \times \mathbf{t}) + K_v B_s (\mathbf{v} \times \mathbf{s}) \end{aligned} \qquad (15)$$

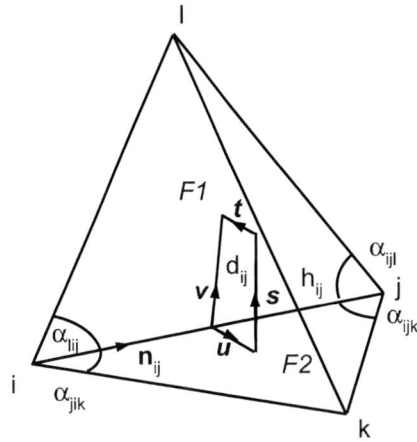

Fig. 1. Mesh element illustrating the ingredients of the decomposition of the Lorentz force vector product. d_{ij} is the (partial) dual area and h_{ij} the distance between node i and j.

As can be seen from Fig. 1, the first two contribution will dominate the result whereas the last two terms vanish for orthogonal faces. For the last two terms it is observed that their contribution seems maximal for acute angles between the adjacent faces, but then the dual area diminishes which results again into a small correction. Therefore, we will implement the first two terms only which definitely suffices for structured grids and which will capture a large portion of the Lorentz force effects on unstructured grids [3]. The magnetic fields perpendicular to the primary grid surfaces $F1$ and $F2$, i.e. B_t and B_s, are obtained from the circulation of the vector potential along these surfaces. The self-induced magnetic field is given by

$$B_\perp = \frac{1}{\Delta S} \sum_k^N A_k h_k \qquad (16)$$

where $A_k \simeq \mathbf{A} \cdot \mathbf{e}_k$ and N is the number of links with lengths h_k around the surface and ΔS is the surface area. For completeness, we mention that the electric field is obtained in the finite-integration method from the voltage differences of the nodes of the discretization grid.

Finally, in order to obtain the variables K_u and K_v, a weighted sum is taken from the current-density projection along $< il >$ and $< jl >$ for K_u and in the same way for K_v, from $< ik >$ and $< jk >$. The weights include the angles α. The motivation for this procedure is found in the requirement that the final current density along the link $< ij >$ should be anti-symmetric in its indices, i.e. $K_{ji} = -K_{ij}$. This can be achieved by using the same K expressions when assembling the contribution of the link $< ij >$ to the nodes i and j. The requirement itself is needed to guarantee current balance.

IV. Static skin effects in conducting wires

The Hall effect induces a skin effect even for static currents. This can be understood by elementary considerations of the

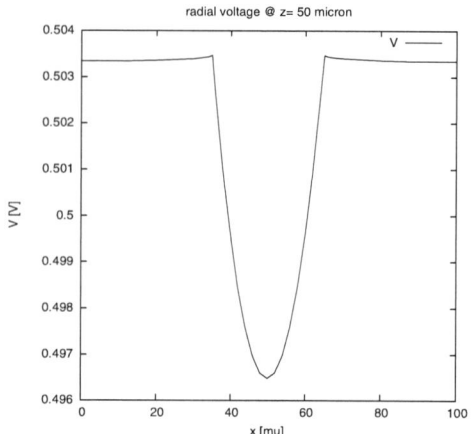

Fig. 3. Radial voltage at mid cut of the conducting wire.

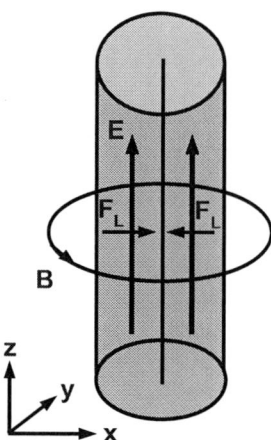

Fig. 2. Illustration of the Lorentz force in a conducting wire.

directions of the electric and magnetic fields. The electric field is in the direction of the current flow. The magnetic field circulates around the wire; moreover, it grows linearly inside the metallic region from the axial center of the wire. Therefore, the Lorentz force acts in the radial direction and the corresponding Lorentz current is compensated by a surface charge distribution that counteracts the Lorentz current (see Fig. 2).

Of course, the net current in the radial direction is vanishing because there are no contacts in this direction to sustain the flow. Qualitatively, one can already conclude that that compensating current is induced by a voltage from the center to the edge that grows quadraticly with the radius, r. This is because inside the wire, the magnetic field is proportional to the radius and therefore the Lorentz force grows linearly with r. The counter force is derived from a potential which is then increasing or decreasing quadratically. A detailed derivation for metals yields

$$V_L(r) = V_0 - k \left(\frac{r}{r_{max}} \right)^2 , \quad k = \frac{\mu_0 R_H}{4\pi^2} \frac{I_{tot}^2}{r_{max}^2} \quad (17)$$

Here μ_0 is the permeability of vacuum, r_{max} the radius of the wire, R_H the Hall coefficient, I_{tot} the current in the wire and V_0 the value of the potential at the center of the wire. This value depends on the position along the wire, i.e. the location with respect to the begin and end point of the wire. The maximum voltage difference ΔV is obtained by setting $r = r_{max}$. Since the parameter k depends on the total current or the applied voltage and the resistivity of the wire material, several equivalent ways exist to represent this parameter. For example, for semiconductor material one obtains

$$k = \mu_{holes} \frac{V_{bias}^2}{R_{wire} \cdot L_{wire}} \times 10^{-7} \ [V] \quad (18)$$

$$k = -\mu_{elec} \frac{V_{bias}^2}{R_{wire} \cdot L_{wire}} \times 10^{-7} \ [V] \quad (19)$$

Note the sign difference between electrons and holes. This reflects the positive or negative value of the Hall coefficient.

A. Self-induced Lorentz force effects in metallic wires

In order to test the numerical implementation, we consider a wire with radius of 15 micron, a length of 100 micron, and a conductance $\sigma = 10^8$ S/m. The Hall coefficient is -10^{-10} Vm/AWb. The applied bias is 1 Volt. In Fig. 3, the voltage is shown along an axis at height of 50 microns (in the middle between to top and bottom of the wire). The center of the wire is located at x= 50 micron. The maximum voltage difference between the voltage at the center and the voltage at the edge of the wire (x=35 , 65 micron) is 0.00699 Volt. The analytic result is obtained by inserting the current, $I_{tot} = 7.0155 \times 10^2$ A into equation (17) which gives $k = -0.00696$ V. The analytic and numerical results are in excellent agreement. The maximum value of the magnetic field being located at the edge of the wire is 9.2 T.

B. Self-induced Lorentz force effects in silicon wires

The implementation for silicon wires is based on the use of the Scharfetter-Gummel current densities as presented in equation (12). Moreover, the mobility represents also the Hall coefficient. For a silicon wire of equal dimensions as above we use a p-type doping of 10^{26} m^{-3} and we use a hole mobility $\mu_{hole} = 0.045$ m^2/Vsec. The resistance of the wire 0.198551 Ω. Using (18) we find that $k = +2.26 \times 10^{-4}$ Volt, whereas the numerical result is 2.264×10^{-4} Volt. Again excellent agreement is found between the numerical and analytic results. It should be noted that the drop in voltage towards the edge implies also a drop in the hole density since the latter is given by $p = n_i \exp (V - \phi_p)$ and ϕ_p is constant at fixed z. The maximum value of the magnetic field is 0.067 T.

V. EXTERNAL FIELDS

The method is as discussed in section 3, can be straight-forwardly extended to include external (constant) magnetic fields. For that purpose the external field is projected on the

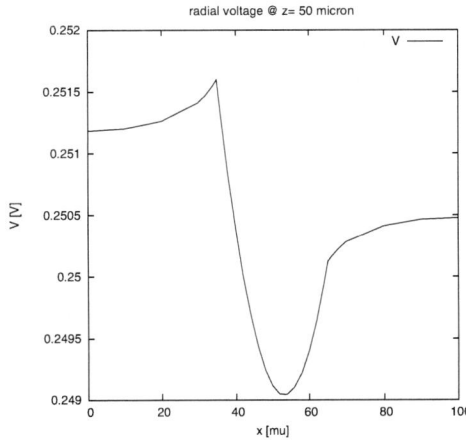

Fig. 6. Radial voltage at mid cut of the silicon wire with an external field in the y -direction. The bias voltage is 0.5 V and the external field is taken 1 T.

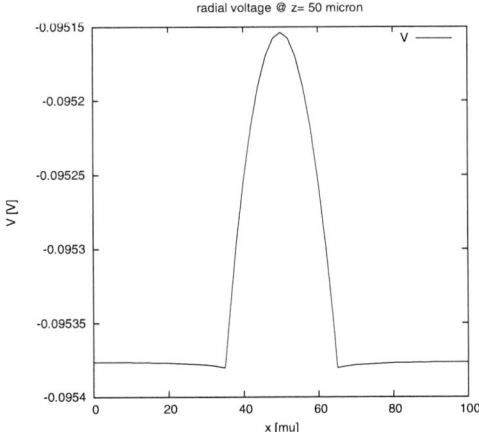

Fig. 4. Radial voltage at mid cut of the silicon wire.

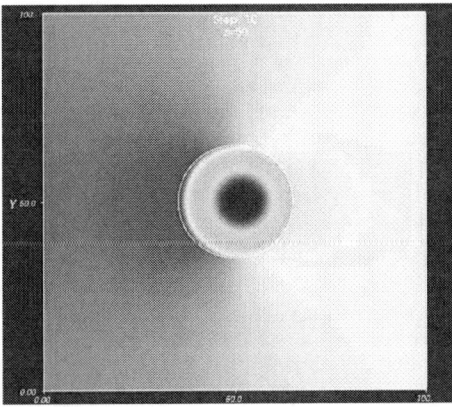

Fig. 5. Radial voltage at mid cut of the silicon wire with an external field in the y -direction (arbitrary units)

local frame base vectors $\{\mathbf{s}, \mathbf{t}\}$. As a test case we consider the silicon wire and consider a magnetic field in the y-direction. This field will break the rotational symmetry. At the left side of the central axis the magnetic field will decrease while at the right side, the field gets enhanced. This should be reflected in the compensating potential that is needed to cancel the Lorentz-force induced radial current. In Fig. 5, the voltage is shown in the xy-plane at z= 50 micron. A cut along the x-axis is shown in Fig. 6. As is seen in Figs. 5 and 6, the radial symmetry is indeed distorted by the external field.

VI. CONCLUSION

We have presented an implementation of the Lorentz force effects in a simulation program which incorporates both external field effects as well as self-induced fields for metallic materials as well as semiconducting materials. The implementation is rooted in the geometrical discretization of the vector potential which exclusively represents the vector potential by scalar variables assigned to the links of the computational grid. Although the method made use of some simplifying assumptions which depend on the use conditions, e.g. $|\mu B| \ll 1$

and on the obtuseness of the grid, these approximations can be easily elevated for completeness but their inclusion will lead to an enhanced filling of the Newton-Raphson matrices and larger CPU cost. The self-induced fields are sufficiently weak such that above approximations are sufficiently accurate to justify their use. Due to the non-linear nature and the self consistency of the self-induced Lorentz force, analytic results are scarce. However, for a simple conducting wire analytic results can be obtained and they serve as an excellent benchmark to test the correctness of the implementation as is illustrated by the comparison of numerical and analytic results. It should be emphasized that the implementation refers solely to local mesh element variables and therefore the corresponding code can be used to analyze the effects of the Lorentz force in arbitrary device layouts in three dimensions. Furthermore, the implementation is *not* a 'post-processing' scheme: the solver arrives at a solution by finding the nodal current balances (eq. (11) in a fully self-consistent manner. Therefore, the sum of the currents of all contacts is equal zero within the numerical noise floor $\sim O(10^{-13})$.

ACKNOWLEDGMENT

This work is partially financially supported by the EU project "ICESTARS" and the Catrene/MEDEA+ project COSIP through the Flemish IWT.

REFERENCES

[1] C. Ricobene, G. Wachutka and H. Baltes, *Operation of Vertical and Lateral Dual Collector Magnetotransistors Studied by Exact 2D-Simulation* proceedings of Simulation of Semiconductor Devices and Processes, SISDEP 1991, Ed. W. Fichtner, D. Aemer, Zurich, September 12-14, 1991-Hartung-Gorre.

[2] W. Schoenmaker and P. Meuris *Electromagnetic Interconnects and Passives Modeling: Software Implementation Issues*, IEEE Trans. on CAD 21, pp. 534-543, 2002

[3] W.J. Schoenmaker, P. Meuris, E. Janssens, K.-J. van der Kolk. N. van der Meijs, W.H.A. Schilders *Maxwell Equations on Unstructured Grids Using Finite-Integration Methods* proceedings of the 12th International Conference in Simulation of Semiconductor Processes and Devices SIS-PAD 2007, Vienna September 2007.

Modeling methodology of high-voltage substrate minority and majority carrier injections

Fabrizio Lo Conte, Jean-Michel Sallese, Maher Kayal
Electronic Laboratory (e-lab.epfl.ch)
EPFL, 1015 Lausanne, Switzerland - fabrizio.loconte@epfl.ch

Abstract—This paper presents a modeling methodology for substrate current coupling mechanisms. An enhanced model of the diode ensuring continuity of minority carriers is used to build an equivalent schematic, accounting for minority and majority carrier propagation in the substrate. For the first time a typical H-bridge structure is simulated with the proposed methodology. The parasitic current injected in the substrate by a high-voltage structure is simulated in a circuit-level simulator as well as with a finite elements method. Both are compared to measurements and show a very good agreement. The simulation resources needed by the proposed equivalent schematics are thus greatly reduced in regard to the finite element approach, offering an efficient tool for substrate modeling in smart power IC's.

I. INTRODUCTION

A new trend in integrated electronics is to concentrate different types of electronic functions on the same substrate, going from sensitive analog circuits to dense and fast digital functions. Moreover, in advanced applications, where actuation of external parts is required, it is common to have high-voltage and high-power circuits integrated in the ASIC.

This kind of technology is called Smart Power IC. The driving capabilities are typically in the order of 40V and can reach several amperes. These integrated circuits are commonly used in consumer electronics (printers, scanners, etc.) as well as in automotive and avionics applications [1].

Typically these technologies are integrated on standard low-voltage CMOS technologies (such as 0.18μm) with some additional steps to implement high-voltage transistors. Electrical isolation between these high-voltage elements can be obtained in different ways, such as self-isolation, trench-dielectric-isolation and junction isolation [1].

In junction isolation topology, isolation is obtained by the reverse biasing of PN junctions, where the P-type part is the substrate and the N-type part is the drain or source of high-voltage NMOS or PMOS transistors, respectively. When inductive or capacitive loads are switched, these junctions can be direct-biased and will inject electrons and holes directly into the substrate. Those electrons and holes will be further coupled by other junctions integrated in the same substrate [1-4].

This electrical coupling noise can severely disturb low voltage analog circuits located in the surrounding area, and may also deeply affect the functionality of digital parts. Such parasitic signals represent the major cause of failure and provoke the need for expensive circuit redesign [1, 3].

Today, this parasitic coupling is modeled using standard bipolar transistors or thyristors models. Typically the coupling between 2 NMOS drains is modeled by a lateral NPN transistor [1-4], where the emitter and collector of the bipolar transistor represents the N-type diffusion of the NMOS drain, and where the P-type base of the lateral bipolar transistor represents the substrate. A cross-section of the high-voltage transistor is shown in Figure 1. The extraction of the parameters for parasitic transistors is highly dependent upon the layout and the coupling effects should be accounted for in a 3-dimensional way [2]. In a classical H-bridge structure, at least four high-voltage transistors are used, leading to a very complex equivalent bipolar schematic with transistor parameters depending on the biasing. Moreover, imposing a bipolar 'meshing' may not be valid in this case.

This lack of design methodology prohibits an efficient design strategy and fails to give clear predictions of electrical perturbations in high voltage integrated circuits.

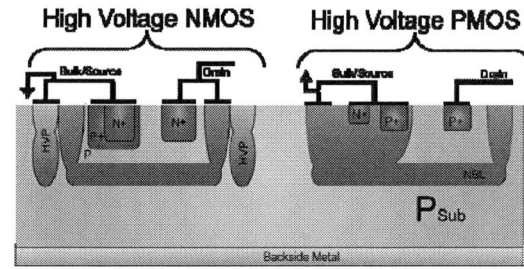

Figure 1. Cross Section of a Smart Power IC LDMOS with bipolar identification and the equivalent diode schematic

A new modeling approach was proposed in [5,6] using an enhanced diode model. The adopted concept consists of providing additional terminals to the diode in order to keep minority carrier continuity at the interconnections. PN junctions are now the elements to be identified, instead of parasitic bipolar transistors. In addition, parameter extraction of the equivalent diode network is greatly simplified: parameters used in the model of the PN junction are mapped to technological and geometrical parameters of the layout.

In this paper we demonstrate that the parasitic substrate current can be modeled by an equivalent schematic of the substrate, followed by a systematic detection of individual PN junctions. This is a major improvement with respect to the tedious and arbitrary process of substrate bipolar

978-1-4244-6658-0/10 $26.00 © 2010 IEEE

transistors identification. The results obtained with this approach are compared to finite element simulations and to the measurements done on an integrated 40V H-Bridge structure.

II. COMPACT MODEL

The substrate equivalent schematic is obtained by interconnecting enhanced PN junctions that account for minority carrier continuity at the boundaries. Minority carrier concentration and gradient are outputted as voltage and current respectively, through an additional terminal. Therefore, minority carriers are not fully recombined at the contact between two diodes, allowing bipolar effect to take place as illustrated in Figure 2.

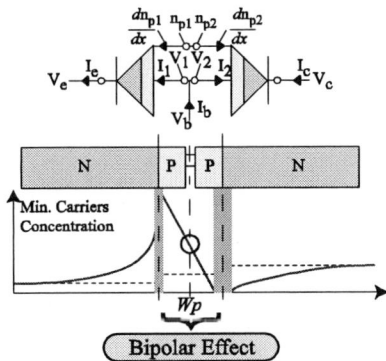

Figure 2. Equivalent bipolar schematic using enhanced diode model.

The parameters used by the enhanced diode model are listed in Table 1. The layout-dependent parameters are extracted from the dimensions of the junction, i.e. the size of the high-voltage NMOS transistor. The additional parameters are technology-dependent and therefore should be extracted from integrated test structures or directly from the components present in the design kit.

Table 1
DIODE PARAMETERS

Type	Parameter Name	Origin
Layout	N-Length	Schematic
	P-Length	Floor-Plan
	Width	Schematic
Technology	N doping	
	P doping	
	Tau$_0$	
	Tau$_n$	
	μ_p	

III. INTEGRATED H-BRIDGE STRUCTURE

Figure 3 shows the integrated structure of a full 40V H-bridge having an ON resistance of 1Ω. Transistors N1, N2, P1 and P2 are classical driving transistors. Diode D1 and D2 are additional free-wheeling diodes used to limit the reverse biasing of the transistor when inductive loads are switched off. The additional transistors N3 and N4 are used during the test to inject or collect charges flowing in the substrate.

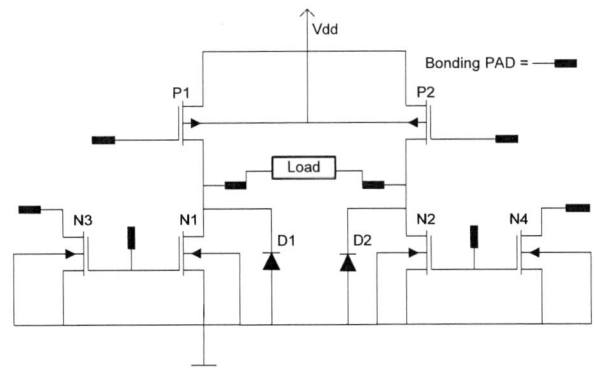

Figure 3. Integrated H-Bridge schematics.

The active dimension of the H-bridge is 760μm width and 700μm height. The placements of the transistors are shown in Figure 4. The two additional NMOS transistors are located in the center of the bridge, which is surrounded by a typical 9μm isolation structure, composed of a deep P diffusion connected to the ground.

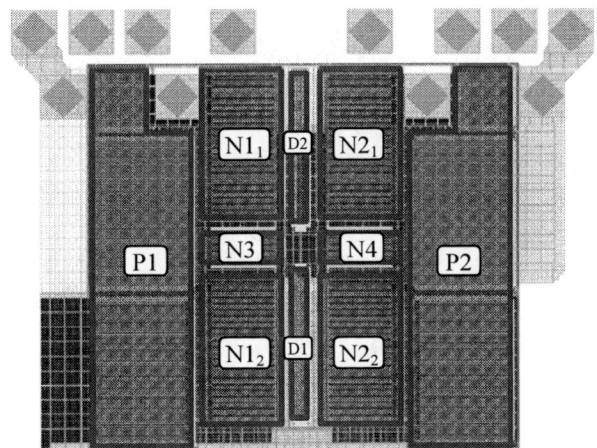

Figure 4. Integrated H-Bridge layout.

IV. FINITE ELEMENT ANALYSYS

In order to optimize the performance and the surface occupied by a transistor, it is essential to layout the transistor in a folded way; this means that the transistor is decomposed in many fingers that are connected together in a parallel.

Figure 5a shows a cross-section of a folded NMOS transistor. It is composed of three fingers that share drain and source terminals.

Simulating such a structure in a finite element simulator is very difficult; thus the structure needs to be simplified in order to reduce its complexity and to preserve carrier injection and collection in the substrate. The discrepancy in the latter is predominantly the result of the N-type buried layer (NBL) [2]. Moreover the current flowing in the drain to bulk diode is already taken into account in the transistor model. We keep only the NBL diffusion as illustrated in figure 5b.

978-1-4244-6658-0/10 $26.00 © 2010 IEEE   115

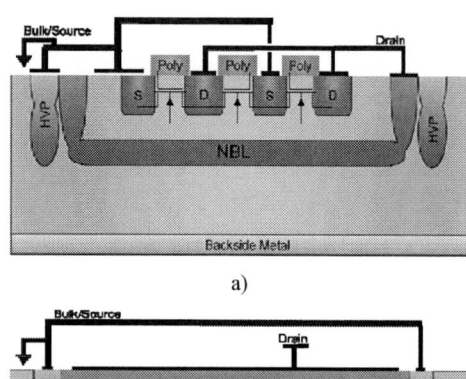

a)

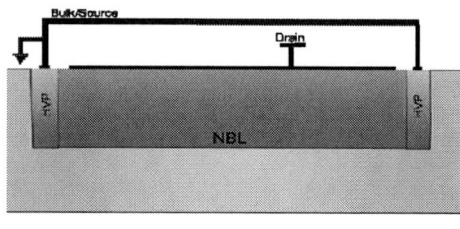

b)

Figure 5. a) Cross-view of an NMOS folded layout, b) simplified structure

The PMOS transistor is different from the NMOS transistor, as shown in Figure 1. In this case, the NBL is connected to the bulk, i.e. the supply voltage.

In addition to the geometrical parameters extracted from the layout, some additional technological parameters are necessary to simulate the structure. These are summarized in Table 2.

Table 2
TECHNOLOGY PARAMETERS

Parameter Name	Value	unit
NBL doping	$8.5e^{18}$	cm^{-3}
HV-P doping	$6.1e^{15}$	cm^{-3}
Substrate doping	$9e^{14}$	cm^{-3}
Junction depth	5	μm
τ_p	$2e^{-6}$	second
τ_n	$5e^{-6}$	second
μ_p	470	$cm^2/V\cdot s$
μ_n	1417	$cm^2/V\cdot s$

In most cases all these technological parameters are confidential. Therefore, we set them to typical values [2, 7, 8] commonly encountered in high-voltage technology.

In the simulation shown in Figure 6 , transistor N4 is biased 0.5V below the substrate voltage, thus injecting a current in the substrate. Other transistors are biased to a positive voltage, thus coupling the injected charge.

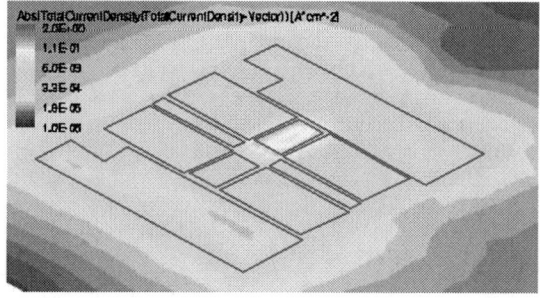

Figure 6. Current density, 10μm below the surface.

The currents coupled by each transistor are shown in Table 3.

Table 3
COUPLED CURRENT

Transistor name	Finite Elements		Measurements	
	Absolute value	Attenuation [dB]	Absolute value	Attenuation [dB]
N1	819nA	-25.26	896nA	-24.48
N2	7.13μA	-6.46	8.45μA	-4.99
N3	162nA	-39.33	244nA	-35.78
N4	−15μA	---	−15μA	---
Vdd	2.02μA	-17.41	2.24μA	-16.51
Gnd	4.85uA	-9.81	---	---

The drain voltage of transistor N4 during measurements was 0.528V

For a fair comparison between measurements and simulations, the injected current was matched to the one obtained from simulations. The results agree, confirming that the technological parameters and the transistor simplification were estimated correctly.

V. EQUIVALENT SCHEMATIC

In a second step, an equivalent schematic was implemented by further interconnecting the enhanced diode components to map the layout. Next the effects arising from the third dimension were obtained by adding a second layer inside the substrate. Additional components were used to model the vertical interconnections.

Figure 7 shows the equivalent schematics used to model the substrate around the transistors N2$_2$-P2-N4. It is composed of two parts. The top-layer models the first 5μm depth of the substrate. The bottom-layer accounts for the depth extending from 5μm to 15μm.

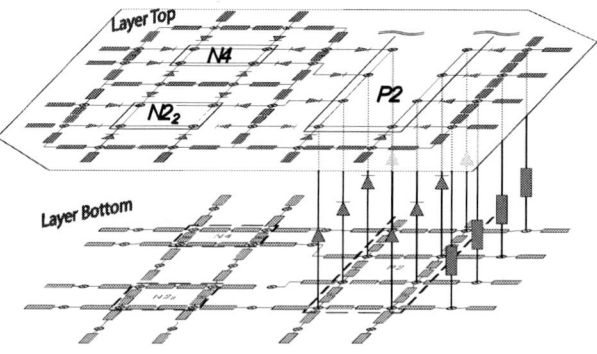

Figure 7. Equivalent schematic for N2$_2$-P2-N4 substrate region. (for readabily purpose some vertical elements are hidden)

All geometrical parameters were extracted from the layout of the H-bridge. The technological parameters are the same as those used for the finite element simulation.

Finally, the full equivalent schematic of the H-bridge and the surrounding substrate was implemented in a SPICE-like simulator. The N4 junction was biased 0.5V below the substrate potential. The results are presented in Table 4.

Table 4
COUPLED CURRENT

Transistor name	Schematic		Measurements	
	Absolute value	Attenuation [dB]	Absolute value	Attenuation [dB]
N1	80.5nA	-35.02	295nA	-24.59
N2	1.59μA	-9.12	2.79μA	-5.06
N3	22.3nA	-46.15	80nA	-35.9
N4	−4.56μA	----	−5μA	---
Vdd	672nA	-16.6	740μA	-16.6
Gnd	---	---	---	---

The drain voltage of transistor N4 during measurements was 0.498V

The parasitic coupled current obtained agrees with the measurements. The error must be analyzed, keeping in mind that this approach is devoted to parasitic substrate simulation, which does not need to be accurate within a few percent. The main difference between finite elements and schematic simulations is the injected current from a junction biased at -0.5V. This mismatch comes from the exponential characteristic of the diode that makes a significant difference in the injected current for a very small voltage difference.

The computational resources needed for these simulations are presented in Table 5, confirming that the SPICE-like simulation is a very good compromise between accuracy and CPU time.

Table 5
SIMULATION RESOURCES

Type	Finite Elements	SPICE Schematics
CPU Time	9h	10s
Memory	2.2GB	30MB

A complete functional simulation accounting for substrate coupled current can be done during the design of the circuit.

VI.   CONCLUSION

This paper presents a general SPICE-compatible modeling methodology to simulate parasitic currents in IC's substrate that are responsible for circuit malfunction or run away, because of latching.

The equivalent schematic is based on an enhanced compact model of the diode, accounting for minority carriers at the component boundary. Classical technological parameters were used as input parameters and their values were validated through finite element simulations. The results obtained by the finite element method and the equivalent schematic agree with the measurement of an integrated high-voltage H-bridge. The simulation resources needed for the proposed modeling methodology are greatly reduced compared to the ones used by the finite element simulator.

This marks the first time that a systematic procedural modeling methodology has been used to build an equivalent schematic of substrate current propagation. It accounts for minority and majority carriers and allows a direct co-simulation of an H-bridge structure that fully includes parasitic substrate current.

ACKNOWLEDGMENT

This work has been supported by the Swiss National Funding Science Foundation, project 200021-125321/1.

The authors thank Advanced Silicon, Lausanne, for the technical and technological support provided during the development of the test circuit.

REFERENCES

[1] B. Murari, F. Bertotti, G. Vignola, *Smart Power ICs 2ⁿᵈ Edition*, pp. 218-220, Springer-Verlag, Berlin, 2002.

[2] M Schenkel, *Substrate Current Effects in Smart Power ICs*, Hartung-Gorre-Verlag, 2003, ISBN 3-89649-848-7.

[3] Ronald R.Troutman, *Latchup in CMOS Technology The problem and its Cure*, Kluwer Academic Publishers second Printing, 1995.

[4] R. J. Widlar, *Controlling substrate currents in junction-isolated ICs*, IEEE J. Solid-State Circuits, vol. 26, pp. 1090–1097, Aug. 1991.

[5] Lo Conte, F.; Sallese, J.-M.; Pastre, M.; Krummenacher, F.; Kayal, M.; "Global Modeling Strategy of Parasitic Coupled Currents Induced by Minority-Carrier Propagation in Semiconductor Substrates," *Electron Devices, IEEE Transactions on*, vol.57, no.1, pp.263-272, Jan. 2010.

[6] Lo Conte, F.; Sallese, J.-M.; Pastre, M.; Krummenacher, F.; Kayal, M.; "Global 2D modeling of minority and majority substrate coupled currents," *Solid State Device Research Conference, 2009. ESSDERC '09. Proceedings of the European*, vol., no., pp.153-156, 14-18 Sept. 2009.

[7] Ikeda et al., "MOS/Bipolar device with stepped buried layer under active regions (Patent style)," U.S. Patent 4 799 098, January 17, 1989.

[8] G. Masetti, M. Severi, and S. Solmi, "*Modeling of Carrier Mobility Against Carrier Concentration in Arsenic-, Phosphorus- and Boron-Doped Silicon*" IEEE Trans. Electron Devices, vol. ED-30, no. 7, pp. 764-769, 1983.

978-1-4244-6658-0/10 $26.00 © 2010 IEEE

Subthreshold FinFET SRAM Cell Optimization Considering Surface-Orientation Dependent Variability

Ming-Long Fan, Vita Pi-Ho Hu, Chien-Yu Hsieh, Pin Su and Ching-Te Chuang

Department of Electronics Engineering & Institute of Electronics, National Chiao Tung University, Hsinchu, Taiwan

E-mail:pinsu@faculty.nctu.edu.tw, ctchuang@mail.nctu.edu.tw

Abstract **-- This work investigates the impact of device intrinsic variation on the stability/variability of subthreshold 6T FinFET SRAM cells with (110)/(100) surface orientations. Due to the difference in the degree of quantum confinement, NFET with (110) orientation shows larger fin Line-Edge-Roughness (LER) induced threshold-voltage variation than the (100) one, while PFET shows the opposite trend. Therefore, the stability of conventional (PU, PD, PG) = (110, 110, 110) cell is inferior and fails to provide sufficient margin. With the optimized orientation, significant μ/σ ratio improvement can be achieved by using (PU, PD, PG) = (110, 100, 100) SRAM cell. Our analysis establishes the potential of 6T FinFET cells with appropriate optimization for emerging subthreshold SRAM applications.**

I. Introduction

For ultra-low power applications in portable devices, implanted medical instruments, and wireless body sensing networks, operating supply voltage V_{dd} below threshold voltage V_{th} is an efficient technique to reduce static and dynamic power consumption. However, the cell stability deteriorates with decreasing V_{dd} and several novel bulk 8T/10T SRAM cells have been proposed to replace conventional 6T SRAM cell for subthreshold applications [1-3]. Due to its better gate control, FinFET exhibits significantly lower subthreshold swing, leakage current and higher I_{on}/I_{off} ratio. In addition, the use of lightly-doped silicon fin improves the immunity to random dopant fluctuation, especially in subthreshold region. Therefore, FinFET emerges as a promising candidate for future subthreshold SRAM applications [4].

Compared with traditional planar structures, the fin height quantization of FinFET limits the option for stability optimization using device sizing. Because of its vertical topology, the conventional sidewall conducting (110) surface orientation of FinFET can be rotated by layout to improve circuit performance [5, 6]. However, due to the difference in effective mass, different surface orientation may result in different device variability caused by quantum confinement. In this work, the cell stability, performance, and variability of various subthreshold FinFET SRAM cells are analyzed.

II. Analysis Framework

Combinations of pull-up (PU), pull-down (PD) and pass-gate (PG) devices with (110) and (100) orientations form 8 types of FinFET SRAM cell. Fig. 1 shows the 6T FinFET SRAM cell layout based on scaled design rules from the published 32-nm technologies [7] and the scaling factor from

the ITRS Roadmap for (PU, PD, PG) = (110, 110, 110) and (110, 100, 100) orientations by rotating the direction of the fin.

To assess the FinFET cell variability, the most important variation source, fin LER [8], is simulated using 3D TCAD mixed-mode simulation [9] together with atomistic Monte Carlo simulation [10]. Fig. 2 shows one of the simulated FinFET samples with fin LER. The quantum confinement model has been calibrated with the exact solution of Schrodinger's equation [11] to consider the impact of surface orientation on the V_{th} variability. In addition, the mobility model has also been calibrated with the published measured data [12]. In the following sections, the analysis is based on FinFET device with $N_a = 10^{17}$ cm^{-3}, $L_{eff} = 25$nm, $W_{fin} = 7$nm, $H_{fin} = 20$nm and EOT = 0.65nm (Fig.2).

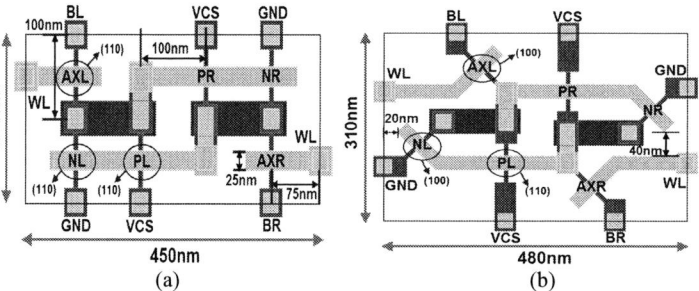

Fig. 1. FinFET cell layouts for (a) devices with all (110) orientation, and (b) (110) pull-up (PL/PR), (100) pull-down (NL/NR), and (100) pass-gate (AXL/AXR) transistors.

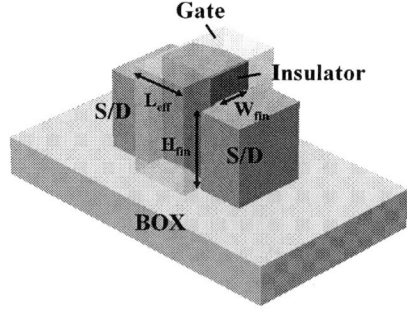

Fig. 2. One of the FinFET samples with fin LER simulated in this work.

III. Effects of Surface Orientation on Cell Stability & Performance

The cell stability for Read and Write is defined in Fig. 3. Cell Read Stability is defined by Read Static Noise Margin (RSNM) calculated from the maximum square that can fit inside the Read butterfly curves. Write-stability (or Write-ability) is quantified by Write Static Noise Margin (WSNM) which is the minimum square spanning between the

978-1-4244-6658-0/10 $26.00 © 2010 IEEE

Write butterfly curves.

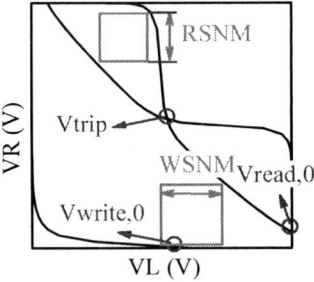

Fig. 3. Butterfly curves used to calculate RSNM and WSNM. The definitions of $V_{read,0}$, V_{trip} and $V_{write,0}$ are shown in the figure.

Also shown on the butterfly curves are voltage levels pertinent/crucial to the cell stability. These are, Read disturb voltage ($V_{read,0}$), inverter trip voltage (V_{trip}) and $V_{write,0}$ (shown in Fig. 3). For Read stability, ($V_{trip} - V_{read,0}$) larger than zero is required. During Write operation, lower $V_{write,0}$ means stronger PG in comparison with PU device and better Write-ability.

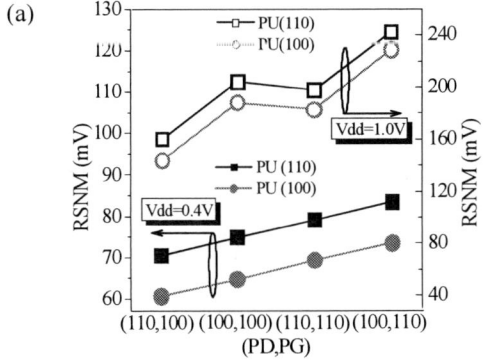

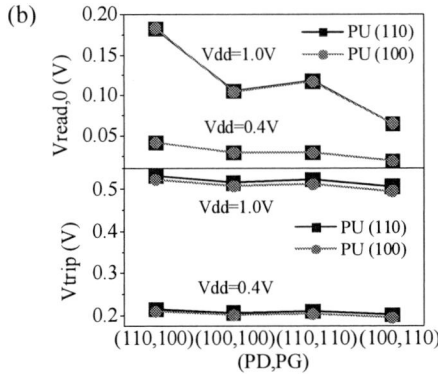

Fig. 4. (a) RSNM and (b) $V_{read,0}$ and V_{trip} comparisons of 8 types of 6T FinFET SRAM cells.

In Fig. 4(a), the nominal RSNM for various cells are compared in subthreshold ($V_{dd}=0.4V$) and superthreshold ($V_{dd}=1.0V$) regimes. As can be seen, FinFET SRAM cells with (PD, PG) = (100, 110) exhibits better RSNM due to the smaller $V_{read,0}$ (Fig. 4(b)). Besides, cells designed with (110) PU transistors are beneficial to increase V_{trip}, thus improving

RSNM as well. Notice that for Read operation in superthreshold region, the PD device is in linear region, thus benefits more from mobility improvement in (100) orientation. On the other hand, the PG device is in saturation, so the benefit of higher mobility in (100) orientation is limited by velocity saturation. Thus, $V_{read,0}$ of (PD, PG) = (100, 100) is smaller than that with (PD, PG) = (110, 110) orientation [6], and cell with (PD, PG) = (100, 100) demonstrates better RSNM than that with (PD, PG) = (110,110) orientation. However, for subthreshold operation, the velocity saturation is negligible. Hence, comparable $V_{read,0}$ for two configurations are observed as shown in Fig. 4 (b), and V_{trip} dominates the RSNM. As a result, cell with (PD, PG) = (110, 110) exhibits better RSNM than that with (PD, PG) = (100, 100) orientation.

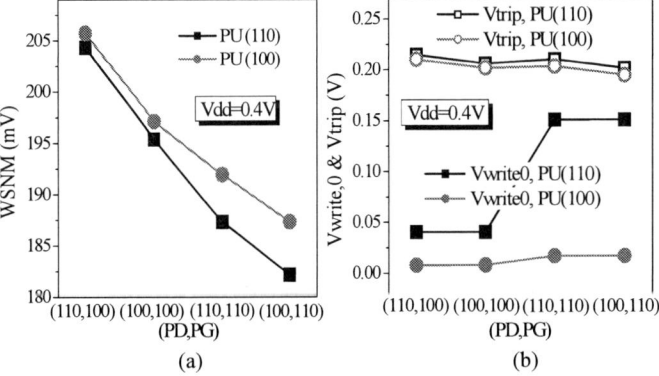

Fig. 5. (a) WSNM and (b) Vwrite,0 and Vtrip comparisons of 8 types 6T FinFET SRAM cells.

Fig. 5(a) compares the nominal WSNM of various cells in subthreshold region. It can be seen that (100) PU device with weaker strength shows lower $V_{write,0}$ (Fig. 5(b)) and better WSNM than that with (110) PU device. Besides, cells with higher V_{trip} help trigger the internal latch effect once the storage node holding "1" is pulled down below the trip voltage of the opposite inverter, thus improving WSNM. Therefore, FinFET SRAM cell designed with (PU, PD, PG) = (100, 110, 100) shows the best WSNM. Compared with WSNM, RSNM is lower in general.

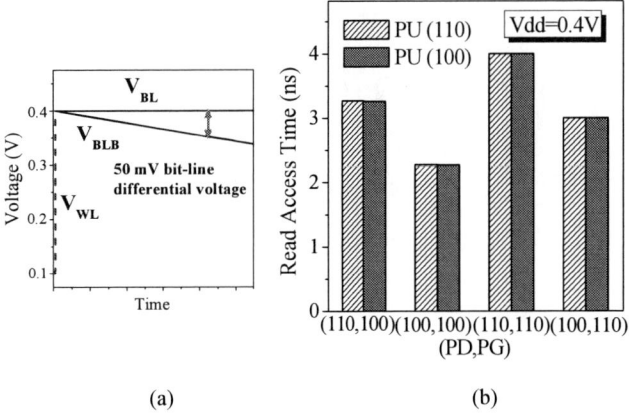

Fig. 6. (a) The definition of "cell" Read access time, and (b) comparison of "cell" Read access time for FinFET cells with various orientations.

For AC performance analysis, the "cell" Read access time is investigated and defined as the time to generate 50 mV bit-line differential voltage (Fig.6 (a)). The "cell" Read access time strongly depends on the Read current through PG and PD transistors. As can be seen in Fig. 6(b), PG and PD devices with faster (100) orientation demonstrates higher Read performance, while there is no dependence on the orientation of PU transistor. For ultra-low power applications, subthreshold SRAMs usually operate from several hundred Hz to several hundred KHz [13]. If we assume the "cell" Read access time contributes to ~ 20% of the total SRAM Read time, the slowest cell of using (110) NFET (~ 4ns) can still meet the requirement.

IV. Effects of Surface Orientation on Variability

In addition to the optimization of nominal cell stability, the impacts of device variations on cell stability are also important. In this section, the impact of local random variation on device variability and optimization of 6T FinFET SRAM cell are analyzed. Among various local random variation sources, Line Edge Roughness induced dimension deviation on fin width (fin LER) has been shown to be the most important one [8]. To assess LER, line edge patterns are generated by Fourier synthesis approach similar to the one in [10] with correlation length = 20 nm and rms amplitude = 1.5nm. Then, the Monte Carlo simulations with 150 samples are performed for each case. Fig. 2 illustrates one of the simulated FinFET samples with fin LER.

In Fig. 7, V_{th} variations induced by fin LER for NFET and PFET of different orientations are compared. Notice that due to the difference in effective mass [14, 15], the effect of quantum confinement varies for different orientations. As such, V_{th} distributions among various orientations considering quantum confinement are significantly different from the classical results. In addition, it is observed that NFET with (110) orientation is more susceptible to fin LER than that with (100) orientation (Fig. 7(a)), while PFET shows the opposite trend (Fig. 7(b)). Therefore, the impact of intrinsic variability difference on FinFET SRAM cell optimization merits investigation.

Fig. 8 compares the butterfly curves of 6T FinFET SRAM cell all with (110) orientation operating in superthreshold and subthreshold regions. 3D atomistic Monte Carlo simulations with 100 samples for each case are performed. As expected, the noise margin decreases with the scaling of V_{dd}. If we target the value of μ/σ ratio to 5-6 for robust SRAM cell design. The calculated μ/σ ratio of cell with all (110) devices reduces from 11.93 ($V_{dd}=1.0V$) to 3.58 ($V_{dd}=0.4V$) which fails to meet the requirement. Therefore, the stability of conventional (110, 110, 110) 6T FinFET SRAM cell is not satisfactory once quantum confinement is considered.

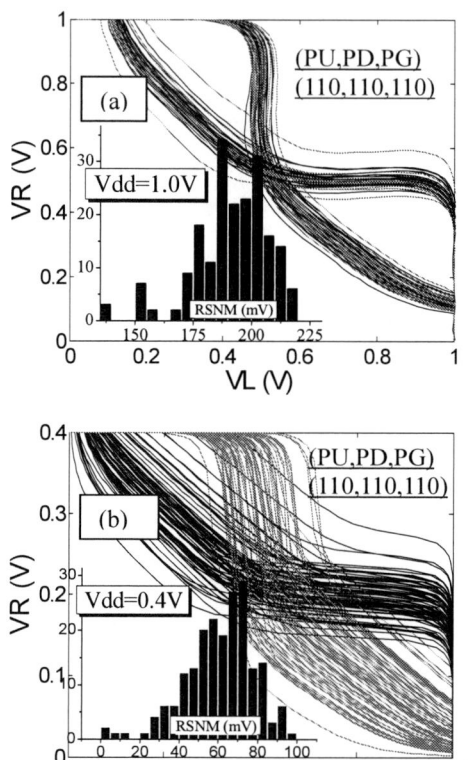

Fig. 8. RSNM variations of conventional FinFET SRAM cell (devices with all (110) direction) operating in (a) V_{dd} = 1.0V and, (b) V_{dd} = 0.4V.

In Fig. 9, the normalized σRSNM and μRSNM/σRSNM of various SRAM cells are compared in subthreshold region. As can be seen in Fig. 9(a), cell designed with all (110) directions shows larger σRSNM than most cells considered (except for the worst cell with (PU, PD, PG) = (100, 110, 110) and is not the optimum choice. This is because NFET oriented in (110) direction suffers larger V_{th} variation (Fig. 7(a)). Among the possible options, (110, 100, 100) SRAM cell with adequate μRSNM and significantly lower σRSNM (~ 40%) exhibits the best μ/σ ratio (6.27) among all cells.

Fig. 10 compares the σRSNM and μRSNM/σRSNM for various cells using classical model and quantum mechanical model (QM). As shown in Fig. 10(a), classical model underestimates the value of σRSNM and can not distinguish the difference of various orientations. The σRSNM of different

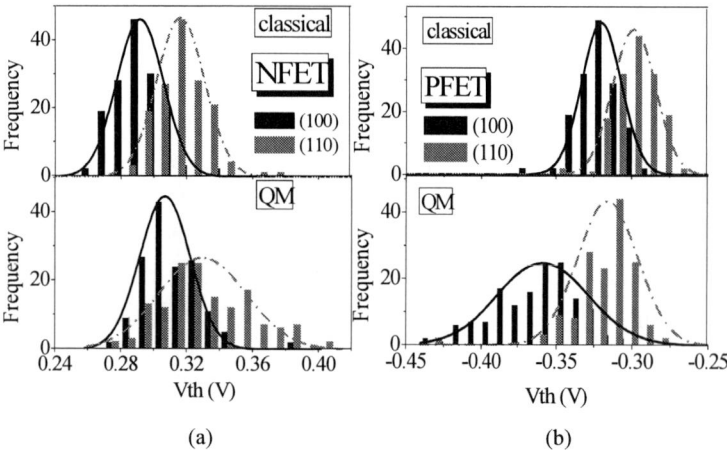

Fig. 7. V_{th} variations induced by fin LER for (a) NFET and (b) PFET (correlation length = 20nm, rms amplitude = 1.5nm [8]).

978-1-4244-6658-0/10 $26.00 © 2010 IEEE

cells are comparable with classical model. Therefore, the μ/σ ratio in the classical case is primarily determined by the value of σRSNM and severely optimistic.

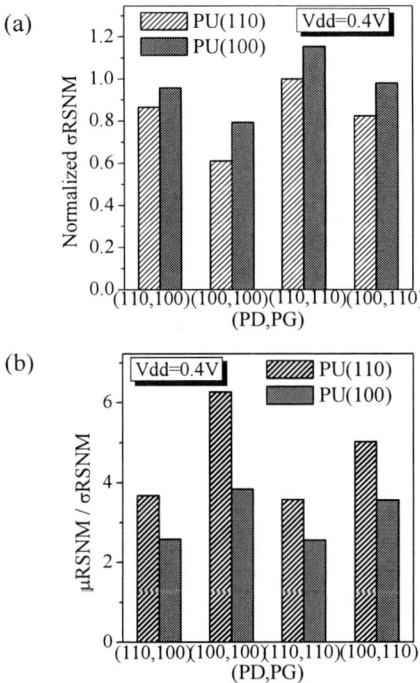

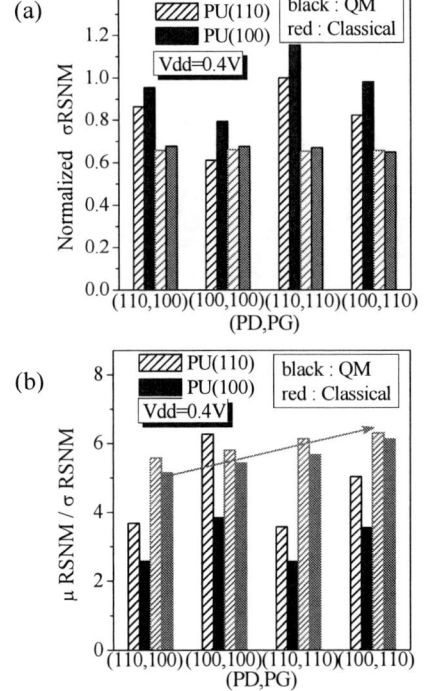

Fig. 9. (a) Comparison of the normalized σRSNM for various cell configurations and, (b) μRSNM/σRSNM comparisons considering fin LER. (110, 100, 100) SRAM cell shows largest μRSNM/σRSNM.

Fig. 10. Comparison of (a) normalized σRSNM and (b) μRSNM/σRSNM for various FinFET cells with and without considering quantum confinement.

Based on the published design rules of 32 nm technologies [7] and the scaling factor from ITRS Roadmap, the cell area of various FinFET SRAM cells are estimated and compared. In Fig. 1(a), we establish a standard 6T thin-cell layout [16] which requires 4.5 fin pitch in horizontal dimension and 2 contacted gate pitch in vertical dimension. Fig. 11 compares the resulting cell area among 8 FinFET SRAM cells. As can be seen, in order to optimize the cell stability, we pay extra area penalty (65% and 17% for (100, 110, 110) and (110, 100, 110) cell, respectively) for rotating the corresponding transistors.

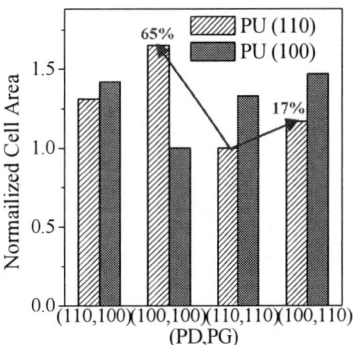

Fig. 11. Comparison of cell area for various FinFET SRAM cells. Cell area is normalized to the cell with (PU, PD, PG) = (110, 110, 110).

V. Conclusion

We have investigated the impact of fin LER on the stability and variability of subthreshold 6T FinFET SRAM cells with (110)/(100) surface orientations. Due to the difference in the degree of quantum confinement, NFET with (110) orientation shows larger fin-LER induced threshold-voltage variation than the (100) one, while PFET shows the opposite trend. Therefore, the stability of conventional (110)-oriented cell is inferior and fails to provide sufficient margin (μ/σ = 3.58). With the optimization of surface orientation, significant μ/σ ratio improvement can be achieved using the (110, 100, 100) cell. As opposed to the necessity of using 8T/10T cells in bulk CMOS technology, this work demonstrates the viability of 6T FinFET SRAM for future ultra-low power applications.

Acknowledgement

This work was supported in part by the National Science Council of Taiwan under Contract NSC 98-2221-E-009-178, in part by the Ministry of Education in Taiwan under ATU Program, and in part by the Ministry of Economic Affairs in Taiwan under Contract 98-EC-17-A-01-S1-124. The authors are grateful to National Center for High-Performance Computing in Taiwan for computational facilities and software.

Reference

[1] T.-H. Kim et al., *IEEE ISSCC*, p. 330, 2007.
[2] I. J. Chang et al., *IEEE ISSCC*, p. 388, 2008.
[3] J. P. Kulkarui et al., *IEEE JSSC*, p. 2303, 2007.
[4] M.-L. Fan et al., *IEEE International SOI Conf.*, Oct. 2009.
[5] K. Shin et al., *IEDM Tech. Dig.*, p. 988, 2005.
[6] S. Gangwal et al., *IEEE CICC*, p. 433, 2006.
[7] X. Chen et al., *Symp. on VLSI Tech.*, p.88, 2008.
[8] E. Baravelli, et al., *IEEE TED*, p. 2466, 2007.
[9] "Sentaurus TCAD, C2009-06 Manual," Sentaurus Device, 2009.
[10] A. Asenov et al., *IEEE TED*, p. 1254, 2003.
[11] *Atlas User's Manual*, SILVACO, Santa Clara, CA, 2008.
[12] C.-Y. Chuang et al., *IEDM Tech. Dig.*, 2009.
[13] J.-J. Kim et al., *IEEE TED*, p. 1468, 2004.
[14] A. Haque et al., *IEEE TED*, p. 1580, 2002.
[15] M. Saitoh et al., *IEDM Tech. Dig.*, p. 711, 2007.
[16] F. Bauer et al., *ESSCIRC*, p. 392, 2007.

Fin-Height Controlled PVD-TiN Gate FinFET SRAM for Enhancing Noise Margin

Y. X. Liu, K. Endo, S. O'uchi, J. Tsukada, H. Yamauchi,
Y. Ishikawa, K. Sakamoto, T. Matsukawa, M. Masahara
Nanoelectronics Research Institute of AIST
1-1-1 Umezono, Tsukuba, Ibaraki, Japan
E-mail: yx-liu@aist.go.jp

T. Kamei, T. Hayashida, A. Ogura
School of Science and Technology
Meiji University
Kawasaki, Kanagawa, Japan

Abstract—**PVD-TiN gate FinFET SRAM half-cells with different β-ratios and fin-height controlled transistors have successfully been fabricated using orientation-dependent wet etching and selective recess RIE. It was found that read static noise margin (SNM) increases significantly by controlling β from 1 to 2. With further increasing β, read SNM increases slightly. On the other hand, write margin shows weak dependence on β. By controlling the fin-heights of pass-gate (PG) and pull-up (PU) transistors to one-half pull-down (PD) transistors, i.e., β = 2, the read SNM was enhanced from 133 to 185 mV at V_{DD} = 1 V. Scaled recess areas down to 103 nm square for low-fins have successfully been fabricated by optimized electron-beam lithography and RIE. The developed fin-height controlled technology is very useful for the fabrication of scaled SRAM without cell area increment.**

I. INTRODUCTION

The FinFET with an undoped channel is well recognized as the most promising candidate for future ultimate-scaled VLSI circuits owing to its superior short-channel-effect (SCE) immunity and the potential robustness against random dopant fluctuations (RDF) [1, 2]. As a promising gate material, PVD-TiN gate has actively been applied in FinFETs because its midgap work function provides well symmetrical threshold voltage (V_{th}) for undoped FinFET CMOS without channel doping [3]. Recently, as the functional CMOS circuits, scaled FinFET SRAMs have actively been developed [4-11]. One of the tough issues in FinFET SRAM design is the β-ratio optimization for obtaining a sufficient static noise margin (SNM) without cell area increment. The β-ratio is defined as β = W_{PD}/W_{PG}, where W_{PD} and W_{PG} are the effective channel widths (W_{eff}) for pull-down (PD) and pass-gate (PG) transistors. In the FinFET SRAM case, however, PD transistors have to be designed as multi-fin channels for increasing β-ratio because the W_{eff} in FinFET is twice of fin-height (H_{fin}), W_{eff} = 2H_{fin}. This inevitably results in a large area penalty even with the sidewall-image-transfer process [9]. To solve such a trade-off relationship, we have proposed a novel FinFET SRAM structure called as Flex-PG SRAM [10, 11] using independent double gate 4-terminal (4T) FinFETs [3, 12] as the PG transistors. As another promising approach, fin-height controlled [3] SRAM should be very important for

enhancing SNM without paying cell area penalty. To set correct fin-height for PG and PD transistors, an optimal β-ratio extraction is essential based on the experimental Si-fin channels because carrier mobility depends strongly on the channel surface orientations. However, very few works regarding the β-ratio dependence of SNM and fin-height controlled SRAMs have been reported [6, 9, 13].

In this paper, we report the detailed experimental results of β-ratio dependence of the noise margins by fabricating PVD-TiN gate FinFET SRAM half-cells with (111)-oriented sidewall channels, and demonstrate fin-height controlled SRAMs with an enhanced SNM without cell area increment.

II. DEVICE FBRICATION

PVD-TiN gate FinFET SRAM half-cells with different β-ratios from 1 to 5 and fin-height controlled transistors were fabricated using orientation-dependent wet etching and selective recess RIE on the same wafer. To evaluate the effective carrier mobility ($μ_{eff}$) in the (111)-oriented sidewall channels, we also fabricated multi-FinFETs on the same wafer. The number of fins (N_{fin}) and gate length (L_g) were designed to be 50 and 20 μm for gate-channel capacitance measurement.

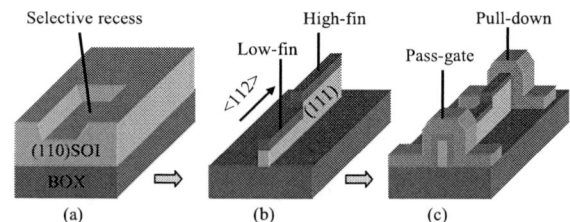

Fig. 1. Device fabrication process flow. (a) selective recess-areas formation for pass-gate and pull-up transistors, (b) fin-patterning in parallel with <112> direction and Si-fin formation using orientation-dependent wet etching, (c) gate oxidation and n+-poly-Si/PVD-TiN gate formation.

Figure 1 shows an abbreviated key process flow of the fin-height controlled SRAM fabrication. At first, undoped (110)-oriented SOI wafers were thermally oxidized to form SiO_2, and recess areas were formed selectively for PG and PU transistors using electron-beam (EB) lithography and RIE,

This work was supported in part by the Development of Nanoelectronics Device Technology of NEDO, Japan then the SiO_2 layer was etched off as shown in Fig. 1(a). Dimension of recess area was 0.6x0.6 μm^2. This recess size can be reduced to sub-100-nm that will be discussed later. After re-oxidation of 36-nm-thick SiO_2, fin patterns were delineated in parallel with <112> direction by EB-lithography, and fin-hard masks were formed RIE. Through the fin-hard masks, SOI layer was etched with a 2.38 % TMAH solution at 50 °C for 30 s as shown in Fig. 1(b). Since the sidewalls of Si-fins have a (111)-oriented plane with an extremely low etch rate in TMAH compared with other planes, very narrow and straight Si-fin channels can be fabricated. After the 2.3-nm-thick gate oxide (t_{ox}) formation at 850 °C, 20-nm-thick PVD-TiN and 100-nm-thick n^+-poly-Si layers were continually deposited as a gate material. After the gate formation by EB-lithography and RIE, ion-implantation (I/I) for source-drain (SD) extension was performed at a fixed low energy of 5-keV [14]. The SD extension I/I dose (D) and tilt-angle (θ) were $D = 4 \times 10^{14}$ cm^{-2}, $\theta = 60°$. After an 80-nm-thick gate sidewall formation, I/I for SD electrode regions was carried out with $D = 1.5 \times 10^{15}$ cm^{-2} and $\theta = 7°$ at a fixed energy of 10-keV, followed by the deposition of 100-nm-thick SiO_2 by TEOS-CVD on the wafers. To activate the implanted impurity, RTA was performed at 915 °C for 2 s. Finally, contact holes and aluminum electrodes were formed, and all wafers were sintered in a forming gas ambient at 450 °C for 30 min.

In order to apply the developed fin-height controlled technology to ultimately scaled FinFET SRAM fabrication, we also optimized EB-lithography and RIE. Figure 2(b) shows the SEM image of the fabricated selective recess areas. Note that 103 nm recess areas were successfully fabricated with a 160 nm pitch. Such scaled fin-height controlled SRAM full-cell is under development and will be reported elsewhere.

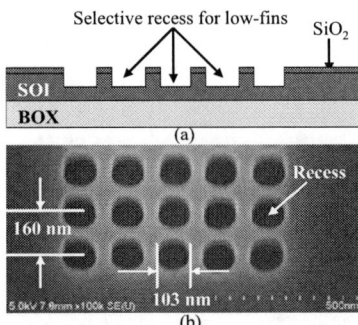

Fig. 2. Nanoscale recess area formation by optimized EB-Lithography and selective RIE. (a) schematic cross-section, and (b) SEM image of the fabricated selective recess areas.

III. EXPERIMENTAL RESULTS AND DISCUSSIONS

A. Mobility

The effective carrier mobilities (μ_{eff}'s) for electron and hole in the fabricated PVD-TiN gate multi-FinFETs with (111) sidewall surface channels were experimentally extracted from differential drain conductance (g_d) at low drain voltage regions (V_d = 0-0.1 V) and surface charge density ($Q_{n,p}$) by split C-V method. Effective transverse electric field (E_{eff}) was estimated from the measured $Q_{n,p}$ using $E_{eff} = \eta Q_{n,p}/\varepsilon_0 \varepsilon_{Si}$,

where ε_0 is the permittivity in vacuum, ε_{Si} is the silicon dielectric constant, and η is a parameter for defining E_{eff}. In the μ_{eff} evaluation in this work, we used $\eta = 1/3$ because the universality is maintained for electron and hole mobilities in a (111) bulk MOSFET when η is chosen to be 1/3 [15].

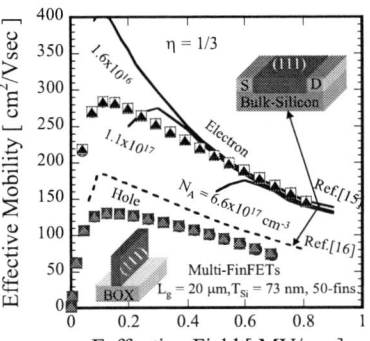

Fig. 3. Comparison between measured μ_{eff} of (111)-oriented multi-FinFETs and reported universal mobility data of (111) bulk MOSFETs.

Figure 3 shows the measured μ_{eff}'s for electron and hole in the fabricated multi-FinFETs with n^+-poly-Si/PVD-TiN gate stack and the reported ones in (111) bulk MOSFETs [15, 16]. It is clear that μ_{eff} values show good agreements with the corresponding universal curves in the (111) bulk MOSFETs thanks to the damage-free Si-fin channels by wet etching. It should be noted that the μ_{eff} values obtained in this work are comparable to those in the n^+-poly-Si gate [17] and better than those in the pure PVD-TiN gate [3] multi-FinFETs we reported previously. This implies that less D_{it} was introduced during a thin PVD-TiN layer deposition and the n^+-poly-Si capping on PVD-TiN gate is very effective to set symmetrical V_{th} for undoped FinFET CMOS without μ_{eff} degradation.

B. β-Ratio Controlled SRAM

Figure 4 shows the schematic circuit diagram of the half-cell of 6T-SRAM and the SEM images of the fabricated SRAM half-cell. The number of fins (N_{fin}) for PD transistors were changed from 1 to 5, but only 1-fin was used for PG and PU transistors, i.e., β = 1-5. The physical L_g was 365 nm.

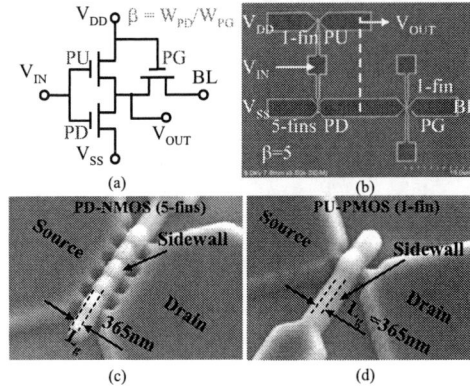

Fig. 4. (a) Schematic circuit diagram of 6T-SRAM half-cell, SEM images of (b) half-cell, (c) 5-fin PD-NMOS, (d) 1-fin PU-PMOS.

Figure 5 shows the typical electrical characteristics of the fabricated PU-PMOS with 1-fin and PD-NMOS devices with different N_{fin}'s from 1 to 5. It is clear that almost ideal S-slopes and symmetric I_d-V_g curves are obtained thanks to a good quality of Si-fins by wet etching and the midgap work function of PVD-TiN gate. Note that V_{th} of PD transistors is almost the same value and to be independent of the N_{fin}. Moreover, it can be clearly seen form Fig. 5(b) that I_d at the same $|V_g$-$V_{th}| = 0.7$ V linearly increases with increasing N_{fin}.

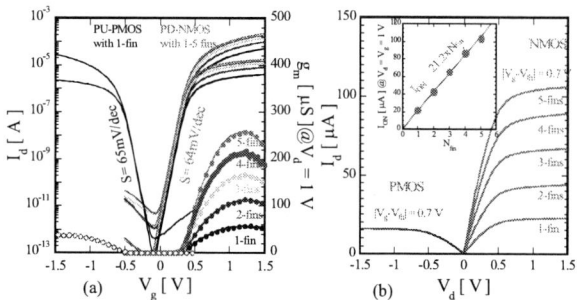

Fig. 5. Typical electrical characteristics of the fabricated PU transistor and PD transistors with different fin numbers (N_{fin}) from 1-5. (a) I_d-V_g & g_m-V_g and (b) I_d-V_d at $|V_g$-$V_{th}| = 0.7$ V.

The butterfly curves of the fabricated SRAM half-cells with different β-ratios from 1 to 5 were measured by changing supply voltage (V_{DD}) form 0.3 to 1 V. Figures 6 shows the typical butterfly curves in the case of β = 2. A high read SNM = 189 mV is obtained at $V_{DD} = 1$ V. Figure 7 summarizes read SNM as a function of V_{DD} with β as a parameter. It is clear that read SNM is greatly improved with increasing β from 1 to 2 at whole range of V_{DD} due to the strength of PD transistor. With further increasing β, the read SNM shows weak dependence on β. However, read SNM and write margin have trade-off relationship. To confirm this fact, write margin was also evaluated by biasing bit-line (BL) to V_{DD} and ground (GND) as shown in Fig. 8. Note that write margin is grater than that of read SNM although it slightly decreases from 379 to 347 mV with increasing β from 1 to 2. Figure 9 summarizes the read SNM and write margin as a function of β at a fixed $V_{DD} = 1$ V. Actually, with increasing β, read SNM increases but write margin decreases, showing trade-off relationship. Considering the area penalty accompanied with introducing multi-fin channels for PD transistor, β = 2 is the most effective way to enhance read SNM.

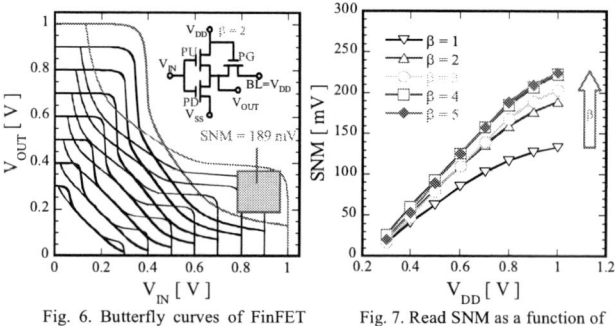

Fig. 6. Butterfly curves of FinFET SRAM half-cell with β = 2.

Fig. 7. Read SNM as a function of V_{DD} with β as a parameter.

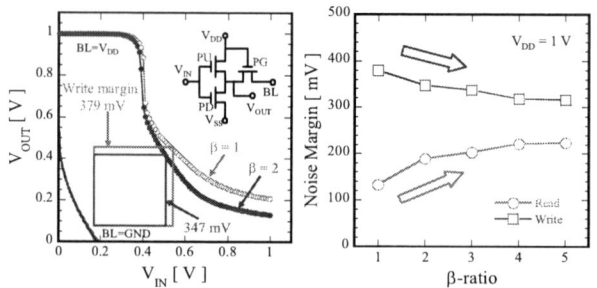

Fig. 8. Write margin extraction by biasing BL = V_{DD} and GND.

Fig. 9. Read and write noise margins as a function of β at $V_{DD} = 1$ V.

C. Fin-Height Controlled SRAM

According to above experimental results of β-ratio dependence of SNM, dramatic improvement in read SNM is achieved when β is chosen to be 2. To realize β = 2 without cell area increment, we fabricated fin-height controlled SRAM half-cells using selective recess RIE. To investigate the influence of fin-heights of PG and PU transistors on the noise margin, three kinds of SRAM half-cells were fabricated. Type A has high-fins for all transistors, type-B has a low-fin for PG transistor only, and type-C has low-fins for both PG & PU transistors. The fin-height for low-fins was controlled to one-half the high-fins by precise control of the selective recess RIE.

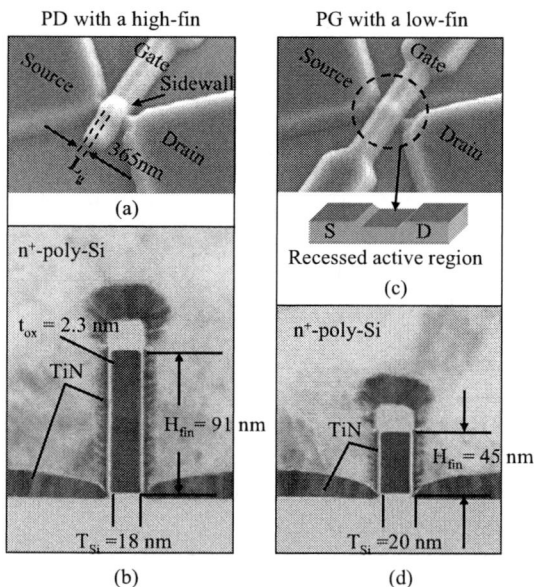

Fig. 10. Device features of the fabricated SRAM half-cell. (a) & (c) SEM images of PD and PG transistors after sidewall formation, and (c) & (d) cross-sectional STEM images of PD and PG transistors.

SEM images of the fabricated PD and PG transistors after gate sidewall formation are shown in Fig. 10(a) and 10(c), respectively. It is clearly confirmed from Fig. 10(c) that fin area is selectively recessed. The corresponded STEM images of PD with a high-fin and PG with a low-fin are shown in Fig. 10(b) and 10(d), respectively. Fin-height (H_{fin}) is 91 and 45 nm for PD and PG transistors, respectively, i.e., β = 2.

978-1-4244-6658-0/10 $26.00 © 2010 IEEE

Moreover, it is noteworthy that an ideal rectangular cross-section Si-fin channel is formed uniformly thanks to the orientation-dependent wet etching. The fin-thickness (T_{Si}) is 18 and 20 nm for PD and PG although they have the same dimension in design. The gate oxide thickness (t_{ox}) is 2.3 nm.

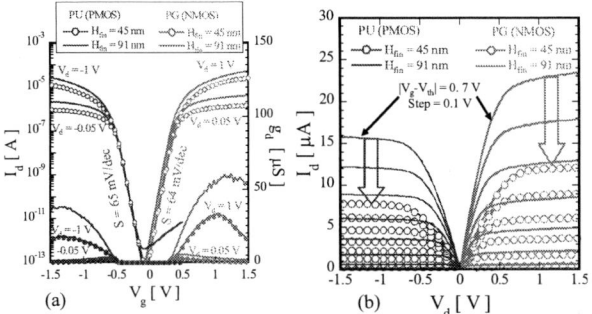

Fig. 11. Electrical characteristics of the fabricated PG and PU transistors with different fin heights of H_{fin} = 45 and 91 nm and with the same L_g = 365 nm. (a) I_d-V_g & g_m-V_g and (b) I_d-V_d.

The measured electrical characteristics of the fabricated PG-NMOS and PU-PMOS with different fin-heights of H_{fin} = 45 and 91 nm are shown in Fig. 11. Note that both PG and PU transistors show almost ideal S-slopes of 64 and 65 mV/decade, and the V_{th} does not depend on H_{fin}. Furthermore, the I_d of low-fin devices is actually reduced about one-half the high-fin devices. This indicates that fin-height control technique is very effective for tuning current drivability.

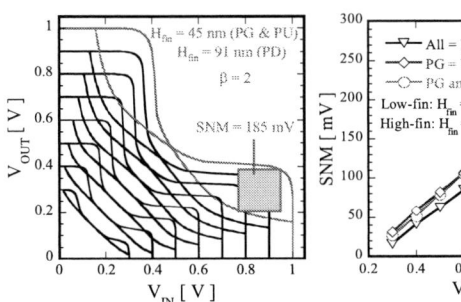

Fig. 12. Butterfly curves of SRAM half-cell with low-fins for both PG & PU, and with a high-fin for PD.

Fig. 13. Comparison between read SNMs for the fabricated three types of SRAM half-cells at V_{DD} = 0.3 -1V.

Butterfly curves were measured at different V_{DD}'s from 0.3 to 1 V. Figure 12 shows the measured typical butterfly curves of the fabricated type-C SRAM with low-fins for both PG & PU transistors. It is noteworthy that its SNMs are almost the same values in the case of 2-fins for PD transistor as shown in Fig. 6 owing to the same β = 2. Figure 13 summarizes the read SNMs for three types of SRAMs as a function of V_{DD}. Note that type-A with all high-fins shows smaller SNMs compared with type-B and type-C due to β = 1. On the other hand, type-B with a low-fin for PG only and type-C with low-fins for both PG & PU show almost the same SNMs at whole range of V_{DD} because read SNM has weak dependence on the current drivability of PU transistor. To clearly compare above three types of SRAM half-cells, the read SNM and write margin at V_{DD} = 1 are summarized in Fig. 14 and Fig. 15. Actually, a significant improvement in SNM is obtained by introducing a low-fin for PG transistor. However, type-B with a low-fin for PG only shows the smallest write margin compared to others because the current drivability ratio of PU over PG transistors (I_{PU}/I_{PG}) becomes the biggest value owing to a low-fin for PG only. Thus, the optimal fin-height controlled SRAM structure is considered to be type-C with low-fins for both PG & PU transistors.

IV. CONCLUSIONS

Fin-height and β-ratio controlled PVD-TiN gate FinFET SRAM half-cells have successfully been fabricated using selective recess RIE and orientation-dependent wet etching. The systematic experimental results show that read SNM increases significantly by adopting β = 2. With further increasing β, read SNM increases slightly, showing a weak dependence on β. The enhanced read SNM has been confirmed by controlling the fin-height of PG to one-half the PD. Write margin of the SRAM half-cell with low-fins for both PG & PU was larger than that in the case of a low-fin for PG only. Thus, it is conclude that the SRAM structure with low-fins for both PG & PU is the optimal structure for enhancing read SNM and write margin simultaneously. The developed fin-height controlled technology is very useful to scaled FinFET SRAM fabrication without cell area penalty.

REFERENCES

[1] S. O'uchi et al., IEDM Tech Dig., p. 709 (2008).
[2] T. Matsukawa et al., VLSI Tech., p. 118 (2009).
[3] Y. X. Liu et al., IEDM Tech Dig., p. 989 (2006).
[4] L. Chang et al., Symp. VLSI Tech., p. 128 (2005).
[5] L. Witters et al., Symp. VLSI Tech., p. 106 (2005).
[6] H. Kawasaki et al., Symp. VLSI Tech., p. 86 (2006).
[7] M. Guillorn et al., IEDM Tech Dig., p. 961 (2009).
[8] H. Y. Chen et al., IEDM Tech Dig., p. 958 (2009).
[9] H. Kawasaki et al., IEDM Tech Dig., p. 289 (2009).
[10] S. O'uchi et al., Custom Integrated Circuits Conf., p. 33 (2007).
[11] K. Endo et al., IEDM Tech Dig., p. 857 (2008).
[12] Y. X. Liu et al., IEDM Tech Dig., p. 986 (2003).
[13] Y. X. Liu et al., IEEE Conf. SNW. p. M0330 (2008).
[14] Y. X. Liu et al., Ext. Abstr. Int. Conf. SSDM. P. 1044 (2009).
[15] S. Takagi et al., IEEE Trans. Electron Devices, 41 (1994), p. 2363.
[16] H. Irie et al., IEDM Tech Dig., p. 225 (2004).
[17] Y. X. Liu et al., Jpn. Appl. Phys. Vol. 45, No. 4B, p. 3084 (2006).

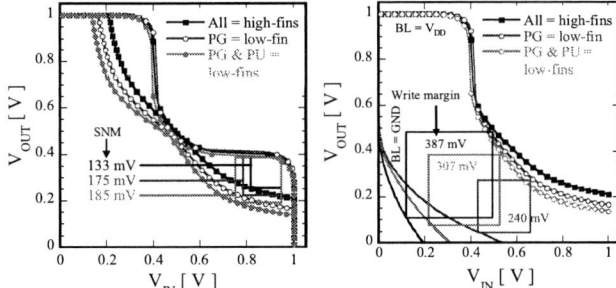

Fig. 14. Comparison between butterfly curves for the fabricated three types of SRAM half-cells at V_{DD} = 1 V.

Fig. 15. Comparison between write margins for the fabricated three types of SRAM half-cells at V_{DD} = 1V.

Dual Channel and Strain for CMOS Co-Integration in FDSOI Device Architecture

C. Le Royer, M. Cassé, F. Andrieu, O. Weber, L. Brevard, P. Perreau, J.-F. Damlencourt, S. Baudot, C. Tabone,
F. Allain, P. Scheiblin, C. Rauer, L. Hutin, C. Figuet*, C. Aulnette*, N. Daval*, B.-Y. Nguyen* and K. K. Bourdelle*

CEA-LETI Minatec, 17 avenue des Martyrs 38054 Grenoble Cedex 9, France

* SOITEC, Parc Technologique des Fontaines, F38190 Bernin, France

cyrille.leroyer@cea.fr

Abstract—We report an original Dual Channel-On-Insulator (DCOI) Fully Depleted CMOS architecture by co-integrating nFETs on sSOI and pFETs on Si/SiGe/(s)SOI with a TiN/HfO$_2$ gate stack (EOT=1.15nm) and down to 40nm gate lengths. We demonstrate for the first time large gains for transconductance (up to +125%) and mobility (+100%) even for short channel pFETs. This enables us to improve the ON(OFF) pFETs trade-off (I_{ON} +23% for a given I_{OFF} =100nA/µm), and thus to obtain similar I_{ON} for n & pFETs (~650µA/µm at V_{DD}=1V). Meanwhile, thanks to a channel material/strain engineering, the threshold voltages are adjusted (V_{th}~+/-0.2V) for high performance (HP) CMOS with a single mid-gap metal gate.

I. INTRODUCTION

With scaling of conventional CMOS devices are suffering from short channel effect (SCE), mobility loss and large access resistance. SOI Single or Multiple Gate architectures are expected to overcome the electrostatics issues. In the search for ON state current improvement, strain engineering has been demonstrated to be efficient for Si bulk CMOS [1], but the mobility gain is degraded for short gate lengths. Tensily Strained (t-Si) Silicon On Insulator (sSOI) substrates combine the advantages of both improved electrostatic integrity and transport properties [2]. The hole mobility which is not modified with t-Si can be improved either by compressive Si (c-Si), or SiGe material (or even by c-SiGe) as channel material. SiGe materials can be implemented by using SGOI wafers (SiGe-On-Insulator [3-5]) or thin SiGe layers on top on Si or SOI [6,7]. But very few studies report on the use of Dual Channel architectures based on thin body (s)Si and SiGe (or Ge) for n and pFETs respectively [3,8,9,10]. In [8] the mobility enhancements using t-Si and c-Ge were investigated in long-channel devices only. In [10] by using sSOI wafers as a template for t-Si nFETs and SiGe/t-SOI pFETs we demonstrated well behaved Dual Channel CMOS with 15nm gate lengths (but with EOT =2nm). We have also observed that the reported +100% mobility gain for long channel devices decrease for sub-100nm gate length devices and issues related to the I_{ON}/I_{OFF} ratio trade-off were not evoked in this previous work.

In this study we investigate a CMOS Dual Channel integration on 300mm sSOI and XsSOI (for nFETS), with Si$_{1-x}$Ge$_x$ (x=20, 30 and 40%) for pFETs. The Si CMOS process features a single high-K and Metal Gate stack, with Si raised S/D and NiPt silicidation. The electrical characterizations demonstrate the potentialities of this DCOI approach for tuning V_{th} (+/-0.2V), improving the electron AND the hole mobility (without significant degradation with L scaling), leading to symmetrical large ON currents.

II. DUAL CHANNEL CMOS PROCESS

A. Electrons and holes high mobility channels

In order to easily introduce SiGe materials in the channel of pFETs, thin Si$_{1-x}$Ge$_x$ layers have been epitaxially grown on 300mm SOI (for x_{Ge}=20 and 30%) and sSOI (for x_{Ge}=40%). Then the surface of the SiGe layers has been covered by an additional ultrathin silicon capping layer (Si cap). The purpose of the Si cap is to avoid a too large interface states density at the high k/SiGe channel interface [5,7]. The different Si cap/SiGe/SOI heterostructures used for pFETs devices integration are summarized in fig. 1:

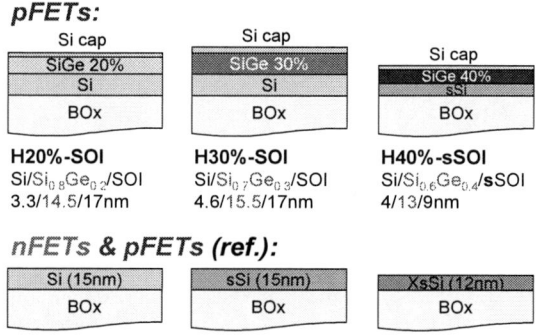

Fig. 1. Schematics of the different heterostructures (H20%-SOI, H30%-SOI, and H40%-sSOI) used for pFETs (+ SOI, sSOI, XsSOI for references).

We can notice that this approach is less complicated than the Ge enrichment technique [11] or the bonding of SiGe layers to produce SGOI structures. Moreover like the Ge enrichment technique it can be applied at the local scale of the (s)SOI wafers in order to obtain SiGe areas for pFETs while keeping (s)Si areas available for nFETs [9]. sSOI and XsSOI wafers have been also processed for comparison purposes.

978-1-4244-6658-0/10 $26.00 © 2010 IEEE

B. CMOS process

The common process for n- and pFETs 300mm wafers is the following:

- Mesa isolation
- Gate stack deposition: PolySi 80nm / TiN 5nm / HfO$_2$ 2.5nm (+patterning)
- Nitride based spacers (7nm)
- Raised Si Source/Drain (10nm)
- Extensions implantations (BF$_2$ and Arsenic)
- 2nd spacers (nitride on SiO2 oxyde)
- S/D implantations (Boron and Arsenic)
- Dopants activation anneal (Spike 1050°C)
- Silicidation (NiPt)
- Forming gas anneal (400°C)

III. ELECTRICAL CHARACTERIZATIONS

A. CV & Threshold voltage

Fig. 2 shows typical CV measurements of the pFET devices based on H40%-sSOI structures (sSOI, XsSOI for references). The Equivalent Oxide thickness (EOT) is the same for all the tested devices (EOT=1.15nm). As for Germanium devices with ultrathin Si capping layers at the high-K/Ge interface [12], a bump occurs due to the accumulation of holes at low gate voltage overdrive at the Sicap/SiGe interface, then at the oxide/Si cap interface.

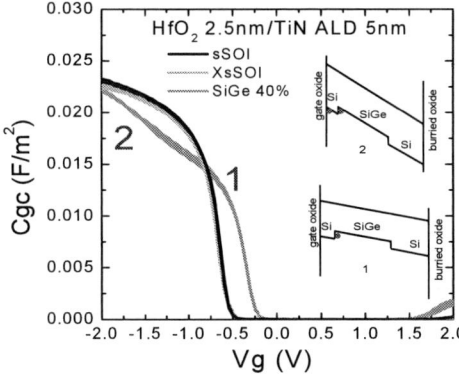

Fig. 2. CV measurements of pFETs devices (L=10μm×W=10μm) on sSOI, XsSOI and H40%-sSOI (heterostructure with Si cap/Si$_{0.6}$/Ge$_{0.4}$/sSOI).

Concerning the threshold voltages, fig. 3 gives an overview of the n and pFETs results: for the Si family (SOI, sSOI, XsSOI) V$_{th,p}$ values are identical but very high (-0.6V), whereas for V$_{th,n}$ XsSOI exhibits lower values (+0.10V) than sSOI (+0.17V) because of the strain-induced conduction band modification. We can observe that TiN does not enable to reach symmetrical (and low) V$_{th}$ values for n and pFETs. But the introduction of SiGe in the channel of pFETs leads to a decrease of the V$_{th,p}$ values (from -0.52 to -0.33V), as illustrated in fig. 4.

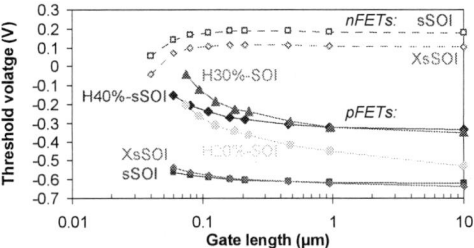

Fig. 3. Impact of gate length on the threshold voltage of n& pFETs for sSOI, XsSOI (n & p) and for the SiGe based heterostructures on (s)SOI (p) at W=10μm and V$_{DD}$=0.05V.

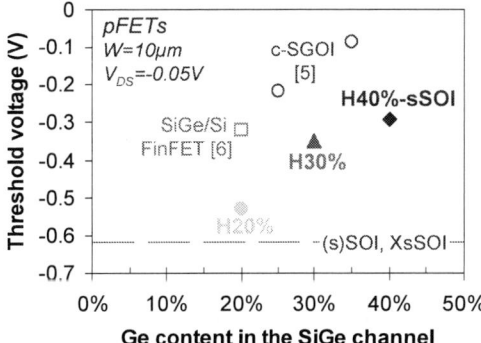

Fig. 4. Impact of Ge content on the pFETs threshold voltage (W=10μm, V$_{DS}$= -0.05V, L=10μm). The c-SGOI data are extracted from [5].

B. Drain Induced Barrier Lowering

From the I$_D$(V$_G$) measurements at low and high V$_{DS}$ (+/- 0.05 & 1V) we have extracted the Drain Induced Barrier Lowering (DIBL) parameter. For the nFETs on sSOI and XsSOI, we obtain a very low DIBL (less than 47mV/V at L=60nm) due to the thin Si layers. For the pFETs, the values depend on the semiconductor(s) thickness under the gate (fig. 5): as the sSOI and XsSOI wafers present a T$_{Si}$ below 18nm, the DIBL is low (less than 25mV/V at L=110nm and 50mV/V at L=60nm). For the SiGe based structures, the stack thickness is larger than 25nm, leading to a DIBL between 85 and 130mV/V at L=60nm.

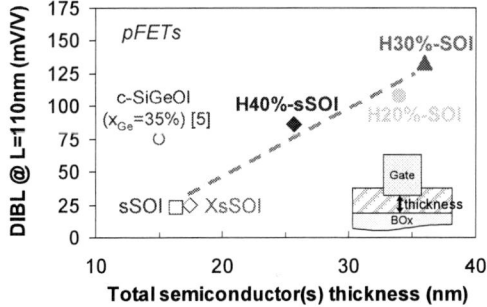

Fig. 5. pFETs DIBL for L=110nm as function of the total semiconductor thickness under the gate (W=0.14μm, V$_{DS}$ = -0.05/-1V).

978-1-4244-6658-0/10 $26.00 © 2010 IEEE

However we can notice that this degradation is only due to a thickness effect and do not rely on the presence of SiGe in the channel. TCAD simulations have been performed in order to investigate the DIBL properties of SOI and Si/SiGe/SOI stacks in the case of scaled comparable total thickness (9nm): it appears that the introduction of a thin SiGe layer does not degrade the electrostatic control, compared to SOI (fig. 6), this regardless of the Ge content in the SiGe layer (for 20 and 40%, all on SOI):

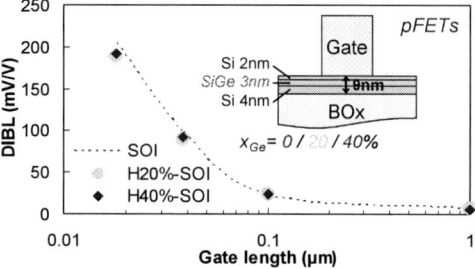

Fig. 6. TCAD simulations results: Impact of gate length on DIBL for pFETs on SOI, H20%-SOI and H40%-SOI with comparable thickness under the gate (9nm).

C. Subthreshold slope

Fig. 7 shows typical subthreshold slope (SS) values measured at low V_{DS} on the different pFETs structures (note that sSOI, XsSOI lead to similar SS results for p and nFETs). We can notice that the lightly degraded SS values for the SiGe based pFETs can be related to the thicker semiconductors stacks as for DIBL. TCAD simulations (not shown here) indicate that SOI and SiGe heterostructures on SOI with comparable thicknesses lead to identical SS values and trends. These good subthreshold characteristics demonstrate that a Si capping layer is efficient to achieve a good quality of the high-K/channel interface, without significantly high interface states densities.

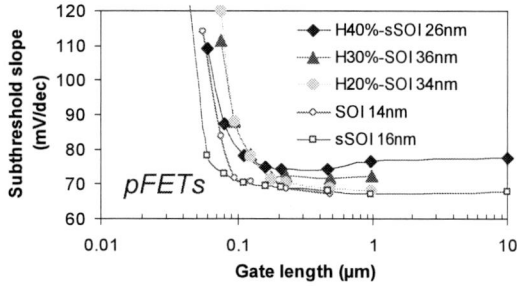

Fig. 7. Measured subthreshold slope values (V_{DS}=-0.05V) for pFETs (W=0.14μm) as a function of gate length.

D. Transport properties

In order to investigate the holes transport properties in the SiGe based heterostructures, we have extracted the transconductance (fig. 8) and the low-field mobility values with the Y-function method (fig. 9), on pFETs. Large $G_{M,max}$ gains are demonstrated for all SiGe based heterostructures as compared to sSOI (up to +110% and 180% for H20% and H30% at L~90nm). The larger transconductance values are due to a smaller access resistance (-40%, not shown here) and

larger low-field hole mobility in the SiGe heterostructures (+30% and +100% for H20% and H30% at L<100nm).

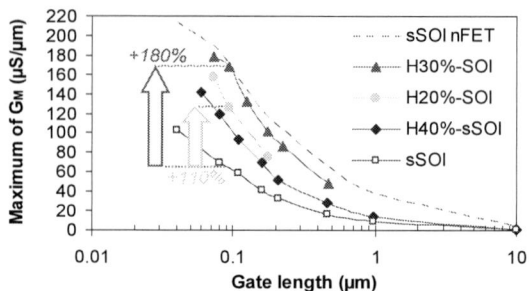

Fig. 8. . Gate length dependence of the maximum of transconductance G_M for pFETs, with a sSOI nFET comparison (dotted line) (W=0.14μm, V_{DS} = +/-0.05V).

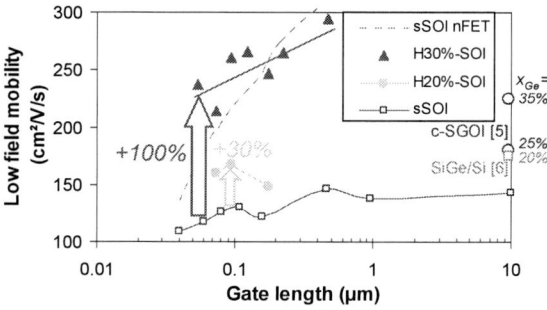

Fig. 9. Hole low-field mobility extracted with the Y function, as a function of gate length with a sSOI nFET comparison for electrons (dotted line) (W=0.14μm, V_{DS} = +/-0.05V).

The H30%-SOI presents better mobility properties compared to H20%-SOI due to i) the larger Ge content in the SiGe layer [3,4,5,7] and ii) probably the larger strain in the SiGe layer due the larger mismatch for x=30% between Si$_{1-x}$Ge$_x$ and Si. The H40%-sSOI case is more difficult to analyze. On the one hand the Ge content is here the largest, but the on the other hand, the mismatch (Si$_{0.6}$Ge$_{0.4}$/sSi with sSi obtained from sSi on relaxed Si$_{0.8}$Ge$_{0.2}$) is lower than for H30%-SOI leading to less strain in the SiGe layer. Moreover, as we used a Si CMOS process flow (including annealing conditions), Ge atoms may have diffused within the H40%-sSOI heterostructure resulting in a final reduced Ge content. As a consequence, the H40%-sSOI exhibit lower Gm,max than the other devices.

Finally the best transport properties are obtained for H30%-SOI; for gate lengths smaller than 100nm, the hole low-field mobility and the transconductance is comparable to, and even exceeds the electron mobility measured on sSOI nFETs.

E. Negative Bias Temperature Instability

NBTI is drastically improved in H40%-SOI compared to sSOI and XsSOI pFETs (fig. 10), in good agreement with prior studies involving Ge or SiGe alloys [3,13,14]. The valence band offset between SiGe and Si cap [13] (band diagram in fig. 2) forms a barrier impeding holes tunnelling into the gate dielectric, which is considered to be the initial step in NBTI degradation.

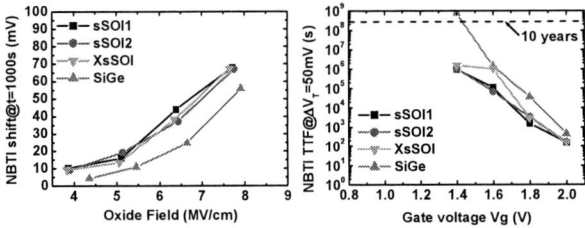

Fig. 10. NBTI shifts at t=1000s for pFETs devices ("SiGe" corresponds to H40%-sSOI), and extrapolated Time To Failure for ΔV_T=50mV.

F. High mobility CMOS & ON/OFF trade-off

Based on the previous analysis, it appears that a promising CMOS integration would involve (s)SOI (or even XsSOI) for nFETS and Sicap / $Si_{1-x}Ge_x$ heterostructures with x~20-30% epitaxially grown on (s)SOI. This Dual Channel approach enables to get well balanced n & pFETs characteristics as far as threshold voltages (fig. 11) and ON currents (fig. 12) are concerned. The symmetrical and high-performance ON-currents are due to the large hole mobility (+30 to 100%, see fig.9), the lowered $V_{T,p}$ and R_{access} improvement (-40%) induced by the SiGe channel, yielding an I_{ON} improvement of +110 to +190% at L=100nm for H20% and H30% with respect to sSOI (Fig. 12).

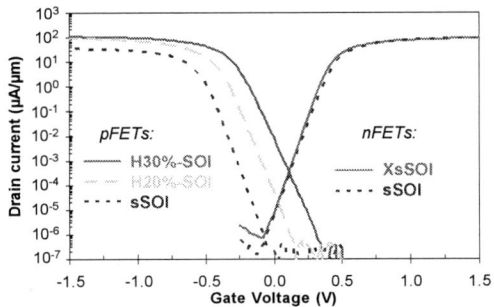

Fig. 11. $I_D(V_G)$ measurements (V_{DS} = +/-0.05V) of ~100nm gate length devices (W=0.14µm).

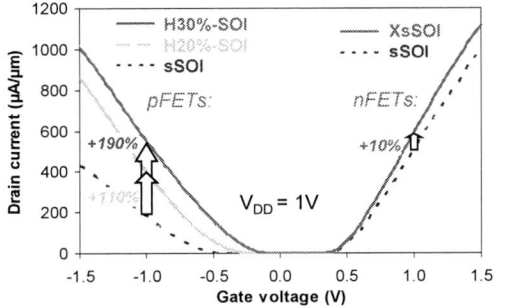

Fig. 12. $I_D(V_G)$ measurements in linear scale for high V_{DS} (= +/-1V) of ~100nm gate length devices (W=0.14µm).

Finally we can notice that the OFF state current values (I_{OFF}) do not degrade excessively the pFETs performances as illustrated in the $I_{ON}(I_{OFF})$ figure of merit (fig. 13): for a fixed I_{OFF} (=100A/µm), a I_{ON} gain remains (+23% for H30%-SOI). Further optimizations of the different thicknesses of the

Si/SiGe/Si heterostructures and of the process flow will enable to optimize the I_{OFF}.

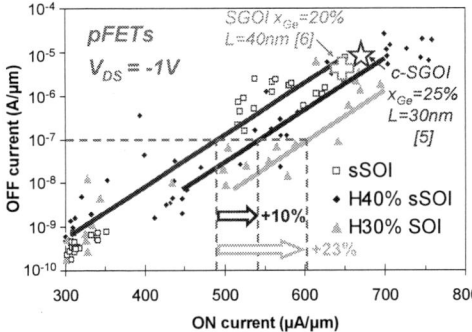

Fig. 13. $I_{ON}(I_{OFF})$ figure of merit of pFETs for H40%-sSOI and H30%-SOI (with sSOI reference) at V_{DS} = -1V (W=0.14µm, L_{min}~40-60nm). For clarity, the H20%-SOI data are not shown here (~ +7% gain in I_{ON}).

IV. CONCLUSION

We present the successful integration of Si cap/$Si_{1-x}Ge_x$ layers (x ranging from 20 to 40%) on 300mm SOI and sSOI with high-K gate dielectric and Metal Gate for pFETs improvements. pFETS devices were fabricated using theses heterostructures (but also n and pFETs references on sSOI and XsSOI). We demonstrate a good quality high-K/channel interfaces thanks to a Si capping layer. All the fabricated devices exhibit excellent Short Channel Effects and DIBL due to the use of thin body structures. The better transport properties of holes in the SiGe heterostructures enable to demonstrate large gains for transconductance (+180%), low field mobility (+100%), access resistance (-40%) and I_{ON} currents (with no significant I_{OFF} degradation). Moreover these gains are maintained with gate down scaling (contrary to many high mobility approaches [1,2]). Finally we have shown that a Dual Channel architecture featuring (s)SOI nFETs and $Si_{1-x}Ge_x$/(s)SOI with x~30% pFETs enables to obtain a high mobility CMOS with symmetrical V_{th} and ON performances.

ACKNOWLEDGMENT

This work was partly funded by the French Public Authorities through the OSEO Nanosmart project, the Medea DECISIF projet and the NANO 2012 program performed in collaboration with IBM and STMicroelectronics.

REFERENCES

[1] S. Natarajan et al., IEDM 2008, pp. 941-943.
[2] V. Barral, et al., IEDM 2007, pp. 61-64.
[3] T. Tezuka, et al., Symposium on VLSI Technology 2005, pp. 80-81.
[4] F. Mayer, et al., IEDM 2008, pp. 163-166.
[5] L. Hutin, et al., Symposium on VLSI Technology, 2010.
[6] C. E. Smith, et al., IEDM 2009, pp. 309-312.
[7] O. Weber et al., IEEE Trans. Electron Devices, 53, 2006, pp. 449-456.
[8] M. L. Lee et al., IEDM 2003, pp. 429-432.
[9] C. Le Royer, et al., EuroSOI conference 2010, pp. 21-22.
[10] F. Andrieu, et al., SOI Conference, p. 223-5, 2005.
[11] J.-F. Damlencourt, et al., ISTDM 2006, pp. 202-208.
[12] H. Grampeix, et al., 39th IEEE SISC, 2008.
[13] B. Kaczer, et al., Microelec. Eng., 86, Issues 7-9, 2009, pp. 1582-1584
[14] S.-H. Lee et al., Symposium on VLSI Technology, 2009, pp. 74-75.

978-1-4244-6658-0/10 $26.00 © 2010 IEEE

UT2B-FDSOI Device Architecture Dedicated to Low Power Design Techniques

J.-P. Noel[1], O. Thomas[1], M.-A. Jaud[1], C. Fenouillet-Beranger[1,2], P. Rivallin[1], P. Scheiblin[1], T. Poiroux[1], F. Boeuf[2], F. Andrieu[1], O. Weber[1], O. Faynot[1] and A. Amara[3]

[1] CEA, LETI, MINATEC, F-38054, Grenoble – jean-philippe.noel@cea.fr
[2] STMicroelectronics, F-38926, Crolles – claire.fenouillet-beranger@st.com
[3] ISEP, F-75006, Paris – amara.amara@isep.fr

Abstract—In this paper, a new ultra-thin body and BOX (UT2B) fully-depleted (FD) silicon-on-insulator (SOI) device architecture based on a stacked back plane (BP) and WELL below the BOX is presented. The proposed device has been developed to boost the gate-to-channel electrostatic control and to be compatible with the adaptive body biasing (ABB) techniques for low power applications. The concept viability and the device electrical characteristics have been demonstrated by TCAD simulations at $L_G=30nm$. The electrical characteristics have been assessed with regular UTB-FDSOI devices for various BOX thicknesses (T_{BOX}). It is shown that the body factor (γ) is for both nMOS and pMOS devices higher than 170mV/V at 10nm of T_{BOX}. On a ring oscillator (RO), the proposed device architecture leads to lower static power dissipation (between a factor 1.2 and 6.7) for a similar propagation delay, compared to devices without BP/WELL.

I. INTRODUCTION

As technologies scale down into the deca-nanometer range, the benefits of linear scaling in terms of speed and power consumption are jeopardized by the increase of leakage currents and V_T variability. To continue increasing the speed of low power applications while keeping adequate static power consumption, the use of multi-V_T CMOS platforms is currently widespread. Indeed, high-V_T (HVT~600mV) transistors are used in non-critical paths to keep low leakage currents, while standard-V_T (SVT~450mV) and low-V_T (LVT~250mV) transistors are commonly used in critical paths to meet timing constraints [1]-[3]. In addition, various techniques of power management and circuit compensation solutions are implemented at the design level. Most of them are based on the body effect, such as reverse source or body biasing techniques [4].

For the 22nm node and below, the ultra-thin body and BOX (UT2B) fully-depleted (FD) silicon-on-insulator (SOI) architecture appears as a promising candidate for low power applications [5]-[7]. In this technology, the V_T depends on the gate material, which requires a processing expertise to finely control the V_T value. Thereby, setting up multi-V_T devices in the undoped FDSOI technology is very challenging. In [8], it was proposed a simple approach to set up multi-V_T UT2B-FDSOI devices based on the combination of different back

plane (BP) doping types and biasing (V_B). The advantage of this approach is the use of a single metal gate (midgap), making it a cost-competitive solution. Nevertheless, it can be further optimized, in particular regarding the SVT device configurations. Indeed, compared to the HVT and LVT device architectures, the SVT one is implemented without BP, which can affect the device electrostatic control. Moreover, the interest of the adaptive body biasing (ABB) techniques to manage the power consumption and to boost the performances of digital circuits is limited by a common substrate for both nMOS and pMOS devices, sharing the same back biasing.

In this paper, a new SVT UT2B-FDSOI architecture based on BP is proposed. It has been developed to boost the gate-to-channel electrostatic control and designed to be compatible with ABB techniques for low power applications. The electrical characteristics have been evaluated by TCAD simulations at $L_G=30nm$. The effectiveness of the proposed device architecture is compared with the SVT device architecture proposed in [8] for different BOX thicknesses (T_{BOX}). The reminder of the paper is organized as follows: In section II, the new SVT device architecture is depicted. In section III and IV, electrical characteristics of devices and ROs are presented and discussed. Finally, conclusions and remarks are drawn in the last section.

II. PROPOSED STANDARD-V_T DEVICE ARCHITECTURE

Table 1 summarizes the different V_T configurations presented in [8] for UT2B-FDSOI nMOS and pMOS devices with regard to the BP doping type and biasing. nMOS or pMOS SVT transistors can be implemented following four configurations. Among them, two are based on BP and two without. SVT devices without BP benefit from a very simple integration and are well dedicated to the high density integrated circuit design. Besides, SVT devices with BP are attractive regarding the use of ABB techniques thanks to the possibility to apply two distinct back biasing on nMOS and pMOS devices due to their complementary BP doping type. However, due to the BP biasing condition, their implementation is more complex. Indeed, for CMOS circuits, it leads to a forward biasing of the resulting PN junction below the BOX, as illustrated on Figure 1. So far, a triple WELL technology becomes mandatory. However, this latter must impact the circuit density.

TABLE I V_T OPTIONS OF UT2B-FDSOI nMOS AND pMOS DEVICES FOR VARIOUS TYPES OF BP DOPING AND BACK BIASING [8]

V_B	0V	V_{DD}	V_B	0V	V_{DD}
n-BP	SVT	LVT	n-BP	SVT	HVT
p-BP	HVT	SVT	p-BP	LVT	SVT
w/o BP	SVT	SVT	w/o BP	SVT	SVT

nMOSFET (V_S=0V) pMOSFET (V_S=V_{DD})

Figure 2 depicts the proposed SVT device architecture taking into account the circuit design environment constraints. It is based on a stack of BP and WELL layers. The BP is enclosed by a deep STI (for example, 300nm from the top of Si-film), which leads to a vertical dielectric isolation of the BP. The WELL doping type used is complementary to the BP one, thus forming a PN junction below the BOX (p-BP/n-WELL or n-BP/p-WELL). Thereby, the BP biasing is controlled by the WELL one. Thanks to this architecture, it becomes possible to apply V_{DD} (resp. 0V) to p-BP (resp. n-BP) without forward biasing the resulting BP/WELL junction. Both nMOS and pMOS devices can be implemented with either n-BP/p-WELL or p-BP/n-WELL stacks. However, to be compatible with BULK CMOS design techniques, the nMOS and pMOS transistors must be based only on a n-BP/p-WELL and p-BP/n-WELL stack, respectively.

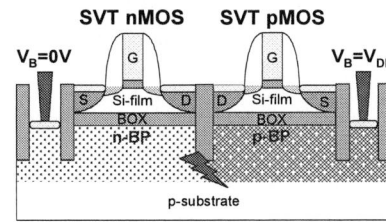

Figure 1. Cross-sectional view of SVT configurations with BP for UT2B-FDSOI nMOS and pMOS devices [8]

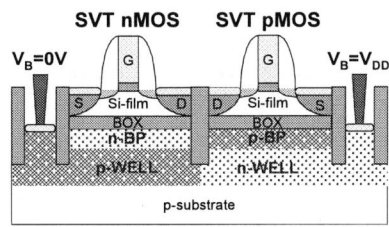

Figure 2. Cross-sectional view of proposed SVT configurations with BP for UT2B-FDSOI nMOS and pMOS devices

To assess the viability and the effectiveness of the proposed device architecture, TCAD simulations have been performed [9]. An improved low field mobility model including surface roughness and remote Coulomb scattering effects calibrated on experimental data is used [10]. Short channel effects have been calibrated on experimental data [11] by adjusting access doping levels and profiles. Simulated devices present a high-κ/SiO$_2$ dielectric gate stack of 1.2nm

equivalent oxide thickness, a Si-film of 6nm with a doping level of 10^{15}cm^{-3}, a BOX thickness of 10nm and BP/WELL junction profiles (n-BP/p-WELL and p-BP/n-WELL) depicted on Figure 3. These junction profiles have been obtained by process simulations [9] based on the doping concentration and energy implantation parameters used in [5]. It results in BP/WELL junctions being roughly 100nm from the BOX (<300nm of deep STI) with a peak of doping concentration of 2.10^{18} cm^{-3} for the BP and 10^{18} cm^{-3} for the WELL.

Figures 4 (a) and 4 (b) show, respectively, the electrostatic potentials of the nMOS and pMOS devices depicted on Figure 2. The resulting electrostatic potentials are compared with the ones of the same devices without WELL, in this case the back biasing is directly tied to the BP and independently for nMOS and pMOS devices. In order to address the most critical case of the BP electrostatic control by the WELL, the devices have been biased in saturated mode. Here, the 1D cuts have been extracted from the Si-film to the substrate near the drain side. Indeed, the drain gives rise to the strongest electrical field which could affect the BP control by the WELL.

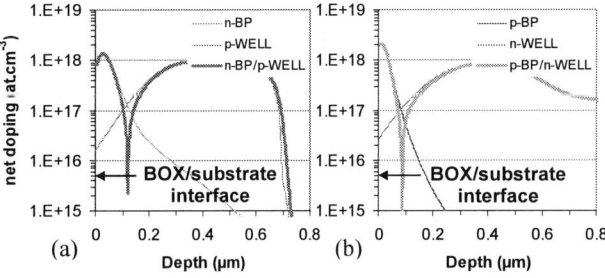

Figure 3. Spike annealing at 1050°C simulated BP and WELL doping profiles for a UT2B-FDSOI device with T_{BOX}=10nm: (a) n-BP/p-WELL and (b) p-BP/n-WELL

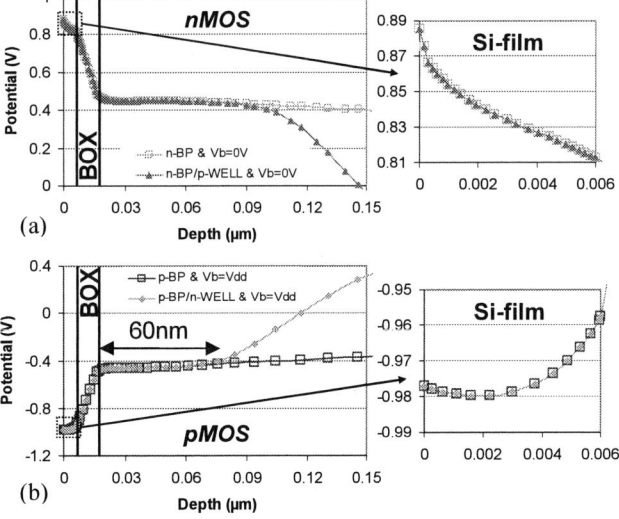

Figure 4. Electrostatic potential for UT2B-FDSOI (a) nMOS and (b) pMOS devices based on various BP/WELL junction configurations with T_{BOX}=10nm at L_G=30nm ($|V_G|$=$|V_{DS}|$=$|V_{DD}|$=0.9V)

TABLE II Electrical Characteristics of UT2B-FDSOI nMOS and pMOS Devices at T_{BOX}=10nm, L_G=30nm and V_{DD}=0.9V

	lin|V_T| (mV)	|DIBL| (mV/V)	I_{ON} (μA/μm)	I_{OFF} (pA/μm)	sat SS (mV/dec)	γ (mV/V)
n-BP	403.1	37	393.1	62.9	77.6	170.4
n-BP/ p-WELL	403.2	37	393.1	62.7	77.7	170.4
p-BP	401.1	48.2	301.5	148.4	79.9	176.8
p-BP/ n-WELL	401.1	48.2	301.5	148.4	79.9	176.8

It is apparent from Figures 4 (a) and 4 (b) that the electrostatic potentials are similar in the Si-film and up to 50nm below the BOX/BP interface with or without WELL. This behaviour allows achieving the same device electrical characteristics, as depicted in Table II. These results show the viability of the proposed device architecture for 10nm of T_{BOX} and higher. Indeed, for thicker BOX, the drain impact on the BP is reduced. Now, regarding the body factor (γ), it is higher than 170mV/V (see Table II) for both nMOS and pMOS devices which shows the suitability of this structure to address power management techniques.

III. DEVICE RESULTS

To assess the performances of the proposed device architecture with a stacked BP and WELL, TCAD simulations have been performed on nMOS and pMOS devices in CMOS circuit biasing conditions. The electrical characteristics have been also compared with the SVT device without BP/WELL [8]. The device models used are the ones defined in section II.

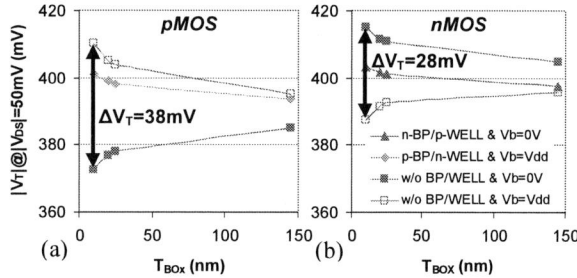

Figure 5. Linear V_T versus T_{BOX} of (a) pMOS and (b) nMOS devices with and without BP/WELL at L_G=30nm

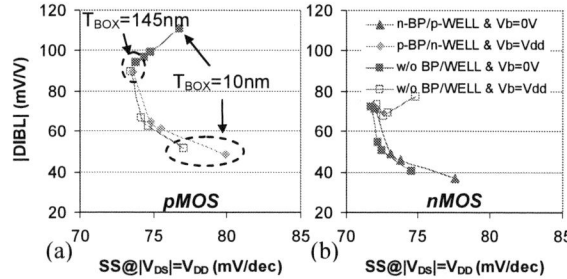

Figure 6. DIBL versus SS (extracted at \|V_{DS}\|=0.9V) of (a) pMOS and (b) nMOS devices with and without BP/WELL for various BOX thicknesses (10nm ≤ T_{BOX} ≤ 145nm) at L_G=30nm

Figure 5 shows the V_T versus T_{BOX}. As expected, the V_T values are roughly similar whatever the BOX thickness. The largest V_T gap between the architectures is lower than 30mV

for nMOS devices (Figure 5 (a)) and 40mV for pMOS devices (Figure 5(b)) at 10nm of T_{BOX}. This is achieved by the devices without BP, reflecting the low body biasing coupling effect due to the BOX/substrate depletion.

Regarding the device electrostatic control, Figure 6 exhibits the DIBL versus the subthreshold slope (SS) for various T_{BOX}. For the nMOS device architectures (Figure 6 (b)) with BP/WELL and the one without BP/WELL at V_B=0V, the DIBL is strongly reduced with T_{BOX}, while SS slightly increases. These two options lead approximately to the same electrostatic control. On the other hand, for the device architecture without BP/WELL at V_B=V_{DD}, there is a back channel conduction for thin BOX, leading to the worst electrostatic configuration in this region. For the pMOS device, the same conclusions can be drawn, as shown in Figure 6 (a). Here, the worst case configuration is the one without BP/WELL at V_B=0 due to the source voltage tied to V_{DD}, which is the CMOS circuit biasing condition.

Figure 7 (a) and 7 (b) show the resulting I_{ON}-I_{OFF} electrical characteristics of the pMOS and nMOS devices, respectively. For the nMOS device architectures with BP/WELL and without BP/WELL at V_B=0V, I_{OFF} is reduced by a factor 2, when reducing T_{BOX} down to 20nm thanks to the DIBL improvement. Below 20nm, the SS degradation becomes higher than the DIBL improvement which leads to a saturation or a slight increase of I_{OFF}. At the same time, the I_{ON} current slightly decreases (by ~6%) due to the DIBL reduction for thin BOX. Regarding the configuration without BP/WELL at V_B=V_{DD}, I_{OFF} increases by a factor 2, especially below 20nm, while I_{ON} is quasi-constant. This is due to both the low DIBL variation and the SS degradation with T_{BOX} in this particular case. Similar behavior is achieved for pMOS devices.

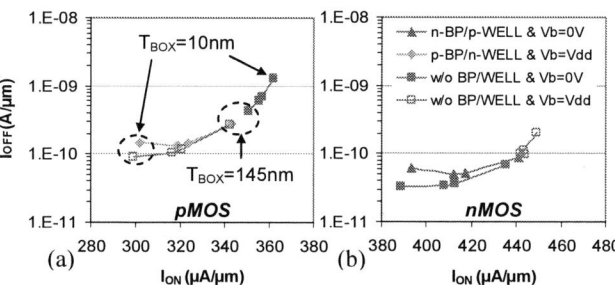

Figure 7. I_{OFF} (extracted at \|V_{GS}\|=0, \|V_{DS}\|=0.9V) versus I_{ON} (extracted at \|V_{GS}\|=\|V_{DS}\|=0.9V) of (a) pMOS and (b) nMOS devices with and without BP/WELL for various BOX thicknesses (10nm ≤ T_{BOX} ≤ 145nm) at L_G=30nm

To conclude, for nMOS at V_B=0V (respectively for pMOS at V_B=V_{DD}), the architectures with n-BP/p-WELL and without BP/WELL (respectively with p-BP/n-WELL and without BP/WELL) yield very similar performances. However, the use of BP isolated by WELL enables a total isolation of the BP. Indeed, it allows applying a back biasing to the nMOS and pMOS transistors independently, giving to the best electrostatic control of the nMOS and pMOS devices at the circuit level. This is not the case without BP/WELL due to the common substrate for both nMOS and pMOS transistors. For the latter, following the back biasing applied (0V or V_{DD}), either pMOS or nMOS devices is in a worst case

978-1-4244-6658-0/10 $26.00 © 2010 IEEE

configuration. In addition, the BP/WELL stack allows using ABB techniques for the power management and process compensations like in BULK technology.

IV. RING OSCILLATOR RESULTS

Once the proposed architecture has been validated on individual MOS devices, the viability in transient mode and the impact on circuit performances have to be demonstrated. To do so, TCAD MixedMode simulations [9] have been performed on ROs with a fan out 1, including a wire load capacitance of 2fF on each gate output. The device models used for the simulations are the same as in the previous sections. The electrical characteristics have been extracted for 10nm of T_{BOX}.

The proposed device architecture sets the BP node in high impedance. If the BP is not efficiently controlled by the WELL, transient signals can affect the BP biasing by coupling capacitance and generate BP noise. To assess the viability of the proposed architecture in transient mode, the propagation delay (τ_P) has been extracted on a RO based on a BP with and without WELL. Figure 8 shows the RO voltage traces (V_{RO}). The results highlight that the τ_P are similar. Indeed, 5.28ps and 5.33ps of τ_P have been respectively extracted, which is less than 1% of difference. This result demonstrates that the transient signals do not affect the electrostatic control integrity of the BP by the WELL.

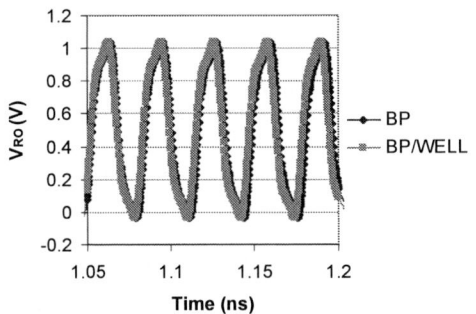

Figure 8. Voltage trace of a RO (V_{RO}) based on devices with or without WELL for T_{BOX}=10nm at V_{DD}=0.9V

For performance assessment, the proposed device architecture has been compared with the one without BP/WELL. Table III gives the different RO configurations defined by the architecture of inverter transistors. The first configuration refers to the proposed architecture and the last two correspond to devices without BP/WELL. For the latter configurations, the same back biasing is applied to the pMOS and nMOS devices due to a common substrate.

TABLE III SIMULATED RO CONFIGURATIONS DEFINED BY THE ARCHITECTURE OF INVERTER TRANSISTORS

	nMOSFET	pMOSFET
(1)	n-BP/p-WELL & V_B=0V	p-BP/n-WELL & V_B=V_{DD}
(2)	w/o BP/WELL & V_B=0V	w/o BP/WELL & V_B=0V
(3)	w/o BP/WELL & V_B=V_{DD}	w/o BP/WELL & V_B=V_{DD}

Figure 9 shows the τ_P versus the static power (P_{STAT}) for the various RO configurations. It is apparent that the τ_P are roughly similar for each configuration. Regarding P_{STAT}, configuration (2) leads to the worst case due to the pMOS leakage currents, which dominate by default and are reinforced in this case. Configuration (1) gives rise to the best case thanks to a better electrostatic control provided by the proposed device architecture.

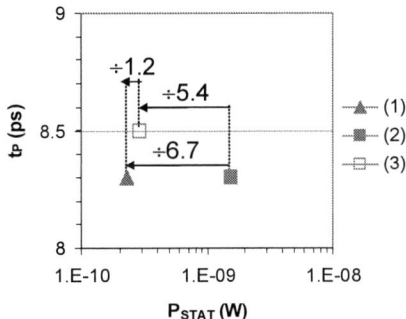

Figure 9. Propagation delay (τ_P) versus the static power (P_{STAT}) of different RO configurations for T_{BOX}=10nm at V_{DD}=0.9V

V. CONCLUSION

A new UT2B-FDSOI device architecture with a BP/WELL stack below the BOX has been presented. This latter has been proposed to optimize the channel electrostatic control, while being compatible with high density CMOS design environment and design techniques for power management (γ>170mV/V). The device viability analysis has shown that the PN junction formed by the BP/WELL stack has no impact on the static and transient device electrical characteristics compared with a device without WELL. The assessment and the comparison of a RO designed with a device without BP/WELL has shown that the proposed device architecture give rises to lower static power consumption with an equivalent propagation delay at 10nm of T_{BOX}.

ACKNOWLEDGMENT

This work has been partly founded by the IBM/STMicroelectronics, DECISIF and Nanosmart R&D projects.

REFERENCES

[1] T. Yamashita *et al.*, ISSCC 2000, pp. 414-416
[2] M. Srivastav *et al.*, IDEAS 2005, pp. 363-367
[3] S. Kunie *et al.*, ASPDAC 2008, pp. 748-753
[4] K. Roy *et al.*, Proc. of the IEEE, vol. 91, no. 2, pp. 305-327, Feb. 2003
[5] C. Fenouillet-Beranger *et al.*, ESSDERC 2009, pp. 89-92
[6] T. Ishigaki *et al.*, ESSDERC 2008, pp. 198-201
[7] O. Weber *et al.*, IEDM 2008, pp. 245-248
[8] J.-P. Noel *et al.*, ESSDERC 2009, pp. 137-140
[9] SILVACO Atlas, Athena, MixedMode, 2010
[10] S. Martinie *et al.*, ESSDERC 2009
[11] F. Andrieu *et al.*, VLSI Technology 2010

978-1-4244-6658-0/10 $26.00 © 2010 IEEE

Ultra-low volume ferromagnetic nanodots for field-coupled computing devices

J. Kiermaier*, S. Breitkreutz*, X. Ju[†], G. Csaba[‡], D. Schmitt-Landsiedel* and M. Becherer*

*Lehrstuhl für Technische Elektronik, Technische Universität München, 80333 Munich, Germany

Email: kiermaier@tum.de

[†]Lehrstuhl für Nanoelektronik, Technische Universität München, 80333 Munich, Germany

[‡]Center for Nano Science and Technology, University of Notre Dame, Notre Dame, IN 46556, USA

Abstract—Focused ion beam irradiation on ferromagnetic Co/Pt films permits controlled modification of the coercivity. This is demonstrated experimentally and mapped to micromagnetic simulations. Temperature measurements prove the thermal stability of films and nanodots in an application relevant temperature range. For the first time, Extraordinary Hall-Effect measurements are performed at a single-domain ferromagnetic nanodot with a target size of $250\,nm$ in a Hall current device. This verifies the thermal stability and the read-out ability of the magnetic bistable states. Thus, the ultra-low volume magnetic Co/Pt dots fulfill the demands for use in field-coupled logic devices.

I. INTRODUCTION

New approaches for alternative signal processing using effects like spin wave interference, domain wall motion and magnetic field-coupling are presented as promising 'beyond CMOS devices' in the ITRS road-map [1]. The field-coupled logic devices, also denoted as MQCA (Magnetic Quantum Cellular Automata) benefit from non-volatility, radiation hardness, low power consumption and simple integration in standard silicon technology by only adding a few technological steps [2]. They are based on deep sub-micron ferromagnetic dots with bistable characteristics given by the magnetization orientation. These single domain nanomagnets interact with their next neighbors by magnetic field-coupling enabling digital signal flow. Boolean logic realized by inverters and majority gates composed of ferromagnetic nanodots is envisioned in [2], [3].

Although the MQCA conception is suggested for general purpose logic in the ITRS road-map, it is particularly suited to complement state-of-the-art CMOS technology for special applications like cryptography or image processing. Therefore, an electrical interface is essential for conversion of the electrical information into magnetic signals, do the profitable magnetic computing and convert it back for further electrical signal processing. We already experimentally demonstrated the functionality of these interface components in [4], [5], however, the output sensor contained a much larger magnetic dot than in a prospective technology.

Fig. 1 demonstrates the assembly of a field-coupled computing system. The bistable magnetic nanodots are realized by patterned Co/Pt multilayer thin films. The electrical information is transmitted to the magnetic system by the magnetic field generated from a current driven wire next to the input dot D_1.

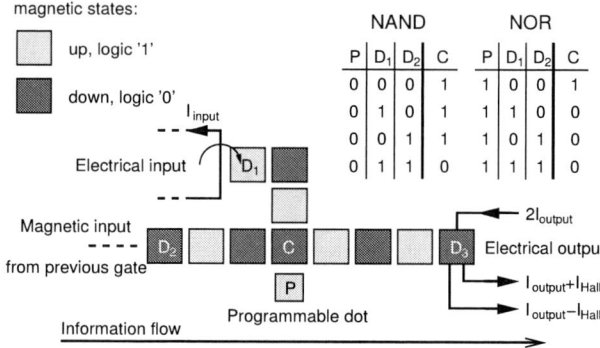

Figure 1. Design of a field-coupled computing system containing an electrical and magnetic input, a majority gate and an electrical output.

The dot D_2 illustrates the interconnect between two majority gates. The magnetic input wires transport the signal to the computing dot C where the logic function is implemented as depicted in the logic table. The majority gate can realize the NAND as well as the NOR function, depending on the logic state of the hard-magnetic programmable dot P. The result is shifted from dot C to the output dot D_3. This nanomagnet is sensed by the extraordinary Hall-Effect and the information is provided as current difference for the consecutive CMOS circuit. A running magnetic field wave is used to clock the system.

In this work, we investigate the magnetic nanodots, the fundamental component of field-coupled systems by magneto-transport measurements, which are an essential tool in the development of nanoscale, low power magnetic logic devices. We explore the characteristics of the magnetic thin films due to predefined ion beam irradiation and temperature variations. Based on the results, a square magnetic nanodot with a edge length of $250\,nm$ is created in this Co/Pt multilayer. The read-out ability and thermal stability are demonstrated for a wide temperature range.

II. EXPERIMENT

A. Sample fabrication

The magnetic nanodots are probed by a sensing structure based on the extraordinary Hall-Effect [6]. First, gold pins

978-1-4244-6658-0/10 $26.00 © 2010 IEEE

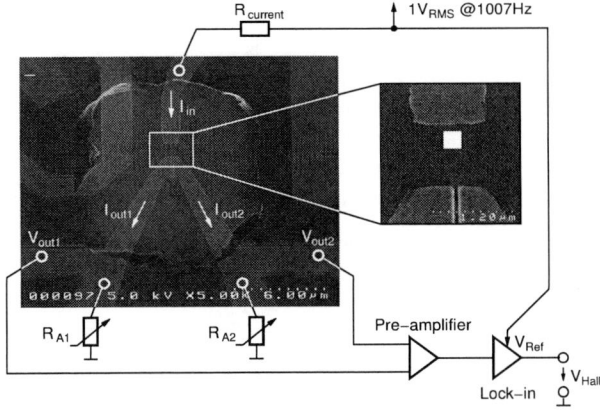

Figure 2. SEM image of a 250 nm by 250 nm ferromagnetic nanodot, schematically visualized as white square in the inset, in a split current Hall-Effect sensing structure with external circuit to determine the binary magnetization state as V_{Hall}.

are deposited by thermal evaporation on the oxidized silicon substrate to contact the magnetic material superimposed afterwards. The 50 nm wide separation between the two current contacts is realized by direct focused ion beam (FIB) milling, using a Micrion 9500 system with a gallium source and an acceleration voltage of 50 kV. The irradiation of the substrate, which occurs inevitably during alignment of the ion beam, roughens the surface. Experiments proved that irradiation up to an ion dose of $7 \cdot 10^{13}$ ions/cm² has no influence on the magnetic properties of the superimposed Co/Pt multilayer.

The $Pt_{3\,nm}$ / [$Co_{0.3\,nm}$ + $Pt_{0.7\,nm}$]$_5$ / $Pt_{2.9\,nm}$ multilayer stack is magnetron sputtered at room temperature in the area of the gold contacts. Fig. 2 shows the Co/Pt multilayer as grey circle with detached material at the border, caused by lift-off structuring. The magnetic properties of the layer stack are defined by the composition and the sputtering parameters. In the final processing step, the 250 nm by 250 nm nanomagnet is defined. Therefore, the perpendicular anisotropy of the entire surrounding Co/Pt multilayer is destructed by FIB irradiation at a dose of $1 \cdot 10^{14}$ ions/cm².

B. Measurement setup

The SEM image in Fig. 2 shows a magnetic nanodot with perpendicular anisotropy located between the current contacts surrounded by soft-magnetic material, whose electric conductivity is preserved after FIB irradiation. The current I_{in} is applied to the sensing area via the upper current contact. There, the extraordinary Hall-Effect causes a deflection of the charge carriers and thereby induces a differential current $\Delta I = I_{out1} - I_{out2}$ in the separated bottom current contacts. This current difference is converted in a voltage difference $\Delta V = V_{out1} - V_{out2}$ by the balancing resistors R_{A1} and R_{A2}. Afterwards ΔV is processed by a lock-in amplifier and is ready for digitization as Hall voltage V_{Hall} at the output of the circuit.

The balancing resistors are utilized to compensate fabrication asymmetries and are adjusted before beginning the

measurements so that there is no voltage difference detectable at the output. An external electromagnet, with a maximum magnetic flux density of ±350 mT perpendicular to the Co/Pt multilayer, is used to magnetize the probe. For temperature dependent experiments, a cooler/heater unit providing a temperature range from 0 °C to 100 °C is implemented in the measurement setup.

In experiments before the definition of the nanodot in the final processing step, that means by utilization of the entire magnetic sensing area of 1 μm by 1 μm, the structure shows an effect in the magnitude of 1 ‰. In detail, by applying a current I_{in} of 100 μA a Hall current of 100 nA is measured. This measurement technique with low voltage noise preamplifier and lock-in amplification allows very narrow-band and therewith low-noise measurements. Analysis of the noise behavior of the sensing structure together with the external circuitry showed a noise voltage of 7 nV/$\sqrt{Hz}$ at a frequency of 1 kHz, which is small enough for detection of the signals from the ultra-low volume Co/Pt dot.

Two different methods are used to probe the samples. On the one hand, a continuous procedure is utilized, where the Hall voltage is measured during an applied external magnetic field sweep. Then, the large FIB irradiated soft-magnetic areas of the Co/Pt multilayer generate a Hall signal, which exceeds the signal of the ferromagnetic nanodot especially at high external magnetic fields. On the other hand in remanence measurements, the magnetization and probing are performed alternatingly in separate time slots. Thus, only the ferromagnetic part of the signal is recorded, the soft-magnetic background is suppressed. In this technique, if the preamplifier input is additionally detached from the sensing structure during the magnetization time slot, the signal generated by the soft-magnetic background cannot saturate the amplifier. This 'input gating' guaranties the highest sensitivity of the measurement setup for probing the Co/Pt nanodot.

III. DISCUSSION

A. Characterization of films

In this section, the characteristic properties of the Co/Pt multilayer stack are investigated with regard to FIB irradiation and temperature dependencies.

1) Focused ion beam modification: By FIB irradiation the magnetic properties of the Co/Pt multilayer, especially the coercivity, can be modified with high accuracy. The bombardment by heavy Ga^+-ions results in atomic displacement in the magnetic layer and atomic blurring occurs in the vicinity of the Co/Pt interfaces, even at low ion doses [7]. Simulations and measurements in this subsection are accomplished at room temperature.

Micro-magnetic simulations with the Object Oriented Micro Magnetic Framework (OOMMF) [8] are calibrated to the used magnetic multilayer as described in [9], which yields an anisotropy constant K_{sat} of $3.25 \cdot 10^5$ J/m³. The Co/Pt multilayer is modeled as a 2-D corresponding magnetic monolayer. If the anisotropy constant K_{var} gets smaller than K_{sat}, the simulation exhibits a decrease in the switching field of the

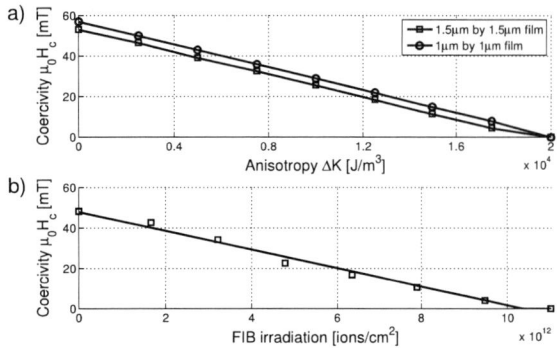

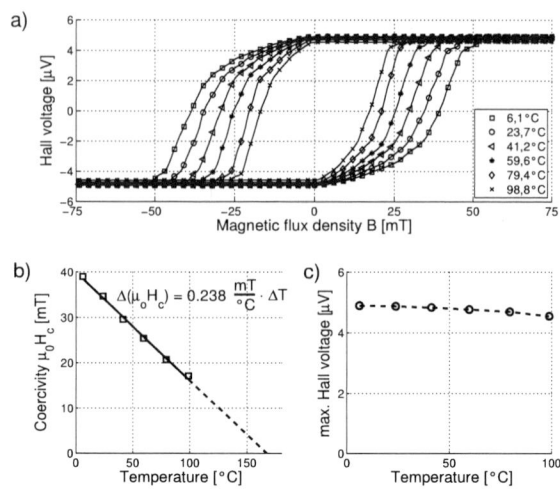

Figure 3. a) Simulation result: The coercivity of two different sized films plotted over the anisotropy variation. b) Magneto optical Kerr-Effect (MOKE) measurement: The coercivity plotted over areal FIB irradiation dose.

Figure 4. a) Hysteresis curves of the Hall voltage plotted over the magnetic flux density at different temperatures. b) Temperature dependence of the coercivity and c) of the maximum Hall voltage, representing the saturation magnetisation.

magnetic layer. Fig. 3a) shows the linear relation between the coercivity and the change in the anisotropy constant $\Delta K = K_{sat} - K_{var}$ for two different sized Co/Pt films. The lager film features a lower coercivity due to stronger magnetostatic force in the nucleation center.

Experimentally, a Co/Pt film is repeatedly irradiated by Ga^+-ions with an acceleration of $50\,kV$ at a dose of $1.6 \cdot 10^{12}$ ions/cm². The polar magneto-optical Kerr-Effect (MOKE) is used for coercivity measurements between the irradiation steps. The Co/Pt multilayer also shows a linear relation between the coercivity and the FIB irradiation dose (Fig. 3b). The coercivity in the starting point differs from the simulation, because in measurement, it depends on the magnetic field sweep rate, which is taken into account in modeling by the Sharrock formalism [10].

The congruence of the simulation and the measurement enables the linkage between the change in the anisotropy ΔK in the simulation and the areal FIB irradiation dose in the experiment. Switching field modification by slight FIB irradiation is essential in many applications that use dots with predefined different switching fields. For example, the programmable dot P in Fig. 1 requires higher switching field than all other nanodots in this design. Now it is possible to model these varied dots in the simulations and predict the behavior of field-coupled magnetic gates and circuits prior to experiments.

2) Temperature dependence: Temperature is a limiting operating parameter for magnetic as well as for charge-based logic. For first temperature explorations, the sensing structure with the magnetic multilayer sputtered on top is used without defining a nanodot by FIB irradiation. The active area between the gold contacts, where the Hall current is generated, is $1\,\mu m$ by $1\,\mu m$. Continuous and remanence measurements display the same hysteresis curve, indicating that the magnetic multilayer only consists of hard-magnetic material.

The as-grown film shows a square hysteresis loop with abrupt switching in the MOKE measurement. The switching signal of the Co/Pt multilayer in the sensing structure, gen-

erated by the extraordinary Hall-Effect, is presented in Fig. 4a). The slower rise of the Hall voltage during the switching process can be ascribed to domain wall pinning and depinning at the intersection of the Co/Pt film and the gold contacts.

On closer examination, the coercivity shows a linear dependence on temperature (Fig. 4b). To change the direction of the magnetization, the sum of the energy contributions from the external magnetic field and the temperature must be higher than the energy barrier of the magnetic multilayer. At elevated temperatures, the Co/Pt film switches already at lower magnetic fields [11]. Raising the temperature by $50\,°C$ corresponds to a reduction of the switching field $\Delta(\mu_0 H_c)$ of $11.9\,mT$.

Fig. 4c) shows the decreasing max. Hall voltage, representing the perpendicular saturation magnetization over increasing temperature. The Hall voltage is reduced because the sensing structure only measures out-of-plain magnetization by the extraordinary Hall-Effect and the magnetization vector of the film includs a partial statistical in-plain component at elevated temperature. As presented in [12] the decrease of perpendicular magnetization rises at higher temperatures.

Extrapolating the curve in Fig. 4b), the explored Co/Pt multilayer shows hysteresis behavior up to a temperature of $165\,°C$. Depending on the necessary coercivity to detect the binary state of a nanodot, the material system is applicable within this temperature range.

B. Characterization of a single nanodot

In this section, a $250\,nm$ by $250\,nm$ Co/Pt nanodot is investigated with regard to temperature stability and readout ability of its two binary states by the electrical sensing structure.

978-1-4244-6658-0/10 $26.00 © 2010 IEEE

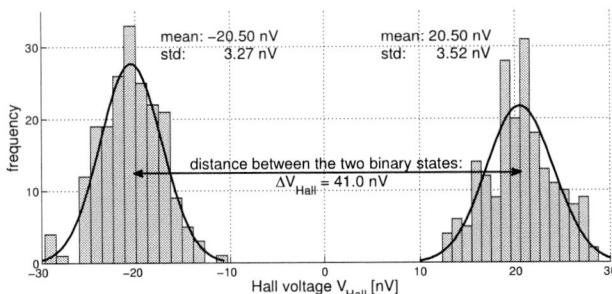

Figure 5. The perpendicular saturation magnetization of a 250 nm by 250 nm nanomagnet is reversed and electrically sensed 400 times at room temperature. The histogram shows the number of measured values within a defined Hall voltage interval resulting in a considerably distance of 41 nV between the two binary states.

Magnetic Force Microscopy (MFM) on the as-grown Co/Pt film shows a natural domain size of about 600 nm after demagnetization in a rotating magnetic field. Because the size of the nanodot is obviously smaller than the natural domain size, the magnetic dot shows single domain behavior with abrupt switching. Domain wall pinning and depinning at the Co/Pt-Au intersection can be neglected after FIB irradiation of the surrounding film of the nanodot due to the soft-magnetic behavior in this area. Considering the beam diameter of the FIB in the layer stack to be 15 nm, the edge length of this dot is reduced to 235 nm.

The nanodot is alternatingly saturated to positive and negative magnetization whereupon in each cycle the Hall voltage, generated by the dot located in the sensing structure, is measured in remanence. Afterwards, a linear long-term drift is eliminated from the 400 measured values.

The frequency of the measurement data is plotted over the saturation Hall voltage in the histogram in Fig. 5. At room temperature, the distance between the mean values of the two binary states ΔV_{Hall} is 41.0 nV. The distributions of the measurements in both states fit well to a Gaussian distribution, depicted as black curve in Fig. 5. The distinct separation of the distribution functions of the two magnetic states demonstrates the stability and read-out ability of this ultra-low volume magnetic nanodot. Measurements are performed from 5 °C to 80 °C. At elevated temperatures, the Hall voltage V_{Hall} decreases, but the binary states of the nanomagnet can be clearly separated at 80 °C by a ΔV_{Hall} of 31.4 nV.

The measurements are repeated on a 300 nm by 300 nm magnetic dot, consisting of the same Co/Pt film. At room temperature, the distance between the binary states ΔV_{Hall} is 58.7 nV. This increase in ΔV_{Hall} can be attributed to the difference in area. The modeled ΔV_{Hall} for a 300 nm square dot is 59.0 nV and fits well to the measurement data.

Scaling nanomagnets is possible as long as the thermal effects in the dot are smaller than its magnetic switching energy. OOMMF simulations show that the ultimate scalability for a stable nanodot is limited to an edge length of approximately 50 nm.

IV. Conclusion

Co/Pt films and ultra-low volume ferromagnetic nanodots are fabricated and characterized with regard to applications in field-coupled magnetic computing devices. The decrease of coercivity by focused ion beam irradiation on films is linked with a reduced anisotropy constant in OOMMF simulations, which enables meaningful predictions of the behavior of field-coupled magnetic logic. Temperature experiments indicate a measurable hysteresis curve of a small Co/Pt film up to 165 °C defining a limiting operating parameter. We presented our smallest single-domain ferromagnetic nanodot at target size with an edge length of 250 nm and proved its electrical read-out ability by the extraordinary Hall-Effect at temperatures up to 80 °C. These results expose the ferromagnetic Co/Pt nanodots as promising fundamental elements and the sensing structure as electrical output for magnetic field-coupled computing devices.

Acknowledgment

The authors would like to thank S. Boche and W. Kraus for valuable input. We also thank H. Mulatz and M. Ganesh for assistance in sample fabrication and the DFG (Grant SCHM 1478/9-1) for financial support.

References

[1] *The International Technology Roadmap for Semiconductors: Emerging Research Devices*, 2009th ed., 2009.

[2] G. Csaba, W. Porod, and A. I. Csurgay, "A computing architecture composed of field-coupled single-domain nanomagnets clocked by magnetic fields," *International Journal of Circuit Theory and Applications*, vol. 31, pp. 67–82, 2003.

[3] A. Imre, G. Csaba, L. Ji, A. Orlov, G. Bernstein, and W. Porod, "Majority Logic Gate for Magnetic Quantum-Dot Cellular Automata," *Science*, vol. 311, pp. 205–208, January 2006.

[4] M. Becherer, G. Csaba, R. Emling, W. Porod, P. Lugli, and D. Schmitt-Landsiedel, "Field-coupled Nanomagnets for Interconnect-Free Non-volatile Computing," in *Digest Technical Papers IEEE International Solid-State Circuits Conference, ISSCC*, 2009, pp. 474–475.

[5] M. Becherer, J. Kiermaier, G. Csaba, J. Rezgani, C. Yilmaz, P. Osswald, P. Lugli, and D. Schmitt-Landsiedel, "Characterizing magnetic field-coupled computing devices by the Extraordinary Hall-Effect," in *IEEE Proceedings of the 39th ESSDERC*, September 2009, pp. 105–108.

[6] M. Becherer, G. Csaba, R. Emling, P. Osswald, W. Porod, P. Lugli, and D. Schmitt-Landsiedel, "Extraordinary Hall-effect sensor in split-current design for readout of magnetic field-coupled logic devices," in *2nd IEEE International Nanoelectronics Conference, INEC08*, Shanghai, 2008.

[7] C. Vieu, J. Gierak, H. Launois, T. Aign, P. Meyer, J. P. Jamet, J. Ferré, C. Chappert, T. Devolder, V. Mathet, and H. Bernas, "Modifications of magnetic properties of Pt/Co/Pt thin layers by focused gallium ion beam irradiation," *Journal of Applied Physics*, vol. 91, no. 5, pp. 3103–3110, March 2002.

[8] http://math.nist.gov/oommf/ (URL).

[9] M. Becherer, G. Csaba, W. Porod, R. Emling, P. Lugli, and D. Schmitt-Landsiedel, "Magnetic Ordering of Focused-Ion-Beam Structured Cobalt-Platinum Dots for Field-Coupled Computing," *IEEE Transactions on Nanotechnology*, vol. 7, no. 3, pp. 316–320, 2008.

[10] M. P. Sharrock, "Time dependence of switching fields in magnetic recording media," *Journal of Applied Physics*, vol. 76, pp. 6413–6418, 1994.

[11] R. Murillo, M. H. Siekman, T. Bolhuis, L. Abelmann, and J. C. Lodder, "Thermal stability and switching field distribution of CoNi/Pt patterned media," *Microsystem Technologies*, vol. 13, pp. 177–180, 2007.

[12] Y. Xiao, J.-H. Xu, and K. V. Rao, "Study of magnetic reorientation phenomenon and magnetic properties of Pd/(Pt/Co/Pt) multilayers," *Journal of Applied Physics*, vol. 79, no. 8, pp. 6267–6269, April 1996.

978-1-4244-6658-0/10 $26.00 © 2010 IEEE

The Curie Temperature as a Key Design Parameter of Ferroelectric Field Effect Transistors

Giovanni A. Salvatore[†], Livio Lattanzio, Didier Bouvet, Adrian M. Ionescu

Nanolab, Ecole Polytechnique Fédérale de Lausanne, Switzerland

[†] giovanni.salvatore@epfl.ch

Abstract- **Interest in Ferroelectric FETs originates from their potential as non-volatile memories and, more recently, as abrupt switches. In this work, we focus on the role of the Curie temperature of the gate stack on the performance of Fe-FETs. The proposed study is based on thin film SOI Fe-FETs using organic ferroelectric gate stacks (45nm P(VDF-TrFE) 70%-30% layer on top of 10nm thermal SiO$_2$). The device static characteristics are investigated from 25 °C up to 155°C, from ferroelectric to paraelectric phase. We report a maximum of the transconductance at the Curie Temperature (T$_c$), where the ferroelectric material loses its spontaneous polarization and the transistor loses its hysteresis. The reported increase of current and transconductance with the temperature up to T$_c$ are in contrast with the behavior in any conventional MOSFET. The experimental results are supported by an appropriate analytical model. We suggest that the Curie Temperature of the gate stack is a key design parameter for the use of Fe-FETs as non-hysteretic switches with increased on-current, close or beyond Tc, or as memory cells, in the hysteretic domain, much below Tc.**

I. INTRODUCTION

Ferroelectrics were first discovered in 1920 by Valasek and, at their early age, they were rather considered as academic curiosities of little application. The discovery of the "robust" BaTiO$_3$ in 1943 [1] was a turnaround and ferroelectric oxides became to be widely used as capacitors in electronic industry. Until the 80's, the main challenges in ferroelectric materials were the modeling of the phase transition and the discovery of novel materials, followed shortly by the integration of ferroelectric thin films on silicon ICs [2]. Since then ferroelectrics have been widely used in electronics for different applications ranging from mobile phone applications to memories. Recently, ferroelectric Field Effect Transistors (Fe-FETs) have attracted great attention from the research community for their promise for both memory and switching applications. One transistor (1T) ferroelectric memory cells have been proposed and experimentally studied [3-5] in order to reduce the size of 1T-1C design with consequent advantages in terms of size, read-out operation and costs. Even more recently, ferroelectrics have been proposed by Salahuddin and Datta [6] as dielectric materials in order to lower the subthreshold slope (SS) and overcome the theoretical limit of the 60mV/dec of the MOS transistor; however, this theory is still under scientific debate. The Salahuddin's assertion is theoretically supported by the Landau-Ginzburg (LG) theory that provides a macroscopic model of the physics of ferroelectrics. The LG description shows that the temperature plays a key role in defining the electrical properties of such materials and in particular the Curie Temperature (T$_c$) is of crucial importance for the proper workability of any ferroelectric transistor.

In this work, we propose a study of ferroelectric FETs static electrical characteristics over a range of temperature going from 25°C to 155°. We particularly focus on the hysteretic non-saturated loops (resulting from small V$_g$ sweeps). As it will be shown later, our fabricated SOI FeFET device shows a memory behavior with stable hysteretic loops up to 100°C. The Fe-FET transconductance shows a unique dependence on temperature that cannot be explained by the threshold voltage shift and by the mobility degradation and needs the introduction of a model in ferroelectric and paraelectric phases based on the Curie temperature. In total contrast with a standard MOSFET transistor, this paper shows that the transconductance improves when increasing the temperature, eventually reaching a maximum near the Curie Temperature of the gate stack and degradates back for higher temperatures. This apparently anomalous behavior could be explained by the dielectric anomaly occurring in the ferroelectric material at T$_c$ modeled by the Landau-Ginzburg theory.

II. MODELLING THE CURRENT AND THE TRANSCONDUCTANCE VERSUS TEMPERATURE IN FE-FET

The dielectric response of a ferroelectric material as function of the temperature can be described by the LG theory [7,8] . According to this approach, the first two terms of the expansion of the Helmholtz free energy F with respect to the macroscopic polarization P read as follows:

$$F = \frac{\alpha}{2} P^2 + \frac{\beta}{4} P^4 \qquad (1)$$

The equation of state dF/dP=E gives a relation between the polarization and electric field E:

$$E = \alpha P + \beta P^3 \qquad (2)$$

Hence, the relative dielectric permittivity of the ferroelectric material can be represented as:

$$\varepsilon = \frac{1}{\varepsilon_0} \frac{\partial P}{\partial E} = \frac{1}{\varepsilon_0} \frac{1}{\alpha + 3\beta P^2} \qquad (3)$$

where ε_0 is the dielectric constant of vacuum. In the Landau theory, the coefficient α is a linear function of temperature vanishing at the Curie-Weiss temperature T$_c$:

$$\alpha = \frac{1}{\varepsilon_0} \frac{T - T_c}{C_{CW}} \qquad (4)$$

where C_{CW} is the Curie-Weiss constant. The sign of the parameter λ is positive in the paraelectric phase and negative in the ferroelectric phase. So one could derive a very simple temperature dependence of the dielectric permittivity, supposed valid for low electric fields:

978-1-4244-6658-0/10 $26.00 © 2010 IEEE

$$\varepsilon_{Fe} = \frac{1}{\varepsilon_0 \alpha} = \lambda \frac{C_{CW}}{T - T_c} \quad with \quad \begin{cases} \lambda = -\frac{1}{2} & T < T_c \\ \lambda = 1 & T > T_c \end{cases} \quad (5)$$

Considering the ferroelectric layer in a physical parallel plate capacitor (as in the gate dielectric stack), the capacitance per unit area is:

$$C_{Ferro} = \frac{\varepsilon_0}{d} \left(\lambda \frac{C_{CW}}{T - T_c} \right) \quad (6)$$

where d is the thickness of the ferroelectric layer.

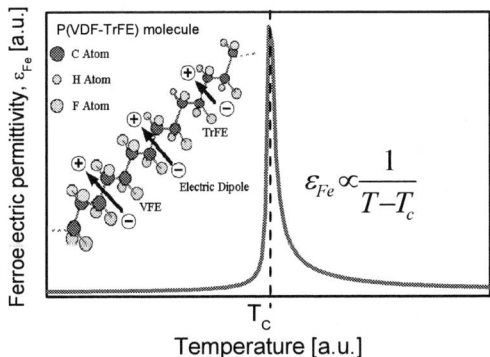

Figure 1- Permittivity of a ferroelectric material as function of the temperature. In a real material, ε_r reaches a large value at Tc, but it does not diverge. The inset shows the molecule of P(VDF-TrFE) used in our work and the orientation of the electric dipoles (black arrows)

It is worth noting that when approaching the transition temperature T_c, the ferroelectric material phase changes from ferroelectric to paraelectric, with a consequent shrinking of the polarization loop and an increase of its permittivity (Figure 3). In this study we are particularly interested in revealing by experiment and modeling how this temperature dependence of the permittivity influences the performance of a ferroelectric transistor. In standard silicon MOSFETs, temperature change mainly impacts the threshold voltage, V_{th}, the carrier mobility in the channel, μ, and the junction leakage, degrading the drain current, I_d, the transconductance, $g_m = dI_d/dV_g$ [9]. The expression of the current of a standard MOSFET transistor in strong inversion is:

$$I_d = \frac{W}{L} \mu(T) C_{ox} \left[\left(V_g - V_{th} \right) V_d - \frac{1}{2} V_d^2 \right] \quad (7)$$

where C_{ox} is the dielectric gate capacitance, Vth is the threshold voltage, Vd is the drain voltage, W/L is the channel width-length ratio and μ is the mobility. If we now consider a ferroelectric transistor such capacitance has to be replaced by:

$$C_{Fe-MOS}(T) = \frac{C_{OX} C_{CW} \lambda \varepsilon_0}{C_{CW} \lambda \varepsilon_0 + d C_{OX}(T - T_c)} \quad (8)$$

that is just the series of C_{ox} and the ferroelectric capacitance (6). Equation (8) shows that in a ferroelectric transistor the gate capacitance is not temperature-independent because of the ferroelectric material contribution. Moreover it is

important to notice that equation (8) suggests the gate capacitance has a maximum at T=Tc, where the C_{Ferro} ideally diverges. Replacing the (8) into the (7) we find out:

$$I_{d,Fe-MOS} = \frac{W}{L} \mu(T) C_{Fe-MOS}(T) \left[\left(V_g - V_{th} \right) V_d - \frac{1}{2} V_d^2 \right] \quad (9)$$

and consequently the transconductance g_m becomes:

$$g_{m,Fe-MOS} = \frac{W}{L} \frac{\mu_0}{\theta_1 (V_g - V_{th})} \left(\frac{T}{300K} \right)^\gamma C_{Fe-MOS}(T) V_d \quad (10)$$

where we have also expressed the mobility dependence versus the temperature and the transversal field. Equation (9) and (10) highlight that the g_m-T degradation due to the mobility reduction can be compensated or dominated by the increase of the C_{Fe-MOS} up to the Curie temperature. Anticipating the experimental results, if the capacitance increase dominates, an exceptional improvement of the transconductance is possible with the temperature, with a maximum near T_c. Consequently, the design of the gate stack capacitance can be optimized taking into account that the Curie Temperature of a stack changes according to its composition:

$$T_c^* = T_c - q \frac{C_{CW}}{\varepsilon_d} \quad (11)$$

where q and ε_d are the volume concentration and the permittivity of the linear dielectric respectively [10]. This effect, originating from the incomplete compensation of the spontaneous polarization, reduces the polarization in such a way that the ferroelectric/dielectric sandwich behaves like a ferroelectric material with a lower transition temperature.

III. DEVICE FABRICATION AND CHARCTERIZATION

A. Fabrication

In order to validate the Curie Temperature based behavior and model previous described, we have fabricated MOSFET transistors on SOI substrates, implementing a gate dielectric stack made of 45nm of P(VDF-TrFE) 70%-30% and 10nm of SiO$_2$. In Figure 2 shows the basic structure of the device.

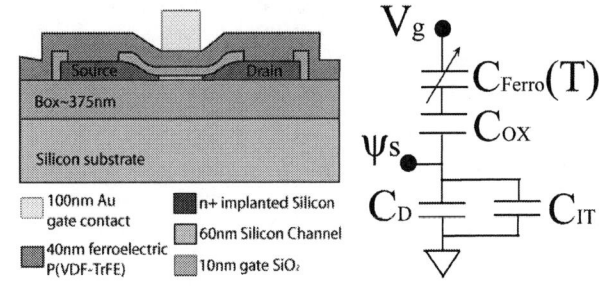

Figure 2- Cross section along the channel of the designed and fabricated ferroelectric P(VDF-TrFE) FET on SOI. The schematic shows the equivalent capacitive model with C_{IT} that takes into account the trap charges at the silicon/oxide interface.

The initial silicon thickness of 100nm is thinned down to 60nm, in order to obtain a recessed-gate structure for a fully-depleted SOI film (p-type doping of $N_A = 1 \times 10^{17}$ cm^{-3}). The thinning is achieved through a LOCOS technique using a 15nm SiO$_2$ hard mask and 150nm of LPCVD Si$_3$N$_4$. The devices are then laterally isolated through a MESA isolation

978-1-4244-6658-0/10 $26.00 © 2010 IEEE   139

and 10nm of silicon oxide are grown for the gate stack. Source and drain contacts are implanted with phosphorous using a photoresist implantation mask. After dopant activation, contact windows are etched with buffered-HF in the 10nm oxide layer to liberate source and drain pads. A 40nm layer of P(VDF-TrFE) 70%-30% is then spun over the whole wafer and annealed at 135°C for 10 minutes. A 100nm layer of gold is finally deposited on the polymer by means of e-beam evaporation, patterned and etched in a KI + I_2 bath to form the gate electrode. Figure 3 shows an AFM image of a fabricated device where the 3D profile highlights the roughness of the coated polymer.

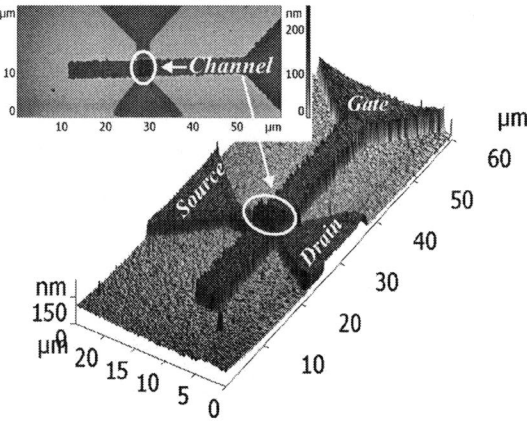

Figure 3- AFM image of a fabricated device showing the drain/source regions, the gate and the channel. The roughness of the spin-coated polymer is also visible in the 3D version.

Note that no metallization has been performed for source and drain contacts. The Fe-FET devices are characterized through the PVDF layer by direct probing; the prober needles are used to scratch out the polymer on source and drain and achieve electrical contacts directly on the implanted silicon pads. A key reason to choose an SOI substrate in our fabrication is the excellent lateral and vertical (from bulk) isolation of the devices, which limits the influence of the leakage currents at high temperatures in the proposed study. The whole characterization that follows has been performed on a Fe-FET geometry with L=10μm, W=20μm.

B. Characterization

We have measured the Id-Vg of the device from 25°C up to 155°C. In order to avoid any contribution from the back gate channel, we set it in accumulation by applying a negative voltage on the back side of the wafer (Vb). Before heating up the device, we measured the *Id-Vb* and the *Id-Vd* curves and verified that no hysteresis was present. This confirmed that the memory effect is exclusively due to the ferroelectric layer and visible only in the front channel conduction. The gate voltage sweep has been limited to low value (-1V<Vg<3V) in order to limit the gate leakages also at high temperatures and in order to fulfill the low electric fields condition under which equation (5) is valid. Figure 4a shows the *Id-Vg* at four different temperatures. The leakage currents is about 100fA at ambient temperature and increases up to 2pA at 155°C. The memory window (Mw) is about 300mV and reaches the zero

at 100°C; the Ion/Ioff ratio (calculated at Vg=1V) is about 50 and decreases to 1 at about 100°C. This analysis demonstrates that our device can be used as memory up to 100°C, where the loop completely closes and the Curie temperature is reached. In Figure 4b we show the threshold voltage shift for the sweeping up branch (*Vth_up* in the plot) of the Id-Vg and for the sweeping down one (*Vth_down* in the plot). The threshold has been calculated with the constant current method for a Id=I_{th}=10nA. The difference between the two thresholds (up and down) is the memory window. Figure 2b interestingly highlights that *Vth_up* decreases almost linearly as for a standard MOSFET, while the *Vth_down* has a non linear behavior below *Tc* (ferroelectric phase) because of the polarization and above *Tc* (paraelectric phase) it becomes linear, i.e. like in a standard MOSFET, because no polarization is anymore present. Moreover, this plot confirms that our gate stack has a Curie temperature: Tc=100°C±5°C.

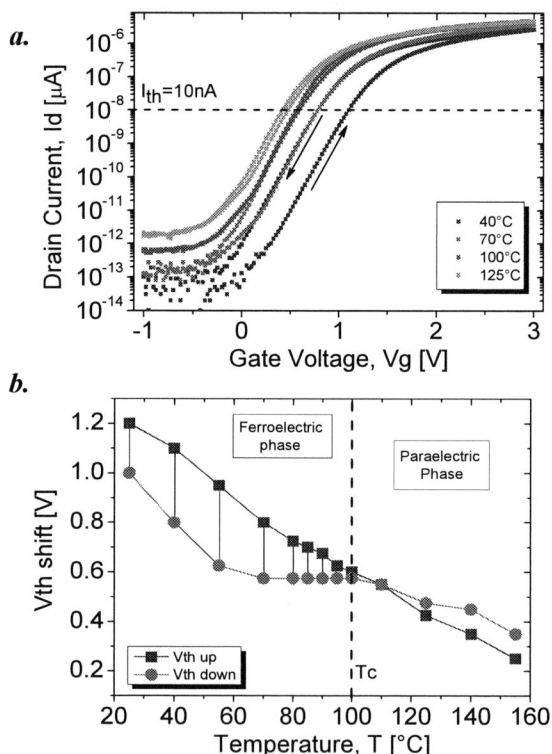

Figure 4- a. Id-Vg measurements at different temperatures. The memory loops closes near T=100°C; *b.* analysis of the threshold shift for both branches of the Id-Vg curve. The memory window (Mw) is equal to the difference between *Vth up* and *Vth down*. The Curie temperature of our gate stack is 100°C±5°C.

As mentioned in the introduction and Section II, the goal of the experiment is to prove prove that the performances, in terms of current and transconductance, of a ferroelectric transistor in strong inversion improve when approaching the Curie temperature. From the model presented in Section II analysis, we expect a current and transcondctance boosting by the temperature improvement up to 100°C, followed by a degradation due to the combined effect of the ferroelectric gate decrease and the mobility reduction. Figure 5 shows the Id-Vg (sweeping up branch in Fig. 5a) and the

978-1-4244-6658-0/10 $26.00 © 2010 IEEE

transconductance a five different temperatures. It is worth noticing that the curves are plotted versus *Vg-Vth* with the threshold voltage extracted at each temperature with the constant current method. This correction is important because in this way we eliminate the influence of the threshold shift induced by the temperature. The inset of figure 5a shows that the drain to source current increases with the temperature till 100°C and then decreases back for higher temperatures (the black arrow in the inset gives the trend of the Id). This behavior is confirmed by the transconductance in figure 5b. The plot clearly shows that the g_m-Vg improves till Tc and then it degrades again. Therefore, the experimental data appear to fully confirm the theoretical analysis developed in Section II. Of course, equation (7) applies to a bulk MOSFET transistor but it is here considered as acceptable estimation for the front channel conduction of the SOI MOSFET when the back channel is accumulated.

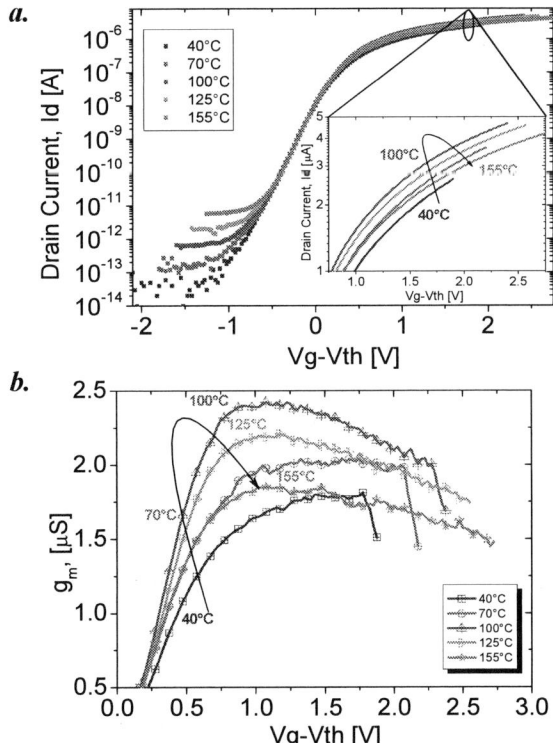

*Figure 5- **a.*** Temperature dependence of the Id current measured for a L=10μ, W=20μm ferroelectric transistor. The current improves up to Tc=100°C and then decreases for higher temperatures; ***b.*** the transconductance trend at different temperatures confirm that there is effectively an improvement till 100°C.

Finally, in Figure 6 we have calculated the transconductance for two different values of Vg-Vth: 1 and 2V, and plotted the improvement, in term of percentage with respect to the ambient temperature value, for all the investigated temperatures. The plot shows clearly that at T=100°C there is an improvement of almost 50% for Vg-Vth=1V and an improvement of 30% for Vg-Vth=2V, which also suggests that the improvement could be dependent on the transverse field. The result reported in Figure 6 greatly

demonstrates that the gate stack capacitance improvement dominates over the mobility degradation from room temperature till the Curie temperature (~100°C, in our case).

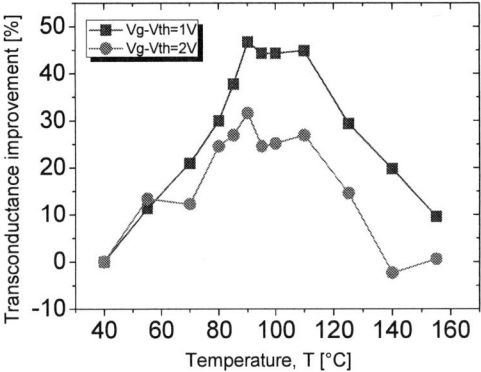

Figure 6- Transconductance improvement (with respect to the ambient temperature value) calculated for two different values of *Vg-Vth* plotted as function of the temperature. The curve for Vg-Vth=1 almost corresponds to the $g_{m,max}$ as visible from Figure 5b. The plot shows that, in the best case, there is an improvement of g_m at Tc of almost 50% with respect to its value at ambient temperature.

IV. CONCLUSION

We have reported on the importance of the Curie temperature for the design of ferroelectric transistors based on the experimental investigation of a SOI Fe-FET with 45nm P(VDF-TrFE) 70%-30% layer on top of 10nm thermal SiO₂ in the gate stack. We theoretically explained based on Landau-Ginzburg theory and experimentally demonstrated (based on measurements conducted from room temperature to 155°C) that a MOSFET with a ferroelectric gate stack could have non-degraded or even improved current and transconductance at high temperature. As a consequence, we suggest that in ferroelectric FETs with appropriately designed Curie temperatures, the performance degradation of logic circuits and/or memories operating near 100°C could be alleviated or even avoided.

ACKNOWLEDGMENT

The technical staff of the Centre of Micro-Nanotechnology (CMI) is thanked for technical support. A particular acknowledgement goes to Michel Schaer and Sara Rigante for AFM images.

REFERENCES

[1] A. von Hippel, "U.S. Nat. Def. Res. Com. Report 300" (NDRC, 1944).

[2] J. F. Scott, C. A. Araujo, *Science* 246, 1400 (1989).

[3] R. C. G. Naber et al, Nature Material, v. 4 n.3 (2005).

[4] G. Salvatore et al., Prof. of ESSDERC 2008, pp. 162-165, (2008).

[5] S. Ducharme et al., IEEE Trans on Dev. And Mater. Reliability, V. 5, No. 4, (2005).

[6] S. Salahuddin, S. Datta, Nano Letters, Vol. 8, pp. 405-410, (2008).

[7] Ginzburg V. L., *Physics-Uspekhi* 44 (10), pp. 1037-1043 (2001).

[8] Tagantsev A.K. et al., *Journal of Electroceramics* 11, pp. 5-66 (003).

[9] Reichert G. et al. *Solid-State Electronics* 39 (9), pp. 1347-1352 (1996).

[10] V. Sherman et Al. *J. Appl. Phys.* 99, 074104 (2006).

Performance trade-offs in polysilicon source-gated transistors

R. A. Sporea, J. M. Shannon, S. R. P. Silva
Advanced Technology Institute
University of Surrey
Guildford, Surrey, GU2 7XH, U.K
r.sporea@surrey.ac.uk

M. J. Trainor
MiPlaza, Philips Research
High Tech Campus 4
5656 AE Eindhoven, The Netherlands
mike.trainor@philips.com

N. D. Young
Philips Research
101 Cambridge Science Park, Milton Road
Cambridge CB4 0FY, U.K
nigel.young@philips.com

Abstract—Self-aligned Schottky-source source-gated transistors (SGTs) have been made in polysilicon. The structures enable a direct comparison to be made between a SGT and a standard thin-film field-effect transistor (FET) on the same device. SGTs having excellent characteristics have been fabricated, with intrinsic gains approaching 10,000. The effects of bulk doping in the polysilicon and of the source barrier modification implant are considered in the context of the electrical output characteristics. It is shown that the choice of source length is a tradeoff between device speed and current uniformity.

I. INTRODUCTION

Source-gated transistors (SGTs) operate using the field dependence of the current through a reverse biased source barrier. As such, their characteristics and properties are very different from those of thin-film field effect transistors (FETs). In this paper we report measurements and modeling on SGTs in polysilicon and show the well known advantages of the SGT compared to standard FETs, namely high output impedance and low saturation voltage [1, 2]. The kink effect [3] due to bipolar amplification of carriers generated in the high field regions around the drain is absent, postulated as a result of minority carrier extraction by the reverse biased source barrier.

In this paper, special attention is given to how the electrical characteristics and intrinsic gain depend on the doping of the polysilicon and on the barrier modification implant. There is some consideration of how source length affects current density and frequency response. We also introduce the hybrid mode of operation in which the current is influenced by the parasitic FET in series with the source barrier. It is suggested that this mode of operation can explain the very low temperature dependence of the drain current measured in some devices. These polysilicon SGTs are particularly relevant to high performance thin-film analog circuits.

Source-gated transistors were fabricated in polysilicon on a glass substrate using back exposure and ion implantation to align a bottom gate with ohmic drain contacts on either side of a Schottky source barrier. A micrograph of one of the devices is shown in Fig. 1. A 40nm layer of polysilicon was formed on top of 200nm SiN_x plus 200nm SiO_2 using a 308nm XeCl excimer laser. The Schottky source barrier was realised via a window in 120nm of SiO_2 and was modified using low energy implants of either P or BF_2. The chromium was extended over the oxide to form a field plate structure to provide field relief at the edge of the source (Fig. 2).

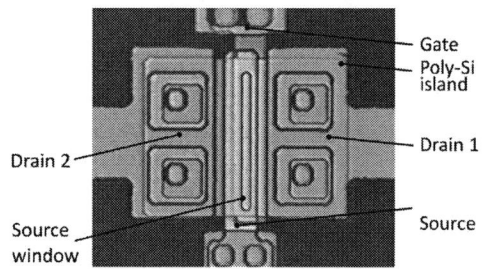

Figure 1. Microphotograph of a self-aligned polysilicon SGT.

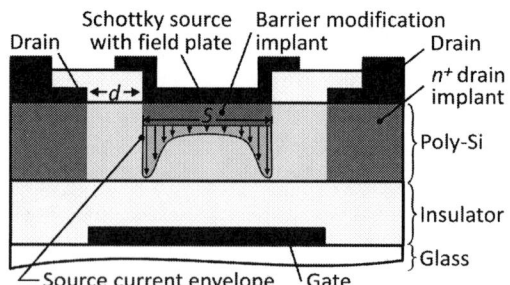

Figure 2. Schematic cross-section of a self-aligned SGT showing current crowding at the edge of the source.

II. RESULTS AND DISCUSSION

A. Effect of bulk doping on SGT and FET characteristics

Since the drain contact of a SGT is forward biased in the on-state, it can be a barrier just like the source. The source and the drain can therefore be reversed to give symmetrical SGT characteristics. However, in the structure made here, the drain contact is ohmic. Therefore, if we make the ohmic contact the source we get an FET and the characteristics of the SGT can be compared directly to those of the FET on the same device. An example is shown in Fig. 3.

The high output impedance and the low saturation voltage of the SGT are apparent, as is the kink effect in the FET. The on-current through the SGT depends on the source barrier height (see Fig 5). It is also seen that the FET does not switch off as well as the SGT. The off-current of the FET increases with n-type doping in the polysilicon (Fig. 4) but we see that the off-current of the SGT is always the same independent of doping. This is because the reverse biased source barrier of the SGT blocks current flow through the polysilicon between source and drain.

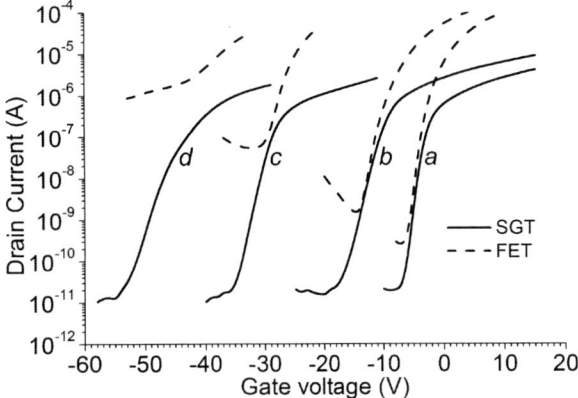

Figure 4. Transfer curves of SGTs and FETs for different substrate dopings. W=50μm, S=8μm, d=10μm, $1 \cdot 10^{13}$ BF$_2$ barrier modification implant; areal doping: a) $0.5 \cdot 10^{12}$; b) $1.5 \cdot 10^{12}$; c) $2.5 \cdot 10^{12}$; d) $3.5 \cdot 10^{12}$ /cm^2 n-type; V$_D$=15V.

B. Effect of barrier modifying implants

Low energy BF$_2$ or P implants into the source window were made before metallization in order to affect the Schottky barrier height. The transfer curves (Fig. 5) are consistent with an increase in barrier height for p-type doping and a decrease for n-type doping, as expected from barrier modification theory [4]. However, the changes in current are small when considering the doses used, which suggests poor electrical activity. This is not surprising, because the annealing temperature was only 500°C to avoid glass compaction. All transistors showed SGT behaviour, even those with the highest n-type implants. As can be seen, the current handling of the SGT can approach that of a FET having the same geometry by tailoring the source barrier. Also, changing the source barrier height has no effect on the off-current; on to off ratios can be more than 6 orders of magnitude.

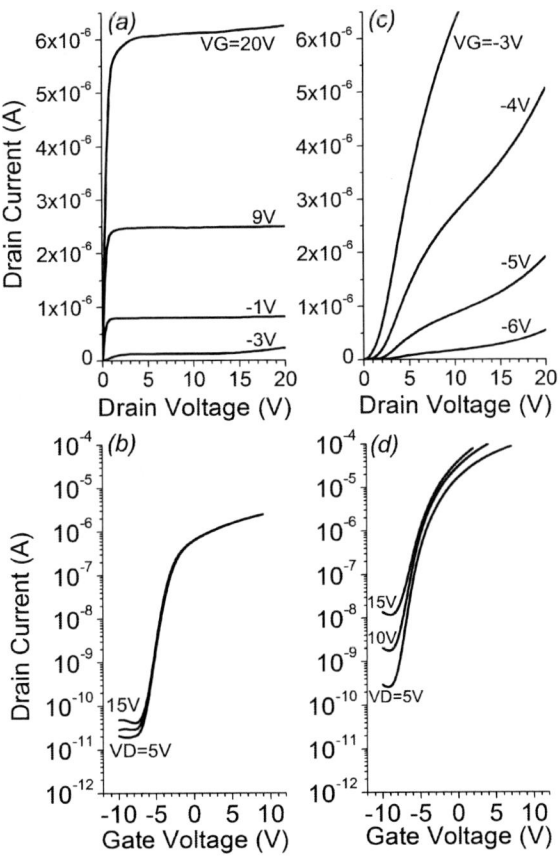

Figure 3. Characteristics of SGT (a), (b) and FET (c), (d) behaviour, at a similar current, obtained by interchanging the source and the drain on the same device. W=50μm, S=8μm, d=10μm, $1 \cdot 10^{13}$/cm^2 BF$_2$ barrier modification implant.

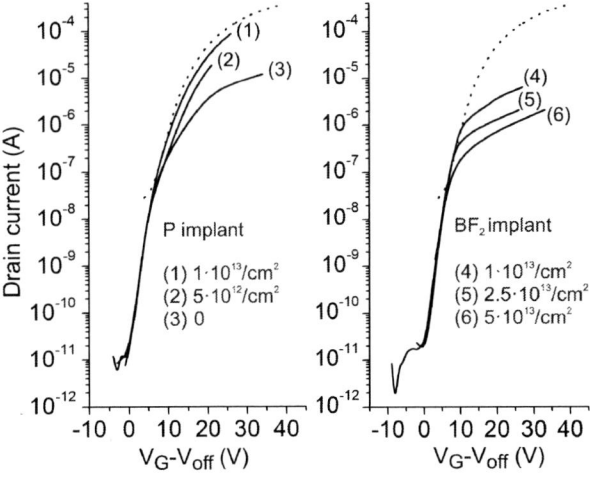

Figure 5. Transfer curves for different barrier implants at V$_D$=5V. Dotted line – FET, continuous line – SGT.

C. Effect of bulk doping and barrier modifying implants on intrinsic gain

The high output impedance of the SGT makes it very useful for analog circuits. A key parameter in this context is the intrinsic gain, g_m/g_d. The high impedance arises because the device first pinches off at the source (at $V_D=V_{SAT1}$), and then at the drain, as in a conventional FET (at $V_D=V_{SAT2}$). [5]

Fig. 6 shows measured intrinsic gains for identical devices but for different n-type doping levels in the polysilicon. Peak gains are very high, approaching 10,000. The peaks in the curves tend to be associated with V_{SAT1} and V_{SAT2} [5] while the falloff in gain at high V_D is related to an increase in g_d due to carrier generation in the high field regions at the drain end of the device.

It is seen that the gain at low V_D around V_{SAT1} increases with decreasing substrate doping levels. This can be explained by an increase in the electric field at the periphery of the source as V_D increases and the depletion layer expands towards the drain. For higher doping levels in the polysilicon, the increase in electric field will be greater, as will the increase in g_d. At higher V_D, however, gains are greatest for higher substrate doping. Since g_m is the same, this effect must be due to carrier generation in the high field regions.

Overall, the gain envelope over V_D is highest for the more lightly doped polysilicon and gain variations with V_D are due to changes in the output conductance.

In Fig. 7, intrinsic gains are shown for two transistors with the same doping level in the polysilicon but with different barrier modification implants. Both are operating at the same current, but gm in (a) is larger than in (b) (see Fig. 5), which explains its higher intrinsic gain. In this case, while the peaks and the dips are due to changes in g_d, the downward shift of (b) relative to (a) results from a change in g_m. For comparison, the intrinsic gain of the FET is ~10.

D. Effect of source length on current density

The change of current with source length (S) has been measured in the past and it has been shown that the

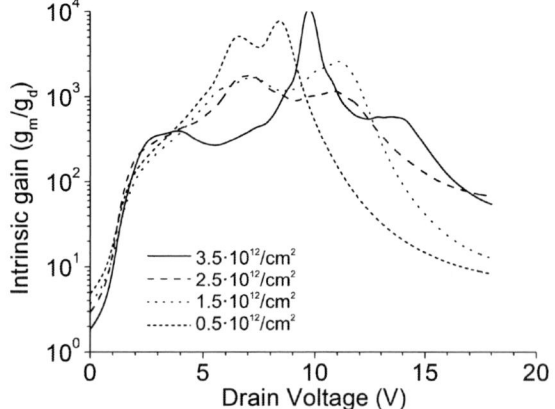

Figure 6. Intrinsic gain as a function of drain voltage for different substrate doping levels.

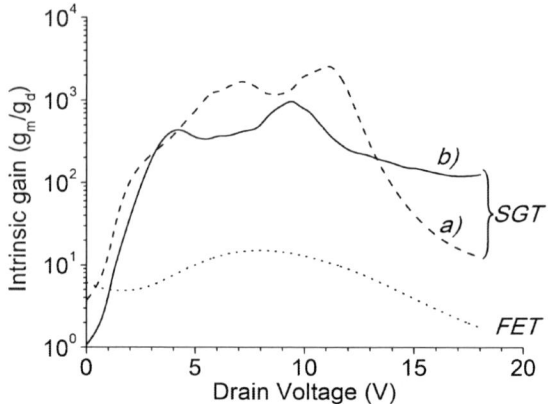

Figure 7. Intrinsic gain for a) $1 \cdot 10^{13}/cm^2$ P and b) $5 \cdot 10^{13}/cm^2$ BF$_2$ measured at 1μA. W=50μm, S=2μm, d=10μm, $0.5 \cdot 10^{12}/cm^2$ n-type bulk doping. Also shown is the gain of an equivalent FET

current is concentrated at the edge of the source as shown in Fig. 2. This has important implications because calculations show [6] that the cutoff frequency (f_T) of the SGT is proportional to the average current density (J_S) passing though the source. In Fig. 8, the results of computer simulations using Silvaco Atlas are plotted for drain current over a range of source lengths. Fig 8(a) shows that, as source length increases, the drain current saturates (S_{SAT}). Therefore, increasing S reduces J_S, since the current becomes almost independent of S. If high f_T is required, the source should be less than S_{SAT}. If current uniformity is required, however, S should be greater than S_{SAT}. The current mismatch between two devices with a change of 0.1μm in source length is shown in Fig. 8(b).

E. Hybrid operation and the temperature dependence of drain current

One parameter that is worse in a Schottky source SGT compared with an FET is the temperature dependence of the current. Since there is a barrier at the source, the current is thermally activated. Although we have been able to pull barriers down to around 150meV, this is still much larger than that measured in a polysilicon FET [7]. One way of reducing the temperature coefficient would be to make a thermionic emission source in which, under bias, carriers would tunnel through the barrier at the Fermi level of the metal [8]. This structure would require careful engineering using thin insulating layers and doping profiles.

The activation energy of current transport against (V_G-V_T) is shown for two SGT devices in Fig. 9. Device (a) is typical of a Schottky source SGT; the source barrier is pulled down by the gate but remains high. However, we have measured SGT devices where the activation energy falls to very low values with increasing V_G, as shown in Fig. 9 (b). Since no deliberate attempt had been made to make a field emission source, it seems unlikely that the low activation energy is due to the source, but it is possible. It was observed that devices with low activation energies operated at the highest currents

978-1-4244-6658-0/10 $26.00 © 2010 IEEE

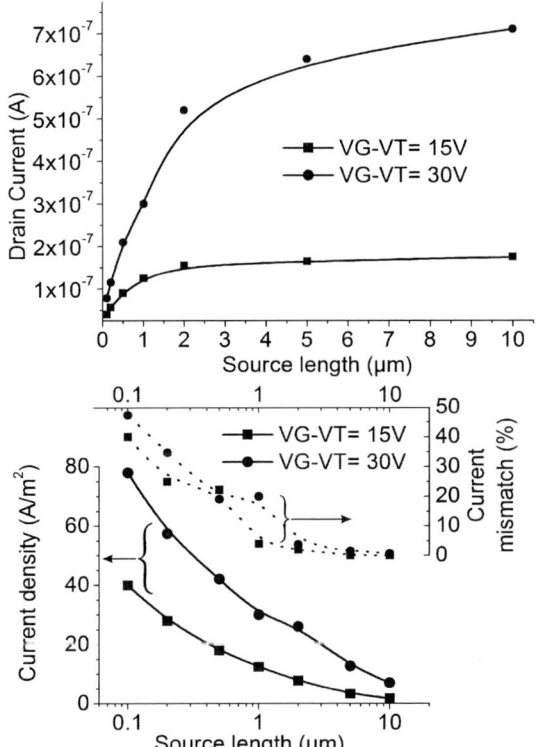

Figure 8. Computed variation of current and current density through the source as a function of source length. Current mismatch was calculatd for a 0.1μm change in source length.

and there was a small increase in V_{SAT1} compared with low current devices, at the same V_G.

We conclude, therefore, that these devices operate in a hybrid mode in which the on-state is partly controlled by the source barrier and partly by the parasitic FET.

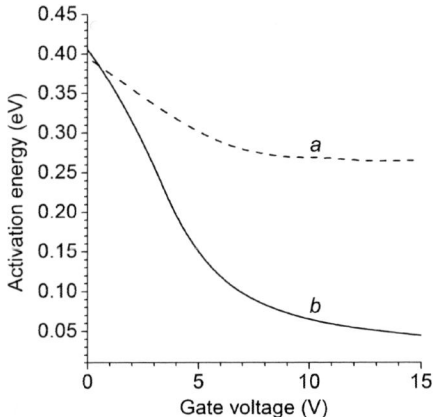

Figure 9. Change of activation energy for current transport for a) a SGT operating at low current and b) a SGT operating at high current.

It seems that there is a negative feedback effect in which the FET, with its very low activation energy of the current restricts the change of current through the source. This hybrid mode could be very important when small changes of current with temperature are required. The tradeoff is a small increase in V_{SAT1}. To fully understand the interaction between the source barrier and the parasitic FET in this situation requires a further 2D analysis.

III. CONCLUSIONS

Self-aligned polysilicon SGTs have been made. Measurements show an absence of the kink effect responsible for degrading the output characteristics of polysilicon FETs. Output impedances are very high, in some cases resulting in intrinsic gains approaching 10,000. While a FET cannot be switched off when the semiconductor is highly doped, an SGT can always be switched off and the off-current is independent of doping level. The intrinsic gain envelope is broader for higher substrate doping but the highest gains were observed with the lowest doping. It is shown that the choice of source length is a tradeoff between speed and current uniformity. Some high current transistors with SGT characteristics have very low activation energies. We attribute this behaviour to a hybrid mode of operation. It has been shown that there are a number of options and tradeoffs which should enable high performance digital and analog circuits to be made in polysilicon.

ACKNOWLEDGMENT

The authors would like to thank EPSRC for the partial funding of this project through IDTA grant number EP/P503982/1 and through the Portfolio Partnership Project.

REFERENCES

[1] J. M. Shannon and E. G. Gerstner, "Source-gated thin-film transistors", IEEE *Electron Dev. Lett.*, 24, no. 6, pp. 405-407, 2003.

[2] J. M. Shannon and E. G. Gerstner, "Source-gated transistors in hydrogenated amorphous silicon", *Solid-State Electronics*, Vol. 48, No. 6, pp. 1155-1161, 2004.

[3] A. Valletta, P. Gaucci, L. Mariucci, G. Fortunato and S. D. Brotherton, "Kink effect in short channel polycrystalline silicon thin film transistors", *Appl Phys Lett*, vol. 85, no. 15, pp. 3113-3115, 2004.

[4] S. M. Sze, "Physics of semiconductor devices", Second Edition, *John Wiley & Sons*, 1981.

[5] R. A. Sporea, M. J. Trainor, N. D. Young, J. M. Shannon, S. R. P. Silva, "Intrinsic Gain in Self-Aligned Polysilicon Source-Gated Transistors", accepted to be published in *IEEE Trans. Electron Devices*, 2010.

[6] J.M. Shannon and F. Balon, "Frequency Response of Source-Gated Transistors", *IEEE Trans. Electron Devices*, Vol. 56, Issue 10, pp. 2354-2356, 2009.

[7] Y. Morimoto, Y. Jinno, K. Hirai, H. Ogata, T. Yamada, and K. Yoneda, "Influence of the Grain Boundaries and Intragrain Defects on the Performance of Poly-Si Thin Film Transistors", *J. Electrochem. Soc.*, Vol. 144, Issue 7, pp. 2495-2501, 1997.

[8] J. M. Shannon and F. Balon, "Source-gated thin-film transistors", *Papers from ITC'07, Solid State Electronics*, Vol. 52, pp. 449 – 454, 2008.

978-1-4244-6658-0/10 $26.00 © 2010 IEEE   145

Analysis and Modeling of Pseudo-Short-Channel Effects in ZnO-Nanoparticle Thin-Film Transistors

Karsten Wolff and Ulrich Hilleringmann

University of Paderborn, Institute for Electrical Engineering and Information Technology

33095 Paderborn, Germany, Email: wolff@sensorik.upb.de

Abstract—Due to the complex nature of the device physics in nanoparticle thin-film transistors (TFT), analytical models for the transistor characteristics are not available for advanced circuit design. The discrepancy between experimental data and the standard MOSFET equations has been neglected up to now, although there are several pseudo-short-channel effects obvious. In this paper, a simple but sufficient model is proposed, which represents the transistor characteristics of ZnO-nanoparticle TFTs by the introduction of two semi-empirical parameters. The model is demonstrated for integrated normally-on and normally-off devices in both linear and saturation regions.

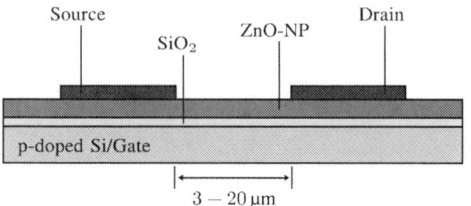

Fig. 1: Schematic crosssection of ZnO-TFTs in Inverted-Staggered-architecture (Backside-Gate/Top-D/S-Electrodes)

I. INTRODUCTION

During the last years, research interest in electronic devices using semiconducting nanoparticles has evolved. Because anorganic nanoparticle transistors show reasonable performance without device degradation, they are considered to be superior to organic electronics. In particular, zinc oxide has been focused on because of its availability, low price and optical transparency. In the meantime many demonstrations of well operating thin-film transistors using ZnO-nanoparticles (ZnO-NP) as semiconductor material can be found in literature [1]–[4]. Nevertheless, modeling of the device characteristics has been disregarded up to now, although lots of nanoparticle transistors cannot be described by the well-known standard equations for conventional MOSFETs [5]:

$$I_D = \beta\mu \left[(V_{GS} - V_{th}) - \frac{V_{DS}}{2} \right] V_{DS} \quad \text{for } V_{GS} - V_{th} > V_{DS}$$

(1)

$$I_D = \frac{\beta\mu}{2} [V_{GS} - V_{th}]^2 \quad \text{for } V_{GS} - V_{th} \leq V_{DS}.$$

(2)

One of the advantages of ZnO-TFTs is the simple integration technique, wherein doping processes can be avoided, because the metal-semiconductor interfaces at the drain- and source-electrodes form Schottky-barriers. Therefore, ZnO-TFTs are classified as Schottky-barrier drain/source field-effect transistors (SB-MOSFET) [6]. The characteristics are highly dependent on the barrier properties, which are not included in the MOSFET theory. Due to the transistor's complex nature, an analytical form of the characteristics is unsuitable for advanced circuit design and simulation.

Based on the standard theory, this approach presents a simple model, which describes the behavior of ZnO-TFTs using nanoparticles. In particular, zinc oxide nanoparticle transistors show severe pseudo-short-channel effects like barrier lowering

and increasing saturation drain currents looking alike channel length modulation, although the transistor dimensions were defined as long-channel. It is demonstrated that these effects can be efficiently modeled by simple, empirical adaptations.

II. TRANSISTOR INTEGRATION

In order to integrate the analyzed transistors, whose cross-section is schematically depicted in Fig. 1, standard 4-inch silicon wafers (boron-doped, <100>-oriented) with specific resistances of about $14\,\Omega\text{cm}$ were cleaned in H_2SO_4/H_2O_2, rinsed and dried, before a thermal oxide (55 nm) was grown. The dispersion of gas phased synthesized zinc oxide nanoparticles (AdNano ZnO 20 DW) was received from Evonik Degussa GmbH. The nanoparticles were well dispersed in water with 34 wt%. Even after a shelf life of 3 months, no phase separation could be observed. The effective agglomerate diameter was in the range of 100 nm, while the primary particle size is even smaller [7]. Due to thinner layers the dispersion was diluted with one part DI-water. In order to crack particle agglomerates before spin-coating, the dispersion was treated in an ultrasonic bath for 30 minutes. The process time and the application of mechanical energy is sufficient because of the excellent suspension stability [8]. Instantly after the redispersion, 2 ml of the solution were spin-coated on the wafers at 2000 revolutions per minute for 30 s. The ZnO-NP formed a transparent film, which was pre-baked on a hot-plate at 120°C for 2 minutes. The baking causes a vaporisation of H_2O and fixed the particle film. To ensure a high degree of adhesion, a further thermal annealing process was performed in oxygen atmosphere at 800°C for 2 hours. As a result the nanoparticles coalesced to larger grains with enhanced electrical parameters. However, this process is not compatible to plastic or glass substrates; the annealing only affects the device performance. Hence, the demonstrated

978-1-4244-6658-0/10 $26.00 © 2010 IEEE

Fig. 2: SEM images of a ZnO-nanoparticle film: (a) as prepared, (b) annealed at $800°C$ in oxygen atmosphere

physical model remains valid for low temperature devices. The nanoparticle films are shown in Fig. 2. The as-prepared film clearly exhibits the primary particle morphology with particle diameters down to $20\,nm$, whereas the annealed film reveals grain sizes up to $0.5\,\mu m$. The annealing process improves the mechanical stability against the quite extensive forces during 'wet' processes, i.e. any wet process can be performed in the following. Therefore, metal drain and source electrodes were deposited by either sputter deposition ($100\,nm$ Au) or evaporation ($150\,nm$ Al) and structured by lift-off technique. Compared to all other approaches, shadow mask evaporation with its approximate $20\,\mu m$ limitation becomes obsolete and the structure dimensions can be vastly scaled due to optical lithography. The minimum and maximum channel length was $3\,\mu m$ and $20\,\mu m$, respectively.

III. Results and Discussion

On-wafer electrical characterization was performed using an HP 4156A precision semiconductor parameter analyzer in ambient air at a temperature of $21°C$. A set of 33 transfer- and output-characteristics of transistors with Al-drain-/source-electrodes and a set of 14 transistors with Au-drain-/source-electrodes were recorded.

I-V-characteristics reveal n-type field-effect-transistors. The transfer and output characteristics of a typical transistor with Al-D/S-electrodes are shown in Fig. 3. The output characteristics exhibit an extensive and well pronounced linear regime because of normally-on behavior. A saturation regime could not be observed before the gate oxide broke down.

Analysis of the transfer characteristics yields field-effect mobilities for Al- and Au-transistors in the range of $4.3 \cdot 10^{-3}\,cm^2(Vs)^{-1}$ to $0.15\,cm^2(Vs)^{-1}$ and $2.7 \cdot 10^{-5}\,cm^2(Vs)^{-1}$ to $1.4 \cdot 10^{-3}\,cm^2(Vs)^{-1}$, respectively. These values are comparable to field-effect mobilities of

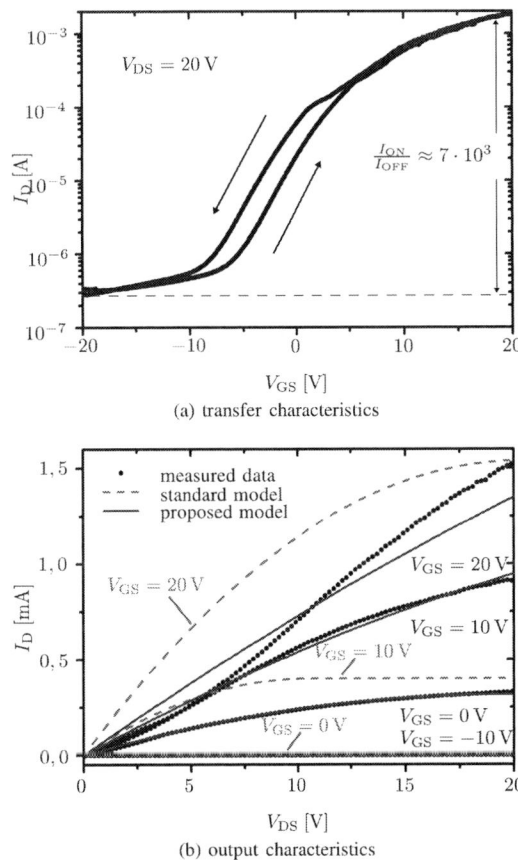

(a) transfer characteristics

(b) output characteristics

Fig. 3: I-V-characteristics of a ZnO-nanoparticle TFT with Al-D/S-electrodes. $L = 8\,\mu m$, $W = 16\,cm$

ZnO-nanoparticle transistors, which are already reported in [1]–[3]. Due to quasi-ohmic contacts between Al and ZnO the mobilities of Al-transistors exceed Au-transistor mobilities [9]. The distributions of the field-effect mobilities are shown in Fig. 4.

The threshold voltages V_{th} are extracted from the $\sqrt{I_{DS}}$-V_{GS}-characteristics. Since the nanoparticle film and the nanoparticle-dielectric interface contain a considerable trap density, charges are trapped during gate bias sweep, which leads to hysteresis and a threshold voltage shift of $\Delta V_{th} = -2.12\,V$ in the case of the device in Fig. 3. Generally, transistors with Au-D/S-electrodes exhibit more pronounced hysteresis of up to $-27\,V$. The distributions of the 'forward' threshold voltages are shown in Fig. 5 for Al-devices. They postulate a dependency on the electric field caused by V_{DS}, which is reasonable, because the device characteristic is dominated by the source contact barrier (Fig. 6). The barrier height and width, however, can be controlled by the drain potential as well due to tilted energy bands and Schottky barrier lowering. Hence, tunneling current and thermionic emission is increased even at low V_{GS}. At first glance, the impact of the mechanism seems to be

978-1-4244-6658-0/10 $26.00 © 2010 IEEE

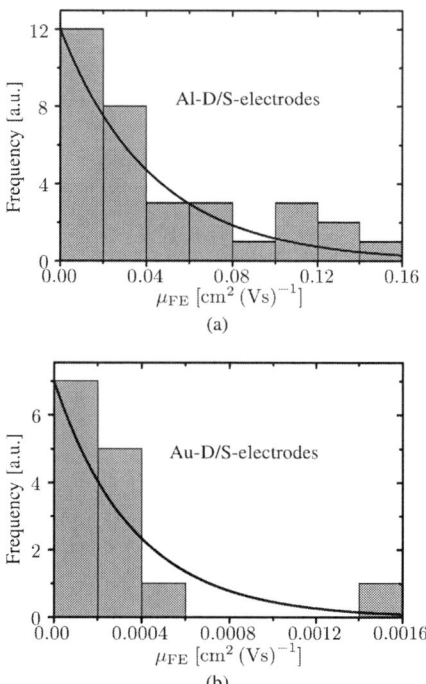

(a)

(b)

Fig. 4: Distribution of the field-effect mobility μ_{FE}: (a) Al-D/S-electrodes, (b) Au-D/S-electrodes

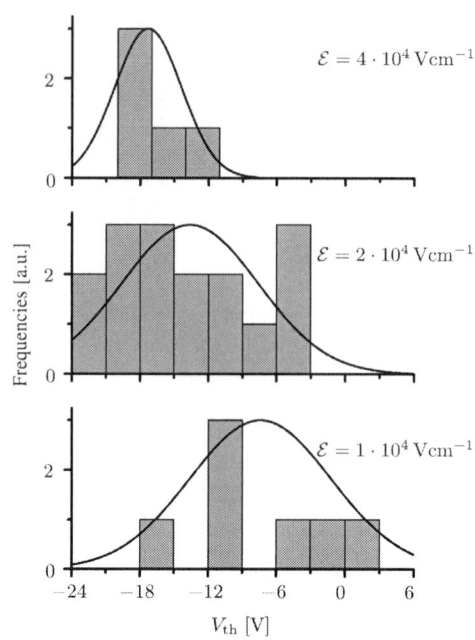

Fig. 5: Distribution of the threshold voltages V_{th} of ZnO-nanoparticle TFTs (Al-D/S-electrodes)

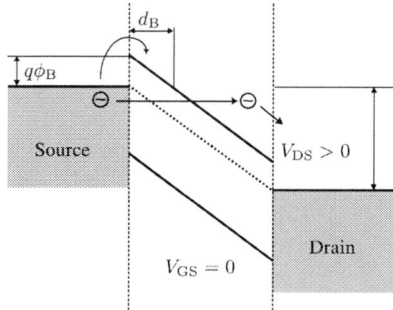

Fig. 6: Band energy diagram of a SB-MOSFET

similar to the drain induced barrier lowering (DIBL) effect in short-channel transistors, but it occurs in a channel length domain, which is untypical for threshold voltage lowering in conventional MOSFETs. Assuming a triangular barrier as sketched in Fig. 6, the barrier width is modulated by

$$d_{\mathrm{B}} = \frac{L\phi_{\mathrm{B}}}{V_{\mathrm{DS}}}. \qquad (3)$$

As known from DIBL, V_{th} is approximated to be linearly dependent on V_{DS}:

$$V_{\mathrm{th}} = \delta V_{\mathrm{DS}} + V_{\mathrm{t0}} \qquad (4)$$

with V_{t0} as the threshold voltage at thermal equilibrium ($V_{\mathrm{DS}} = 0, V_{\mathrm{GS}} = 0$).

Furthermore, a serious degradation of the charge mobility is expected, which is caused by accumulation of charges in the nanoparticles at the dielectric interface. With $V_{\mathrm{GS}} > 0$ channel charges are accumulated (Fig. 7). When the electric field becomes so strong that the Debeye length L_{D} is smaller than the nanoparticle radius, a great amount of channel charges are trapped and therefore unavailable for the lateral drain current. As a result the effective mobility decreases [4]. Smaller nanoparticles lead to attenuation of the 'trapping' indeed, but mobility is decreased in general due to a higher density of grain boundaries. For the proposed model a mobility degradation term is introduced following FU ET AL. [10]:

$$\mu = \frac{\mu_0}{1 + \theta(V_{\mathrm{GS}} - V_{\mathrm{t0}})}. \qquad (5)$$

Then equation (1) for the linear regime becomes

$$I_{\mathrm{D}} = \frac{\beta\mu_0}{1 + \theta(V_{\mathrm{GS}} - V_{\mathrm{t0}})} \left[(V_{\mathrm{GS}} - V_{\mathrm{t0}}) - \left(\delta + \frac{1}{2} \right) V_{\mathrm{DS}} \right] V_{\mathrm{DS}} \qquad (6)$$

within the same limits as in Eq. (1). Comparing the standard equations to the proposed model, equation (1) is certainly inappropriate, although the device-defining electrical parameters V_{th} and μ_{FE}, extracted from the transfer characteristic, have been used. Plotting the modeled output characteristics with $\delta = -0.316$ and $\theta = 0.036\,\mathrm{V}^{-1}$ (Fig. 3b), the simple model fits the measured data quite well (blue graphs). The upward bending of the I-V-curves in the low-V_{DS}-portion, which is typically observed in SB-MOSFETs and caused by the contact properties, is balanced sufficiently by the proposed model.

Since a saturation regime was not observed in Al-devices, saturation behavior of Au-devices is analyzed instead. These

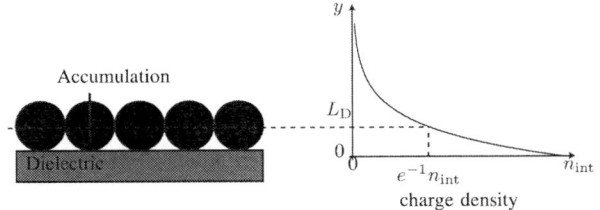

Fig. 7: Charge accumulation at the nanoparticle/dielectric interface.

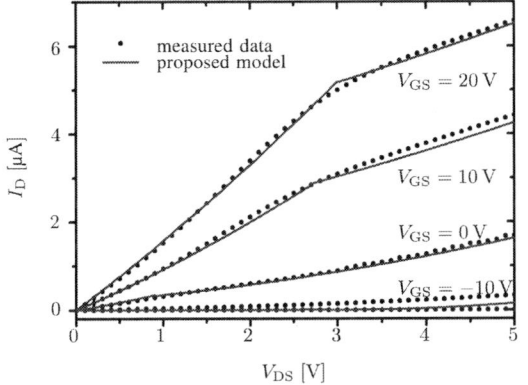

Fig. 8: Output characteristics of a ZnO-nanoparticle TFT with Au-D/S-electrodes. $L = 20\,\mu\text{m}$, $W = 45.1\,\text{cm}$

devices usually show worse performance regarding field-effect mobility and hysteresis, but they require much lower drain voltages than Al-devices. Yet, the demonstrated model for the linear regime is applicable, too, even though the device is a normally-off transistor and the linear portion in the output characteristics is therefore less pronounced.

Analyzing the measured output characteristics of a $20\,\mu\text{m}$ transistor in Fig. 8, a steady slope of the drain current can be seen in the saturation region. Although supposable, the slope is neither caused by channel length modulation nor by any other short-channel effect like SCLC dominated punch-through. In fact, the same source barrier modulation and mobility degradation effects as before influence the drain current, leading to a non-constant saturation behavior. Using the assumptions of (4) and (5), the drain current then becomes

$$I_{\text{D}} = \frac{\beta \mu_0}{2\left[1 + \theta\left(V_{\text{GS}} - V_{\text{t0}}\right)\right]} \left[V_{\text{GS}} - \delta V_{\text{DS}} + V_{\text{t0}}\right]^2 . \quad (7)$$

The curves of the proposed model are plotted in Fig. 8 as blue graphs with $\delta = -1.83$ and $\theta = 0.018\,\text{V}^{-1}$. As one can see, the model is an adequate fit to the measured data in the saturation region as well as in the linear regime.

IV. Conclusion

Thin-film transistors with ZnO-nanoparticles as semiconductor material were integrated and analyzed concerning pseudo-short-channel effects. It was shown, that the transistors, which belong to the class of SB-MOSFETs, are seriously subject to threshold voltage lowering and charge mobility degradation. Whereas the threshold voltage lowering is mainly due to source barrier modulation caused by the drain influenced electric field, the mobility degradation worsens with increasing gate bias and accumulation of charges near the semiconductor/dielectric-interface.

It was shown that the standard MOSFET equations are not valid, when using transistor parameters extracted from the transfer characteristics. However, further circuit design demands sufficient modeling concerning its simulation. The complex nature of the device physics forbids analytical derivation of the transistor equations. Therefore, a model for nanoparticle ZnO-TFTs was proposed, which represents the transistor characteristics in the triode region as well as in the saturation region. Introducing two empirical parameters, the model sufficiently approximates the measured data. The parameters can be extracted easily from measured data and adjusted to both normally-on and normally-off transistors.

Acknowledgement

The authors would like to thank the German DFG for funding (Hi551/24-1) and Evonik Degussa GmbH, Creavis Technologies & Innovation for the nanoparticle supply.

References

[1] S. Lee, Y. Jeong, S. Jeong, J. Lee, M. Jeon, and J. Moon, "Solution-processed ZnO nanoparticle based semiconductor oxide thin film transistors," *Superlattices and Microstructures*, vol. 44, no. 6, pp. 761–769, 2008.

[2] K. S. Volkman, A. B. Mattis, Molesa, S.E., Lee, J.B., Fuente Vornbrock, A. de la, T. Bakhishev, and V. Subramanian, "A novel transparent air-stable printable n-type semiconductor technology using ZnO nanoparticles," 2005.

[3] H. Faber, M. Burkhardt, A. Jedaa, D. Kälblein, H. Klauk, and M. Halik, "Low-Temperature Solution-Processed Memory Transistors Based on Zinc Oxide Nanoparticles," *Advanced Materials*, vol. 21, no. 30, pp. 3099–3104, 2009.

[4] K. Okamura, N. Mechau, D. Nikolova, and H. Hahn, "Influence of interface roughness on the performance of nanoparticulate zinc oxide field-effect transistors," *Appl. Phys. Lett.*, vol. 93, no. 8, pp. 083 105–1 – 3, 2008.

[5] S.M. Sze, *Physics of semiconductor devices*. New York, USA: John Wiley & Sons, Inc., 1981.

[6] R. J. Tucker, C. Wang, and S. P. Carney, "Silicon field-effect transistor based on quantum tunneling," *Appl. Phys. Lett.*, vol. 65, no. 5, pp. 618–620, 1994.

[7] Degussa AG (Advanced Nanomaterials), "Datasheet "AdNano Zinc Oxide"," www.advancednanomaterials.com, 2006.

[8] J. S. Chung, P. J. Leonard, I. Nettleship, K. J. Lee, Y. Soong, V. D. Martello, and K. M. Chyu, "Characterization of ZnO nanoparticle suspension in water: Effectiveness of ultrasonic dispersion," *Powder Technology*, vol. 194, no. 1-2, pp. 75–80, 2009.

[9] K. Ip, T. G. Thaler, H. Yang, S. Youn Han, Y. Li, P. D. Norton, J. S. Pearton, S. Jang, and F. Ren, "Contacts to ZnO: Proceedings of the International Conference on Materials for Advanced Technologies (ICMAT 2005) Symposium N - ZnO and Related Materials," *Journal of Crystal Growth*, vol. 287, no. 1, pp. 149–156, 2006.

[10] Y. K. Fu, "Mobility degradation due to the gate field in the inversion layer of MOSFET's," *Electron Device Letters, IEEE*, vol. 3, no. 10, pp. 292–293, 1982.

978-1-4244-6658-0/10 $26.00 © 2010 IEEE

Comparison of strained SiGe heterostructure-on-insulator (001) and (110) PMOSFETs: $C - V$ characteristics, mobility, and ON current

A. T. Pham [a], C. Jungemann [b], and B. Meinerzhagen [a]

[a] BST, TU Braunschweig, 38023 Braunschweig, Germany
e-mail: pham@nst.ing.tu-bs.de
[b] EIT4, Universität der Bundeswehr München, 85577 Neubiberg, Germany

I. INTRODUCTION

Strained SiGe heterostructure-on-insulator (HOI) stacks in combination with non-traditional surface and channel orientations are promising concepts for improving PMOS performance. On the research level, transport and electrostatics of strained SiGe HOI PMOSFETs with the traditional (001) surface orientation have previously been studied both experimentally [1] and by simulation [2]–[4]. Advanced HOI PMOS structures contain a heterostructure of Si/SiGe/Si in the channel as shown in Fig. 1. The inclusion of a SiGe layer in the channel improves the mobility of the hole inversion layer, but at the same time increases the gate leakage current due to the lower band gap between the conduction band and the valence subbands of the SiGe alloy (e.g. 0.661eV for Ge) compared to the band gap of a Si layer (1.12eV). As a consequence the drain leakage current is reduced by the Si layers sandwiched between the SiGe layer and the oxide layers. The channel layers are strained in order to boost channel mobility and ON current. During the fabrication of HOI PMOS structures [1] the semiconductor layers are first grown pseudomorphically on a relaxed $Si_{1-y}Ge_y$ substrate with a Ge content y and subsequently transferred to oxide, such that the strain state of the original epitaxial layer stack is retained after the transfer. The difference $z - y$ between the Ge contents in the thin $Si_{1-z}Ge_z$ layer and the relaxed $Si_{1-y}Ge_y$ substrate determines the compressive strain in the $Si_{1-z}Ge_z$ layer.

In this work, for the first time, the simulation of strained SiGe HOI PMOSFETs is extended to the (110) surface orientation in order to provide a better understanding of the differences between the (001) and (110) surface orientations for strained SiGe HOI PMOSFETs. The comparison is focused on many important aspects including $C - V$ characteristics, mobility, and ON current. The (110) surface orientation is of special interest because it was observed both experimentally [5] and using simulations [6], [7] that the (110) surface orientation enhances the effective hole mobility for unstrained Si bulk PMOSFETs compared to the (100) surface.

II. SIMULATION METHODS

The simulations are based on the self-consistent solution of the 6×6 $\vec{k} \cdot \vec{p}$ Schrödinger Equation (SE), the multi subband Boltzmann Transport Equation (BTE), and Poissons Equation (PE) [8], [9]. This simulation method captures the influence of size quantization, strain, surface/channel orientations, and SiGe alloys on a solid physical basis. Each device is partitioned into many vertical slices along the channel direction [10]. For each slice the 1D SE is solved. The biaxial strain tensor is determined based on the Ge contents z and y in the channel layer and the virtual substrate, respectively

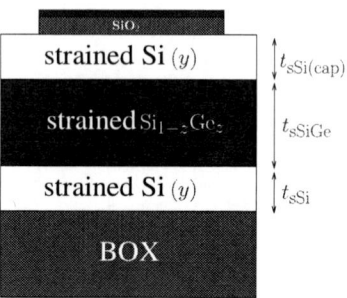

Fig. 1. HOI sSi(cap)/sSi$_{1-z}$Ge$_z$/sSi PMOS structure. The biaxial strain in the Si and SiGe layers is assumed to be consistent with a pseudomorphic growth of each layer on a relaxed Si$_{1-y}$Ge$_y$ virtual substrate.

[7]. The Pikus-Bir strain Hamiltonian is calculated using the resultant strain tensor and added to the 6×6 $\vec{k} \cdot \vec{p}$ Hamiltonian. The valence band offset energies due to the Si/SiGe/Si heterostructure are calculated based on the average valence band energy of each alloy layer following [11]. The diagonal Hamiltonian due to these valence band offsets is also added to the total Hamiltonian. Spin-orbit coupling is considered by the additional spin-orbit coupling Hamiltonian. Size quantization effects and the electrostatic potential are considered by solving the 1D 6×6 $\vec{k} \cdot \vec{p}$ envelop function SE which results from the total Hamiltonian. The solution of SE provides the valence subband structure and allows to calculate the scattering overlap factors. A rotation of the $\vec{k} \cdot \vec{p}$ Hamiltonian and the strain Hamiltonian is necessary in order to handle different surface orientations [7], [12]. The stationary multi subband BTE is solved with a deterministic method based on a Fourier harmonic expansion of the distribution function up to high order (13th). This method is well documented in [9]. The transport simulations include scattering due to phonon, surface roughness and alloy disorder. The calculation of the transition rates of these scattering mechanisms is described in [3]. The hole density resulting from the subband structure and the distribution function is considered in the 2D PE in order to obtain a self-consistent solution. The electron density and the hole density in the bulk layer below the BOX layer are calculated classically based on the electrostatic potential and a constant quasi-Fermi level. The three equations are solved self-consistently using an efficient quasi Newton iteration scheme [9]. For a given gate bias the typical CPU times for obtaining the mobility and the ON current are 20-30 minutes and 3-4 hours, respectively.

978-1-4244-6658-0/10 $26.00 © 2010 IEEE

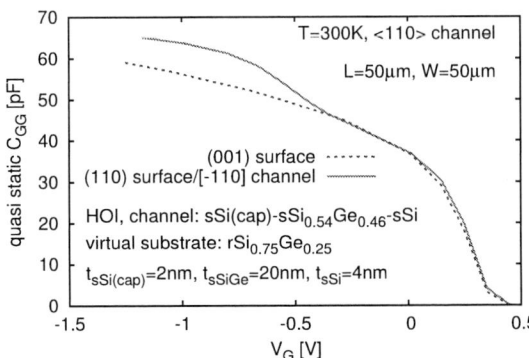

Fig. 2. $C - V$ characteristics for the two strained SiGe HOI (001) and (110) PMOS structures. $t_{ox} = 1nm$, $L = 50\mu m$, $W = 50\mu m$. The $C - V$ curve for the (001) surface orientation shows no plateau , whereas a plateau is visible for the (110) orientation.

III. RESULTS

For the simulations of $C - V$ characteristics and mobilities a homogenous channel HOI PMOS structure is simulated. The thicknesses of the top and bottom oxide are 1nm and 10nm, respectively. The thicknesses of the channel stack layers are $t_{sSi(cap)} = 2nm$, $t_{sSiGe} = 20nm$, and $t_{sSi} = 4nm$. The thickness of the bulk layer below the bottom oxide is 800nm. The Ge contents in the strained $Si_{1-z}Ge_z$ layer and in the relaxed $Si_{1-y}Ge_y$ virtual substrate are $z = 0.46$ and $y = 0.25$, respectively. A uniform donor doping concentration of $10^{15}cm^{-3}$ is assumed in in both the channel layers and the bulk Si region. A metal gate contact and an ohmic contact at the bottom of the bulk region are assumed. A bulk bias V_B of 0V and a constant lattice temperature of 300K are assumed in all simulations. The simulations are performed for the two interface orientations (001) and (110). The channel directions for the (001) and (110) interface orientations are [110] and [−110], respectively.

The quasi static gate capacitance defined as $C_{GG} = \partial Q_G / \partial V_G$ is evaluated based on the variation of the gate charge Q_G w.r.t. a small variation of the gate bias V_G. Fig. 2 shows the $C - V$ characteristics for the two different strained SiGe HOI PMOS structures over a wide range of V_G. Please note that C_{GG} refers to a gate length L and width W of $50\mu m$. The $C - V$ curves for the two interface orientations differ substantially. The difference is small for $|V_G| < 0.5V$ resulting in a very small threshold voltage difference of a few mV only. However for $|V_G| > 0.5V$ the gate capacitance of the (110) PMOS structure is much larger than the gate capacitance of the (001) alternative. For the strained SiGe HOI (001) PMOS structure experimental results [1] show that the $C - V$ curves have a plateau in a certain range of gate biases when the strained Si cap thickness is larger than 4nm. For a Si cap thickness smaller than 2nm this plateau vanishes. Consistent with [1] the simulated $C - V$ curve (see Fig. 2) for the HOI (001) PMOS structure with 2nm Si cap thickness shows no plateau. However this plateau occurs for the $C - V$ curve of the HOI (110) PMOS structure even for the small Si cap thickness of 2nm as shown in Fig. 2.

The difference between the CV curves for the two surface orientations is due to the difference of the spatial hole density distributions for the two orientations as shown in Fig. 3. For low inversion charge ($N_{inv} = 2 \times 10^{12}cm^{-2}$) the spatial hole density distributions within the two structures show peaks only in the SiGe layer near the Si cap/SiGe interface (Fig. 3(top)). Moreover, it is shown in the same figure that the sum of the electrostatic potential energy and the valance band offset energy is lowest within the SiGe layer at the Si cap/SiGe interface. This is on one hand due to the large valence band offset potential between the Si cap and the SiGe layer (≈ 0.2 eV for the Ge content of $z = 0.46$ considered here) and on the other hand due to the small variation of the electrostatic potential in these two layers. Consequently, the hole density is moved away from the SiO_2/Si interface and confined mainly in the SiGe layer near the Si cap/SiGe interface. Moreover, the distance between the SiO_2/Si interface and the hole density peak is slightly shorter for the (110) surface orientation. This is consistent with the observation that the gate capacitance for the (110) structure is slightly larger than the gate capacitance for the (001) structure (Fig. 2). For higher inversion charges (here $N_{inv} = 1.5 \times 10^{13}cm^{-2}$) the spatial hole density distributions for the two structures are completely different as shown in Fig. 3(bottom). For the (001) orientation the holes are still mainly confined close to the Si cap/SiGe interface, although the first quantum well near the SiO_2/Si interface has now a much smaller energy than the second quantum well in the SiGe layer near the Si cap/SiGe interface. The spatial hole density distribution for the (110) structure however shows two distinct peaks indicating that the holes in this case move more and more from the SiGe to the Si quantum well with increasing gate voltage. Moreover, the dominant peak of the hole density distribution for the (110) structure is much closer to the SiO_2/Si interface compared to peak position for the (001) structure. This is consistent with the higher gate capacitance for (110) structure (Fig. 2).

Equi energy lines in multiples of 100meV are shown in Fig. 4 for the first HH subband and the two surface orientations. The equi energy lines for the two cases differ substantially. The typical 4- and 2- fold symmetries of the subband structures for the (001) and (110) surface orientations, respectively, are clearly visible. In the optimum channel direction ([−110]) for the (110) structure the distance from the Γ point ($\vec{k} = \vec{0}$) to the point on the first 100meV equi energy line is much shorter for the (110) surface orientation than for the (001) case irrespectively of the channel direction chosen for the (001) case. This indicates that the curvature and consequently the effective mass resulting from the first HH energy dispersion relation near the Γ point for this direction is much smaller for the (110) structure than for (001) alternative. Please note that both the ohmic mobility tensor as well as the effective mass tensor for the (001) case are isotropic. The effective mass for the [−110] channel direction and the (110) surface as calculated directly based on the subband structure is shown in Fig. 5 as a function of inversion charge in comparison to isotropic (001) case. It can be clearly seen that the effective mass for the (110) case is about a factor of two smaller compared to the (001) case. This is consistent with the the different curvatures of the first HH energy dispersion relations near the Γ point mentioned above for the two surface orientations.

For the (001) surface orientation the effective ohmic mobility is calculated and shown in Fig. 6 for an unstrained Si bulk PMOS structure and the strained SiGe HOI PMOS structure. The phonon and surface roughness scattering parameters have been calibrated to reproduce the experimental mobility by Takagi et al. [13] for unstrained bulk PMOS.

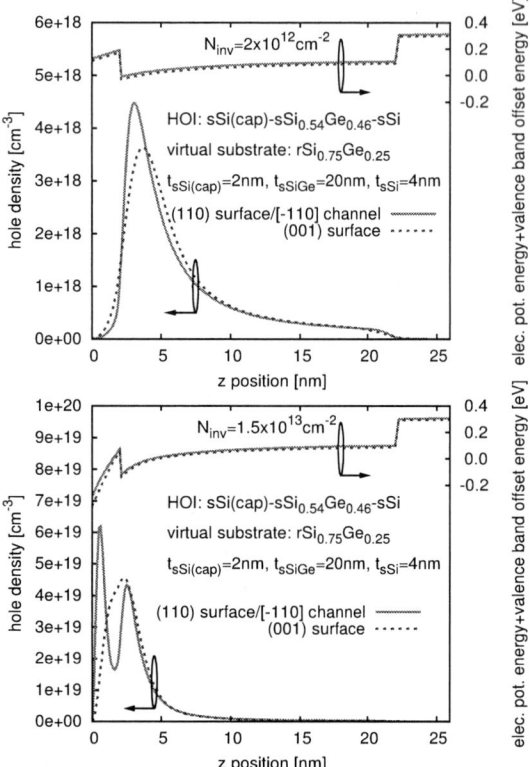

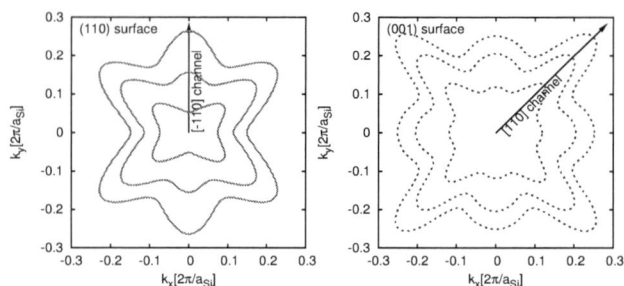

Fig. 4. Equi energy lines in multiples of 100meV of the first HH subband for the strained SiGe HOI (001) and (110) PMOS structures. For both devices the bottom of the first HH subband is shifted to 0meV. $N_{inv} = 1.5 \times 10^{13} \mathrm{cm}^{-2}$.

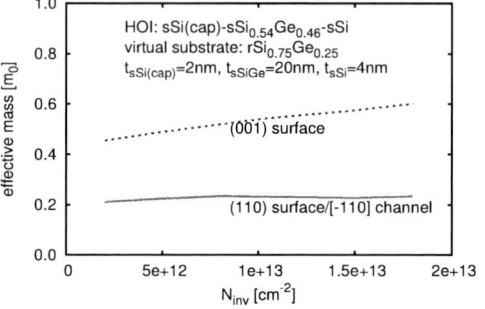

Fig. 3. Spatial distributions of hole density and electrostatic potential + valence band offset energy in the strained SiGe HOI (001) and (110) PMOS structures. Top: $N_{inv} = 2 \times 10^{12} \mathrm{cm}^{-2}$ ($V_G = 0.014$V for the (001) surface and $V_G = 0.015$V for the (110) surface). Bottom: $N_{inv} = 1.5 \times 10^{13} \mathrm{cm}^{-2}$ ($V_G = -1.059$V for the (001) surface and $V_G = -0.990$V for the (110) surface).

Fig. 5. Effective mass in [-110] channel direction for the strained SiGe HOI (110) PMOS structure compared to the (001) case.

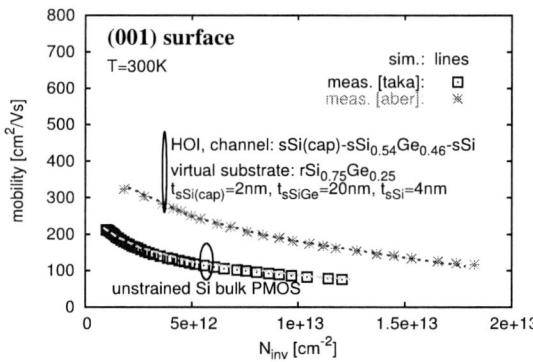

For SiGe in addition alloy scattering is important. With an alloy scattering deformation potential of 0.445eV an excellent agreement between the simulated mobilities and the measured low-field mobility data of Aberg [1] is achieved (see Fig. 6). All simulations in this paper are based on the same scattering parameters extracted only based on the two key experiments shown in Fig. 6 as just described. For the (110) surface orientation and the [-110] channel direction, the simulation results reproduce well the measured mobility data of Irie [5] for an unstrained Si bulk PMOS structure (Fig. 7). Moreover, as can be seen in the same figure the simulation results predict that the mobility in the strained SiGe HOI (110) PMOS structure is much higher than the mobility for the unstrained Si bulk (110) PMOS case. For the same strained SiGe HOI PMOS structure and the optimum [-110] channel direction for the (110) surface case the mobility for the (110) surface orientation is much higher than the mobility for the (001) surface orientation as shown in Fig. 7 and 6. This is consistent with the reduction of the effective mass for the (110) case compared to (001) shown in Fig. 5. The dependence of the mobility on strained Si cap thickness is shown in Fig. 8. For the (001) surface orientation the simulation results agree well with measured data from [1]. For all cap thicknesses the mobility of the (110) structure is higher. Moreover, in both cases the maximum mobility results for the Si cap thickness of 1nm.

Short channel strained SiGe HOI (001) and (110) PMOS-FETs are simulated next. The vertical PMOS stack is the same as for the previous simulations. The channel length is 22nm

Fig. 6. Mobility characteristics of strained SiGe HOI (001) PMOS compared to unstrained Si (001) bulk PMOS. Measured data for HOI and bulk PMOS are taken from [1] and [13], respectively.

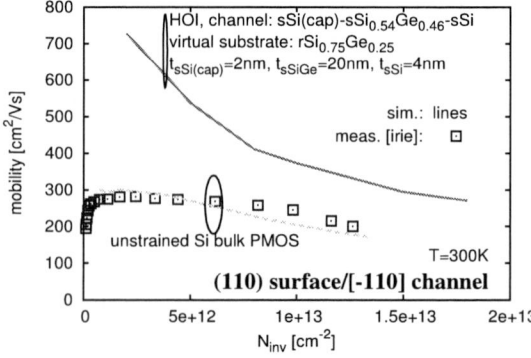

Fig. 7. Mobility characteristics of strained SiGe HOI (110) PMOS compared to unstrained Si (110) bulk PMOS. Measured data for bulk PMOS are taken from [5].

978-1-4244-6658-0/10 $26.00 © 2010 IEEE   152

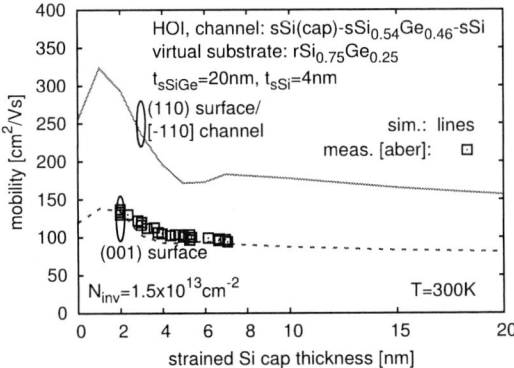

Fig. 8. Mobility vs strained Si cap thickness in strained SiGe HOI (001) and (110) PMOS structures. The measured data for strained SiGe HOI (001) PMOS are taken from [1].

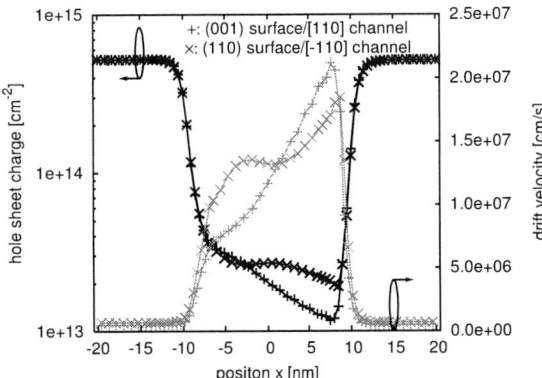

Fig. 9. Hole sheet charge and drift velocity along the channel in the 22nm strained SiGe HOI (001) and (110) PMOSFETs. $V_{GS} = -1$V and $V_{DS} = -0.7$V.

and the acceptor doping concentration in the source and drain regions is 2×10^{20} cm^{-3}. The hole sheet charge and drift velocity along the channel are shown in Fig. 9 for the two cases and the ON state bias $V_{GS} = -1$V and $V_{DS} = -0.7$V. At about -3nm (near the source side) the hole charge in the 2 devices is nearly the same, but the drift velocity for the (110) device is much larger than the drift velocity for the (001) device. This results in a much higher ON current for the (110) orientation compared to the traditional (001) orientation. As shown in Fig. 10 the ON current for the short channel device with (110) surface orientation is enhanced by about a factor of 1.5 compared to the (001) case. This ON current enhancement factor is smaller than the respective enhancement factor for the homogenous channel low-field mobility (about 2.0). This difference has two main reasons: short channel effects and the non-linear device characteristics for higher $|V_{DS}|$.

IV. CONCLUSIONS

A comprehensive comparison between strained SiGe HOI (001) and (110) PMOSFETs including important aspects like $C-V$ characteristics, mobility, and ON current has been performed by simulations. The simulations are based on the solution of 6×6 $\vec{k} \cdot \vec{p}$ SE, multi subband BTE, and PE and capture size quantization, strain, crystallographic orientation, and SiGe alloy effects on a solid physical basis. The simulation results show that the strained SiGe HOI PMOSFET with (110) surface orientation has a higher gate capacitance and a much higher mobility and ON current compared to the analog device with the traditional (001) surface orientation.

V. ACKNOWLEDGMENT

The authors thank Prof. S. Mantl and Dr. S. Fest at the Forschungszentrum Jülich for helpful discussions. This work has been partially supported by the EU through the NANOSIL NoE (No. IST–216171).

REFERENCES

[1] I. Aberg, C. N. Chleirigh, and J. L. Hoyt, "Ultrathin-Body Strained-Si and SiGe Heterostrcucture-on-Insulator MOSFETs," *IEEE Trans. Electron Devices*, vol. 53, no. 5, pp. 1021–1029, 2006.

[2] C. D. Nguyen, A. T. Pham, C. Jungemann, and B. Meinerzhagen, "TCAD ready density gradient calculation of channel charge for Strained Si/Strained Si$_{1-x}$Ge$_x$ dual channel pMOSFETs on (001) Relaxed Si$_{1-y}$Ge$_y$," *J. Compu. Electr.*, vol. 3, pp. 193–197, 2004.

[3] A. T. Pham, C. Jungemann, and B. Meinerzhagen, "Physics-based modeling of hole inversion layer mobility in strained SiGe on insulator," *IEEE Trans. Electron Devices*, vol. 54, no. 9, pp. 2174–2182, 2007.

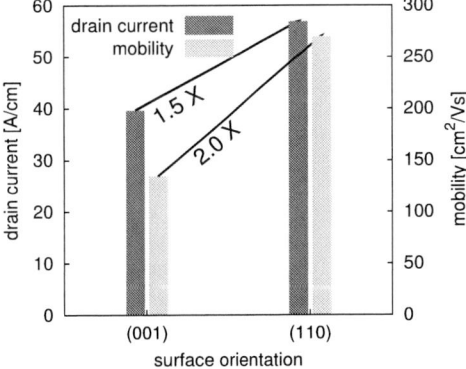

Fig. 10. ON current and homogenous channel low-field mobility for the strained SiGe HOI (001) and (110) PMOSFETs. $V_{GS} = -1$V for all cases. $V_{DS} = -0.7$V for the short channel PMOSFETs. The channel directions for (001) and (110) surface orientation are [110] and [−110], respectively.

[4] T. Krishnamohan, A. T. Pham, C. Jungemann, B. Meinerzhagen, and K. Saraswat, "Mobility modeling in ultra-thin (UT) strained Germanium (s-Ge) quantum well (QW) heterostructure pMOSFETs," in *SiGe and Ge; Processing, and Devices Symposium*, 2008.

[5] H. Irie, K. Kita, K. Kyuno, and A. Toriumi, "In-Plane Mobility Anisotropy and Universality Under Uni-axial Strains in n- and p-MOS Inversion Layers on (100), (110), and (111) Si," in *IEDM Tech. Dig.*, 2004.

[6] M. V. Fischetti, Z. Ren, P. M. Solomon, M. Yang, and K. Rim, "Six-band $k \cdot p$ calculation of the hole mobility in silicon inversion layers: Dependence on surface orientation, strain, and silicon thickness," *J. Appl. Phys.*, vol. 94, pp. 1079–1095, 2003.

[7] A. T. Pham, C. Jungemann, and B. Meinerzhagen, "Microscopic modeling of hole inversion layer mobility in unstrained and uniaxially stressed Si on arbitrarily oriented substrates," *Solid–State Electron.*, vol. 52, pp. 1437–1442, 2008.

[8] A. T. Pham, C. Jungemann, and B. Meinerzhagen, "Deterministic multisubband device simulations for strained double gate PMOSFETs including magnetotransport," in *IEDM Tech. Dig.*, 2008.

[9] A. T. Pham, C. Jungemann, and B. Meinerzhagen, "On the numerical aspects of deterministic multisubband device simulations for strained double gate PMOSFETs," *J. Compu. Electr.*, vol. 8, pp. 242–266, 2009.

[10] D. J. Widiger, I. C. Kizilyalli, and J. J. Coleman, "Two-dimensional transient simulation of an idealized high electron mobility transistor," *IEEE Trans. Electron Devices*, vol. 32, no. 6, pp. 1092–1102, 1985.

[11] M. M. Rieger and P. Vogl, "Electronic-band parameters in strained Si$_{1-x}$Ge$_x$ alloys on Si$_{1-y}$Ge$_y$ substrates," *Phys. Rev. B*, vol. 50, p. 8138, 1994. Erratum.

[12] W. H. Seo and J. F. Donegan, "6x6 effective mass Hamiltonian for heterostructures grown on (11N)-oriented substrates," *Phys. Rev. B*, vol. 68, pp. 0753181–0753188, 2003.

[13] S. Takagi, A. Toriumi, M. Iwase, and H. Tango, "On the universality of inversion layer mobility in Si MOSFET's: Part I–Effects of substrate impurity concentration," *IEEE Trans. Electron Devices*, vol. 41, pp. 2357–2362, 1994.

978-1-4244-6658-0/10 $26.00 © 2010 IEEE

A New Model for the Backscatter Coefficient in Nanoscale MOSFETs

J.-L.P.J. van der Steen[†], P. Palestri[*], D. Esseni[*] and R.J.E. Hueting[†]

[†]MESA+, University of Twente, 7500AE Enschede, The Netherlands
[*]DIEGM, University of Udine, Via delle Scienze 208, 33100 Udine, Italy
Email: j.l.p.j.vandersteen@utwente.nl

Abstract—In this work, we present a new model for the backscatter coefficient in nanoscale MOSFETs. The model assumes that only few backscattering events occur, which is likely to hold for devices with channel length in the order of the carrier mean free path. Both elastic and inelastic scattering mechanisms are accounted for. Moreover, the model naturally captures the effect of degeneracy. The model is compared with Monte–Carlo simulations for a broad range of channel lengths, temperatures, and electric fields, obtaining in general a very good agreement.

I. INTRODUCTION

In today's extremely scaled MOSFETs, the channel length is in the order of the carrier mean free path (λ). Consequently, carriers encounter only few scattering events when moving from source to drain, commonly referred to as quasi-ballistic transport. Some of the carriers do not reach the drain since they are backscattered to the source. The so-called backscatter coefficient (r) is given by the ratio of fluxes of respectively the negatively and positively directed carriers at the virtual source (VS), i.e. the top of the source/channel barrier. Modelling of r is an active area of research [1]–[3] since r provides an estimate of how close to the ballistic limit a device operates. Since carriers experience only few scattering events, the carrier distribution will deviate significantly from the equilibrium one. To illustrate this point, Fig. 1 shows the velocity distribution along the channel of a 32 nm Single–Gate SOI MOSFET obtained from a Multi-Subband Monte Carlo simulator (MSMC) [4]. The device parameters are reported in [5]. Up to a distance of approximately L_{kT}, i.e. where the voltage drop from the VS equals the thermal voltage, the positive velocity distribution essentially retains its initial Maxwellian shape, slightly displaced towards higher v_x. However, as we move further along the channel towards the drain, the distribution starts to deviate significantly from a Maxwellian ('thermal') distribution. The well-known model for r, originally proposed in [6] and elaborated upon in [7], relates r to L_{kT}

$$r = \frac{L_{kT}(1-\beta)}{L_{kT}(1-\beta) + \lambda} \quad (1)$$

with $\beta = \exp(-L/L_{kT})$, $\lambda = 2\mu K_B T/q\nu_{th}$ and $\nu_{th} = \sqrt{2K_B T/\pi m_x}$. However, as shown in [7], (1) implicitly assumes a linear potential profile, and a thermalized carrier distribution. Furthermore, λ is related to the low-field mobility (μ), so it is not directly linked to the individual scattering mechanisms and does not separate the effects of elastic and inelastic scattering.

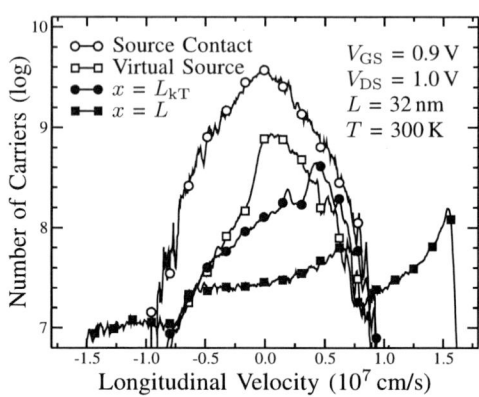

Fig. 1. Velocity distribution at several points along the channel of a 32 nm Single–Gate SOI MOSFET [5]: at the source contact, at the virtual source ($x = 0$), at L_{kT} (i.e. where the voltage drop from the virtual source equals the thermal voltage, $L_{kT} \approx 5$ nm), and at the drain end of the channel (L).

In this work, we introduce a new model for r which, differently from [6], does not require the fluxes to be close to equilibrium nor does it pose any restriction on the shape of the backscattered distribution. Furthermore, this model treats elastic and inelastic scattering separately, and can handle situations with a strong carrier degeneracy (e.g. such as at the VS of MOSFETs operating in the ON-condition). Using a simple test case, the model results are compared with values obtained from MSMC simulations [4], [8], for a wide range of electric field, channel length and temperature values.

II. MODEL FRAMEWORK

Before proceeding to the derivation of the model, we briefly discuss the assumptions which are central to the model framework. First, we assume that only few scattering events occur in the channel and that, in particular, those carriers which are scattered back to the VS, have encountered on average just a single scattering event. Further, we consider a single subband with parabolic energy dispersion.

A. Derivation of the Single–Scattering model

We assume that the flux of particles F^+ moving from the VS to the drain can be considered as a ballistic one, i.e. back-scattering events are very rare and do not significantly affect F^+. We denote as I_0^+ the current associated with F^+. The energy distribution of the charges belonging to F^+ is

978-1-4244-6658-0/10 $26.00 © 2010 IEEE

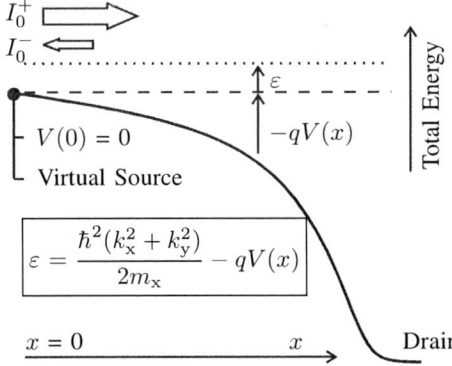

Fig. 2. Schematic of the model framework. I_0^+ and I_0^- are resp. the injected ballistic flux and the backscattered flux, at the virtual source; ε is the total energy and $V(x)$ is the potential.

indicated as $n^+(x,\varepsilon)$. The integral of $n^+(x,\varepsilon)$ over the total energy ε gives the inversion density of carriers moving inside the channel with positive group velocity. The total energy ε includes the kinetic energy and the subband energy. As sketched in Fig. 2, $\varepsilon = 0$ is taken at the bottom of the subband at the VS ($x = 0$). Since we assume F^+ to be ballistic, $n^+(x,\varepsilon)$ is null for $\varepsilon < 0$ at any x along the channel. The number of carriers belonging to F^+ suffering a scattering event per unit time, unit distance and unit energy is given by $n^+(x,\varepsilon)/\tau$, where $1/\tau$ is the scattering rate. At this stage we assume τ not to depend on energy and assume the scattering events to be elastic and isotropic. The backscattered flux I_0^- at the VS can thus be expressed in terms of $n^+(x,\varepsilon)$, τ, and the probability $\alpha(x,\varepsilon)$ that a carrier, once scattered, reaches the virtual source again. We write

$$I_0^- = q \int_0^L \int_0^\infty \frac{n^+(x,\varepsilon)}{\tau} \alpha(x,\varepsilon)\, d\varepsilon\, dx \quad (2)$$

with L the channel length. The probability $\alpha(x,\varepsilon)$ that a carrier with total energy ε, after scattering at position x, has sufficient longitudinal energy to surmount the barrier and to arrive at the source, is given by [6]

$$\alpha(x,\varepsilon) = \frac{1}{\pi} \arccos \sqrt{\frac{qV(x)}{\varepsilon + qV(x)}} \quad (3)$$

We note that $\alpha(x,\varepsilon) = 0.5$ at the source and that it gradually decreases towards the drain. So, even if we do not embrace any concept of K_BT-layer or similar, the scattering events occurring close to the VS are the most effective in back-scattering carriers and, consequently, give the main contribution to r. $n^+(x,\varepsilon)$ can be determined by considering that, under ballistic transport, the carrier energy distribution can be obtained following the approach in [4]. However, here we consider the *total* energy, rather than the longitudinal energy.

Assuming Fermi-Dirac (FD) statistics, we obtain:

$$F(x,\varepsilon) = \frac{N_0^+}{K_BT \ln\left[1 + \exp(\eta)\right]}$$
$$\times \frac{1}{\exp\left(\dfrac{\varepsilon}{K_BT} - \eta\right) + 1} \frac{2}{\pi} \arccos \sqrt{\frac{qV(x)}{\varepsilon + qV(x)}} \quad (4)$$

with N_0^+ the inversion density of carriers with positive group velocity at the VS; η is the degeneracy level, defined as $\eta = [E_F - E_0]/K_BT$ at the VS.
Since the ballistic current is $I_0^+ = qN_0^+ v_{inj}$ with $v_{inj} = \nu_{th} \cdot \left(\mathcal{F}_{1/2}/\mathcal{F}_0\right)$ [9], $\mathcal{F}_0 = \ln\left[1 + \exp(\eta)\right]$ and $\mathcal{F}_{1/2}(\eta)$ the FD integral of order $1/2$, we can express N_0^+ in (4) as a function of I_0^+ and then write

$$n^+(x,\varepsilon) = \gamma(x,\varepsilon) \cdot \frac{I_0^+}{q} \quad (5)$$

with

$$\gamma(x,\varepsilon) = \frac{1}{K_BT \ln\left[1 + \exp(\eta)\right] v_{inj}}$$
$$\times \frac{1}{\exp\left(\dfrac{\varepsilon}{K_BT} - \eta\right) + 1} \frac{2}{\pi} \arccos \sqrt{\frac{qV(x)}{\varepsilon + qV(x)}} \quad (6)$$

Note that $V(x)$ can have an arbitrary profile with $V(x) \geq 0$. Substituting (5) in (2) and assuming no injection from the drain, yields

$$r \equiv \frac{I_0^-}{I_0^+} = \int_0^L \int_0^\infty \frac{\gamma(x,\varepsilon)}{\tau} \alpha(x,\varepsilon)\, d\varepsilon\, dx \quad (7)$$

B. Reduction of the Positive Flux

In presence of non-negligible scattering, the positive flux $I^+(x)$ will significantly decrease along the channel due to the carriers which change momentum from positive to negative. In this section, we discuss how scattering modifies the energy distribution of this flux, denoted with $I^+(x,\varepsilon)$. The reduction of $I^+(x,\varepsilon)$ is proportional to $n^+(x,\varepsilon)$ and to $1/\tau$. Since the scattering is assumed to be isotropic, the probability that a carrier's momentum is redirected towards the source equals $1/2$. Thus, we can write

$$\frac{dI^+(x,\varepsilon)}{dx} = -\frac{qn^+(x,\varepsilon)}{2\tau} = -\frac{1}{2\tau}\frac{I^+(x,\varepsilon)}{v_x^+(x,\varepsilon)} \quad (8)$$

where we have used $I^+(x,\varepsilon) = qn^+(x,\varepsilon)v_x^+(x,\varepsilon)$. The velocity $v_x^+(x,\varepsilon)$ is assumed to be equal to the velocity of a ballistically moving flux, which can be shown to be

$$v_x^+(x,\varepsilon) = \sqrt{\frac{2\varepsilon}{m_x}} \frac{1}{\arccos \sqrt{\frac{qV(x)}{\varepsilon + qV(x)}}} \quad (9)$$

Then, we find for $I^+(x,\varepsilon)$

$$I^+(x,\varepsilon) = I^+(0,\varepsilon) \exp\left(-\int_0^x \frac{1}{2\tau v_x^+(x',\varepsilon)}\, dx'\right) \quad (10)$$

978-1-4244-6658-0/10 $26.00 © 2010 IEEE

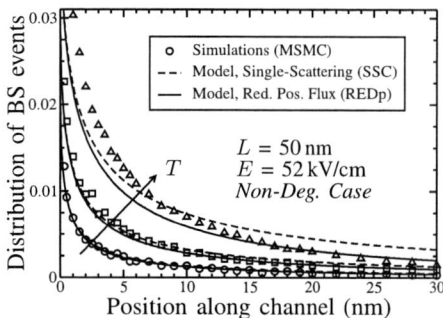

Fig. 3. Distribution of backscattering events contributing to r [10]. The plot shows both the model and MSMC values for T = 100–400 K. The curves are normalized such that integrating yields r, the backscatter coefficient.

The carrier concentration is then given by

$$n^+(x,\varepsilon) = N_0^+ v_{\text{inj}} \gamma(x,\varepsilon) \exp\left(-\int_0^x \frac{1}{2\tau v_x^+(x',\varepsilon)}\, dx'\right)$$

We thus see that r can be obtained from (7) by replacing $\gamma(x,\varepsilon)$ with

$$\gamma^*(x,\varepsilon) = \gamma(x,\varepsilon) \exp\left(-\int_0^x \frac{1}{2\tau v_x^+(x',\varepsilon)}\, dx'\right) \quad (11)$$

III. RESULTS AND DISCUSSION

In this section we compare the model results with the MSMC backscatter characteristics, for varying L, temperature T and longitudinal field E. To facilitate a clear and direct comparison, we employ, unless stated otherwise, a constant accelerating field, a template material (as in [8]) and no injection from the drain. The template material features a single spherical valley ($m_{\text{x}} = 1.0 m_0$), and $\mu = 400$ cm²/Vs at 300 K, obtained by adjusting the coupling constant of acoustic phonons. Since we consider a single circular subband with parabolic energy dispersion, the scattering rate with elastic acoustic phonons, $1/\tau$ in the model, is constant over energy and has been taken from the MSMC simulator without any adjustment in all forthcoming figures.

A. Elastic Scattering

Initially, we use Boltzmann's approximation, which in our model is naturally obtained by setting $\eta \ll 0$ in (6). The impact of degeneracy will be discussed in the second part of this section. Fig. 3 shows, for several temperatures, the distribution of carriers backscattered to the VS versus the position at which the momentum was redirected towards the source. Most scattering events which contribute to r occur close to the VS. Clearly, the single–scattering model [(6)–(3), labeled SSC] is able to capture the general features of the distribution produced by the MSMC. The quantitative agreement in the tail of the distribution can be improved by accounting for reduction of the positive flux [(7) with (11), REDp], particularly at high temperature and under low-field conditions, i.e. conditions which induce enhanced scattering.

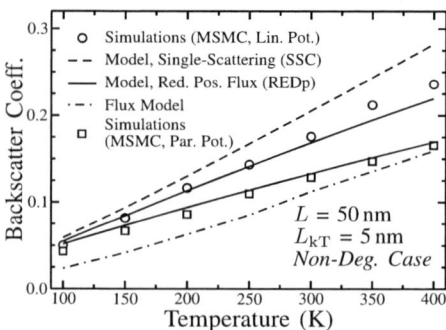

Fig. 4. r vs. temperature, along with the Flux model values from (1). The MSMC and REDp results are shown for a linear potential profile (corresponding to $E = 52$ kV/cm) and a parabolic potential. The SSC and Flux model values are shown for the linear profile only.

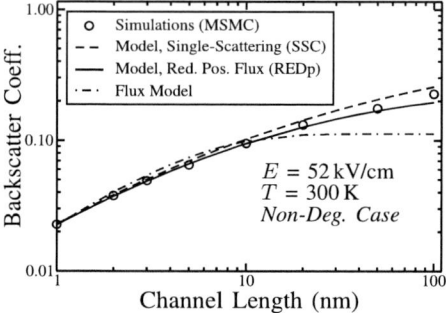

Fig. 5. r vs. channel length; $E = 52$ kV/cm, T = 300 K. The SSC model starts to deviate from the MSMC values for the longer channel lengths. The agreement can be improved by accounting for reduction of the positive flux.

Fig. 4 shows the values of r, corresponding to the curves in Fig. 3, along with the values predicted by the conventional model based on the thermal fluxes ["Flux model", (1)]. The value of λ, to be used in (1), is obtained by noting that the low-field mobility μ is related to the scattering rate ($1/\tau$ in the SSC model) through $\mu = q\tau/m_{\text{x}}$, hence $\lambda = \tau\sqrt{2\pi K_B T/m_{\text{x}}}$. By accounting for the reduction of the positive flux, the model values essentially coincide with the MSMC results for both the linear and parabolic potential profile. The Flux model systematically underestimates r, which is attributed to the non-thermal nature of the negative flux [11].

Fig. 5 depicts r versus L, showing that the SSC tracks fairly well the MSMC results; again, the agreement can be improved by accounting for the positive flux reduction. The r predicted by the Flux model, instead, saturates for the longer channels. The findings from Figs. 4 and 5 are concisely summarized in Fig. 6, which shows r for different values of E. Clearly, the entire range of $r(E)$ is captured by the REDp model. For the higher fields (corresponding to $r < 0.2$), the single–scattering assumption of the SSC model turns out to be a good approximation, judging from the agreement in the SSC and MSMC results. Eq. (1) shows good agreement for the lowest fields only, which can be explained by noting that it assumes close to equilibrium (hence low-field) conditions.

In Figs. 3–6 we have assumed a non-degenerate electron gas. Fig. 7, instead, reports r as a function of the degree of

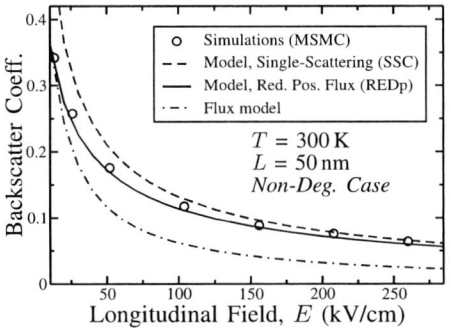

Fig. 6. r vs. longitudinal field, showing the model with and without reduction of the positive flux, along with the MSMC and the flux model results.

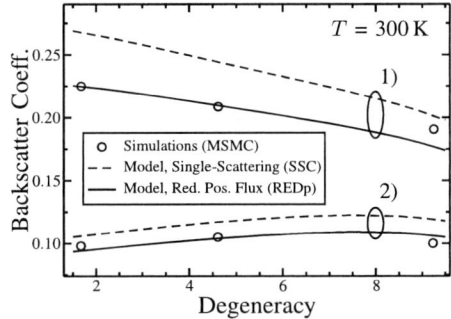

Fig. 7. r for different values of the degeneracy level η, showing two sets of curves: set 1) uses $L = 20\,\mathrm{nm}$, $E = 13\,\mathrm{kV/cm}$; set 2) is obtained with $L = 50\,\mathrm{nm}$, $E = 156\,\mathrm{kV/cm}$. In either case, r is found to be only weakly dependent on η.

degeneracy at the VS, for both a low [set 1)] and a high-field case [set 2)]. Although r decreases [set 1)] or reaches a maximum [set 2)], the changes in absolute terms are very modest, both in the model and in the simulations.

B. Inelastic Scattering

So far, we have only considered backscattering due to elastic scattering (e.g. acoustic phonons). In order to include also optical phonon (OP) absorption and emission, (3) can be extended to

$$
\alpha^*(x, \varepsilon) = \left(\frac{1}{\tau_{AP}} \frac{1}{\pi} \arccos \sqrt{\frac{qV(x)}{\varepsilon + qV(x)}} \right.
$$
$$
+ \frac{1}{\tau_{OE}} \frac{1}{\pi} \arccos \sqrt{\frac{qV(x)}{\varepsilon - E_{OP} + qV(x)}} \theta\left(\varepsilon - E_{OP}\right)
$$
$$
\left. + \frac{1}{\tau_{OA}} \frac{1}{\pi} \arccos \sqrt{\frac{qV(x)}{\varepsilon + E_{OP} + qV(x)}} \right) \tag{12}
$$

in which τ_{OA}^{-1} and τ_{OE}^{-1} are the scattering rates of OP absorption and emission, respectively; E_{OP} is the OP energy, and $\theta(x)$ is the unit step function which ensures that OP emission can occur only if $\varepsilon \geq E_{OP}$. To illustrate the impact of OP scattering, Fig. 8 depicts r versus E_{OP}. The OP coupling constant D_{OP} in the MSMC was increased to have significant OP scattering. D_{OP} used in Figs. 3–7 was lower ($2 \times 10^8\,\mathrm{eV/cm}$), causing scattering to be dominated by acoustic rather than

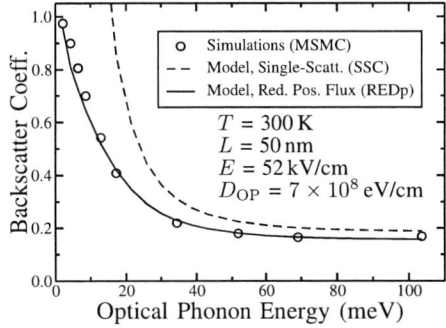

Fig. 8. r vs. the optical phonon energy E_{OP}. For each E_{OP}, the scattering rates have been directly obtained from the scattering rates in the MSMC simulator.

optical phonons. The model values are obtained from (7) and (12). To include the positive flux reduction, (11) is modified by adding, for each scattering mechanism, an exponential term with the corresponding τ. For each E_{OP}, the values of τ_{OA} and τ_{OE} have been extracted from the scattering rates calculated in the MSMC simulator, which renders the model and MSMC results directly comparable.

Both the model and the MSMC predict r to be constant for E_{OP} above a given energy, since elastic scattering becomes dominant. For smaller values of E_{OP}, r quickly increases due to the onset of OP emission. In fact, for very small E_{OP}, the OP scattering rate greatly exceeds the acoustic scattering rate ($>10\times$). As a result, r approaches unity.

IV. CONCLUSION

In this work we present a new model for the backscatter coefficient, which is based on the assumption of only few scattering events in the channel and does not pose any restriction on the distribution of backscattered carriers. The model accounts for the effect of degeneracy and separately takes into account both elastic and inelastic scattering. Using a simple test case, the model results have been compared with Monte–Carlo simulations, showing generally a very good agreement.

ACKNOWLEDGMENT

The authors would like to thank Prof. L. Selmi for his constant support and encouragement during this work.

This work was supported by NXP Semiconductors, Eindhoven, The Netherlands and in part by the ENIAC project MODERN.

REFERENCES

[1] V. Barral et al., IEEE Trans. El. Dev., vol. 56, no. 3, pp. 408–419, 2009.
[2] C. Jeong et al., IEEE Trans. El. Dev., vol. 56, no. 11, pp. 2762–2769, 2009.
[3] M. Fischetti et al., J. Comput. Electr., vol. 8, no. 2, pp. 60–77, 2009.
[4] L. Lucci et al., IEEE Trans. El. Dev., vol. 54, no. 5, pp. 1156–1164, 2007.
[5] P. Palestri et al., Solid-State Electr., vol. 53, pp. 1293–1302, 2009.
[6] M. Lundstrom et al., IEEE Trans. El. Dev., vol. 49, no. 1, pp. 133–141, 2002.
[7] R. Clerc et al., IEEE Trans. El. Dev., vol. 53, no. 7, pp. 1634–1640, 2006.
[8] P. Palestri et al., in Proc. IEDM, 2006, pp. 1–4.
[9] F. Assad et al., IEEE Trans. El. Dev., vol. 47, no. 1, pp. 232–240, 2000.
[10] P. Palestri et al., IEEE Trans. El. Dev., vol. 52, no. 12, pp. 2727–2735, 2005.
[11] R. Clerc et al., in Proc. ULIS, 2008, pp. 125–128.

978-1-4244-6658-0/10 $26.00 © 2010 IEEE

Multi-Subband Monte Carlo simulation of bulk MOSFETs for the 32nm-node and beyond

Carlos Sampedro[*], Francisco Gámiz[*], Andres Godoy[*], Raul Valín[†], Antonio García-Loureiro[†], and Noel Rodríguez[*]

[*]Departamento de Electrónica y Tecnología de Computadores
Universidad de Granada, 18071 Granada (SPAIN). Email: csampe@ugr.es
[†]Departamento de Electrónica y Computación
Universidade de Santiago de Compostela, Santiago de Compostela 15782 (SPAIN)

Abstract— With the 32 nm node in mass production, simulations tools have to include quantum effects to correctly describe the behavior of the devices. The Multi-Subband Monte Carlo (MSB-MC) approach constitutes today's most accurate method for the study of nanodevices with important applications to SOI devices. This work presents an MSB-MC study of 32 nm node and beyond bulk-nMOSFETs which still are the mainstream in the semiconductor industry.

I. Introduction

As the semiconductor industry has developed new fabrication techniques and device concepts during the last 45 years to obey Moore's Law, the simulation community has been improving in parallel the computational models in order to predict and optimize the performance of next generation devices. These efforts allow an important reduction in design time and cost. Nowadays, an aggressive scaling of the devices is required in order to fit the requirements of the International Technology Roadmap for the Semiconductor Industry (ITRS) [1]. The study of such structures, from a simulation point of view, demands the inclusion of quantum effects to correctly catch the complex electrostatics of the device and an accurate transport model to take into account the effects of non-uniform transport. This is a very challenging task for the simulation community and different solutions mainly based in full quantum approaches or Monte Carlo methods are preferred. Concerning full quantum approaches, the most extended are based on numerical solutions of the Schrödinger equation or the Non-Equilibrium Green's Functions theory (NEGF) [2]. In a quantum model, the transport of charged particles is treated coherently according to a quantum wave equation. In the simplest case of a single-particle Hamiltonian, carriers are considered as non-interacting waves described by the Schrödinger equation. The introduction of scattering mechanisms in the simulations involves a very high computational cost and for this reason, only simplified models can be used in practical quantum simulations. Ensemble Monte Carlo (EMC) simulators are widely used since they present several advantages compared to full quantum approximations. Among them, it can be highlighted a reduced computational cost, the possibility of considering a wide variety of scattering mechanisms and high accuracy for devices with silicon thicknesses as low as a few nanometers [3]. The inclusion of quantum

effects can be afforded from two points of view. In one hand, a correction term to the electrostatic potential can be included to mimic the carrier concentration profile obtained from the solution of the Schrödinger equation. Different models have been developed following this philosophy, giving a good accuracy-computational cost ratio [4]–[6]. In the other hand, the Multi-Subband Ensemble Monte Carlo approach (MSB-EMC) couples the solution of the Schrödinger equation solved in the confinement direction of the device for different slices and the Boltzmann Transport equation (BTE) solved by the MC method from source to drain. This method has been widely used in the last years [7]–[11] and provides what is to date the most detailed description of carrier transport in semiconductor devices, since the scattering rates are obtained from a quantum solution. However, this method is difficult to implement in systems where quantum confinement is not geometrical but electro-statically induced. As a consequence, previous works are mainly devoted to SOI devices and only a few of them to bulk MOSFETs which still constitute the mainstream in Integrated Circuit (IC) production [12].

This paper presents, a thorough study of ultrashort channel bulk-nMOSFETs addressed for the sub-45 nm technological nodes using a MSB-EMC. The main differences with SOI MSB-EMC simulators are depicted and the obtained results are compared to equivalent fully depleted single gate SOI devices (FD-SGSOI) addressed for the same technological nodes.

II. Description of the Bulk-MSB simulator

The Multi-Subband method has been developed from the mode-space approach of quantum transport [13] and considers a decoupling between the transport and the confinement problems. In this way, the Schrödinger equation is only solved in the confinement direction whereas the transport problem can be dealt in a semiclasical way. The main limitation of the method arise from the decoupling method itself, avoiding the inclusion of coherence phenomena in a direct way that could be of special interest for the study of some structures. This is not the case for standard MOSFET devices where quantum transport effects can be neglected or included separately, i.e. tunneling mechanisms as band-to-band [14]. From the simulation point of view, our transistor is considered as a stack of slices along the confinement direction z (Figure 1). The elec-

978-1-4244-6658-0/10 $26.00 © 2010 IEEE

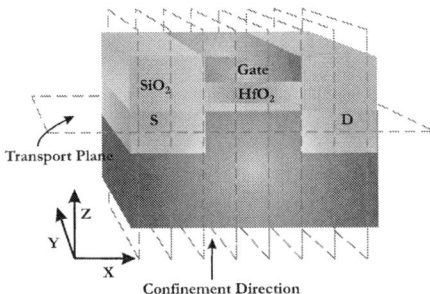

Fig. 1. Simulated structure for the considered bulk-nMOSFET. 1D Schrödinger equation is solved for each grid point in the transport direction whereas Boltzmann Transport equation is solved by the Monte Carlo method in the transport plane.

trostatics of the system is calculated from the self-consistent solution of the 2D Poisson's equation and the 1D Schrödinger equation solved for each considered slice and conduction band valley. As a result, the evolution along the transport direction, x, for the i-th valley and the ν-th subband of the eigen-energies, $E_{i,\nu}(x)$, and the wave functions, $\xi_{i,\nu}(x,z)$ are obtained. The transport along the Source/Drain axis is calculated from the MC solution of the Boltzmann Transport equation (BTE) in the transport plane. The conduction band is considered as non-parabolic, and the effects on the device characteristics are included in both confinement following [7] and transport approaches. In opposition to standard EMC simulators and according to the space-mode approach, the driving field undergone by a simulated super-particle is calculated from the derivative of $E_{i,\nu}(x)$, i.e. the driving force is different for each of the subbands corresponding to a given valley. To obtain the subband population in each grid point in the transport direction, a re-sampling of the super-particles using the cloud-in-cell method is performed. This population is used to weight the corresponding density of probability $|\xi_{i,\nu}(x,z)|^2$ to calculate the electron density, $n(x,z)$. Finally, the electrostatic potential is updated by solving the 2D Poisson equation using the previous $n(x,z)$ as input. This approach is especially appropriate for the study of 1D confinement in nanoscale devices. The drawback compared with semi-classical MC codes, is an important increase of the computational effort. This issue is partially overcome thanks to the parallel implementation of the code.

Concerning scattering models, acoustic and intervalley phonons are included [7] whereas surface roughness scattering is described in [15], [16] and Coulomb interaction in [17]. All the models consider non parabolic and ellipsoidal bands. There is an important increase in the computational effort of the code because it is necessary to calculate a scattering table for each slice and update them for every new solution of the Schrödinger equation to keep the self-consistence of the simulation. The reason comes from the evolution of the eigen-energies and wave functions along the transport direction which makes vary the scattering rates associated for a given mechanism in different positions of the device. In one hand,

the different distribution of the eigen-energies changes the allowed inter-subband energies and, in the other hand, the variation of the wave functions whith the position affects the form factor even in intra-band processes. Figure 2 shows the wavefunction corresponding to the ground level for a 22 nm bulk-nMOSFET, following the NANOSIL templates, when $V_{GS} = V_{DS} = 1\ V$. It can be observed an important variation in the shape, and therefore in the form factor, for source/drain and channel. Moreover, in the end of the channel, close to the drain edge, the high lateral field applied strongly affects the wave form.

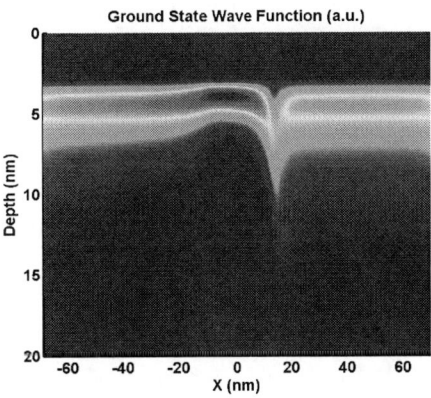

Fig. 2. Wavefunction evolution for the fundamental subband in a 22 nm channel length bulk-nMOSFET with $V_{GS} = V_{DS} = 1\ V$. The strong field near the drain edge affects the wavefunction in an important manner.

The main issue when the MSB method is applied to bulk MOSFETs comes from the physical structure of the device. In SOI devices, specially in MuGFETs and FD-SGSOI, the potential well necessary to induce the quantum confinement has an important geometrical component [18]. The existence of more than one gate in MuGFETs or a buried oxide, BOX, in the case of SGSOI, creates a geometrically defined quantum well which is modulated by the applied voltage. In a bulk device, the quantum well is induced in two main ways. In the channel area, as the gate bias is increased, the conduction band in the substrate bends and the inversion region is created in a electrically induced well. In the source and drain regions, the quantum well is induced by the potential barrier in the PN junction formed by the highly doped region region and the substrate. The semiconductor there, is in accumulation near the interface with the field oxide and the carrier concentration fades as the junction region is reached. In such situation, the quantum-well profile slowly changes (specially when Gaussian doping profiles are considered) and, therefore, is far from the step-like approximation. The eigenvalues seeking must be then carefully carried out. Moreover, to perform a realistic simulation of bulk-MOSFETs it is necessary to consider an important part of the substrate which implies a big number of grid points in the confinement direction. It is mandatory then, to use a numerical scheme for the solution of the Schrödinger equation that can handle these issues. One of the fastest,

978-1-4244-6658-0/10 $26.00 © 2010 IEEE

easiest to implement and most commonly used methods is the Numerov's method [19]. However to be applied in a correct way the grid spacing must be uniform with the direct implication of a huge number of grid points. It has been also shown [20] that in some conditions, specially for slow variations of the confining potential and close eigenstates, the standard method can skip eigenvalues. Again, this situation can be observed in the highly doped source and drain areas. In such scenario, the method implemented for the bulk-MSB simulator is based in the shooting algorithm [21] which has been widely used for mobility extraction in 2D electron gas systems [22], [23]. Furthermore, the method has been also successfully applied for highly doped accumulated or depleted systems like polysilicon gates [24]. Finally, the computation time penalty compared to Numerov's scheme is compensated by the fact that more efficient non-uniform grids can be used.

III. SIMULATION RESULTS

A set of bulk-nMOSFETs with different channel length (from 32 to 16 nm) have been considered to show the performance of the MSB-MC simulator applied to bulk devices. For the sake of comparison, the results will be compared with FD-SGSOI devices. In all the cases a gate stack consisting of 2.4 nm of HfO_2 on top of 0.7 nm of SiO_2 (EOT = 1.1 nm) and midgap metal gate is considered. Source and Drain regions are doped with $N_{Dpeak} = 5.2 \times 10^{19}$ cm^{-3} and Gaussian decays following the templates proposed in the NANOSIL group [10]. The substrate is doped uniformly with $N_A = 3 \times 10^{18}$ cm^{-3}. For the FD-SGSOI device, a lightly doping is considered in the channel with uniform doping in source and drain regions and a lateral doping decay as its bulk counterpart to model the source-channel-drain transitions. The channel thickness is 7 nm and the BOX is 20 nm thick.

In order to compare the confinement effects in both devices, Figure 3 shows the electron concentration profile in the midpoint of the channel for the 32 nm devices at low drain voltage and equivalent inversion charge (3.5×10^{12} cm^{-3}).

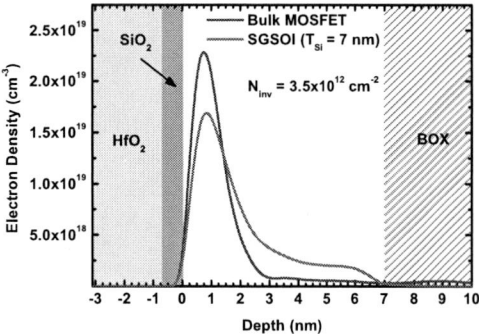

Fig. 3. Electron concentration profile in the midpoint of the channel for the 32 nm devices.

As can be observed, the geometrical confinement induced by the BOX spreads the charge in the Si slab whereas in the bulk device the charge is mainly located only in the

region affected by the gate. It is also shown how the charge vanishes inside the oxide since the penetration of the wave functions are considered in the calculations. Figure 4 shows the inversion charge in the channel at high drain conditions. Short channel effects (SCEs) are more clearly noticed in the bulk case as expected (top). Thanks to the BOX and the geometrical confinement, the electrostatic control of the gate is stronger in the FD-SGSOI device (bottom).

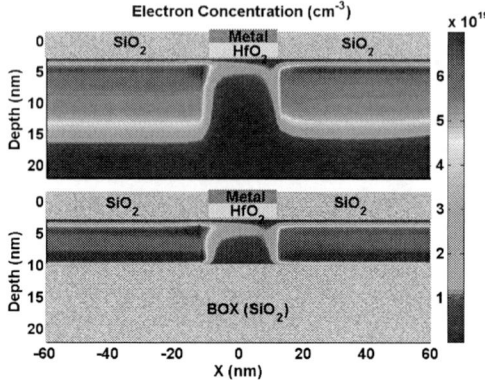

Fig. 4. Electron concentration profile for the 22 nm length bulk (top) and FD-SGSOI (bottom) devices at $V_{GS} = V_{DS} = 1$ V. The gate stack is also shown for sake of clarity.

The differences between both devices are also clearly shown in the subband distribution. Figure 5 shows the first subbands for the situation depicted in Figure 4. Due to the effect of geometrical confinement, the separation among the subbands is bigger in the FD-SGSOI case (bottom) than in bulk devices (top). This fact is specially relevant in the source and drain areas. The effect of geometrical confinement is also important in the tranpost properties. Due to the redistribution of the subbands, Δ_2 valleys in the SOI device are more populated than Δ_4 valleys, with the consequent decrease in the effective conduction mass.

This fact combined with a less effective injection in the channel of carriers belonging to excited subbands due to the higher value of the barrier, lead to a higher current level for the FD-SGSOI devices compared to its bulk counterpart as shown in Figure 6.

IV. CONCLUSION

This work presents a thorough study of state of the art bulk-nMOSFETs which still constitute the mainstream technology in commercial ICs. A MSB-EMC simulator has been used to forecast the behavior of the bulk devices addressed for the 32nm and beyond nodes using the most accurate approach for the study of electron transport in nanodevices which, up to now, has been mainly applied to SOI devices.

ACKNOWLEDGMENT

The authors would like to thanks the support given by the European Union (EUROSOI+ CA and NANOSIL), the

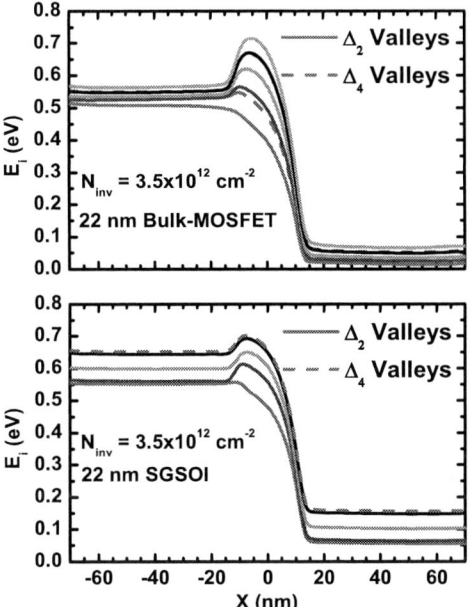

Fig. 5. Subband distribution for the 22 nm bulk-nMOSFET (top) and FD-SGSOI (bottom) at V_{DS}=1V. The differences on the confinement are clearly shown in the different separation ordering among the subbands for the two cases.

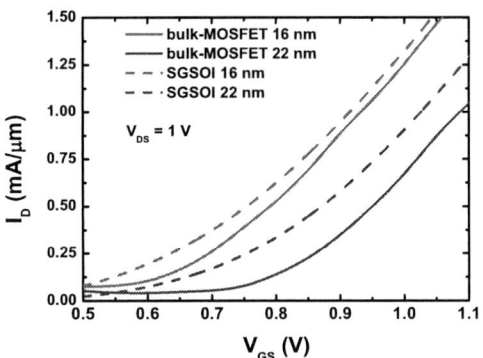

Fig. 6. I_D vs. V_{GS} curves for bulk-nMOSFETs (solid) and FD-SGSOI (dashed) for 22 (blue) and 16 nm (red) channel lengths.

Spanish Government (FIS2008-05805, TEC2008-06758-C02-01), and Junta de Andalucia (P06-TIC1899).

REFERENCES

[1] International Roadmap for the Semiconductor Industry, "http://www.itrs.net/," 2009.

[2] H. Tsuchiya, A. Svizhenko, M. P. Anatram, M. Ogawa, and T. Miyoshi, "Comparison of Non-Equilibrium Green's Function and Quantum-corrected Monte Carlo Approaches in Nano-MOS Simulation," *J. Comp. Elec*, vol. 4, pp. 35–38, 2005.

[3] R. Ravishankar, G. Kathawala, U. Ravaioli, S. Hasan, and M. Lund-strom, "Comparison of Monte Carlo and NEGF Simulation of Double Gate MOSFETs," *J. Comp. Elec.*, vol. 4, pp. 39–43, 2005.

[4] A. Asenov, A. R. Brown, and J. R. Watling, "Quantum Corrections in the Simulation of Decanano MOSFETs," *Solid State Elec.*, vol. 47, no. 7, pp. 1141–45, 2003.

[5] D. Ferry, S. Ramey, L. Shifren, and R. Akis, "The Effective Potential in Device Modeling: the Good, the Bad and the Ugly," *J. Comp. Elec.*, vol. 1, pp. 59–65, 2002.

[6] C. Sampedro-Matarin, F. J. Gamiz, A. Godoy, and F. Garcia-Ruiz, "The Multivalley Effective Conduction Band-Edge Method for Monte Carlo Simulation of Nanoscale Structures," *Electron Devices, IEEE Transactions on*, vol. 53, no. 11, pp. 2703–2710, 2006.

[7] M. V. Fischetti and S. E. Laux, "Monte Carlo Study of Electron Transport in Silicon Inversion Layers," *Physical Rev. B*, vol. 48, no. 4, pp. 2244–2274, 1993.

[8] J. Saint-Martin, A. Bournel, F. Monsef, C. Chassat, and P. Dollfus, "Multi Sub-Band Monte Carlo Simulation of an Ultra-Thin Double Gate MOSFET with 2D Electron Gas," *Sem. Sci. and Tech.*, vol. 21, pp. 29–31, Apr. 2006.

[9] E. Sangiorgi, P. Palestri, D. Esseni, C. Fiegna, and L. Selmi, "The Monte Carlo approach to transport modeling in deca-nanometer MOSFETs," *Solid-State Electronics*, vol. 52, no. 9, pp. 1414 – 1423, 2008.

[10] P. Palestri *et al.*, "A comparison of advanced transport models for the computation of the drain current in nanoscale nmosfets," *Solid-State Electronics*, vol. 53, no. 12, pp. 1293 – 1302, 2009. Papers Selected from the Ultimate Integration on Silicon Conference 2009, ULIS 2009.

[11] C. Sampedro, F. Gámiz, A. Godoy, R. Valín, A. García-Loureiro, and F. Ruiz, "Multi-Subband Monte Carlo study of device orientation effects in ultra-short channel DGSOI," *Solid-State Electronics*, vol. 54, no. 2, pp. 131 – 136, 2010. Selected Full-Length Extended Papers from the EUROSOI 2009 Conference.

[12] M. Fischetti, T. O'Regan, S. Narayanan, C. Sachs, S. Jin, J. Kim, and Y. Zhang, "Theoretical Study of Some Physical Aspects of Electronic Transport in nMOSFETs at the 10-nm Gate-Length," *Electron Devices, IEEE Transactions on*, vol. 54, pp. 2116 –2136, sept. 2007.

[13] R. Venugopal, Z. Ren, S. Datta, M. S. Lundstrom, and D. Jovanovic, "Simulating Quantum Transport in Nanoscale Transistors: Real Versus Mode-Space Approaches," *Journal of Applied Physics*, vol. 92, pp. 3730–3739, Oct. 2002.

[14] M. V. Fischetti and S. E. Laux, "Monte Carlo analysis of electron transport in small semiconductor devices including band-structure and space-charge effects," *Phys. Rev. B*, vol. 38, pp. 9721–9745, Nov 1988.

[15] F. J. Gamiz, J. B. Roldan, J. A. Lopez-Villanueva, P. Cartujo-Cassinello, and J. E. Carceller, "Surface Roughness at the Si–SiO$_2$ Interfaces in Fully Depleted Silicon–On–Insulator Inversion Layers," *J. Appl. Phys.*, vol. 86, pp. 6854–6863, Dec. 1999.

[16] F. Gámiz, P. Cartujo-Cassinello, J. B. Roldán, and F. Jiménez-Molinos, "Electron transport in strained si inversion layers grown on sige-on-insulator substrates," *Journal of Applied Physics*, vol. 92, no. 1, pp. 288–295, 2002.

[17] F. Jiménez-Molinos, F. Gámiz, and L. Donetti, "Coulomb scattering in high-κ gate stack silicon-on-insulator metal-oxide-semiconductor field effect transistors," *Journal of Applied Physics*, vol. 104, no. 6, p. 063704, 2008.

[18] J.-P. Colinge, *FinFETs and Other Multi-Gate Transistors.* Springer Publishing Company, Incorporated, 2007.

[19] J. M. Blatt, "Practical points concerning the solution of the Schrödinger equation," *Journal of Computational Physics*, vol. 1, no. 3, pp. 382 – 396, 1967.

[20] J. Kim, C. Chen, and R. Dutton, "An effective algorithm for numerical Schrödinger solver of quantum well structures," *Journal of Computational Electronics*, vol. 7, pp. 1–5, march 2008.

[21] J. Lopez-Villanueva, I. Melchor, and J. Jimnez-Tejada, "Modified Scroedinger Equation Including Nonparabolicity for the Study of a Two Dimensional Electron Gas," *J. App. Phys.*, vol. 75, no. 8, pp. 4267–4269, 1993.

[22] F. Gamiz, J. Lopez-Villanueva, J. Banqueri, J. Carceller, and P. Cartujo, "Universality of electron mobility curves in mosfets: a monte carlo study," *IEEE Trans. Elec. Dev.*, vol. 42, pp. 258–265, Feb. 1995.

[23] F. J. Gamiz and M. V. Fischetti, "Monte Carlo Simulation of Double-Gate Silicon-On-Insulator Inversion Layers: The Role of Volume Inversion," *J. Appl. Phys.*, vol. 89, pp. 5478–5487, May 2001.

[24] N. Rodriguez, F. Gamiz, R. Clerc, G. Ghibaudo, and S. Cristoloveanu, "The Quantization Impact of Accumulated Carriers in Silicide-Gated MOSFETs," *Electron Device Letters, IEEE*, vol. 29, pp. 628 –631, june 2008.

Modeling Study on Carrier Mobility in Ultra-Thin Body FinFETs with Circuit-Level Implications

Mirko Poljak[*] and Tomislav Suligoj
FER-ZEMRIS, University of Zagreb
Unska 3, HR-10000 Zagreb, Croatia
[*]Corresponding author: mirko.poljak@fer.hr

Vladimir Jovanović[1]
DIMES, Delft University of Technology
Feldmannweg 17, 2628 CT Delft, The Netherlands
[1]also working at FER-ZEMRIS

Abstract—**Influence of different active surface orientations and fin-width scaling on electron and hole mobility in ultra-thin body FinFETs is examined. Results of mobility modeling are validated on experimental data, including (111)-oriented FinFETs from our previous work which are here proven to exhibit no mobility degradation caused by fin-width fluctuations. We show that (111)-oriented FinFETs are the optimum solution when performance and layout area of 6T SRAM cell are concerned, since they enable SRAM with a minimum number of fins per cell.**

I. Introduction

FinFET is the most promising candidate to succeed bulk MOSFET due to fabrication compatibility with CMOS process [1]. It is believed that in System-on-Chip (SoC) integration FinFET would be used for circuits which are highly sensitive to short channel effects (SCE) or threshold voltage variations, such as SRAM cells [2]. If FinFETs are to be introduced into future CMOS, its fin-width will be scaled from 12 nm at the 22-nm node down to 6 nm at the 11-nm node, in order to maintain immunity to SCEs [2]. At this scale, quantum confinement plays a crucial role in carrier transport and it is therefore necessary to investigate electron and hole mobility behavior with the downscaling of fin-width for all standard active-surface orientations, i.e. (100), (110) and (111). Moreover, an optimum solution is needed for the current-matching of n-channel and p-channel devices in CMOS inverters [3].

In this work, physics-based and computationally light effective-mass approximation (EMA) and momentum-relaxation-time approximation (MRTA) are used to calculate carrier mobilities. FinFETs are simulated as double-gate devices which is a reasonable assumption [4] having in mind FinFET dimensions for advanced CMOS nodes [2]. Electron mobility in double-gate structures has been extensively studied in the past, both experimentally and numerically, in particular for (100)- and (110)-oriented devices [5,6]. Therefore, the emphasis in our work is placed on hole mobility behavior in highly scaled double-gate devices. Experimental results exist for holes on (100) and (110) surfaces [7,8] whereas the modeling lags behind. Recently,

hole transport in double-gate devices with ultra-thin body (UTB) has been investigated, but without comparison to the experimental data [9]. In the following sections, we present modeling results on hole and electron mobility behavior with the downscaling of body thickness (T_{Si}). Results of hole mobility modeling are validated by comparison with existing experimental data for (100), (110) and (111) orientations. Furthermore, we discuss circuit-level implications for CMOS inverter and 6T SRAM cell design using FinFETs with different active-surface orientations.

II. Physics-Based Mobility Modeling

Effective mass approximation (EMA) is used to calculate phonon-limited electron and hole mobility. EMA is computationally most effective physics-based approach for carrier transport modeling, especially when compared to state-of-the-art multi-subband Monte Carlo simulators [10]. However, it has been argued that EMA for electrons fails when (110) surface is considered due to strong non-parabolicity of silicon Δ-valley in (110)-direction [11]. To include this effect, we used a modified quantization mass for Δ_2-electrons on (110) surface from [12], which was obtained by fitting EMA to results obtained using the method of linear combination of bulk bands [13]. The modeling of hole transport is more challenging compared to electron transport due to nonparabolic and anisotropic valence bands. Only recently, $k \cdot p$ method was used to simulate hole transport properties [9]. Although good agreement was achieved with experimental results for (100)-oriented MOSFET [14], this method has a great computational burden. De Michielis et al. [15] have developed a semianalytical anisotropic nonparabolic model for the 2D hole energy dispersion on (100) and (110) surfaces. This model is used in our work to calculate the effective masses for the calculation of hole mobility using EMA. Hole confinement masses for (111) surface were taken from [16], whereas the effective masses in transport direction were fitted to obtain the inversion charge densities and distributions as in [9], where a 6-band $k \cdot p$ method is employed. The resulting hole distributions between the gates at a given inversion charge are shown in Fig. 1 and a good agreement with [9] is achieved.

This work was sponsored by the Ministry of Science, Education and Sports of the Republic of Croatia under contracts 036-0361566-1567 and 036-0982904-1642.

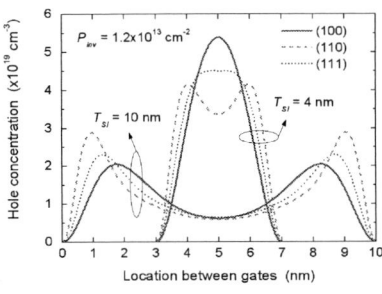

Fig. 1. Hole distribution between gates for T_{Si} of 4 nm and 10 nm, for all surface orientations.

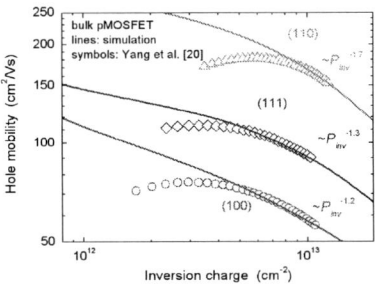

Fig. 2. P_{inv}-dependence of surface-limited hole mobility for different surfaces.

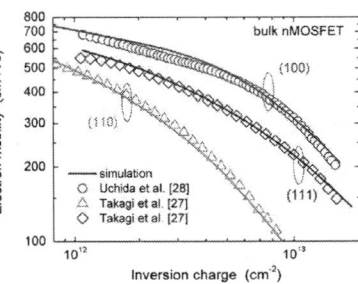

Fig. 3. Electron mobility model calibrated on experimental data for nMOSFETs.

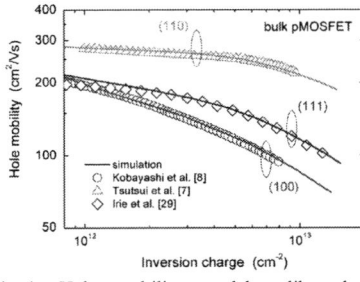

Fig. 4. Hole mobility model calibrated on experimental data for bulk pMOSFETs.

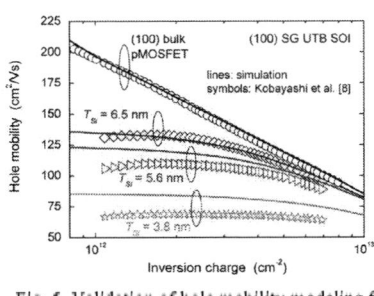

Fig. 5. Validation of hole mobility modeling for (100)-oriented single-gate UTB SOI devices.

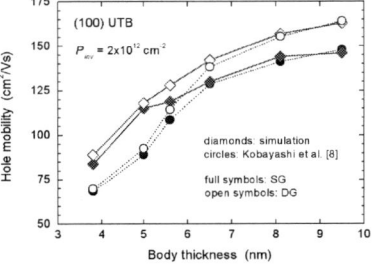

Fig. 6. Model validation for (100)-oriented single- and double-gate UTB MOSFETs.

Using the EMA, the Schrödinger-Poisson system is solved self-consistently to obtain the envelope functions of eigenstates and eigenenergies. Momentum-relaxation time approximation (MRTA) is used to calculate carrier mobilities. Acoustic and optical phonon scattering is taken into account, including intra- and intervalley transitions for electrons, and intra- and intersubband transitions for holes. Calculation procedures are taken from [17] and adapted for 2DEG (as in [18]) and 2DHG. Parameters needed to calculate the phonon-limited mobility were calibrated on experimental data for bulk electron and hole mobility on (100), (110) and (111) surface. The model assumes standard channel directions, <110> for (100) and (110), and <112> for (111) surface. Surface-roughness-limited mobility is included analytically as a power-law dependence on the inversion charge density. We have previously reported on the extraction of these dependencies for electrons [19]. The extraction for holes using data from [20] is shown in Fig. 2. Since we are interested in low-doped body FinFETs with high-quality interfaces assumed, Coulomb scattering is not included in our model. The ΔT_{Si}-induced scattering [21] is neglected because we investigate FinFETs with body thickness down to 6 nm, as this is a targeted fin-width at the 11-nm technology node [2]. The effect of acoustic phonon confinement is not included as it affects the mobility behavior only quantitatively and is significant only in n-type devices [22,23].

Our model is calibrated on bulk MOSFET mobility data for (100), (110) and (111) surfaces. Different sources of experimental data are used for calibration. Mobility data at high inversion densities is needed for surface-limited mobility extraction, whereas phonon-limited mobility model is calibrated on measurements which contain results for both bulk and UTB devices. Dependences of surface-limited hole mobility on inversion charge density (Fig. 2) are in accordance with hole distributions in Fig. 1. Surface-scattering matrix elements depend on wavefunction derivatives at the Si-SiO$_2$ interface [24]. As can be seen in Fig. 1, (110)-oriented devices have the steepest hole distribution near the interface which is linked to stronger P_{inv} dependence compared to (100) and (111) double-gate transistors. Simulation results shown in Fig. 3 and 4 for bulk electron and hole mobility confirm that good agreement to experimental data is achieved.

Next, we validate the hole-mobility model through comparison with the available experimental data. The comparison for (100)-oriented single-gate UTB silicon-on-insulator (SOI) with body thickness of 6.5 nm, 5.6 nm and 3.8 nm from [8] is shown in Fig. 5. Simulation results, obtained using the same model parameters as for bulk mobility, match experimental data for devices with body thickness down to 6.5 nm. For thinner body devices, our modeling overestimates the hole mobility due to ΔT_{Si}-induced scattering which is clearly present in devices from [8]. Nevertheless, the shape of the mobility dependence on inversion charge density closely follows experimental data which verifies the calculations of phonon-limited hole mobility. Hole mobility dependence on body-thickness is shown in Fig. 6, for single- and double-gate UTB MOSFETs. Hole mobility is extracted at P_{inv} of $2 \cdot 10^{12}$ cm^{-2}, as in [8], to make a proper comparison. An excellent agreement is achieved both for single- and double-gate devices with body thickness down to 6 nm. Double-gate MOSFETs from [8] also suffer from ΔT_{Si}-induced scattering which reduces hole mobility in extremely thin channels and is not covered by our model. In p-type devices with (100) active surface, monotonic mobility degradation with body thickness downscaling is observed both in experimental and in simulation results. As

978-1-4244-6658-0/10 $26.00 © 2010 IEEE

(110)-oriented p-type transistors are concerned, experimental data is available only for single-gate UTB pMOSFETs. Mobility measurements from [7] are used for comparison with simulations in Fig. 7. All fabricated (110) p-type devices exhibit stronger Coulomb scattering as evident from the mobility drop at lower inversion carrier densities [7], which is mostly caused by lower quality of Si-SiO$_2$ interfaces at (110) silicon surfaces [3]. Since Coulomb scattering is not included in our model, there is a stronger discrepancy between simulations and experimental data at mid-inversion charge densities. However, the comparison in Fig. 7 demonstrates correct overall physical behavior of simulated hole mobility, including the mobility enhancement around T_{Si} of 4 nm.

III. RESULTS AND DISCUSSION

Carrier mobility dependence on fin-width of nFinFETs and pFinFETs is shown in Fig. 8. NFinFETs with (100) sidewalls exhibit mobility enhancement for body thickness around 3 nm. This is caused by two interfering mechanisms with T_{Si} downscaling: the overall increase in form-factors due to stronger confinement in thinner devices, and the repopulation effect because of the increasing occupancy of Δ_2-levels and lower transport mass of Δ_2-electrons. This enhancement is masked in experimental data for (100) double-gate MOSFETs because of strong ΔT_{Si}-induced scattering [21]. (110) and (111)-oriented nFinFETs experience a monotonic mobility degradation with the downscaling of T_{Si}. Nevertheless, electron mobility is rather constant down to T_{Si} of 8 nm and 6 nm for nFinFETs with (110) and (111) sidewalls, respectively. Figure 8e also contains experimental results for electron mobility in (111) FinFETs from our previous work [25], which agree well with the simulations (measured mobility values were taken at mid-effective field value, corresponding roughly to P_{inv} of 10^{12} cm^{-2}). This agreement indicates very low carrier scattering induced by fin-width-fluctuations and consequently high fin-width uniformity along the channel in [25]. PFinFETs show qualitatively opposite

mobility behavior when compared to their n-type counterparts. Namely, (110) and (111)-oriented devices exhibit a weak mobility enhancement, whereas pFinFETs with (100) sidewalls show no mobility enhancement. (100) pFinFETs experience a strong mobility degradation for T_{Si} under 6 nm, while (110) and (111) transistors have a mobility peak at T_{Si} around 4 nm. The observed qualitative hole mobility behavior both at low and high inversion charge densities matches results from [9], where a 6-band $k \cdot p$ – Poisson solver is employed, for all simulated surfaces. The reason for mobility enhancement in (110) and (111) pFinFETs comes from the same effect as for electrons in (100) nFinFETs, i.e. an interplay between form-factor increase and repopulation of hole bands. Donetti et al. [9] have demonstrated dependence of heavy-hole transport mass on body thickness as an additional mechanism in hole mobility enhancement. However, as shown here, mobility enhancement effect is reproduced by using EMA which reaffirms the suitability of the well-calibrated EMA for hole mobility calculations. Experimental hole mobility of a pFinFET device with T_{Si} of 1.9 nm is added to the plot in Fig. 8f. Because of the high uniformity of the fin-width in these devices the simulation

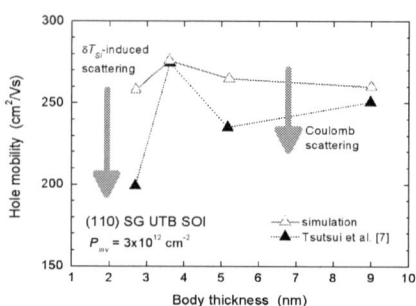

Fig. 7. Comparison of simulation and experimental data for (110)-oriented single-gate UTB MOSFETs.

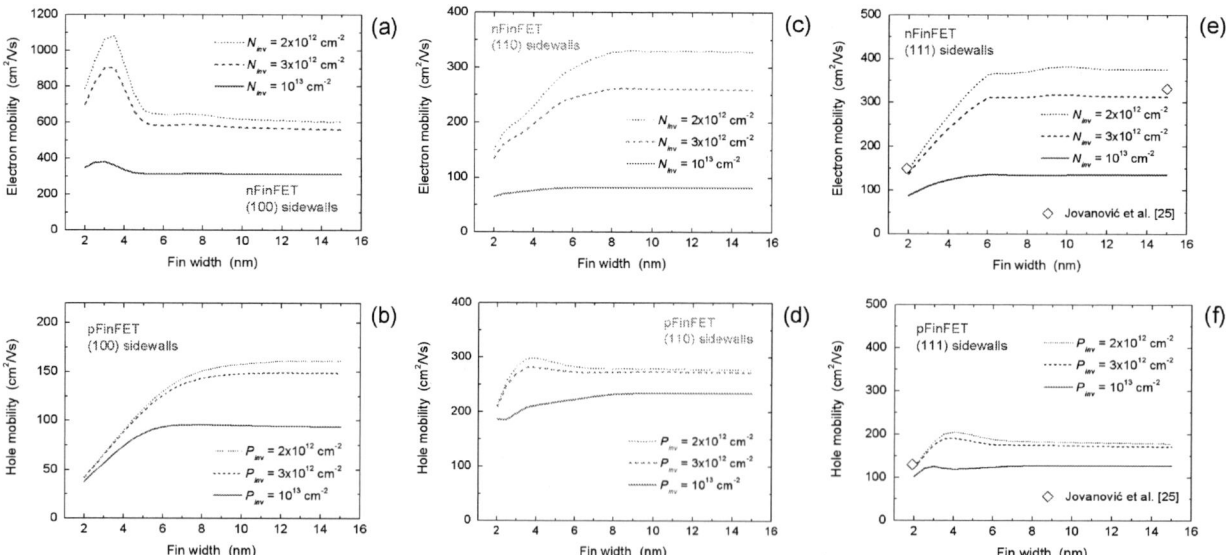

Fig. 8. Carrier mobility dependence on fin-width for nFinFETs is shown in a-c-e, while for pFinFETs the results are presented in b-d-f.

978-1-4244-6658-0/10 $26.00 © 2010 IEEE

results match well with the experimental data. We note that nFinFET and pFinFET with (111) active surface have approximately equal carrier mobilities for T_{Si} of 6 nm and thicker (135 cm^2/Vs and 128 cm^2/Vs, respectively) at high inversion charge density, i.e. in the on-state. Also, 2-nm thick (111) FinFETs have approximately matching carrier mobilities independently of inversion charge density.

Current matching (i.e. symmetrical currents) of pull-up (PU) and pull-down (PD) transistors is the most important condition for good performance of a CMOS inverter. The symmetry implies low static noise margin (SNM) which is inherently important for SRAM also [3]. In standard bulk MOSFET technology, CMOS inverter is designed with wider pMOSFET than nMOSFET due to higher electron mobility. On the other hand, the effective channel width of FinFETs is determined by the fin-height. Therefore, channel width adjustment can be done either by adjusting of the fin-height or the number of fins. The first proposal is impractical because a complicated process must be used to fabricate FinFETs with different fin-heights on the same chip [26]. Figure 9 shows electron-to-hole mobility ratio (μ_e/μ_h) for different fin-widths in three different FinFETs. The mobility values are taken at $N_{inv} = P_{inv} = 10^{13}$ cm^{-2}. Current-matching between nFinFET and pFinFET can be achieved by using different number of parallel fins having in mind the ratio between carrier mobilities. With the symmetric V_{th}, the fin-number ratio must be equal to inverse of mobility ratio for current matching ($N_{p\text{-}fins}/N_{n\text{-}fins} = \mu_e/\mu_h$). FinFETs with (100) sidewalls exhibit an increasing μ_e/μ_h ratio with T_{Si} downscaling, whereas the ratio in (110) and (111) oriented FinFETs is almost independent of T_{Si}. The results for FinFETs with (111) sidewalls ($\mu_e/\mu_h \approx 1$) clearly show that a matched CMOS inverter nFinFET and pFinFET can be constructed using devices with the same number of fins. On the contrary, inverter with (100) devices would demand 3x higher number of fins for pFinFETs (4 fins total), and inverter with (110) devices would demand 3x higher number of fins for nFinFETs (4 fins total). Therefore, FinFET with (111) sidewalls are an optimum solution for efficient use of silicon real-estate (2 fins total). This is illustrated in Fig. 10 where layouts of 6T SRAM cell using different FinFETs are shown. SRAM cells using FinFETs with (100) or (110) sidewalls need 8 fins, while SRAM cell with (111) FinFETs demands only 4 fins, including the pass-gate (PG) transistor. Reduction of layout area is obvious.

IV. CONCLUSION

Hole and electron mobility behavior with downscaling of fin-width is investigated in FinFETs with (100), (110) and (111) active surface orientations. Phonon scattering rates are calculated within MRTA based on EMA. The experimental results of hole mobility in ultra-scaled double-gate devices were accurately reproduced using the physics-based and computationally light approach using EMA. Excellent matching was achieved for devices which do not exhibit strong ΔT_{Si}-induced scattering, typically down to T_{Si} of 6 nm. For FinFETs with (111) sidewalls with extremely-high fin-width uniformity, the presented model matches the experimental results down to the fin-width of 1.9 nm. For circuit-level implications, it is demonstrated that (111)-

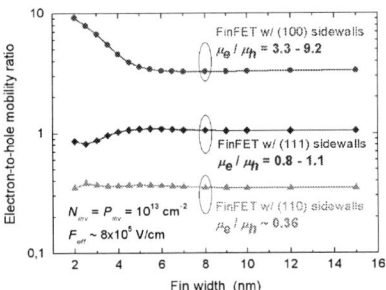

Fig. 9. Dependence of electron-to-hole mobility ratio on fin-width.

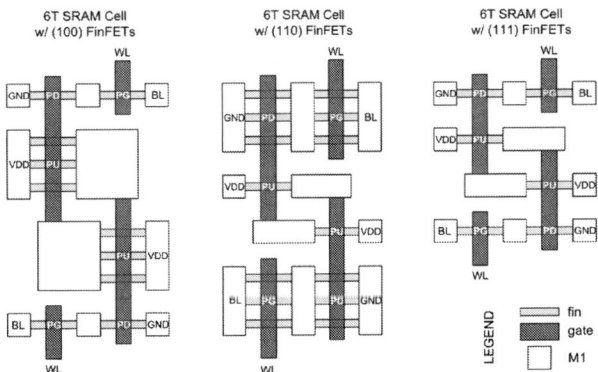

Fig. 10. Layouts of 6T SRAM cells using FinFETs with different active-surface orientations. Fin-separation in multiple-fin transistors is reduced in this schematic 6T SRAM layout for space saving purposes.

oriented FinFETs have matching carrier mobilities at high inversion densities, which can lead to the more efficient use of the silicon real-estate in 6T SRAM cells compared to (100) or (110)-oriented FinFETs.

REFERENCES

[1] D. Hisamoto et al., IEEE Trans. Electron Dev. 47 (12), p. 2320, 2000.
[2] H. Kawasaki et al., IEDM Tech. Dig., p. 1-4, 2009. [3] Y. Taur and T.H. Ning, Fundamentals of Modern VLSI Devices, Cambridge University Press, 1998. [4] N. Serra et al., IEDM Tech. Dig., pp. 1-4, 2009. [5] D. Esseni et al., IEEE Trans. Electron Dev. 50 (12), p. 2445, 2003. [6] G. Tsutsui et al., IEDM Tech. Dig., p. 729, 2005. [7] G. Tsutsui et al., IEEE Electron Dev. Lett. 26 (11), p. 836, 2005. [8] S. Kobayashi et al., IEDM Tech. Dig., p. 707, 2007. [9] L. Donetti et al., Solid-State Electronics 54 (2), p. 191, 2010. [10] M. De Michielis et al., IEEE Trans. Electron Dev. 56 (9), p. 2081, 2009. [11] K. Uchida et al., IEDM Tech. Dig., pp. 1-4, 2008. [12] N. Serra et al., Proc. ULIS, p. 113, 2009. [13] D. Esseni et al., Phys. Rev. B, Condens. Matter 72 (16), p. 165342, 2005. [14] L. Donetti et al., Semicond. Sci. Technol. 24, p. 035016, 2009. [15] De Michielis et al., IEEE Trans. Electron Dev. 54 (9), p. 2081, 2007. [16] E.X. Wang et al., IEEE Trans. Electron Dev. 53 (8), p. 1840, 2006. [17] M. Lundstrom, Fundamentals of Carrier Transport, 2nd edition, Cambridge University Press, 2000. [18] S. Takagi et al., J. App. Phys. 80 (3), p. 1567, 1996. [19] M. Poljak et al., Proc. ULIS, p. 21, 2010. [20] M. Yang et al., IEEE Electron Dev. Lett. 24 (5), p. 339, 2003. [21] K. Uchida et al., IEDM Tech. Dig., p. 33.5.1, 2003. [22] L. Donetti et al., IEEE Electron Dev. Lett. 30 (12), p. 1338, 2009. [23] L. Donetti et al., App. Phys. Lett. 88 (12), p. 122108, 2006. [24] D. Esseni et al., IEEE Trans. Electron Dev. 50 (7), p. 1665, 2003. [25] V. Jovanović et al., Proc. ESSDERC, p. 241, 2009. [26] Y.X. Liu et al., Microelectron. Eng. 84 (9), p. 2101, 2007. [27] S. Takagi et al., IEEE Trans. Electron Dev. 41 (12), p. 2363, 1994. [28] K. Uchida et al., IEDM Tech. Dig., p. 47, 2002. [29] H. Irie et al., IEDM Tech. Dig., p. 225, 2004.

978-1-4244-6658-0/10 $26.00 © 2010 IEEE

TCAD Based Device Architecture Exploration Towards Half-Terahertz Silicon/Germanium Heterojunction Bipolar Technology

Arturo Sibaja-Hernandez, Shuzhen You[1], Stefaan Van Huylenbroeck,
Rafael Venegas, Kristin De Meyer[1], Stefaan Decoutere
IMEC, Kapeldreef 75, B-3001 Leuven, Belgium
Tel: (+) 32 16 288 566, e-mail: Arturo.Sibaya-Hernandez@imec.be
[1]Also with KULeuven-ESAT, Kasteelpark Arenberg 10, B-3001 Leuven, Belgium

Abstract— **A 2D TCAD based device architecture exploration of SiGe:C NPN HBTs is presented. Two novel and one conventional self-aligned architecture are explored by process and device simulation. All these three architectures show their capability of achieving maximum oscillation frequency (f_{max}) of 500 GHz for scaled layout rules.**

I. INTRODUCTION

SiGe:C HBTs have continuously pushed the usable operating frequency towards sub-millimeter or so-called terahertz frequency range, defined as extending from 300 GHz to 10 THz, and have become appealing for applications in the field of high-speed communication, radar, and THz imaging and sensing [1-3]. The European joint research project DOTFIVE [2] has set its goal at 500 GHz f_{max} at room temperature. Partners of this project have achieved SiGe:C HBTs of 400 GHz f_{max}, which are today's state-of-the-art SiGe:C HBTs [4-5]. These partners achieve this milestone through completely different architectures. Reference [4] summarizes the improvement of the conventional double-polysilicon fully self-aligned selective epitaxial growth HBTs (FSA-SEG), while [5] demonstrates a novel self-aligned SiGe:C HBT architecture featuring a single-step epitaxial collector-base process or grow-in-one-go process (G1G) [6].

In order to further approach the next milestone of 500 GHz f_{max}, it is important to scale down the devices. Scalability is explored in this work for three kinds of architectures:
(1) G1G architecture [5, 6], which is a novel self-aligned architecture.
(2) Low-parasitic collector (LPC) construction for high-speed SiGe:C HBTs [7], which is also a novel self-aligned architecture.
(3) Fully self-aligned (FSA) architecture, a conventional self-aligned architecture [4, 8].

To bring confidence in the TCAD based device architectures exploration the reference TCAD platform is firstly validated with recent silicon material from IMEC, which is presented in section II. The updated TCAD platform is used then in the exploration of the three self-aligned

architectures, which is demonstrated in section III. Sensitivity study on the emitter widths, emitter-base spacers, and poly-to-emitter overlap is performed to explore the limits of the different architectures. It will be shown that all these three architectures are capable of achieving 500 GHz f_{max}.

II. VERIFICATION OF TCAD PLATFORM

In this work, a 2D-TCAD process and device simulation platform for high speed bipolar HBT devices with f_T/f_{max} values exceeding 200 GHz is used as a reference TCAD platform [9, 10]. This 2D-TCAD platform is implemented in the process and device simulators TSUPREM-4 [11] and MEDICI [12] respectively. The 2D-TCAD process simulation, calibrated with SIMS and TEM pictures, has been successfully complemented with the use of SSRM measurements, which provide emitter-base and collector-base junction delineation and two-dimensional carrier distributions. An excellent agreement in vertical and lateral electrical junction depths between SSRM measurement and the 2D-TCAD simulation validates the 2D process calibration [10]. Accurate 2D process simulations enable excellent correspondence between measured and simulated electrical characteristics. Confidence has been brought by the simulation for updated IMEC G1G processed devices. Fig. 1 compares the simulated electrical characteristics to the measured results of the devices featuring 220/400 GHz f_T/f_{max} reported in [5]. The accuracy of the calibrated reference platform provides the baseline for further architecture exploration.

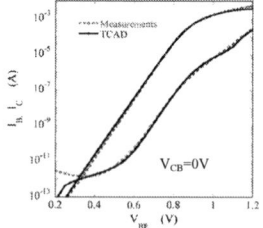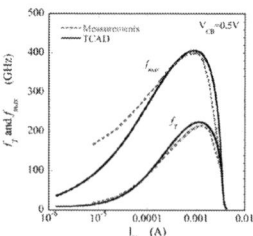

Fig.1 Gummel plot (left) and f_T/f_{max} (right) overlays between measurements and simulations for the G1G reference device.

This work is supported by European joint research project DOTFIVE.

978-1-4244-6658-0/10 $26.00 © 2010 IEEE

III. ARCHITECTURE EXPLORATION

In this section, the potential of the G1G, LPC and FSA architectures to deliver half-terahertz f_{max} will be explored. The simulated 2D half-structure of these three architectures are shown in Fig. 2. The important parameters to scale are described in the figures as well. They are effective emitter window width (W_{emi}), width of L-shaped spacer (L_{spa}), width of cavity (W_{cav}), and poly-to-emitter enclosure (PE). In order to assure the consistency in the denotation among different architectures, PE is defined as the enclosure of poly emitter to effective emitter window.

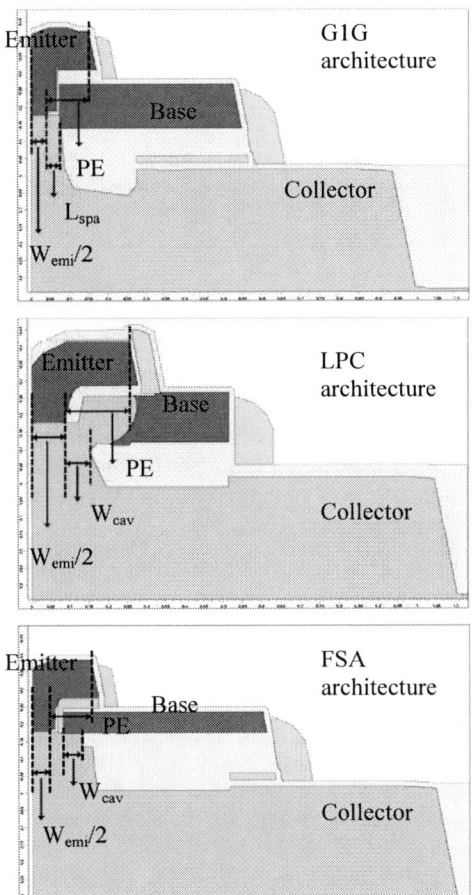

Fig. 2. Simulated 2D half-structures of the G1G, LPC, and FSA architectures.

For a fair comparison between architectures, the simulation used identical as-grown collector-base epitaxial layers (Fig. 3). However, the different architectures imply different thermal budgets for the diffusion of the dopants, such that the final profiles are slightly different. This approach allows us to separate the impact of the different architectures from the vertical profile optimization. The drawback is that it is more difficult to compare the simulated results with experimental results. As the simulator was calibrated based on a 2-step germanium profile, this composition of the epitaxial base was used for the present architecture study. The germanium and doping profiles for the as-grown profiles are shown in Fig. 3. Compared to the profile in the reference platform [5], the improved profile in Fig. 3 uses a two step in-situ As-doped collector layer instead of phosphorous to better control the end-of-line collector profile resulting in better RF performances. This updated platform is used in further simulation in the G1G, LPC and FSA architectures.

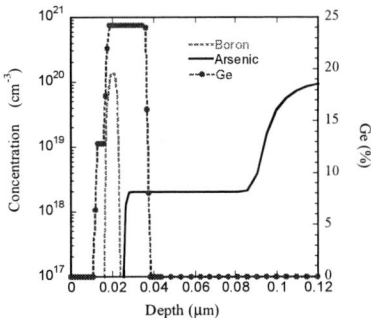

Fig. 3. Profiles of the as-grown base epitaxial layer used in the G1G, LPC and FSA architectures study.

The simulations for these three architectures were performed with the same set of coefficients as in the calibrated reference platform. The active concentration of boron in polysilicon was set to 2e19cm⁻³. The emitter contact on top of the polysilicon emitter presents a surface recombination velocity for holes (VSURFP) 1.3e5cm/s. External resistances are lumped at the emitter (15Ω-μm) and base (10Ω-μm) contacts to represent the contact parasitics.

A. G1G architecture exploration

A sensitivity study on the PE enclosure, L_{spa} width and effective emitter window width is performed to explore the potential of the G1G architecture. The initial structure presents the reference values of PE=112nm, L_{spa}=37nm and W_{emi}=80nm. Figs. 4-6 show on the left side the RF performance of explored G1G devices and on the right side the normalized values of emitter-base capacitance (C_{BE}), base-collector capacitance (C_{BC}), emitter resistance (R_E), base resistance (R_B), product of $R_E C_{BE}$ and product of $R_B C_{BC}$. The electrical characteristics shown in Figs. 4-6 are normalized to the electrical characteristics of the initial structure.

Fig. 4(left) shows the impact of PE on RF performance and Fig. 4(right) gives the normalized parameters afore mentioned. When PE is scaled down, the polysilicon-base link resistance is decreased resulting in a reduced base resistance. In addition to the reduction of base resistance, scaled PE reduces the couple capacitance between the emitter and external poly base, which benefits f_T. However R_E increases as PE is scaled down, which brings a trade-off between R_E and C_{BE} in improving RF performance. As a result, the simulated peak f_T decreases as PE decreases due to a higher emitter resistance. f_{max} increases as PE decreases as a result of the reduction in R_B. The next parameter to be varied is L_{spa}. As shown in Fig. 5, f_T decreases when L_{spa} decreases due to an increase of collector resistance while decreasing the collector active area of the device. f_{max} benefits from the reduction of R_B

978-1-4244-6658-0/10 $26.00 © 2010 IEEE

and C_{BC} with decreasing L_{spa}. The scaling of L_{spa} however should be limited to avoid tunneling current that occurs when the two highly doped regions, the emitter and the external-base regions, are brought close to each other. f_{max} is significantly impacted by the dimension of the emitter window width through the internal base resistance reduction [14]. As expected, f_{max} increases as W_{emi} decreases, shown in Fig. 6, due to the reduction of R_B and C_{BC}. f_T decreases because of increase in R_E and R_C. Figs. 4-6 show that the most important parameter boosting f_{max} is the reduction in the emitter window width (W_{emi}), and that for the smallest simulated emitter window width of 40nm, the G1G architecture is capable of achieving 500 GHz f_{max}.

B. LPC architecture exploration

The LPC architecture presents a novel collector construction for high-speed SiGe:C HBTs that substantially reduces the parasitic base collector capacitance by selectively under-etching of the collector region [7, 13]. A sensitivity study on PE, W_{emi} and W_{cav} is performed to explore the potential of the LPC architecture. Combined with proper scaling of the PE and cavity dimensions, the LPC architecture is pushed towards 500 GHz f_{max} as W_{emi} is scaled down. As shown in Fig. 7, for the scaled device with PE=62nm, W_{cav}=13nm, f_{max} increases as W_{emi} decreases due to the reduction in R_B and C_{BC}. f_{max} reaches 500 GHz for the smallest simulated dimension.

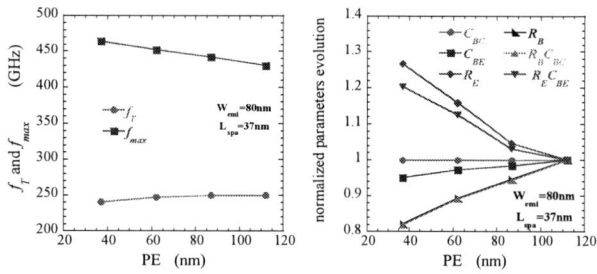

Fig. 4. Impact of the poly to emitter enclosure (PE) on (left) RF performance and (right) normalized parameters C_{BC}, C_{BE}, R_E, R_B, R_BC_{BC}, and R_EC_{BE}.

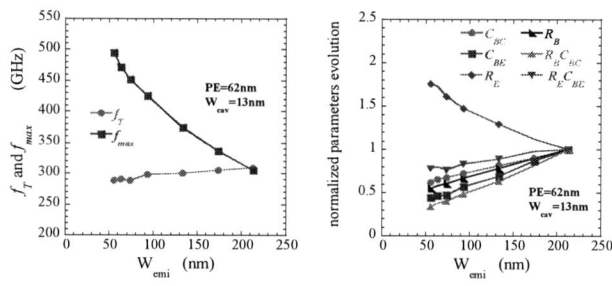

Fig. 7. Impact of effective emitter window width on (left) RF performance and (right) normalized parameters C_{BC}, C_{BE}, R_E, R_B, R_BC_{BC}, and R_EC_{BE}.

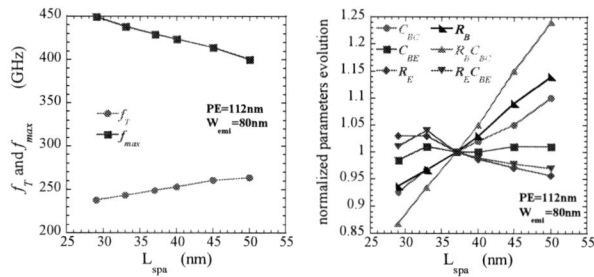

Fig. 5. Impact of the width of the L-shaped spacer (L_{spa}) on (left) RF performance and (right) normalized parameters C_{BC}, C_{BE}, R_E, R_B, R_BC_{BC}, and R_EC_{BE}.

C. FSA architecture exploration

The FSA architecture is a conventional self aligned architecture widely used in industry [4, 8]. A sensitivity study on PE, W_{emi} and W_{cav} is performed to explore the potential of the FSA architecture. As expected, the devices with smaller PE and W_{emi} give higher f_{max}. In this architecture the thickness of the external poly base layer was chosen not as thick as in the G1G and LPC architectures to: i) keep low topography, ii) maintain the parasitic C_{BE} low, and iii) have better comparison with published material [4, 15]. With thinner external poly base, the sheet resistance of poly base is higher compared to the G1G and LPC architectures. In addition to the higher external base resistance the FSA architecture presents higher C_{BC} which results in lower f_{max} for the same scaled layout rule. As shown in Fig. 8, the smallest device gives approximately 400 GHz f_{max}. Improving the active boron concentration in the external poly base can effectively reduce the base resistance hence improve f_{max}. In [15], a 300Ω/sq poly base sheet resistance is obtained for a 1080°C spike anneal in the FSA architecture. This is translated in an increase of the boron activation level to 6e19cm^{-3} for the simulations of the FSA architecture. It brings the simulated f_{max} values to 500GHz performance for the smallest device (Fig. 9).

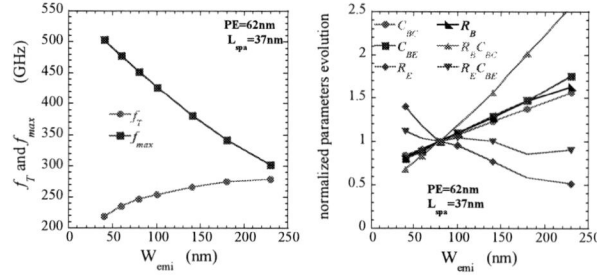

Fig. 6. Impact of effective emitter window width (W_{emi}) on (left) RF performance and (right) normalized parameters C_{BC}, C_{BE}, R_E, R_B, R_BC_{BC}, and R_EC_{BE}.

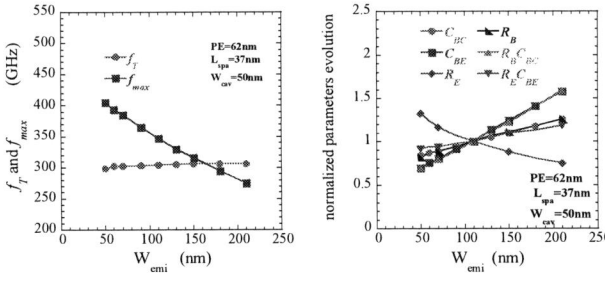

Fig. 8. The impact of effective emitter window width on (left) RF performance and (right) normalized parameters C_{BC}, C_{BE}, R_E, R_B, R_BC_{BC}, and R_EC_{BE}.

D. Comparison of the G1G, LPC and FSA architectures.

Fig. 9 overlays the RF performances of the G1G, LPC and FSA architectures, which have the capability of reaching 500GHz f_{max} as the effective emitter window is scaled down. Some published results are positioned in the plot as well, which shows that the simulation can predict the RF performance with reasonable accuracy. Compared to [5] with non-optimized collector, the G1G simulation in this work uses the improved two-step As-doped collector that gives higher f_T and f_{max} because of the sharper low-to-high dopant transition in the collector and a lower R_C. The simulated profile of the FSA architecture is completely different with respect to [4], where an optimized standard collector module with selectively implanted collector has been used.

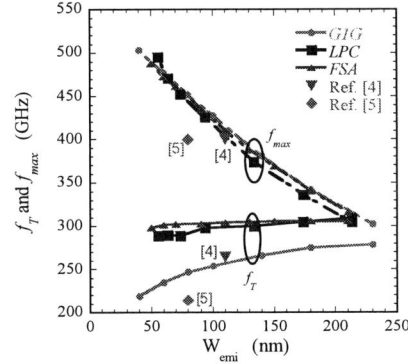

Fig. 9. Comparison of RF performance between the G1G, LPC and FSA architectures

Tab. 1. Scaled parameters used to obtain the results in Fig. 9.

	PE	L_{spa}	W_{cav}
G1G architecture	62	37	
LPC architecture	62	37	13
FSA architecture	62	37	50

For a common set of process simulation models, the sensitivity study shows the advantage of the G1G and LPC architectures over the conventional FSA architecture in

pursuing 500GHz f_{max} due to a reduction of parasitic base-collector capacitance. The difference in f_T between the different architectures comes from the difference in thermal budget after the growth of the SiGe:C layer. The intrinsic base in the G1G architecture sees the thermal budget of the formation of the external poly base, while this is not the case for the other two architectures. This difference results in lower f_T values for the G1G architecture compared to the LPC and FSA architectures. Despite the lower f_T values of the G1G architecture it reaches f_{max} values similar to the LPC architecture due to a lower base resistance. Note that in this work the performed TCAD exploration has not considered vertical profile engineering to further boost the RF performance.

IV. CONCLUSION

2D TCAD based device architecture exploration has been presented. A sensitivity study of the G1G, LPC and FSA architectures shows that these three architectures are capable of achieving 500 GHz f_{max} with aggressively scaled down of the emitter window width.

ACKNOWLEDGMENT

The authors wish to acknowledge the support of the European Commission in the frame of the FP7 IST project DOTFIVE (IST-216110).

References

[1] U. Pfeiffer, et al., "Opportunities for silicon at mmWave and terahertz frequencies", Proc. IEEE BCTM, pp. 149-156, 2008.

[2] http://www.dotfive.eu

[3] S. Decoutere, et al., "Advanced process modules and architectures for half-terahertz SiGe:C HBTs", Proc. IEEE BCTM, pp. 9-16, 2009.

[4] P. Chevalier, et al., "A conventional double-polysilicon FSA-SEG Si/SiGe:C HBT reaching 400 GHz f_{max}", Proc. IEEE BCTM, pp. 1-4, 2009.

[5] S. Van Huylenbroeck, et al., "A 400 GHz f_{max} fully self-aligned SiGe:C HBT architecture", Proc. IEEE BCTM, pp. 5-8, 2009.

[6] J. Donkers, et al., "A novel fully self-aligned SiGe:C HBT architecture featuring a single step epitaxial collector-base process", Proc. IEDM, pp. 655-658, 2007.

[7] B. Heinemann, et al., "A low-parasitic collector construction for high-speed SiGe:C HBTs", Proc. IEEE IEDM, pp. 251-254, 2004.

[8] T. F. Meister, et al., "SiGe bipolar technology with 3.9 ps gate delay", Proc. IEEE BCTM, pp. 103-106, 2003.

[9] A. Sibaja-Hermandez, "Optimization of the RF performance of SiGe:C HBTs for BiCMOS technologies", Ph. D theis, K. U. Leuven, Belgium, 2007.

[10] A. Sibaja-Hermandez, et al., "2D-TCAD process calibration for a high speed QSA SiGe:C HBT verified with SSRM", ECS Transactions, Vol. 3, pp. 387-395, 2006.

[11] TSUPREM-4 computer code from Synopsys, Inc..

[12] MEDICI computer code from Synopsys, Inc..

[13] H. Rücker, et al., "SiGe:C BiCMOS Technology with 3.6 ps Gate Delay", Proc. IEEE IEDM, pp. 121-124, 2003.

[14] S. M. Sze, Physics of semiconductor devices, Second edition, John Wiley & Sons, New York, 1981.

[15] B. Geynet, et al., "SiGe HBTs Featuring $f_T>$400GHz at Room Temperature", Proc. IEEE BCTM, pp. 121-124, 2008

978-1-4244-6658-0/10 $26.00 © 2010 IEEE

Integrated Phototransistors in a CMOS Process for Optoelectronic Integrated Circuits

Plamen Kostov, Wolfgang Gaberl and Horst Zimmermann

Institute of Electrodynamics, Microwave and Circuit Engineering, Vienna University of Technology

Gusshausstr. 25/354, 1040 Vienna, Austria

[plamen.kostov | wolfgang.gaberl | horst.zimmermann]@tuwien.ac.at

Abstract — This work presents integrated pnp phototransistors built in a 0.6 µm OPTO ASIC CMOS process using a low doped epi wafer as starting material. Several phototransistors with different designs of the base and emitter area were realized and characterized. For these novel photodetectors responsivities up to 65 A/W for DC light and up to 37.2 A/W for modulated light were achieved. Other transistors reach bandwidths up to 14 MHz. Due to the used standard silicon CMOS process low-cost integration is possible. Analog and digital circuitry can be implemented together with active optical detectors. This paves the way for high performance optical sensors and cost efficient SoC devices. Typical application examples include highly sensitive optical sensors, active pixel image sensors, light barriers and optical distance measurement sensors as well as 3D cameras.

I. INTRODUCTION

Photodetectors convert optical signals into electrical ones. Different types of them can be built in a standard CMOS process. Often used photodetectors are e.g. PN-, PIN-diodes, avalanche photodiodes (APD) and phototransistors (PT).

PN diodes can be considered as the most common photodetectors. Integrated PN detectors are typically very shallow (only a few µm), which can be a problem for near-infrared light, since e.g. 850 nm has a 1/e penetration depth of around 16.6 µm [1]. Common PN diodes can be designed in two different structures which give either high speed or high responsivity. First, a PN diode might use a common well/substrate structure. This structure will convert charges generated in any depth, even deep inside the substrate. Since the charges in the field free substrate are only carried by diffusion (very slow mechanism) these detectors are rather slow. Therefore, the electrical performance shows a good responsivity but huge diffusion tails. Second, a PN diode can be designed to avoid charges generated deep in the semiconductor. This can be done by means of a p+/n-well structure, which leads to a rather shallow (~1 µm) remaining effective active thickness area. All charges generated below this area (slow charges) can be drained. Drained charges do not contribute to the electrical output signal. This leads to fast but very inefficient detectors.

PIN diodes avoid the limitations of the ordinary PN diodes, because of their additional low-doped intrinsic layer.

PIN diodes are integrated in CMOS processes using special starting materials, where a 15 µm low doped epitaxial (epi-) layer is applied on the top of the wafer [1]. This epi-layer forms the intrinsic zone between the p- and n-area and leads to a thick space-charge-region (SCR). Inside this thick intrinsic zone an electrical field can be generated easily by a low voltage. The applied voltage leads to a thick drift zone. In that zone most of the charges are generated. By this mean, the detector gains speed and high responsivity at the same time. Therefore the PIN diode is the mostly used photodetector in applications where high speed is desired. Nevertheless, the responsivity is limited for optimum quantum efficiency ($\eta = 1$) to 0.65 A/W @ 850 nm and 0.55 A/W @ 650 nm [1]. In the case of optimum quantum efficiency all photons are used to generate charges inside the detector. Typical applications for PIN diodes are for example in data communication and optical distance measurement circuits [2].

Photodiodes have no inherent current amplification. Therefore their maximum quantum efficiency is limited to 1. This limit can be exceeded by the use of APDs or PTs. Both use mechanisms for amplifying the received photocurrent. This amplification is desirable and important for detecting weak optical signals. Amplification and very high gain in APDs is achieved by the avalanche multiplication process. However, this process needs voltages of several tens of volts [3]. High voltages for avalanche photodiodes are hard to handle in integrated circuits. This is a big drawback for systems on chip (SoC) applications. Other drawbacks of APDs are a rather high dark count rate in Geiger mode and the very narrow bias voltage range for linear operation. Any change in the bias voltage can lead to a nonlinear behavior. Such a behavior is therefore not feasible for image sensor or optical distance measurement applications, especially when measurements are taken in daylight conditions. Due to all these effects an APD needs a complex control circuitry for many application fields. In [4] a shallow APD with a responsivity of 4.6 A/W at 430 nm with a reverse bias of 19.5 V is reported.

PTs do not need high voltages for amplifying the current. This is a major benefit compared to APDs. A PT consists of a photodiode (base-collector junction) with an internal bipolar transistor for current amplification. In the photodiode (base-

---

Funding from the Austrian Science Fund (FWF) in the project P21373-N22 is acknowledged.

collector) region generated charges are separated. For a pnp PT electrons are swept into the base area and holes into the collector. The electron accumulation in the base makes the base potential more negative. Therefore the p+ emitter starts to inject holes into the base. This mechanism amplifies the generated (primary) photocurrent. PTs in standard-burried-collector (SBC) technology with a responsivity of 2.7 A/W at 850 nm were reported in [5].

In this paper, we present several silicon integrated PTs with different layouts. Cheap implementation in a CMOS process makes them easy to use as single chip solutions with silicon-based optoelectronic integrated circuits (OEICs). OEICs have many advantages, e.g. no bond wires are necessary, a simple mounting and packaging process can be used and a high mechanical reliability is reached. According to these benefits as well as the low operating voltage and high current amplification, these PTs are appropriate for image sensors, active pixels, optocouplers, light barriers and other SoC applications.

II. DEVICE STRUCTURES

In our work, we implement several 100×100 μm pnp-type PTs in a 0.6 μm OPTO ASIC CMOS technology. The OPTO ASIC process used implements PIN photodiodes. The only difference compared to the PIN OPTO ASIC process is the starting material. All PTs use a wafer starting material with a thick (~15 μm) low doped (2×10^{13} cm^{-3}) p-epi layer and a shallow (~1 μm) n-epi layer (10^{14} cm^{-3}). The p-epi layer ensu-

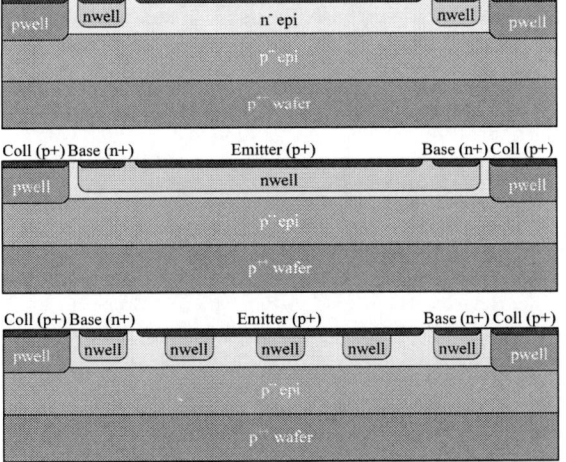

Figure 1. Cross-sections of the different base designs:
n-epi (top), n-well (middle) and modulated n-well (bottom)

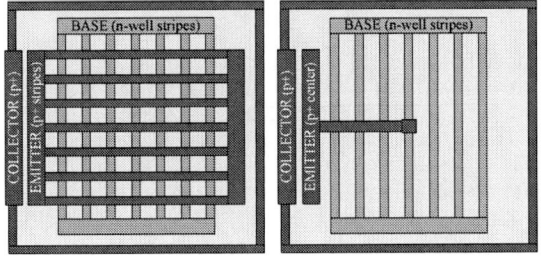

Figure 2. Top view of the different base designs

res a thick collector-base SCR even for low voltages. Such a thick SCR device is well suited for deep penetration light (e.g. 850 nm). Different designs of the base (n-type) and emitter (p-type) region lead to varying characteristics for each PT. These designs are illustrated in fig. 1 and 2. These designs as well as the collector area are explained in the following three subsections.

A. Collector Area

The p-type substrate forms the collector. It is connected via a large-area ring of substrate contacts on the border of the PT. Due to the fact that the collector is tied to substrate potential, the pnp PT can only be used in emitter follower setup.

B. Base Area

The base is formed by the 1 μm thick low doped n-epi layer. To further adjust the base doping additional n-well areas can be implanted. Three different base doping layouts are investigated: a low doped epi-layer base, a full n-well base and a base with n-well stripes.

First, the topmost picture of fig. 1 shows the lowest doped base layer. This 1 μm thick design leads to an increased thickness of both SCRs (BC, BE). The SCR increase results in a reduced effective base width W_B. The reduction of W_B leads to a decrease of the base-collector and base-emitter capacitances C_{BC} and C_{BE}. Accordingly the PT's bandwidth increases due to faster transport of the minority carriers through the (thinner effective) base region. C_{BC} and C_{BE} are the reason why PTs are slower than photodiodes. Both capacitances and the base transit time τ_B are the main parameters defining the –3 dB bandwidth of PTs [6]:

$$f_{-3dB} = \frac{1}{2\pi\beta \cdot \left(\tau_B + \frac{k_B T}{q I_E} \left(C_{BE} + C_{BC} \right) \right)}. \quad (1)$$

The increase of both SRCs causes the effective base width to shrink, which may become a problem when both SCRs touch each other. In this case a reach-through current between emitter and collector will arise. This current is dependent on the collector emitter voltage.

Second, the middle picture of fig. 1 shows the highest doped base. It uses a full n-well. The higher doped base leads to larger capacitances, which results in a slower device. Furthermore, the higher doping increases also the effective base width W_B. Therefore the current gain β decreases as

$$\beta = \frac{1}{\frac{W_B^2}{2\tau_b D_p} + \frac{D_n}{D_p} \frac{W_B}{L_n} \frac{N_D}{N_A}} \quad (2)$$

where τ_b the minority carrier lifetime in the base, D_p and D_n the carrier diffusion coefficients in the base, L_n the diffusion length of electrons in the emitter, and N_D and N_A the donor and acceptor densities in the emitter and base are [7].

Third, to achieve a good tradeoff between gain and speed, n-well stripes with three different width-to-spacing ratios were designed:

- NW$_{33}$: 1 µm stripe with 2 µm space,
- NW$_{50}$: 1 µm stripe with 1 µm space and
- NW$_{66}$: 2 µm stripe with 1 µm space.

Due to the thermal budget of the process, the different stripes diffuse into an inhomogeneous doped base layer. Thereby it is possible to adjust the effective base doping and the effective base width even in a standard ASIC process.

C. Emitter Area

Different emitter designs have been realized. The largest emitter is a full emitter plane. Other designs with smaller emitters are shown in fig. 2.

The 97×97 µm full plane emitter leads to a low current density resulting in a reduced cut-off frequency. The main advantage of the full plane emitter is a very homogenous electric field inside the PT so generated charges only need to transit vertically. However, the full plane emitter area also increases the base-emitter capacitance C_{BE}. This increase dominates and causes according to (1) a reduced cut-off frequency.

The left structure of fig. 2 shows an emitter layout with stripes (1.4 µm stripes, 8.4 µm gap), whereas the right one shows a small emitter dot (1.4×1.4 µm). These structures have a higher current density and lower capacitance due to the reduced emitter area. Thereby the cut-off frequency is increased. However, generated charges have to travel longer distances for smaller emitter areas which results in a reduced gain due to recombination. Furthermore a device with a small emitter area (7.0×1.4 µm) in one corner of the PT was produced. This structure has no emitter inside the optical area and also no metal structures. Therefore this structure might be combined with an opto-window etching process step to reduce attenuation and reflection effects of the oxide stack. However, to compare the different device layouts against each other, the corner emitter design was not realized with an opto-window. With an additional opto-window the optical performance of the corner layout might be increased by around 2-3 dB depending on the used wavelength.

III. RESULTS AND MEASUREMENTS

Three different measurement setups were used for characterizing the PTs. The electrical current amplification was verified through Gummel measurements. The DC responsivity was measured by sweeping the power of the light source at 675 nm and 850 nm. Finally the AC responsivity respectively the bandwidth was measured at an average optical power of –10 dBm and –21.2 dBm (@ 850 nm). For these measurements the extinction ratio of the light source was chosen at 2:1.

A. Gummel measurements

All PTs with only the low doped epi-layer base show a reach-through effect due to the above mentioned reason. All devices with higher doped base show the expected current amplification β between 57 (for the full n-well base and corner emitter PT) and 176 (for the NW$_{33}$ base and small center emitter PT – fig. 3). A different performance is measured with the NW$_{33}$ base and full emitter PT, which is at the edge of a reach-through scenario. It has an abnormal Gummel plot,

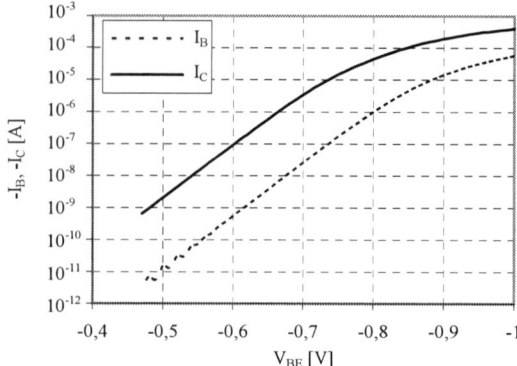

Figure 3.   Gummel plot of the NW$_{33}$ base and center emitter transistor

wherein the I_B and I_C graphs run nonparallel. For low currents the gain reaches up to 10^5. Nevertheless, the dark current was only about 4.56 µA at a collector-emitter voltage of –4 V. For all other (non-reach-trough) devices dark currents below 60 pA were measured.

B. DC measurements

The DC responsivity was measured at 675 nm and 850 nm by changing the optical power with an optical attenuator. The optical light power was varied from –55 dBm to –11 dBm and monitored by using an optical 50/50 beam splitter and optical power meter. Increasing the light power shows an exponential decrease of the responsivity due to the operating point variation. Furthermore, the applied collector-emitter voltage was also varied from –1 V to –8 V.

For 675 nm three PTs were measured. The devices show only a small dependence on the collector-emitter voltage. A maximum responsivity of about 65 A/W for the NW$_{50}$ and full emitter PT was achieved.

Since 850 nm light has a larger penetration depth, the effect of the SCR is of higher influence. Values between 20 A/W and 40 A/W were achieved at –55 dBm for small emitter devices. The responsivity for these devices at –11 dBm was about 2 A/W. The smallest responsivity change was measured at the full and striped emitter devices. The responsivity change is below 20 % over the whole light power range. A totally different result shows the near to reach-through PT (NW$_{33}$ base and full emitter). For this PT the corrected responsivity (with subtracted dark current) is in the range of 23000 A/W at –55 dBm optical light power. Moreover at –11 dBm optical light power the device still shows a responsivity of 80 A/W.

C. AC measurements

AC light was used for measuring the dynamic responsivity and the bandwidth of the PTs. The measurements were done at a low optical power level (–21.2 dBm) as well as at a high one (–10 dBm). Eight devices were characterized at three different collector-emitter voltages: V_{CE} = –2 V, –5 V and –10 V. The operating points of the PTs were adjusted by sinking a current I_B from the base (via an on-chip 1 MΩ resistor). Five different operating points, including floating base were investigated. Figure 4 depicts the results of the measurements for 850 nm.

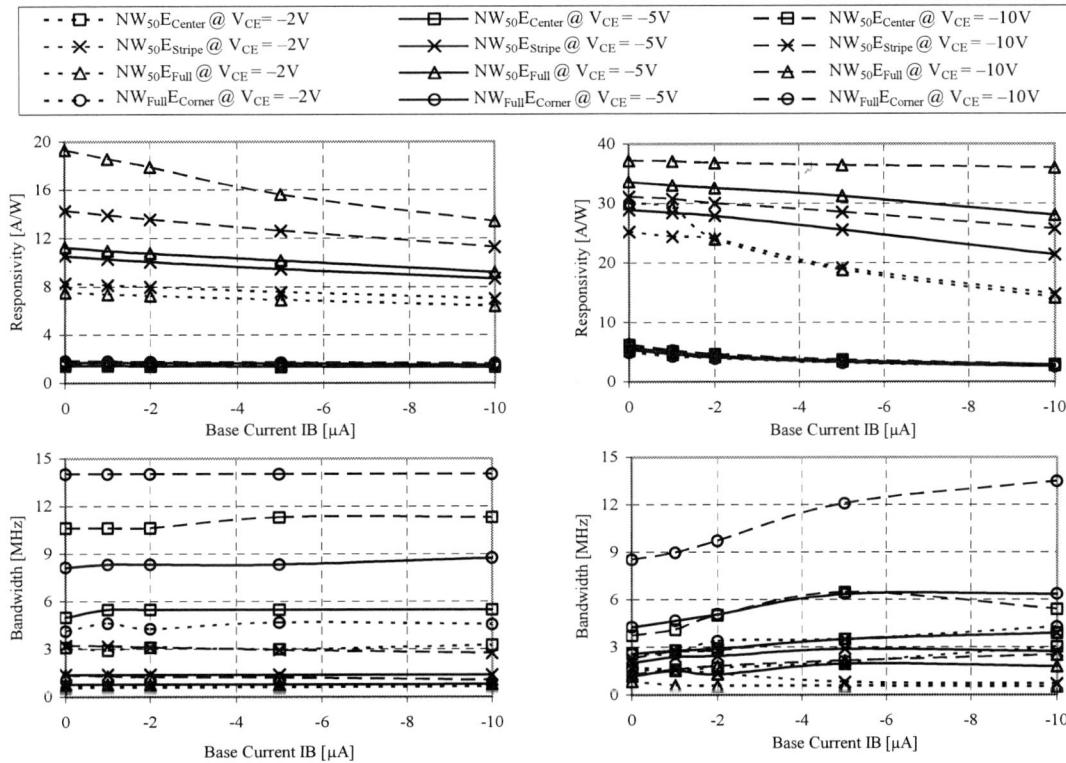

Figure 4. The two left diagrams show responsivity (top) and bandwidth (bottom) for high optical power (−10 dBm). The two right diagrams are respectively for low light conditions (−21.2 dBm).

The measurements at −10 dBm show a decrease of the responsivity for larger base currents. The small-emitter devices have a small but rather constant responsivity. The larger-emitter devices show a collector-emitter voltage depending responsivity which is generally higher. The bandwidth is nearly constant for all operating points, since the incident optical power already ensures a certain operating condition. However, the bandwidth is considerably influenced by the collector-emitter voltages mainly due to increased SCRs, which leads to a thinner effective base.

The measurements at −21.2 dBm show a decrease of the responsivity for higher base currents for all PTs. The responsivity of the small emitter size devices does not depend much on the collector-emitter voltage. The devices with larger emitter sizes (striped and full emitter) show a strong increase of the responsivity for higher collector-emitter voltages. The collector-emitter voltage mainly affects the bandwidth of the devices. Smaller emitter layouts show higher bandwidths than larger.

IV. CONCLUSION

New fully integrated silicon phototransistors in a CMOS OPTO ASIC technology are presented by using special starting material wafers. Several devices with different designs of base and emitter area have been produced and characterized. Electrical Gummel plot measurements showed current gains up to 176. Responsivities up to 65 A/W (about

25 times better than that of SBC npn phototransistors in SiGe BiCMOS) and bandwidths up to 14 MHz were measured. A close-to-reach-through phototransistor showed operating point dependent current amplification up to 50.000 for low optical input power. The responsivity and bandwidth can be adjusted by the layout (n-well stripe widths and spacing as well as emitter area and position). Fully integrated custom tailored phototransistors therefore are well suited for many optical sensing applications, SoCs and also imaging systems.

REFERENCES

[1] H. Zimmermann, "Integrated Silicon Optoelectronics", 2nd ed., Springer-Verlag, Berlin, Heidelberg, 2010.

[2] G. Zach, M. Davidovic and H. Zimmermann, "Extraneous-Light Resistant Multipixel Range Sensor Based on a Low-Power Correlating Pixel-Circuit", IEEE ESSCIRC 2009, pp. 236-239.

[3] S. Cova, M. Ghioni, A. Lacaita, C. Samori and F.Zappa, "Avalanche photodiodes and quenching circuits for single-photon detection", Applied Optics, vol. 35, no. 12, pp. 1956-1976, 1996.

[4] A. Pauchard, A. Rochas, Z. Randjelovic, P.A. Besse and R.S. Popovic, "Ultraviolet Avalanche Photodiode in CMOS Technology", IEEE IEDM 2000, pp. 709-712.

[5] T. Yin, A.M. Pappu, A.B. Apsel, "Low-cost, high-efficiency, and high-speed SiGe phototransistors in commercial BiCMOS", IEEE Photonics Technology Letters, vol. 18, no. 1, 2006, pp.55-57.

[6] G. Winstel and C. Weyrich, "Optoelektronik II", p. 97, Springer, Berlin, Heidelberg, 1986

[7] P. Gray, P. Hurst, S. Lewis and R. Meyer, "Analysis and Design of analog integrated Circuits", p. 8 ff., Wiley, 2008

978-1-4244-6658-0/10 $26.00 © 2010 IEEE

CMOS Process Enhancement for High Precision Narrow Linewidth Applications

Frank Hochschulz, Uwe Paschen and Holger Vogt
Fraunhofer Institute for Microelectronic Circuits and Systems IMS
Duisburg, Germany
Email: frank.hochschulz@ims.fraunhofer.de

Abstract— During the last years CMOS technologies found widespread use in the development and fabrication of optical sensors and imagers [1][5]. However, during the development of CMOS photo diodes for special applications requiring the detection of radiation with a small spectral linewidth various aspects have to be considered that are negligible for photo diodes employed in common imaging applications. One very important aspect is the influence of interference effects due to the dielectric stack that covers the photo diodes in every CMOS process. This results in a dramatic modulation of the electronic signal as a function of the wavelength, stack thickness and spectral width of the impinging radiation. The uncertainties introduced by these oscillations disfavor standard CMOS imaging solutions for small spectral width applications. In common imaging applications with a larger spectral linewidth these modulations are not visible due to the convolution of the spectral sensitivity with the incoming spectral profile. This paper describes a CMOS process addition that can be applied to photo diodes to significantly reduce the sensitivity modulations by etching the dielectric stack in photo active areas. Using this approach the sensitivity modulations due to interference effects have been nearly eliminated for wavelengths above 300 nm. This enables the use of standard CMOS processes for spectroscopy or special imaging applications like laser illumination.

I. Introduction

In standard CMOS processes photo diodes are buried under several layers of dielectric materials insulating the metal layers. In order to prevent mechanical damage and as a diffusion barrier for humidity and ions in many cases an additional passivation layer made of silicon nitride is deposited on top of the final chip. When analysing the ability of a photo diode to convert impinging photons into electron hole pairs and to generate a photo current these coating layers have to be considered. From the generated photo current I_{Photo} and the impinging radiant flux Φ the spectral sensitivity or quantum efficiency $Q.E.$ can be derived according to:

$$Q.E. = \frac{hc}{e\lambda} \frac{I_{photo}}{A\Phi} \qquad (1)$$

with h being the Boltzmann constant, c the speed of light, e the elementary charge, λ the wavelength and A being the area of the photo diode.

The radiant flux entering the silicon through the dielectric layers depends not only on reflection and transmittance at every material boundary according to the Fresnel equations but also on the interference of reflected waves. This effect

modulates the transmittance depending on the phase difference $\Delta\phi$ of the waves. For a single layer with a refractive index of $n_{Silicon} > n > n_{Air}$ this phase difference is:

$$\Delta\phi = 2\frac{2\pi}{\lambda}nd\cos\theta \qquad (2)$$

with d being the thickness of the layer and θ being the angle of propagation inside the layer (for the following considerations $\theta = 0$ is assumed). The period of the resulting modulation is proportional to $\cos(\Delta\phi)$ and the distance between two adjacent maxima is $\frac{\lambda^2}{2nd-\lambda}$ resulting in a period of 1-50 nm in the wavelength region between 200 and 1100 nm for the present stack thickness of several micro meters (Fig. 1).

The amplitude of the modulation is determined by the refractive indices yielding up to 60 % for the materials used in the present process.

A stack thickness variation of $\Delta d = \frac{\lambda}{4n}$ leads to a phase difference of π. This means for example that a process induced thickness variation of 112 nm leads to a quantum efficiency change of 55 % for light with a wavelength of 650 nm. For a typical stack this thickness variation corresponds to a relative change of only about 2 %. Since imagers are comparably large this thickness variation can become significant even across a single die yielding a huge variation of sensitivity for different pixels of a single imager.

Finally the spectral width of the incident radiation has to be taken into account by convoluting the spectral sensitivity of the photo diode with the spectral profile of the incident radiation. For a large spectral linewidth many oscillations are averaged resulting in a smooth spectral sensitivity. However, for a small spectral linewidth only a small fraction of an oscillation is averaged keeping the oscillations dominant in the spectral sensitivity. A simulation showing this effect for radiation with a linewidth of 6.4 nm impinging on a standard photo diode is shown in figure 1. The value of 6.4 nm is chosen to take the spectral linewidth of the monochromator into account, that is employed in the measurement of diodes.

This means that if only one of the parameters stack thickness, spectral width or wavelength is not known with the required precision a large uncertainty is introduced when measuring the intensity of small spectral width visible radiation with a standard CMOS photo diode.

978-1-4244-6658-0/10 $26.00 © 2010 IEEE

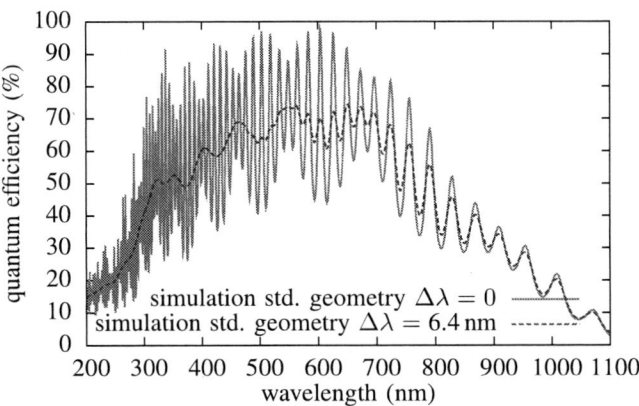

Fig. 1. Simulation of standard photo diode for two different spectral widths $\Delta\lambda$ of the impinging radiation. The oscillations are mitigated partially because of the convolution of the impinging spectral profile with the monochromatic sensitivity.

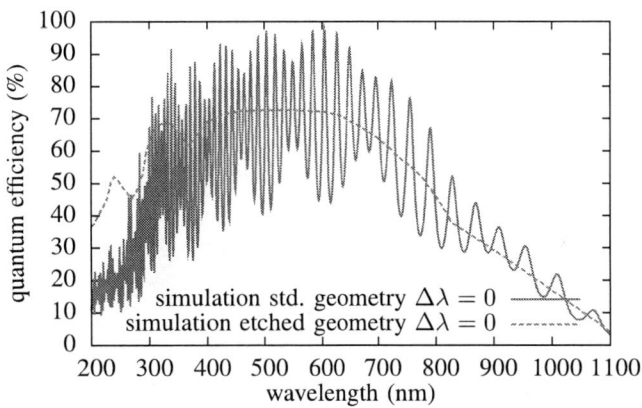

Fig. 2. Comparison of the simulation of the quantum efficiency. Solid line standard dielectric stack, dashed line stack removed to a residual oxide thickness of 50 nm. Both simulations assume the incident light to be monochromatic.

II. STUDIED GEOMETRY

A. Studied CMOS process

The underlying CMOS-process studied in this paper is a four metal layer double poly 0.35 µm LOCOS CMOS process with a silicon nitride passivation layer. Three of the four metal layers are planarized via chemical mechanical polishing (CMP). The dielectric stack on top of the diode consists of approx. 5.65 µm silicon oxide and approx. 750 nm silicon nitride.

B. Deep Dielectric Stack Etching

As described earlier (eq. 2) the modulation of the quantum efficiency of a photo diode due to interference effects in the dielectric stack is proportional to $\cos\left(\frac{d}{\lambda}\ldots\right)$. Thus, by reducing the thickness d of the dielectric stack in photo active regions the period of the oscillations can be increased and for a very small thickness the oscillations can be suppressed completely. This is achieved by introducing an additional process step to the CMOS process featuring an additional lithography step followed by silicon nitride and deep oxide etching.

After the deposition of the silicon nitride passivation a lithography step is performed to open the photo active and pad regions. The pad etching removes the silicon nitride and a fraction of the silicon oxide on the photo active areas. An additional lithography step opening only photo active regions allows the etching of the dielectric stack down to a remaining oxide thickness of less than 50 nm. For wavelengths smaller than 300 nm this thickness is too small to allow a phase difference of π and thus completely removes oscillations due to interference effects for larger wavelengths as shown in figure 2.

Due to the missing oscillations different spectral widths $\Delta\lambda$ have nearly no impact on the Q.E.

C. Simulation Setup

In order to estimate the sensitivity of a photo diode covered by a multi layer stack simulations are conducted based on the transfer matrix formalism which allows convenient handling of reflection, transmittance and interference for multi layer systems. The refractive index of silicon is taken from [3], the refractive indices of the other layers have been determined by spectroscopic ellipsometry.

After entering the silicon the absorption within the first 10 µm is calculated. The generation of electron hole pairs upon absorption is set to 1 for the wavelength region of interest (200-1100 nm). The wavelength step of the simulation is 1 nm. Finally a convolution of the spectral sensitivity with a Gaussian distribution is performed to incorporate the spectral width of the impinging radiation. The resulting parameter is the wavelength dependent quantum efficiency of a photo diode.

D. Measurement Setup

To measure the quantum efficiency of the photo diodes a test system for wavelengths between 450 and 1100 nm is available. The measurements are conducted with a wavelength step of 1 nm. The system is using a 250 W halogen bulb and a Oriel Instruments Cornerstone 260 Monochromator. Using a chopper in the light path the photo current is measured using a lock in amplifier. The impinging radiant flux is determined by a measurement with a calibrated Hamamatsu silicon photo diode. A second calibrated Hamamatsu silicon photo diode used during recalibration and measurement steps allows the compensation of flux variations of the halogen bulb.

III. RESULTS

N well/p sub [2] photo diodes with a size of $(300\,\mu m)^2$ and $(7\,\mu m)^2$ have been fabricated using the CMOS process mentioned earlier in a standard version and with the additional deep stack etching. Figure 3 shows a SEM image of an array of $(7\,\mu m)^2$ diodes. After packaging and bonding the spectral response of the diodes has been measured using the

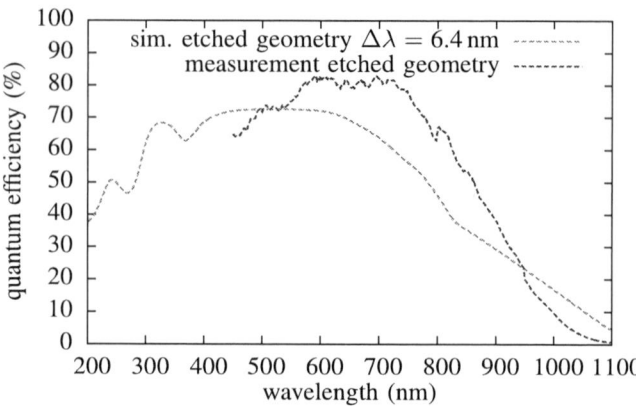

Fig. 5. Comparison of simulated quantum efficiency (spectral width $\Delta\lambda = 6.4$ nm) with a measurement of the $(300\,\mu m)^2$ deep stack etched photo diode.

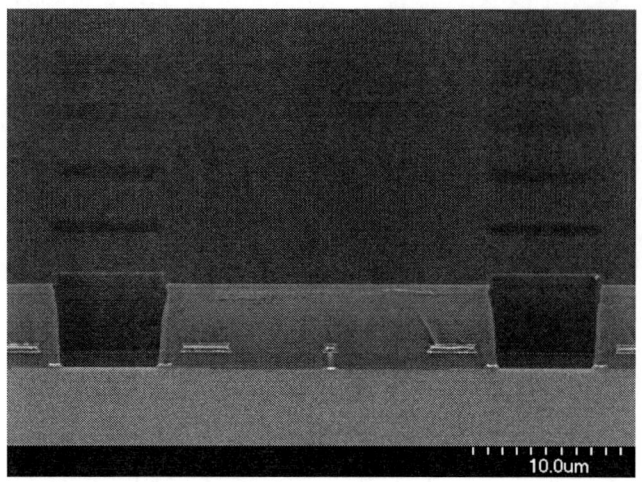

Fig. 3. SEM cross section of an array of $(7\,\mu m)^2$ diodes processed with the deep stack etching.

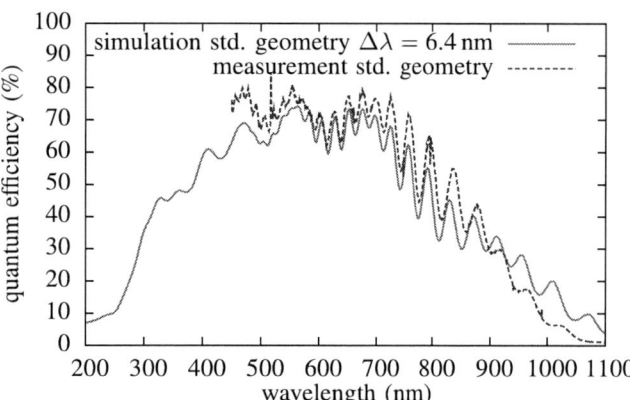

Fig. 4. Comparison of simulated quantum efficiency (spectral width $\Delta\lambda = 6.4$ nm) with a measurement of the $(300\,\mu m)^2$ standard photo diode.

test system described earlier. Following the measurements the photo diodes have been analysed with a SEM cross section in order to measure the thicknesses of the involved layers. These thicknesses are used for the simulations presented in this paper.

A comparison of the measured quantum efficiency of the standard photo diode with the simulation is shown in figure 4. Despite the simple simulation approach a good agreement is achieved. For the deep stack etched photo diode shown in figure 5 a smooth response as predicted by the simulation is measured. However, there is a quantitative difference in the absolute values between the simulation and the measurement of the quantum efficiency. This might be due to effects induced by the more complex geometry of the edges of the photo diode or the missing simulation of effects influencing the generated charge carriers. A more sophisticated simulation approach taking these effects into account is subject of further research.

IV. CONCLUSION

Detecting visible radiation with a small spectral linewidth using a standard CMOS photo diode introduces large uncertainties in intensity measurements which are not visible in common imaging applications with a large spectral linewidth. Measurement and simulation of the quantum efficiency of a standard photo diode are in very good agreement using a simple simulation approach based on the transfer matrix formalism both showing an oscillation caused by interference effects in the dielectric stack covering the photo diode. Removing the dielectric stack in photo active areas by a CMOS process addition both simulation and measurement show that the oscillations could be completely suppressed, yielding a rather smooth spectral sensitivity. Thus, the presented CMOS process addition enables the use of CMOS imagers for applications with a small spectral linewidth like spectroscopy or laser light illumination. The exact shape of the quantum efficiency of the improved photo diodes could not be simulated precisely and is subject to further research.

REFERENCES

[1] Magnan, P. *Detection of visible photons in CCD and CMOS: A comparative view* Nuclear Instruments and Methods in Physics Research Section A: Accelerators, Spectrometers, Detectors and Associated Equipment, 2003, 504, 199 - 212
[2] Murari, K. and Etienne-Cummings, R. and Thakor, N. and Cauwenberghs, G. *Which Photodiode to Use: A Comparison of CMOS-Compatible Structures* Sensors Journal, IEEE, 2009, 9, 752-760
[3] Lide, D. R. (ed.) *CRC Handbook of Chemistry and Physics* CRC Press, Inc., 1995
[4] Durini, D. and Hosticka, B. *Photodetector structures for standard CMOS imaging applications* Research in Microelectronics and Electronics Conference, 2007. PRIME 2007. Ph.D., 2007, 193-19
[5] Theuwissen, A. *CMOS image sensors: State-Of-the-art and future perspectives* Solid State Device Research Conference, 2007. ESSDERC 2007. 37th European, 2007, 21-27

A 2μm Diameter, 9Hz Dark Count, Single Photon Avalanche Diode in 130nm CMOS Technology

Justin A. Richardson, Lindsay A. Grant
ST Microelectronics Imaging Division
33 Pinkhill, Edinburgh
EH12 7BF, Scotland, UK
justin.richardson@st.com

Eric A. G. Webster, Robert K. Henderson
Institute for Integrated Micro and Nano Systems
School of Engineering, The University of Edinburgh,
King's Buildings, Mayfield Road, Edinburgh, EH9 3JL, UK

Abstract — **We report a CMOS single photon avalanche diode (SPAD) with a 2μm active diameter, 9Hz dark count rate at 20°C, photon detection efficiency peak of 14% at 500nm, implemented in a 130nm process technology. The implicit guard ring structure relies on retrograde well doping and overcomes a key problem in scaling SPAD devices to small dimensions. TCAD-based device simulations point the way to high resolution, high fill-factor single photon imaging.**

I. INTRODUCTION

Smart sensors comprising arrays of single photon detectors integrated with fast electronic processing on a single CMOS die have recently become commercially available [1]. These much-heralded replacements for photomultiplier tubes and silicon photomultipliers are targeted at positron emission tomography (PET) systems and large-scale nuclear science instrumentation. However, much research activity has been devoted to other time correlated or photon counting applications such as fluorescence lifetime imaging microscopy (FLIM), fluorescence correlation spectroscopy (FCS), wavefront-sensing in astronomy, as well as ranging and 3D cameras. A number of compact, single chip imaging solutions have been proposed based on CMOS SPADs with increasing levels of in-pixel and on-chip signal processing.

An impediment to further scaling of CMOS SPAD arrays to resolutions much above several 10's of kilo pixels is the size of the detector itself. The smallest reported SPAD device with low dark count rate (DCR) (below 100Hz) has an active diameter around 8μm [2-5]. Although 2μm active diameter devices have been reported using STI guard rings, the dark count was extremely high (100's kHz) [6]. Indeed, the authors of [5] conclude that SPADs with conventional p-well guard rings cannot be scaled much below 5μm because the depletion regions around the p-well implants expand and merge such that the active area of the SPAD is almost fully depleted. The SPAD then performs like a p-well n-well diode as the p+ n-well breakdown junction no longer operates [5].

The incentives for reduction of SPAD active area go beyond the desire to increase spatial sampling resolution. A number of device characteristics are improved; small devices have lower jitter and the mean DCR across a population of devices is reduced [4]. Moreover, the yield of an array of SPAD pixels, defined as the proportion of devices exhibiting DCR below an acceptable threshold, is improved. A fixed density of traps or defects per unit area will be localized within fewer small SPADs than in an array of larger devices occupying the same total area. Higher resolution arrays allow defective pixels to be less noticeable when masked in subsequent digital processing, as is conventionally done in CMOS image sensors. Currently, SPAD device dimensions have not allowed the pixel pitch to be reduced much below 20-30μm. At such dimensions, it is difficult to apply wafer-scale microlenses to recover fill-factor. SPAD arrays have so-far employed chip-scale microlenses with difficult alignment tolerances which are not amenable to volume manufacturing [7]. N-well sharing has been successfully applied to linear arrays to improve fill-factor but is difficult to apply to 2-dimensional arrays [8].

In this paper we demonstrate that the SPAD structure proposed in [9] removes the area limitation imposed by explicit guard rings. Measured results of a CMOS SPAD device which simultaneously achieves low dark count and the smallest published active diameter of 2μm are presented. TCAD modeling results indicate that continued scaling of the device to even smaller dimensions should be possible, making this structure a promising candidate for future 'jot' imaging [10].

II. SPAD DESIGN

Fig. 1.shows a 2μm active region diameter SPAD derived from the 8μm device published in [9]. The p+ and p-well dimensions defining the active area have been scaled, whilst maintaining the spacings of the p-well periphery with respect to the STI boundary.

978-1-4244-6658-0/10 $26.00 © 2010 IEEE

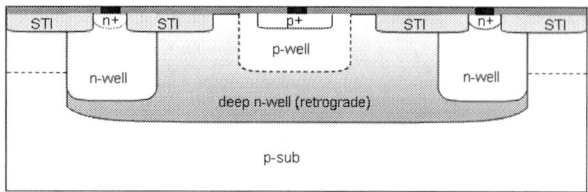

Figure 1. SPAD cross section

As shown in the micrograph of the device in Fig. 2, the active area has been scaled to 2μm diameter, leaving the exterior ring of n-well and STI extending to the same diameter as the original 8μm device. Clearly, in order to realise a small SPAD pixel pitch, the complete device periphery also requires to be reduced. However, it is our intention in this work to demonstrate that the guard ring is effective at small dimensions.

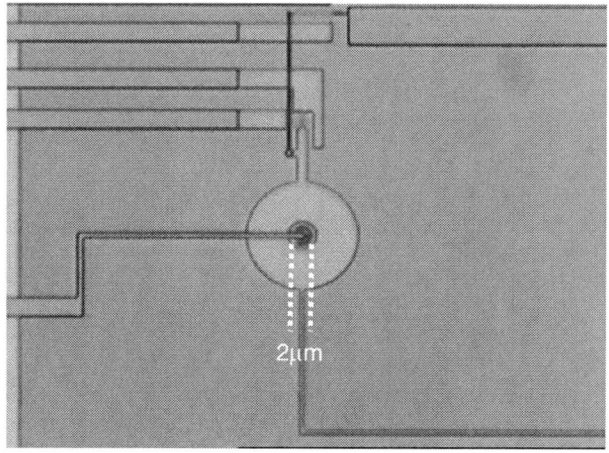

Figure 2. Device micrograph

The key to low DCR Geiger mode operation at small active diameter is the implicit guard ring formed by inhibiting n-well in the periphery of the p-well active area with a ring of p-well blocking layer. The entire device is enclosed in a deep n-well contacted with a ring of n-well and n+ contacts. The guard ring thus receives no n-well or p-well implants and only the deep retrograde n-well doping.

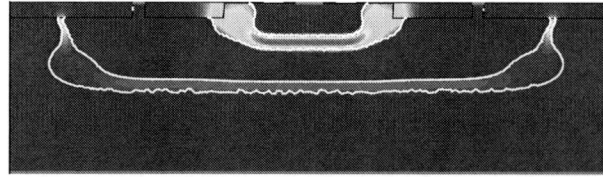

Figure 3. TCAD field simulation (arbitrary scale; red indicates high field strength and blue low field strength)

Fig. 3 shows an electric field plot produced by a breakdown voltage simulation with Synopsys Sentaurus TCAD using a Delaunay-conforming meshing strategy. The white lines indicate the extent of the depletion regions around the anode and cathode. This simulation clearly shows the high field region (red) confined uniformly within the active area of the device. The guard ring shows the natural reduction in field strength from medium field strength (green) to low field strength (blue) at the surface. which is related to the graded doping of the deep n-well implants. The STI is spaced by around 0.75μm from the edge of p-well and plays no particular role in the operation of the guard ring.

It is clear from Fig. 3 that the high field region of the device is simply defined by the region where p-well overlaps the deep n-well. A single geometric parameter (the drawn area of p-well) can therefore be adjusted to vary the device active area. The guard ring is formed implicitly around the edge by the graded doping of the deep n-well implant. This leads us to believe the scaling of the remaining area of the detector to even smaller dimensions poses few difficulties.

III. MEASURED RESULTS

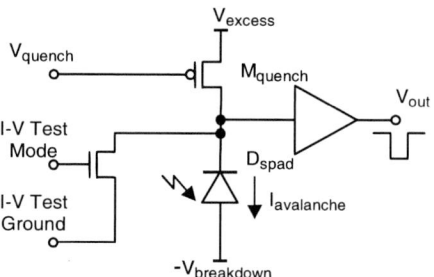

Figure 4. SPAD test circuit

Fig. 4 shows the circuit configuration employed to test the SPAD device. A PMOS passive quench transistor is employed, the voltage V_{quench} can be adjusted to vary the quench impedance although a bias voltage of 0V was used in the experimental work presented in this paper. An additional NMOS transistor is employed to allow I-V DC characterization of the SPAD, in which case V_{quench} and I-V Test Mode are both set to the positive power supply rail. The excess bias supply V_{excess} and power supply of the buffer are separated to distinguish current flowing into the SPAD from that driving off-chip load. The buffer supply voltage is 1.2V for all measurements. A minimum dead time of around 20ns is observed, and can be configured as desired dependent on bias voltages.

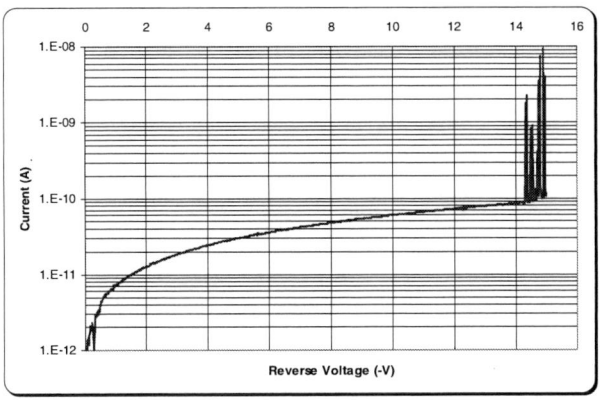

Figure 5.   SPAD I-V characteristic

The SPAD reverse I-V characteristic is shown in Fig. 5. The breakdown voltage is approximately 14.2V. The flickering around the breakdown knee demonstrates a low ignition rate under dark conditions. The breakdown voltage of the comparable 8μm active-area diameter device of [9] was 14.36V. The similarity of these breakdown voltages indicating that the same p-well deep n-well junction is responsible for avalanche. This is contrary to the explicit p-well guard ring where scaling of the active area to 5μm caused a large shift in breakdown voltage due a different junction participating in the avalanche [3].

The jitter of the SPAD was measured using a 470nm Picoquant pulsed laser diode. Care was taken to attenuate the laser such that only around 1 in every 100 laser pulses causes the SPAD to trigger in order to avoid photon pile-up distortion of the jitter histogram. A LeCroy Wavemaster 40Gs/s oscilloscope was used to digitise the time interval between the SPAD pulse and the laser trigger. A histogram of 20,000 such time intervals is plotted in Fig. 6 and the measurement was repeated at different excess bias voltages ranging from 0.4V to 1.2V beyond breakdown.

A typical FWHM jitter of 140ps is observed at an excess bias of 0.8V beyond breakdown. The 8μm active-diameter SPAD with similar guard-ring structure [9] exhibited a jitter of 200ps using an identical measurement procedure. An increase of excess bias to 1.2V reduces the FWHM jitter to around 136ps while a reduction in excess bias to 0.4V increases the FWHM jitter to 150ps. The laser jitter of 70ps is included in these measurements as well as the jitter of the on-chip buffer. The jitter tail extends for only 500ps beyond the peak and has a very distinct exponential characteristic.

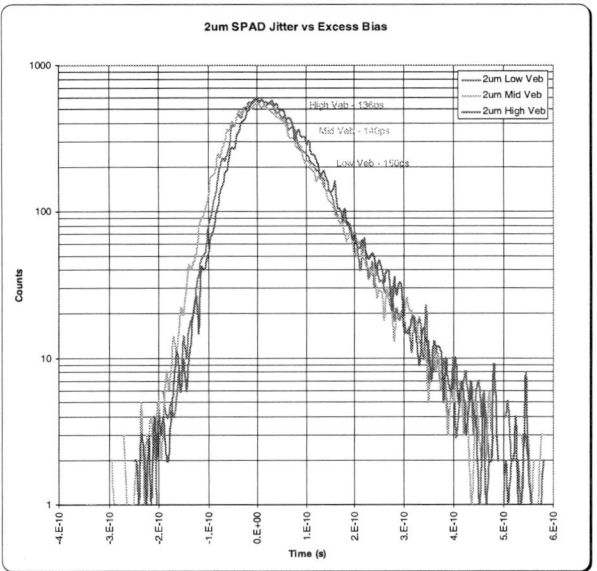

Figure 6.   SPAD jitter at different excess bias voltages

The photon detection efficiency (PDE) of the device is plotted in Fig. 7 at three different excess bias voltages ranging from 0.4V to 1.2V. A peak PDE of 14% at 500nm is observed at an excess bias voltage of 1.2V. The PDE is enhanced by the CMOS image sensor process which provides optimized dielectric layers above the detector. A cavity etch step reduces the number of metal interconnect layers to only two and significantly reduces the dielectric stack height above the detector. The active area obscured by the metal contact and routing has been taken into account in the calculation of PDE. Nevertheless the peak PDE of 14% is considerably lower than the PDE of 28% for the 8μm device described in [9] for reasons that are still under investigation.

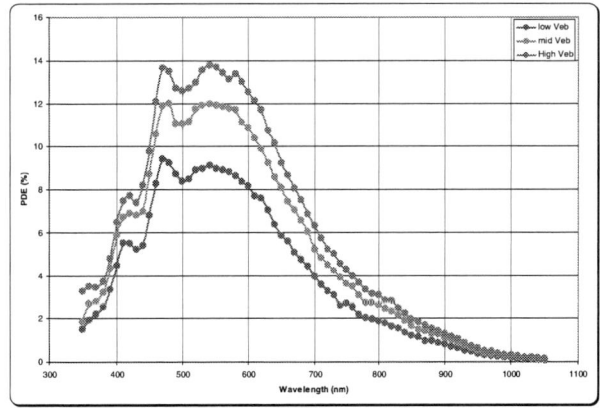

Figure 7.   SPAD photon detection efficiency at different excess bias voltages

978-1-4244-6658-0/10 $26.00 © 2010 IEEE

In Fig. 8 the DCR of the device is plotted against temperature over the range -20°C to +40°C. The dark count at room temperature is around 9Hz which is typical of measurements made on other 2μm SPADs amongst the bonded samples from the batch of test die. The variation of DCR for the 8μm device from [9] over a comparable temperature range is much larger, around two orders of magnitude, whereas for the 2μm device only a single order of magnitude is observed. This would suggest that the dominant source of dark count has switched from thermal generation to band-to-band tunneling associated with the junction's low electric field. The low activation energy of the DCR appears to support this conclusion.

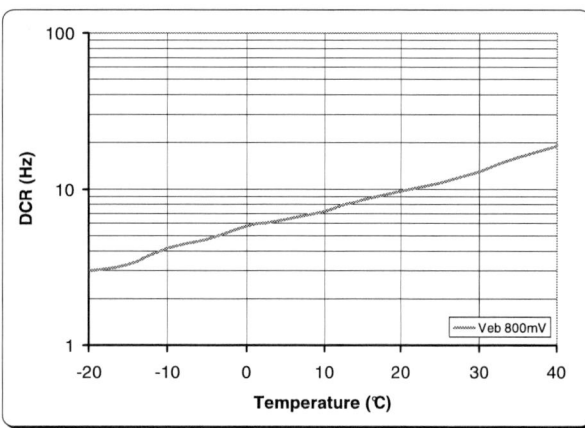

Figure 8. SPAD DCR versus temperature at excess bias voltage of 0.8V

IV. CONCLUSIONS

This paper demonstrates the successful scaling of our retrograde well SPAD design without compromising any detector performance parameters. Timing performance and dark count have improved in line with expectations although the photon detection efficiency has reduced.. The resultant device is a strong candidate for high spatial and temporal resolution array based single photon detection applications.

REFERENCES

[1] T. Frach, G. Prescher, C. Degenhardt, R. de Gruyter, A. Schmitz, R. Ballizany, "The Digital Silicon Photomultiplier – Principle of Operation and Intrinsic Detector Performance", IEEE Nuclear Science Symposium Conference Record, 2009, p1959-1965.

[2] M. Gersbach, C. Niclass, E. Charbon, J. Richardson, R.K. Henderson, L. Grant. 'A Single Photon Detector Implemented in a 130nm CMOS Imaging Process', Solid-State Device Research Conference, ESSDERC 2008, 38th European, p270-273.

[3] Faramarzpour, N., Deen, M.J., Shirani. S., Fang, Q.: 'Fully Integrated Single Photon Avalanche Diode Detector in Standard CMOS 0.18um Technology', Electron Devices, IEEE Transactions on, Vol. 55, Issue 3, 2008, p760-767.

[4] L. Pancheri, D. Stoppa "Low-Noise CMOS Single-Photon Avalanche Diodes with 32 ns Dead Time", Proc. IEEE European Solid-State Devices Conference, Munich, 2007, p362-365.

[5] Marwick, M.A., Andreou, A.G.: 'Single photon avalanche photodetector with integrated quenching fabricated in TSMC 0.18um 1.8V CMOS process', Elect. Letters, Vol. 44, No. 10, 2008, p643-644.

[6] H. Finkelstein, M.J Hsu, S.C. Esener, 'STI-Bounded Single Photon Avalanche Diode in a Deep-Submicrometer CMOS Technology', IEEE Electron Device Letters, Vol. 27, Issue 11, 2006, p887-889.

[7] S. Donati, G. Martini, M. Norgia: "Microconcentrators to recover fill-factor in image photodetectors with pixel on-board processing circuits", Optics Express, 15, (2007), p18066-18074.

[8] L. Pancheri, D. Stoppa, "A SPAD-based Pixel Linear Array for High-Speed Time-Gated Fluorescence Lifetime Imaging", European Solid-State Circuits Conference, ESSCIRC 2009, p428-431.

[9] J. Richardson, L. Grant, R. K. Henderson, "A Low Dark Count Single Photon Avalanche Diode Structure Compatible with Standard Nanometer Scale CMOS Technology", IEEE Photonics Technology Letters, Vol. 21, No. 14, July 2009, p1020-1022,

[10] E. R. Fossum, "What to do with sub-diffraction-limit (SDL) pixels? – a proposal for a gigapixel digital film sensor", IEEE Workshop on Charge-Coupled Devices and Advanced Image Sensors, Nagano, Japan, 2005, p214-217.

Buried Finger Concept for a Correlating Double Cathode Photodetector in BiCMOS

A. Nemecek[a,b], H. Zimmermann[a]

[a] Vienna University of Technology, Institute of Electrodynamics, Microwave and Circuit Design
Gusshausstrasse 25/354, 1040 Vienna, Austria
[b] present address: University of Applied Sciences Wiener Neustadt,
Johannes Gutenberg-Strasse 3, 2700 Wiener Neustadt, Austria
alexander.nemecek@fhwn.ac.at, horst.zimmermann@ieee.org

Abstract — **A new and efficient Double Cathode Photodetector (DCP) for the correlation of incident light and an electrical modulation signal is presented. The concept of buried fingers using a PIN setup augments the separation efficiency up to η=67% {54%}, results in a bandwidth of up to f_{-3dB}~300MHz {80MHz} and ensures high responsivity of R=0.47A/W {0.36A/W} at λ=660nm {850nm}. The integrated photodetector is realized in a 0.6µm OPTO ASIC BiCMOS process.**

I. INTRODUCTION

Correlating photodetectors are required for e.g. time of flight (TOF) based distance measurement. In contrast to pulsed runtime TOF, correlation based systems compare the delayed modulated optical reception signal, which is backscattered from a general object surface, with the transmitted signal. This correlation can be performed either in a consecutive (on-chip) readout circuit or already within the photodiode thus easing readout in every pixel of a possible multipixel integrated distance measurement sensor of an optoelectronic integrated circuit (OEIC). Different correlating photodetector concepts have been developed in order to perform optimized signal mixing of the optical and electrical modulated signals using either structured polysilicon modulation gates [1, 2, 3, 4], interdigitated fingers with doped implants [5] or Schottky contacts [6] respectively. The mode of operation of [1, 2, 3 4] is similar, as it is based on a cross field caused by the external modulation voltage applied to the polysilicon gates. The impact of this modulation voltage drops with increasing depth of silicon, thus reducing the yield of correlating photodetectors especially for light in the near infrared (NIR). As the use of invisible light is a key requirement for range finding applications, the demand for an efficient correlating photodetector in the NIR range is strong. The fast Double-Anode Photodetector (DAP) in [5] performs signal correlation using interdigitated finger structures of surface-near implant regions which means reduced correlation ability in the NIR. The very high bandwidth and separation

efficiency of the Metal-Semiconductor-Metal (MSM) detector in [6] rely on GaAs-Schottky barriers requiring floating bulk potential and thus excluding this photodetector for the use for integrated multipixel sensors. Regarding spectral absorption in silicon the photo generation of electron-hole pairs is described by Lambert-Beer's law $P(z)=P_0\,e^{-\alpha z}$, where P_0 is the optical power at the surface, decreasing with depth z depending on the absorption coefficient α, or the penetration depth $1/\alpha$ respectively (see fig. 1).

II. BURIED FINGER CONCEPT

The proposed correlating photodetector consists of two buried readout finger cathodes and a backside anode as shown in fig. 2 (right) and is built up as follows: The P+ doped wafer features a 10µm thick P-- epitaxial layer with a low doping concentration N_A<10^{14}cm^{-3}. Exploiting the possibilities of a BiCMOS process, where an N+ buried layer is processed for vertical NPN transistors, structured N+ buried cathode fingers are foreseen. In the next process step another P- epitaxial layer (N_A>10^{14}cm^{-3}) covers this buried layer. Aside P well regions with P+ buried layers below enclose the photodetector and

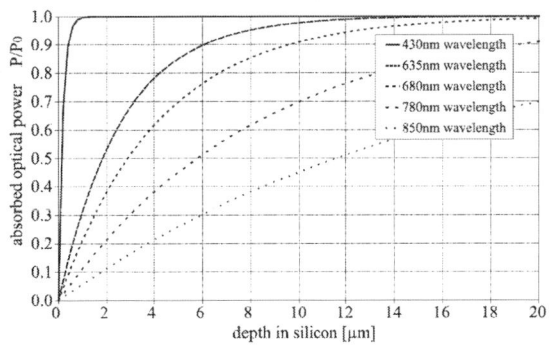

Figure 1. Absorbed optical power versus depth in silicon.

This work was partially supported by the Austrian Science Found (FWF) within project # P17801.

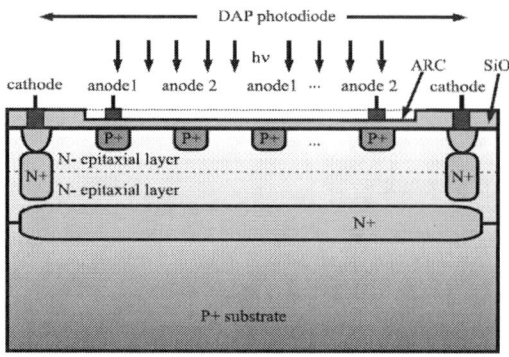 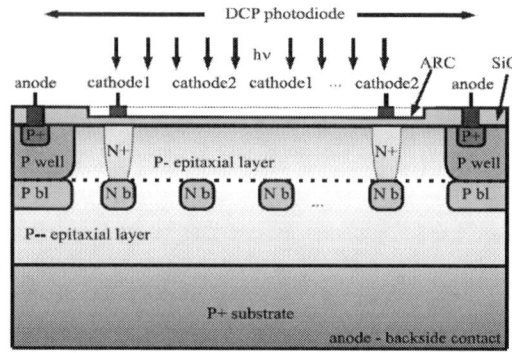

Figure 2. Cross section: comparison of a DAP from [5] (left) and the new DCP concept (right).

also provide a suitable topside connection to the backside anode along nearly the whole chip area with $R''=0.16\Omega cm^2$. For a chip area of e.g. 0.5mm × 0.5mm a serial resistance from the backside anode to the top contact of $R=64\Omega$ results which is a sufficiently low value. The buried finger concept is of advantage because of two reasons: First a higher impact on the separation efficiency also in deeper semiconductor regions is achieved especially for higher penetration depth of light up to NIR. Second, the problem of a relevant reach through current between the cathodes is avoided, as a realization of implanted finger cathodes next to the oxide would matter a conducting channel under the oxide, due to segregation of positive charges in the oxide. Furthermore also threshold implants do not increase the relevant doping concentration for the space charge regions (SCR) around the cathodes. In other words a low doping concentration in combination with the buried finger concept ensures a wide extended SCR which is necessary for high separation efficiency.

III. FUNCTION PRINCIPLE

Simulations regarding reliable doping profiles were done using the 2D semiconductor simulation tool MEDICI. Fig. 3 shows a simulated potential plot of the photodetector setup for cathode potentials of $V_{C1}=0V$ (low), $V_{C2}=3V$ (high). Hence most photo generated electrons are forced to the more positive cathode C2, while holes are directed to the negative anode. Switching the external modulation voltage changes the cathode potentials and leads to the dual case, where electrons are attracted by cathode C1, the now more positive one. As can be seen from fig. 3, the potential of the buried cathode directly impacts a depth region down to nearly $z\sim 5\mu m$ then starting to decay versus deeper regions. This is a better behavior in terms of correlation ability for NIR compared to the DAP in [5], where this depth impact is lower than $z<1\mu m$ due to the surface near implant finger setup. As could be shown, the suspected local spread of the buried cathode doping profile due to the exposed total thermal budget of this early process step in chip manufacturing did not reduce the functionality of the DCP in a critical extend.

IV. CHARACTERIZATION PARAMETERS

A photodetector can be described by the following parameters: the responsivity R is the yield of photocurrent I_{photo} compared to the incident optical power P_{opt}. If every photon with the wavelength λ (c speed of light, q elementary charge and h Planck's constant) generates one electron-hole pair, the quantum efficiency QE is 100%. In order to ensure high quantum efficiency, a stack of silicon oxide Si_2O and silicon nitride Si_3N_4, the Anti Reflection Coating (ARC), is processed to minimize reflection loss at the detector surface.

$$R = I_{photo} / P_{opt} = (QE\, q\, \lambda) / (h\, c) \tag{1}$$

Describing correlating photodetectors the separation efficiency η is necessary to express the demodulation contrast from the high i_{high} and low i_{low} values of the photocurrents, respectively.

$$\eta = (i_{high} - i_{low}) / (i_{high} + i_{low}) \tag{2}$$

Furthermore the rise time t_{rise} is related to the bandwidth f_{-3dB} according to the first order approximation $f_{-3dB} = 0.35 / t_{rise}$.

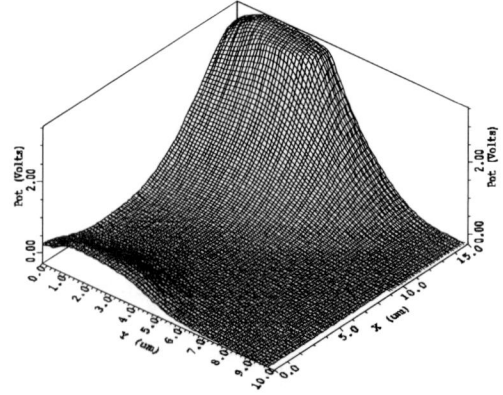

Figure 3. Simulated potential plot ($V_{C1}=0V$, $V_{C2}=3V$, $V_A=0V$).

978-1-4244-6658-0/10 $26.00 © 2010 IEEE

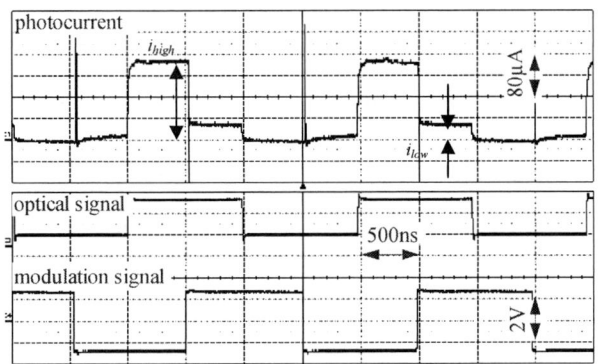

Figure 4.   Measured transient step response: top-photocurrent, middle-optical signal and bottom-electrical modulation signal (inverted).

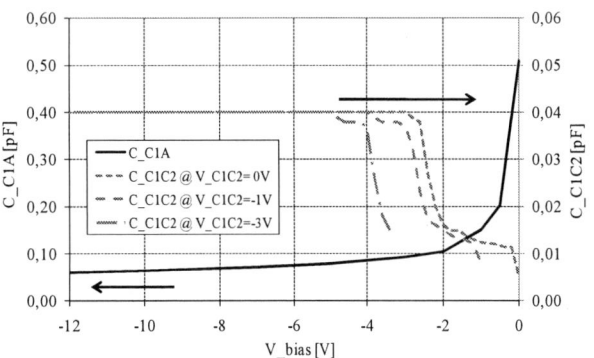

Figure 5.   Measured parasitic capacitances between both cathodes C_{C1C2} (left) and cathode anode C_{C1A} (right).

V.   RESULTS

The photodetector is realized in a 0.6μm BiCMOS process with an active optical area of 100μm × 100μm and a cathode finger width of 1μm. As the buried fingers are exposed to the whole thermal budget of chip fabrication, also a DCP with a finger width of 2μm was realized as a reference, but no significant difference could be realized. A chip photograph of two realized DCPs is shown in fig. 8. For transient measurements of the photodiode, cathode C1 is connected to the modulation voltage of a pulse generator $V_{mod}=\pm1.5V$, cathode C2 is used for read out at a 1GHz oscilloscope and a negative bias voltage is applied to the anode contact. The optical signal is digitally modulated with a delayed modulation signal and fed to the photodiode via multimode fiber on a wafer prober. Fig. 4 shows the voltage across the 50Ω input impedance of the oscilloscope of the photocurrent, the electrical modulation signal, and the delayed modulation signal representing the optical signal. From the difference of the high and low states of the photocurrent from fig. 4, values of the separation efficiency of η=67%, 61%, 35% {56%, 42%, 33%} have been measured for a cathode finger pitch of 12μm, 15μm, and 20μm respectively at a wavelength of λ=660nm {850nm} at a low bias voltage of V_{bias}=-1.5V. Compared to (2) it must be mentioned, that e.g. η=67% means a five times higher current i_{high} compared to i_{low}. Though the measured rise time of t_{rise}=20ns (fig. 4) is relatively high due to dominating carrier diffusion, the switching time of the photocurrent due to the change of the electrical modulation, which is more relevant for signal correlation, is t_{switch}<5ns. A responsivity of R=0.47A/W {0.36A/W} at λ=660nm {850nm} is measured. As can be seen from fig. 5, the measured capacitance for depleted operation between cathode and anode is below $C_{C1,A}=C_{C2,A}$<0.2pF, the capacitance between the two cathodes is about $C_{C1,C2}$~$0.2C_{C1,A}$. Bandwidth measurements are done with a network analyzer: A laser is arranged in the transmission path, the modulated light is coupled into the DCP and the photocurrent of both connected cathodes is feed back to the return path of the network analyzer, calibrated with a 12GHz photoreceiver. As shown in fig. 6, a bandwidth of f_{-3dB}=284MHz / 322MHz, {59 MHz / 75MHz} is measured for DCPs with 15μm finger pitch and Field OXide (FOX) on

top of the diode at V_{bias}=-5V/-12V at λ=660nm {850nm}. For DCPs without FOX a bandwidth of f_{-3dB}=40MHz / 43MHz {33MHz /32MHz} is measured. This significant reduction of the achieved bandwidth obviously dates from a reduced pureness of the intrinsic region. Due to the etching process stopping at substrate level, the original intrinsic junction must have been blurred in terms of doping concentration. Hence the difference in bandwidth with and without FOX is comparable to the speed difference of PIN and PN photodiodes: Whilst slow diffusion is a main carrier mechanism at common PN-junctions resulting in a bandwidth of some 10MHz, drift is the dominating effect for intrinsic PIN-photodiodes resulting in an increased bandwidth by a factor of up to 100 in [7]. Low dark current of below I_{dark}<5pA {10pA} is measured at an operating bias voltage of V_{bias}<10V {17V} at room temperature (see fig. 7).Tab. 1 gives a summary of the measurement results.

VI.   APPLICATION

The presented DCP has been implemented in an integrated rangefinder sensor containing a simple pixel readout using a capacitor for integration of the already correlated photocurrents by the DCP. Furthermore a quasi-differential

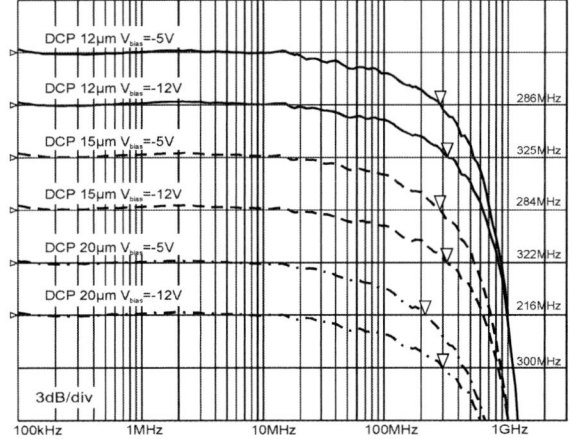

Figure 6.   Measured frequency response of DCP at λ=660nm with cathode pitch of 12μm/15μm/20μm each at V_{bias}=-5V/-12V (from top to bottom).

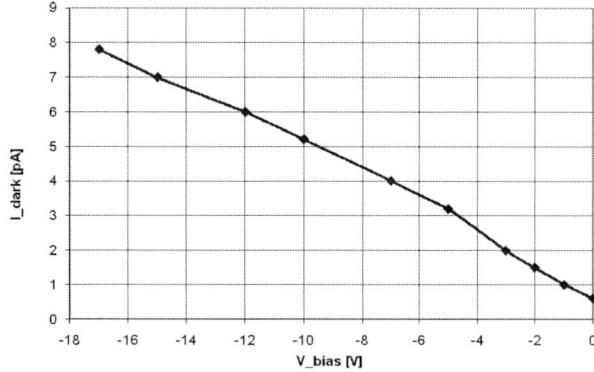

Figure 7. Measured dark current I_{dark} versus bias voltage V_{bias}.

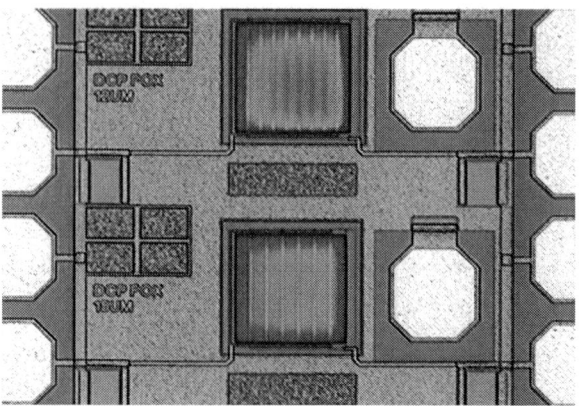

Figure 8. Chip photograph: DCP 12μm (top) and DCP 15μm (bottom).

output driver and a timing unit synchronizing the 10MHz clock complete the first demonstrator single pixel sensor [8] ending up in an optical fill factor of ~67%. Using a laser source with an optical power of P_{laser}=1mW at a wavelength of λ=650nm, distance measurements up to x=3.4m {6.2m} at a standard deviation of σ<1cm {5.7cm} for a measurement time of t_{meas}=20ms targeting diffuse white paper was achieved.

VII. CONCLUSION

A new buried finger concept for correlating photodetectors is presented using BiCMOS technology. Due to this buried cathode fingers a higher correlation capability also for longer wavelengths is achieved and legitimates this PIN setup in terms of: high responsivity R=0.47A/W {0.38A/W}, low parasitic capacitance C_{CIA}<0.1pF and high bandwidth up to f_{-3dB}~300MHz {80MHz} together with the achieved separation efficiency of up to η=67% {56%} at λ=660nm {850nm} which is better compared to the results η=20% at a bandwidth of 1MHz at λ=850nm in [3], the η=62% at 29MHz in [4] and the η=57% at λ=660nm in [5].

ACKNOWLEDGMENT

The authors would like to thank M. Hofbauer and W. Gaberl for support with device simulations. Furthermore A. Nemecek wants to thank G. Zach and K. Oberhauser for valuable discussions leading him to the idea of the DCP.

REFERENCES

[1] A. Izhal, T. Ushinaga, T. Sawada, M. Homma, Y. Maeda, and S. Kawahito, 'A CMOS time-of-flight range image sensor with gates on field oxide structure," *in Proc. of the 5th IEEE Int. Conf. on Sensors*, 2005, pp. 141-144, Irvine, USA.

[2] T. Oggier, R. Kaufmann, et al., "3D-imaging in real-time with miniaturized optical range camera," *in Proc. of the 6th International Conference for Optical Technologies, Optical Sensors and Measuring Techniques OPTO 2004*, pp. 89- 94, 2004.

[3] R. Lange, P. Seitz, A. Biber, and S.C. Lauxtermann „Demodulation pixels in CCD and CMOS technologies for time-of-flight ranging," in Proc. of SPIE Sensors and Camera Systems for Scientific, Industrial, and Digital Photography Applications, vol. 3965, pp.177-188, 2000.

[4] A. Nemecek, G. Zach, and H. Zimmermann; "Correlating Photodetector with Current Carrying Photogate for Time-of-Flight Distance Measurements", SPIE Photonics Europe 2008 - Proc. of Optical Sensors, vol. 7003-21, pp. 70030L-1–8, 2008.

[5] K. Oberhauser, A. Nemecek, Ch. Sünder, H. Zimmermann: 'Universal Integrated PIN Photodetector'; *Proceedings of the 34th European Solid-State Device Research Conference ESSDERC*, pp. 349-352, Leuven, Belgium, 2004.

[6] Ultrafast MSM photodetectors g4176 series, preliminary data sheet. Hamamatsu Photonics Co. [Online]. Available: www.hamamatsu.com.

[7] A. Nemecek, G. Zach, R. Swoboda, K. Oberhauser, H. Zimmermann, "Integrated BiCMOS PIN photodetectors with high bandwidth and high responsivity," *IEEE Journal of Selected Topics in Quantum Electronics*, vol. 12, num. 6, pp. 1469 - 1475, 2006.

[8] G. Zach, A. Nemecek, K. Oberhauser, and H. Zimmermann: "Correlating Buried-Finger Photodetector for Time-Of-Flight Applications," *in Proceedings of SPIE 19th International Congress on Photonics in Europe – Optical Measurement Systems for Industrial Inspection VI*, vol. 7389, S. 738936-1 - 738936-9, Germany, 2009.

TABLE I.   MEASUREMENT RESULTS OF DCP A.) COVERED WITH FOX AND B.) WITHOUT FOX FOR DIFFERENT BIAS VOLTAGES AND WAVELENGTHS.

		660nm				850nm			
	V_{bias}	<-17V...0V>	-1V	-5V	-12V	<-17V...0V>	-1V	-5V	-12V
	finger pitch / width	R [A/W]	η [%]	f_{-3dB} [MHz]		R [A/W]	η [%]	f_{-3dB} [MHz]	
FOX	12μm / 1μm	0.47	45	286	325	0.36	38	60	80
	15μm / 1μm	0.47	38	284	322	0.36	33	59	75
	20μm / 1μm	0.47	33	216	300	0.36	33	42	60
	15μm / 2μm	0.48	29	260	300	0.36	26	75	76
no FOX	12μm / 1μm	0.47	67	46	48	0.38	56	35	38
	15μm / 1μm	0.47	61	40	43	0.38	42	33	35
	20μm / 1μm	0.47	35	47	42	0.37	33	37	36

Understanding dark current in pixels of silicon photomultipliers

R. Pagano, S. Lombardo and S. Libertino
CNR-IMM
Strada VIII Zona Industriale, 5, 95121, Catania, Italy
roberto.pagano@imm.cnr.it

G. Valvo, G. Condorelli, B. Carbone, D.N. Sanfilippo
and G. Fallica
IMS-R&D STMicroelectronics,
Stradale Primosole, 50 95121 Catania, ITALY

Abstract— **Silicon photomultipliers are nowadays considered a promising alternative to conventional vacuum tube photomultipliers. The physical mechanisms operating in the device need to be fully explored and modeled to understand the device operational limits and possibilities. In this work we study the dark current behavior of the pixels forming the Si photomultiplier as a function of the applied overvoltage and operation temperature. The data are well modeled by assuming that dark current is caused by current pulses triggered by events of diffusion of single minority carriers (mostly electrons) injected from the boundaries of the active area depletion layer (dominating at temperatures above 0°C) and by thermal emission of carriers from Shockley-Read-Hall defects in the depletion layer (dominating at temperatures below 0°C).**

I. INTRODUCTION

Silicon Photomultipliers (SiPMs) are a very promising alternative to conventional photomultipliers (PM) thanks to some interesting characteristics: they are insensitive to magnetic fields, hence can be used in environments with high fields; their operation voltage is far lower, and they ensure better robustness and reliability than PM; they are much cheaper than their traditional counterpart [1-2].

SiPM structure consists in a parallel array of equal single pixels, each one made of a silicon p-n junction avalanche photodetector with an integrated resistor. The SiPM is biased above the breakdown voltage, that is, each pixel is operated in Geiger mode, above the breakdown voltage (BV) of the p-n junction. The junction is carefully doped in order to have breakdown only in the central active area of the pixel, used for the photon detection, and by the avalanche mechanism (not by Zener). To understand the photon detection concept, let us assume to bias such junction above breakdown with a fast voltage step. In this condition, if no carrier is present in the depletion region the junction is highly sensitive to the detection of single photons. In fact, if the photon is absorbed by creating an electron-hole pair, both carriers will start to drift in the high field region of the depletion layer and, being the voltage above breakdown, this drift will result with a 100% probability in the impact generation of a second e-h pair, and so on, up to the build-up of the junction avalanche.

The avalanche is limited by the buildup of a limiting space charge in the depletion layer which decreases the field [3]. Moreover, since the photodector has a resistor in series, when the avalanche current flows through the resistor, the voltage applied to the junction drops below BV. It quenches the avalanche, the current decreases to zero, and the voltage across the p-n junction increases again above BV. The pixel is ready again to detect the arrival of a new photon. Clearly, all the transients recorded are the result of both capacitive effects and (generally faster) avalanche build-up characteristic times.

Such ideal picture is strongly modified by the occurrence of phenomena leading to dark current, generally attributed to generation effects from Shockley-Read-Hall (SRH) defects in the depletion layer, afterpulsing effects, and diffusion of carriers from the quasi-neutral boundaries of the p-n junction [4].

The purpose of this work is to understand the behavior of dark current in single pixels of SiPMs, by separately taking into account the contribution given by the avalanche build-up and quenching, and the effect of generation / diffusion of carriers in the depletion layer in order to provide a detailed understanding of the current-voltage (IV) curves. We propose a physical model of the I-V above breakdown voltage able to reproduce the voltage and temperature dependence of the current for the studied devices.

II. DEVICE STRUCTURE

Devices were realized by STMicroelectronics on silicon epitaxial n-type wafers and formed from planar microcells. An implanted p-layer forms an enrichment region which defines both the active area and the breakdown voltage (BV) of the junction. The anode is contacted by sinkers created around the photodiode active area by means of a high-dose boron implantation. The cathode is given by the diffusion of arsenic from a doped in-situ thin polysilicon layer deposited on the top of the structure. The quenching resistor, made from low-doped polysilicon, is integrated on the cathode of the cell itself. Thin optical trenches filled with oxide and metal surround the pixel active area in order to reduce electro-optical coupling effects (crosstalk) between adjacent microcells. A double-layer antireflective coating made of silicon oxide and silicon nitride enhances the spectral response of the device in the blue and near ultraviolet wavelength ranges. The pixels have a square geometry with an active area side of 40x40 μm^2 [5]. Figure 1 shows a schematic drawing of the pixel, with the structure of the p-n junction. The depletion region relevant for photon detection is the one below the enrichment layer, where the field is higher, being larger the doping level.

III. RESULTS AND DISCUSSION

The current-voltage (I-V) curves of single cell reverse biased in the region 24V – 36V, in dark, as a function of the device temperature, from -25°C to 65°C are reported in Fig. 2. Breakdown is clearly visible at voltages of 27-29 V, with the well known increase of the BV with the temperature.

Though the measurements show steady-state I-V curves, the time resolved analysis of the current at the oscilloscope at a fixed bias above BV reveals that the time averaged breakdown current of Fig. 2 is indeed a random sequence of current spikes. Fig. 3 shows such spikes in a semilog time scale at various bias levels at room temperature.

Each I-t curve is indeed the average of 1000 traces. It is evident that after an initial spike the current has an exponential decrease with time, with the same characteristic time as the voltage level is changed. These dark counts are attributed to generation and / or diffusion from quasi neutral boundaries of a single free carrier which initiate the avalanche in a short time scale.

The current, however, does not go immediately to zero since there is the displacement current due to diode capacitance recharge to the pristine voltage level. In such a picture the integrated current signal, usually referred to as gain, is approximately equal to:

$$\overline{G} = \frac{\overline{Q}}{q} = \frac{2C}{q}\left(V_{POL} - BV\right) \qquad (1)$$

where C is the effective pixel capacitance, V_{POL} the applied

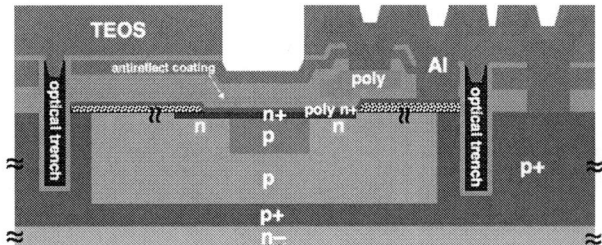

Figure 1.   Schematic of SiPM cell.

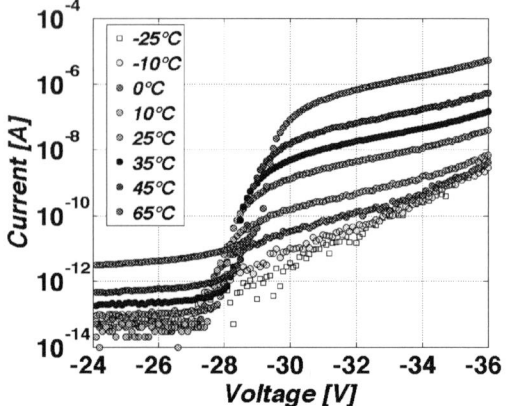

Figure 2.   I-V curves in reverse voltage as a function of the device temperature from -25°C (white square) up to 65 °C (magenta circles).

bias, and q the elementary charge. The factor 2 is needed since we detect both the initial current spike due to avalanche build-up and quenching, followed by the recharge of the diode effective capacitance. According to this picture, the time constant of the exponential I vs. t trace after the initial current spike should simply be equal to $\tau_C = R_{quench} \times C$, where R_{quench} is the value of the quenching resistance. Such interpretation is confirmed by the excellent agreement between the experimental time constants and the τ_C values. The agreement is also found when temperature is changed. In such case R_{quench} varies because of the temperature dependence of the resistance value of the integrated resistor [6], but still the measured time constants are perfectly consistent with the τ_C values.

According to this picture the measured DC current (Fig. 2) can then simply evaluated as:

$$I\left(V,T\right) = qG\left(V,T\right)\cdot \tilde{f}_{DC}\left(V,T\right)\cdot A_d \qquad (2)$$

where $\tilde{f}_{DC}$ is the frequency of "dark" events per unit area, i.e., events of generation and / or diffusion from quasi neutral boundaries of single free carriers into the active detection volume of the photodetector, and A_d is the corresponding detection area. G is the product of $\overline{G}$ of Eq. (1) times the probability P_a that an injected free carrier actually initiates the avalanche.

The gain $\overline{G}$ can be evaluated as the integral of signals such as those of Fig. 3 from 0 to 3-4 times τ_C. In particular Fig. 4 reports the measured gain evaluated by integration from 0 to 160 ns as a function of voltage for a number of temperatures. In the same figure we also report the theoretical gain $2 \times C(V\text{-}BV)/q$ evaluated at a single temperature (-25°C). Only one model curve is calculated for clarity, and the others corresponding to the higher temperatures are simply obtained by shifting the first one to the right because of the temperature dependence of BV.

978-1-4244-6658-0/10 $26.00 © 2010 IEEE

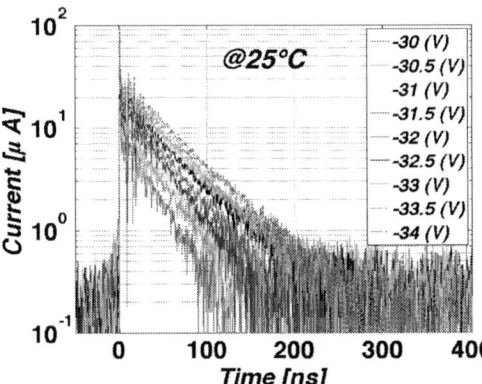

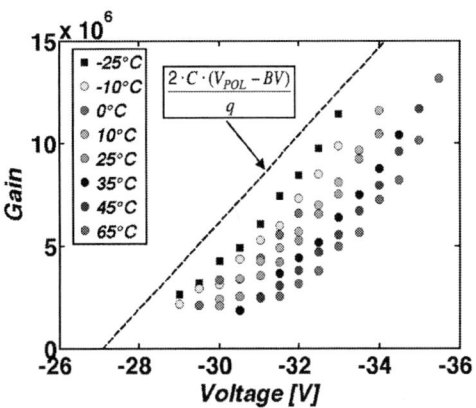

Figure 3. Dark Current as a function of time for biases ranging from -30V (lowest line) up to 34 V (highest line) acquired at room temperature.

Figure 4. Gain as a function of voltage. The dashed line is the theoretical gain evaluated at -25°C.

We first note that the model is quite close to the data (curve at -25 °C) but the experimental curve is non linear, with an approximately quadratic trend with voltage. The super-linear behavior is also observed at higher temperatures, without any particular change of trend, except for the well known shift of BV as temperature increases. The non linear behavior of gain is an important feature of SiPMs and we have investigated this issue in further detail. In particular we have measured the gain by an alternative, independent method, hereafter proposed. From Eq. (2) it is easy to estimate the photodetector current under illumination I_{light}. In fact one expects that:

$$I_{light} = q \cdot G \cdot \left(\tilde{f}_{DC} + QE \cdot \tilde{f}_{photon} \right) \cdot A_d \approx \\ \approx q \cdot G \cdot QE \cdot \tilde{f}_{photon} \cdot A_d \qquad (3)$$

where f_{Phot} is the photon flux incident on the pixel and QE is the corresponding external quantum efficiency. If we are in a condition where $f_{Phot} \gg f_{DC}$, G can be evaluated as:

$$G = \frac{I_{Light}}{\tilde{f}_{photon} \cdot QE \cdot A_d}. \qquad (4)$$

As already underlined G in Eqs. (2)-(4) is the product of $\overline{G}$ (Eq.(1)) times P_a, so the two parameters coincide only if P_a is one, i.e., 100% probability to trigger the avalanche.

Figure 5 reports an example of I-V characteristics of a SiPM pixel under illumination with laser light at 659 nm at flux levels ranging from 2.2 nW/cm^2 up to 22 μW/cm^2. Above BV the pixel operates linearly up to about 200 nW/cm^2, and a tendency to signal saturation is evident above such intensity. The saturation above 200 nW/cm^2 is well explained by dead time effects, of the order of 200 ns as shown in Fig. 3.

Data such as those of Fig. 5 allow to evaluate G by using Eq. (4). By assuming a QE value of 0.15 at the 659 nm laser wavelength [7] we determine G and the results are shown in Fig. 6. In particular the figure shows the comparison between

the $\overline{G}$ values with those of G and the linear model of Eq. (1). It is evident a surprisingly good match between G and $\overline{G}$ at high voltage, while a small difference is observed at low voltages.

We now proceed in our analysis by discussing the dark count frequency f_{DC}. Ideally with no SRH center generating free carriers in the detection volume, f_{DC} should at least be equal to the frequency of free carrier injection from the quasi-neutral boundaries, given by the well known expression:

$$\tilde{f}_{DIF} = \frac{n_i^2 D_n}{N_a L_n} = \sqrt{\frac{D_n}{\tau_n}} \cdot \frac{n_i^2}{N_a} \qquad (5)$$

where n_i is intrinsic carrier concentration, N_a is the dopant concentration at the depletion layer boundary of the enrichment, D_n is the electron diffusivity and L_n is the diffusion length.

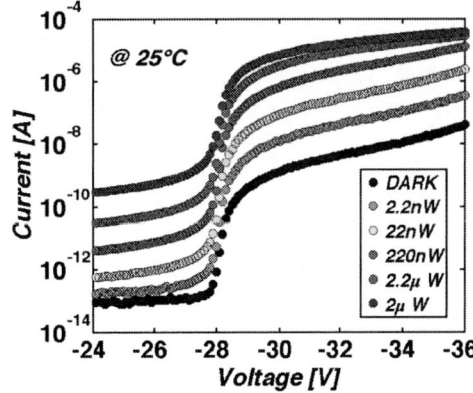

Figure 5. IV characteristics in reverse voltage in dark (dark) and under illumination from 2.2 nW/cm^2 (green) to 22 μW/cm^2 (blue)

If we also assume the presence of defects, the related emission frequency is given by the well known SRH expression:

$$\tilde{f}_{TH} = N_{dif} \cdot W \cdot \gamma_n \cdot \sigma_n \cdot T^2 \cdot \exp\left(-\frac{E_c - E_T}{kT}\right) \quad (6)$$

where N_{dif} is the defect concentration, W the depletion layer width, γ_n an universal constant, σ_n the defect cross-section, E_C-E_T the defect ionization energy, T the temperature, and k the Boltzmann constant.

Figure 7 shows the comparison between the experimental dark I-V characteristics and the model, by assuming the G values of Fig. 6. The agreement between data and model is extremely good. We fit the data both as a function of voltage and as a function of temperature by assuming the well known relationship between carrier diffusivity and mobility, and $L_n = 10$ μm, $\mu_n = 1100$ cm^2/Vs, and $N_a = 1.5e16$ cm^{-3} in Eq. (5), while for thermal diffusion (Eq. (6)) we have assumed $N_{dif} = 1e9$ cm-3, E_C-$E_T = 0.54$ eV, $\sigma_n = 1.6e\text{-}15$ cm^2, with the universal constant $\gamma_n = 1.78e21$ cm^{-2}s^{-2}K^{-2} as reported in [8]. The remarkable agreement between data and model is obtained by assuming quite reasonable values of the fit parameters, and this suggests that the present model catches quite well the behavior of the device. We also note that these devices present a dark current only limited by carrier diffusion already at quite low temperatures, essentially almost at 0 °C, indicating a remarkably low SRH defect concentration (of the order of 1e9 cm^{-3}).

IV. CONCLUSIONS

In this paper we have reported on the realization of Silicon Photomultipliers, we have described a physical model on the dark count rate of SiPM single pixels, and we have compared this model to experimental data taken on SiPM realized by

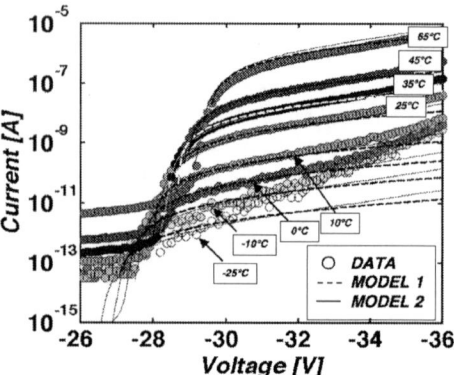

Figure 7. Comparison of the experimental I-V curves (circle) in reverse voltage and the physical model as (lines) a function of the device temperature from -25°C (white) up to 65 °C (magenta)

STMicroelectronics. The model fits nicely the data and demonstrates that state-of-the-art SiPM can have at room temperature a dark current rate limited only by carrier diffusion.

ACKNOWLEDGMENTS

CNR authors gratefully acknowledge partial grant support by IMS R&D, STMicroelectronics.

REFERENCES

[1] G. Bondarenko, B. Dolgoshein, V. Golovin, Ilyin, R. Klanner, E. Popova, "Limited Geiger-mode silicon photodiode with very high gain", Nucl. Phys. B Proc. Suppl., vol.61 B, 1998, pp. 347-352.

[2] N. Otte, B. Dolgoshein, J. Hose, S. Kleminin, E. Lorentz, R. Mirzoyan, E. Popova, M. Teshina, "The Potential of SiPM as Photo Detector in Astropaticle Physics Experiments like MAGIC and EUSO", Nucl. Phys. B Proc. Suppl.,vol 150, 2006, pp. 144-149.

[3] S. Cova, M. Ghioni, A. Lacaita, C. Samori, F. Zappa, "Avalanche photodiodes and quenching circuits for single photon detection," Appl. Opt., vol 35, 1996, pp. 1956–1976.

[4] E. Sciacca, A.C. Giudice, D. Sanfilippo, F. Zappa, S. Lombardo, R. Cosentino, C. Di Franco, M. Ghioni, G. Fallica, G. Bonanno, S. Cova, E. Rimini, "Silicon Planar Technology for Single-Photon Optical Detectors", IEEE Trans. on Elect. Dev., vol 50, 2003, pp. 918-925.

[5] M. Mazzillo, G. Condorelli, D. Sanfilippo, G. Valvo, B. Carbone, G. Fallica, S. Billotta, M. Belluso, G. Bonanno, L. Cosentino, A. Pappalardo, P. Finocchiaro, "Silicon Photomultiplier Technology at STMicroelectronics" IEEE Trans. on Nucl. Sci., vol 56, 2009, pp. 2434-2442.

[6] R. Pagano, S. Libertino, G. Valvo, G. Condorelli, B. Carbone, A. Piana, M. Mazzillo, D. N.Sanfilippo, P.G. Fallica, G. Falci, S. Lombardo, "Dark Count in Single Photon Avalanche Si Detectors", in Optical Components and Materials VII, ed. by S. Jiang, M.J.F. Digonnet, J.W. Glesener, J.C. Dries, Proc. of SPIE vol. 7598, 2010, pp. 75980Z.

[7] M. Mazzillo,G. Condorelli,A. Piazza,D. Sanfilippo,G. Valvo,B. Carbone,G. Fallica,S. Billotta,M. Belluso,G. Bonanno, A. Pappalardo, L. Cosentino, P. Finocchiaro, "Single-photon avalanche photodiodes with integrated quenching resistor", Nucl. Instr. & Met. In Phys. Res., vol. 591, pp 367-373, 2008.

[8] K. Schroder, Semiconductor Material and Device Characterization,3rd ed.,Wiley-Interscience,2006,pp 262.

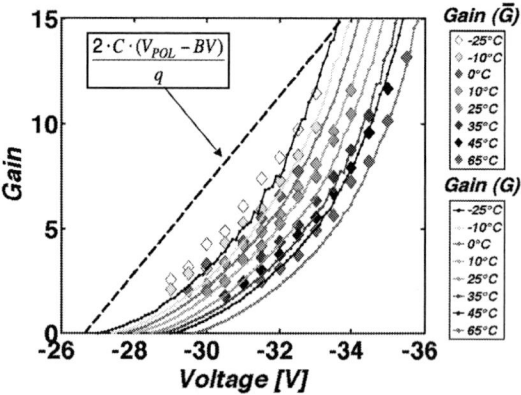

Figure 6. Gain as a function of voltage for temperatures from -25°C up to 65°C: dots are data reported in Fig.4, lines the gain determined from eq. (4) as described in text.. The results are compared with the model of eq. (1)

Hot-carrier Stress induced degradation in Multi-STI-Finger LDMOS: an experimental and numerical insight

S. Poli, A. Loi, S. Reggiani, G. Baccarani,
E. Gnani, A. Gnudi,
ARCES and DEIS, University of Bologna
Bologna, Italy, email: spoli@arces.unibo.it

M. Denison, S. Pendharkar, R. Wise, S. Seetharaman
Texas Instruments, Inc.
Dallas, Texas

Abstract— Degradation induced by hot-carrier stress (HCS) in a Multi-STI-Finger (MF) LDMOS is analyzed through both electrical measurements and TCAD simulations. The critical HCS issues have been first addressed on a conventional STI-based LDMOS. Then, the detrimental effect of extended Si/SiO$_2$ interfaces along the silicon fingers in the MF-LDMOS has been widely investigated. Experimental results are analyzed and discussed on the basis of numerical simulations. The application of a time-dependent HCS degradation model is successfully proved for the first time on the conventional and MF-LDMOS devices at different stress biases and ambient temperatures.

I. INTRODUCTION

The lateral double-diffused DMOS (LDMOS) transistors are widely used in mixed-signal technologies due to their compatibility with standard CMOS devices. Recently, different architectures have been proposed to optimize the performance of LDMOS devices for high voltage applications, namely, those based on the use of dielectric RESURF [1] and those based on interleaved silicon/STI fingers with lateral field plates [2, 3, 4]. Here a Multi-STI-finger (MF) LDMOS with pulled-back gate, tapered field plate, and drain-sided STI comb is studied. This concept represents a cost efficient approach to improve the specific on-resistance (R_{SP}), as it is based on layout design variations. Even though experimental demonstration of their applicability has been already proven, just little information is given on electrical reliability concerns [5], and a TCAD-based study of hot-carrier stress (HCS) induced degradation is still lacking for such devices. Furthermore, the STI-based LDMOS, used as a starting reference in this work, shows a rugged behavior extending the safe operating area to very large drain/gate voltages [6]. This was achieved with a buried-body implant introduced to suppress the parasitic bipolar transistor, thus eliminating the snapback effect. Hot-carrier characterization at such stress biases still requires a comprehensive analysis [7].

In this work an experimental analysis of the HCS-induced degradation of a MF-LDMOS is presented, extended to different stress biases and ambient temperatures. Results have been compared with the parameter drifts of a standard STI-based LDMOS. Moreover, the extensive use of TCAD simulation in both DC and transient conditions allowed us to

Work supported by the SRC Research Contract No. 2007-VJ-1667

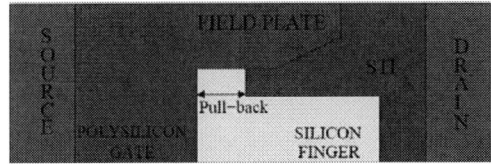

Figure 1. Schematic top view of the multi-STI-finger LDMOS representing the device unit cell used for TCAD simulations. The gate edge is pulled back with respect to the STI source-side edge.

thoroughly investigate the mechanisms responsible for the HCS degradation, providing a physical insight on the device behavior. Drift curves of the linear drain current as functions of the stress time are calculated by TCAD simulations which nicely predict the experimental data over a range of drain biases and ambient temperatures, showing for the first time that the TCAD tool can accurately solve the kinetics of the trap formation induced by hot carriers at the STI interface in standard and MF-LDMOS devices.

II. CALIBRATION OF THE SIMULATION DECK

Fig. 1 reports the schematic top view of the n-channel MF-LDMOS transistor investigated in this paper. The MF structure has been designed as a layout variation of the reference rugged LDMOS device, introducing laterally interleaved STI and silicon fingers along the drift region. The polysilicon gate is pulled back with respect to the STI, whereas the lateral capacitive coupling between adjacent field plates (realized by means of polysilicon fingers) ensures the depletion of the silicon-finger regions at high drain voltages, giving the advantage of an extra active area without affecting the breakdown voltage (V_{BD}). MF-LDMOS devices with different field-plate geometrical features (i.e., field-plate length, with a maximum relative variation of 25%, and with/without tapering) have been characterized. Very limited variations of both V_{BD} and R_{SP} have been observed for the different samples, indicating that the length and shape of the field plate cannot significantly affect the dielectric RESURF induced by the lateral capacitances. An overall reduction of R_{SP} of about 10% was obtained at the same V_{BD} for the MF-LDMOS when compared to the reference device.

978-1-4244-6658-0/10 $26.00 © 2010 IEEE

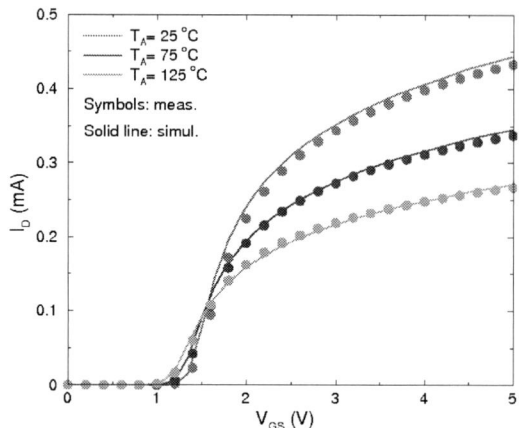

Figure 2. Turn-on characteristics at $V_{DS} = 0.1$ V with $T_A = 25$, 75, and 125 °C. (Symbols) Experiments and (lines) simulation results.

3D simulations have been carried out on the device unit cell (half pitch, from source to drain, and half width, from the center of the silicon finger to the center of the STI finger). Simulations have been performed using the electro-thermal model available in the Sentaurus-Device tool by Synopsys [8], which couples the drift-diffusion transport equations with the heat transfer equation. The definition of the analytical doping profiles has been derived from SIMS measurements and process simulations on the reference LDMOS device [9], and a fine tuning of the deck was carried out through the comparison with measured turn-on and output I-V curves.

In Figs. 2 and 3, the simulated characteristics are compared with experiments. The numerical simulation nicely predicts the experimental turn-on curves at different ambient temperatures (T_A) perfectly capturing the thermal behaviour. Being instrumental for the following investigations, the output characteristics are reported at $V_{GS} = 2$ and 4 V and for V_{DS} up to values well above the nominal voltage ($V_{DD} = 30$ V). At $V_{DS} > 30$ V, the output curves show a clear current increase, which is due to the on-set of the impact-ionization generation. This is confirmed by the body current (I_B) vs V_{GS} (inset of Fig. 3), showing a significant increase of I_B for the MF-LDMOS around $V_{GS} = 2$ V for both the V_{DS} biases corresponding to the low- and high-drain stress conditions: at such operating regimes, the impact-ionization generation takes place mainly at the p-body/n-well junction, and, compared to the conventional LDMOS, a higher I_B is observed for the MF structure due to a slight (about 15%) layout reduction of the accumulation region length inducing an increase of the electric field peak at the junction. The high field regime in the proximity of the source-side STI corner is the main cause of strong HCS effects: the presence of a large density of "hot" electrons flowing close to the Si/SiO$_2$ interface leads to a fast trap generation and to the consequent trapping of charges in the on-current condition which causes large shifts (degradation) of the drain current itself.

Even though I_B increases with V_{GS} (inset of Fig. 3), the stress biases at $V_{GS} = 2$ V represent a worst case condition for the HCS-induced degradation of the linear drain current $I_{D,lin}$ for both devices. Indeed, at higher V_{GS}, I_B is sustained by

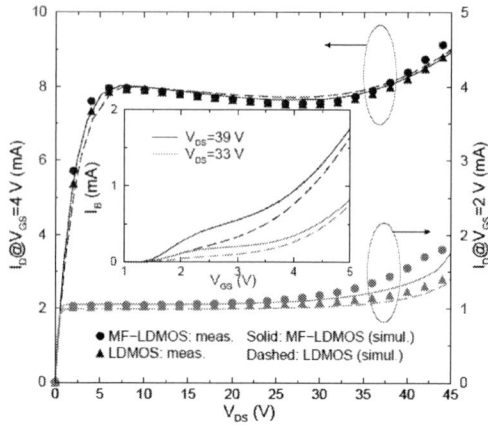

Figure 3. Output characteristics at (right scale) $V_{GS} = 2$ V and (left scale) 4 V with $T_A = 25$ °C. (Symbols) Experiments and (lines) simulation results. An extended range of V_{DS} has been considered as of interest for the HCS analysis. Inset: simulated I_B as a function of V_{GS} at low ($V_{DS} = 33$ V) and high ($V_{DS} = 39$ V) stress biases ($T_A = 25$ °C) for the (solid lines) MF and the (dashed lines) conventional LDMOS.

impact-ionization generation caused by the Kirk effect taking place at n/n+ region close to the drain contact, while no critical electric field is found at the source-side STI corner, leading to a much more limited trap generation along the Si/SiO$_2$ interface and to a significantly reduced HCS degradation of the $I_{D,lin}$. This is confirmed by the experimental analysis reported in the following.

III. HOT-CARRIER STRESS ANALYSIS

The analysis of the HCS-induced degradation has been carried out by exploring different stress biases (varying both drain and gate voltages) and different ambient temperatures. Both the MF-LDMOS and the conventional STI-based LDMOS have been characterized. The considered devices, characterized by the same channel geometry and doping profiles, show very close operating regimes and electric field distributions, and thus can be fairly compared, gaining an insight on the new features of the multi-finger structure. Identical and negligible threshold-voltage shifts have been measured in the bias and temperature ranges under analysis and a small impact of the channel degradation on the transconductance has been found as well. Thus, the investigation has been focused mainly on the $I_{D,lin}$ degradation, capturing the reliability concerns related to the presence of the additional silicon fingers and the 3D effects on the STI corners.

A. Dependence on the drain voltage

In Fig. 4 (top), the measured relative $\Delta I_{D,lin}$ (absolute value) as a function of stress time is reported for both devices at a fixed gate voltage ($V_{GS} = 2$ V) and different V_{DS}. First, a reduced dependence of the $\Delta I_{D,lin}$ shifts on V_{DS} is observed for the MF-LDMOS compared to the reference one. This is mainly due to the enhancement of the critical electric field across the p-body/n-well junction described in the previous section, which causes the MF-LDMOS to experience a higher impact-ionization regime in the investigated V_{DS} range.

978-1-4244-6658-0/10 $26.00 © 2010 IEEE

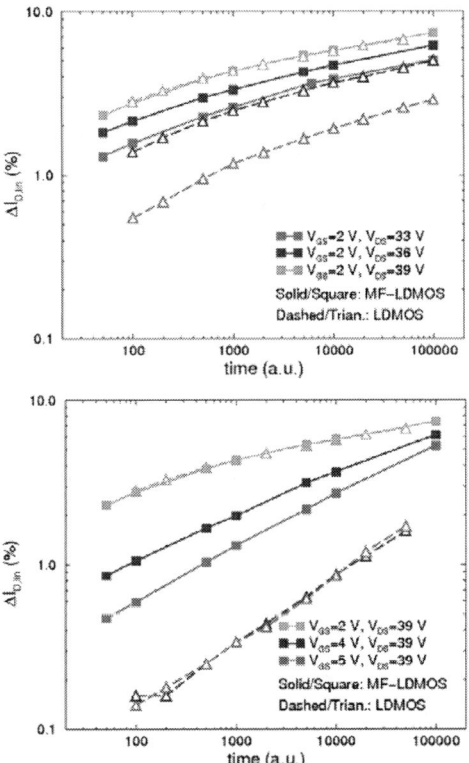

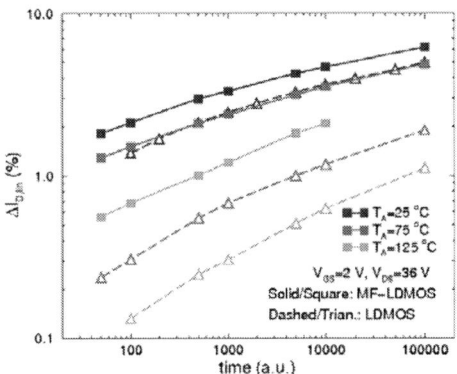

Figure 5. Measured $I_{D,lin}$ shift vs. stress time at V_{GS}=2 V and V_{DS}=36 V for different ambient temperatures (T_A = 25, 75, and 125 °C).

Figure 4. Measured $I_{D,lin}$ drift as a function of stress time for (top) different V_{DS} and (bottom) different V_{GS}. T_A = 25 °C. Solid lines: MF-LDMOS. Dashed lines: conventional LDMOS.

Additionally, provided that the STI effective width is reduced by the presence of the silicon fingers and that the electric field distribution at the source-side STI corner is no more uniform and less effective, a beneficial effect on the HCS degradation is expected. Vice versa, no advantage is observed in experimental curves of the new device at any considered V_{DS}. The reported larger shifts can be explained only by additional contributions to the trap formation and charge trapping, which should be localized not only at the STI corner along the drift region, but also at the top and sidewall Si/SiO$_2$ interfaces of the silicon fingers. It is worth noting that the numerical analysis shows up to 20% of the on-current flowing across the finger, and that a strong $I_{D,lin}$ sensitivity to charge trapped at the finger interfaces has been also verified. Moreover, the TCAD analysis of the electric field distribution, electron density, and electron temperature in stressed conditions reports critical hot-carrier spots along the finger with high carrier densities which can induce significant trap formation.

B. Dependence on the gate voltage

The previous analysis has been further extended to different V_{GS} biases and fixed V_{DS} = 39 V. The characterization of the $\Delta I_{D,lin}$ is reported in Fig. 4 (bottom). As V_{GS} is increased, and the intrinsic MOSFET is driven from saturation towards the linear regime [8], a drastic reduction of the electric field peak close to the source-side STI corner is experienced. Moreover, the effect of self-heating is expected to play a relevant role in reducing the hot-electron concentration and HCS degradation. In such conditions, the standard LDMOS shows a reduction of the $\Delta I_{D,lin}$ values of almost one order of magnitude and the shifts are found to be independent of V_{GS}, as the gate modulation of the critical electric field peak at the drain-side of the drift region has little effect on the current degradation. On the contrary, remarkably higher shifts are reported for the MF-LDMOS and with a clear dependence on V_{GS}. Once more, the physical explanation for such differences needs to be ascribed to the trap formation at the oxide interfaces along the silicon fingers, as no significant differences are observed in the MF-LDMOS field distribution at the STI corner and under the STI with respect to the reference structure. Along the fingers, the top and sidewall surfaces experience a strongly non-uniform distribution of the current density and hot-electron spot. The concurrent increase of the carrier density with V_{GS} and decrease of the electric field (and, consequently, of the hot spot) with decreasing V_{GD} are clearly observed. The modulation of the $\Delta I_{D,lin}$ with V_{GS} can be thus explained by considering that the dependence on the electron temperature is expected to play the major role in the trap formation.

C. Dependence on the ambient temperature

A relevant reduction of the critical field peaks with T_A is expected, which is consistent with the decrease of I_D and I_B, and with the increase of V_{BD} (about 9% for the standard and 3% for the MF-LDMOS at T_A = 125 °C). Hence, a strong reduction of the HCS effects is expected as well [9]. The measured $\Delta I_{D,lin}$ curves at different T_A are reported in Fig. 5. Due to the strong correlation between T_A and the longitudinal electric field, the different $\Delta I_{D,lin}$ modulation with temperature reported for the reference and the MF-LDMOS can be explained by the same arguments used for the dependence on V_{DS}.

IV. SIMULATION OF THE HOT-CARRIER DEGRADATION

Beside the experimental characterization of the $\Delta I_{D,lin}$ dependencies and their TCAD analysis based on DC results, transient simulations have been carried out by using the

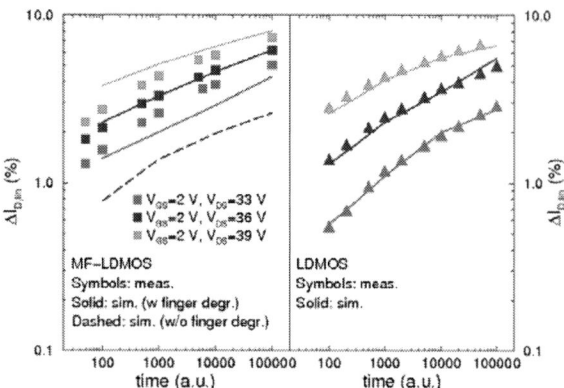

Figure 6. Measured and simulated $I_{D,lin}$ shifts vs. stress time for (left) MF-LDMOS and (right) standard LDMOS. Dashed line: simulated $\Delta I_{D,lin}$ for the MF-LDMOS with no traps at the finger interfaces.

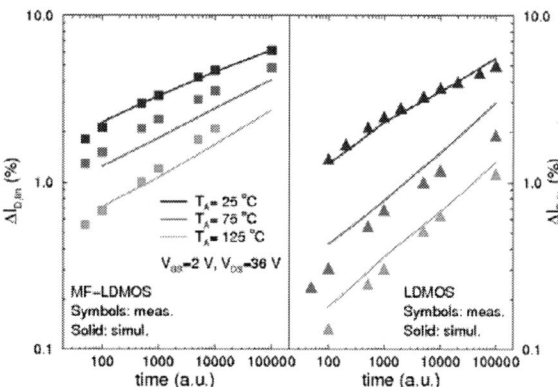

Figure 7. Comparison between measured and simulated $\Delta I_{D,lin}$ shifts vs. stress time at different T_A for (left) MF-LDMOS and (right) standard LDMOS. V_{GS} = 2 V, V_{DS} = 36 V.

kinetic equation of trap formation at the Si/SiO$_2$ interface [11]. The HCS model is based on the solution of the Si-H defect kinetics equation, and takes into account the interface disorder and the Si-H bond activation energy evolution as the bonds are broken. The HCS parameters have been calibrated on the experimental $\Delta I_{D,lin}$ shifts of the reference LDMOS [12]. Here, we extend the analysis to the 3D MF-LDMOS device and to different T_A. It is important to point out that the same set of parameters used for the calibration of the standard LDMOS has been used for the MF-LDMOS, extending its application to the additional oxide interfaces along the finger.

Fig. 6 reports the comparison between the simulated and the experimental $\Delta I_{D,lin}$ at different V_{DS} and T_A = 25 °C. The implemented HCS model correctly takes into account the dependence on a uniformly distributed T_A, but is not suited for local temperature variations. For this reason, TCAD analyses

have been carried out at low V_{GS} to avoid self-heating effects. The electro-thermal simulations nicely predict the experimental results. In particular, the different dependencies on V_{DS} shown by the two structures are well captured. In addition, TCAD simulations allowed us to prove the important role played by the degradation in the finger region: the blue dashed line in Fig. 6, left, has been obtained by artificially switching off the trap formation at the finger interfaces in the MF-LDMOS, showing that the $\Delta I_{D,lin}$ values drastically drop below those of the standard one.

Finally, simulation results at different T_A are reported in Fig. 7 for both devices. The measured $\Delta I_{D,lin}$ are nicely reproduced by TCAD results revealing that the localization of interface traps in the finger interfaces play a major role in degrading the MF structure. The nice agreement with experiments allowed us to confirm the HCS-degradation analysis reported in the previous Section.

V. CONCLUSIONS

An experimental and TCAD analysis of the HCS degradation has been carried out on a multi-STI-finger LDMOS device compared with a standard one. Different bias and ambient temperatures have been addressed to the purpose of gaining an insight on the different HCS mechanisms. The TCAD degradation model, calibrated on the conventional device, nicely predicts the degradation features of the 3D multi-STI-finger structure, clearly revealing the role played by the interfaces along the fingers.

ACKNOWLEDGMENT

The authors would like to thank Phil Hower and John Lin at Texas Instruments.

REFERENCES

[1] A. Heringa and J. Šonský, *Proc. of the ISPSD'06*, Napoli (Italy), June 4-8, 2006, pp. 1-4.

[2] J. Šonský and A Heringa, *Proc. of the ISPSD'07*, Jeju (Korea), May 27-30, 2007, pp. 77-80.

[3] US patent application #20090256199.

[4] A. Heringa et al., *Proc. of the ISPSD'08*, Orlando (FL), May 18-22, 2008, pp. 271-274.

[5] J. Perez-Gonzalez et al., *Proc. of the ISPSD'09*, Barcelona (Spain), June 14-18, 2009, pp. 61-64.

[6] P. Hower et al., *Proc.of the ISPSD'05*, Santa Barbara (CA), May 23-26, 2005, pp. 327-330.

[7] S. Poli et al., Proc. of the *ISPSD'10*, Hiroshima (Japan), June 6-10, 2010, pp. 311-314.

[8] Synopsys Inc., "Sentaurus device simulator (release A-2008.09)," 2008.

[9] S. Reggiani et al., IEEE Trans. on Electron Devices 56, p. 2811, 2009.

[10] T. Nitta et al., *Proc. of the ISPSD'09*, Barcelona (Spain), June 14-18, 2009, pp. 84-87.

[11] O. Penzin et al., IEEE Trans. On Electron Devices 50, p. 1445, 2003.

[12] S. Reggiani et al., Proc. of the *IRPS'10*, Anaheim (California), May 2-6, 2010.

Repetitive Avalanche Cycling of Low-Voltage Power Trench n-MOSFETs

Olayiwola Alatise*, Ian Kennedy, George Petkos, Keith Heppenstall, Khalid Khan, Jim Parkin Adrian Koh and Phil Rutter

Innovation R&D, NXP Semiconductors, Manchester, UK. SK7 5BJ. *olayiwola.alatise@nxp.com

ABSTRACT

Low voltage discrete power trench n-MOSFETs in TO-220 packages have been subjected to over 200 million cycles of repetitive unclamped inductive switching (UIS) at a mounting base temperature of 150° C and at different avalanche currents. Hot-hole injection into the gate dielectric during avalanche conduction causes a reduction in the threshold voltage as the number of avalanche cycles increase. The relationship between the change in the threshold voltage and the number of avalanche cycles is shown to be a power-law with the pre-factor dependent on the test conditions and the MOSFET technology. Experiments show that the power law pre-factor is proportional to the avalanche current in agreement with the predictions of the "lucky-electron" model. Interestingly, the pre-factor also responds proportionally to the MOSFETs cell pitch. A 40% increase in the avalanche current caused a 30% increase in the power law pre-factor while a 37.5% reduction in the MOSFETs cell pitch caused a 40% reduction in the power law pre-factor. Smaller cell pitch MOSFETs also exhibit improved on-state resistance stability with avalanche cycling however with higher drain-source leakage.

1. INTRODUCTION

Power MOSFETs are increasingly being used in automotive applications where they drive inductive loads like electric motors and alternators. These MOSFETs can be subjected to energy spikes from unclamped inductive switching (UIS) capable of causing MOSFET thermal failure either by parasitic bipolar latch-up and/or intrinsic temperature limitations of the semiconductor [1, 2]. To this end, manufacturers quote the maximum avalanche power densities and durations otherwise known as the safe-operating-area (SOA) on the datasheets so that designers can guarantee the reliability of the automotive system. However, other failure mechanisms arise due to the repetitive avalanche cycling of energy spikes within the MOSFET SOA. Hot-carrier injection (HCI) during

repetitive avalanche constitutes a major reliability challenge to power MOSFETs because of its impact on the threshold voltage, off-state drain leakage and on-state resistance. In this paper, the impact of HCI from repetitive UIS is assessed on power MOSFETs with different cell pitches.

2. EXPERIMENTAL SET-UP

The MOSFETs under investigation are automotive grade devices rated to 20 V and 30 V source-drain breakdown-voltage (B_{VDSS}). 21 mm^2 active area MOSFETs with 2.5 µm and 4 µm cell pitch and 76 nm thick gate oxides are fabricated with 1.5 µm trench depth and room temperature threshold voltage (V_{GSTX}) centered at 3.5 V and on-state resistance (R_{DSON}) at 2 mΩ. A cross-sectional SEM of the MOSFET's active area is shown in Fig. 1. The MOSFETs are packaged in standard TO-220 packages comprising of 3 source wires, 1 gate wire and the drain connected to the lead frame.

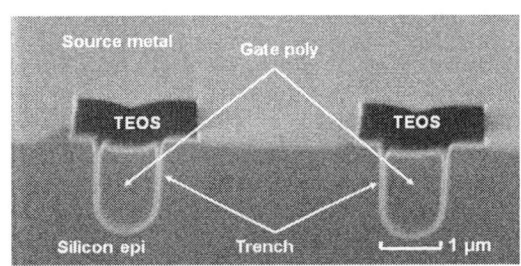

Fig. 1. Cross-sectional SEM of the power MOSFET showing 2 trenches, gate poly, source metal and TEOS passivation.

A custom-built avalanche test equipment is used for repetitive ruggedness testing. Fig. 2 shows the circuit diagram of the test equipment capable of testing 16 devices simultaneously. The equipment comprises of 3 high voltage charging MOSFETs (B_{VDSS} of 100 V) connected in parallel to the power supply through the avalanche inductor. The device under test (DUT) is connected in parallel with the high voltage MOSFETs to the power supply through the

978-1-4244-6658-0/10 $26.00 © 2010 IEEE

inductor. High voltage MOSFETs are used to charge the inductor so as to ensure that the DUT is avalanched alone since the MOSFET with the lower B_{VDSS} always enters avalanche first i.e. an avalanche current (I_{AV}) always flow through the MOSFET with the lowest B_{VDSS}.

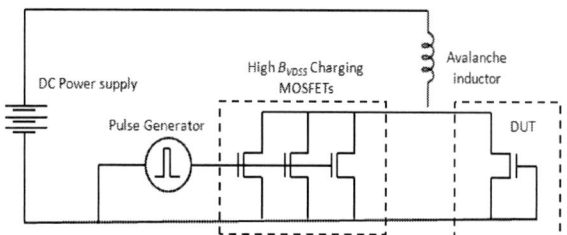

Fig. 2. The circuit diagram of the custom-built repetitive avalanche test equipment showing the power supply, pulse generator, high B_{VDSS} charging MOSFETs, the avalanche inductor and the DUT.

The gate terminals of the 100 V charging MOSFETs are connected to a pulse generator whereas the gate of the DUT is grounded as shown in the circuit diagram of Fig 2. The period of the pulse generator is used to modulate the mounting base temperature (T_{MB}) at which the MOSFET is tested and is varied until T_{MB} is 150 °C. The magnitude of the voltage pulse is used to set I_{AV} which will be proportional to the gate drive in the charging MOSFETs whereas the avalanche inductor is used to set the avalanche duration. When the input voltage of the pulse generator is high, the 100 V MOSFETs conduct current hence charging the inductor and when it is low, the inductor dissipates the energy into the DUT which is driven into avalanche. Since the gate of the DUT is grounded, the MOSFET never switches on and hence only conducts in avalanche. Fig. 3 illustrates the typical avalanche behavior of one of the DUTs by showing I_{AV} and V_{DS} as functions of time.

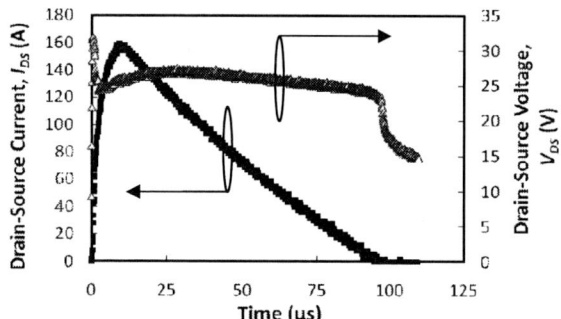

Fig. 3. The drain-source avalanche current and drain source voltage shown as functions of the avalanche time. As the avalanche current flows through, the drain-source voltage rises to the B_{VDSS} until the current reaches zero. The avalanche duration is determined by the avalanche current, MOSFET B_{VDSS} and the magnitude of inductance.

The avalanche duration in Fig. 3 is approximately 100 μs, the peak I_{AV} is 160 A and the measured B_{VDSS} averages at 25

V. Fig. 3 shows an increase in B_{VDSS} within the first 25 μs because of the positive temperature coefficient of B_{VDSS}. This is due to avalanche mean-free-path reduction from increased phonon scattering caused by rising temperature.

3. RESULTS AND DISCUSSION

Repetitive UIS is performed at different I_{AV} so as to investigate the impact of I_{AV} on the magnitude of ΔV_{GSTX}. The experiments are performed with an I_{AV} of 160 A and 225 A at a T_{MB} of 150 °C. Experimental measurements of the magnitude of the change in threshold voltage, ΔV_{GSTX}, show a power law relation to the number of avalanche cycles.

$$\Delta V_{GSTX} = A \cdot N^n \qquad (1)$$

Fig. 4 shows the log-log plot of the magnitude of ΔV_{GSTX} as a function of N where it can be seen that A is 1.9×10^{-13} and 2.5×10^{-13} for I_{AV} of 160 A and 225 A respectively.

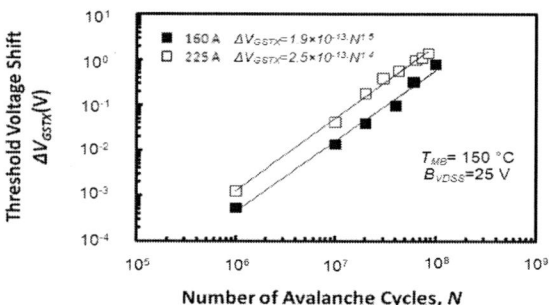

Fig. 4. The log-log plot showing the power law relationship between the magnitude of the change in threshold voltage (ΔV_{GSTX}) and the number of avalanche cycles. A 40% increase in the avalanche current resulted in a 30% increase in the power law pre-factor.

The V_{GSTX} reduction is due to hot-hole injection into the gate dielectric during impact ionization. Since the MOSFET gate is grounded (V_{GS}=0 V) in repetitive avalanche mode conduction, electrons generated from impact ionization are swept into the drain by the electric field whereas the holes generated from impact ionization are injected either into the gate oxide, can cause parasitic bipolar latch-up and/or recombine in the body of the MOSFET (in a MOSFET with a separate body contact, the holes will flow out as substrate current). If the hole current resulting from impact ionization is large enough, the MOSFET would undergo bipolar thermal runaway in a single pulse of UIS, hence the SOA on the MOSFET's datasheet would reflect the limitations of the device. The observation of hot-hole injection is in agreement with the work of Doyle et al [3, 4] which showed that hole injection dominates the degradation mechanism at low gate voltages (V_{GS}) whereas electron injection dominates the degradation mechanism at high gate voltages. For power MOSFETs, this translates to hot-hole injection being the active degradation mechanism during avalanche operation (V_{GS}=0 V and V_{DS}=B_{VDSS}) and hot-electron degradation being the active degradation mechanism during linear mode operation (V_{GS} and V_{DS} = High). It is expected that an increase in I_{AV} would accelerate the change in V_{GSTX} since higher I_{AV} will cause more impact ionization and hence, more hot-hole injection. It can be seen from Fig. 4 that the power law pre-factor (A) increases approximately linearly with I_{AV} i.e. as I_{AV} is increased by 40%, A increases by 30%.

The fundamental theory of HCI can explain the relationship between the power-law pre-factor, A and I_{AV} and why ΔV_{GSTX} is accelerated by increased I_{AV}. The "lucky-electron" model introduced by Hu et al [5] showed that the rate of supply of hot electrons into the gate dielectric and the substrate current was proportional to $I_D \cdot exp(-\Phi_B/q\lambda E_m)$ and $I_D \cdot exp(-\Phi_i/q\lambda E_m)$ respectively where I_D is the drain current, Φ_B is the barrier energy at the Si/SiO$_2$ interface, Φ_i is the minimum energy for impact ionization, λ is the hot-electron mean free path and E_m is the maximum channel electric field [6]. It can be seen from the equations that both the rate of carrier injection into the gate dielectric and the substrate current depend on the concentration of carriers in the channel i.e. I_{AV}. This is in agreement with the experimental observation that the power law pre-factor is proportional to I_{AV} i.e. the rate of ΔV_{GSTX} shifting is proportional to I_{AV}.

The impact of the cell pitch on HCI during repetitive UIS is investigated by avalanche cycling 4 μm and 2.5 μm cell pitch 30 V rated MOSFETs. The results of the experiments are illustrated in Fig. 5 where the ΔV_{GSTX} is shown as a function of N for the different cell pitch devices.

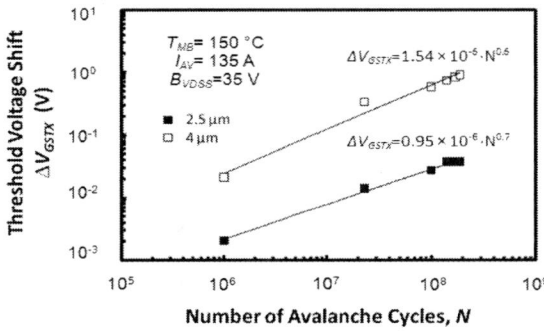

Fig. 5. The log-log plot showing the power law relationship between ΔV_{GSTX} and the number of avalanche cycles for the 4 μm and 2.5 μm cell pitch trench devices tested under repetitive UIS. A 37.5% increase in channel density resulted in a 40% decrease in the power law pre-factor.

It can be seen from the results in Fig. 5 that the smaller cell-pitch MOSFETs exhibit better V_{GSTX} stability. This is due lower avalanche current per unit cell in the smaller cell pitch devices which according to the "lucky-electron" model reduces HCI because the rate of charged carrier supply into the gate dielectric is proportional to the current in the channel. Since power MOSFETs are essentially many

trench FETs connected in parallel to common terminals, reducing the cell pitch is in essence, adding more FETs into the same silicon area. A 2.5 μm cell pitch power MOSFET has 37.5% more trench FETs than a 4 μm cell pitch power MOSFET, hence will be more resistant to ΔV_{GSTX} shifting under the same I_{AV} since increasing the channel density by 37.5% has the same effect as reducing the current per unit channel by the same amount. As in the case of Fig. 4 where the power law pre-factor (A) responded linearly to a change in I_{AV}, A in this experiment responded linearly to the change in the MOSFET's cell pitch. As the cell pitch is reduced from 4 μm to 2.5 μm (channel density is increased by 37.5%), A is reduced by 40%. Hence, cell pitch scaling is recommended for MOSFETs that are to be repetitively avalanched in automotive applications.

The I_{DSS} is also monitored as a function of avalanche cycling for the 4 μm and 2.5 μm cell pitch MOSFETs. Fig. 6 shows the I_{DSS} as a function of N where it can be seen that the smaller cell pitch device exhibits over 100% more leakage on average.

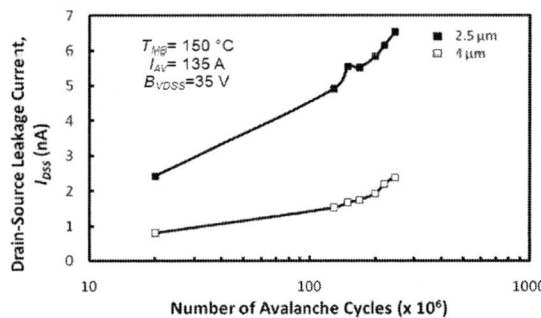

Fig. 6. The I_{DSS} of the MOSFETs shown as functions of the number of avalanche cycles for the 4 μm and 2.5 μm cell pitch trench devices tested under repetitive UIS. The 2.5 μm cell pitch devices exhibit 100% higher due to the greater channel density hence more source-drain conductive paths.

This is due to the higher channel density, hence more leakage paths as the cell pitch is reduced. After 240 million cycles, I_{DSS} increased by 100% and 160% for the 4 μm and 2.5 μm cell pitch MOSFETs respectively. Hence, not only does the smaller cell pitch device exhibit more drain leakage, but the drain leakage appears to be more sensitive to avalanche cycling.

The R_{DSON} is also monitored as a function of avalanche cycling. Fig. 7 shows the R_{DSON} as a function of N where it can be seen that the 2.5 μm cell pitch MOSFET exhibited 25% smaller R_{DSON} compared with the 4 μm cell pitch MOSFET. The smaller R_{DSON} with the smaller cell pitch device is expected since there are more source-drain conduction paths in the same area of silicon as the cell pitch is reduced. However Fig. 7 shows that R_{DSON} is less sensitive to avalanche cycling in the 2.5 μm cell pitch MOSFET i.e.

978-1-4244-6658-0/10 $26.00 © 2010 IEEE   195

the R_{DSON} after 240 million cycles increased by 13% and 2% for the 4 µm and 2.5 µm cell pitch MOSFETs respectively.

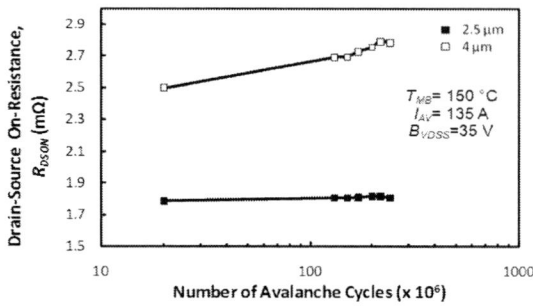

Fig. 7. The R_{DSON} of the MOSFETs shown as functions of the number of avalanche cycles for the 4 µm and 2.5 µm cell pitch trench devices tested under repetitive UIS. The 2.5 µm cell pitch devices show 25% improvement in R_{DSON} and better R_{DSON} stability over the number of avalanche cycles.

Thermo-mechanical stresses from temperature excursions are known degrade the contact resistance between the source metal and the bond wires (as well as the Si back contact and lead frame solder). This causes the R_{DSON} of the MOSFETs to increase with the number of cycles [7, 8]. Thermal cycling experiments have shown continuous degradation in the integrity of the wire bond (due to differences in the thermal expansion coefficients) with an increasing number of cycles [8]. Fig. 7 shows that the 2.5 µm cell pitch MOSFETs exhibit better R_{DSON} stability but worse I_{DSS} stability in comparison to the 4 µm cell pitch MOSFETs.

The subthreshold slopes of the 2.5 µm and 4 µm cell pitch MOSFETs are also compared. Fig. 8 shows the measured subthreshold slopes at the different intervals of repetitive UIS.

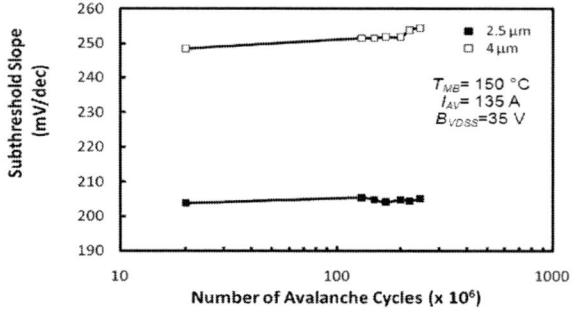

Fig. 8. The subthreshold slopes of the 4 µm and 2.5 µm cell pitch MOSFETs showing 25% improvement in the smaller cell pitch device. This improvement is due to higher gate capacitance from higher channel density. The stability of the subthreshold slope indicates that interface trap generation is not an active degradation mechanism during repetitive UIS.

Both MOSFETs show subthreshold slope stability during repetitive UIS however the 2.5 µm cell pitch MOSFET

exhibits 20% less subthreshold slope. This can be attributed to a higher gate capacitance since there are more trench-FETs in the smaller cell pitch device.

4. CONCLUSIONS

It has been shown that hot-hole injection into the gate dielectric is responsible for the reduction of the threshold voltage during repetitive UIS. The relationship between the change in the threshold voltage and the number of avalanche cycles has been shown to be governed by a power law. Experimental repetitive UIS measurements show that the power law pre-factor responds linearly to changes in the avalanche current. Repetitive UIS experiments were performed on 4 µm and 2.5 µm cell pitch 30 V MOSFETs which showed that smaller cell pitch devices exhibit better threshold voltage stability. The better threshold voltage stability of the smaller cell pitch devices is due to reduced avalanche current per unit cell, which according to the "lucky-electron" model results in less HCI. The results show that the power law pre-factor also responds linearly to changes in the cell pitch (37.5% reduction in cell pitch caused 40% reduction in avalanche pre-factor) and that reducing the cell pitch has the same effect as reducing the avalanche current. The 2.5 µm cell pitch MOSFET also exhibited 25% improved R_{DSON} with better R_{DSON} stability and 20% improved subthreshold slope although with 100% higher I_{DSS} and worse I_{DSS} stability.

REFERENCES

[1] I. Pawel, R. Siemieniec, M. Rosch, F. Hirler, and R. Herzer, "Experimental study and simulations on two different avalanche modes in trench power MOSFETs," *IET Circuits Devices Syst.,* vol. 1, pp. 341-346, 2007.

[2] P. Rutter, K. Heppenstall, A. Koh, G. Petkos, and G. Blondel, "High Current Repetitive Avalanche of Low Voltage Trench Power MOSFETs," *International Symposium of Power Semiconductor Devices & ICs,* pp. 112-115, 2009.

[3] B. Doyle, M. Bourcerie, J. Marchetaux, and A. Boudou, "Interface State Creation and Charge Trapping in the Medium-to-High Gate Voltage Range (Vd/2 > Vg > Vd) During Hot-Carrier Sressing of n-MOS Transistors," *IEEE Trans. Electron Devices,* vol. 37, pp. 744-754, 1990.

[4] B. Doyle, M. Bourcerie, C. Bergonzon, R. Benecchi, A. Bravis, K. Mistry, and A. Boudou, "The Generation and Characterization of Electron and Hole Traps Created by Hole Injection During Low Gate Voltage Hot-Carrier Stressing of n-MOS Transistors," *IEEE Trans. Electron Devices,* vol. 37, pp. 1869-1876, 1990

[5] C. Hu, "Lucky Electron Model of Channel Hot Electron Emission," *In IEDM Tech Dig,* pp. 22-25, 1979.

[6] C. Hu, S. Tam, F. Hsu, P. Ko, T. Chan, and K. Terrill, "Hot-Electron-Induced MOSFET Degradation-Model, Monitor and Improvement," *IEEE Trans. Electron Devices,* vol. 32, pp. 375-385, 1985.

[7] T. Matsunaga and S. Sudo, "Evaluation of fatigue life reliability and new lead bonding technology for power modules," *Mitsubishi Electr. Adv.,* vol. 113, pp. 13–16, 2006, Technical Reports.

[8] W. Loh, M. Corfield, H. Lu, S. Hogg, T. Tilford, and C. Johnson, "Wire bond for power electronic modules—Effect of bonding temperature," in *Proc. Int. Conf. Thermal, Mech. Multi-Phys. Simul. Experiments Microelectron. Micro-Syst., EuroSime,* 2007, pp. 1–6

Investigation of a Dual Channel N/P-LDMOS and Application to LDO Linear Voltage Regulation

Marie Denison, Yizhong Xie and Hannes Estl
Texas Instruments, Inc.
13121 TI Boulevard, Dallas, TX 75243, USA
Tel. +1-214-567-6575, e-mail: mdenison@ti.com

Abstract — **A simulation analysis of a configurable dual channel n/p LDMOS is presented. The n, p and combined n/p channel regimes are analyzed, showing the transistors potential for enhanced saturation current density without high-temperature Safe Operating Area (SOA) degradation. Possible circuit applications are proposed. In particular, it is shown that the device is a good candidate for implementation in Low-Dropout (LDO) voltage regulators, in which it allows reducing the need for a charge pump circuit, and thus increasing performances in terms of power consumption, noise and Electromagnetic Interferences (EMI).**

I. INTRODUCTION

Lateral double-diffused DMOS (LDMOS) and Drain Extended MOS (DEMOS) transistors are widely used as high-voltage switches and drivers in mixed-signal technologies such as Linear BiCMOS (LBC) technologies. N-channel (n-ch) devices are mainly used for low-side and high-power applications due to their lower specific on-resistance (Rsp). P-channel (p-ch) devices are mostly used for high-side or pull-up operation although n-channel devices combined with a charge pump circuit are preferred when the area penalty of a p-ch DEMOS or LDMOS is too large. A charge pump circuit used to pull up the gate of an n-ch transistor above the supply voltage may consume a considerable area, however in some cases, like for instance when the charge pump can be shared between different circuit parts, utilization of n-channel high-side devices can be more cost effective than p-channel transistors. Recently, a dual n/p channel device over SOI was proposed [1, 2]. It exploits the alternation of n-type and p-type stripes in a superjunction lateral MOSFET design, which, in the dual channel device, are used both as RESURF regions and as drain extension for the channel of respective type. To the knowledge of the authors, details on the device operation in saturation and circuit applications have not been reported yet.

In this work, the electrical characteristics of an n/p LDMOS based on an n-buffered superjunction n-ch LDMOS in bulk silicon [3] are studied by TCAD simulations with particular focus on the saturation regime. Possible circuit implementations are discussed, such as configurable low-side/high-side switching and Low-Dropout (LDO) voltage regulator. For the latter, an analysis is provided on how the device could improve circuit area efficiency and performance by reducing or even avoiding the need for a charge pump circuit.

II. N/P LDMOS STUDY

Fig. 1 shows a sketch of the transistor design. On the low-side of the device, a p-doped well forms the backgate of the n-channel and the drain of the p-channel device, and contains a heavily n-doped (n+) source diffusion for the n-channel component of the device. On the high-side, an n-well forms the backgate of the p-channel and the drain of the n-channel device, and contains a heavily p-doped (p+) source diffusion for the p-channel transistor portion. The device is comprised in a deep n-well over a low-doped p-substrate. In the following, we consider a device with 65V breakdown voltage and a gate oxide rated to 5V, simulated in 3D using the Synopsys tool S-Device [4]. The threshold voltage of the n-channel component of the device (when the gate-source voltage Vgs of the p-ch component is equal to 0) is 1V, whereas the threshold voltage of the p-ch component (when the gate-source voltage of the n-ch component is equal to 0) is 0.6V. As usual in superjunction devices, the doses of the n- and p- drift regions are approximately balanced along the drain extension. Because of this, the ratio of the p- to n-drift region on-state resistance is expected to follow the electron and hole mobility ratio. In our particular example, the total on-state resistance factor between p-ch and n-ch components is 4.1, as a result of different charge geometries and shorter n-channel length.

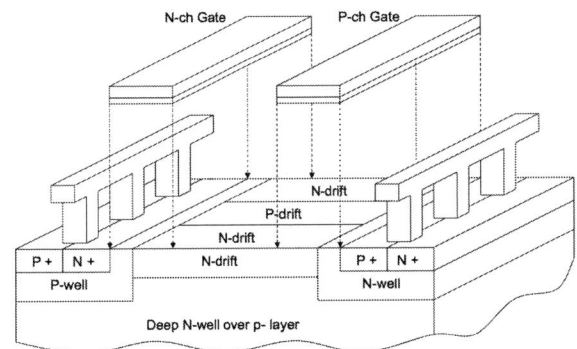

Fig. 1: Sketch of an n/p LDMOS transistor in deep n-well buffer (the p-substrate is not shown).

Fig. 2 shows the transistor output characteristics obtained at 2V gate-source bias with the n-channel activated, the p-channel activated (gate-source voltage given in absolute value), or both n- and p-channels simultaneously activated. As can be seen, the drain current of the device when both channels are active is roughly the sum of the current of the individual channel components as observed in [2] in linear mode.

Fig. 3 shows the transistor output characteristics obtained at 5V gate-source bias with the n-channel activated, the p-channel activated, or both n- and p-channels simultaneously activated. This time, the linear current of the device when both channels are on is more than 10% higher and the saturation current is more than three times larger than the sum of the individual n-ch and p-ch currents. This effect is due to the activation of the parasitic bipolar components of the device [1, 2]. Under sufficient current density, the parasitic NPN (Emitter = n+ source, Base = p-well, collector = n-well) and PNP (Emitter = p+ source, Base = n-well, collector = p-well) transistors couple to each other in a thyristor-like system, as sketched in Fig. 4 which also shows the electron density across the p-drift region. At high temperature, the saturation current of the device in dual mode does not significantly decrease, while the onset of current instability is shifted to higher drain voltage, see also Fig. 2 for low gate bias. This advantageous behavior results from the negative temperature coefficient of the MOS saturation current, which counter-acts the positive temperature coefficient of the parasitic bipolar currents [5].

Comparing Fig. 2 to Fig. 3, one can notice a degradation of the SOA at high Vgs under p-ch and combined p-ch and n-ch modes. This degradation is due to the relatively low n-well doping used in this particular simulation, such that avalanche electron current from the p-ch component activates the device parasitic PNP component. Higher n-well doping under the p-ch source (as reported in [6] for an n-ch LDMOS) is observed to remedy to this SOA limitation (not shown).

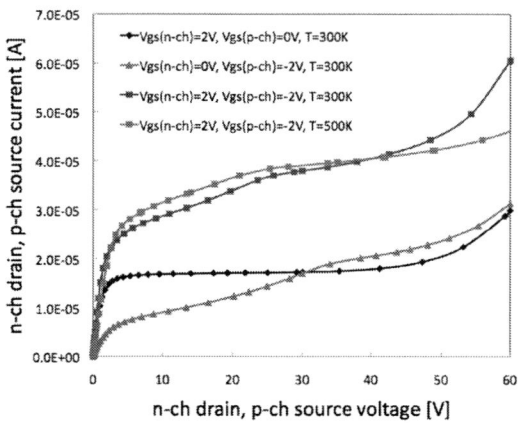

Fig. 2: Output characteristics of an n/p LDMOS transistor at gate-source bias |Vgs|=2V.

The n/p LDMOS can thus be used as n-ch LDMOS, p-ch LDMOS or dual n/p-channel LDMOS with, under sufficient biasing of the n-ch and p-ch gates, lower specific on-resistance and higher saturation current than a single channel LDMOS.

III. N/P LDMOS APPLICATIONS

A possible application of the device makes use of its configurability in n- or p-mode. ICs are sometimes designed to switch loads which may be low-sided or high-sided (in combination with a charge pump circuit) depending on customer needs. With an n/p LDMOS, the same device can be configured as low-side (Fig. 5a) or as high-side (Fig. 5b) without combination with a charge pump circuit, with the

constraint that the on-state resistance is at least about 3 times higher in high-side mode than in low-side mode. Charge pump circuits can be the source of significant Electromagnetic Interferences (EMI), and consume a certain amount of quiescent current. Reducing or avoiding their usage may therefore improve the related performance of the IC. On the other hand, complexity is added for gate driving, which now has to handle both low- and high-side operation modes.

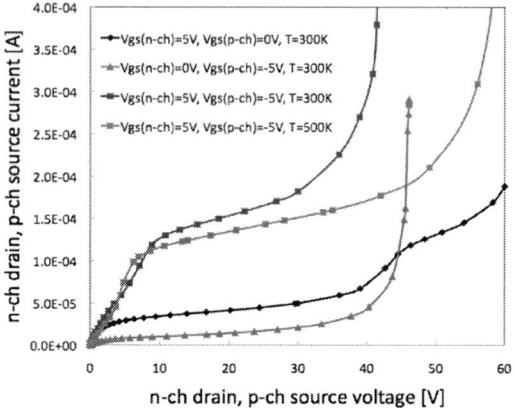

Fig. 3: Output characteristics of an n/p LDMOS transistor at gate-source bias |Vgs|= 5V.

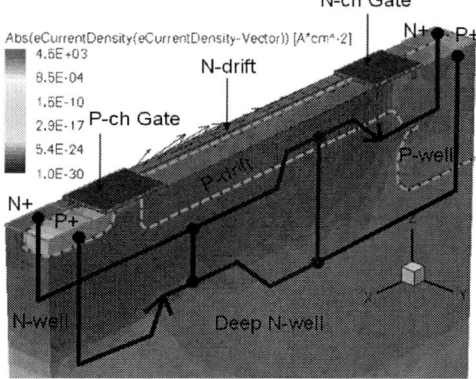

Fig. 4: 3D TCAD view of an n/p LDMOS transistor in saturation with both channels active (Gate-source bias |Vgs| on both sides = 5V). The parasitic NPN and PNP components of the devices are sketched.

In a configurable low-/high-side switch, the LDMOS works either as n- or as p-channel, but not as combined n/p-LDMOS. An application which can greatly benefit from the n/p LDMOS unique dual feature is the low-dropout linear voltage regulator [7]. This circuit provides a fixed output voltage from an unregulated input supply. When an n-channel LDMOS is used as high-side regulating device, its gate-source voltage is limited by the voltage difference between the unregulated input and the regulated output, unless a charge pump is used to supply the gate when this difference is too low to maintain the output voltage. A p-channel LDMOS, on the other hand, has its gate-source voltage referred to the unregulated supply and therefore need not any charge pumping. However, its on-state resistance can be significantly lower than the one of an n-channel device so that it cannot drive enough current within a reasonable area, even when the gate is fully activated. The diagram in Fig. 6 shows how an n/p LDMOS can improve the efficiency of an LDO voltage regulator, in the example of a 14V battery unregulated supply, 5V regulated output and a maximum load current of 500mA.

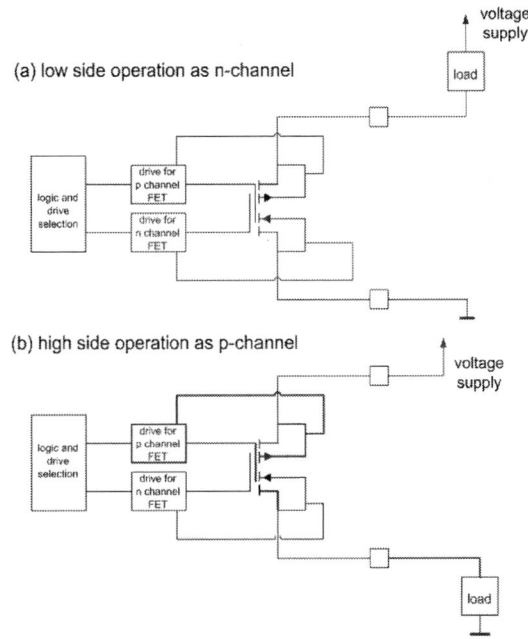

Fig. 5: Configurable (a) low-side n-ch and (b) high-side p-ch LDMOS.

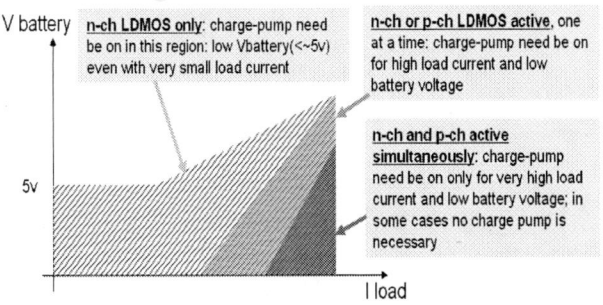

Fig. 6: Charge pump utilization reduction in an LDO with 5V output voltage by means of an n/p LDMOS: the hacked area shows battery voltage-load current range where charge pumping is needed when regulation is achieved by an n-ch device only; the light gray area shows the charge pump operation area when a p-ch component is added as an alternative to n-ch regulation; the dark gray area shows further charge pump regime reduction when both n-ch and p-ch components can be active at a time.

Under n-channel regulation, the charge pump need be on in a wide current and battery voltage range, when the latter is low and/or the load current is high, as represented by the hacked area in Fig. 6 and in the diagram in Fig. 7. If a p-ch could be activated when the n-ch gate-source bias becomes insufficient (Fig. 8), the hacked area could be reduced to the light gray area, where the p-ch device has reached its maximum current. If both n-ch and p-ch could be activated at the same time, the highest current / lowest battery voltage regime could also be served without charge pump (see dark gray area in Fig. 6). If both n-ch and p-ch current components would not only add up, but also amplify each other, as when

the parasitic npn-pnp system of the n/p LDMOS is active, the entire operating range could possibly be covered without any charge pump.

In this work, the influence of an alternate p-ch LDMOS or of a parallel n/p LDMOS on an LDO is studied by means of the four test circuits represented in Fig. 9, for 3.3V output regulation and 5.5V nominal input voltage. The top circuit represents the n/p LDMOS as an n-ch device in parallel with a p-ch device. The bottom left circuit is driven by an n-ch LDMOS; the bottom center one by an n-ch LDMOS with charge pump; the bottom right one by a p-ch LDMOS. All n-ch and p-ch devices have the same W/L ratio. The coupled bipolar mode with higher maximum current capability is not covered in this example. Two different cases are considered in the following:

(i) load current (I-load) sweep at 5.5V battery voltage,
(ii) battery voltage (Vbattery) sweep at two different fixed load current (500mA and 100mA).

The output voltage is plotted versus swept parameter for each sub-circuit in Figs. 10-12. The circuit limit is reached when the output voltage drops away from the specified 3.3V.

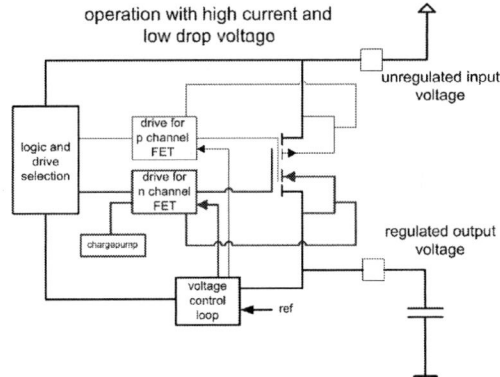

Fig. 7: n/p LDMOS in LDO configured as n-ch LDMOS with charge pump.

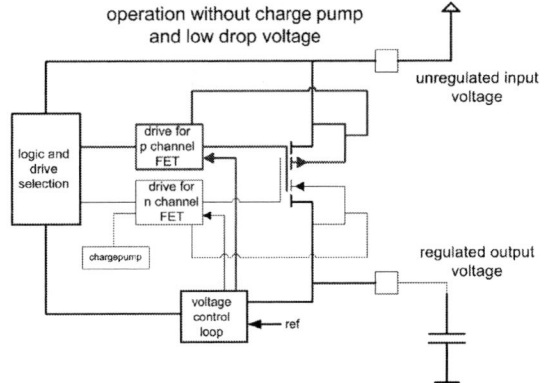

Fig. 8: n/p LDMOS in LDO configured as p-ch LDMOS.

As shown in Fig. 10, at 5.5V battery voltage, the worst performer is the n-ch device without charge pump, followed by the p-ch LDMOS. The parallel n-ch and p-ch device performs as good as the n-ch device with charge pump circuit.

978-1-4244-6658-0/10 $26.00 © 2010 IEEE

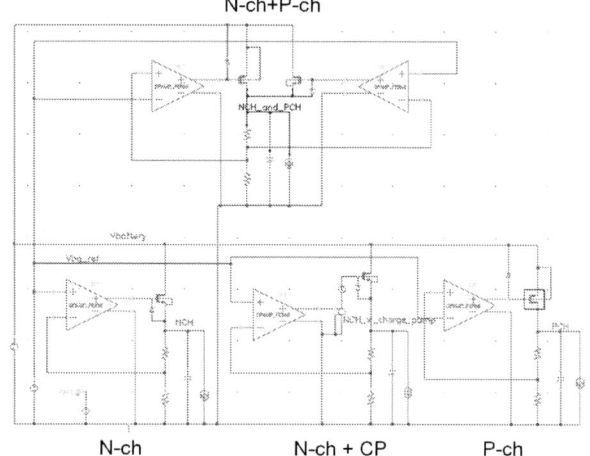

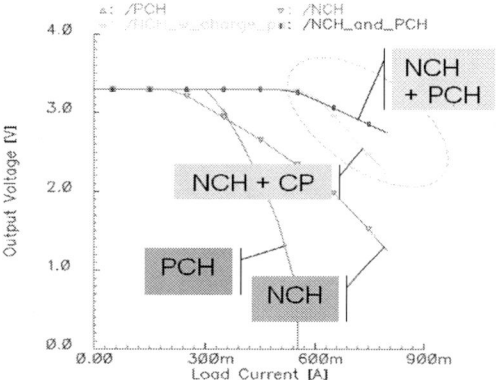

Fig. 9: Four test circuits to model an LDO response to different LDMOS implementations.

Fig. 10: Output voltage resulting from I-load sweep @ Vbattery=5.5V in circuit of Fig. 9.

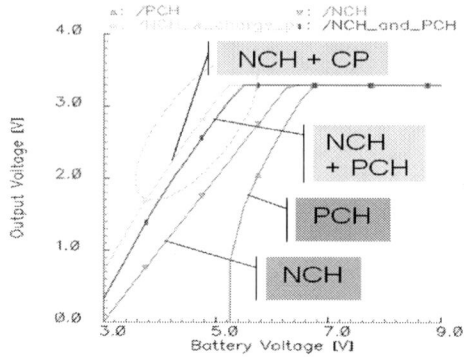

Fig. 11: Output voltage resulting from I-load sweep @ Vbattery=5.5V.

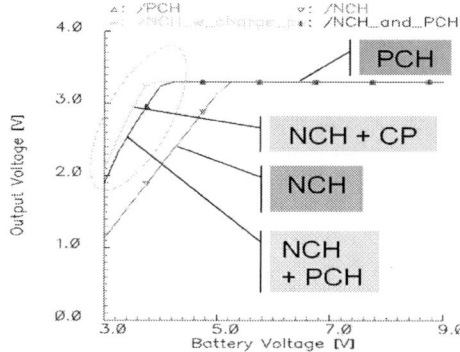

Fig. 12: Output voltage resulting from Vbattery sweep @ I-load=500mA

need for a charge pump circuit, improving the LDO quiescent power consumption, noise and EMI performances. Extended device and circuit studies including fabrication and spice modeling would be desirable to assess the full potential of this interesting component.

At high load current, Fig. 11 shows that the output voltage drops first in the p-ch sub-circuit due to the limited conductivity of the device. The sub-circuit with the parallel n-ch and p-ch devices performs nearly as well as the n-ch transistor with charge pump. Similar result is obtained at lower current, with a wider operating battery voltage window, see Fig. 12. As mentioned earlier, this is even not considering any current enhancement resulting from the internal bipolar coupling discussed in Section II.

IV. CONCLUSIONS

Electrical characteristics of an n/p LDMOS transistor in bulk-silicon were investigated by simulations. The studied device was showed to benefit from strong saturation current enhancement when both channel components are biased well above their respective threshold voltage. At high temperature, the saturation current shows little degradation and widened SOA limit. Configurable low-side/high-side switch and driver for low-dropout voltage regulator are possible applications of this transistor. In an LDO circuit, it was shown that the n/p LDMOS promises to significantly reduce or even avoid the

ACKNOWLEDGMENT

The authors would like to thank Allen Bowling, Sameer Pendharkar and Frank Tsai at Texas Instruments for their support to this study.

REFERENCES

[1] F. Udrea, A. Popescu and W. Milne, "The 3D RESURF junction", Proc. International Semiconductor Conf. (CAS) 1998, pp. 141–144.

[2] F. Udrea, A. Popescu and W. Milne, "3D RESURF double-gate MOSFET: A revolutionary power device concept", Electronics Letters, Vol. 34, No. 8, 1998, pp. 808-809.

[3] I.Y. Park and C.A.T. Salama, "CMOS compatible super junction LDMOST with n-buffer layer", Proc. of the ISPSD 2005, pp. 159-162.

[4] Sentaurus Device™, Synopsys, Inc., Mountain View, CA, USA.

[5] S.M. Sze, "Physics of semiconductor devices", John Wiley and Sons, 1981, ISBN 0-471-05661-8.

[6] S. Reggiani, G. Baccarani, E. Gnani, A. Gnudi, M. Denison and S. Pendharkar, "Explanation of the rugged LDMOS behavior by means of numerical analysis", IEEE Transactions on Electron devices, Vol. 56, No. 11, Nov. 2009.

[7] Dave Heisley and Bert Wank, "DMOS delivers dramatic performance gains to LDO regulators", Electronics Design, Strategy, News (EDN) magazine, June 22nd, 2000, pp. 141-151.

High Transconductance AlGaN/GaN HEMT with Thin Barrier on Si(111) Substrate

F. Lecourt, Y. Douvry, N. Defrance, V. Hoel,
J.C. De Jaeger
IEMN (Institut d'Electronique, de Microélectronique et de
Nanotechnologie), UMR CNRS 8520
Villeneuve d'Ascq, France
francois.lecourt@ed.univ-lille1.fr

S. Bouzid, M. Renvoise, D. Smith and H. Maher
OMMIC
Limeil-Brévannes, France
h.maher@ommic.com

Abstract—The fabrication of high transconductance AlGaN/GaN high electron mobility transistors (HEMTs) grown on high-resistivity silicon substrate is reported with an AlGaN barrier thickness of only 12.5 nm. A maximum DC current density of 655 mA/mm, a current gain cutoff frequency (F_T) of 75 GHz and a power-gain cutoff frequency (F_{MAX}) of 125 GHz are obtained for a 0.125 μm gate length transistor. The device provides a record peak extrinsic transconductance of 332 mS/mm and an intrinsic value of 509 mS/mm. To the authors' knowledge, the obtained transconductances are the highest reported values from AlGaN/GaN devices grown on a Si(111) substrate. This performance demonstrates the potential of GaN transistors on silicon for low-cost microwave power applications.

I. INTRODUCTION

GaN based high electron mobility transistors (HEMTs) have emerged as very attractive candidates for high temperature, high-voltage, and high-power operation in microwave as well as lower frequencies [1], [2]. They are more and more being used for both civilian and military applications.

This study focuses on AlGaN/GaN-HEMTs on a Si(111) substrate. Though silicon substrates provides less thermal dissipation and more lattice constraints with GaN than SiC, the interest of this technology is mainly due to its lower fabrication cost. The quality of the epitaxys on such substrates has been optimised in order to improve the 2DEG carrier concentration and mobility. The motivation of this work is in being able to manufacture low cost transistors and to reach the electrical performances that are obtained on other epi-materials.

GaN grown on Si substrate received a lot of attention in the past few years and now, very good material parameters have been obtained for AlGaN/GaN HEMTs on Si(111) [3], [4]. The state of the art on Si(111) exhibits a power density of 5.1 W/mm at 18 GHz and 12 W/mm at 2.14 GHz [5], [6]. Moreover, the state of the art for F_T and F_{Max} are 107 GHz and 150 GHz respectively [7].

In this paper, we present $Al_{0.26}Ga_{0.74}N$/GaN HEMT grown on high-resistivity Si(111) substrates with a record-setting 332 mS/mm extrinsic transconductance. The originality of the epitaxy is a barrier thickness of only 12.5 nm. This permits to reduce the gate length mitigating short channel effect. Moreover, a decrease of the barrier thickness gives a low pinch-off voltage value and an increase of the transconductance.

The device structure and fabrication are first described. Then, DC and small signal results are reported and analyzed. Finally, pulsed current-voltage measurements are presented and discussed.

II. DEVICE STRUCTURE AND FABRICATION

A. Device Structure

The AlGaN/GaN transistor epitaxy was grown on a high-resistivity Si(111) substrate by metal-organic chemical vapor deposition (MOCVD). Device structure consists of 12.5 nm $Al_{0.26}Ga_{0.74}N$ barrier and 2 nm cap layer of GaN. From Van Der Pauw measurements performed on different areas of the wafer, an average sheet carrier density of 7.05×10^{12} cm^{-2} (±0.3), an electron mobility of 2160 cm^2/V.s (± 35) and a sheet square resistance of 410 $\Omega/\square$ (± 20) are obtained.

B. Device Fabrication

Transistor fabrication starts with Ti/Al/Ni/Au (12/200/40/100 nm) ohmic metallization followed by a rapid thermal annealing (RTA) at 850 °C during 30 s under a nitrogen atmosphere. The device isolation is obtained from He^+ ion multiple implantation based on different energies and doses. With these parameters, the implanted ion range is as deep as 760 nm in the GaN buffer. Contact resistance and specific contact resistivity, measured by the transmission line model (TLM) on different patterns, are 0.28 Ω.mm (± 0.3) and 2.3×10^{-6} $\Omega.cm^2$ (± 0.5) respectively. T-shaped gate based on Ni/Au (40/300 nm) metallization with 0.125 μm footprint is defined by electron-beam lithography using a tri-layer resist stack (PMMA/Copolymer/PMMA). The 125 nm physical gate

978-1-4244-6658-0/10 $26.00 © 2010 IEEE

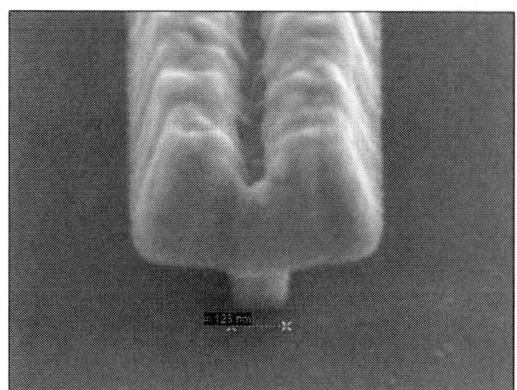

Figure 1. A cross-sectional SEM image of the 125 nm T-shaped gate.

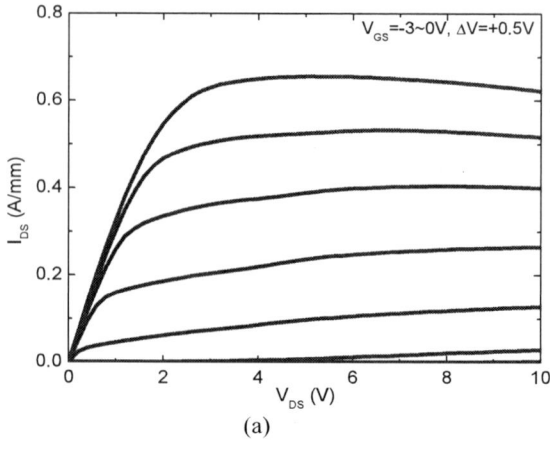

(a)

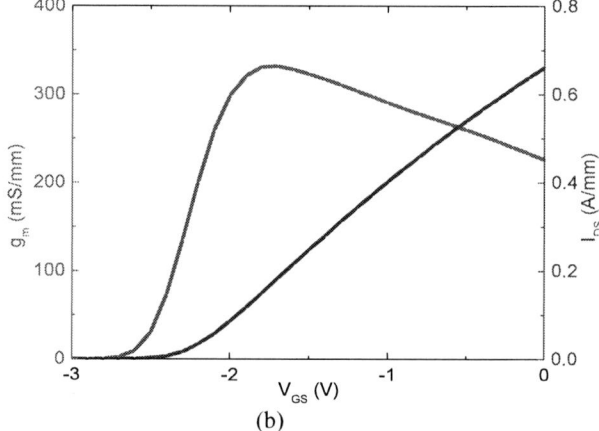

(b)

Figure 2. DC characteristics of the 120 nm gate AlGaN/GaN HEMT.
(a) I_{DS} (V_{DS}) characteristics measured for V_{GS} = -3 to 0 V, step +0.5 V.
(b) Transfer characteristics at V_{DS} = 4 V with a record peak extrinsic of transconductance of 332 mS/mm at V_{GS} = -1.7 V.

length was observed by scanning electron microscope (SEM), as shown in Fig. 1. Finally, devices are passivated with silicon nitride (240 nm) deposited by plasma-enhanced chemical vapor deposition (PECVD) at 300 °C without pre-treatment.

III. DC AND SMALL SIGNAL MICROWAVE RESULTS

All DC and small signal measurements were carried out on $2 \times 50 \times 0.125$ µm² HEMT devices using microwave probes.

A. DC Characteristics

DC measurements are performed on a HP4142B static modular source and monitor. From this DC characterization, a maximum drain current density of 655 mA/mm is obtained at V_{GS} = 0 V with a pinch-off voltage of -2.5 V (Fig. 2(a)). The knee voltage is less than 3 V indicating excellent ohmic contact fabrication. The transfer characteristics at V_{DS} = 4 V are represented in Fig. 2(b). The peak extrinsic transconductance is 332 mS/mm at V_{GS} = -1.7 V constituting to our knowledge the best reported value for AlGaN/GaN on high resistivity silicon substrate.

It can be noted that the short channel effects are reduced due to the low barrier layer thickness; which is illustrated by the good ratio of I_{ON}/I_{OFF} (> 1×10^4). This can be explained by a better gate control due to an increase aspect ratio between the gate length and the barrier layer thickness.

To determine the further physical parameters of the devices, "end" resistance measurements were performed [8], [9]. The following set of equations can be written (1), where R_s is the static source resistance, R_d the drain resistance and R_i the equivalent resistance occurring under the gate metallization.

$$R_s + \left(\frac{R_i}{2} \right) = 18.5 \ \Omega$$

$$R_d + \left(\frac{R_i}{2} \right) = 22.5 \ \Omega. \tag{1}$$

To extract R_s and R_d, the whole on-resistance R_{ON} is measured at low V_{DS} and its evolution versus the gate bias is expressed following (2) where G_0 is the full channel conductance.

$$R_{ON} = R_s + R_d + 1/[G_0 * (1 - \eta)] \tag{2}$$

with

$$\eta = (V_{bi} - V_{GS})/(V_{bi} - V_t). \tag{3}$$

V_{bi} and V_t correspond to the built-in and threshold voltages respectively in (3). The used approach is similar to the technique proposed by Hower and Bechtel [10], but modified and adapted to the HEMT technology [11]. By this method, it is found $R_s + R_d$ = 24.8 Ω, which in turn gives R_s = 10.4 Ω and R_d = 14.4 Ω.

The extraction of the intrinsic transconductance g_{mi} is then possible, assuming (4) where g_{me} is the already known extrinsic transconductance value.

$$g_{mi} = g_{me}/(1 - R_s * g_{me}). \tag{4}$$

978-1-4244-6658-0/10 $26.00 © 2010 IEEE   202

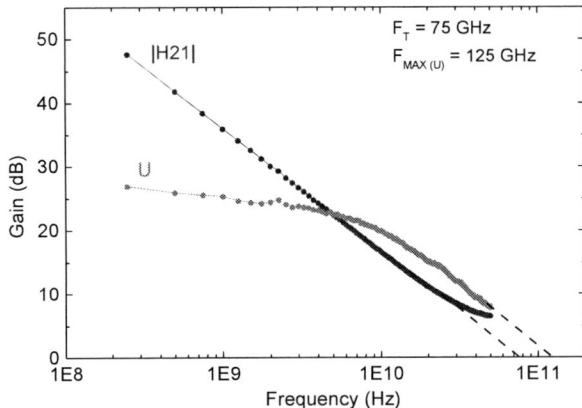

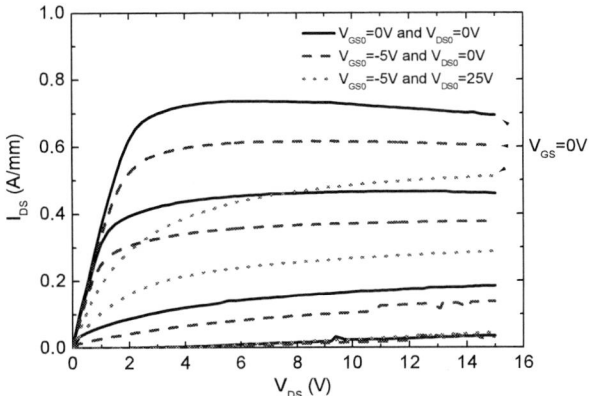

Figure 3. Microwave characteristics of T-shaped 0.12 μm gate length AlGaN/GaN HEMT at V_{DS} = 4 V and V_{GS} = -1.8 V. Extrapolation at -20dB/dec of $|H_{21}|$ and U yield F_T = 75 GHz and F_{MAX} = 125 GHz respectively.

Figure 4. DC pulsed $I_{DS}(V_{DS})$ characteristics of T-shaped 0.12 μm gate length AlGaN/GaN HEMT at three differents bias points. (V_{DS0} = 0 V, V_{GS0} = 0 V), (V_{DS0} = 0 V, V_{GS0} = -5 V) and (V_{DS0} = 25 V, V_{GS0} = -5 V) for V_{GS} from -3 to 0 V by step of 1 V.

An intrinsic transconductance of g_{mi} = 509 mS/mm is obtained which is an outstanding value for AlGaN/GaN technology on silicon substrate.

B. RF Characteristics

The scattering parameters were measured from 0.25 to 50 GHz using an Agilent Technologies N5245A vector network analyzer and line-reflect-reflect-match (LRRM) calibration. Pad capacitances were extracted using an on-wafer open calibration structure.

Fig. 3 shows the current gain modulus ($|H_{21}|$) and the Mason's maximum unilateral gain (U) derived from S-parameters measurement versus the frequency. At V_{DS} = 4 V and V_{GS} = -1.8 V, an extrinsic current gain cutoff frequency (F_T) of 75 GHz and a maximum power gain cutoff frequency (F_{MAX}) of 125 GHz are obtained. These values are very good for AlGaN/GaN HEMT on high resistivity silicon with a 0.125 μm gate length. Moreover, a good F_{MAX}/F_T ratio value of 125/75 GHz is observed.

C. Pulse Characteristics

Pulse measurements were carried out in order to study the trap response to an applied electrical field. To perform these measurements, the devices are characterized with a measurement setup providing 500 ns pulse lengths with a 0.3% duty cycle. This method is usually used to de-correlate the contribution of the traps located within the AlGaN barrier (mainly under the gate metallization and at the material surface) or within the GaN buffer. It relies on the exclusive use of cold quiescent bias points, permitting to neglect current drop due to thermal effects.

As an example, Fig. 4 shows the DC pulsed characteristics wherein all quiescent bias points are chosen to reveal the gate and drain lag effects. The pulse $I_{DS}(V_{DS})$ characteristics determined at the quiescent point V_{DS0} = 0 V, V_{GS0} = -5 V (beyond pinch-off voltage) are compared to the $I_{DS}(V_{DS})$ reference determined at the quiescent point V_{DS0} = 0 V, V_{GS0} = 0 V in order to analyze the gate lag effect. On the same figure,

the pulse $I_{DS}(V_{DS})$ characteristics determined at V_{DS0} = 25 V, V_{GS0} = -5 V are presented in order to show the drain lag effect. Even with high electrical stress upon the gate (about twice the pinch-off voltage), the lag remains rather low with a current drop of only 15% at V_{DS} = 6 V. Regarding the drain lag contribution, it has been evaluated to be about 23% at the same V_{DS}.

These results are very promising considering that the transistor process is based on epi-material with only 12.5 nm barrier thickness. Associated with the high values of F_T and F_{MAX} (75 and 125 GHz respectively), good microwave power performance can be expected in Ka-band.

CONCLUSION

0.125 μm gate length AlGaN/GaN HEMT grown on a (111) high resistivity silicon substrate exhibits record extrinsic and intrinsic transconductances g_{me} = 332 mS/mm and g_{mi} = 509 mS/mm respectively. The device features very interesting extrinsic peak cutoff frequencies of F_T = 75 GHz and F_{MAX} = 125 GHz. This performance indicates that AlGaN/GaN on silicon technology is a good alternative to silicon carbide (SiC) to obtain low-cost microwave power transistor fabrication. On-going work is being performed to further shrink the gate length in order to achieve even higher RF performance.

ACKNOWLEDGMENT

The authors would like to thank the DGA for the financial support.

REFERENCES

[1] T. Palacios, A. Chakraborty, S. Rajan, C. Poblenz, S. Keller, S. P. DenBaars, J. S. Speck, and U.K. Mishra, "High-power AlGaN/GaN HEMTs for Ka-band applications," IEEE Electron Device Letters, vol. 26, no. 11, pp. 781–783, November 2005.

[2] M. Micovic, A. Kurdoghlian, P. Hashimoto, M. Hu, M. Antcliffe, P. J. Willadsen, W. S. Wong, R. Bowen, I. Milosavljelic, A. Schmitz, M. Wetzel, and D. H. Chow, "GaN HFET for W-band power applications," IEDM Tech. Dig., 2006, pp. 425–427.

[3] A. Minko, V. Hoël, E. Morvan, B. Grimbert, A. Soltani, E. Delos, D. Ducatteau, C. Gaquière, D. Théron, J. C. De Jaeger, H. Lahreche, L. Wedzikowski, R. Langer, and P. Bove, "AlGaN/GaN HEMTs on Si With Power Density Performance of 1.9 W/mm at 10 GHz," IEEE Electron Device Letters, vol. 25, no. 7, pp. 453–455, July 2004.

[4] K. Cheng, M. Leys, S. Degroote, J. Derluyn, B. Sijmus, P. Favia, O. Richard, H. Bender, M. Germain, and G. Borghs, "AlGaN/GaN High Electron Mobility Transistors Grown on 150mm Si(111) Substrates with High Uniformity," Japanese Journal of Applied Physics, vol. 47, no. 3, pp. 1553–1555, March 2008.

[5] D. Ducatteau, A. Minko, V. Hoël, E. Morvan, E. Delos, B. Grimbert, H. Lahreche, P. Bove, C. Gaquière, J. C. De Jaeger, and S. Delage, "Output Power Density of 5.1W/mm at 18 GHz With an AlGaN/GaN HEMT on Si Substrate," IEEE Electron Device Letters, vol. 27, no. 1, pp. 7–9, January 2006.

[6] J. W. Johnson, E. L. Piner, A. Vescan, R. Therrien, P. Rajagopal, J. C. Roberts, J. D. Brown, S. Singhal, and K. J. Linthicum, "12 W/mm AlGaN/GaN HFETs on Silicon Substrates," IEEE Electron Device Letters, vol. 25, no. 7, pp. 459–461, January 2004.

[7] S. Tirelli, D. Marti, H. Sun, A. R. Alt, H. Benedickter, E. L. Piner, and C. R. Bolognesi, "107-GHz (Al,Ga)N/GaN HEMTs on Silicon With Improved Maximum Oscillation Frequencies," IEEE Electron Device Letters, vol. 31, no. 4, pp. 296–299, April 2010.

[8] K. Lee, M. Shur, K.W. Lee, T. Vu, P. Roberts, and M. Helix, "A new interpretation of "End" resistance measurements," IEEE Electron Device Letters, Vol. 5, Issue 1, pp. 5-7, 1984.

[9] S.-M.J. Liu, S.-T. Fu, M. Thurairaj, and M.B. Das, "Determination of source and drain series resistances of ultra-short gate-length MODFETs", IEEE Electron Device Letters, Vol. 10, Issue 2, pp. 85-87, 1989.

[10] P. L. Hower, N. G. Bechtel, "Current Saturation and small-signal characteristics of GaAs FET's", IEEE Trans. on Electron Device, Vol. 20, Issue 3, Mar. 1973.

[11] M. Berroth, R. Bosch, "Broad-band determination of the FET small-signal equivalent circuit", IEEE. Trans. on Microwave Theory and Techniques, Vol. 38, Issue 7, pp.891-895, 1990.

978-1-4244-6658-0/10 $26.00 © 2010 IEEE

Study of GaN HEMTs Electrical Degradation by Means of Numerical Simulations

Valerio Di Lecce*, Michele Esposto, Matteo Bonaiuti, Fausto Fantini, Alessandro Chini*

Department of Information Engineering
University of Modena and Reggio Emilia
Modena, ITALY
*{valerio.dilecce, alessandro.chini}@unimore.it

Abstract—**In this paper, we investigate the effects of dc stress on GaN high-electron mobility transistors (HEMTs) by means of numerical simulations. Following stress tests showing a degradation of static characteristics (dc), the formation of an electron trap in the AlGaN barrier layer was related to the observed degradation according to the results obtained from numerical simulations carried out by introducing a trapping region underneath the gate edge. The worsening of the device dc performance is evaluated by changing the extension of the degraded region and the trap concentration while studying the variation of parameters like the saturated drain current I_{DSS}, the output conductance g_O, and the device transconductance g_M. An increase in the trap concentration induces a worsening of any of the abovementioned parameters; an increase in the extension of the degraded region induces a degradation of I_{DSS} and g_M, but can reduce g_O.**

I. INTRODUCTION

The need for high-speed and high-power electron devices is constantly increasing, following the demand of the market for solutions capable of supporting the technological advances in the field of Information and Communication Technology. Very interesting perspectives are offered by compound semiconductors, in particular by Gallium Nitride: its wide bandgap makes it an exceptional solution for sustaining high voltages, and its ability to form heterojunctions makes it very useful for the fabrication of high-electron mobility transistors (HEMTs) compared to other wide bandgap semiconductors like Silicon Carbide. Its high electron velocity completes the frame, making it one of the best choices for high-power and high-frequency applications.

Being expression of a very recent technology, GaN-based devices still suffer from various reliability problems both in dc [1] and RF [2] operation. In the last decade, a number of technological solutions have been developed in order to overcome such problems: among them are surface passivation, field-plate structures, and gate recessing [3].

Several studies have analyzed the failure mechanisms of

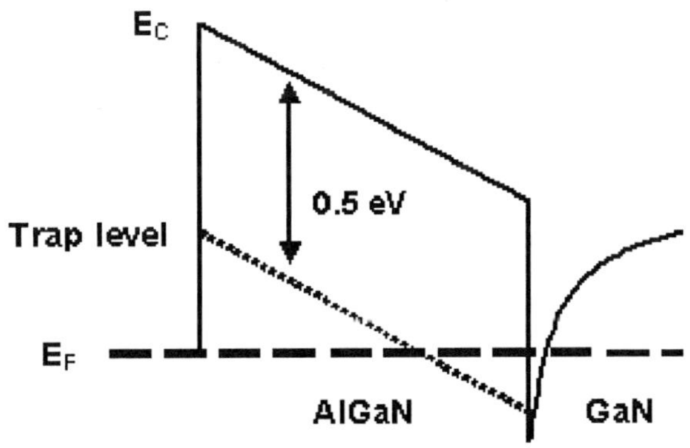

Fig. 1. Simplified band diagram of an AlGaN/GaN HEMT structure showing an acceptor trap density located 0.5 eV below the conduction band edge. Traps close to the AlGaN/GaN interface are filled (negatively charged) even in open-channel conditions.

GaN HEMTs during reverse gate bias stresses. A common feature in many of them is the identification of a reverse "critical voltage" beyond which the gate current increases severely or, more generally, a serious degradation takes place. In [4] Joh and Del Alamo found a critical $|V_G| = 20\text{-}25$ V in the $V_{DS} = 0$ V condition beyond which I_G increased abruptly. Similar results were found in [5] and [6]: in both cases the occurrence of such degradation phenomena was explained by a defect formation in the AlGaN barrier at the gate edge through the inverse piezoelectric effect.

In one of our previous works a reverse gate-drain bias stress with the source terminal floating was found to cause both a decrease of the saturated drain current and an increase of the dc-to-RF dispersion [7]. In our case, the presence of trapping region centered at the gate edge toward the drain contact with an acceptor trap level located 0.5 eV from the conduction band in the AlGaN barrier correctly predicted the experimentally observed degradation of both dynamic and static drain current levels (Fig. 1).

This work was supported in part by the Italian Ministry of Education, University and Research (MIUR) under the FIRB 2006 project "Enabling technologies, characterization, and modeling for wideband reconfigurable integrated electronic components for high frequencies applications", and under the PRIN 2007 project "GaN wideband microwave and mm-wave integrated circuits for low-noise and high-power subsystems".

978-1-4244-6658-0/10 $26.00 © 2010 IEEE

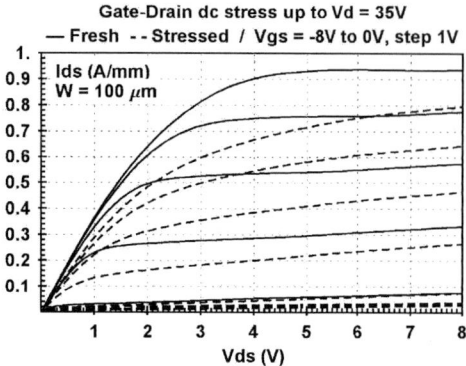

Fig. 2. I_D–V_D characteristics for a device stressed by applying a reverse gate-drain bias up to 35 V, in step of 2.5 V with a duration of 5 min. The source terminal was floating. Saturated drain current I_{DSS} has significantly decreased, and output conductance g_O has increased.

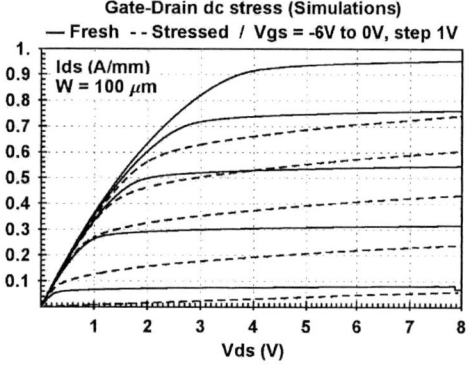

Fig. 3. I_D–V_D characteristics for the device structures simulated in the fresh condition and with 100 nm-long acceptor trap region in the AlGaN barrier layer, centered at the drain edge of the gate terminal. Trap concentration was 200×10^{18} cm^{-3}.

The aim of this paper is to study the effects induced by the variation of the trapping region extension and of the trap concentration on the device dc characteristics, in particular on the saturated drain current I_{DSS}, the output conductance g_O, and the device transconductance g_M, in order to gain a better insight of the degradation mechanisms observed in [8] and [9].

II. EXPERIMENTAL RESULTS

We carried out dc stresses on single heterojunction Al$_{0.28}$Ga$_{0.72}$N/GaN HEMTs grown by MOCVD on SiC substrate with a 25 nm-thick barrier layer. They were fabricated including Ti/Al/Ni/Au ohmic contacts, Ni/Au Schottky gate, and a SiN passivation layer. The tested devices had a gate length $L_G = 0.25$ μm and a total gate periphery of 4 × 25 μm. Nominal gate–drain and gate–source spacings were 2.05 and 1.4 μm, respectively. The devices were step-stressed by applying a drain-gate voltage up to 35 V in 2.5 V/step with a step duration of 5 min, keeping the source terminal floating. At the end of each stress step, I_D–V_D and I_D–V_G characteristics were acquired in order to monitor the device degradation; reverse gate current measurements were performed too.

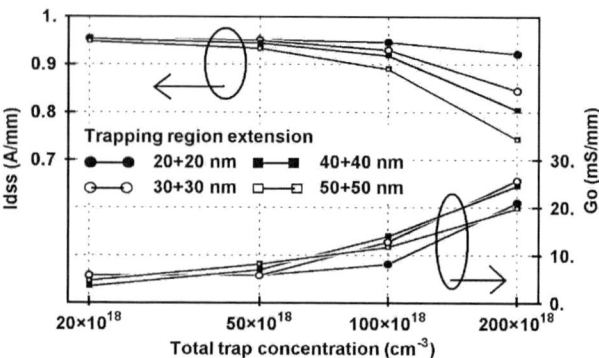

Fig. 4. I_{DSS} and output conductance values obtained by numerical simulations for different trap concentrations and trapping region extensions. The defective region equally extends underneath gate terminal and drain access region. Increasing trap concentration degrades device performance. Defective region is short: increasing its extension does not affect g_O.

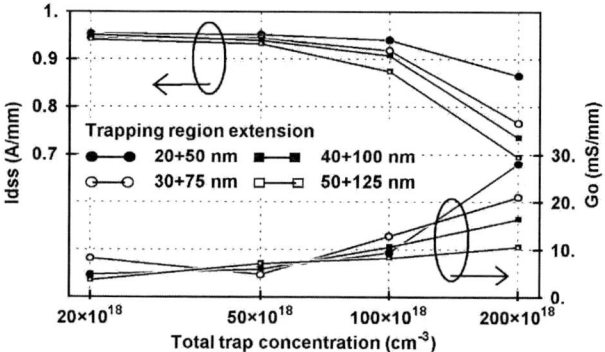

Fig. 5. I_{DSS} and output conductance values obtained by numerical simulations for different trap concentrations and trapping region extensions. Defective region is longer and extends mainly underneath the drain access region. Increasing its extension degrades I_{DSS} but eventually reduces the output conductance.

The saturated drain current I_{DSS}, evaluated at $V_{GS} = 0$ V and $V_{DS} = 8$ V, decreased by 15% at the end of the stress campaign. On the other hand, the output conductance increased significantly (see Fig. 2). The peak of the device transconductance g_M also decreased by 21%, from 260 to 205 mS/mm (not shown). During the step stress carried out, the device experienced also an increase in the reverse gate current up to 15 mA/mm, as previously observed in [9] (not shown).

III. NUMERICAL SIMULATIONS

Two sets of structures were considered for numerical simulations. In the first one, the degraded region is centered at the gate edge and its overall extension spans in the source-drain direction from 40 to 100 nm in 20 nm/step. In the second set, the degraded region extends for 2/7 underneath the gate terminal and for the remaining 5/7 underneath the gate-drain access region, its overall extension varying from 70 to 175 nm in 35 nm/step. The degraded region extends for the whole barrier thickness, in accordance with the results

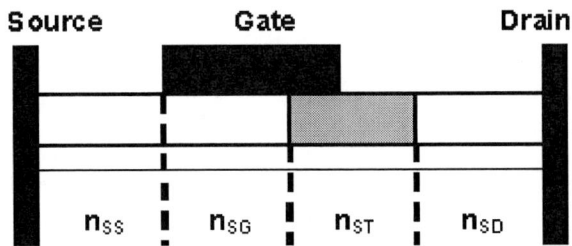

Fig. 6. Cross-section of the device structure showing the defective region in the AlGaN barrier in grey. The carrier sheet concentration in the channel n_{ST} corresponding to such region is smaller than in the rest of the channel, explaining the decrease in I_{DSS}.

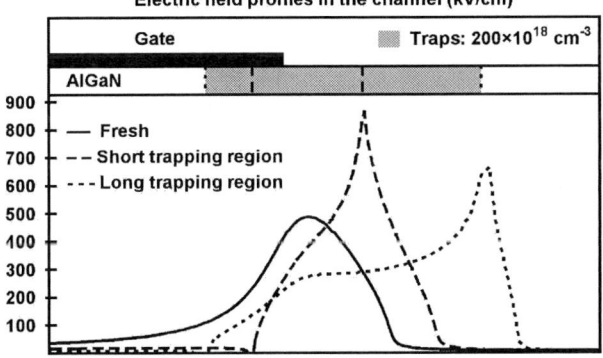

Fig. 7. Simulated electric field along the channel for a fresh device, and a structure with a short defective region or a long defective region: in the first case the electric field shows a higher peak and its distribution is narrow, whereas in the second case the peak is lower and the distribution broadens.

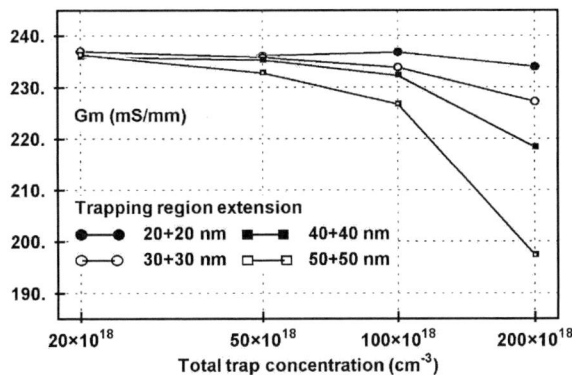

Fig. 8. Peak transconductance g_M values as obtained by numerical simulations. The defective region equally extends underneath gate terminal and drain access region. Increasing the trap concentration or the defective region extension degrades the device transconductance.

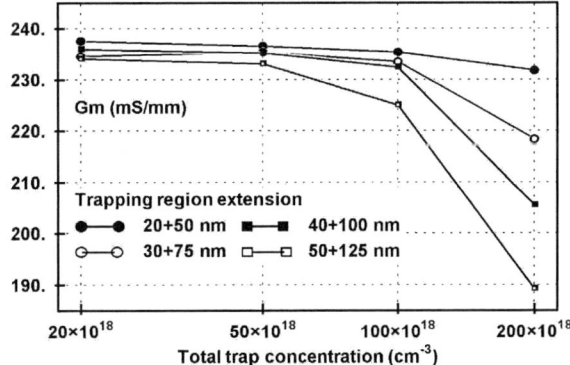

Fig. 9. Peak transconductance g_M values as obtained by numerical simulations. The defective region is longer and extends mainly underneath the drain access region. Increasing the trap concentration or the defective region extension degrades the device transconductance anyway.

obtained in [7], with traps uniformly distributed throughout it. Total trap concentrations of 20, 50, 100, and 200×10^{18} cm^{-3} were considered. The I_D–V_D characteristics were simulated with a drift-diffusion model by biasing the device and solving then the currents under stationary conditions, so that all transients can be considered to have expired. As can be seen in Fig. 3, the introduction of a trap concentration at the edge of the gate contact leads to both a degradation of the saturated drain current I_{DSS} and to an increase in the output conductance g_O, as experimentally observed: it appears therefore as an appropriate model to describe the occurring degradation phenomena.

In the first set of structures (trapping region extension 40–100 nm), I_{DSS} degrades significantly at the increasing of the trap concentration, the decrease being proportional to the lateral extension of the trapping region. In the same way, the output conductance increases uniformly at the increasing of the trap concentration (Fig. 4). In Fig. 5 the same results are shown for the second set of structures (trapping region extension 70–175 nm). The degradation of I_{DSS} is even more serious, whereas the output conductance tends to decrease for the longer degraded regions.

The reason for the decrease in I_{DSS} has been explained in [7]: The acceptor trap is located 0.5 eV from the conduction band: the trapping region extends for the whole AlGaN barrier thickness, so the traps closer to the AlGaN/GaN interface approach the Fermi level and thus a fraction of them are filled also in open-channel condition: a negative charge is therefore present in the barrier layer, close to the interface. Such negative charge decreases the effectiveness of the positive piezoelectric charge, inducing a reduction in the 2-DEG concentration. From the results obtained in [7], we can state that traps are filled only within very few nanometers from the interface: this statement is further confirmed by our simulations, which show that a total trap concentration of 20–200×10^{18} cm^{-3} constitutes in the end an actual sheet charge density of 1.4–6.7×10^{12} cm^{-2}, compared to a 2-DEG sheet charge density in the channel of 12.4×10^{12} cm^{-2}. In Fig. 6 a cross-section of the device structure is reported: the value of the carrier sheet concentration in the channel under the trapping region (n_{ST}), is lower than in the other regions. On one hand, this limits the saturation current I_{DSS}, on the other hand an increase in the extension of the degraded region increases the channel resistivity anyway.

978-1-4244-6658-0/10 $26.00 © 2010 IEEE

The change in the output conductance can be explained as follows: The portion of the channel underneath the trapping region has a lower carrier concentration, acting therefore as a bottleneck for the flow of electrons. However, the same portion of the channel is subject to a high horizontal electric field at the drain end of the gate. At low V_{DS} such electric field is low, and electrons are provided with less energy, thus enhancing the bottleneck effect, leading to a significant current degradation compared to the fresh device. However, the electric field at the drain end of the gate increases with V_{DS}: if the degraded region is sufficiently short, electrons are provided with more energy and are significantly accelerated through the bottleneck, managing to pass through it, so the current increases. Our simulations show in fact that for a short degraded region the electric field peak at the drain end of the gate is much higher compared to a fresh device. Fig. 7 depicts the simulated electric field profiles in the channel for a fresh device and two with a trap concentration of 200×10^{18} cm^{-3} located in a degraded region either 70 or 175 nm long. As can be seen, if traps are present in a shorter degraded region the electric field peak is much higher and its distribution narrows: this enhances short-channel effects such as the increase in the output conductance g_O, as reported in [10]. Fig. 7 shows also that if the degraded region becomes longer, the electric field distribution broadens and its peak decreases, thus reducing short-channel effects and leading to a smaller output conductance: as a matter of fact, as the trapping region extension increases, the bottleneck region becomes longer too, so electrons require more energy (i.e., higher V_{DS}) to pass through it.

By simulating the I_D–V_G characteristics in the saturation region we extracted the peak device transconductance g_M for the different structures analyzed. The dependence of the transconductance on the extension of the degraded region and on the trap concentration showed a trend almost identical to the one found for the I_{DSS} current: a decrease in g_M is induced both by an increase in the trapping region extension or by an increase in the trap concentration. This happens for any of the degraded region extensions taken into account, as shown by Fig. 8 and 9.

IV. CONCLUSION

The effects of dc stresses on AlGaN/GaN HEMTs were investigated by means of experimental measurements followed by numerical simulations. Tested devices subject to reverse gate-drain bias with the source terminal floating experienced a degradation in dc drain current, an increase in the output conductance and a collapse of the transconductance.

Following the results obtained in [7], we have verified that the presence of an acceptor trap level located 0.5 eV from the conduction band in the AlGaN barrier at the edge of the gate contact reasonably predicts the observed degradation phenomena: simulations have shown that both the decrease of the drain current and the degradation of the transconductance can be caused by either an increase in the trap concentration or in the extension of the trapping region. However, the output conductance still increases with the trap concentration, but it tends to decrease as the degraded region becomes longer, because of the broadening of the electric field distribution at the drain edge of the gate and the consequent weakening of short-channel effects.

The study carried out suggests that an increase in the trap concentration might have occurred in our devices during the stress, accompanied probably by a modest increase in the trapping region extension, as suggested by the high output conductance at the end of the stress campaign.

REFERENCES

[1] M. Faqir, G. Verzellesi, G. Meneghesso, E. Zanoni, and F. Fantini, "Investigation of High-Electric-Field Degradation Effects in AlGaN/GaN HEMTs", *IEEE Trans. Electron Devices*, Vol. 55, No. 7, Jul. 2008, pp. 1592–1602.

[2] M. Faqir, G. Verzellesi, A. Chini, F. Fantini, F. Danesin, G. Meneghesso, E. Zanoni, and C. Dua, "Mechanisms of RF Current Collapse in AlGaN–GaN High Electron Mobility Transistors", *IEEE Trans. Device Mater. Rel.*, Vol. 8, No. 2, Jun. 2008, pp. 240–247.

[3] U. K. Mishra, L. Shen, T. E. Kazior, and Y.-F. Wu, "GaN-based RF Power Devices and Amplifiers", *Proc. IEEE*, Vol. 96, No. 2, Feb. 2008, pp. 287–305.

[4] J. Joh and J. A. del Alamo, "Critical Voltage for Electrical Degradation of GaN High-Electron Mobility Transistors", *IEEE Electron Device Lett.*, Vol. 29, No. 4, Apr. 2008, pp. 287–289.

[5] J. Joh, L. Xia, and J. A. del Alamo, "Gate Current Degradation Mechanisms of GaN High-Electron Mobility Transistors," in *IEDM Tech. Dig.*, 2007, pp. 385–388.

[6] J. Joh and J. A. del Alamo, "Mechanisms for Electrical Degradation of GaN High-Electron Mobility Transistors", in IEDM Tech. Dig., 2006, pp. 415–418.

[7] A. Chini, V. Di Lecce, M. Esposto, G. Meneghesso, and E. Zanoni, "Evaluation and Numerical Simulations of GaN HEMTs Electrical Degradation", *IEEE Electron Device Lett.*, Vol. 30, No. 10, Oct. 2009, pp. 1021–1023.

[8] E. Zanoni, F. Danesin, M. Meneghini, A.Cetronio, C. Lanzieri, M. Peroni, and G. Meneghesso, "Localized Damage in AlGaN-GaN HEMTs Induced by Reverse-Bias Testing", *IEEE Electron Device Lett.*, Vol. 30, No. 5, May 2009, pp. 427–429.

[9] A. Chini, M. Esposto, G. Meneghesso, and E. Zanoni, "Evaluation of GaN HEMT degradation by means of pulsed, I–V, leakage and DLTS measurements", *IET Electronics Lett.*, Vol. 45, No. 8, pp. 426–427, Apr. 2009.

[10] Y. Awano, K. Tomizawa, N. Hashizume, M. Kawashima, "Monte Carlo Particle Simulation of a GaAs Short-Channel MESFET", *IET Electronics Lett.*, Vol. 19, No. 1, Jan 1983, pp. 20–21.

On the Influence of Flash Peak Temperature Variations on Schottky Contact Resistances of 6-T SRAM Cells

C. Kampen, A. Burenkov, and J. Lorenz

Fraunhofer Institute for Integrated Systems and Device Technology IISB

Schottkystrasse 10, 91058 Erlangen, Germany, Email: christian.kampen@iisb.fraunhofer.de

Abstract— The influence of temperature variations during flash annealing on contact resistances in 6-T SRAM cells was studied. TCAD simulations of 32 nm single gate FD SOI devices were carried out. The active regions of a 6-T SRAM cell were simulated by 3D process simulations to calculate the Schottky contact resistances. A coupled spike and flash annealing scheme was used to anneal the devices. Flash annealing temperature fluctuations were modeled in TCAD simulations and the resulting contact resistance values were calculated. SPICE parameters of the FD SOI devices were extracted and used in circuit simulations. The dependence of contact resistances on temperature fluctuations were taken into account in the SPICE simulation by analytical models.

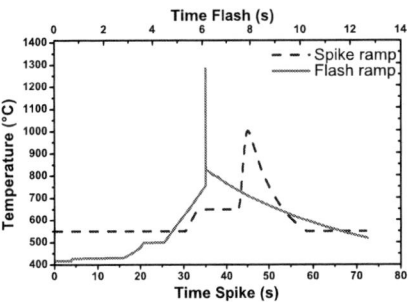

Fig. 1. Temperature profiles of the RTA and MSA annealing schemes

I. INTRODUCTION

Flash annealing has become a commonly used method for achieving high active doping concentrations at low diffusivity in todays CMOS devices. Due to its short time scales (about one millisecond) and high temperature (e.g. 1300 °C), the activation rate is very high while the redistribution of dopants is small. As Schottky contact resistances mainly depend on the active contact area and the active surface doping concentration at the metal/silicon interface, flash annealing is beneficial for achieving low contact series resistances. Due to short annealing times and layout dependent optical properties significant temperature variations on the wafer [1] can occur during the flash annealing, which mainly impact the doping activation level at high doping concentrations. As a consequence, the main influence of temperature variations in flash annealings is expected on source/drain resistances and especially on contact resistances. The effect of contact resistance variation due to variation of the local annealing temperatures during flash annealing is studied in this paper. To our knowledge, this is the first study on variations of contact resistances in silicon CMOS transistors.

II. SIMULATION METHODOLOGY

The simulation in this work was split into five parts. At first, single gate FD SOI MOSFETs with a physical gate length of 32 nm, a gate oxide thickness of 1.2 nm, a silicon film thickness of 10 nm and a buried oxide thickness of 20 nm were simulated by process and device simulations. The annealing of the devices was done by a coupled spike and flash annealing scheme we already presented in [2]. Special models for diffusion and activation of arsenic [3] and boron [4], calibrated for spike and flash annealings, were used in the process simulations. Furthermore, heavy doped ground plane [5] layers were used to suppress short channel effects. Following the TCAD simulations, SPICE parameters of the MOSFET devices were extracted using the EPFL-EKV model [6]. The method of SPICE parameter extraction is presented elsewhere [7].

A. Threshold voltage dependence on flash peak temperature variations

Secondly, the influence of flash peak temperature fluctuations on the threshold voltages of the NMOS and PMOS devices was studied. Figure 1 shows the temperature profiles of the spike annealing and the flash annealing, used in this work. The flash profile is denoted by the solid line, the spike profile by the dashed line. As can be seen, the peak temperature of the flash annealing amounts to approximately 1300 °C and lasts about one millisecond.

To take the effect of peak temperature fluctuations into account in process simulations, the peak temperature profile of the flash annealing was fitted by the following function:

$$T = T_0 + A \cdot exp(-exp(-z) - z + 1) \qquad (1)$$

with

$$z = \frac{t - t_c}{\alpha}. \qquad (2)$$

T_0 in Eq. 1 is the start temperature and A is the amplitude to describe the maximum peak temperature. t in Eq. 2 is the time, t_c the point in time the peak of the flash annealing starts, and α is a fit parameter to characterize the peak duration.

Equations 1 and 2 were used in the process simulations to modify the temperature ramp of the flash annealing, regarding to the maximum temperature peak. At least, the variation of the peak temperature profile was done by changing the amplitude parameter A. In this work, A was assumed to be Gaussian distributed. Thereby, 3σ was assumed to be 50 °C [1]. Figure 2 shows exemplary a set of flash peak temperature profiles, calculated by Eqs. 1 and 2.

In the next step, the impact of flash peak temperature variations on the threshold voltage of SG FDSOI MOSFETs was investigated. Therefore, 11 values for A were chosen. The resulting temperatures are in the range of 1230 °C to 1330 °C. After process simulations, device simulations for each temperature step were performed and the threshold voltages at low drain voltages and at high drain voltages were extracted from the transfer characteristics. Figure 3 shows the threshold voltage values for the NMOS (circles (V_{drain}=50 mV) and squares (V_{drain}=1.1 V)) and for the PMOS (stars (V_{drain}=-50 mV) and triangles (V_{drain}=-1.1 V)) plotted versus the peak temperature of the flash annealing. Due to the low diffusivity caused by the flash annealing, the threshold voltages stays on a constant level of approximately 420 mV in case of low drain voltages and 320 mV at high drain voltages for both devices. Hence, peak temperature fluctuations do not affect the threshold voltage of the SG FDSOI devices, investigated in this work and have, therefore, not to be taken into account in the later SPICE simulations.

B. Contact Resistance Calculation

Thirdly, a closer look to the design of a 6 transistor SRAM cell [8] was taken, to localize contact resistances. Figure 4 shows the typical layout of a 6-T SRAM cell. The active n+/p+ regions are denoted by the horizontal rectangles, the gate electrodes by the thin solid color vertical rectangles, and metal-1 layer is given by the striped structures. The corresponding circuit scheme is displayed in Fig. 5. Ten contact resistances were localized in total: Resistances R1 and R5 resulting from the contact pads at the sources of the NMOS devices M1 and M2, R4 and R8 from the supply voltage contact pads at the sources of the PMOS devices M2 and M4. R3 and R7 can be traced back to the direct contact of metal-1 at the drain region of the PMOS transistors M2 and M4, well as R9 and R10 of

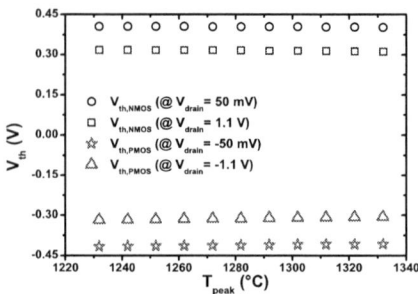

Fig. 3.   Threshold voltage behavior at increasing flash annealing temperature

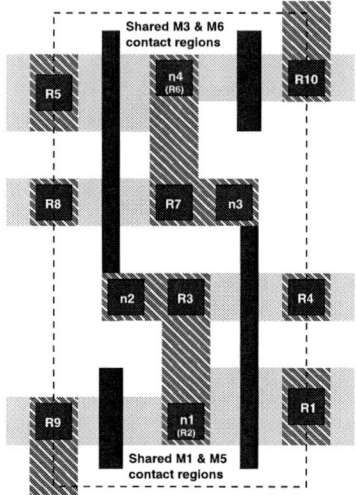

Fig. 4.   Layout of one 6-T SRAM cell used in this work

the access transistors M5 and M6, which are connected to the bitlines. As the active regions of the access transistors and the flip-flop NMOS transistors are shared by both devices, as seen in Fig. 4, no contact resistances have to be taken into account between transistors M9 and M1, as well as between M10 and M3.

Fourthly, the active source/drain regions of the respective transistors were simulated separately by 3D process and device simulations to calculate the Schottky contact resistances. The length of the contact regions was assumed to be four times the gate length, as well for the single parts of the respective transistors, as for the shared drain regions of transistors M9, M1, M10, and M3. The widths of the active regions were chosen to be 80 nm in case of M2, M4, M9, and M10, while 130 nm were assumed for M1 and M3. The silicon thickness of the contact regions was set on 10 nm, corresponding to the film thickness of the SG FDSOI MOSFETs. Figures 6(a) and 6(b) show the structures of the shared drain region of M9 and M1 and the active drain region of transistor M2, respectively. As can be seen, the isolation, which surrounds the MOSFETs, was taken into account in the simulation to ensure the correct calculation of the dopant segregation

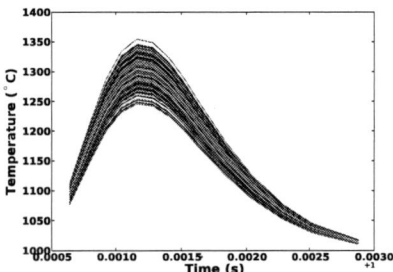

Fig. 2.   Peak temperature variations

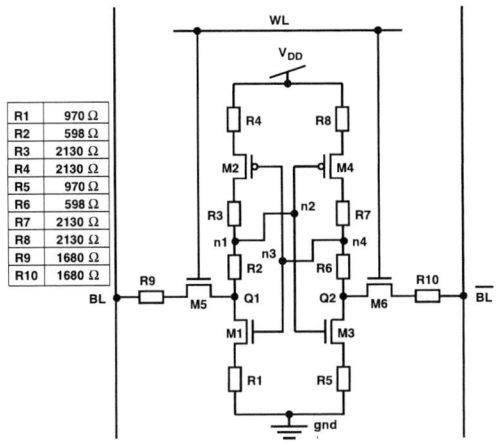

R1	970 Ω
R2	598 Ω
R3	2130 Ω
R4	2130 Ω
R5	970 Ω
R6	598 Ω
R7	2130 Ω
R8	2130 Ω
R9	1680 Ω
R10	1680 Ω

Fig. 5.   Contact resistance network in a 6-T STRAM cell

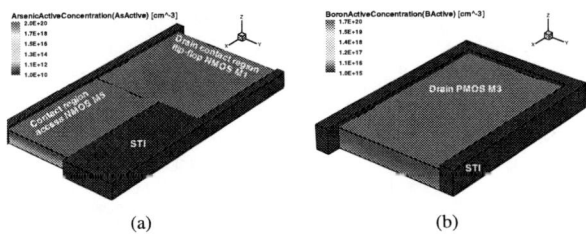

(a)   (b)

Fig. 6.   Contact area of the access NMOS and flip-flop NMOS transistors

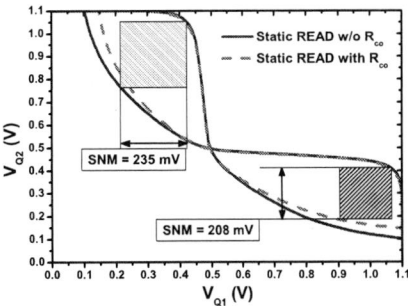

Fig. 7.   Influence of contact resistances on the static noise margin

behavior at the silicon/silicon-oxide interface. Ion implantation was simulated, followed by the annealing scheme, presented in Fig. 1. Device simulations were used to calculate the Schottky contact resistances. The Schottky contact resistances were calculated, assuming a Schottky barrier height of 0.6 eV for electrons and 0.51 eV for holes. The resulting contact resistances R1 to R10 for the nominal annealing are listed in Fig. 5. Due to the large area of the access transistors M9 and M10 and the flip-flop NMOS transistors M1 and M3, the resistances R2 and R6 were calculated to be the smallest with 598 Ω. The contact resistances of the PMOS source/drain contacts on the other hand were calculate to be the largest with a value of 2.13 kΩ, which results from the smaller area and the slightly lower surface doping concentration ($1.1 \cdot 10^{20}$ cm^{-3}) compared to the NMOS ($1.4 \cdot 10^{20}$ cm^{-3}).

Finally, the 11 temperature values, already used for calculating the threshold voltage dependence of the CMOS devices, were used to calculate numerically the dependence of Schottky contact resistances on the peak temperature. A simple parabolic function was found to be a good approximation to describe the temperature dependence of R1 to R10. Fit parameters for each contact pad were extracted, to be used in SPICE simulations.

III. SIMULATION RESULTS

First of all, the influence of the nominal contact resistances on the static and dynamic performance of the 6-T SRAM cell

was investigated. Therefore, one simulation run was performed without taking contact resistances into account and a second one using the network of Fig. 5. Figure 7 shows the static read butterfly characteristics of the two simulation runs. The solid line denotes the case without taking contact resistances into account, while the dashed curve shows the results of the second simulation run, including contact resistances. Only a small impact of R_{co} was observed for the static case. The static noise margin (SNM) amounts to 235 mV in case of no contact resistances were assumed. If R_{co} were taken into account, the SNM is reduced by 13 %.

A stronger impact can be observed in the dynamic READ and WRITE case. Figure 8 shows the results for the READ (Fig. 8(a)) and the WRITE (Fig. 8(b)) operation. It has to be mentioned that a fan-out capacitance of 6 fF was assumed for the bitlines during the simulation of the READ operation [9]. A rise and fall time of 1 ps was used for the wordline signal. As can be seen in Fig. 8(a), the READ delay, calculated at 0.55 mV, is enlarged from 50 ps (without contact resistances) by 28 % if contact resistances were taken into account. The strongest effect of contact resistances is seen, if the WRITE operation is investigated. Here, the nominal SRAM cell (without R_{co}) completes the WRITE operation afer 11.5 ps. If the network of Fig. 5 is used, the WRITE delay is increased to 17.5 ps, which is more than 50 % of the nominal delay.

Finally, temperature variations in the range of 100 °C [1] around the nominal temperature of 1282 °C were accounted for. Here, a Gaussian distribution with a standard deviation of 16.6 °C (1.3 % from mean value) was assumed. The mean value was set on the nominal peak temperature. Then, 1000 simulations were undertaken to observe the influence of peak temperature variation on the SRAM cell. First of all, the impact of the peak temperature fluctuations on the contact resistances was observed. Figure 9 shows the distribution of the source contact resistances R1 and R5 of transistors M1 and M3. As can be seen, a wide spread in the range of 1400 Ω to 2100 Ω was calculated. The mean values amounts to 1692 Ω and the standard deviation to 122 Ω. For all contact resistances, a relative standard deviation of 7 % from the mean value was calculated. Secondly, the impact of the contact resistance variations on the electrical performance of the SRAM cell was investigated. Figure 10 displays the distribution of the

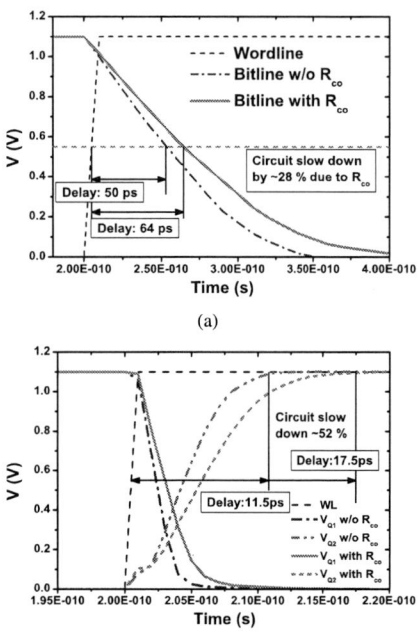

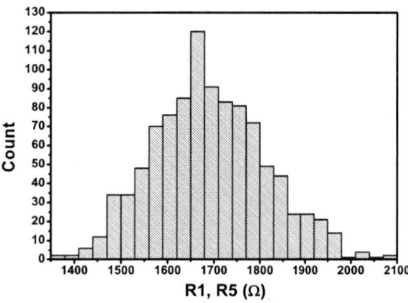

Fig. 8. Influence of contact resistances on the dynamic performance: (a) READ delay; (b) WRITE delay

Fig. 9. Probability density function of the source contact resistances of transistors M1 and M3

SNM (Fig. 10(a)) and the READ delay (Fig. 10(b)). As can be seen, even temperature variations of 100 °C have only a moderate impact, as well on the static behavior as on the dynamic performance. The SNM is changed in the range of approximately 211 mV to 203 mV. The 3σ fluctuation amounts to about 3 %. Comparable small impact of peak temperature fluctuations was found in case of the dynamic performance (Fig. 10(b)). The mean value was calculated to be at 63.8 ps with a 3σ fluctuation of 6 %. As well as the READ delay, 3σ of the WRITE delay amounts to approximately 6 %.

IV. CONCLUSION

The impact of flash annealing peak temperature fluctuations on the electrical performance of 6-T SRAM cells was studied in this work. A significant impact of contact resistances on the dynamic performance was found, which amounts to

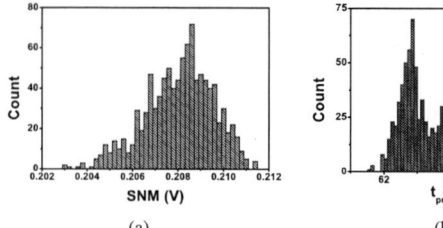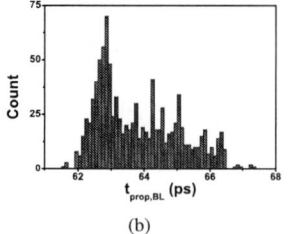

Fig. 10. Probability density function of the static noise margin (a) and the READ delay (b)

50 % circuit speed loss. 3D TCAD simulations were used to calculate the dependence of Schottky contact resistances on the peak annealing temperature of flash annealing schemes. The dependence of contact resistances on the annealing temperature was taken into account in SPICE simulations. 1000 SPICE SRAM simulations for randomly changed peak temperature of flash annealings were undertaken. A significant sensitivity of contact resistances on annealing fluctuations was found ($3\sigma=21$ %). However, only a moderate impact on the overall electrical SRAM cell performance was found, when flash annealing peak temperature variations were taken into account.

ACKNOWLEDGMENT

This research has been supported by the Fraunhofer Internal Program under Grant No. MAVO 817 759.

REFERENCES

[1] T. Kubo, T. Sukegawa, E. Takii, T. Yamamoto, S. Satoh, and M. Kase, "First quantitative observation of local temperature fluctuations in millisecond annealing," in *15th IEEE International Conference on Advanced Thermal Processing of Semiconductors, RTP2007*, October 2007, pp. 321–326.

[2] C. Kampen, A. Martinez-Limia, P. Pichler, J. Lorenz, and H. Ryssel, "Advanced annealing schemes for the 32nm node," in *13th Internaltional Conference on Simulation of Semiconductor and Decices SISPAD 2008*, 2008.

[3] A. Martinez-Limia, P. Pichler, C. Steen, S. Paul, and W. Lerch, "Modelling and simulation of advanced annealing processes," *Materials Science Forum*, vol. 279, pp. 573–574, 2008.

[4] J. Schermer, A. Martinez-Limia, P. Pichler, C. Zechner, W. Lerch, and S. Paul, "On a computationally efficient approach to boron-interstical clustering," 2008, accepted to be published.

[5] T. Ernst and S. Cristoloveanu, "Buried oxide fringing capacitance: A new physical model and its implication on SOI device scaling and architecture," in *SOI Conference*, 1999, pp. 38–39.

[6] *The EPFL-EKV MOSFET Model Equations for Simulations*, Version 2.6 ed., Electronics Laboratories, Swiss Federal Institute of Technology (EPFL), 1998.

[7] C. Kampen, A. Burenkov, J. Lorenz, and H. Ryssel, "FD SOI MOSFET compact modeling including process variations," in *11th International Conference on Ultimate Integration of Silicon, ULIS 2010*. Institute of Electrical and Electronics Engineers IEEE, 2010, pp. 173–176.

[8] A. Pouydebasque, B. Dumont, S. Denorme, F. Wacquant, M. Bidaud, C. Laviron, A. Halimaoui, C. Chaton, J. D. Chapon, P. Gouraud, F. Leverd, H. Bernard, S. Warrick, D. Delille, K. Romanjek, R. Gwoziecki, N. Planes, S. Vadot, I. Pouilloux, F. Arnaud, F. Boeuf, and T. Skotnicki, "High density and high speed SRAM bit-cells and ring oscillators due to laser annealing for 45nm bulk CMOS," in *Electron Device Meeting, IEDM Technical Digest. IEEE International.* IEEE, December 2005, pp. 663–666.

[9] A. C. J. M. Rabaey and B. Nicolic, *Digital Integrated Circuits*. Pearson Education International, 2003.

Temperature Dependent Dielectric Absorption Characterization and Modeling for SiN, Al₂O₃ and Ta₂O₅

H. Muminovic, P. Riess, P. Baumgartner, P. Klein*

Infineon Technologies Munich, *University of Applied Sciences Munich

Abstract--This paper reports temperature dependant dielectric absorption measurements of SiN, Al₂O₃ and Ta₂O₅ MIM capacitors between 100Hz and 10GHz over a broad temperature range. The SiN dielectric clearly shows a temperature dependent dielectric absorption whereas for Al₂O₃ and Ta₂O₅ the effect is nearly temperature independent. A model is proposed for the 3 dielectric materials which takes into account dielectric absorption and temperature dependence. Depending on the dielectric material the extraction strategy has to be modified. The temperature dependence of the capacitance has to be split between temperature coefficient and temperature dependent dielectric absorption.

I. INTRODUCTION

Dielectric absorption (DA) is known as the intrinsic property of a capacitor to recharge itself after being shortly discharged. This effect is due to dipoles and or trapping detrapping mechanism in the dielectric material [1]. As a consequence the precision of analogue digital converters (ADC) and voltage controlled oscillators (VCO) is limited [2-4]. When measured in the time domain the capacitor exhibits a discharge that is stretched over time. When measured in the frequency domain, the dielectric absorption is extracted by measuring the frequency dependence of the capacitance. In this paper we characterize the dielectric absorption for frequencies between 100Hz and 10GHz focusing on its temperature dependency for SiN, Al₂O₃ and Ta₂O₅ dielectrics. In a second step we apply the extraction strategy of temperature dependent dielectric absorption model used for SiN capacitors [5] to the two other dielectrics. It is discussed under which condition this method can be applied to Al₂O₃ and Ta₂O₅. Finally the impact of the temperature dependency of the dielectric absorption on the MIM cap RF performance is discussed for the three dielectric materials.

II. DIELECTRIC ABSORPTION CHARACTERIZATION

The capacitors used in this study are metal insulator metal (MIM) capacitors which are integrated in a standard 0.13μm CMOS process. The dielectric materials used are SiN, Al₂O₃ and Ta₂O₅. The dielectric characteristics are listed in table 1. The bottom and top electrode of the MIM capacitor are made of TiN.

	SiN	Ta₂O₅	Al₂O₃	
tox [nm]	29	50	20	50
area cap [fF/um2]	2,10	4,50	3,50	1,40
Dielectric loss between 100Hz and 1GHz [% of total cap]	~4%	~1,5%	~4%	~4%
Tc1 @ 10kHz [ppm/K]	25	115	250	250
Tc1 @ 100MHz[ppm/K]	140	115	250	250

Table 1. Dielectric characteristics of SiN, Al₂O₃ and Ta₂O₅

In order to cover the complete frequency range capacitances between 70pF down to 300fF are needed to be measured. The low frequency measurements are done using a Precision LCR Meter (Agilent 4284E 20Hz to 10MHz). For higher frequency RF equipment is necessary in order to cover a frequency range between 100MHz and 10GHz. Using a standard Network analyser (Agilent 8753E, 30MHz-6GHz) and capacitances between 70pF and 300fF the frequency dependency of the capacitance is extracted from S parameter measurements. A two step open/thru deembedding is used to eliminate the undesired parasitics. At high frequency, for large capacitors the measurement is distorted due to the self resonance frequency of the capacitance (cf. Figure 1).

The capacitance of the 3 dielectrics was measured over a broad frequency range. Figure 2 compares the absorption rate of SiN, Al₂O₃ and Ta₂O₅ at 25°C. Figure 2 clearly shows that the dielectric absorption of SiN is strong frequency dependent whereas Al₂O₃ and Ta₂O₅ have a constant absorption rate over the full frequency range (cf. Figure 2).

978-1-4244-6658-0/10 $26.00 © 2010 IEEE

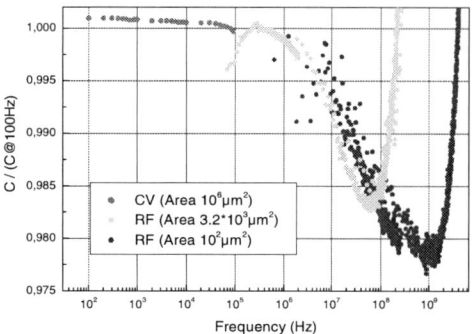

Fig. 1. Measurement of the MIM capacitance over a broad frequency range. Below 1MHz an LCR meter is used and at high frequencies the capacitance is extracted from S parameter measurements. Different sized test structures are used to cover the complete frequency range.

SiN exhibits two different absorption regimes. At low frequency below 100kHz SiN has the lowest dielectric absorption in the range of 100 ppm/decade whereas Al_2O_3 shows the highest absorption rate of 6000 ppm/decade. At high frequency above 10MHz Ta_2O_5 shows the lowest dielectric absorption of 2500 ppm/decade.

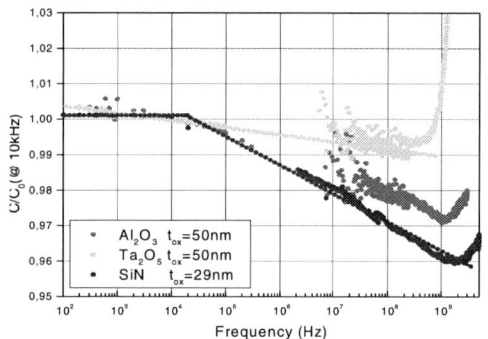

Fig. 2. Frequency dependence of the SiN, Al_2O_3 and Ta_2O_5 normalized MIM capacitances. Comparison of the dielectric absorption rates at room temperature.

Moreover it can be seen for Al_2O_3 that the dielectric absorption is independent of the thickness of the dielectric and the area capacitance indicating that intrinsic properties of the materials are characterized (cf. Figure 3).

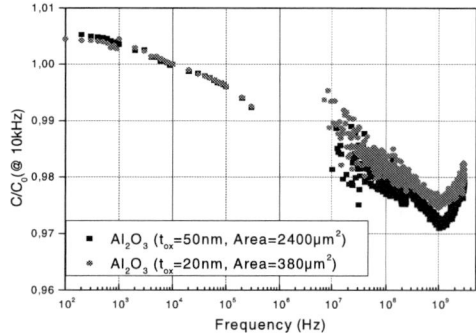

Fig. 3. Frequency dependence of the Al_2O_3. Capacitaces with thickness 20nm and 50nm. The dielectric absorption is independent of the thickness and area capacitance.

The capacitances are measured from -40°C up to 150°C. Figure 4 shows very characteristic temperature behaviour of dielectric absorption for SiN capacitance (cf. Figure 4). At high temperature the dielectric absorption onset occurs only at 10MHz whereas at -40°C the onset occurs at 10kHz. In addition it can be clearly seen that the dielectric absorption at high temperature above 10MHz is greater (higher C(f) slope) then at low temperature. This very specific signature leads to a temperature dependent dielectric absorption extraction and modeling of SiN which was described in detail by Riess and Baumgartner [5]. The key assumption being that the dipoles/or trapping detrapping mechanisms at the origin of the dielectric absorption has an activation energy which follows the Arrhenius law and that a distribution of those extracted activation energies is necessary to model the temperature dependent dielectric absorption.

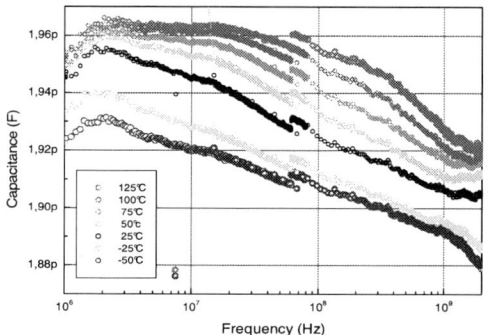

Fig. 4. SiN MIM capacitance frequency dependent measurements for temperatures between -50°C and 125°C.

Compared to SiN we determined another physical behaviour for Al_2O_3 and Ta_2O_5 concerning the temperature dependency of the dielectric absorption. The dielectric absorption of Al_2O_3 and Ta_2O_5 is temperature independent with constant absorption rate over a broad temperature range from -40°C up to 150°C (cf. Figure 5-6).

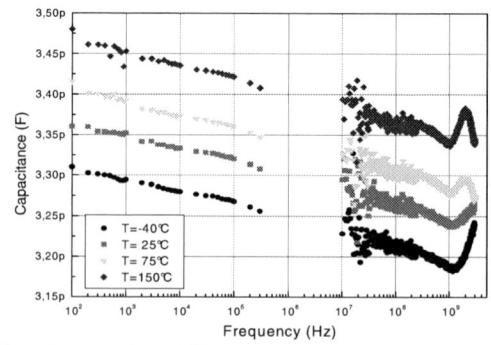

Fig. 5. Al_2O_3 MIM capacitance frequency dependent measurements for temperatures between -40°C and 150°C. The dielectric absorption of Al_2O_3 is temperature independent.

978-1-4244-6658-0/10 $26.00 © 2010 IEEE   214

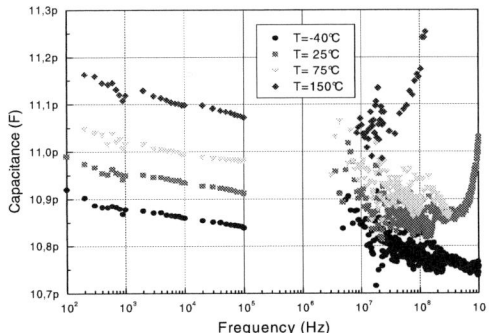

Fig. 6. Ta$_2$O$_5$ MIM capacitance frequency dependent measurements for temperatures between -40°C and 150°C.

Nevertheless a very strong temperature dependency of the measured capacitance, commonly modeled by the temperature coefficient T$_{c1}$ (cf. Equation 1) at a fixed frequency, is observed over the full frequency range (cf. Figure 7).

$$C_{MIM}(T) = C_0 \left(1 + T_{c1}(T - T_0)\right) \quad (1)$$

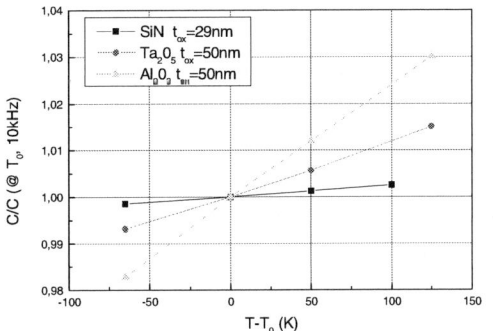

Fig. 7. Temperature dependency of the MIM capacitance between -40°C and 150°C.

For these two dielectrics the temperature coefficient T$_{c1}$ is constant over the full frequency range whereas it was frequency dependent for SiN (cf. Figure 8). Based on this measurement data it can be identified two different classes of dielectrics: The first one where the temperature dependency of the dielectrics strongly modifies the dielectric absorption rate and the second where the dielectric absorption rate is temperature independent with a strong temperature dependency of the capacitance.

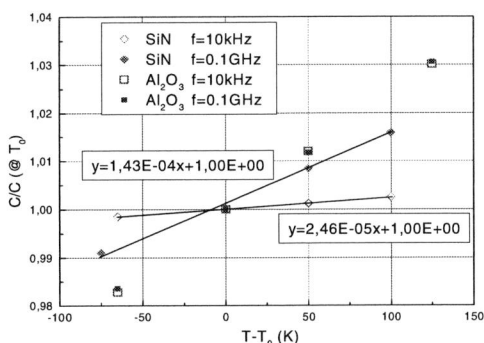

Figure 8. Comparison of the temperature dependency of the SiN and Al$_2$O$_3$ MIM capacitance for 10kHz and 0.1GHz.

In the next step we apply the extraction strategy developed in [5] to Al$_2$O$_3$ and Ta$_2$O$_5$ in order to verify in which extent the model extraction methodology can be applied. We apply the activation energy extraction method in order to determine the resistance values R$_n$ (cf. Equation 2) for the RC network which is needed for the modeling of the dielectric absorption (cf. Figure 9). The number of parallel RC circuits is chosen in order to have 2 RC corner frequencies per frequency decade.

$$R_n(T) = K^{-1} \exp\left(\frac{E_{an}}{K_b \, T}\right) \quad (2)$$

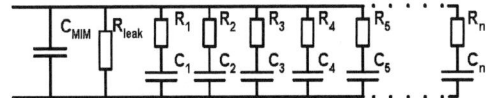

Figure 9. Dielectric absorption modeling network.

Also regarding the extracted activation energy we observed differences between the 3 dielectrics. Figure 10 shows the activation energy extracted for the 3 dielectrics. For SiN we get values from 0.2-0.3 eV with a constant K factor whereas for Al$_2$O$_3$ and Ta$_2$O$_5$ the K factor is spread over several decades and the extracted activation energy is much higher then with SiN. The extracted E$_a$ values of Al$_2$O$_3$ and Ta$_2$O$_5$ are then introduced in the dielectric absorption network choosing an average value for K.

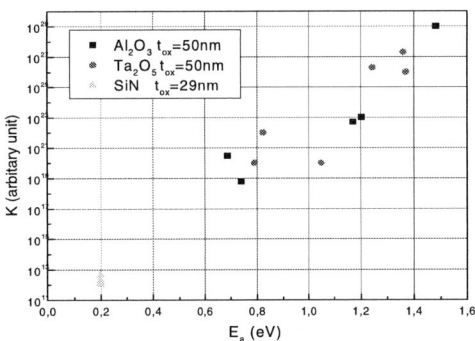

Figure 10. Extracted activation energy and K factor for SiN, Al$_2$O$_3$ and Ta$_2$O$_5$.

In Figure 11 we see a good fit achieved for SiN using the dielectric absorption network.

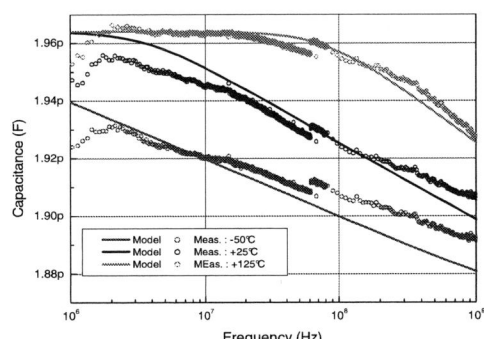

Figure 11. Modeling of the SiN MIM cap temperature and frequency dependency using the dielectric absorption temperature model [5].

On the other hand figure 12 shows the model hardware correlation for Al_2O_3. It can be seen that the extracted model overestimates the temperature activation and shows a temperature dependent slope of the dielectric absorption which is not measured experimentally. This is due to the fact that the temperature activation is extracted twice with T_{c1} and a second time with the activation energy needed for the temperature dependent dielectric absorption model.

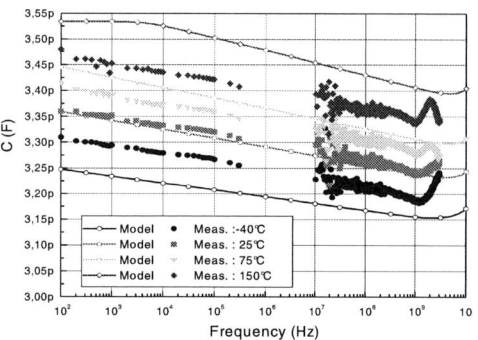

Figure 12. Modeling of the Al_2O_3 MIM capacitance temperature and frequency dependent using the dielectric absorption temperature model [5].

III. TEMPERATURE DEPENDENT DIELECTRIC ABSORPTION MODELING

Due to the different physical behaviour of the temperature independent dielectric absorption and the strong temperature dependence of the capacitance for Al_2O_3 and Ta_2O_5, compared to SiN where the dielectric absorption is strongly temperature dependent, the MIM capacitances of the two dielectrics can't be modeled with the temperature model of the dielectric absorption extracted for SiN [5]. These physical properties of Al_2O_3 and Ta_2O_5 led to a different model approach. From a modeling point of view it is possible to switch of the temperature dependency in the dielectric absorption model, which accounts for the temperature dependent dielectric absorption rates. At the moment the physical mechanisms behind this effect are not fully understood. R_n is defined as temperature independent (T is set to T_0 for all temperatures) therefore there is no longer a temperature dependent dielectric absorption. In this case the temperature dependency of the MIM capacitances is modeled with the temperature coefficient T_{c1}. Figure 13 shows the improved model hardware correlation when the temperature activation is only included in the temperature coefficient T_{c1}.

Depending on the dielectric material the dielectric absorption depends on temperature or is not affected by it. Therefore we propose two modeling strategies for dielectric absorption. In the case of temperature dependent dielectric absorption where the temperature coefficient T_{c1} strongly varies with the frequency the temperature model of the dielectric absorption [5] can be applied. For capacitances with a temperature independent dielectric absorption where the temperature coefficient is frequency independent there is no need to extract the activation energy E_a. The temperature dependency of the capacitances can be modeled by T_{c1} only as shown in figure 13.

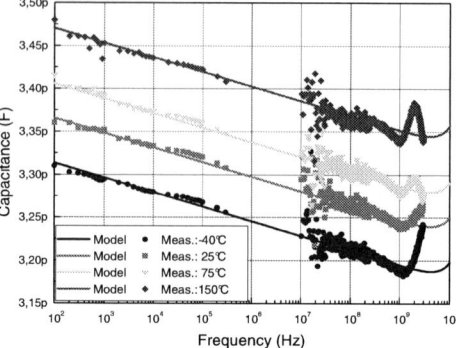

Figure.13. Modeling of the MIM capacitance temperature and frequency dependency using the dielectric absorption temperature model [5] whereas R_n is temperature independent.

IV. CONCLUSION

In this work we have characterized temperature dependency of the dielectric absorption of SiN, Al_2O_3 and Ta_2O_5 MIM capacitor for RF applications. Based on experimental results we indentify two classes of dielectrics where the temperature dependence of the capacitance has to be divided between the linear temperature coefficient and the temperature dependent dielectric absorption. We propose a modelling solution for the 3 dielectrics which takes into account the dielectric absorption and its temperature dependency. These models can directly be used for circuit simulation.

The work also clearly shows that depending on the dielectric material different temperature dependencies of the dielectric absorption exist and different modeling approaches are necessary.

REFERENCES

[1] A. K. Jonscher, J. Phys. D: Appl. Phys. 32 (1999) R57-R70.

[2] J. W. Fattaruso, et al. , IEEE Journal of Solid-State Circuits, Vol. 25, No. 6, December 1990, pp 1550-1561.

[3] A. Zanchi, et al, IEEE Journal of Solid-State Circuits, Vol. 38, No. 12, December 2003, pp 2077-2086.

[4] J.C. Kuenen, et al, IEEE Transaction on Instrumentation and Measurement, Vol. 45, No. 1, February 1996, pp 89-97.

[5] P. Riess, et al, ESSDERC., 2006, pp 459-462.

[6] J-P. Manceau, et al, IEEE ICMTS Conference, march 6-9 2006, pp 199-204

[7] H. Reisinger, et al, IEDM Tech. Dig. 2001, pp 267-270.

[8] P.C. Dow, IRE Trans. Electr. Comp., pp. 17-22, Mar. 1958.

[9] Ken Kundert, www.designers-guide.com, July 2004.

[10] Z. Lu, et al, J. Vac. Sci. Technol. A 13(3), May/Jun 1995, pp 607-612.

[11] C. H. Ng, et al, IEEE Transaction on Electron Devices, Vol. 52, No. 7, July 2005, pp 1399-1409.

[12] J. A. Babcock, et al, IEEE Electron Device Letters, Vol. 22, No. 5, May 2001, pp230-232.

[13] M. Kropfitsch, et al, ISCAS 2006, pp

978-1-4244-6658-0/10 $26.00 © 2010 IEEE

Strained MOSFETs on Ordered SiGe Dots

Johann Cervenka, Hans Kosina,
and Siegfried Selberherr
Institute for Microelectronics, TU Wien
Wien, Austria
Email: cervenka@iue.tuwien.ac.at

Jianjun Zhang, Nina Hrauda,
Julian Stangl, and Guenther Bauer
Johannes Kepler University
Linz, Austria

Guglielmo Vastola,
Anna Marzegalli, and Leo Miglio
University of Milano-Bicocca
Milano, Italy

Abstract—The potential of strained DOTFET technology is demonstrated. This technology uses a SiGe island as a stressor for a Si capping layer, into which the transistor channel is integrated. The structure information is extracted from AFM measurements of fabricated samples. Strain on the upper surface of a $30\,\mathrm{nm}$ thick Si layer is in the range of $0.7\,\%$, as supported by finite element calculations. The Ge content in the SiGe island is $30\,\%$ on average, showing an increase towards the top of the island. Based on realistic structure information, three-dimensional strain profiles are calculated and device simulations are performed. Up to $15\,\%$ enhancement of the NMOS saturation current is predicted.

I. INTRODUCTION

Strain engineering as a means to enhance electronic device performance has become an integral part of contemporary CMOS technology. In addition, novel device architectures have been proposed to improve the way in which strain is induced in the device channel. Schmidt and Eberl proposed to use self-assembled SiGe islands as stressors for Si capping layers [1]. In this way higher strain values can be reached as compared to strained Si grown pseudomorphically on relaxed SiGe buffers.

In this work, the process for growing self-organized SiGe islands is briefly described, followed by an experimental and theoretical assessment of the strain in the capping layer, and a prediction of performance enhancement of n-type FETs integrated in the strained capping layer by means of three-dimensional device simulation.

II. GROWTH OF REGULAR ARRAYS OF SIGE ISLANDS

The growth procedures are described in detail in [2] and [3]. The samples were grown the on a Si(001) substrate, on which a square pattern of pits with a period of $800\,\mathrm{nm}$ had been defined by e-beam lithography, and transferred by reactive ion etching to form pits with a width of $170\,\mathrm{nm}$ and a depth of $75\,\mathrm{nm}$. A $36\,\mathrm{nm}$ thick Si buffer layer was deposited on the pit-patterned substrates by molecular beam epitaxy, while the substrate temperature was ramped from $450\,^\circ\mathrm{C}$ to $550\,^\circ\mathrm{C}$. Then the substrates were heated to a growth temperature of $720\,^\circ\mathrm{C}$, at which 6 mono-layers of Ge were deposited to form one dome-shaped SiGe island per pit. A Si capping layer of $30\,\mathrm{nm}$ thickness was deposited after cooling the substrate to $360\,^\circ\mathrm{C}$, to avoid intermixing between the SiGe island and the Si cap. The surface morphology was investigated after each growth step by atomic force microscopy (AFM). Figure 1 shows the AFM image of the final surface of the Si cap, which actually

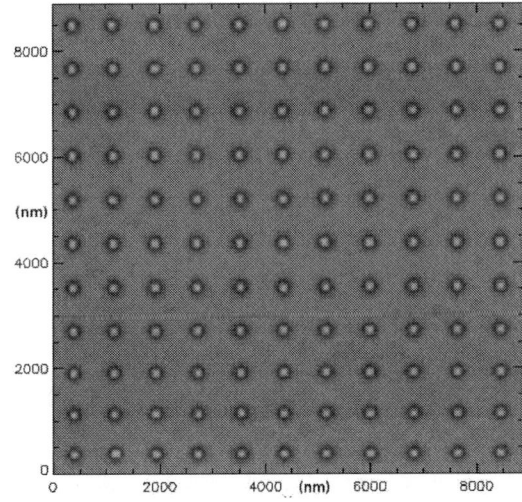

Fig. 1. AFM micrograph of the sample surface after $30\,\mathrm{nm}$ of Si cap has been deposited onto SiGe islands grown in pits of a prepatterned Si substrate.

conformally replicates the surface after formation of the SiGe islands. Line scans across several pits and across a single pit are shown in Figure 2, crossing the center of the pits and buried SiGe islands.

III. MODELING OF THE STRAIN FIELD

The processes of substrate patterning and SiGe island growth have been optimized with respect to subsequent device fabrication. Excellent island ordering, as well as island shape, size, and composition uniformity have been achieved. This optimization has been supported by simulations. Three-dimensional AFM data have been directly imported into a finite element code for strain calculation. In particular, AFM data were taken both after the Si buffer growth (surface of the pit) and after Ge depositions. As compared to the ideal equilibrium shape of islands as reconstructed from facet plots[4], the procedure used here includes more details of the actual structure such as edge rounding and the trench surrounding the island. In the elastic field calculations a Ge content profile in the island has been taken into account with an average Ge content of 0.3. In an iterative procedure the elastic energy is minimized. At the top of the island a higher

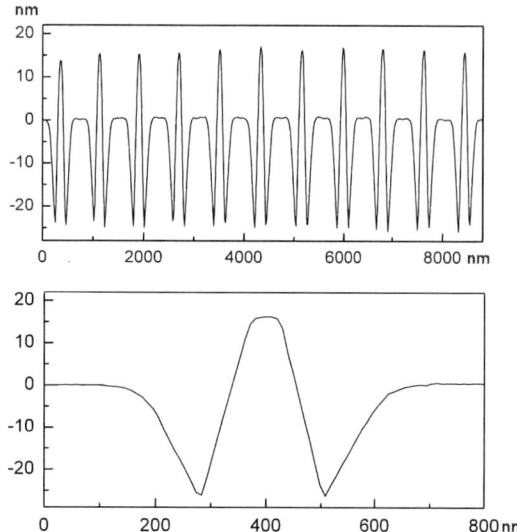

Fig. 2. AFM linescan across the islands indicates the excellent homogeneity of the island array (upper panel). The shape of the cap is shown with high resolution in the lower panel.

Ge content is found, whereas at the lower interface the Ge content is reduced by intermixing with the Si buffer layer. The Si capping layer is deposited at sufficiently low temperature to prevent any significant intermixing of the deposited Si with the SiGe island, so that Ge content profiles can be estimated prior to capping. Then in the whole structure including the capping layer the strain field is calculated using a finite-element code. The sample considered here shows at the upper surface of the 30 nm Si capping layer, where the transistor channel will be integrated, up to 0.7 % biaxial, tensile strain (Fig. 3, Fig. 4, and Fig. 4). This value has also been confirmed by high-resolution X-ray diffraction [5] and Raman spectroscopy in conjunction with simulations [6].

IV. THE DOTFET PROCESS

Removing the SiGe island at some stage of the process, a silicon on nothing (SON) device architecture can be realized. In this work, however, we consider a process in which the island is preserved. In this case the thermal budget must be kept sufficiently low to prevent intermixing of Si and Ge between the Si capping layer and the SiGe island. This can be achieved by state-of-the art low-temperature processing, including low-temperature formation of the gate stack and laser annealing of the source/drain implants. A low-complexity, dedicated n-channel MOSFET process has been developed [7].

V. THREE-DIMENSIONAL DEVICE SIMULATION

With the AFM data of the buffer and the SiGe island surfaces, the geometrical structure has been built. The Si capping layer is treated in the simulation by a conformal deposition. The correct representation of the Si cap is very important since the simulated current is quite sensitive to the strain distribution at the surface. An unstructured mesh is used,

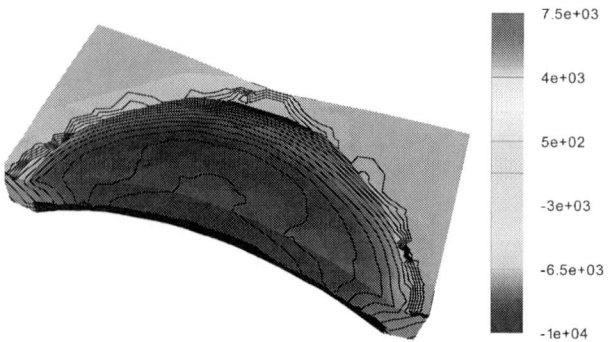

Fig. 3. Strain component e_{xx} (channel length direction) in the Si capping layer.

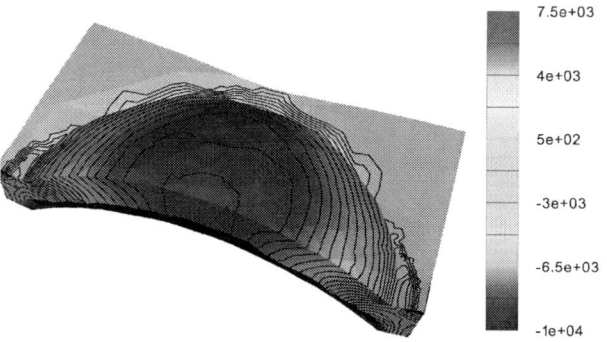

Fig. 4. Strain component e_{yy} (channel width direction) in the Si capping layer.

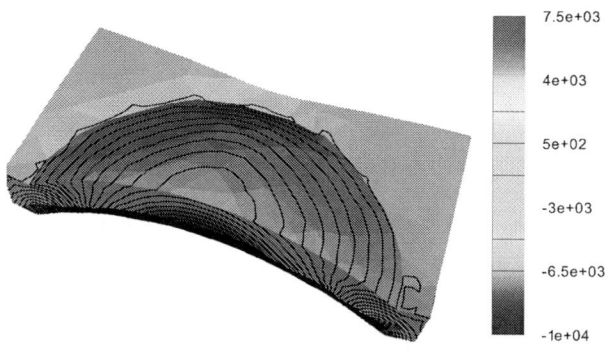

Fig. 5. Strain component e_{zz} (vertical direction) in the Si capping layer.

which has to be sufficiently dense near the surface to resolve the surface inversion layer. On the other hand, in the SiGe island and the underlying Si buffer layer, lower point densities can be used.

A. Structure definition

First the strained Si surface is defined. In a an initial step the set of measured xyz-data points has to be converted to a triangular surface mesh. Due to the high point density of the input data, a smoothing stage with proper element elimination is applied to these surfaces. Then the three-dimensional mesh generator Netgen [8] is used, which is well suited because of its robust mesh generation technique able to handle the high aspect ratio of gate length/width to gate oxide thickness.

To achieve the desired high mesh density in the channel region a maximum tetrahedron height for the mesh elements in the oxide region is assigned. Neighboring regions will start with these small elements growing towards the outer boundaries. With this technique an appropriate resolution in the channel region is achieved.

The gate stack consists of a 1.5 nm thick oxide and a 60 nm × 60 nm polysilicon gate. A bulk contact is attached to the Si buffer layer. Source and drain regions are 60 nm wide and approximated by analytical profiles. In Figure 6 the final structure with the simulation mesh is shown. The transistor is cut along its symmetry axis and only one half of the whole structure is analyzed.

In the Si cap layer a constant boron doping of 4×10^{18} cm^{-3} is assumed. Into the access regions from source/drain to the channel an arsenic profile with a maximum doping of 5×10^{20} cm^{-3} is implanted (Figure 7). Finally, the strain profile is interpolated to the device simulation mesh.

B. Electrical Characterization

Three-dimensional device simulations are performed using MINIMOS-NT [9]. The physical models invoked include the strain-dependent low-field mobility model described in [10] and the IMLDA quantum correction model [11]. To determine the strain-induced current enhancement, comparative simulations of the same structure are performed with and without taking the strain effect on the mobility into account.

In Figure 8 the output characteristics for the unstrained and strained device are depicted. The transfer characteristics for drain voltages of 0.05 V and 1.5 V can be seen in Figure 9. Fig. 10 shows the achieved drain current enhancement as a function of the drain voltage. The current enhancement shrinks with growing gate voltage. Due to saturation of the electron velocity this value also shrinks towards higher drain voltages. At $V_{GS} = 1.5$ V, $V_{DS} = 1.5$ V an enhancement of 15 % can be evaluated.

For comparison, the two-dimensional case shows a current enhancement of 16 % along the axis of the device. Due to the lowered strain values in the peripheral regions under the gate the achieved current enhancement will be reduced. However, as the strain variation under the gate is not too high this difference is marginal. Enlarging the gate width will degrade this situation.

VI. DISCUSSION

A similar technology which utilizes local SiGe stressors has been reported by IBM in 2006 [12]. In this case strain is induced by elastic relaxation of a Si/SiGe bilayer structure. Uniaxial strain of 0.24 % in the Si channel results in a drive current improvement of 15 % [12]. In comparison, the DOTFET process induces a more uniform strain up to 0.7 % in the Si channel. The whole device is integrated in the coherently

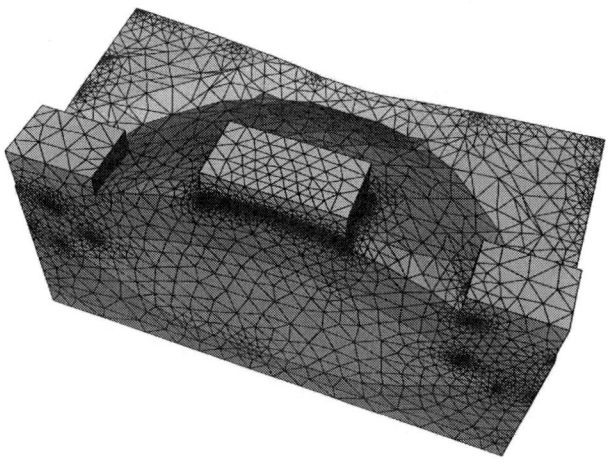

Fig. 6. The geometry of the generated transistor structure. Only one half of the transistor is simulated.

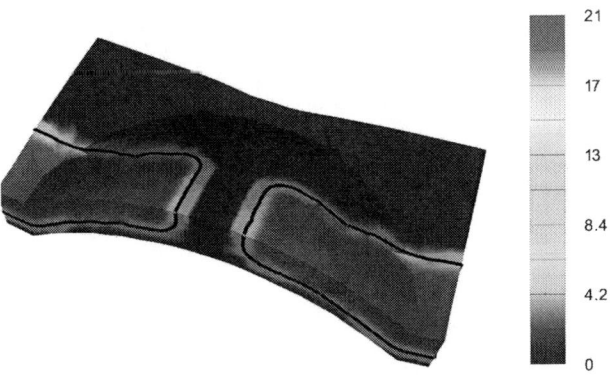

Fig. 7. The Arsenic Doping with shown pn-junction in the capping layer.

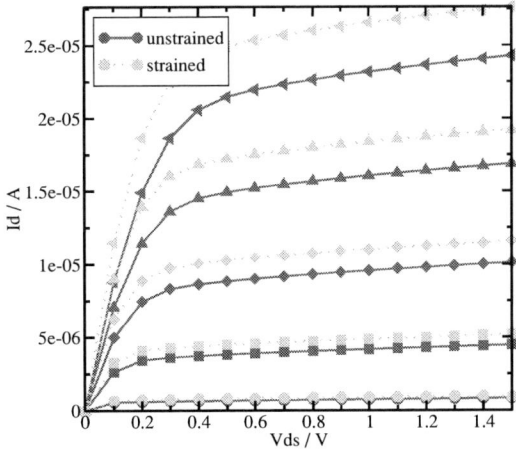

Fig. 8. The unstrained and strained output characteristics for gate voltages of 0.7 V, 0.9 V, 1.1 V, 1.3 V, and 1.5 V are shown.

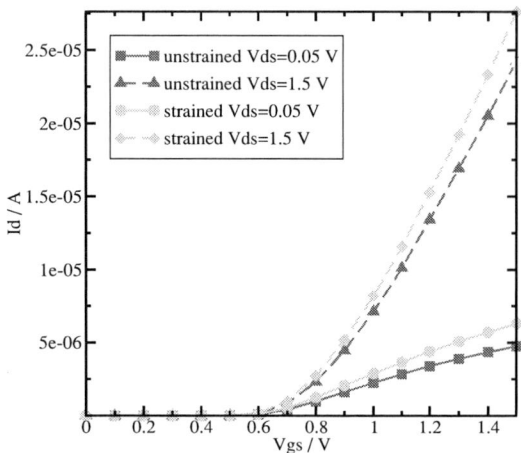

Fig. 9. The unstrained and strained transfer characteristics for drain voltages of 0.05 V and 1.5 V.

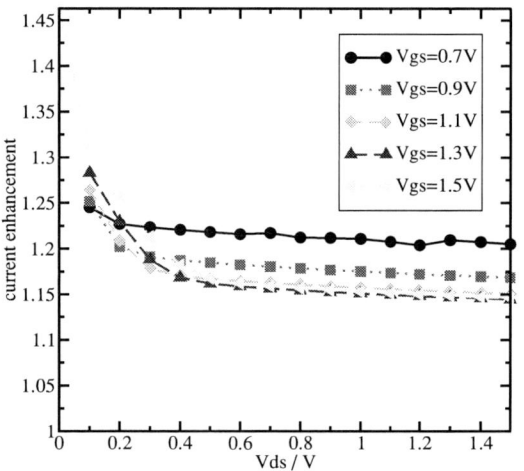

Fig. 10. The current enhancement $I_{D,strained}/I_{D,unstrained}$ for several gate voltages.

grown Si capping layer, wheres in the IBM process, source and drain are grown by selective epitaxy. Our simulations predict for the DOTFET the same drive current enhancement as has been reported for the IBM process, despite the strain in the latter is about three times lower (0.24 % versus 0.7 %). This can be partly attributed to different extraction methods for the current enhancement used in [12] and in our simulations. This comparison also indicates that the parameters used in our simulations give a conservative estimate of the current enhancement of the DOTFET.

VII. CONCLUSION

In this work the potential of the DOTFET has been studied by three-dimensional simulations. The geometry has been extracted from actually fabricated samples. The strain field is obtained by comprehensive simulations verified by strain measurements on the strained Si layer. A conservative estimate for the NMOS drive current enhancement of about 15 % is obtained.

ACKNOWLEDGMENT

This work has been supported by the EC, project No. 012150-2 (d-DOTFET) and the Austrian Science Fund FWF, projects F2507 and F2509 in the SFB 025 (IR-ON). We thank J. Moers and D. Grützmacher (FZ Jülich, Germany) for supplying the patterned Si wafers.

REFERENCES

[1] O. Schmidt and K. Eberl, "Self-assembled Ge/Si dots for faster field-effect transistors," *IEEE Transactions on Electron Devices*, vol. 48, no. 6, pp. 1175–1179, 2001.

[2] Z. Zhong and G. Bauer, "Site-controlled and size-homogeneous Ge islands on prepatterned Si (001) substrates," *Applied Physics Letters*, vol. 84, no. 1, pp. 1922–1924, 2004.

[3] J. Zhang, M. Stoffel, A. Rastelli, O. Schmidt, V. Jovanovic, L. Nanver, and G. Bauer, "SiGe growth on patterned Si(001) substrates: Surface evolution and evidence of modified island coarsening," *Applied Physics Letters*, vol. 91, p. 173115, 2007.

[4] M. Stoffel, A. Rastelli, J. Tersoff, T. Merdzhanova, and O. Schmidt, "Local equilibrium and global relaxation of strained SiGe/Si(001) layers," *Physical Review B*, vol. 74, p. 155326, 2006.

[5] N. Hrauda, J. Zhang, J.Stangl, A. Rehman-Khan, G. Bauer, M. Stoffel, O. Schmidt, V. Jovanovich, and L. Nanver, "X-ray investigation of buried SiGe islands for devices with strain-enhanced mobility," *Journal of Vacuum Science and Technology B*, vol. 27, no. 2, pp. 912–918, 2009.

[6] E. Bonera, F. Pezzoli, A. Picco, G. Vastola, M. Stoffel, E. Grilli, M. Guzzi, A. Rastelli, O. Schmidt, and L. Miglio, "Strain in a single ultrathin silicon layer on top of SiGe islands: Raman spectroscopy and simulations," *Physical Review B*, vol. 79, p. 075321, 2009.

[7] L. Nanver, C. Biasotto, V. Jovanovic, J. Moers, D. Grützmacher, J. Zhang, G. Bauer, O. Schmidt, L. Miglio, H. Kosina, A. Marzegalli, G. Vastola, G. Mussler, N. Hrauda, J. Stangl, J. Cingel, and E. Bonera, "SiGe dots as stressor material for strained Si devices," in *Proc. 5th International SiGeTechnology and Device Meeting (ISTDM)*, May 2010.

[8] J. Schöberl, "NETGEN – Automatic Mesh Generator," *http://www.hpfem.jku.at/netgen/index.html*.

[9] *MINIMOS-NT 2.1 User's Guide*, Institut für Mikroelektronik, Technische Universität Wien, Austria, 2004.

[10] S. Dhar, H. Kosina, V. Palankovski, E. Ungersböck, and S. Selberherr, "Electron mobility model for strained-Si devices," *IEEE Transactions on Electron Devices*, vol. 52, no. 4, pp. 527–533, 2005.

[11] C. Jungemann, C. Nguyen, B. Neinhüs, S. Decker, and B. Meinerzhagen, "Improved modified local density approximation for modeling of size quantization in NMOSFETs," *Proc. Intl. Conf. Modeling and Simulation of Microsystems 2001*, vol. 1, pp. 458–461, 2001.

[12] R. Donaton, D. Chidambarrao, J. Johnson, P. Chang, Y. Liu, W. Henson, J. Holt, X. Li, J. Li, A. Domenicucci, A. Madan, K. Rim, and C. Wann, "Design and fabrication of MOSFETs with a reverse embedded SiGe (rev. e-SiGe) structure," in *IEDM Technical Digest*, 2006, pp. 465–468.

978-1-4244-6658-0/10 $26.00 © 2010 IEEE

Optimized Oxygen Annealing Process for V_{th} Tuning of p-MOSFET with High-k/Metal Gate Stacks

T. Kawanago[1], Y. Lee[1], K. Kakushima[2], P. Ahmet[1], K. Tsutsui[2], A. Nishiyama[2], N. Sugii[2], K. Natori[1], T. Hattori[1] and H. Iwai[1]

[1] Frontier Research Center, Tokyo Institute of Technology, 4259, Nagatsuta, Midori-ku, Yokohama 226-8502, Japan,
[2] Interdisciplinary Graduate School of Science and Engineering, Tokyo Institute of Technology,
Tel.: +81-45-924-5847, E-mail: kawanago.t.ab@m.titech.ac.jp

Abstract— A demonstration of V_{FB}/V_{th} tuning has been conducted by optimized annealing in oxygen ambient for direct contact of high-k with Si gate stacks. The amount of oxygen atoms has been controlled by optimized annealing temperature and the thickness of the gate electrode. The shift in V_{FB} has been confirmed irrespective of gate dielectric materials and the thickness. The V_{th} of pMOSFET can be controlled to positive direction by 520 mV without any EOT penalty. Once a shift in V_{FB}/V_{th} is obtained, the values are found to be stable even after following forming gas annealing.

I. INTRODUCTION

Continuous scaling in the equivalent oxide thickness (EOT) with high-k/metal gate stacks is crucially important to suppress not only gate leakage current but also the severe short-channel effect for highly scaled FETs[1]. Several groups have reported that an EOT below 0.5 nm can be achieved by using a structure with Hf-based oxides directly in contact with Si by careful high-k deposition and process techniques to inhibit or scavenge the SiO_x-based interfacial layer growth [2, 3]. Besides, it has been reported that a direct contact structure can also achieved using rare earth such as La_2O_3 for gate dielectrics owing to the material nature to form La-silicate at the interface and fairly nice nMOSFET operation has been demonstrated with extremely scaled EOT [4].

One the prominent features of La_2O_3 is that a lowering of the effective work function (EWF) of the gate electrode and low threshold voltage (V_{th}) nMOSFET is successfully achieved with W/La_2O_3 gate stacks as shown in **Figure 1 (a)** and **(b)**. However, for pMOSFET, the V_{th} becomes so high (to negative direction) so that one must consider an effective way to reduce and tune the V_{th}. Many studies have been reported on the V_{FB}/V_{th} modulation of high-k gate stack including Al incorporation for Hf-based oxide [5]. It has been reported that metals with high work function behave like metals of midgap after recovery annealing in forming gas ambient [6, 7] due to the oxygen deficiency in the high-k. Therefore, careful processes optimization to compensate the defects in the high-k for the tuning in V_{FB}/V_{th} without any cost in the EOT should be conducted [8, 9]. One way to compensate the oxygen deficiency is to incorporate oxygen atoms in the metal electrode as a reservoir and indeed oxygen doped W has

showed a fairly nice tuning of the V_{FB}/V_{th} by 400 mV toward the Si valence band edge [10]. This experimental result indicates the possibility of V_{FB}/V_{th} tuning not only by addition of elements into high-k but also by process such as annealing after gate electrode formation. In this paper, we experimentally investigate the effect of oxygen atoms in the gate electrode on V_{FB}/V_{th} shift. Optimization of the processes including the annealing temperature has been carefully conducted to achieve a V_{FB}/V_{th} tuning and small state density with low EOT at the same time. Finally, the impact of the process will be discussed through fabrication of pMOSFET.

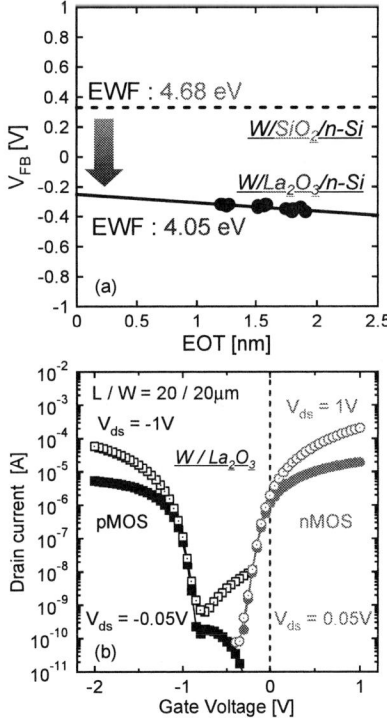

Figure 1.   (a) EOT-V_{FB} plots of $W/La_2O_3/n$-Si capacitors. The effective work function is extracted form the intercept of y-axis. (b) I_d-V_g characteristics of $W/La_2O_3/Si$ n&pMOSFETs.

This study was supported by New Energy and Industrial Technology Development Organization (NEDO), and Grant-in-Aid for research fellows of Japan Society for the Promotion of Science (JSPS).

II. DEVICE FABRICATION AND EXPERIMENT

La_2O_3 was deposited on HF-last n-Si wafer for MOS capacitors by e-beam evaporation in an ultra-high vacuum chamber, followed by *in-situ* W (tungsten) metal deposition by RF sputtering. The metal was patterned by reactive ion etching (RIE) with SF_6 chemistry to form gate electrodes. Thermally-grown SiO_2 MOS capacitors were also fabricated as a reference. The substrate impurity concentration of MOS capacitors is $3 \times 10^{15} cm^{-3}$. The samples were post-metallization annealed in forming gas ambient ($H_2:N_2$=3:97%) at 800 °C for 30min. Source and Drain pre-formed n-Si (100) substrates were also utilized to fabricate pMOSFETs with a substrate doping concentration of $3 \times 10^{16} cm^{-3}$ measured using reverse substrate bias technique [11]. Al was deposited on the source/drain region and back side of the substrate as a contact. Oxygen atmospheric annealing ($O_2:N_2$=5:95%) was applied to supply additional oxygen into gate stacks. Finally, recovery annealing (F.G.A) was performed. Process flow is summarized in **Figure 2**. EOT and V_{FB} were estimated by NCSU CVC program [12]. Split-CV method was employed to measure an effective hole mobility of pMOSFET [13].

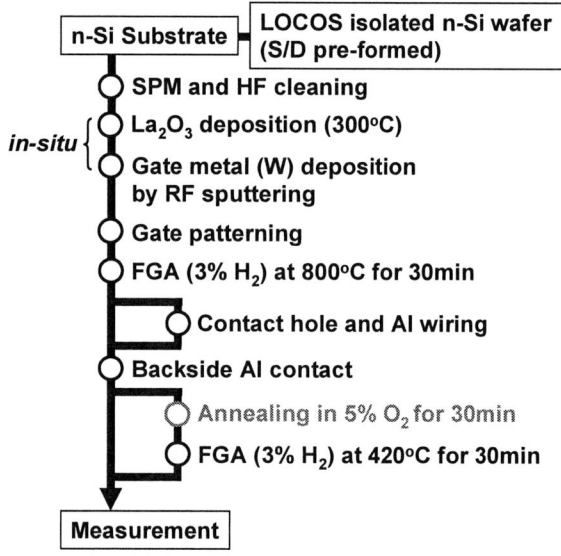

Figure 2. Fabrication flow of high-k gate MOSCAP and pMOSFET.

III. RESULTS AND DISCUSSION

Figure 3 (a) shows the C-V characteristics of W/SiO_2 MOS capacitors. The C-V curves shifted to the positive direction while increasing the annealing temperature. Compared to the report with Ar/O_2 mixture deposition in ref [10], this experimental result suggests that the oxygen atoms can be successfully introduced into the high-k layer during the oxidant ambient annealing. The positive V_{FB} shift on high-k gate stack was also confirmed as shown in **Figure 3 (b)**. The amount of positive V_{FB} shift by 440 mV at a temperature of 340 °C is slightly higher than that of the SiO_2 case. The penalty on the EOT was less than 1 Å, therefore, it can be concluded that the V_{FB} shift by oxygen annealing process can be obtained used for V_{FB} tuning in metal/high gate stack.

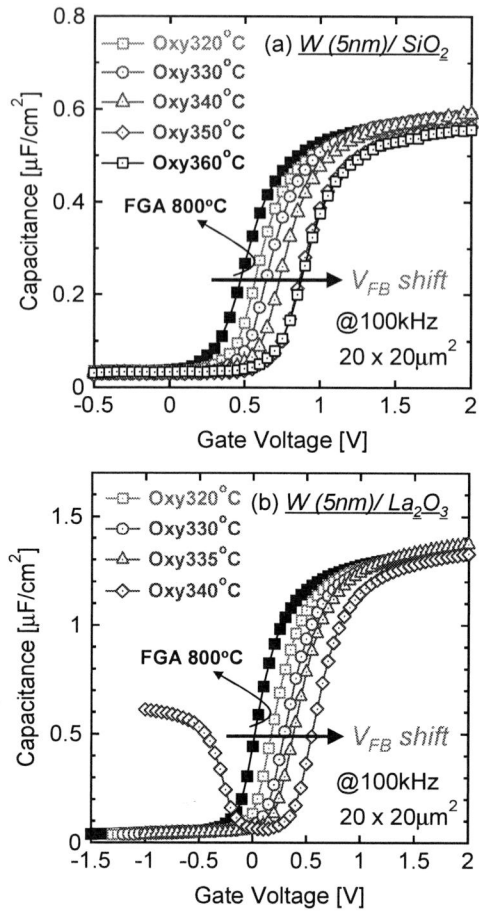

Figure 3. C-V characteristics of MOS capacitors with (a) SiO_2 and (b) high-k. Positive V_{FB} shift can be observed by increasing the annealing temperature in dilute oxygen atmospheric.

To investigate the influence of the oxygen annealing, the relationship between the V_{FB} shift and the annealing temperature is plotted in **Figure 4 (a)**. Although V_{FB} shift monotonically increases at a temperature below 350 °C, the V_{FB} shift starts to saturate above 360 °C (shown in the dashed line). This behavior can be observed in both gate dielectrics, however, the amount of increase in the V_{FB} shift on high-k is larger than that on SiO_2. **Figure 4 (b)** shows the comparison of the V_{FB} shift on SiO_2 and high-k. The slope of the experimental results is slightly larger than the ideal slope (S=1). Recent theoretical study explains that the work functions of the metals on Hf-based dielectrics [14] are determined by the atomic bonding dependent interface structure at metal/high-k interface. It is considered that these experimental results shown in **Figure 4** reflect the difference of interface structure. One of the possible origins of V_{FB} shift associated with oxygen incorporation is the potential shift at metal/dielectrics interface proposed by refs [7, 10].

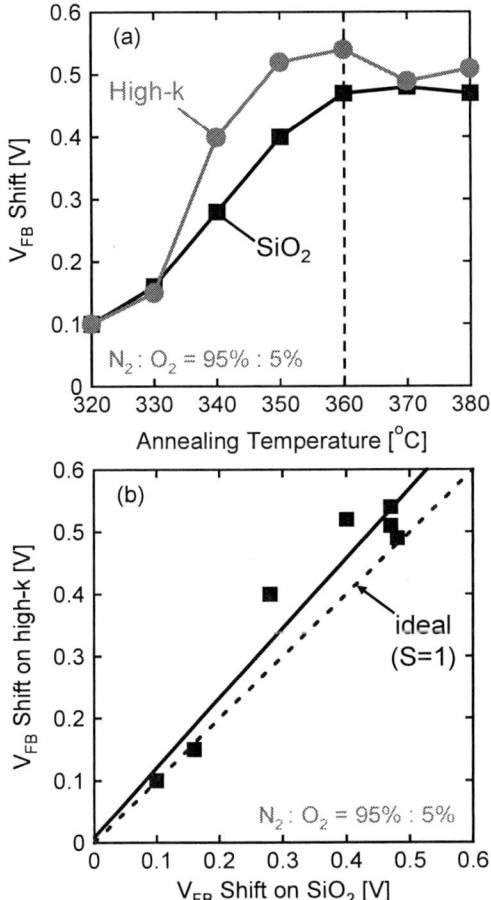

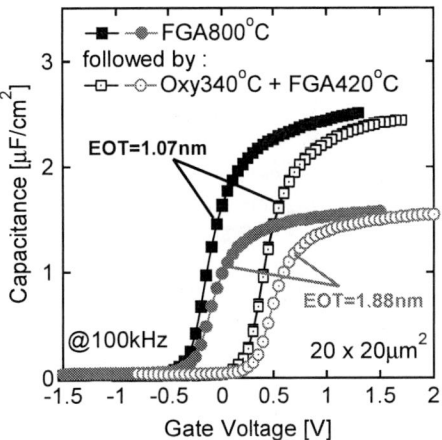

Figure 4. (a) V_{FB} shift on both SiO_2 and high-k as a function of oxygen annealing temperature. (b) Comparison of the V_{FB} shift on SiO_2 and high-k. Slope is larger than 1. (The dashed line represents the S=1)

Figure 5. C-V caracteristics with different EOT. F.G.A at 420 °C was performed after oxygen annealing. Same amout of the V_{FB} shift can be observed irrspective of EOT.

The C-V curves of MOS capacitors turn up in inversion region as shown in **Figure 3 (b)**, which implies de-passivation of dangling bonds or newly created minority carrier generation center after oxygen annealing [7]. It is well known that interface property strongly affects the inversion carrier mobility of MOSFET and F.G.A is effective to terminate the dangling bonds at SiO_2/Si interface [11]. The subsequent F.G.A showed improvement in the C-V characteristics at inversion region, while maintaining the V_{FB} at positive value even with the annealing in reductant ambient, indicating that the supplied oxygen atoms to compensate the defects in the high-k layer is preserved (**Figure 5**). Although, the V_{FB} behavior of metals with high work functions is reported to be unstable and the negative V_{FB} shift occurs after F.G.A. at temperatures below 400 °C [6, 7], our results are completely different and suggest that pMOSFET characteristics may be improved by F.G.A. with V_{th} tunning.

As the metal gate electrode is supposed to be the reservoir for the oxygen atoms, the effect of metal gate thickness on the electrical V_{FB} shift is conducted. **Figure 6** show the C-V characteristics as a function of the thickness of W gate electrode after oxidation annealing with followed by F.G.A. Contrary to the expectation, a positive V_{FB} shift can be observed with thinner W thickness. Therefore, we can speculate that the storage of oxygen atoms supplied during oxidant ambient to compensate the oxygen deficiency in the high-k is effective within the surface of W layer with a thickness of the order of 10 nm. On this account, one must prepare an oxygen reservoir adjacent to the high-k.

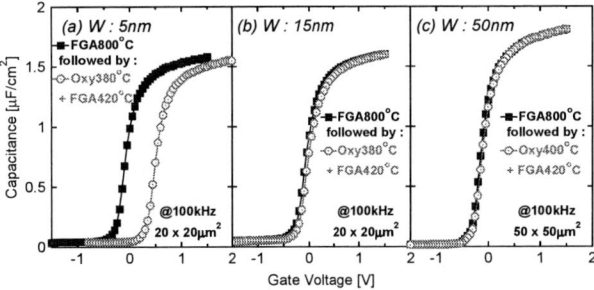

Figure 6. W thickness dependence of C-V characteristics. W thickness is (a) 5nm, (b) 15nm, and (c) 50nm, respectivery.

IV. MOSFET CHARACTERIZATION

The effect of positive shift in V_{FB} on the electrical properties, especially on effective hole mobility, was conducted with pMOSFET. **Figure 7 (a)** shows the I_d-V_g characteristics. A V_{th} lowered by 520 mV is in good agreement with the shift in the capacitors. The sub-threshold slope (SS) showed a slightly degradation, however, the F.G.A process recovers the SS. In **Figure 7 (b)**, the gate-to-channel capacitance (C_{gc}-V) curves confirm little difference in the

EOT. We also confirmed the equal amount of V_{th} shift at N_s of 5×10^{11} cm^{-2} as shown in **Figure 7 (b)**.

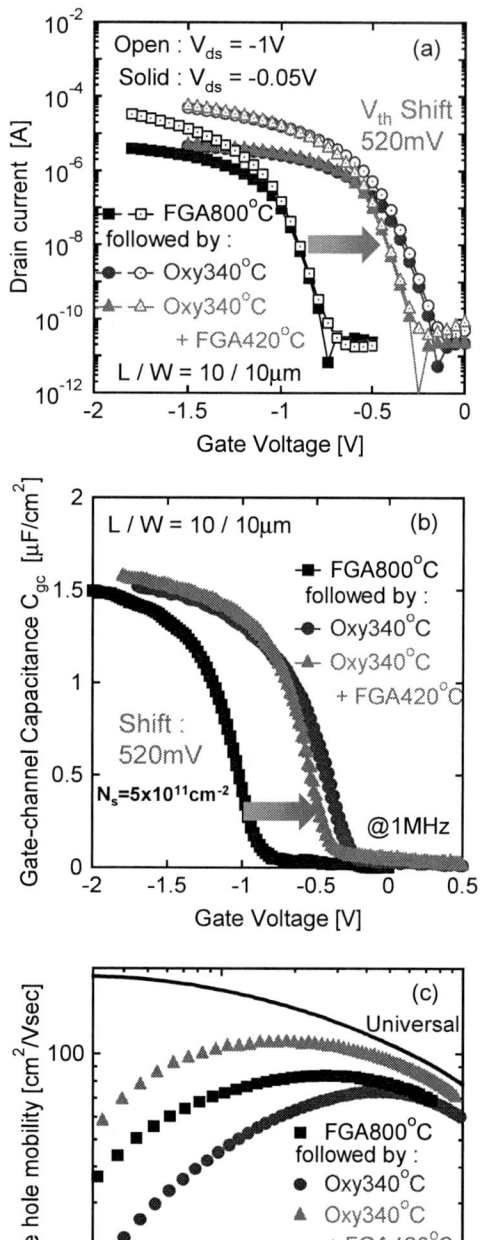

Figure 7. (a) I_d-V_g characteristics of pMOSFETs, (b) gate-to-channel capacitonce, and (c) effective hole mobility, respectively.

Figure 7 (c) shows the effect of annealing on the effective hole mobility. The reduced mobility can be clearly observed after oxygen annealing at low surface carrier region, indicating the increase in the Coulomb scattering. However, with the combination of F.G.A, the effective hole mobility showed a dramatically improvement with peak mobility of 100 cm^2/Vs.

V. CONCLUSIONS

We have experimentally demonstrated a V_{FB}/V_{th} tuning method by the incorporation of oxygen atoms into the metal gate electrode. The thickness dependent metal electrode on the V_{FB} shift has indicated that the position of the oxygen atom reservoir, favorable at adjacent to the high-k layer, is important to compensate the defect in the high-k. The electrical characteristics of pMOSFET have showed a Vth shift of 520 mV accompanied by an improvement in the mobility, indicating the compensation of defects in the high-k.

REFERENCES

[1] T. Skotnicki et al., "Innovative Materials, Devices, and CMOS Technologies for Low-Power Mobile Multimedia," IEEE Trans. Electron Devices, vol. 55, no. 1, pp. 96, January (2008).

[2] M. Takahashi et al., "Gate-first processd FUSI/HfO2/HfSiOx/Si MOSFETs with EOT=0.5 nm", in IEDM Tech. Dig., pp.523, (2007).

[3] J. Huang et al., "Gate First High-k/Metal Gate Stacks with Zero SiOx Interface Achiving EOT=0.59nm for 16nm Application," in VLSI Symp. Tech. Dig., pp. 34, (2009).

[4] K. Kakushima et al., "Further EOT scaling below 0.4 nm for high-k gated MOSFET", Technical digest of International Workshop on Dielectric Thin Films for Future ULSI Devices: Science and Technology, pp.9, (2008).

[5] T. Ando et al., "Understanding Mobility Mechanisms in Extremely Scaled HfO2 (EOT 0.42 nm) Using Remote Interfacial Layer Scavenging Technique and Vt-tuning Dipoles with Gate-First Process," in IEDM Tech. Dig., pp.423, (2009).

[6] C. Choi et al., "Quasi-damascene metal gate/high-k CMOS using oxygenation through gate electrodes," Microelectronic Engineering, 86, pp.1737, (2009).

[7] E. Cartier et al., "Role of Oxygen Vacancies in V_{FB}/V_t stability of pFET metals on HfO2," in VLSI Symp. Tech. Dig., pp. 230, (2005).

[8] E. Cartier, M. Hopstaken and M. Copel, "Oxygen passivation of vacancy defects in metal-nitride gated HfO2 /SiO2 /Si devices," Appl. Phys. Lett. 95, 042901, (2009).

[9] E. Cartier et al., "pFET V_t Control with HfO2/TiN/poly-Si Gate Stack Using a Lateral Oxygenation Process," in VLSI Symp. Tech. Dig., pp. 42, (2009).

[10] M. E. Grubbs, M. Deal, Y. Nishi and B. M. Clemens, "The Effect of Oxygen on the Work Function of Tungsten Gate Electrodes in MOS Devices," IEEE Electron Device Lett., vol. 30, no. 9, pp. 925, September (2009).

[11] Y. Taur and T. H. Ning, Fundamentals of Modern VLSI Devices. New York: Cambridge Univ. Press, 1998.

[12] J. R. Hauser and K. Ahmed, "Characterization of Ultra-Thin Oxides Using Electrical C-V and I-V Measurements," Proc. AIP Conf., pp.235, (1998).

[13] J. R. Hauser, "Extraction of Experimental Mobility Data for MOS Devices," IEEE Trans. Electron Devices, Vol. 43, no. 11, pp. 1981, (1996).

[14] K. Shiraishi et al., "Universal theory of workfunctions at metal/Hf-based high-k dielectrics interfaces-Guiding principles for gate metal selection-," in IEDM Tech. Dig., pp.43, (2005).

Experimental Analysis of Surface Roughness Scattering in FinFET devices

Jae Woo Lee[1,2], Doyoung Jang[1,2], Mireille Mouis[1], Gyu Tae Kim[2], Thomas Chiarella[3], Thomas Hoffmann[3] and Gérard Ghibaudo[2]

[1]IMEP-LAHC, INPG/MINATEC, 3 parvis Louis Néel, BP 257, 38016 Grenoble France
[2]School of Electrical engineering, Korea University, Seoul 136-701, Korea
[3] IMEC, Kapeldreef 75, B-xxx Leuven, Belgium
email: orion627@hanmail.net

Abstract—**The surface roughness scattering in n-type FinFET devices is accurately analyzed based on mobility measurements at low temperatures. Using the top/side wall current separation technique, the effective mobility for each surface has been extracted. Considering the second order mobility degradation factor, the surface roughness scattering of top and sidewall conduction has been quantitatively compared.**

I. INTRODUCTION

It is well known that bulk CMOS technologies are foreseen to face severe integration difficulties for sub 32nm node generations. The triple-gate field-effect transistor like FinFET constitutes a possible option as well as FD-SOI or planar DG-MOSFETs due to their immunity to the short channel effects (SCE) and the proximity problem in the standard bulk planar CMOS processing [1-2]. Nevertheless, the FinFET architecture could suffer from some specific technological limitations related to Fin patterning, gate stack and junction conformality, as well as spacer and source/drain engineering.

In this paper, we characterized the conduction properties of n-type FinFETs. Notably we separated each contribution of the interface depending on the orientation (the top surface and the sidewall of the fin) based on the analysis of the drain current variation with the fin width. The scattering properties at each interface were deduced from the low temperature measurements. From the second order mobility degradation factor, contributions of the surface roughness scattering are analyzed in details.

II. EXPERIMENTAL DETAILS

A. Device Fabrication

The n-channel triple-gate FinFETs were fabricated at IMEC (Leuven) on SOI wafers with 145 nm buried oxide thickness, following the process described elsewhere [3]. The channel of the transistors is undoped with background boron doping of 10^{15} cm^{-3}.

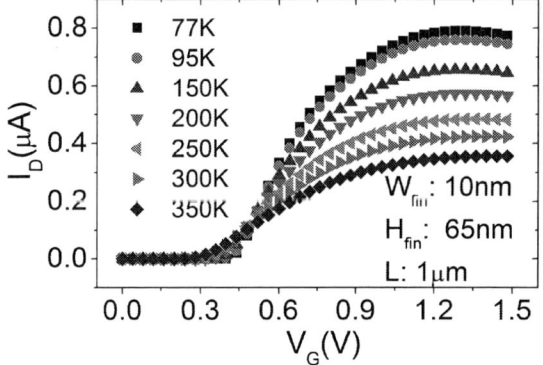

Figure 1. I_D-V_G characteristics in the various temperature with 10mV V_D.

As the gate insulator, HfSiO was deposited by metalorganic chemical vapour deposition (MOCVD) with an equivalent gate oxide thickness of 1.7 nm, whereas 5 nm thick of Physical vapor deposition (PVD) TiN, capped with 100nm poly was deposited for gate metallization. The dimensions of the devices are defined as following: the distance between the gate and the large source/drain pads is $f_s = 0.2$ μm with doping about 2×10^{20} cm^{-3} in the source/drain areas, the spacer width is about 50 nm and the extensions underneath the spacer have a doping of 5×10^{19} cm^{-3}, the fin height H_{fin} is kept constant at 65 nm, the fin width W_{fin} is varied from 10 to 1000 nm and the gate length L is fixed at 1 μm.

B. Measurements

SussMicroTec LT probe station and HP 4155A semiconductor parameter analyzer were used to measure static I_D-V_G characteristics. Every FinFET device was measured in the linear operation regime (V_D=10mV) and on the wide temperature range 77K-350K.

978-1-4244-6658-0/10 $26.00 © 2010 IEEE

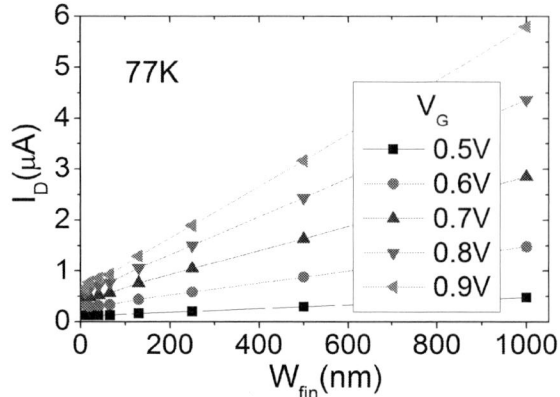

Figure 2. I_D-W_{fin} for the top/sidewall current separation at 77K.

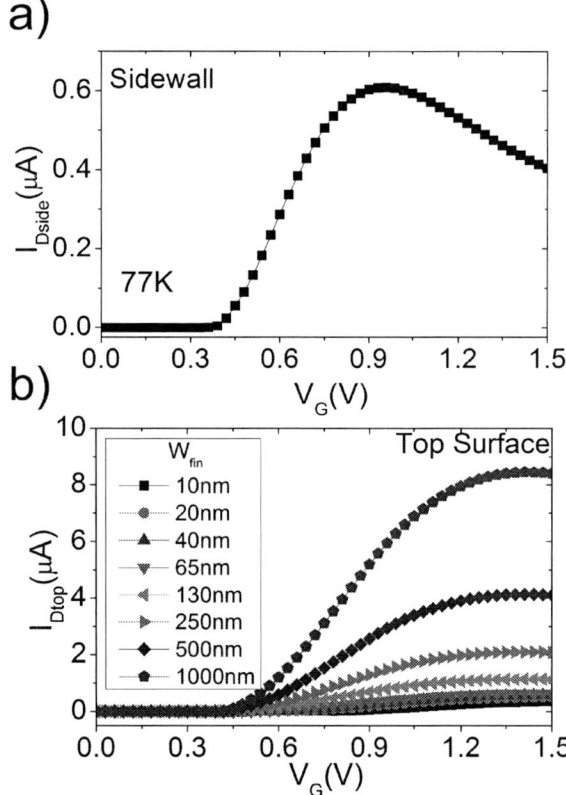

Figure 3. Separated drain current of a) side wall and b) top surface at 77K.

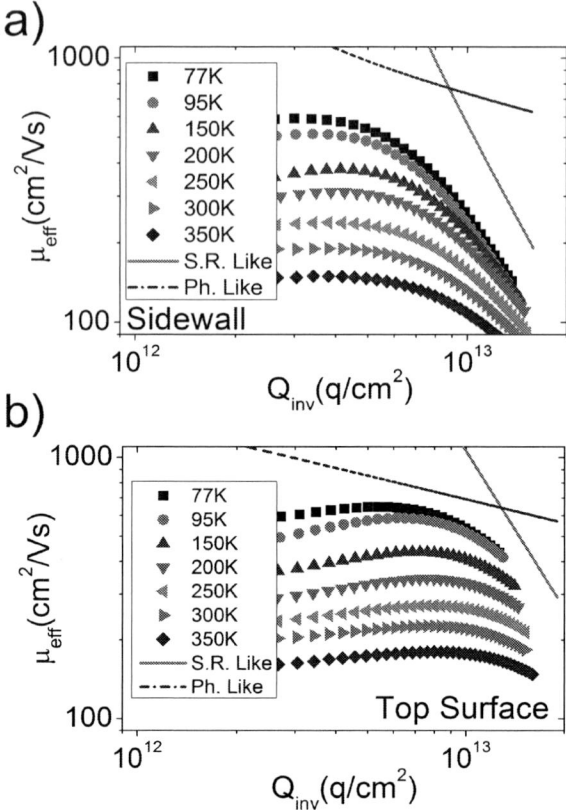

Figure 4. Mobility characteristics of each separated surface at various temparatures. a) Sidewall and b) top surface of 130nm W_{fin}. Lines in figures indicate suface roughness limited mobility (red line) and phonon scattering limited (blue line-dot-line).

III. RESULTS AND DUSCUSSION

Figure 1 shows the I_D-V_G characteristics obtained for given fin width (W_{fin}=10nm) and fin height (H_{fin}=65nm) with temperature variation.

For this analysis, a long gate length (L=1µm) was selected to avoid short channel effects.

At low temperatures, the current level increases due to the reduction of lattice scattering [4-5]. However, drain current degradations appear at high electric field (V_G >1.2V) because the surface roughness scattering becomes dominant for the effective mobility [6-7].

To analyze the effective mobility μ_{eff}, the standard model of the drain current I_D in linear regime is used [8]:

$$\mu_{eff} = \frac{L}{W C_{ox}(V_G - V_{th})V_D} I \; , \tag{1}$$

where W is the effective channel width (W_{fin}+2H_{fin}), C_{ox} the gate oxide capacitance, V_{th} the threshold voltage.

To study the surface roughness scattering of each wall, the top/sidewall current separation technique was carried out [9]. The drain currents were plotted versus W_{fin} for each V_G in Fig. 2. According to the good linearity of the I_D as a function of W_{fin}, the sidewall current I_{Dside}, corresponding to the 2xH_{fin} width, can be extrapolated at W_{fin}=0. Then, the top surface current is easily calculated as:

$$I_{Dtop} = I_D - I_{Dside} \; . \tag{2}$$

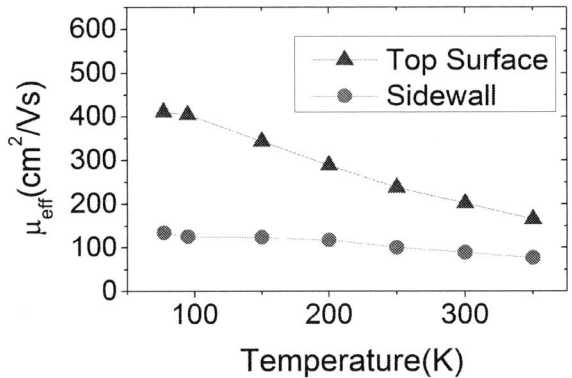

Figure 5. Temperature variation of μ_{eff} for side wall and top surface at 77K ($Q_{inv} = 1.3 \times 10^{13}$ q/cm^2).

Figure 3 shows the separated current of top/sidewall. Note that this degradation indicates a huge surface roughness for the side wall (Fig. 3 a). It is not due to self heating effect because of the low V_D.

The side wall current dramatically decrease at high electric field comparing to the total current (Fig. 1) and top surface current (Fig. 3 b).

The effective mobilities at the sidewall and the top surface calculated by using Eq. 1 are shown in Fig. 4. The inversion charge density was calculated by $Q_{inv} \approx C_{ox}(V_G-V_{th})$.

It can be noted that the mobility on the side wall decreases dramatically at high Q_{inv}. The effective mobility of the inversion layer usually consists of three dominant scattering mechanisms with Coulomb centres, phonons and surface roughness. All three mechanisms show a different dependence with the effective field and correspondent with Q_{inv}. Especially, the mobility derived from the surface roughness scattering decreases with the effective electric field E_{eff} as [10-12]:

$$\mu_{SR} \propto \frac{E_{eff}^{-2}}{\Delta^2 \lambda^2} . \quad (3)$$

where Δ is the rms value of the surface roughness and λ its correlation length. Note also that the slope of the log-log plot of the surface roughness scattering mobility follows -2 power law in agreement with Eq. (3) and well matches at strong electric field (high Q_{inv} in Fig. 4 a). On the top surface, however, the effective mobility decreases slowly with the increase of Q_{inv}, indicating that surface roughness is not so efficient. It means that the other scattering mechanisms dominate the mobility.

Moreover, note that the pure surface roughness scattering mobility does not depend on the temperature as in Eq. (3). Fig. 5 shows the temperature dependence of the effective mobility at each interface for fixed Q_{inv} values of 1.3×10^{13} q/cm^2. The effective mobility degradation at the top surface as a function of temperature is observed. In contrast, in the case of the sidewall, the effective mobility shows constant behaviour, which means that the surface roughness scattering is dominant

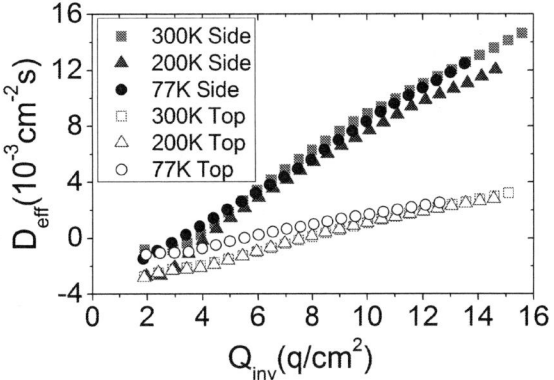

Figure 6. Derivative of reciprocal effective mobility D_{eff} of the top surface and sidewall versus effective charge density.

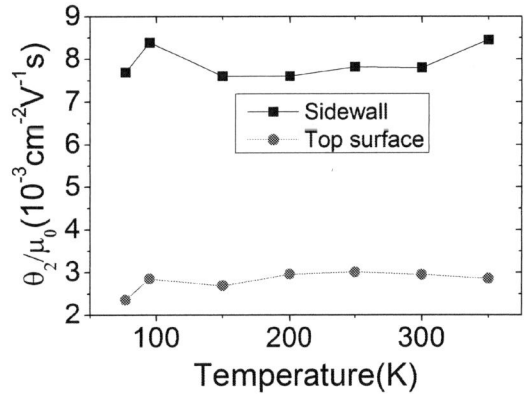

Figure 7. SR mobility amplitude θ_2/μ_0 of sidewall and top surface as a function of Temparature.

for the sidewall conduction, whereas several scattering mechanisms are simultaneously occurring at the top interface.

The effective mobility can be approximated as [7, 13]:

$$\mu_{eff} = \frac{\mu_0}{1 + \theta_1(V_G - V_{th}) + \theta_2(V_G - V_{th})^2} , \quad (4)$$

where μ_0 is the low field mobility, θ_1 and θ_2 mobility degradation parameters. According to the Eq. (3), the θ_2 correlates to the surface roughness scattering.

This model can be investigated from the derivative of the reciprocal effective mobility D_{eff} [8]:

$$D_{eff} = \frac{d(1/\mu_{eff})}{dV_G} = \frac{\theta_1 + 2\theta_2(V_G - V_{th})}{\mu_0} . \quad (5)$$

Figure 6 shows the variations of the derivative D_{eff} at the top surface and the sidewall with effective charge density. One should note the good linearity of the $D_{eff}(V_G)$ characteristics, emphasizing the validity of the mobility law of Eq. 4. As can be seen from Fig. 6, θ_1 values, deduced from y-axis intercept,

978-1-4244-6658-0/10 $26.00 © 2010 IEEE

are found to be slightly negative. This feature, which corresponds to the increase of mobility (Fig. 4), is indicative of a Coulomb scattering contribution at low field. The SR mobility amplitude θ_2/μ_0 can be extracted from the slope of D_{eff} vs V_G plot. Figure 7 gives the variations of θ_2/μ_0 with temperature. As can be seen, the SR mobility term θ_2/μ_0 is found nearly independent of temperature, inferring the SR scattering rate extraction.

At 77K, θ_2/μ_0 is around $7.7 \times 10^{-3} cm^{-2}V^{-1}s$ for the sidewall and $2.4 \times 10^{-3} cm^{-2}V^{-1}s$ for the top surface. The surface roughness scattering is therefore about 3 times higher on the sidewall surface than on the top interface. From Eq. 3, it can be stated that the geometrical factor $\Delta \cdot \lambda$ of the sidewall is 1.7 times higher than for the top surface.

IV. Conclusion

In this paper, a full separation of top surface versus sidewall conduction in n-type FinFETs has been performed, for the first time, as a function of temperature. The effective mobility associated to each surface has thus been extracted as a function of inversion charge for different temperatures. This allowed us to separately calculate the surface roughness mobility amplitudes for each surface, demonstrating that the sidewall SR scattering rate is about 3 times higher than for the top surface. The enhanced scattering on the side wall should correspond to a 1.7 increase of the rms roughness amplitude, which can be attributed to sidewall surface roughness induced by the etching process during fin patterning. The improvement of the etching process would allow taking full advantage of ultra thin FinFET architectures in the next CMOS generations.

Acknowledgment

This work has been supported by the Nanoscience foundation, Grenoble, France, by European Network of Excellence FP6/IST/NoE/Nanosil under contract no. 216171, by the National Research Foundation (NRF, 2009-0083380), by the World Class University program (WCU, R322009000100820) and (NRF, Nano 2009-0082826).

References

[1] X. Sun, Q. Lu, V. Moroz, H. Takeuchi, G. Gebara, J. Wetzel, S. Ikeda, C. Shin, and T. Liu, "Tri-gate bulk MOSFET design for CMOS scaling to the end of the roadmap," IEEE Electron Device Lett., vol. 29, pp. 491-493, 2008.

[2] J. P. Colinge, "Multi-gate SOI MOSFETs," Microelectron. Eng., vol. 84, pp. 2071-2076, 2007.

[3] N. Collaert, M. Demand, I. Ferain, J. Lisoni, R. Singanamalla, P. Zimmerman, Y.-S. Yim, T. Schram, G. Mannaert, M. Goodwin, et al., "Tall triple gate devices with TiN/HfO2 gate stack", VLSI Symp. Tech. Dig., p. 108-109, 2005.

[4] D. S. Jeon and D. E. Burk, "MOSFET ELECTRON INVERSION LAYER MOBILITIES - A PHYSICALLY BASED SEMI-EMPIRICAL MODEL FOR A WIDE TEMPERATURE-RANGE," IEEE Trans. Electron Devices, vol. 36, pp. 1456-1463, 1989.

[5] B. Cheng and J. Woo, "Measurement and modeling of the n-channel and p-channel MOSFET's inversion layer mobility at room and low temperature operation," J. Phys. IV, vol. 6, pp. 43-47, 1996.

[6] A. Hartstein, A. B. Fowler, and M. Albert, "TEMPERATURE-DEPENDENCE OF SCATTERING IN THE INVERSION LAYER," Surf. Sci., vol. 98, pp. 181-190, 1980.

[7] T. Ong, P. Ko, and C. Hu, "50-A gate-oxide MOSFET's at 77 K," IEEE Trans. Electron Devices, vol. 34, pp. 2129-2135, 1987.

[8] C. Dupre, T. Ernst, S. Borel, Y. Morand, S. Descombes, B. Guillaumot, X. Garros, S. Becu, X. Mescot, G. Ghibaudo, and S. Deleonibus, "Impact of isotropic plasma etching on channel Si surface roughness measured by AFM and on NMOS inversion layer mobility," Proc. ULIS. 2008., pp. 133-136, 2008.

[9] K. Bennamane, T. Boutchacha, G. Ghibaudo, M. Mouis, and N. Collaert, "DC and low frequency noise characterization of FinFET devices," Solid State Electron., vol. 53, pp. 1263-1267, 2009.

[10] S. Takagi, A. Toriumi, M. Iwase, and H. Tango, "On the universality of inversion layer mobility in Si MOSFET's: Part I-effects of substrate impurity concentration," IEEE Trans. Electron Devices, vol. 41, pp. 2357-2362, 1994.

[11] G. Reichert and T. Ouisse, "Relationship between empirical and theoretical mobility models in silicon inversion layers," IEEE Trans. Electron Devices, vol. 43, pp. 1394-1398, 1996.

[12] F. Gamiz and J. B. Roldan, "Scattering of electrons in silicon inversion layers by remote surface roughness," J. Appl. Phys., vol. 94, pp. 392-399, 2003.

[13] P. K. McLarty, S.Cristoloveanu, O. Faynot, V. Misra, J. R. Hauser, and J. J. Wortman, "A SIMPLE PARAMETER EXTRACTION METHOD FOR ULTRA-THIN OXIDE MOSFETS," Solid State Electron., vol. 38, pp. 1175-1177, 1995.

Parameter Extraction of Nano-Scale MOSFETs Using Modified Y Function Method

Subramanian N, G. Ghibaudo, M. Mouis

IMEP-LAHC, MINATEC, Grenoble-INP, 3 Parvis Louis Néel

38000 Grenoble, France

subraman@minatec.inpg.fr, ghibaudo@minatec.inpg.fr

mouis@minatec.inpg.fr

Abstract—This paper presents a new modified Y function method for parameter extraction of nano scale MOSFETs, taking into account the gate voltage dependent source-drain series resistance R_{SD} in the linear operation regime. Two approaches have been considered. The first one explores Y function without any prior assumption on $R_{SD}(V_g)$ while the second one assumes a linear $R_{SD}(V_g)$ for simpler extraction. Both methods were applied to real devices and successfully extracted the parameters.

I. INTRODUCTION

As MOSFET channel length is reduced, the influence of series source/drain (R_{SD}) resistance and the error induced by the same in the extraction of other MOSFET parameters becomes more and more important and has to be properly addressed. The Shift and Ratio method and others [1-3] cannot be used now, since the assumption of constant mobility with channel length is no longer valid.

Constant mobility method [4], which does not need prior information on gate oxide capacitance C_{ox}, effective channel length L_{eff}, and channel width W, takes advantage of the universality of mobility at high electric field to extract a constant R_{SD} neglecting its gate voltage (V_g) dependence. The total resistance R_{tot} vs $1/\beta$ method [5] does not need any prior knowledge of L_{eff} and mobility-L_{eff} dependence. The R_{tot} is defined as the ratio V_d/I_{DS} in the linear regime and β is same as in Eq 1. This method, however, does not take into account the threshold voltage (V_t)-L_{eff} dependence and also limits the extraction up to linear $R_{SD}(V_g)$ relationship. The extraction also depends on the Mc Larty method [6] for extracting β, which requires a second derivative function (which takes into account the effect of R_{SD} and corrects up to linear $R_{SD}(V_g)$). This, however, adds additional numerical error to the extraction process.

This paper presents a new method to extract the gate voltage dependent R_{SD} and other MOSFET parameters. This is based on Y-Function method [7, 11], which is modified here to accommodate the $R_{SD}(V_g)$. The new method takes into account the L_{eff} dependence of V_t and mobility and does not need any prior knowledge of C_{ox}, L_{eff}, and W for R_{SD} extraction. Two methods were explored and are explained in the next two sections.

This work has been partially supported by French national project NANO2012 and by European Network of Excellence FP6/IST/NoE/Nanosil under contract no. 216171.

II. METHOD 1

In this method, no assumption is made on R_{SD}–V_g relationship. Therefore it is a very general extraction procedure. The I_{DS} model used for linear regime (small V_d, $V_g > V_t$) is given in Eq 1 [8, 10].

$$I_{DS} = \frac{\beta V_{gt} V_d}{1 + \left(\Theta_{10} + \beta R_{SD}(V_g)\right)V_{gt} + \Theta_2 V_{gt}^2}, \beta = \frac{\mu_0 C_{ox} W}{L_{eff}}, V_{gt} = V_g - V_t \quad (1)$$

Where V_{gt} is the gate over drive, V_t is the threshold voltage and μ_0 is the low field mobility. The Θ_{10}, Θ_2 are the first order and second order mobility attenuation factors. The $\Theta_{10} + \beta R_{SD}(V_g)$ term is generally grouped together as Θ_1. The function, involving Θ_1 and Θ_2, models the mobility attenuation with gate over drive. Generally Θ_{10} and Θ_2 have to be re-interpreted to obtain physically meaningful parameters (surface roughness, phonon scattering etc) [10]. This is however outside the scope of this paper. The Y-function is defined as $I_{DS}/\sqrt{gm}$ and it is easy to show from Eq (1) that the measured Y-function Y_{mes} is,

$$Y_{mes} = \frac{I_{DS}}{\sqrt{gm}} = \frac{\sqrt{\beta V_d} V_{gt}}{\sqrt{1 - (\Theta_2 + R'_{SD}\beta)V_{gt}^2}} \quad (2)$$

The prime in the Eq 2 for R_{SD} term denotes derivation with respect to V_g. Therefore we can build a new function Y_0, the Y-function without the effect of Θ_2 and R'_{SD}, given by

$$Y_0 = Y_{mes}\sqrt{1 - (\Theta_2 + R'_{SD}\beta)V_{gt}^2},$$

The slope of the Y_0 vs V_g plot gives β and the x-axis intercept gives V_t. The denominator of Eq 2 is found from the following equation,

$$\Theta_{eff}(V_g) \equiv \frac{\beta V_d}{I_{DS}} - \frac{1}{V_g - V_t} = \Theta_{10} + R_{SD}(V_g)\beta + \Theta_2 V_{gt}$$

Therefore

$$R'_{SD}(V_g)\beta + \Theta_2 = \Theta'_{eff}(V_g) \quad (3)$$

$$\Theta_{10} + R_{SD}(V_g)\beta = \Theta_{eff}(V_g) - \Theta_2 V_{gt} \quad (4)$$

The extraction procedure is shown in Fig 1. The terms β

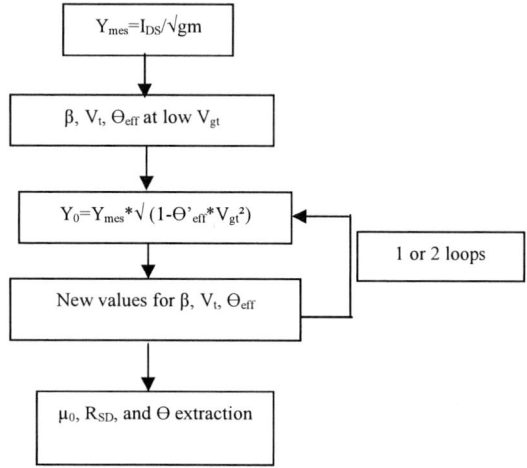

Figure 1. The Flow chart showing the Y – function correction steps

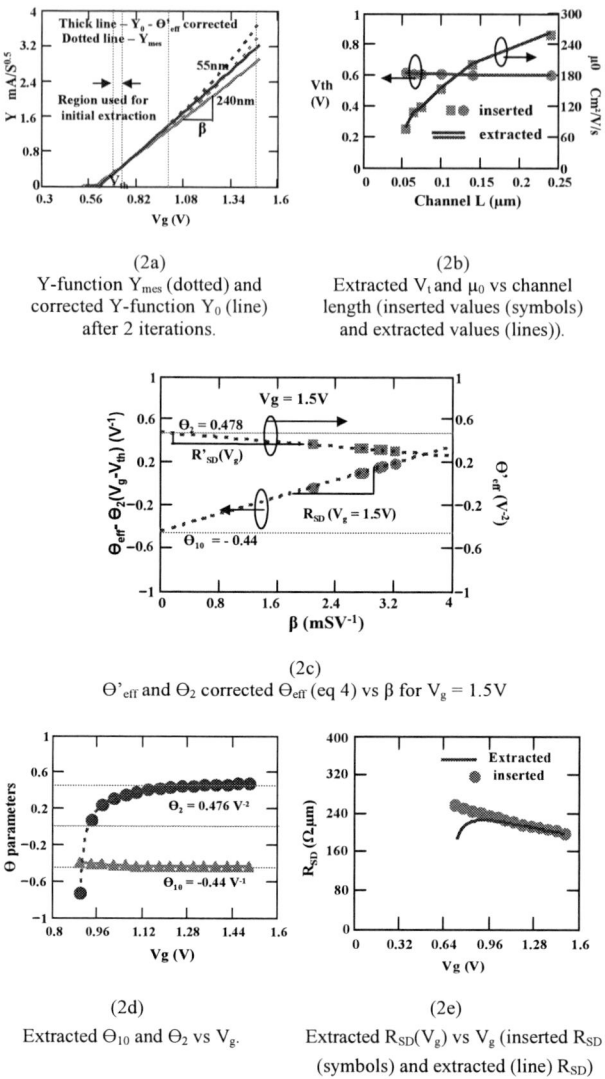

(2a) Y-function Y_{mes} (dotted) and corrected Y-function Y_0 (line) after 2 iterations.

(2b) Extracted V_t and μ_0 vs channel length (inserted values (symbols) and extracted values (lines)).

(2c) Θ'_{eff} and Θ_2 corrected Θ_{eff} (eq 4) vs β for $V_g = 1.5V$

(2d) Extracted Θ_{10} and Θ_2 vs V_g.

(2e) Extracted $R_{SD}(V_g)$ vs V_g (inserted R_{SD} (symbols) and extracted (line) R_{SD})

Figure 2. Extraction results using Method 1

and V_t are extracted from Y_{mes} at low V_{gt} where the high field effect is negligible. The region for extracting the initial values from Y_{mes} for the iteration in a real device must be carefully chosen. The region should not be too close to the threshold voltage while keeping the criteria of low V_{gt}. The β and V_t are then used to calculate $\Theta_{eff}(V_g)$. The Θ'_{eff} (Eq 3) is used to correct the Y function, Y_{mes}, for large V_g and deduce Y_0. The terms β, V_t and $\Theta_{eff}(V_g)$ are re-calculated from the corrected region of the Y function. After one or two iteration steps as shown in Fig 1, a good linearity is achieved for $Y_0(V_g)$ from which the final values of β, V_t and Θ_{eff} are finally obtained. This has to be done for different channel lengths. From the Θ'_{eff} vs β plot for different V_g, we get Θ_2 (Eq 3) from the y-axis intercept and R'_{SD} from the slope. The mean Θ_2 from the plateau of Θ_2 - V_g plot is used to correct Θ_{eff} for each gate voltage (Eq 4) and by plotting the RHS of Eq 4 vs β for each V_g, Θ_{10} (y-axis intercept) and $R_{SD}(V_g)$ (slope) are extracted. The plateau for Θ_{10} vs V_g plot is taken as the final value of Θ_{10}. See Fig 2c, 2d and 2e.

In order to validate the extraction procedure, I_{DS}-V_g characteristics were simulated using Eq 1 for different channel lengths. I_{DS} was assumed to be exponentially small for $V_g<V_t$. The values of V_t, μ_0, Θ_{10} and Θ_2 were arbitrarily chosen. R_{SD} was simulated according to Eq 5 and the values of 'R_0' and 'α' were arbitrarily chosen (350 and 0.5 respectively in this example).

$$R_{SD} = \frac{R_0}{1 + \alpha V_g} \quad (5)$$

The extraction method should be able to recover the values of the parameters with acceptable error. The Y functions, Y_{mes} and Y_0 are shown in Fig 2a. It is to be noted that here, only 2 iterations were needed before obtaining a notably linear Y-function Y_0. The final values of β and V_t were then extracted from the linear portion of Y_0. The initial values of V_t and μ_0 were faithfully regained as shown in Fig 2b. Knowing β and V_t, Θ_{eff} and Θ'_{eff} were calculated. The Fig 2d shows the extraction result for Θ_2 and Θ_{10}. We can also see that Θ_{10} and Θ_2 reach a plateau for high V_g. The mean values for Θ_{10} and

Θ_2 were -0.44 and 0.476 compared to the initial values -0.5 and 0.5. As seen from Fig 2e, R_{SD} has been successfully extracted for high V_g. Therefore all the parameters were extracted with very good agreement with the initial values. Even though this is a simulation exercise, there still remains an ambiguity near V_t (due to numerical calculations). This limits the extraction to high V_g. In a real device also we can extract only in the high inversion regime where this model is valid and therefore this simulation gives a good image of how the extraction is going to behave for a real device. Section 4 will discuss the results obtained on real sample devices using this method. The first method can be simplified by assuming a simple R_{SD}-V_g relationship. We can extract a quadratic fit to R_{SD}-V_g of the form $A+BV_g+CV_g^2$. In this case the $R'_{SD}(V_g)$ function extracted from the Θ'_{eff} vs β plot for different V_g will give the terms 'B' and 'C' (by plotting $R'_{SD}(V_g)$ vs V_g) and can be used with $R_{SD}(V_g)$ to extract the term 'A'. In section 3, a second method is explained where a linear R_{SD}-V_g is assumed for simpler extraction.

III. METHOD 2

In this method R'_{SD} is assumed constant. The iteration proceeds by taking a mean value of Θ'_{eff} in strong inversion. The governing equations for extracting R_{SD}, Θ are as given below.

$$\Theta_{eff}(V_g) = \Theta_{10} + R_0\beta + R'_{SD}V_g\beta + \Theta_2 V_{gt} \qquad (6)$$

$$\Theta'_{eff}(V_g) = R'_{SD}\beta + \Theta_2 \qquad (7)$$

$$\Theta_{10} + R_0\beta = \Theta_{eff}(V_g) - R'_{SD}V_g\beta - \Theta_2 V_{gt} \qquad (8)$$

We can also follow method 1 and use the same procedure as for quadratic fit explained in the previous section (only here, a mean R'_{SD} will be taken). However by taking a constant Θ'_{eff}, the loop operation and further extraction steps become simpler.

The result of the extraction on the same simulated device is shown in Fig 3. Figure 3a gives the extracted V_t and μ_0. Figure 3c shows the Θ'_{eff} vs β relationship, from which Θ_2 and R'_{SD} are extracted from y-axis intercept and slope respectively. Equation 8 was then used to extract Θ_{10} and R_0 from the y-axis intercept and slope from the plot of RHS vs β (Fig 3c). The result for R_{SD} using R_0 and R'_{SD} is shown in Fig 3b. The extracted value for Θ_{10} and Θ_2 were -0.424 and 0.483 respectively against the initial values -0.5 and 0.5. All the parameters were successfully extracted with good agreement with their initial values. However the extracted R_{SD} is an approximate function compared to the real R_{SD}. The error in the extraction of parameters is then dependent on how far the linear approximation is valid.

IV. EXTRACTION ON REAL DEVICES

Extraction results for real devices are shown in Fig 4. Devices used were Bulk+/Fully Depleted Silicon on Nothing devices [9 and references there in] provided by ST Microelectronics with channel length varying from 240nm to 55nm and 1µm wide. The result for Θ'_{eff} vs V_g is shown in Fig 4a. The slope for Θ'_{eff} at high V_g shows that R'_{SD} is not a constant (Eq 3) and steeper slope for 55nm curve shows that it is proportional to β (Eq 3). For method 1 Θ'_{eff} was used as such while for method 2, a mean value from high V_g plateau was taken. Here for both methods only 2 iterations were needed. Figure 4c shows Θ'_{eff} vs β and RHS of Eq 4 vs β at V_g 1.5V showing a good correlation for both functions with β. Figure 4b is the RHS of Eq 8 vs V_g for method 2 after extracting Θ_2 and R'_{SD} from Θ'_{eff} vs β shown in Fig 4d. As can be seen, the function is fairly constant with V_g. It should also be noted here that the function is larger for shorter channel length showing that it is proportional to β. This can be seen from Fig 4d which shows a good correlation for this function with β. Therefore the linear approximation for R_{SD} holds good in this case. Figure 4g gives Θ_{10} and Θ_2 extracted using method 1. At high V_g both parameters are reaching a plateau. The value of Θ_{10} is -0.403 from method 1 and -0.346 from method 2. The value of Θ_2 is 0.51 from method 1 and

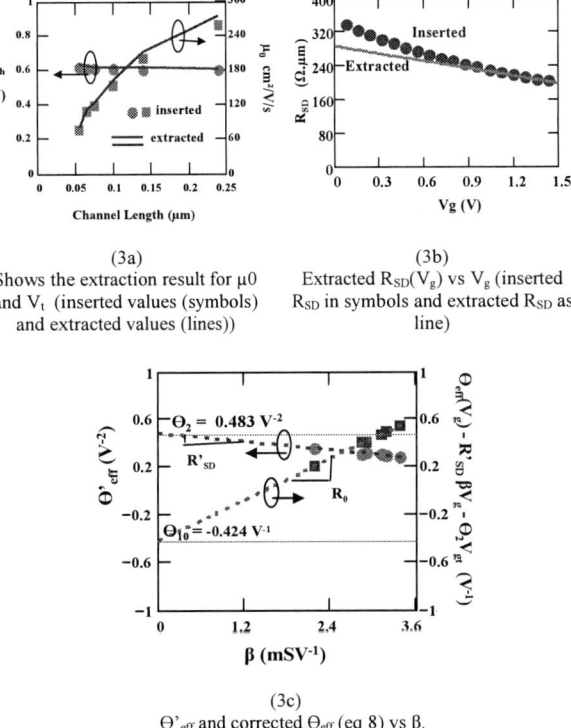

(3a)
Shows the extraction result for $\mu 0$ and V_t (inserted values (symbols) and extracted values (lines))

(3b)
Extracted $R_{SD}(V_g)$ vs V_g (inserted R_{SD} in symbols and extracted R_{SD} as line)

(3c)
Θ'_{eff} and corrected Θ_{eff} (eq 8) vs β.

Figure 3. Extraction results using method 2.

0.58 from method 2. A comparison of the final results obtained using methods 1 and 2 is given for V_t, μ_0, and R_{SD} in Fig 4e, f and h respectively. The V_t and μ_0 values obtained from methods 1 and 2 show a good agreement. Comparison of R_{SD} curves shows that R_{SD} is not exactly linear against V_g but can still be linearly approximated without much error, in this case, in spite of Fig 4a. This is also consistent with Fig 4b and 4d. Figures 4i and 4j show the extracted model fit for I_{DS} (symbols) vs measured I_{DS} (lines) for long channel (240nm) and short channel (55nm) devices using methods 1 and 2. In this case method 2 also gives a good approximation to I_{DS}-V_g. The excellent fit obtained (above 0.9V) shows the validity of the extraction procedure for advanced MOSFET devices.

V. CONCLUSIONS

We have presented a new Y-function based parameter extraction method for nano-scaled MOSFETs which takes into account the V_g dependence of R_{SD}. The extraction of parameters μ_0, V_t, $R_{SD}(V_g)$ and Θ has been carried out on simulated MOSFET devices with an arbitrary R_{SD} function in order to validate the theory. The methodology was successfully validated on simulated data. Both methods were also applied on experimental data of real nano scale devices and successfully extracted the parameters, proving their validity for real devices. This method is more accurate compared to others [1-5] as it takes into account the L_{eff} dependence of mobility and threshold voltage and doesn't need values of C_{ox}, W, L_{eff} or any prior assumption on R_{SD}-V_g relationship for the extraction.

REFERENCES

[1] Y.Taur et.al "New shift and ratio method for MOSFET Channel-Length Extraction", IEEE EDL v13, n5, p267, 1992.

[2] William P.N. Chen et.al, "A New series Resistance and Mobility Extraction Method by BSIM Model for Nano-Scale MOSFETs" VLSI TSA, pp 143-144, 2006

[3] Clifford Y.Hwang, Tsung-Chia, and Jason C. S. Woo "Extraction of Gate Dependent Source/Drain Resistance and Effective Channel Length in MOS Device at 77K" IEEE TED, vol42, n10, p1863, 1995

[4] Da-Wen Lin, M.L. Cheng, S.W. Wang, C.C. Wu and M.J. Chen "A Constant-Mobility Method to Enable MOSFET Series-Resistance Extraction" IEEE EDL, vol28, n12, p1132, 2007

[5] D.Fleury, A.Cros, G.Bidal, J Rosa and G.Ghibaudo "A New technique to Exact the Source/Drain Series Resistance of MOSFETs" IEEE EDL vol32, n9, p975, 2009

[6] P K McLarty, S. Cristoloveanu, O Faynot, V Misra, J R Hauser and J J Wortman "A Simple Parameter Extraction Method for Ultra Thin Oxide MOSFETs" Sol Stat Elec, vol 38, n6, p1175, 1995

[7] G.Ghibaudo, "New Method for the extraction of MOSFET Parameters" Elec Letters, vol 24 n9, p543 (1988)

[8] Tong-Chern Ong, Ping K Ko, and Chenming Hu, "50 A° Gate-Oxide MOSFET's at 77K" IEEE TED, vol34, n10, 1987

[9] F.Boeuf et al, "Optimization of Bulk+/SON Integration for Low stand-by Power (LstP) Application" Proc.SSDM (Ext Abs), p1030, 2009

[10] D. Fleury, A. Cros, H. Brut and G. Ghibaudo " New Y-Function Based Methodology for Accurate Extraction of electrical Parameters on Nano-Scaled MOSFETs" ICMTS, p160-165, 2008

[11] C.Mourrain, B.Cretu, G.Ghibaudo and P.Cottin "New Method For Parameter Extraction in Deep Submicrometer MOSFETs " ICMTS, p181-186, 2000

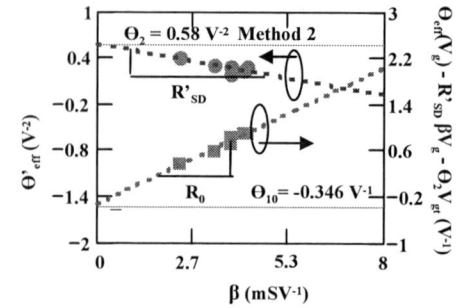

(4d)
Eq 7 and 8 (method 2)

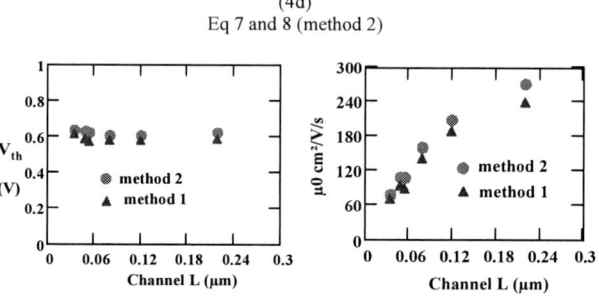

(4e)
Extraction result for V_t vs Channel length (μ)

(4f)
$\mu 0$ vs channel length (μ)

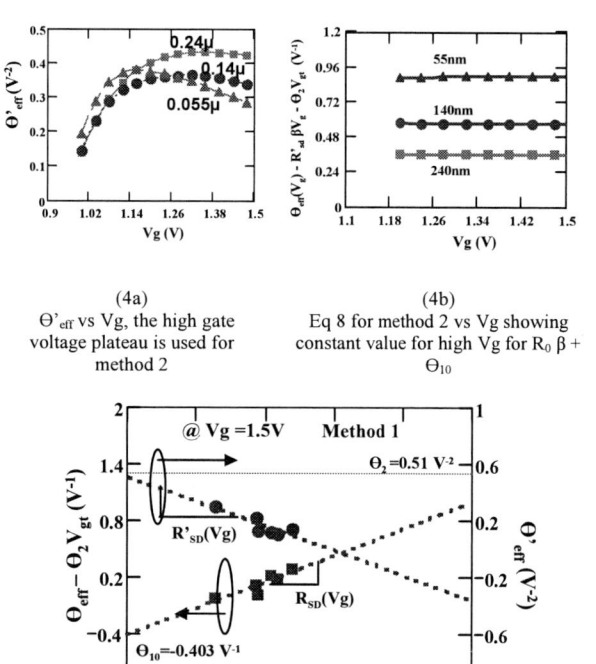

(4a)
Θ'_{eff} vs Vg, the high gate voltage plateau is used for method 2

(4b)
Eq 8 for method 2 vs Vg showing constant value for high Vg for $R_0 \beta + \Theta_{10}$

(4c)
Eq 3 and 4 result for $V_g = 1.5V$. (method 1)

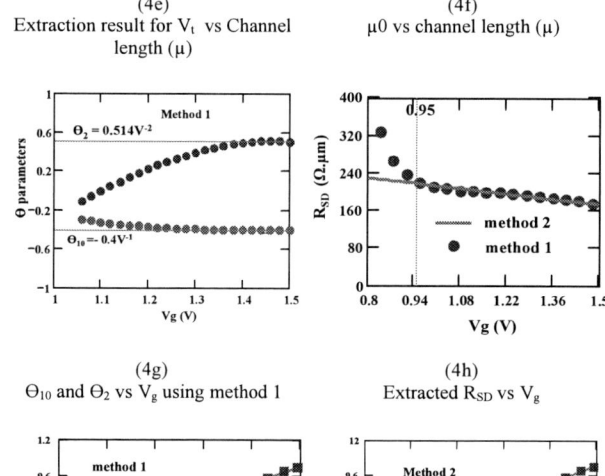

(4g)
Θ_{10} and Θ_2 vs V_g using method 1

(4h)
Extracted R_{SD} vs V_g

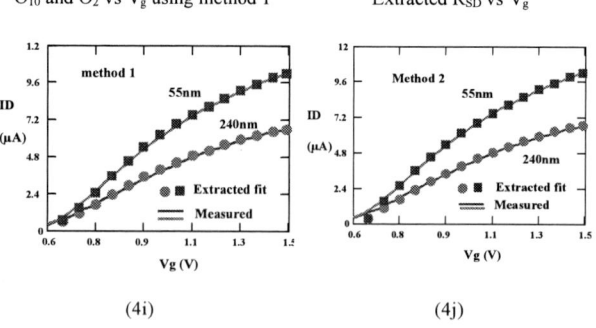

(4i)
I_{DS}-V_g measured vs extracted using method 1

(4j)
I_{DS}-V_g measured vs extracted using method 2

Figure 4.
Shows extraction results for real devices (channel lengths from 240nm to 55nm and 1µm wide).

978-1-4244-6658-0/10 $26.00 © 2010 IEEE

Carbon-doped GeTe Phase-Change Memory featuring remarkable RESET current reduction

G. Betti Beneventi[*][§], L. Perniola[*], A. Fantini[*], D. Blachier[*], A. Toffoli[*], E. Gourvest[‡||], S. Maitrejean[*],
V. Sousa[*], C. Jahan[*], J.F. Nodin[*], A. Persico[*], S. Loubriat[*], A. Roule[*], S. Lhostis[*‡], H. Feldis[*], G. Reimbold[*],
T. Billon[*], B. De Salvo[*], L. Larcher[§], P. Pavan[§], D. Bensahel[‡], P. Mazoyer[‡], R. Annunziata[‡], and F. Boulanger[*].

[*] CEA-LETI, MINATEC, 17 rue des Martyrs, F-38054 Grenoble Cedex 9, France
[§] Università degli Studi di Modena e Reggio Emilia, via G.Amendola 2, Pad. Morselli 42100 Reggio Emilia, Italy
[‡] STMicroelectronics, Central R&D, 850 rue Jean Monnet, F-38926 Crolles Cedex, France
mail to: *giovanni.bettibeneventi@cea.fr*

Abstract—In this paper we present a study of Phase-Change non-volatile Memory (PCM) devices integrating carbon-doped GeTe as chalcogenide material. Carbon-doped GeTe, named GeTeC, remarkably lowers the RESET current and features very good data retention properties as well. In particular, GeTe PCM with 10% carbon inclusions (named GeTeC10%) yields about 30% of RESET current reduction with respect to pure GeTe and GST. Furthermore, our GeTeC10% memory cells are expected to guarantee a 10-years-lifetime-temperature of about 127°C, which is one of the highest ever reported for PCM. The outstanding properties of GeTeC make this material promising for non-volatile memory technologies.

I. INTRODUCTION

Phase-Change Memory (PCM) technology is today considered as one of the most promising candidates for next generation non-volatile memories [1]. Compared to the Flash mainstream and to other emerging technologies, PCM offers better scalability (up to few nanometers [2]), faster programming (in the order of few nanoseconds [3]) and improved endurance (up to 10^9 programming cycles [4]).
The PCM device is basically a thin film chalcogenide resistor, which resistance can be tuned through suitable electrical pulses leading to reversible phase-change. Indeed, the chalcogenide material can exist in two different phases: a low-resistive, ordered, polycrystalline configuration (named SET state) and a high-resistive, disordered, amorphous one (named RESET) [5]. One of the key issues of the PCM technology is RESET current reduction. In fact, to program the cell in RESET state, the current has to be large enough to melt the chalcogenide active region. The high RESET current limits the minimum size of the selection device in series to the cell (hence the maximum memory density) and the capability of parallel programming at array level (hence write bandwidth or throughput) [6]. Moreover, in order to make the PCM technology suitable not only for consumer applications but also for automotive memory products, the 10-years-lifetime-temperature must increase above the actual limit, that records about 85°C for the most known and used chalcogenide material, i.e. $Ge_2Sb_2Te_5$ (GST) [7]. In this work, we propose an experimental study of carbon-doped GeTe chalcogenide alloy (GeTeC) as possible solution to the previous mentioned issues. Carbon can be a contaminant during Chemical Vapor

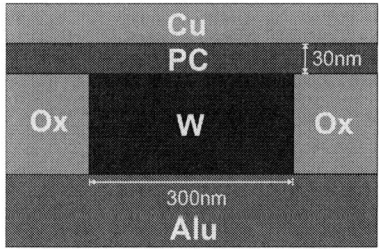

Fig. 1. Schematic of the cross section of our pillar-type PCM device. The core of the memory cell is a 30-nm thick chalcogenide phase-change layer, indicated as PC in the figure, placed on a 300-nm wide pillar tungsten plug, W, and insulating material, Ox. Top and bottom electrodes are Cu and Alu, respectively.

Deposition (CVD) of phase change materials because of the use of organic precursors, hence the importance of studying the electrical impact of carbon inclusions in GeTe. In order to do it, two different carbon impurity fractions were obtained by plasma-assisted co-sputtering. Their percentages, revealed by Rutherford Back-Scattering (RBS) and Nuclear Reaction Analysis (NRA) measurements, are: GeTe with 4% carbon and GeTe with 10% carbon (named GeTeC4% and GeTeC10%, respectively).
The paper is organized as follows. In section II we report experimental data on RESET-to-SET programming characteristics, focusing in particular on SET programming time. Furthermore, we discuss GeTeC data retention properties showing a 10-years-lifetime-temperature of 127°C. Section III is dedicated to SET-to-RESET transition. Experimental data on RESET current and RESET power are presented. A possible explanation concerning the physical origin of our findings is finally proposed.

II. GeTeC PCM DATA RETENTION AND RESET-TO-SET TRANSITION

To characterize the electrical behavior of GeTeC, simple pillar-type PCM devices where fabricated. In our cells, a 300 nm wide W pillar is in direct contact with a 30 nm thick phase-change layer, see Fig. 1. The GeTeC material has been deposited by plasma-assisted co-sputtering from 2 targets (stoichiometric GeTe and Carbon) in Ar atmosphere,

978-1-4244-6658-0/10 $26.00 © 2010 IEEE   233

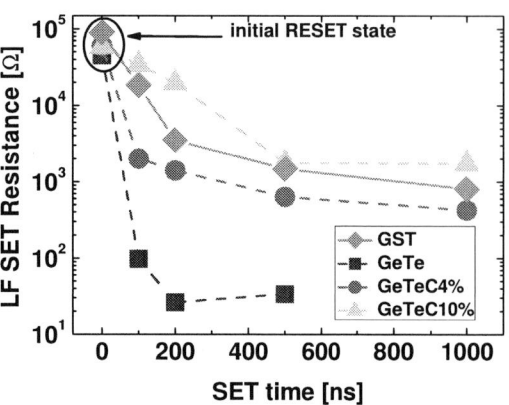

Fig. 2. Low-field (LF) SET state resistance as a function of SET time for GST, GeTe, GeTeC4% and GeTeC10% PCM. 100 ns, 200 ns, 500 ns and, for GST and GeTeC only, 1 μs SET pulses are applied to the cells and then resistance is measured with LF readout pulses. The point correspondent to 0 ns reports the initial resistance value of the PCM in RESET state. Each point is obtained by averaging on about 30 cells.

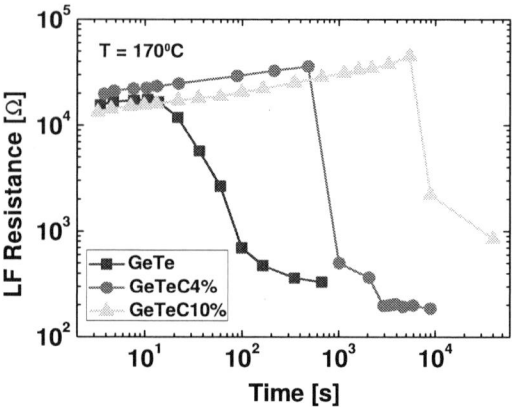

Fig. 3. Data retention measurements on PCM cells integrating GeTe, GeTeC4% and GeTeC10% at a temperature of 170°C. Every cell has been programmed with the same amorphization pulse and then LF resistance has been monitored. Each curve is representative of tests performed on 30 cells, on average.

at a pressure of 0.005 mbar and at room temperature.

A. RESET-to-SET programming time

An important parameter describing the phase-change properties of a chalcogenide alloy for PCM applications is the time required to crystallize the melt-quenched amorphous spot (i.e. RESET-to-SET programming time). In Fig. 2 we plot the PCM resistance value obtained by applying a crystallization pulse (SET pulse) to amorphous cells of each alloy composition. The device resistance is read through a low-field (LF) signal following the SET programming pulse. Note that the SET pulses are always applied starting from an identical amorphous RESET state (corresponding to the point at 0 ns in the figure). The data refer to pulses of the same amplitude but of different widths: 100 ns, 200 ns, 500 ns and 1 μs. While for GeTe-based memories a 200 ns pulse is already sufficient to bring the cell to a minimum resistance value, GeTeC4% and GeTeC10% require more time to reach a full SET state. This is in agreement with material characterization on GeTeC as-deposited full-sheet depositions: the amorphous phase of GeTeC is more disordered with respect to the GeTe one, requiring more energy to be switched in an ordered configuration [8]. As we will see in the next section, this property has, on the other hand, a beneficial effect on data retention. Nevertheless, the time required to program GeTeC-based devices in the SET state remains in the order of hundreds of ns, resulting comparable to the time needed for the reference PCM material GST. Besides, we note also that the resistance of the SET state increases with carbon doping, being in the order of kΩ for GeTeC, while few Ω for GeTe.

B. Data retention properties

While the polycrystalline state of the PCM is inherently stable, being the lowest possible energy configuration of the system, the amorphous material is affected by spontaneous crystallization, which represents the most important concern

for data retention performance of PCM. Moreover, temperature plays a major role in crystallization phenomena, according to Arrhenius law [9]. Recent works point out that GeTe and carbon-doped GeTe thin films have higher crystallization temperatures than the chalcogenide material reference GST, leading to superior data retention performances when integrated in memory cells [10]- [11]. Furthermore, carbon doping largely enhances data retention properties of pure GeTe [8]. In Fig. 3 data retention measurements performed on PCM cells with amorphous GeTe, GeTeC4% and GeTeC10% are displayed. The PCM cells are initially programmed in the high-resistive amorphous phase, then the resistance increases because of the *drift* phenomenon [12], and, finally, the resistance falls down owing to spontaneous crystallization. Note that the resistance drop significantly shifts at higher times as the carbon concentration is increased. In particular GeTeC10% reaches the characteristics SET resistance value almost 2 decades after pure GeTe. In Fig. 4 the dispersion of fail times of GeTeC10% and GeTe PCM is compared, where fail time is considered to occur when the resistance is reduced of 50% with respect to the initially programmed value. The graph displays the ratio between fail time standard deviation (σ) and mean value (μ) for three different temperatures: 160°C, 170°C and 180°C. In each measurements the PCM cells are programmed in the amorphous state and then LF resistance is monitored. It appears clear that carbon doping helps in reducing the large dispersion of fail times, being the ratio σ/μ always smaller for GeTeC10% than for GeTe. In Fig. 5 the GeTeC10% mean fail times are recorded for five different temperatures (155°C, 160°C, 170°C, 175°C, and 180°C) and then 10 years extrapolation is obtained simply applying Arrhenius law. The E_A extracted value, 4.33 eV, is in good agreement with optical characterization on fullsheet material depositions (i.e. 4.16 eV [8]). It is worth noting that the 10-years-lifetime-temperature extrapolated for GeTeC10%-based PCM devices is about 127°C suggesting that GeTeC10% could satisfy the specifications of embedded memories.

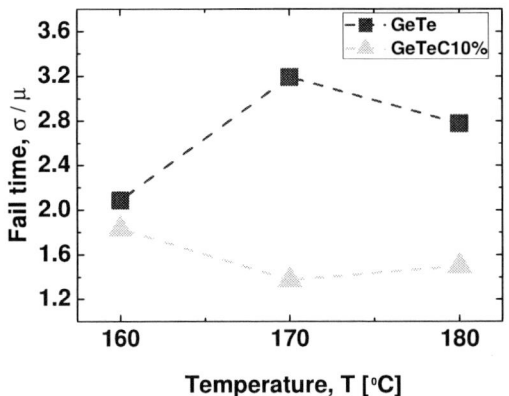

Fig. 4. Fail time standard deviation (σ) / mean value (μ) for GeTe and GeTeC10%-based PCM devices for three different temperatures. Each point has been obtained averaging measurements on about 30 cells.

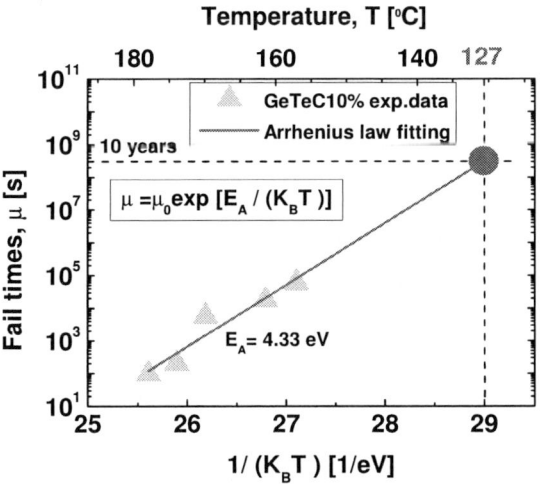

Fig. 5. Fail time mean value μ as a function of $1/(K_B T)$ for GeTeC10% PCM-based devices (K_B is the Boltzmann constant). Symbols are data, line is fitting based on the Arrhenius law annotated in the figure. The E_A value, extracted from the fitting, is indicated in the graph. Each point has been obtained averaging measurements on about 30 cells.

III. GeTeC SET-TO-RESET OPERATION

A. RESET current and RESET power

Fig. 6(a) shows data on SET-to-RESET transition for GST, GeTe, GeTeC4% and GeTeC10%-based PCM. The electrical setup employed for these measurements is described in detail in [10]. The PCM cell is programmed with 50 ns width pulses of increasing current amplitudes and very fast 10 ns trailing edge (I_{PROG} pulses). After each pulse the LF resistance is measured. Each programming pulse is preceded by a SET pulse (width > 1 μs), in order to analyze the effect of the increasing programming signal starting from a fully-polycrystalline SET state We define the current to RESET the device, I_{RESET}, as the I_{PROG} needed to obtain the 90%

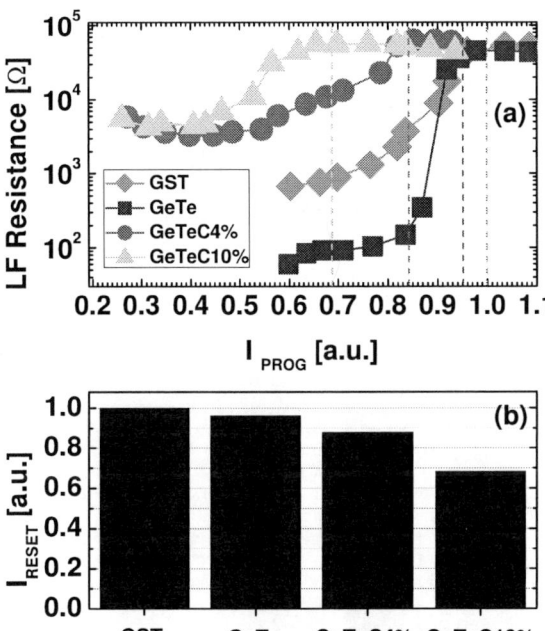

Fig. 6. (a) SET-to-RESET transition for GST, GeTe, GeTeC4% and GeTeC10% PCM. LF device resistance is plotted against programming current I_{PROG}. Before each I_{PROG} pulse a SET pulse (width > 1 μs, not shown in the figure) is applied to the cell in order to re-initialize the PCM on a full SET state. Vertical dash-type lines trace the correspondence $I_{PROG} = I_{RESET}$ for each alloy. Each curve is representative of about 10 devices. (b) RESET current mean values for GST, GeTe, GeTeC4% and GeTeC10% devices. The I_{RESET} mean values are normalized with respect to the GST ones. Each data is obtained averaging on about 10 PCM devices.

of the maximum LF resistance value of the whole SET-to-RESET transition. Interestingly, I_{RESET} lowers as the carbon percentage rises. Note also that the RESET resistances are similar for each material, suggesting that the resistivity of the four different alloys in the amorphous melt-quenched state should be practically the same. Consider now Fig. 6(b): the mean values of I_{RESET} are plotted for each material. In first approximation GST and GeTe are characterized by a comparable RESET current (I_{RESET} GeTe $\sim$ 95% I_{RESET} GST), while a reduction of more than 10% is obtained for GeTeC4%, and of more than 30% for GeTeC10%.

Furthermore, using the I_{RESET} value previously obtained, it is possible to calculate the power associated to the RESET operation P_{RESET} as $P_{RESET} = I_{RESET}^2 \cdot R_{ON}$, where R_{ON} is the resistance associated to the PCM device during programming, when the phase-change material is melt and highly conductive [13]. In our cells R_{ON} is mainly determined by the series resistance of the metal lines and of the W plug, being the ON resistance of the 30-nm thick phase-change layer very low. As R_{ON} is measured constant among the different alloys (see Fig. 7), the RESET power decreases with the square of I_{RESET}; this leads to a P_{RESET} reduction of more than 20% for GeTeC4% and of more than 50% for GeTeC10% (see Fig. 8).

978-1-4244-6658-0/10 $26.00 © 2010 IEEE   235

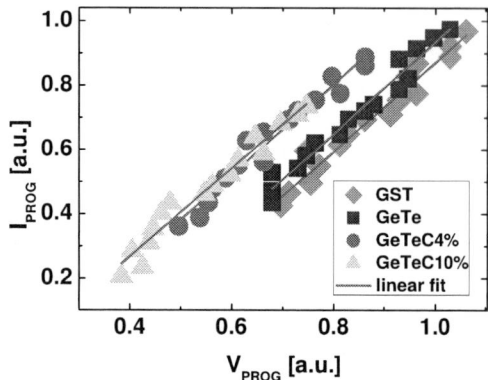

Fig. 7. I_{PROG}-V_{PROG} characteristics of PCM cells in the *ON* region for each alloy, where V_{PROG} is the voltage drop on the PCM device during programming. R_{ON}, extracted from the slope of the curves, is about 28-29 Ω for each composition (sum of the resistance of metal lines and tungsten plug). Slightly differences of Ω fractions between the curves stay within experimental accuracy limitation.

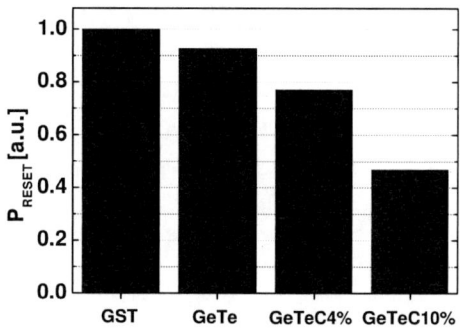

Fig. 8. RESET power mean values for GeTe, GeTeC4% and GeTeC10% devices. The P_{RESET} mean values are normalized with respect to the GST ones. Each data is obtained averaging on 10 PCM devices.

B. Discussion

The electro-thermal properties of the materials, along with the geometrical shape of the PCM cell, are the parameters that determine the I_{RESET} value. For pillar-type PCM device, I_{RESET} can be analytically expressed through a simple formula accounting for Joule heating [6]. In our structure, the heater plug is made of tungsten, featuring a relatively high thermal conductivity, thus the thermal resistance of the heater plug results negligible with respect to the chalcogenide one. Indeed, we verified by simulation that the temperature peak during program is placed in the middle of the chalcogenide layer. For these reasons, the equation given in [6] becomes:

$$ I_{RESET} = \sqrt{\frac{4(T_{RESET} - T_0)}{R_{ON,c}\, R_c^{TH}}}, \qquad (1) $$

where T_{RESET} is the RESET temperature, intrinsic properties of the chalcogenide material, defined as the maximum temperature generated in the device during the I_{RESET} pulse, T_0 is the room temperature, $R_{ON,c}$ is the chalcogenide contribution to the total R_{ON} of the PCM device, and R_c^{TH} is the thermal resistance of the chalcogenide active material. In this formula, the chalcogenide parameters are T_{RESET}

and the thermal conductivity of the chalcogenide k_c^{TH}, being $R_c^{TH} = th_c/(k_c^{TH} A)$, where th_c and A are the chalcogenide layer thickness and the plug-chalcogenide contact area. Therefore, carbon doping could either a) decrease T_{RESET}, or b) decrease k_c^{TH}, or c) both. In previous works focusing on other systems always made by a chalcogenide alloy with dielectric co-sputtering inclusions, i.e. GST carbon-doped and GST with SiO_x or SiN_x dopants, the RESET current diminution has been interpreted as a consequence of the *effective* k_c^{TH} reduction [14] [15]. In particular, the decrease of the *effective* thermal conductivity of the active layer has been correlated with the actual reduction of the chalcogenide programmable volume caused by the formation of nanoclusters of immiscible chalcogenide-dielectric mixtures. It is worth noting that the electrical properties of carbon-doped GST, GST-SiO_x and GST-SiN_x systems have many analogies with the ones of GeTeC, namely: a) an important reduction of the RESET current with doping, b) the increase of the SET state cell resistance for the doped material with respect to the undoped one and c) the evidence that the resistivity of the melt-quenched amorphous phase does not change with doping. Further analysis are in progress in order to deeper understand our results and determinate which is the dominant effect in our system.

IV. CONCLUSIONS

In this paper we have presented novel experimental findings about carbon-doped GeTe PCM devices. carbon doping has a beneficial effect on PCM data retention: PCM devices integrating GeTeC10% can guarantee a 10-years-lifetime-temperature of about 127°C. Moreover, carbon doping reduces fail time dispersion. SET operation for GeTeC devices results slower with respect to GeTe ones, although it remains in the hundreds of ns range, featuring GST-like SET program times. Our electrical characterization data highlight a reduction of both RESET current and power when carbon is added. In particular, GeTeC10% PCM devices yield about 30% of RESET current reduction in comparison to GST and GeTe ones, which translates in about 50% RESET power reduction. Both data retention up to 127°C and 30% RESET current reduction indicate GeTeC10% as a promising candidate for both embedded and stand-alone PCM applications.

REFERENCES

[1] R. Bez, IEDM Tech. Dig., 89-92, 2009.
[2] S. Raoux et al., IBM J. Res. & Dev., Vol. 52, No.4/5, 465-479 (2008).
[3] G. Bruns et al., Appl. Phys. Lett., Vol.95, 043108 (2009).
[4] G. Servalli, IEDM Tech. Dig., 113-116, 2009.
[5] M. Wuttig, Nature Materials, Vol. 4 (2005).
[6] U. Russo et al., IEEE Trans. El. Dev., Vol. 55, No.2, (2008).
[7] J.H. Ho et al., IEDM Tech. Dig., 49-52, 2006.
[8] G. Betti Beneventi et al., Proc. of IMW, 21-24, 2010.
[9] A.L. Lacaita et al., Phys.Stat.Sol.(a), Vol. 205, No.10, 2281-2297 (2008).
[10] A. Fantini et al., Proc. of IMW, 66-67, 2009.
[11] L. Perniola et al., Electron Dev. Lett., Vol. 31, 5, 488-490 (2010).
[12] D. Ielmini et al., IEDM Tech. Dig., 939-942, 2007.
[13] A. Pirovano et al., IEEE Trans. El. Dev., Vol. 51, No.3, (2004).
[14] W. Czubatyj et al., Proc. EPCOS Conf., 143-152, 2006.
[15] T.Y. Lee et al., Appl. Phys. Lett., Vol. 94, 243103 (2009).

Current distributions of BJT-based decoding array for Phase Change Memory

Domenico Ventrice, Alessandro Calderoni and Paolo Fantini
Technology Development
Numonyx
Via Olivetti, 2 – 20041 Agrate Brianza, Italy
domenico.ventrice@numonyx.com

Abstract—**Phase Change Memory (PCM) is today demonstrating to be one of the mainstream memories for the next decade [1]. In particular, the PCM technology based on BJT array for the storage cells decoding appears the mainstream architectural approach in order to keep both compact the cell layout and provide the proper amount of programming current. In this scenario, the present work investigates the electrical spreads of the BJT decoding arrays in 45 nm technology presented in ref. [1] for both the readout and programming conditions. A detailed experimental investigation, together with SPICE-like Monte Carlo simulations, are used to assess the sources of current variability allowing a better control of the BJT-based decoding current distributions.**

I. INTRODUCTION

Among the new concepts that challenged the conventional Non-Volatile Memories (NVM) during the last decade, the $GST(Ge_2Sb_2Te_5)$-based Phase Change Memory (PCM) has been demonstrated a mature technology with a set of new interesting features for novel applications in the memory market [1]. The recent development of the 45-nm technology node on 1Gb PCM product represents a tremendous progress in this way [2]. PCM technology relies on a resistor (R) built with a chalcogenide alloy (GST) able to reversibly phase change from the ordered-crystalline state (set) to the amorphous-insulating one (reset), under a proper electrical pulse [3]. The decoding of large density PCM array through BJT transistors (BJT-PCM) appears the mainstream architectural approach leading to the 1T/1R structure [4]. Thus, apart from the current density requirements to program cells, that are expected to increase with the scaling nodes [5], also the electrical spread of the decoding arrays needs to be well investigated in order to better control both the readout and the programming current distributions. More precisely, besides the BJT selector, also the heater resistor is a fundamental element for the programming operation of the GST active region since it modulates the Joule heating effect promoting the phase transition. Thus, the BJT+heater block plays an important role in the PCM cell decoding and

functionality and must be statistically analyzed. Furthermore, both of these auxiliary devices for the PCM cell operations are involved in the continuous downscaling that recently led the NVM technology to dimensions of few tens of nanometers. One of the consequences of downscaling is the increasing of the variability effects at single-device level, potentially affecting their performance and reliability. These phenomena can be addressed by 3D TCAD simulations [6], but the huge computational load makes this approach unpractical when extensive statistical analyses under different operating conditions are required, whereas Spice-like simulations appear viable. *In this work, the statistical analysis of the BJT+heater decoding array of 45 nm technology node is carried out through both a detailed experimental characterization and Monte Carlo (MC) Spice-like simulations. Their comparison allows to single out the sources of variability of the readout and programming current distributions for the PCM bits, then to assess their impact on the real operating conditions of the device.*

II. EXPERIMENTAL

In order to evaluate the decoding spreads excluding any impact deriving from the GST layer, the 45 nm BJT-PCM

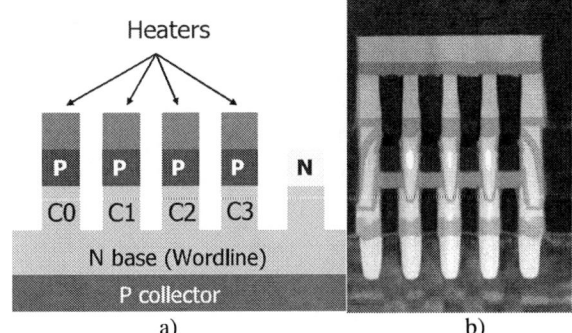

a)   b)

Figure 1. a) Schematic layout of the four BJTs mini-string for the PCM cells decoding. b) TEM cross-section of a 1Gb PCM cell array [1].

978-1-4244-6658-0/10 $26.00 © 2010 IEEE

array [1] without GST deposition (no-GST) has been fabricated and experimentally characterized. In the following we will use the term cell to identify the bit decoding even if the GST switching layer was not present. Fig. 1 shows the schematic layout of the mini-string of four PCM cells sharing the same base contact with the corresponding TEM picture of our 45-nm PCM technology [1]. We call internal and external BJTs those ones are farer (C1, C2) or closer (C0, C3) to the base contact, respectively (see Fig. 1a). We collected through direct memory access (DMA) the current distributions of 12k cells captured from a homogeneous central part among the 256 tiles organizing our 1 Gbits devices [2]. In this way we annihilated any *IR* drop difference and, in the frame of our work, we can lamp the memory access with a single series resistance. The DMA acquisition allows to group the experimental data according to they come from the internal or external BJTs of the mini-string sketched in Fig 1. So, we can decouple the base resistance effect, which value is for construction different for internal and external cells, for both the readout and program condition. Also possible noise sources of such scaled devices have been investigated through the noise statistical measurement when the ΔI distribution between two consecutive current samplings is performed.

III. STATISTICAL SPREADS OF I_{READ}

Statistical spread of decoding array must be correctly evaluated, in particular, when the set (high conductive state) distribution is considered [7]. On the contrary, heater and BJT selector do not contribute significantly to the width and shape of the reset currents distribution, since its own resistive effect [8]. Fig. 2 shows the experimental current distributions when the Bit Line (BL) bias was close to the reading condition. We separated four BJTs current distributions. We notice the perfect distribution overlap between the two inner and the two external BJTs, thus demonstrating a well controlled process

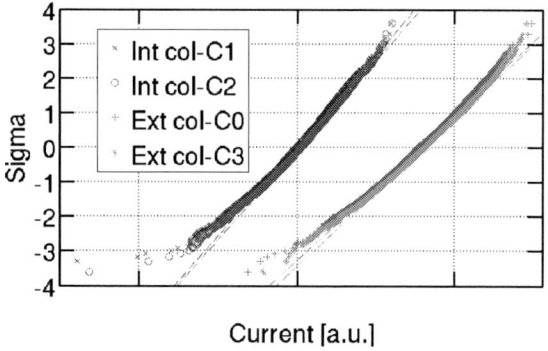

Figure 2. Cumulative current distributions obtained from a 12K decoding array (no-GST) biased in the readout condition. Currents are separated accordingly to their coming from the BJTs C0, C1, C2, C3 of the quadruplet of Fig. 1.

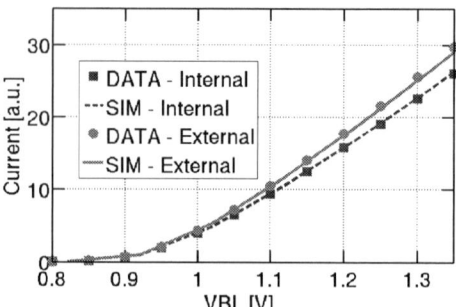

Figure 3. Average I-V characteristics on 12k BJT+heater decoding array for internal (C1-C2) and external (C0-C3) cells.

able to guarantee an excellent matching between equally designed decoding structures in 45 nm technology. We also observe a quasi-Gaussian shape (dashed lines) of the currents distribution apart from a slight enlargement opening at lower currents. Fig. 3 shows the average *I-V* curves obtained from the 12k cells correspondingly to the inner and external BJTs (squares and dots, respectively). It also reports (lines) the corresponding Spice simulation of the BJT with the heater resistance in series. The internal and external models only differ for the base resistance (R_B) value of the BJT that is the only building difference in agreement with Fig. 1. In order to study the sources of variability determining the current distributions of Fig. 2, statistical Spice simulations were run by using the Monte-Carlo (MC) method. As a first step, Gaussian spreads of some key parameters are considered to investigate the sensitivity on the width and shape of current distributions. The intra-die dispersions for technology induced variability (heater area (A) and thickness (t_H)) as well as due to the doping fluctuations affecting the BJT parameters (V_{BE} and R_B) were derived from in-line process control measurements. Fig. 4 shows the SPICE simulated cumulative distribution for

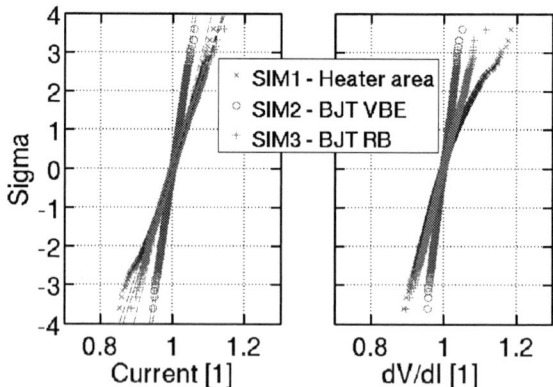

Figure 4. MC Spice simulations of the decoding array obtained with a Gaussian spread either of the heater area, or of the VBE or RB parameters. Both the view of the normalized current and normalized resistance spreads are reported in the left and right side, respectively.

both the currents and the differential resistances (*dV/dI*) in the readout condition, when only the internal BJT is considered. The three simulated cumulative distributions are obtained with the introduction in the model of a Gaussian spread of: 1) heater area (pluses); 2) V_{BE} (circles); 3) R_B (crosses). Although the main distribution is always Gaussian, we can

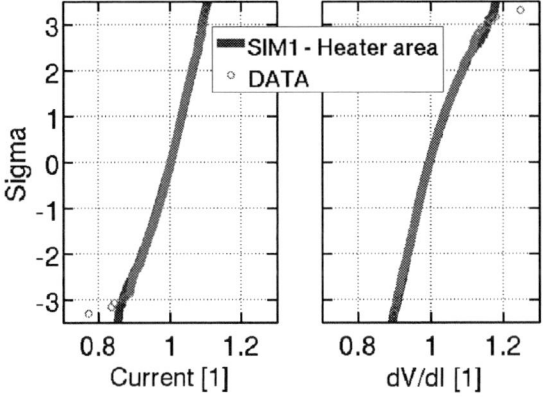

Figure 5. Normalized cumulative distributions for the readout current (left) or differential resistance (right) of the internal BJT decoding array. Also the MC Spice-like simulations are reported (continuous line).

capture the role played by each parameter in determining the shape of the distribution tails when they are seen both in current and mirrored in the differential resistance plot. In particular, the spread of the heater contact area appears the only parameter able to account for the low-currents tail and the convex resistances distribution. Fig. 5 shows that these features well agree with the experimentally collected distribution related with the internal BJT and heater. Statistical Spice simulations, that well capture the experimental distributions, are obtained with the heater area spread as the main source of variability.

This result indicates that heater needs to be well controlled in order to keep the distribution tight. In fact, when we focus on the low current tail, we observe that an equivalent failure probability (without GST) versus the verify current is well captured when the heater area fluctuates in the model (Fig. 6).

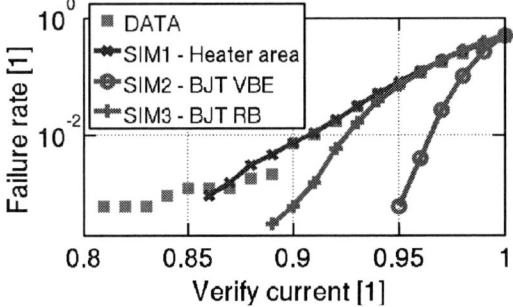

Figure 6. Failure probability as a function of the verify current (normalized). The experimental curve (squares) and the calculated ones considering the different main spread sources (coherently with Figure 4) are reported.

IV. NOISE

As a consequence of the impressive scaling down of dimensions the low-frequency noise, typically critical for analog applications, has been recognized to be a strong concern also for nanoscaled NVM devices, since it produces current fluctuations that can cause errors in the memory

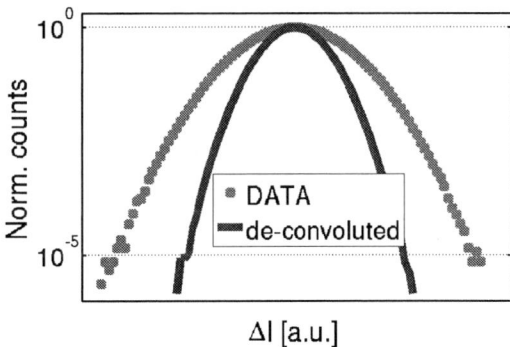

Figure 7. Experimental noise collected through ΔI measurements showing a Gaussian distribution. Noise estimation of the decoding array when that one induced by the experimental set-up is de-convolved (continuous line).

readout operation [11, 12]. In this scenario, also possible noise sources and the related weight of the nanoscaled cell selection devices worth noting to be investigated. To this aim, with a similar approach followed to statistically characterize noise in FG-based memory cells [12], we measured the ΔI distribution between two consecutive current samplings on 12k cells of no-GST wafer. The experimental measurement is reported in Fig. 7 showing a Gaussian noise distribution. The Gaussian noise is in agreement with the *1/f*–like noise that can affect the decoding array and recently observed in PCM cells [9, 10]. Being the measured current fluctuations comparable with those ones of the experimental set up resolution, we also accounted for it. So after the de-convolution of the noise due to the experimental apparatus, that one related with the decoding array can be computed (continuous line in Fig. 7). The extracted normalized noise intensity is negligible (ΔI/I < 0.05%) with respect to that one captured by the cell [9].

V. STATISTICAL SPREADS OF $I_{PROGRAM}$

The programming currents distribution is a key figure of the capability to handle the programming operation in PCM technology. Since heater is responsible of the *dV/dI* slope in the high current programming region also when the GST switching layer is deposited [7], the statistical spread of the heater resistance is fundamental. Again, the no-GST wafer allows extracting this information. Fig. 8 shows the experimental distribution of the *dV/dI* slope captured at a fixed bias driving a current around the programming one (~200uA). The different average current between external/internal BJT (not reported in Figure) is essentially due to the different value of the base resistance that is again systematic, demonstrating a

978-1-4244-6658-0/10 $26.00 © 2010 IEEE

good process control. Fig. 8 also reports the comparison with the Spice simulation obtained by using the same model and spread of the readout condition. The satisfactory agreement (slightly overestimating the tail, probably as a consequence of the high temperature heater *dV/dI* collected in the readout and programming region is shown. A strong correlation in between can be observed, thus demonstrating that the heater resistance that played a non negligible role in the readout region, as pointed out in section III, is also the key parameter in the high current region.

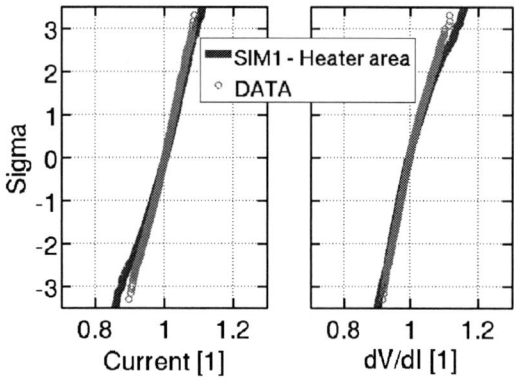

Figure 8. Experimental and calculated cumulative distributions of the normalized current and differential resistance in programming condition.

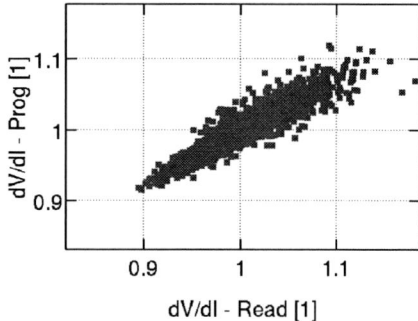

Figure 9. Correlation plot between the differential normalized resistances in read and program conditions.

VI. CONCLUSIONS

A detailed analysis on the statistical spreads of the BJT-based decoding array for PCM technology has been presented on the basis of the four BJTs mini-string selection scheme adopted in the 45 nm technology. Both our experimental investigation of no-GST array and the simulation analysis show that BJT decoding is a robust architectural choice. The

heater element gives rise to non-Gaussian tails in the lower currents of the readout region. The correlation between the readout and programming resistances has been shown. Furthermore, we have shown that the decoding array noise does not represent a critical issue. This work addresses the first statistical analysis of the BJT-based array decoding, being precious for the next PCM technology nodes.

ACKNOWLEDGMENT

Authors would like to thank P. Cappelletti, R. Bez, G. Servalli and A. Marmiroli for helpful discussions and support. This work has been also partially supported by ENIAC under the MODERN project 120003.

REFERENCES

[1] G. Servalli , "*A 45nm Generation Phase Change Memory Technology*", in IEDM Tech. Dig. pp. 113-116, 2009, Baltimora (ML).

[2] C. Villa, D. Mills, G. Barkley, H. Giduturi, S. Schippers and D. Vimercati, "*A 45nm 1Gbit 1.8V Phase-Change Memory*"in ISSCC Tech Dig. pp.270-273 (2010).

[3] Agostino Pirovano, Andrea L. Lacaita, Fabio Pellizzer, Sergey A. Kostylev, Augusto Benvenuti, Roberto Bez, "*Low-Field Amorphous State Resistance and Threshold Voltage Drift in Chalcogenide Materials*", IEEE Transactions on Electron Devices, vol. 51, no. 5, 0018-9383 (2004).

[4] A. Redaelli, A. Pirovano, I. Tortorelli, F. Ottogalli, A. Ghetti, L. Laurin and A. Benvenuti "*Impact of the Current Density Increase on Reliability in Scaled BJT-selected PCM for High-Density Applications*", to be published in IRPS Tech Dig. (2010).

[5] http://www.itrs.net/Links/2008ITRS/Home2008.htm.

[6] A. Asenov, A.R. Brown, J.H. Davies, S. Kaya and G. Slavcheva, "*Simulation of intrinsic parameter fluctuations in decananometer and nanometer scale*", IEEE Trans. El. Dev., vol. 50, pp. 1837-1852 (2003).

[7] D. Ventrice, A. Calderoni, A. Spessot, P. Fantini, A. Sanasi, S. Braga, A. Cabrini and G. Torelli, "*Statistical Modeling of Bit Distributions in Phase-Change Memories*", in Proc. ESSDERC pp. 157-160 (2009).

[8] A. Calderoni, M. Ferro, D. Ventrice, D. Ielmini and P. Fantini, "Reset current distributions in Phase Change Memories", to be published in IRPS Proc. (2010).

[9] P. Fantini, G. Betti Beneventi, A. Calderoni, L. Larcher, P. Pavan and F. Pellizzer, "*Characterization and Modeling of Low-Frequency Noise in PCM device*", in IEDM Tech. Dig. pp. 219-222, 2008, San Francisco (CA).

[10] G. Betti Beneventi, A. Calderoni, P. Fantini, L. Larcher, P. Pavan, "*Analytical model for low-frequency noise in amorphous chalcogenide-based phase-change memory devices*", Journal of Applied Physics, Vol. 106, Issue 5, pp. 54506-54513 (2009).

[11] D. Fugazza, D. Ielmini, S. Lavizzari, and A. L. Lacaita "*Distributed-Poole-Frenkel modeling of anomalous resistance scaling and fluctuations in phase-change memory (PCM) devices*" in IEDM Tech. Dig. pp. 723-726, 2009, Baltimore (ML).

[12] R. Gusmeroli, C. Monzio Compagnoni, A. Riva, A. Spinelli, A. L. Lacaita, M. Bonanomi and A. Visconti, "Defects spectroscopy in SiO2 by statistical random telegraph noise analysis" in IEDM Tech. Dig. pp. 439-442, 2006, San Francisco (CA).

978-1-4244-6658-0/10 $26.00 © 2010 IEEE

SET switching effects on PCM endurance

Vincenzo Della Marca, Francesca Carboni, Luca Larcher*, Andrea Padovani*, and Paolo Pavan

Dipartimento di Ingegneria dell'Informazione, Università di Modena e Reggio Emilia,
Via Vignolese 905, 41125 Modena, Italy, paolo.pavan@unimore.it
*Dipartimento di Scienze e Metodi dell'Ingegneria, Università di Modena e Reggio Emilia,
Viale G. Amendola 2, Pad. Tamburini, 42100 Reggio Emilia, Italy

Abstract — **In this paper we report results on PCM endurance failure characterization. We show that endurance failure is related to SET pulse features and we analyze and model SET operation to obtain a better understanding and improve endurance performance. Results give interesting insights on the crystallization process of GST material. SET obeys to a constant energy law. Fast SET pulses require high power; slow SET pulses can be implemented in low power applications. Results may be used for optimized SET/RESET operation to achieve better endurance.**

I. INTRODUCTION

Phase Change Memory (PCM) device based on resistive switching of materials like $Ge_2Sb_2Te_5$ (GST) is a concept proposed for future Flash memory technologies. Studies on chalcogenide materials like GST have been carried out since the '70 [1], but more recently their possible use in PCM has been claimed as they show fast phase transformation, and can undergo high number of cycles. High endurance and long retention time along with compatibility with CMOS technology make PCM a candidate for future generation of both stand-alone and embedded memories [2][3].

GST characteristics concerning SET and RESET states are related to the kinetics of phase transition, particularly for the amorphous (RESET) to crystalline (SET) phase transition of GST, which involves a wide variety of material related phenomena, thus representing an arena of important technical and scientific interest in understanding and developing optimized materials and structures for phase change memory devices [4].

The switching endurance between SET and RESET states leads to two possible failures: 1) the stuck-reset failure (SRF), where the cell can no longer switch to SET, and 2) the stuck-set failure (SSF), where the cell can no longer switch to RESET. However there is no clear understanding of the mechanisms involved in degradation and, eventually, in failure [3].

In this work, we show results on PCM endurance and relate the failure events to SET switching characteristics. We monitor cell characteristics during degradation and associate endurance failure to SET conditions. We also characterize in detail SET operation and interpret it adopting two different models presented in the literature.

II. CELL STRUCTURE AND EXPERIMENTAL SET UP

μ-trench PCM cells have been presented and processed in different technology nodes. Details on process, cross section and layout of cells used in this work are reported elsewhere [5].

The experimental set-up is schematically reported in Fig. 1 [5]. *PROGRAM* pulses, with amplitude V_{SET} and duration PW, were applied by a pulse generator HP 8110A. During the pulse, the voltage drop across the cell (V_{CELL}) was sampled by a LeCroy W44Xs scope. The corresponding current I_{PROG} is then calculated as $(V_{SET} - V_{CELL})/R_{LOAD}$.

The *READ* operation was performed using HP 4142B parameter analyzer, applying a constant current I_{READ} and measuring the corresponding voltage V_{READ}. The cell resistance was then evaluated as $R = V_{READ}/I_{READ}$.

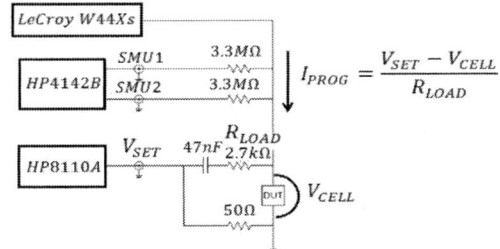

Fig. 1. Experimental set-up used in this work.

III. RESULTS AND DISCUSSION

a) Endurance characterization

Devices have been cycled applying this sequence of operations: READ, RESET, READ, SET. READ and RESET are always the same (*READ* as in Section II, *RESET*: V_{RESET} = 5.5V, PW_R = 200 ns). RESET and SET resistances are plotted against cycle number. Results are shown in Figs. 2 and 3, obtained with SET pulses having different duration. In Fig. 2, T_{SET} is kept constant to 300 ns and V_{SET} is 2.2V, 2.8V and

---

This work was supported in part by Italian MUR with the PNR 2005, RBIP06YSJJ, "Characterization and Optimization of Phase Change Memories"

978-1-4244-6658-0/10 $26.00 © 2010 IEEE

3.4V to give a resistance value around 10kΩ. In Fig. 3 T_{SET} is kept at 600 ns and V_{SET} 2.4V, 2.5V and 3.4V to give a resistance value around 10kΩ, as well.

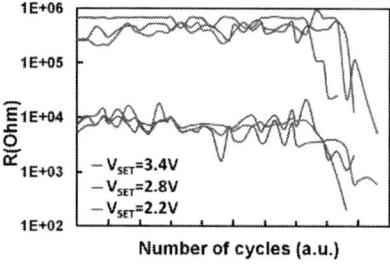

Fig. 2. W/E cycling test of a PCM device for PWs=300ns; V_{SET} is 2.2 (red), 2.8 (blue) and 3.4 (green) V.

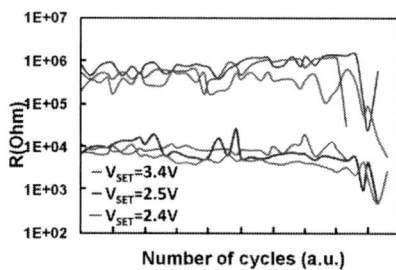

Fig. 3. W/E cycling test of a PCM device for PWs=600ns V_{SET} is 2.4 (red), 2.5 (blue) and 3.4 (green) V.

Cells fail after many cycles, and the observed failure is always the same: first, the SET resistance decreases, then RESET is not complete, making the RESET resistance to decrease as well, then the cell suddenly goes in OPEN (high resistance) condition. This final OPEN condition can be due to heater failure, or to void formation or delamination that catastrophically severs the electrical path through the device, typically at material interface such as the heater-to-GST contact in mushroom cells [3]. We note that the maximum number of cycles depends both on SET PWs and power delivered by the cell, contrarily of [6] where SET conditions are reported to have no influence on degradation. This dependence is shown in Fig 4. The number of cycles increases on decreasing the power delivered for the SET operation, for both SET pulse duration.

We monitored cell characteristics (I-V and R-I) on increasing number of cycles. From I-V, Fig. 5, we can see that the heater resistance decreases. But, from Fig. 6, we can see that also SET resistance decreases and RESET triggers at higher I_{PROG}. Thus, we can suppose that failure starts with the reduction of the SET resistance, making the RESET operation less effective: from Fig. 6 we can see that the cell can still be RESET but at very high current values, since a larger current has to be dissipated on a lower resistance to achieve the same temperature for crystal amorphization.

A detailed analysis of the SET operation is therefore needed.

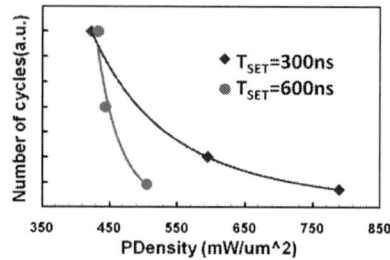

Fig. 4 Maximum number of cycles vs. power density delivered for different T_{SET}.

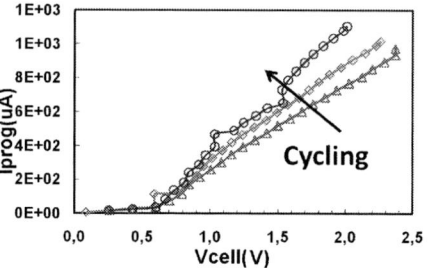

Fig. 5. I-V curve of a PCM cell fresh and after W/E cycles

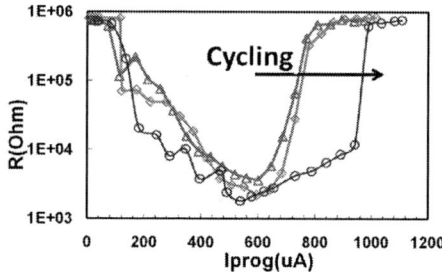

Fig. 6. R-I_{PROG} of a PCM cell fresh and after W/E cycles.

b) SET characterization

Devices have been programmed into SET states by applying programming pulses with fixed V_{SET} and increasing PW_S, see Fig. 7.

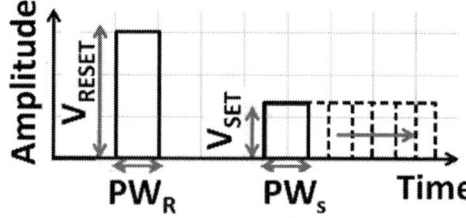

Fig. 7. RESET-SET characterization procedure. After a RESET pulse is applied, a SET pulse is applied. For a fixed V_{SET}, PW_S is increased.

978-1-4244-6658-0/10 $26.00 © 2010 IEEE

After every SET pulse, the cell resistance was measured and a new RESET pulse was applied (V_{RESET} = 5.5V, PW_R = 200 ns). Thus, every SET operation was performed starting from the same initial condition, i.e. with cells in the full RESET state.

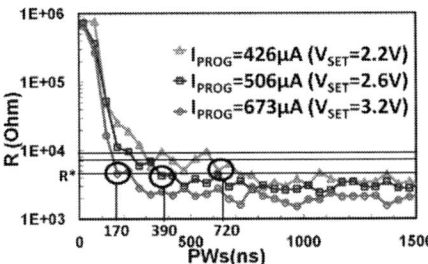

Fig. 8. Typical RESET to SET transition measurements. R is plotted vs. PWs.

Typical results are shown in Fig. 8 [5]. Experimental data point out that: (i) resistance drops approximately with an exponential law, and the slope increases with I_{PROG}; (ii) resistance fluctuations are noticeable, although data were averaged on several measurements and smoothed; this is even more remarkable at low I_{PROG}; (iii) R-PW_S characteristics tend to saturate to a unique low resistance value (corresponding to the heater resistance, 2 kΩ) and the saturation time increases on decreasing I_{PROG}.

During the SET operation and after a short delay time, V_{CELL} and I_{PROG} keep constant and, consequently, the electrical power $P_{SET} = V_{CELL} \cdot I_{PROG}$ provided to the cell is constant, as shown in [5]. This delay time is not related to the GST phase transition but it depends on the electrical switching time and measurement set–up parasitic effects [4]. Thus, P_{SET} can be used as a characteristic parameter of the SET operation once the pulse amplitude is fixed, while the energy E_{SET} provided to the cell during the same operation linearly increases with PW_S:

$$E_{SET} = P_{SET} \cdot \left(PW_S - T_0 \right) \tag{1}$$

where T_0 is derived as 35 ns [5]. On the basis of the previous considerations, we define as SET transition time T_{SET} the pulse width value yielding a resistance approximately R*= 5 kΩ. The value of R* is chosen as the indicator of RESET to SET phase transition completeness: for example R* = 10 and 7 kΩ indicate a partial SET, while we consider R*= 5 kΩ an indication of a complete SET process (see Fig. 8), which is actually completed when R reaches the heater resistance 2 kΩ.

We can study T_{SET} as a function of P_{SET}, and we did it also for cells with different geometries. Data represent the power delivered for a complete SET and they obey to the following law:

$$\left(P_{SET} - P_0 \right) \left(T_{SET} - T_0 \right)\Big|_W = K \cdot A \tag{2}$$

where A is the contact area. If we divide both terms by A, the characteristics for different A superimpose and we can plot T_{SET} as a function of power density P_{SET}/A, see Fig. 9.

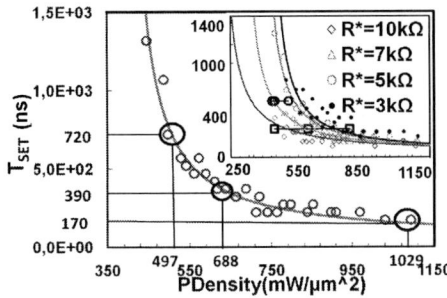

Fig. 9 T_{SET} vs. Power Density. Symbols are experimental data, lines results from eqn. (2). In the inset, the same data are plotted for different R*. Marks correspond to the cycling conditions used in Figs. 2 and 3, for PW_S = 600 and 300 ns.

K is a constant independent of area and represents the energy density (per unit area) needed to achieve a complete change of phase of the amorphous GST material. P_0 is the power at melting point, defined as the point where the low resistance curve changes slope, as in [7], for these cells is 390 mW/μm². Since in (2) all the quantities are known, apart from K, the energy density provided to the heater-GST stack during programming can be evaluated. In our case this quantity lies in the range of $8 \cdot 10^{-7}$ J/μm².

The inset in Fig. 9 shows that a similar law can be obtained for partial SET (for example, choosing R* = 10, 7, 5, 3 kΩ), as well. Moreover, it also justifies the choice of R*=5 kΩ as a complete SET indicator, since the curves obtained for R* < 5 kΩ superimpose.

IV. MODELS

A better understanding of SET can be achieved from the analysis of results obtained using two analytical models presented in the literature able to reproduce the phase transition dynamics [9], [10].

1. Model A [9]

With this numerical model for chalcogenide glasses [9], assuming that only holes are involved in conduction [11], we obtained the results shown in Fig. 10 (line). The slope of the line is 2kΩ and it is the heater value. In the same plot we superimpose the I-V values that correspond to the experimental data associated to R*=5kΩ in Fig. 8.

The slope in Fig. 10 gives R= 2kΩ, different from the I-V values associated with the symbols, derived when R* is 5kΩ in Fig. 8. Indeed, this data, together with data from the inset in Fig. 9, confirm that R* = 5kΩ can be considered the threshold of a complete SET operation, and used to characterize the SET process.

We used the model in [9], [11] to calculate the electrical power P_{SET} and result show the same trend of (2) see Fig. 11.

978-1-4244-6658-0/10 $26.00 © 2010 IEEE

Parameters T_0, P_0, k, have the same values derived experimentally.

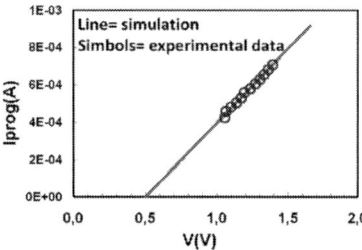

Fig. 10. IV characteristics result from model A. The red line corresponds to the analytical simulation, the circles correspond to experimental data associated to R*=5kΩ defined in Fig. 8.

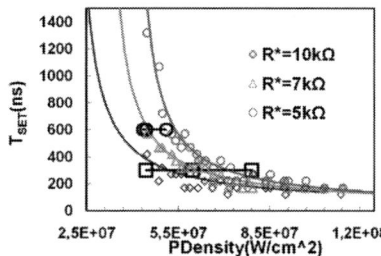

Fig. 11. T_{SET} vs Power density, Symbols are experimental data, lines results from model A and Model B. Marks correspond to the cycling conditions in Figs. 2 and 3.

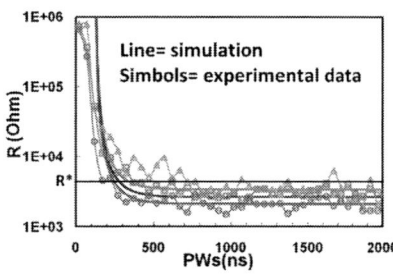

Fig. 12. R_{SET} vs. PW_S The lines represent the simulations and the symbols correspond to experimental data, for different I_{PROG}.

2. Model B [10]

This is a simple physical model of threshold switching in phase change memory cells based on the field induced nucleation of conductive cylindrical crystallites [10]. We implemented it and obtained the I_{PROG} and R_{SET} vs. PW_S. In Fig. 12 we show R vs. PW_S, and the agreement with experimental data is excellent. Following the same procedure as in Fig. 8, we calculate the electrical power density needed to obtain R*= 10, 7 and 5kΩ. Results are shown in Fig 10, and they superimpose exactly to the results obtained with Model A, with the same parameters T_0, P_0, and k.

Fig. 11 shows also T_{SET} and Power conditions used for endurance experiments in Figs. 2 and 3. If we add the information from Fig. 4, we can conclude that cycling with partial SET increases the number of cycles before failure. The curve obtained for R*=5kΩ is representative for a complete SET and will give the worst conditions for cycling.

V. CONCLUSIONS

In this paper we present results on endurance of PCM cells. We show that SET operation determines the maximum number of cycles. We also investigate SET operation and propose a new characterization methodology confirming that the SET operation is a constant energy operation, thus suggesting that GST crystallization and amorphization have different underlying physics. However, microscopic aspects related to GST phase change (as discussed, for example, in [8]) cannot be analyzed with the proposed measurement technique, which mainly aims to support an optimized design rule of memory device operating conditions in terms of SET time and power.

We have also used two models for phase change transition proposed in the literature to confirm the constant power relation for a correct SET. Increasing SET power is therefore a possible endurance failure trigger.

REFERENCES

[1]. S.Ovshinsky, "Reversible electrical swirching phenomena in disordered structures," *Phys. Rev. Lett.*, vol. 21, no. 20, pp. 1450-1453, Nov 1968.

[2]. B.C. Lee et al., "Phase-change Technology and the future of main memory," *IEEE Micro*, vol. 30, pp. 131-141, Jan/Feb 2010.

[3]. G.W.Burr et al., "Phase change memory technology," *J.Vac.Sci.Technol.*, pp. 223-263, Mar 2009.

[4]. H.-S. P. Wong S. Kim, "Analysis of Temperature in Phase Change Memory Scaling," *IEEE Electron Dev. Letters.*, vol. 28, no. 8, pp. 697-699, 2007.

[5]. G. Puzzilli et al., "On the RESET-SET Transition in Phase Change Memories," *ESSDERC*, pp. 158-161, 2008.

[6]. L.Goux et al., "Degradation of the reset switching during endurance testing of a phase-change line cell," *IEEE Trans. Electron Dev*, vol. 56, pp. 354-358, 2009.

[7]. I. V. Karpov et al., "SET to RESET Programming in Phase Change Memories," *IEEE Elec. Device Lett*, vol. 27, no. 10, Oct 2006.

[8]. A. Lacaita et al., "Electrothermal and phase-change dynamics in chalcogenide-based memories," *IEDM Tech. Digest*, pp. 911-914, 2004.

[9]. A.Redaelli et al., "Threshold switching and pahse transitino numerical models for phase change memory simulation," *J. Appl Phys*, vol. 103, no. 11, p. 1101, 2008.

[10]. A. Chimenton et al., "A new analytical model of erase operation in phase change memories," *IEEE Electron Dev. Letters*, 2009.

[11]. M. Boniardi et al., "A physics-based model of electrical conduction decrease with time in amorphous Ge2Sb2Te5," *J. Appl Phys*, vol. 105, no. 08, p. 4506, 2009

Study of N-induced traps due to nitrided metal gate in HK/MG nMOSFETs

M. Cassé, X. Garros, O. Weber, F. Andrieu, G. Reimbold, F. Boulanger

[1] CEA-LETI Minatec, 17 rue des Martyrs 38054 Grenoble Cedex 9, France

Abstract— We report an experimental study of the defects induced by the TiN metal gate. N-induced defects are evidenced and energy profile through the Si band gap is measured by original spectroscopic charge pumping measurements. The density of defects is then correlated to the electron mobility degradation and compared to a theoretical model.

I. INTRODUCTION

Today large efforts are made to integrate high-k dielectrics and metal gates in CMOS devices. The threshold voltage can be adjusted by using different metal gate materials, deposition processes or by varying the metallic layer thickness [1,2]. However, such advanced gate stacks have also an impact on the other electrical characteristics, like the inversion carrier mobility or the reliability [3]. Correlation between mobility and NBTI has been evidenced in HK/MG transistors due to nitrogen incorporation originating from the metal nitride and/or the nitridation of the high-k oxide [4]. The nitrided gate (TiN, TaN,…) acts as a nitrogen reservoir which diffuses through the oxide toward the Si channel interface and creates some defects (fig.1), at the origin of degraded performance.

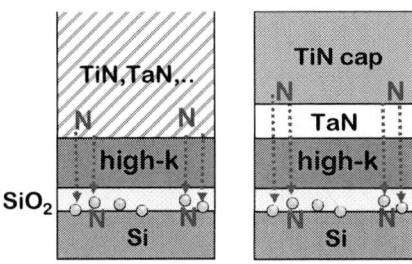

Fig.1: Model of degradation induced by N diffusion from nitrided metal gates (from Ref.[3]). Defects are created at the SiO₂/Si interface.

In this work, we present an experimental study of the effect of a TiN gate. Spectroscopic charge pumping has been performed to evidence the energy profile within the Si band gap of the interface traps created by N diffusion. The effect of these additional interface defects on electron mobility has then been investigated by low temperature measurements. Mobility results have finally been confronted to a theoretical model including additional Coulomb scattering induced by charges located at the Si interface, and discussed with regard to the D_{it} energy profile.

II. DEVICES DESCRIPTION AND EXPERIMENTAL SET-UP

For the purpose of our study, FD-SOI MOSFETs with varying PVD TiN metal gate thicknesses (from 3nm up to 15nm) were fabricated. A 3 nm thick HfO₂ dielectric deposited by ALD on a ~0.8nm interfacial SiO₂ oxide completes the gate stack [5]. Mobility and charge pumping

(CP) measurements were performed from room temperature down to 20K on long n-channel MOSFETs. For CP experiments on SOI devices, gated diodes with a N+ and a P+ contact are required (fig.2) in order to collect the charge pumping current [6].

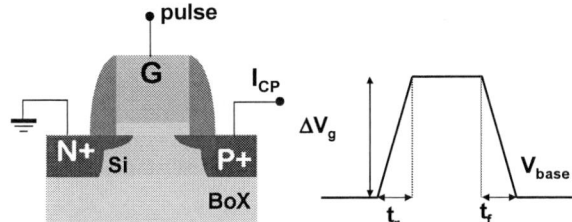

Fig.2: Schematics of the experimental set-up used for charge pumping spectroscopy. A trapezoidal pulse (amplitude ΔVg, rise and fall time t_r,t_f), with constant amplitude and varying base level V_{base}, is applied to the gate and CP current is measured on the P+ contact.

In the CP technique, a trapezoidal voltage pulse is applied to the transistor gate with a varying base level and a constant amplitude ΔV_g greater than the Si band gap (fig.2)[7,8]. This pulse alternatively fills the interface traps with electrons and holes, thereby causing a recombination current to flow in the P+ and N+ regions of the gated diode. By varying the base level V_{base} from accumulation to inversion, the measured charge pumping current I_{CP} has a typical "hat" shape. The maximum of the two-level CP current can be expressed as:

$$I_{CP} = qfA \int_{E_{em,h}}^{E_{em,e}} D_{it}(E)\, dE \qquad (1)$$

where A is the gate area, f the frequency of the pulse signal, q the electron charge, and $D_{it}(E)$ the density of interface traps. The integration boundaries are the hole and the electron emission levels $E_{em,h}$ and $E_{em,e}$, which are function of the temperature (T), the rise or fall time (t_r, t_f), and the capture cross-section of electrons or holes ($\sigma_{n,p}$). The mean value integrated over the band gap can be measured by sweeping the frequency of the pulsed signal using:

$$\overline{D_{it}} = \frac{1}{qA\Delta E_{em}} \frac{dI_{CP}}{df} \qquad (2)$$

The emission levels can be modulated by varying either the fall or rise time, or the temperature. In particular, varying t_r while keeping t_f constant, and reversely, allows to extract the D_{it} energy profile in the forbidden band gap [7,9] using:

$$D_{it}(E_{em}) = \frac{1}{qAfk_BT} \frac{dI_{CP}}{d\ln(t_{r,f})} \qquad (3)$$

By repeating the measurements at different temperatures from 20K to 300K, we can scan the energies E_{em} within the gap from conduction or valence band edge toward the midgap [10,11].

III. RESULTS AND DISCUSSION

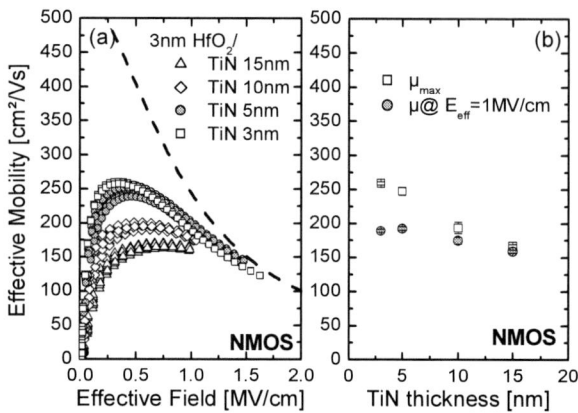

Fig.3: (a) Effective mobility as a function of electric field measured for different TiN layer thicknesses from 3nm up to 15nm. (b) Maximum mobility and mobility extracted at 1MV/cm as a function of TiN thickness. Clear mobility degradation is observed as the TiN thickness increases, mostly at low-medium electric field.

Fig.3 shows the electron mobility measured by standard split-CV at room temperature as a function of electric field on devices with different TiN thicknesses. Clear mobility degradation is observed as the metallic layer thickness increases, mostly at low-medium electric field, up to 1MV/cm (fig.3b). For thin metallic layer (below 5nm) the effective mobility saturates to its maximum value. These results show that a non negligible part of the mobility degradation observed in HK/MG MOSFETs can actually be due to the effect of the nitrided metal electrode. Indeed the TiN layer acts as a nitrogen reservoir; the thicker the layer is, the more N is released and can diffuse toward the Si channel. As a consequence, a higher density of interface traps is generated as the TiN thickness increases. In order to investigate further the nature of these N-induced defects, spectroscopic charge pumping has been performed on corresponding gated diodes (fig.4). Sweeping the temperature from 300K down to 20K has allowed us to scan the Si band gap in the E-E_i=±[0.3;0.58]eV energy range (E_i is the intrinsic Fermi level) [10]. The different spectra are plotted in fig.4 for 5nm, 10nm and 15 nm thick TiN.

As expected the mean D_{it} obtained from Eq.2 increases from $1.5\times10^{11}cm^{-2}eV^{-1}$ for the 5nm TiN up to $3.6\times10^{11}cm^{-2}eV^{-1}$ for the thicker TiN. The $D_{it}(E)$ energy profiles give a more detailed view of this TiN effect: a strong D_{it} peak appears in the upper half-part of the Si band gap for the thickest layers, at E-E_i~0.45eV, reaching $3\times10^{12}cm^{-2}eV^{-1}$ for 15nm TiN. In the lower half part of the Si band gap, a lower D_{it} peak is also observed symmetrically near E-E_i=-0.45eV. The energy profiles are gathered and re-plotted in linear scale in fig.5. They are compared with the energy distribution of Pb-centers usually observed on Si/SiO2 interface due to dangling bonds [12]. The defects observed in our devices clearly have a

different energy signature, attributed to the different nature of the N-induced defects.

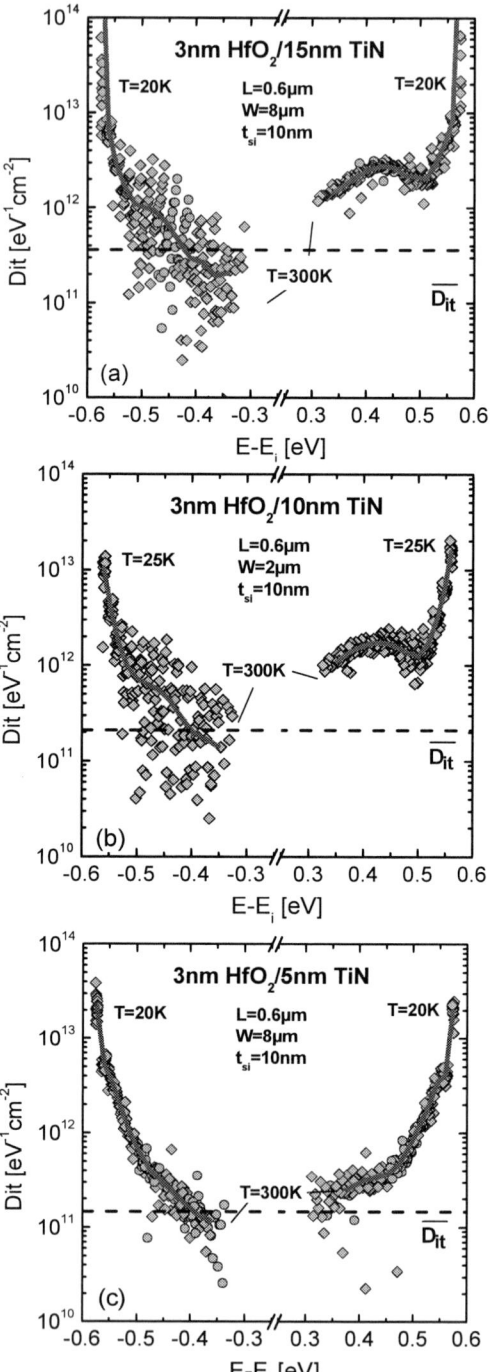

Fig.4: Energy distribution of interface traps obtained by spectroscopic charge pumping for SOI devices with: (a) 15nm, (b) 10nm, and (c) 5nm TiN metal gate. Profiles are obtained through low temperature measurements from 20K to 300K (see text). The bold line represents the mean value of $D_{it}(E)$. The dashed line is the measured mean value of interface trap density $\overline{D_{it}}$ over the full energy range at 300 K (see Eq.2). The intrinsic Fermi level E_i is used as energy reference.

Similar results have already been obtained on oxynitride SiON devices which strengthen our conclusions. Previous works have shown that oxide nitridation reduces the density of Pb-centers, but causes at the same time structural changes at the interface [13]. Thus electrical measurements have shown that the energy distribution across the Si band gap of NBTI-induced interface states varies with nitridation. Nitrided oxides exhibit a higher interface state density near the Si conduction band edge [14,15], whereas pure SiO_2 exhibits more states at mid-gap and close to the Si valence band edge [12]. Ab-initio calculations have also demonstrated energy levels for the N-related defects close to the bottom of the conduction band [16].

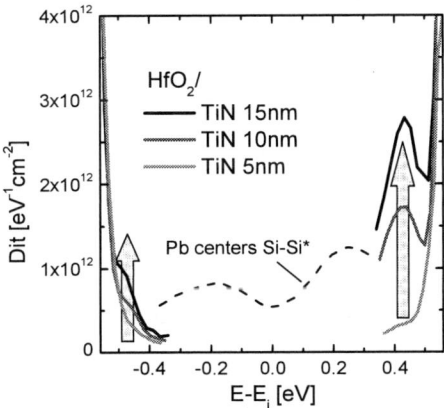

Fig.5: Linear plot of Dit energy profile obtained for the three TiN thicknesses, compared with a typical energy profile of Pb centers (dashed line) as usually observed at (100) Si/SiO_2 interface (see for example ref.[12]).

Now that the defects generated by N diffusion from the TiN layer have been clearly identified, we have then studied their impact on electron mobility. Fig.6 shows the effective mobility extracted by split-CV at different temperatures for 5nm, 10nm and 15nm thick TiN devices. Lower temperature dependence of the mobility is clearly observed with increasing TiN thickness. This suggests the contribution of an additional scattering mechanism. At very low temperature (typically below 20K), phonon scattering disappears and the effective mobility is only limited by Coulomb scattering (CS) at low electric field and surface roughness scattering (SRS) at high field [17]. The linear dependence with inversion carrier density N_{inv} at low electric field (dashed line in fig.6) reveals a stronger Coulomb scattering contribution μ_{CS} for thicker TiN. These results are well correlated with the D_{it} peak in the upper part of the Si band gap previously observed. If we consider that the N-induced defects are amphoteric as it is generally admitted for Pb-centers, then all the states above E_i are acceptorlike, and donorlike below E_i. Consequently, for NMOS in strong inversion, the Fermi level E_F lies above E_i, and all the traps between E_F and E_i are negatively charged (occupied acceptors), whereas those below E_i are neutral. These negative charges are located at the Si/SiO_2 interface and also probably in the oxide near that interface. For NMOS, the sharp D_{it} peak observed in figs. 4 and 5 is most probably responsible for the higher CS evidenced on low temperature mobility of thick TiN devices.

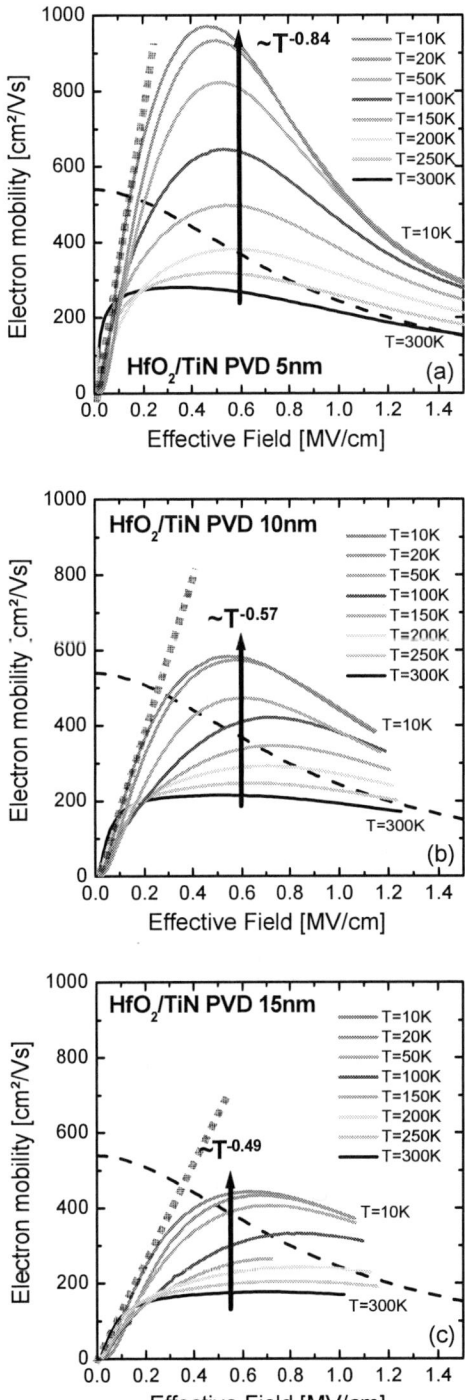

Fig.6: Electron mobility as a function of electric field measured at different temperatures, from 300K down to 10K, on long channel NMOS (L=10µm), with (a) 5nm TiN, (b) 10nm TiN and (c) 15nm TiN. At low temperature, the electron mobility is limited by Coulomb scattering (CS) and surface roughness scattering only. As the TiN thickness increases, the low temperature mobility saturates to lower values, showing the additional CS contribution of increasing Nit. Bold dashed lines are just a guide for the eyes to show the CS contribution μ_{CS}.

978-1-4244-6658-0/10 $26.00 © 2010 IEEE

Finally we have calculated the effect of interface traps on mobility using momentum relaxation time approach [18,19]. The scattering potential induced by a point charge at Si/SiO$_2$ interface has been evaluated by solving the Poisson equation through the gate stack, and corresponding $\mu_{CS,Nit}$ has then been calculated. This additional CS contribution has been added to the experimental TiN 5nm curve, used as reference here, and using the Mathiessen's rule:

$$\frac{1}{\mu_{tot}} = \frac{1}{\mu_{TiN\,5nm}} + \frac{1}{\mu_{CS,N_{it}}} \qquad (4)$$

The total mobility μ_{tot} obtained by this way is plotted for the TiN 10nm and TiN 15nm, using the amount of interface charges as a fitting parameter to experimental curves.

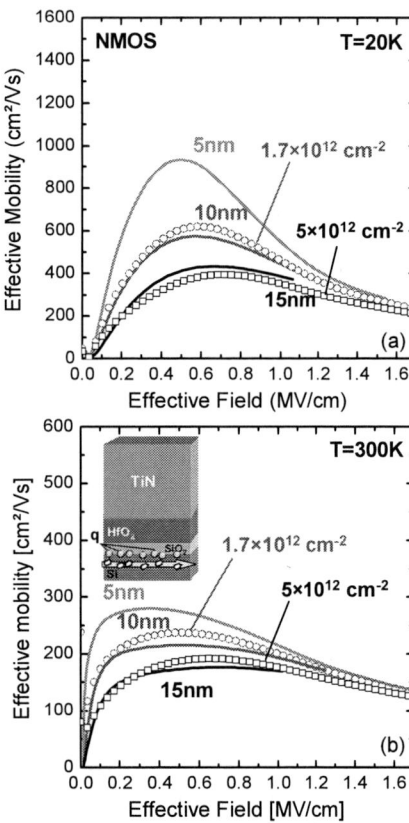

Fig.7: Calculated electron mobility for N_{it} =1.7× 10^{12} cm^{-2} and 5× 10^{12} cm^{-2} at SiO$_2$/Si interface (symbols) compared with experimental data of fig.6 (lines), at (a) 20K, and (b) 300K. Insert: schematic of the model used for N_{it}. The total mobility has been reconstructed using the calculated $\mu_{CS,Nit}$ added to the TiN 5nm curve following Eq.4.

Good agreement has been obtained between calculations and experimental curves with N_{it}=1.7×10^{12} cm^{-2} and N_{it}=5×10^{12} cm^{-2} for TiN 10nm and TiN 15nm respectively. The density of interface traps needed to fit the data seems higher – even if in a reasonable order ~10^{12}cm^{-2} – than the density measured by spectroscopic charge pumping (the Dit(E) profile in fig.4 must be integrated from E$_i$ to E$_F$ which lies in the CB). This "discrepancy" may suggest a high density of traps near/above the Si conduction band edge that can not be evaluated by CP. Furthermore in our calculation we have made the assumption

of surface charges at the Si/SiO$_2$ interface, which are the most effective to degrade mobility. A more likely distributed charge within few Å of SiO$_2$ could also lead to the same μ_{eff} degradation, reducing the amount at the interface needed to be introduced in our calculation.

IV. CONCLUSION

N-induced defects due to nitrided metal gate have been studied by spectroscopic charge pumping. The spectra reveal a sharp peak of D$_{it}$ near the conduction band edge (E-E$_i$=0.45eV), and a much lower one near the valence band edge in a symmetrical position, as a signature of these defects. The electron mobility is degraded accordingly, due to additional CS evidenced by low temperature measurements. Theoretical calculations well reproduce the experimental data, but by including slightly higher density of traps than the one measured by CP.

ACKNOWLEDGMENT

This work was partly funded by European PULLNANO and FOREMOST projects.

REFERENCES

[1] L. Colombo, A.L.P. Rotorandaro, M.R. Visokay, and J.J. Chambers, in "High Dielectric Constant Materials", chap.15, Springer (2005).

[2] K. Choi, H.-C. Wen, H. Alshareef, R. Harris, P. Lysaght, H. Luan, P. Majhi, and B.H. Lee, in *ESSDERC proceedings*, 101 (2005).

[3] G. Reimbold, J.Mitard, M.Cassé, X.Garros, C.Leroux, L.Thevenod, and F. Martin, *Proc. 207th ECS meeting: Silicon nitride, silicon dioxide thin insulating films, and other emerging dielectrics VIII*, 437 (2005).

[4] X. Garros, M. Cassé, G. Reimbold, F. Martin, C. Leroux, A. Fanton, O. Renault, V. Cosnier, and F. Boulanger., in *VLSI Symp. Tech. Dig.*, pp 68-69 (2008).

[5] F. Andrieu, C. Dupré, F. Rochette, O. Faynot, L. Tosti, C. Buj, E. Rouchouze, M. Cassé, B. Ghyselen, I. Cayrefourcq, *et al.*, in *VLSI Symp. Tech. Dig.*, pp. 221–224 (2006).

[6] T. Elewa, H. Haddara, S. Cristoloveanu, and M. Bruel, *J. Phys. Colloq.* **49**, 137 (1988).

[7] J.Brugler and P. Jespers, *IEEE Trans. Electron Devices* **16**, 297 (1969).

[8] G. Groeseneken, H. Maes, N. Beltran, and R. de Keersmaecker, *IEEE Trans. Electron Devices* **31**, 42 (1984).

[9] G. Van den bosch, G. Groeseneken, P. Heremans, and H. Maes,, *IEEE Trans. Electron Devices* **38**, 1820 (1991).

[10] M. Cassé, S. Thiele, K. Tachi, and T. Ernst, *Appl. Phys. Lett.* **96**, 123506 (2010).

[11] W. Wang, J. Deng, J.C.M. Hwang, Y. Xuan, Y. Wu, and P.D. Ye, *Appl. Phys. Lett.* **96**, 072102 (2010).

[12] G.J. Gerardi, E.H. Poindexter, P.J. Caplan, and N.M. Johnson, *Appl. Phys. Lett.* **49**, 348 (1986).

[13] J. Stathis and S. Zafar, *Microelectr. and Reliab.* **46**, pp.270-286 (2006).

[14] J. Stathis, G. LaRosa, and A. Chou, in *IRPS proceedings*, p1-7 (2004).

[15] S. Fujieda, Y. Miura, M. Saitoh, E. Hasegawa, S. Koyama, and K. Ando, *Appl. Phys. Lett.* **82**, 3677 (2003).

[16] E.-C. Lee , *Phys. Rev. B* **77**, 104108 (2008).

[17] S. Takagi, A. Toriumi, M. Iwase, and H. Tango, *IEEE Trans. Electron Devices* **41**, 2357 (1994).

[18] D. Esseni and A. Abramo, *IEEE Trans. Electron Devices* **50**, 1665 (2003).

[19] M. Cassé, L. Thevenod, B. Guillaumot, L. Tosti, F. Martin, J. Mitard, O. Weber, F. Andrieu, T. Ernst, G. Reimbold, T. Billon, M. Mouis, and F. Boulanger , *IEEE Trans. Electron Devices* **53**, 759 (2006).

978-1-4244-6658-0/10 $26.00 © 2010 IEEE

Carbon Junction Implant: Effect on Leakage Currents and Defect Distribution

Guntrade Roll[1], Stefan Jakschik[1], Matthias Goldbach[2], Thomas Mikolajick[1,3], Lothar Frey[4]

[1]Namlab gGmbH, Dresden, Germany
[2]Qimonda Dresden GmbH&Co OHG, Dresden, Germany
[3]Institut of Semiconductor and Microsystems Technology, University of Dresden, Germany
[4]Fraunhofer IISB, Erlangen, Germany
Corresponding author: email: Guntrade.Roll@namlab.com; phone 049 351 212499033

Abstract— **In this paper we present a detailed investigation on the influence of carbon co-implantation in the source/drain extension on leakage current and defect density in PFET transistors. Carbon is used to reduce the transient enhanced boron diffusion, to decrease short channel effects and to control the overlap length of the transistor. Leakage currents are measured and separated in order to analyze the influence of the carbon on the different MOSFET regions. An increase in carbon dose by a factor of 1.15 leads to an enhanced source/drain extension leakage which is caused by carbon induced defects. No effect of the carbon implantation on source/drain junction leakage was found as the co-implant is located above the source/drain depletion region. In addition an increase of gate induced drain leakage with carbon dose was observed. This increase is further analyzed by charge pumping technique.**

I. INTRODUCTION

The continued improvement in CMOS technology requires transistor scaling while maintaining acceptable leakage current. Halo and source/drain extension implants (SDE) are used to reduce short channel effects. This includes carbon co-implantation to minimize the boron transient enhanced diffusion in PFETs. There are numerous studies on the effect of carbon co-implants on dopant diffusion [1, 2]. It was also suggested using carbon for phosphorous halo diffusion control [3], and carbon cluster implants are in discussion for 26nm CMOSFET [4]. It has been shown that C interstitial defects increase junction leakage currents [5]. In this paper we present the missing detailed study on the effect of the SDE carbon co-implant on the leakage current of the different PFET regions. The leakage current contributions are separated and identified, and leakage mechanisms are analyzed to distinguish between the influence of the electric field and the change in defect density for samples with different C doses. This allows retrieving information on the influence of the C on electrically active traps which are further analyzed by charge pumping (CP) measurements. The transistors used in this study are DRAM peripheral MOSFETs. To increase retention, the devices are subject to a long term high temperature anneal. This step is the key to reduce defect density and junction leakage [6], for low standby power DRAM targets.

II. PROCESSING

Silicon oxide was grown and followed by a nitridation step. Samples with two different gate oxide thicknesses with EOTs of 2.3nm (Fig. 1) and 5.2nm were investigated. The gate electrode is a polysilicon (B-doped) - tungsten stack. An oxide spacer was deposited to protect the gate edge and the source/drain extensions were implanted with 20keV and $3 \cdot 10^{14}$atm/cm² Ge, 3keV and $4 \cdot 10^{14}$atm/cm² BF$_2$, and C. The wafer with high C dose had an implant of 4keV and $4 \cdot 10^{14}$atm/cm². For the wafer with the low C dose $3.5 \cdot 10^{14}$atm/cm² was used. The halo, a region of increased doping at the gate edge, was implanted in this step using 40keV and $2.3 \cdot 10^{13}$atm/cm² P. Afterwards the silicon substrate was recrystallized and a second spacer was deposited prior to the formation of the source/drain region. The source/drain implants contained Ge with an energy of 20keV and a dose of $3 \cdot 10^{14}$atm/cm² and BF$_2$ with 2.2keV and $2.5 \cdot 10^{15}$atm/cm². Subsequently, the substrate was recrystallized using a 1020°C rapid thermal anneal step. Finally, the defect concentration was reduced with an anneal of 800°C for 30min. The resulting junction profile as measured by SIMS is presented in Fig. 2.

III. MEASUREMENTS AND RESULTS

A. Measurements

MOSFET chains connected in series with an overall area of

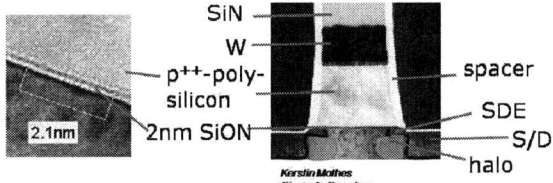

Fig. 1 Overview of the fabricated PFET samples. The PFET has a silicon oxynitride gate oxide with a thickness of about 2.1nm.

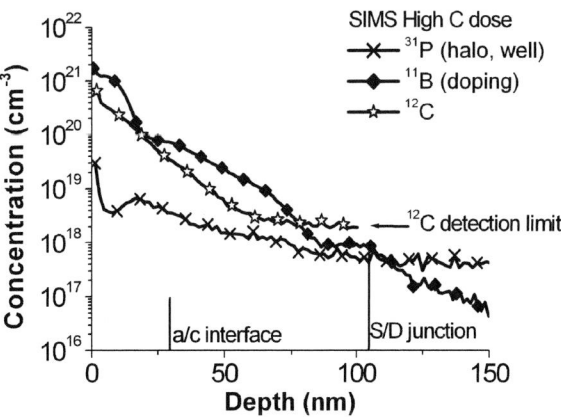

Fig. 2 Source/drain junction profile of the fabricated PFETs. The indicated a/c interface was estimated from the Ge implantation depth calculated by SRIM.

978-1-4244-6658-0/10 $26.00 © 2010 IEEE

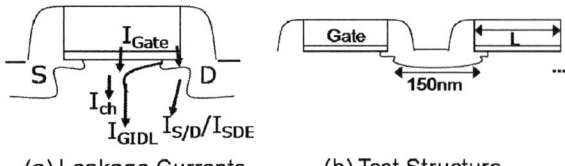

(a) Leakage Currents   **(b) Test Structure**

Fig. 2(a) Overview of the different leakage current paths in the transistor.
Fig. 2(b) Schematic sketch of the test structure.

$5000\mu m^2$ and gate lengths from 65nm to 5μm were used for electrical characterization. The test structure is presented in Fig. 3b. The number of connected devices varies for each gate length. The gate width is either 5μm or 148μm. The area to perimeter ratio is between 0.0325 and 2.5.

Temperature dependent current voltage - and capacitance voltage measurements were performed at the Keithley 4200 measurement unit with fA- and pF resolution. For off current measurements the gate was biased at 0V and the source and drain were connected together to measure the bulk-, gate- and source/drain leakage currents simultaneously. Fig. 3a presents a schematic of the leakage current. In order to analyze the defect density in the different space charge regions of the transistor the leakage currents have to be separated.

B. Source/Drain-, SDE- and Channel leakage

The bulk is biased and the current is measured at the source/drain region (Fig. 4a). Test structures with different area perimeter ratios were measured to distinguish between the area dependent channel current and the perimeter dependent source/drain (S/D) - and SDE leakages. The area- and perimeter current for a certain test structure can then be calculated from the current density as shown in Fig. 5. This procedure is similar to the calculation of area and perimeter currents in pn-junctions [6]. The current to the side-walls of the transistor can be neglected due to the shared source/drain of the transistors (Fig. 3b). A diode test structure was used to distinguish between the current from the SDE and S/D regions. The leakage current in the long channel device of Fig. 5 is due to the SDE in the high voltage regime. For the short channel samples, with a high width to area ratio the overall leakage is controlled by the SDE contribution.

Leakage current mechanisms are analyzed to distinguish between the influence of the electric field and the change in defect density on the current. Thermal generation, direct band to band tunneling, thermal assisted tunneling via defects, and Frenkel-Poole current are possible leakage mechanisms. References [7], and [8] present methods to separate the different mechanisms using the activation energy.

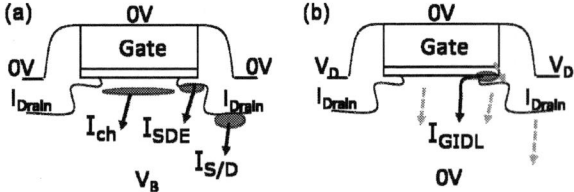

Fig. 4(a) Different leakage current contributions when the bulk is biased.
Fig. 4(b) Different leakage current contributions when S/D is biased. The GIDL leakage (black line) can be calculated subtracting the current measured in 4(a) and subtracting the gate overlap leakage.

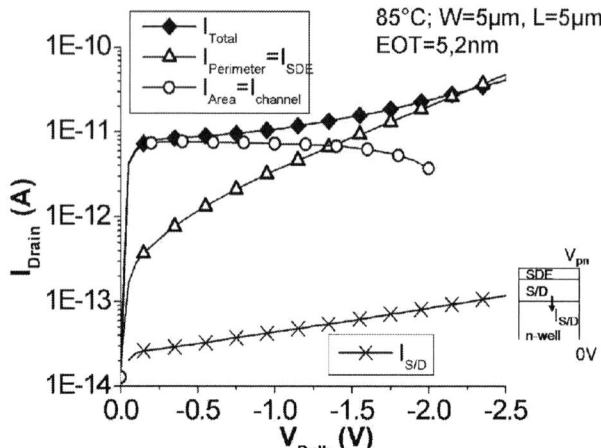

Fig. 5 Separation of the area dependent channel leakage (I_{ch}), and the perimeter dependent source/drain leakage ($I_{S/D}$) and the source drain extension leakage (I_{SDE}) for a long channel device with an EOT of 5.2nm. $I_{S/D}$ can be neglected for the overall leakage (I_{Total}).

Channel Leakage

The channel current is a generation current which occurs in the depletion region beneath the channel (V_{fb}=0.8V). The temperature dependent measurements reveal a thermal generation current with trap energy between 0.3eV and 0.45eV for all the samples. As expected the channel leakage or trap density is not affected by the carbon implant.

Source/Drain Leakage

The S/D leakage reveals thermal assisted tunneling via defects with a trap energy of 0.61eV, fitted for an effective mass of 0.16 of the electron mass [7, 8]. Previous studies [7] indicate that the S/D leakage is not affected by the carbon implant, because the carbon is located above the S/D depletion region. The pn-junction is placed at approximately 110nm below the Si surface (Fig. 2). The C concentration at this point is lower than the SIMS detection limit.

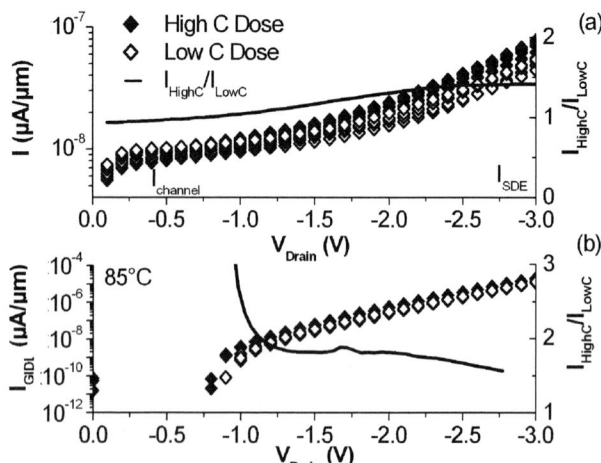

Fig.6 (a) Dependence of the channel- and the SDE leakage on the C dose. The current ratio for the high C -, and low C dose samples is 1 for the channel - and 1.4 for the SDE leakage for the devices with an EOT of 5.2nm.
Fig 6 (b) Influence of the carbon dose on the GIDL current. The current ratio is approximately 2 in the trap assisted tunneling region for the device with an EOT of 5.2nm.

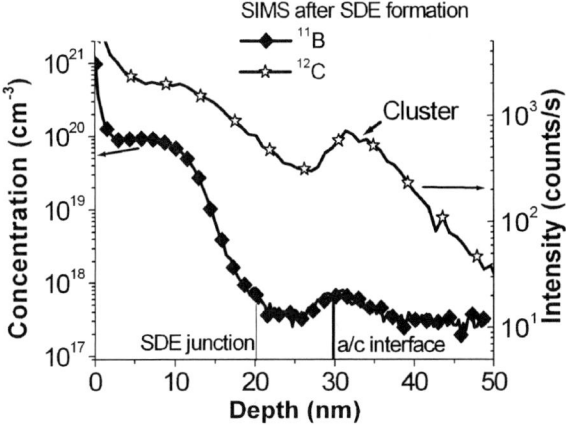

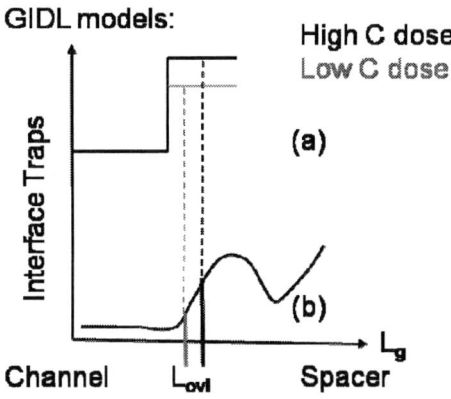

Fig. 7 Source/drain extension SIMS profiles prior to annealing. A clustering of C and B close to the amorphous/crystalline interface was found.

Fig. 8 Two different models that could explain the increase GIDL:
(a)   An increase of carbon related defects.
(b)   A steep increase of interface defects with overlap length.

SDE Leakage

For the SDE leakage a decrease of the activation energy with increasing voltage is observed (not shown here). A thermally assisted tunneling or Frenkel-Poole mechanism is proposed. The higher carbon dose increases the leakage by a factor of 1.4 (Fig. 6a). This increase is probably due to excess carbon interstitials in the SDE depletion region. The SDE junction is approximately 20nm from the Si interface (Fig. 7). The SIMS results measured after SDE implantation prior to annealing reveal a clustering of boron and carbon at a depth of about 30nm. This is the estimated amorphous/crystalline interface determined from SRIM calculations (Fig. 7). It is known that end of range defects form at the amorphous/crystalline interface [9]. A leakage current increase due to carbon interstitial defects in the pn-junction depletion region was observed in earlier studies [5]. The clustering is dissolved during annealing. However, the authors suggest that small amounts of interstitial defects remain and lead to increased

defect density.

C.   GIDL Leakage

The GIDL leakage is especially sensitive to defects at the interface between the gate oxide and silicon at the gate edge. The density of defects at the gate edge is expected to be higher than in the gate channel area and influences the reliability [10]. GIDL was measured as part of the overall off-current (I_{Drain}) with the gate and bulk contact biased to 0V for increasing source/drain voltage (Fig. 4b). The GIDL current is calculated by subtracting the gate leakage (I_{Gate}) and the leakage ($I_{SDE+channel}$) that is measured at the drain in the configuration of Fig. 4a (1) [11].

$$I_{GIDL} = I_{Drain} - I_{Gate} - I_{SDE+channel} \qquad (1)$$

The temperature dependent GIDL reveals thermal assisted tunneling via defects in the low voltage region and band to band tunneling in the high voltage region (not shown).

The interface trap density is sensitive to the thermal assisted tunneling current. An increase in GIDL by a factor of 2 for the samples with high carbon dose is found (Fig. 6b). Taking into account the overlap length the current should decrease with increasing C dose, because the effective area and electric field are reduced. The difference in overlap length between the two samples is small and within the statistical variations of the split capacitance voltage measurements. We propose two possible models for this GIDL increase (Fig. 8). The first model indicates an increase of the interface defect concentration due to the carbon implantation (Fig. 8a). The second model suggests a rapid change in defect concentration at the gate edge so that even a small change of the overlap length would lead to an enhancement of the leakage current (Fig. 8b). Charge pumping measurements were performed to analyze the source of this GIDL increase.

D.   Charge Pumping

Charge Pumping (CP) is a powerful and well known method to characterize defects in the gate oxide to silicon interface [12]. However, lateral profiling is not easily done. To analyze the interface defects in the gate area a frequency sweep,

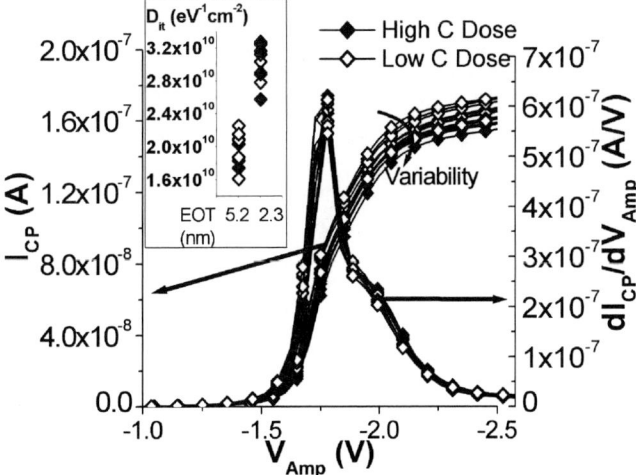

Fig. 9 Results of the charge pumping amplitude sweep, revealing a side peak caused by the increased V_{th} at the gate edge due to halo implantation. No difference between high - and low C doses can be found within the statistical variation. The inset shows the average D_{it} in the gate area for the thick oxide sample $(1.96\pm0.63) \cdot 10^{10}\text{cm}^{-2}\text{eV}^{-1}$ and the thin oxide $(3.02\pm0.70) \cdot 10^{10}\text{cm}^{-2}\text{eV}^{-1}$ for a amplitude of 1,5V.

978-1-4244-6658-0/10 $26.00 © 2010 IEEE   251

whereby the rise time and fall time were 20% of the pulse period was performed. From these measurements the interface trap density (D_{it}) can be calculated [12].

We adapted the method developed in reference [10] to compare the GIDL current results with the CP measurements. The halo implant leads to an increased doping at the gate edge and, therefore, a threshold voltage shift. By analyzing the variable amplitude measurements a side peak was found in the derivation of the CP current with the amplitude (Fig. 9). The maximum of this side peak should be proportional to the defect density at the gate edge.

No change of this side peak in the derivation curve was found for the different C doses within the accuracy of the measurement (Fig. 9). This leads to the conclusion that if there is a change of interface defect concentration at the gate edge due to the carbon implants it is either very small or highly localized. The average D_{it} is given in the inset of Fig. 9. A comparison of the thin oxide sample with an EOT of 2.2nm to the thick oxide sample shows that defect density increases by a factor of 1.5 for the thin oxide sample. This could be due to B penetration in the silicon oxynitride from the polysilicon.

IV. SUMMARY AND CONCLUSION

This study presents methods to separate and analyze the components of the MOSFET leakage current which are used to determine the defect concentration change in different transistor regions.

We focused on the impact of carbon co-implantation in the source/drain extension (SDE). The ratio between the different carbon implantation doses was only 1.15. The change in the overlap length of the SDE is below the split capacitance voltage detection limit. The results are summarized in Fig. 9. As expected, no defect density change occurs underneath the channel. The authors found the source/drain leakage current to remain unchanged because the carbon is located above the space charge region. This is not the case for the shallower SDE where a leakage/defect ratio between high - to low

carbon dose of 1.35 was determined. SIMS measurements reveal a carbon clustering below the amorphous/crystalline interface prior to annealing. We conclude that carbon interstitial defects in this region increase the leakage current. An increase in GIDL leakage due to interface defects was also found. Further work is required to clarify this effect.

Carbon co-implants are suitable to produce shallow junctions, avoiding advanced anneals such as laser - or flash anneal. It has been shown how the leakage currents are influenced by the amorphization depth and carbon implant energy. Leakage currents can be reduced significantly by the chosen annealing conditions.

ACKNOWLEDGMENT

The author would like to thank Heiko Hortenbach for SIMS measurements, Kerstin Mothes for the TEM pictures. This work is supported by the German Federal Ministry of Education and Research in the framework of the project MEGAEPOS.

REFERENCES

[1] Graoui, and Foad, "A comparative study on ultra-shallow junction formation using c-implantation with fluorine and carbon in pre-amorphized silicon," Materials Science and Engineering B, vol. 124-125,pp. 188-191, 2005.

[2] Pawlak, Janssens, Brijs, Vandervorst, Collart, Felch, and Cowern, "Effect of amorphization and carbon co-doping on activation and diffusion of boron in silicon,"Applied Physical Letters, vol. 89, 062110, 2006.

[3] Pawlak, Duffy, Augendre, Severi, Janssens, Absil, Vandervorst, Collart, Felch, Schreutelkamp, Cowern, "The carbon co-implant with spike RTA solution for phosphorous extension," Proceedings of the Material Research Society Spring Meeting 2006 Symposium C, vol. 912, 0912-C01-06, April 2006.

[4] Yako, Yamamoto, Uejima, Hase, and Hane, "26nm gate length CMOSFET with aggresively reduced silicon position by using carbon cluster co-implanted raised source/drain extension structure," Proceedings of Symposium on VLSI Technology 2009, pp.160–161, June 2009.

[5] Tan, Chor, Lee, Quek, and Chan, "Enhancing leakage supression in carbon-rich silicon," Electron Device Letters, vol. 27, no. 6, pp. 442-444, 2006

[6] Poyai et al.," High purity silicon VI: Lifetime and leakage current studies in shallow p-n junctions fabricated in a deep high-energy boron implanted well", Proceedings of the Electrochemical Society, vol. 2000-17, p. 404, 2000.

[7] Guntrade Roll, Matthias Goldbach, and Lothar Frey, "Leakage current and defect characterization of p+n source/drain diodes," submitted to IEEE Transactions on Electron Devices.

[8] Hurkx, Graaff, Klosterman, and Knuvers,"A new analytical diode model including tunneling and avalanche breakdown," IEEE Transactions on Electron Devices, vol. 39, no. 9, pp. 2090, 1992

[9] Pichler, and Stiebel, "Current status of models for transient phenomena in dopant diffusion and activation," Nuclear Instruments, and Methods in Physics Research B, vol. 186, pp. 256-274, 2002.

[10] Chen, Balasinski, and Ma, " Lateral profiling of oxide charge and interface traps near the MOSFET junctions," IEEE Transactions on Electron Devices, vol. 40, no. 1, pp. 187, 1993.

[11] Gilbert, Rideau, Dray, Agut, Minondo, Juge, Masson, and Bouchakour, "Characterization and modeling of gate-induced-drain-leakage," IEICE Transactions on Electron Devices, vol. E88, no. 5, pp.829, 2005.

[12] Groeseneken, Maes, Beltran, and Keersmaecker, " A reliable approach to charge-pumping measurement in MOS transistors," Transactions on Electron Devices, vol. ED31, no.1, pp.42-53, 198

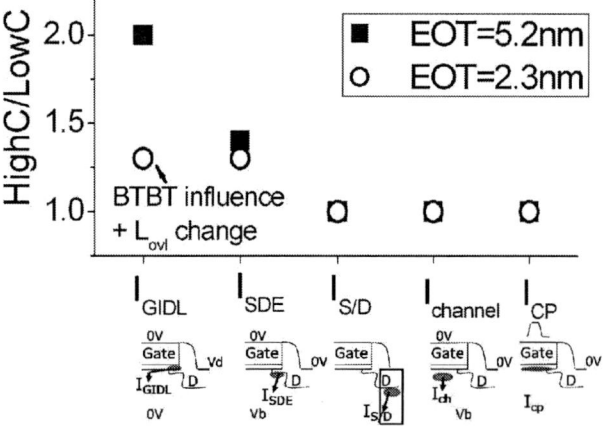

Fig 9 Measurement results for the thick and thin oxide samples are summarized. No influence of the C dose on to the channel-, source/drain- and CP current is found. The SDE leakage is increased by a factor of 1.3-1.4 by the higher carbon dose. The GIDL leakage is also increased with increased C dose. The thin oxide sample shows less sensitivity in GIDL to carbon dose. The overlap length is approximately 7.3nm for the thick and 6.8nm for the thin oxide sample.

Grain boundary-driven leakage path formation in HfO$_2$ dielectrics

G. Bersuker, J. Yum, V. Iglesias[1], M. Porti[1], M. Nafría[1], K. McKenna[2], A. Shluger[2], P. Kirsch and R. Jammy

SEMATECH, Austin, TX, USA

[1]Universitat Autònoma de Barcelona, Spain

[2] University College London, United Kingdom, and Tohoku University, Japan

Abstract— **The time evolution of the leakage current in HfO$_2$ based MIM capacitors under constant voltage stress applied either continuously or with periodic interruptions was studied for a range of stress voltages and temperatures. The data analysis was performed based on the results of the conductive AFM measurements demonstrating preferential current flow along grain boundaries (GB) in the HfO$_2$ dielectric and *ab initio* calculations, which show the formation of a conductive sub-band due to oxygen vacancy precipitation at GBs. The proposed model suggests that the observed reversible increase in leakage current is caused by the defects segregation at GBs and electron trapping/detrapping at these defects. The energy characteristics of the electrically active defects extracted from the electrical measurements match well with those calculated for neutral oxygen vacancies in hafnia.**

INTRODUCTION

Metal oxides are finding ever widening applications in microelectronics, specifically in the gate stacks of advanced devices (high-k dielectrics) and resistive memory (ReRAM). Consequently, understanding the mechanisms controlling the electrical characteristics and reliability of this class of materials is of primary importance. Towards this goal, the nature of electrically active defects in metal oxides must be addressed up front. Since the characteristic feature of dielectric degradation under electrical stress is an increase in leakage current and its subsequent breakdown, we need, more specifically, to identify the defects contributing to the current through these dielectrics and the specific structural features of these materials, which might facilitate stress-induced defect generation.

Until recently, reliability studies of these dielectrics focused mainly on MIS gate stacks used in transistor structures. However, the electrical characteristics of MIS structures are strongly affected by the SiO$_2$ layer, usually present at the interface between the substrate and the high-k gate dielectric that complicates evaluation of the intrinsic properties of the latter. On the other hand, MIM structures (directly relevant to ReRAM, RF, etc. applications), which do not generally include a "parasitic" dielectric film, allow measurements of only a few electrical characteristics (primarily, leakage current), which limits the ability to obtain and analyze characteristics of electrically active dielectric defects. As a consequence, analysis of the MIM electrical data must rely on a physical model of the process describing the leakage current.

Because of extensive experience in the processing, treatment, and structural and electrical properties of high-k HfO$_2$ dielectric, it is a convenient model material with which to investigate reliability mechanisms. In this study, by analyzing the time-dependent characteristics of the leakage current in TiN/HfO$_2$/TiN capacitors subjected to constant voltage stress, we extracted characteristics of the defects in HfO$_2$ that control the leakage current and identify their possible atomic structure using the results of *ab initio* calculations.

PROPERTIES OF THE HFO$_2$ GRAIN BOUNDARIES

The MIM structures were fabricated on 200 mm p-epi wafers using TiN metal electrodes and 7 nm ALD HfO$_2$ dielectric. Leakage current was measured on 10^{-5} cm^2 capacitors subjected to a constant voltage stress (CVS) in the voltage and temperature ranges of V_g = 1.4~2.0 V and T = 25°C~100°C, respectively. For each of these conditions, CVS was applied for 3000 sec either continuously or interrupted for various durations after every 1000sec of stress and then resumed, Fig. 1. During stress interruptions, both bottom and top electrodes were either grounded or floated; either condition resulted in identical subsequently measured current

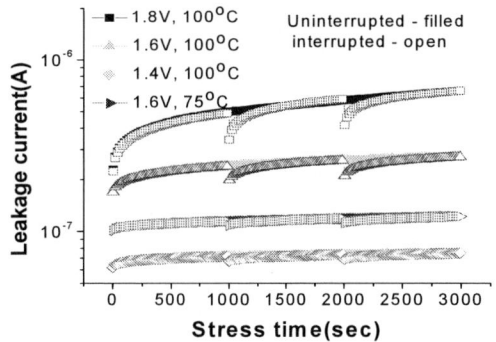

Figure 1. Examples of the stress current time dependencies in TiN/7 nm HfO$_2$/TiN MIM capacitor under CVS. At each stress condition, the current was measured both continuously (filled symbols) and with stress interruption every 1000s when the bias was switched to 0V for 300s and then resumed (open symbols).

978-1-4244-6658-0/10 $26.00 © 2010 IEEE

values. The post-interruption stress currents were lower than those pre-interruption; the current decrease depended on the interuption time. The 300sec interruption was determined to be sufficient for the current drop to reach near-saturation under all stress conditions.

Because of good device-to-device uniformity, we can compare the uninterrupted and interrupted stress current data (collected on different fresh devices) by matching adjacent devices, which exhibited nearly identical currents during the initial 1000 sec of stress. The characteristic feature observed within the entire range of the CVS conditions is that after the stress is resumed, the current increases much faster until it merges with the uninterrupted current.

To interpret the above electrical data, we focused on the structural properties of the dielectric. As-deposited sufficiently thick (> 4 nm) ALD HfO$_2$ films are expected to be crystallized [1]. Conductive atomic force microscopy (CAFM) data collected on a crystallized 5 nm HfO$_2$ film deposited using the same ALD process, Fig.2, exhibits perfect correlation between the (a) topographical and (b) current image obtained at 6.5V under ultrahigh vacuum (UHV) conditions using diamond-coated Si tips. CAFM measurements can distinguish individual nanocrystals and allow a magnitude of the current both along and across the grain boundary (GB) (depression in the topographical profile) to be profiled. Results demonstrate that the current through the crystalline dielectric preferentially flows along the grain boundaries, consistent with scanning tunneling microscopy (STM) measurements [2].

To identify the structural properties of the GB responsible for the current flow, we performed *ab initio* calculations (see the methodology in [3, 4]) of the GB structures in monoclinic HfO$_2$. Several possible GB

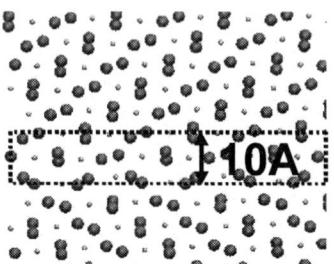

Figure 3. Ab initio model of GB in monoclinic HfO$_2$. O sub-lattice are the large red balls, Hf atoms are shown by small green balls. Boxed area outlines a GB region

configurations were explored; here we show the (101) twin boundary in m-HfO$_2$, which is constructed by mirroring the (101) surface plane, Fig.3. It has a low formation energy of 0.60 Jm^{-2} because there are no undercoordinated ions and open voids as are sometimes seen in other boundaries. Therefore, the defect segregation is less profound in this boundary than in others, making it a useful model system as more general boundary configurations might be expected to exhibit even stronger effects.

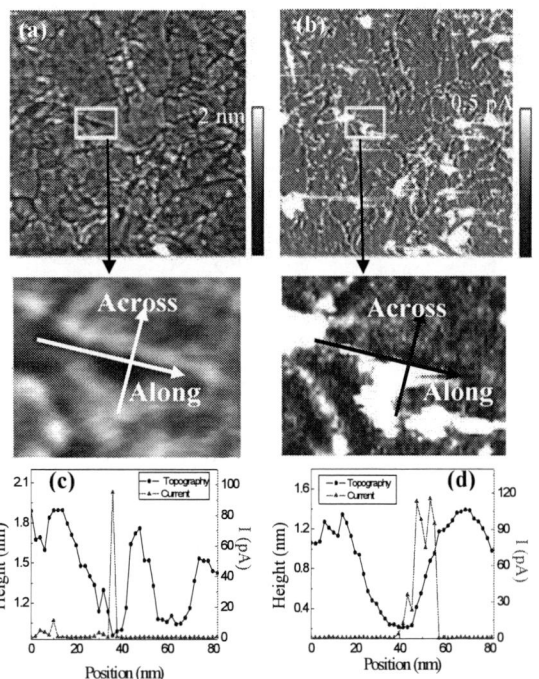

Figure 2. (a) Topographical and (b) current CAFM images obtained at 6.5V. (c) and (d) correspond to topographical and current profiles obtained across and along the GB, respectively

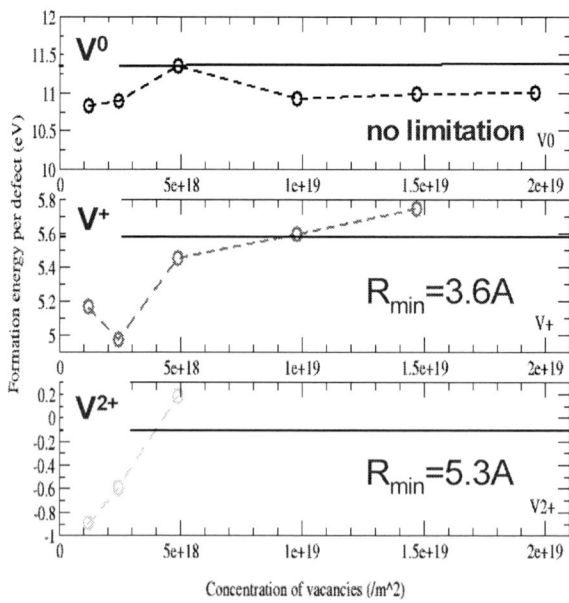

Figure 4. Formation energy vs. concentration of neutral and positively charged vacancies at GB. R$_{min}$ indicates the minimum achievable distance between the vacancies in GB

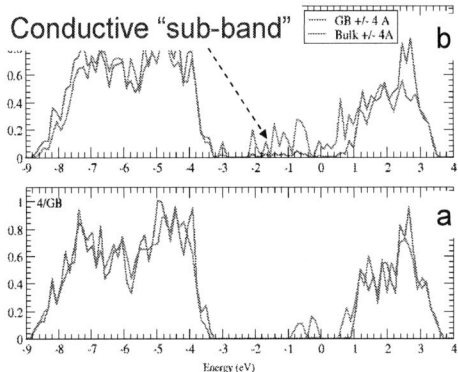

Figure 5. Density of states associated with the bulk of the monoclinic HfO₂ grain and GB. (a) and (b) correspond to low and high concentration of V^0 at GB, respectively.

When diffusing neutral or positively charged O-vacancies (randomly or field-driven, respectively) encounter a GB region, the width of which in this model is approximately 1 nm, they tend to segregate there with an energy gain of 0.40eV and 0.65eV, respectively. The activation energy for a positive O-vacancy diffusion is calculated to be 0.65–0.70 eV, while for a neutral vacancy it is ≥2 eV. As expected, neutral O-vacancies can achieve a much higher density at the GB, Fig.4. Positively charged vacancies may reach a similar density when neutralized during stress by, for instance, capturing the injected electrons.

At a high enough density of the O-vacancies at the GB, a localized conductive "sub-band" is formed along the GB, as demonstrated by the high density of the localized gap states, Fig.5, which effectively constitute a percolation path for the current flow. Electrons injected into the dielectric may be trapped by the neutral vacancies at the GB, which represent shallow electron traps with a thermal ionization energy of 0.3–0.5 eV. Electron hopping between the vacancies along the GB (i.e., between the localized states in the gap) is responsible for the leakage current.

REVERSIBLE LEAKAGE CURRENT MODEL AND PARAMETER EXTRACTION

The above results suggest the explanation for the reversible current growth observed under stress. An increasing concentration of O-vacancies at the GB forms a percolation path during stress. Since there is no barrier for vacancy segregation at the GB, this process is controlled by vacancy diffusion through the grains. The injected electrons are trapped at these GB vacancies and transported between the vacancy sites. When bias is removed, the electrons occupying the gap states associated with the GB vacancies are released back into the electrode (with significant energy gains), thus reversing O-vacancies to their neutral charge state and reducing the leakage current towards its pre-stress value. After the stress is resumed, the current increases due to repopulation of the already present GB vacancy-induced gap states. The current growth rate during stress is determined by the slower of these processes (exhibiting higher activation energy), while a higher rate of current increase after the resumption of stress (see Fig. 1) is governed by the process with lower activation energy. Therefore, the reversible electron trapping/detrapping by the O-vacancies at the GB results in a repeatable current-switching sequence during each applied/removed bias cycle, as observed in the TiN/HfO₂/TiN capacitors, Fig.1.

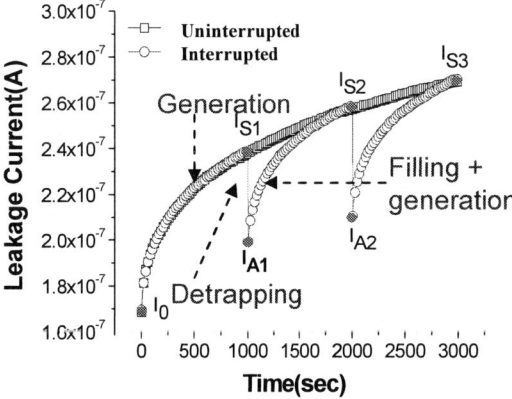

Figure 6. Current time dependency under the CVS at Vg = 1.6V/100°C. Stress sequence is described in the caption of Fig.1.

To verify this model and extract characteristics of the O-vacancy participating in conduction through the dielectric, we analyzed leakage current time dependency data for the entire range of the stress conditions. An example of this analysis for a specific stress condition is shown in Fig. 6. Two processes exhibiting different characteristic times can be identified: a slow process, which follows the uninterrupted stress current, and a faster one, observed after the stress is resumed, during which the current increases much faster. As a working hypothesis, we assumed that the rate of the slower process is dominated by "generation" of the defects contributing to the observed current increase. The faster process, on the other hand, primarily represents reversible and repeatable electron trapping and detrapping (during stress interruption) at these defects.

In analogy with the adiabatic approximation used to separate the electron (fast) and nuclei (slow) subsystems, we first modeled the faster process of the electron capture/emission at the available traps and then extracted the characteristics of the slower process of trap generation. The former process can be described by the kinetic equation

$$\frac{\partial n}{\partial t} = P\left(n_0 - n\right) - P_d n \qquad (1)$$

where n is the density of the filled traps, n_o is the initial available trap density at the start of each stress cycle (points I_0, I_{A1}, I_{A2} in Fig.6), and P and P_d are the probabilities of an electron trapping and detrapping, respectively. P can be approximated by $P \equiv \sigma J/q$ for which σ is the defect capture

978-1-4244-6658-0/10 $26.00 © 2010 IEEE

cross-section, J is the injection current density (determining the density of carriers reaching the traps), and q is the electron charge. The solution of this equation is:

$$ n = A\left(1 - e^{-Bt}\right),\; A = \frac{n_0}{\left(1 + \dfrac{P_d}{P}\right)},\; B = P + P_d $$

In Fig.7, the interrupted stress I(t) dependencies after the first (A1→S2) and second (A2→S3) stress cycles were fitted using Eq. (2) (after subtracting the uninterrupted current S1-S2 and S2-S3, respectively, to correct for the additional O-vacancies generated at GB during these stress periods). The fitting parameters A and B were extracted for each stress voltage and temperature, and then P_d values were calculated based on the obtained B values, measured J, and $\sigma = 10^{-15} \mathrm{cm}^2$ (assumed to be temperature independent, $\sigma = $Const.). Considering a temperature-activated detrapping process, $P_d \propto \exp(-E_t / kT)$, with the activation energy E_t, k is the Boltzmann constant, we extracted the E_t values from the slopes of the $\ln(P_d)$ vs. $1/kT$ dependency at various stress voltages.

Using the Frenkel-Poole correction for the electric field-induced barrier lowering effect, $E_t = E_t^0 - p_0 \cdot E_{ox}$, where p_0 is the

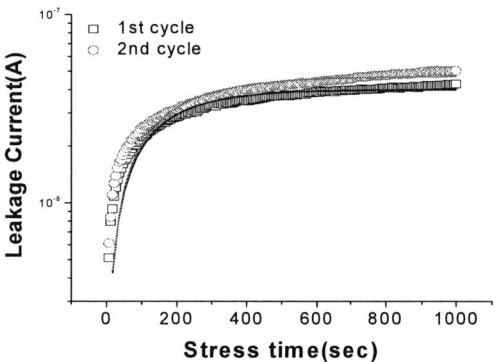

Figure 7. Experimental (symbols) and fitted using Eq. (1) (lines) time dependencies of the current after resumption of the stress presented in Fig. 6. 1st and 2nd cycles refer to two subsequent stress interruptions.

molecular dipole moment associated with the polarization of the chemical bonds, which value for HfO_2, $p_0 = 10.2$ e·cm, was estimated in [5], we extracted the intrinsic activation energy E_t^0, Fig.8. This procedure, performed for each stress interruption cycle (shown in Fig. 1), yielded a practically identical value $E_t^0 = 0.35$ eV for both cycles; this value matches with the above calculated ionization energy of the negatively charged O-vacancy at GB. These results indicate that the process of filling of the available GB traps with the injected electrons may be responsible for the fast current increase observed after resumption of the stress.

The n_0 values calculated from the A parameter by Eq. (2) were found to be temperature- and voltage-dependent.

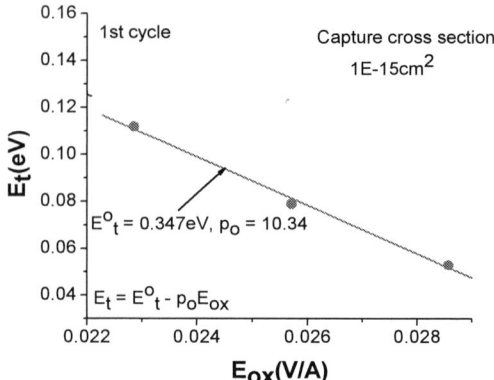

Figure 8. Extraction of the activation energy for the trap filling process.

Performing a similar analysis procedure for the n_0 values ($n_0 \sim \exp(-E_{ac}/kT)$) and including the barrier lowing correction using the same p_0 value, the activation energy E_{ac}^0 for the new trap formation process can be extracted: $E_{ac}^0 = E_{ac} + p_0 \cdot E_{ox} \approx 1.6\text{-}2.0$ eV. This value is close to the calculated activation energy for the neutral oxygen vacancy diffusion along the GB suggesting that this process may be the limiting factor in leakage path formation.

SUMMARY

We have proposed a model describing the formation of the leakage path in crystallized HfO_2. The model suggests that migration of the O-vacancies and their eventual segregation at the GB lead to the formation of a conductive sub-band within the dielectric energy gap. The injected electrons captured by the GB defects are effectively transported through the percolation path along the GB. Electron trapping/detrapping of the GB vacancies is responsible for the current high-low switching observed when stress is interrupted. The trap energy characteristics extracted from the electrical measurement data match with those calculated for the neutral oxygen vacancies precipitated at the GB. The leakage current model proposed here suggests that the breakdown filament in metal oxides may preferentially form along the grain boundaries.

REFERENCES

[1] P. Lysaght, J.Woicik, A. Sahiner, B.-H. Lee, R. Jammy, "Characterizing crystalline polymorph transitions in HfO_2 by extended x-ray absorption fine-structure spectroscopy," Appl.Phys.Lett 91, 122910, 2007

[2] K. S. Yew, Y. C. Ong, D. S. Ang, K. L. Pey, G. Bersuker, P. S. Lysaght, and D. Heh, "Nanoscale Characterization of HfO_2/SiO_x Gate Stack Degradation by Scanning Tunneling Microscopy," Proc. SSDM, 2009

[3] D. M. Ramo et al, "Theoretical Prediction of Intrinsic Self-Trapping of Electrons and Holes in Monoclinic HfO_2," Phys. Rev. Lett. 99, 155504, 2007

[4] K. P. McKenna, et al., "The interaction of oxygen vacancies with grain boundaries in monoclinic HfO_2", Appl. Phys. Lett. 95, 222111, 2009

[5] J. McPherson, J-Y. Kim, A. Shanware, and H. Mogul, "Thermochemical description of dielectric breakdown in high dielectric constant materials," Appl.Phys.Lett. 82, 2121, 2003

Extracting Accurate Position and Energy Level of Oxide Trap Generating Random Telegraph Noise(RTN) in Recessed Channel MOSFET's

Sunyoung Park[1], Sanghoon Lee[1], Yeonsung Kang[1], Byung-Gook Park[1], Jong-Ho Lee[1], Jooyoung Lee[2], Gyoyoung Jin[2], and Hyungcheol Shin[1]

[1]ISRC and School of Electrical Engineering, Seoul National University
San 56-1, Shinlim-dong, Kwanak-gu, Seoul 151-742, Korea
[2]DRAM Core Technology Lab, Samsung Electronics Co., Ltd.,
San #16 Banwol-Dong, Hwaseong-Si, Gyeonggi-Do 445-701, Korea
email : hwtkjs@snu.ac.kr

Abstract—**In this paper, we have proposed an extraction method to find accurate oxide trap locations and energy level in recessed channel structure such as SRCAT. Analytical models for poly depletion effect and the surface potential variation in the cylindrical coordinate were derived and applied to DRAM SRCAT.**

I. INTRODUCTION.

In modern VLSI technology, metal-oxide-semiconductor field effect transistors (MOSFETs) have been getting smaller and smaller for its performance and cost. As effective channel length (L_{eff}) is getting shorter, the short-channel effect is becoming more severe [1]-[2]. Among various 3-D transistors, recessed channel MOSFETs such as sphere-shaped recess-channel-array transistor (SRCAT) have attracted great attention for high speed and low power DRAM and flash memory device applications due to their strong immunity over short-channel effects [3]-[6]. However, even in these recessed channel transistors, degradation of device performance due to RTS noise is also unavoidable like in the other 3-D devices. What is worse, it is more difficult to characterize an oxide trap causing RTN in recessed channel MOSFETs (RC-MOSFET) compared to other planar-type MOSFETs because of the cylindrical shape. Even though extensive researches on RTN in conventional MOSFETs have been performed, RTN in SRCAT has not been reported yet. In this study, we have measured RTN in SRCAT and extracted accurate trap locations and energy level with the newly derived analytical equations for poly depletion and surface potential variation in cylindrical shape of recessed channel transistor with cylindrical shape.

II. MODELING

The RTS noise has been measured in SRCAT fabricated in DRAM 60nm technology. The drain current was observed in time domain varying bias voltages for characterization of the oxide trap.

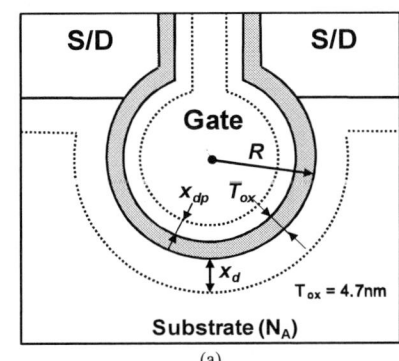

(a)

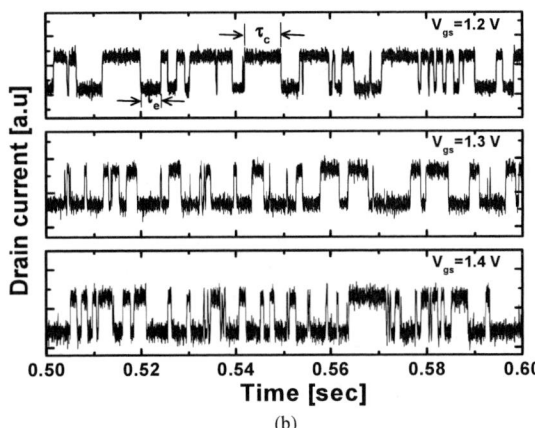

(b)

Figure 1. (a) The cross section of the SRCAT device. (b) Random telegraph noise in the SRCAT measured at the indicated gate voltages.

The structure of SRCAT device is presented in Fig. 1 (a). The SRCAT devices have more strong immunity over SCE than the planar MOSFET because it can have longer L_{eff}.

978-1-4244-6658-0/10 $26.00 © 2010 IEEE

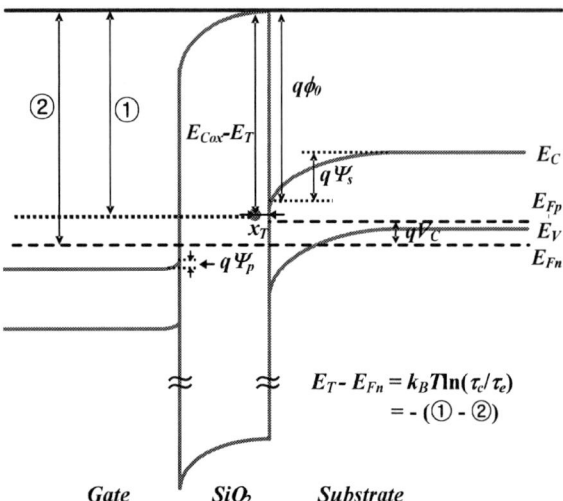

Figure 2. The energy band diagram of SRCAT. Oxide band get bent like natural logarithm function. x_T and E_{Cox}-E_T is shown.

Therefore, as the device is shrunk down, SRCAT is being used instead of Planar MOSFET in DRAM technology.

A. Extraction equation for the vertical position and energy level of oxide trap

The RTN is measured in SRCAT devices as shown in Fig. 1 (b). As soon as an electron is captured into an oxide trap, the current level is decreased because of the carrier number fluctuation and surface mobility fluctuation [7]. Therefore, the time during the high current level is capture time (τ_c) and the time during the low current level is emission time (τ_e).

We noticed that the decrease in τ_c is observed with respect to V_{gs}. If V_{gs} is increased, the trap energy level moves down relative to the Fermi level of substrate and the probability of the capture process is increased.

Previously, the accurate extraction equation for the trap position and energy level in planar MOSFET was reported [8]-[9]. In the case of traps in SRCAT devices, the extraction equation should be revised because the energy band structure of the SRCAT is different from that of the planar MOSFET as shown in Fig. 2 because the recessed channel transistor has cylindrical shape [10]. The oxide is bent with the form of natural logarithm function. From the energy band diagram of the SRCAT, we have derived the relationship between trap energy and time constants as:

$$\ln\frac{\tau_c}{\tau_e} = \frac{1}{kT}\left[\begin{array}{l}\left\{q\phi_0 - q\psi_s + (E_C - E_{Fp} + qV_C)\right\} \\ -\left\{(E_{Cox} - E_T) + q\dfrac{\ln(1-\dfrac{x_T}{R})}{\ln(1-\dfrac{T_{ox}}{R})}(V_{gs} - V_{FB} - \psi_s - \psi_p)\right\}\end{array}\right] \quad (1)$$

where E_{Cox} is the conduction band edge of oxide, E_{Fp} is the Fermi level of Si bulk, E_T is the oxide trap energy level, E_c is the conduction band edge of silicon, ψ_s is surface potential, ψ_p is the potential drop in the poly gate, V_c is the channel

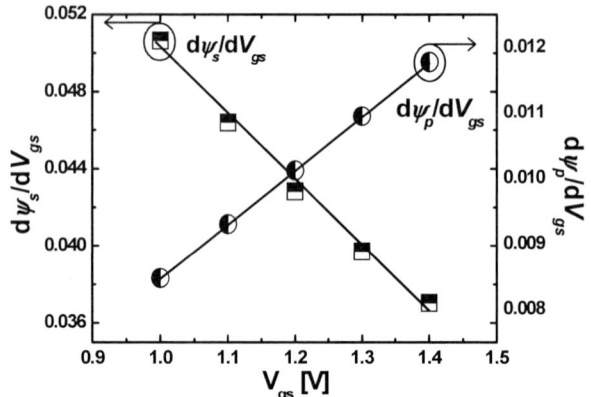

Figure 3. The derivative term of ψ_s and ψ_p. It will be used for extracting the vertical trap position(x_T).

potential at which the trap is located ($V_c \approx y_T V_{ds}/L_{eff}$), V_{FB} is the flat band voltage, ϕ_0 is the barrier height between Si and SiO$_2$, R is the radius of recessed channel and T_{ox} is the oxide thickness.

By differentiating eq. (1) with respect to gate voltage, the equation for extracting trap depth in the SRCAT is derived as:

$$x_T = \left(1 - \left(1 - \frac{T_{ox}}{R}\right)^{\left(\frac{k_BT}{q}\frac{d\ln(\tau_c/\tau_e)}{dV_{gs}} + \frac{d\psi_s}{dV_{gs}}\right)\left/\left(\frac{d\psi_s}{dV_{gs}} + \frac{d\psi_p}{dV_{gs}} - 1\right)\right.}\right) \cdot R \quad (2)$$

To extract the accurate trap position, the derivative term of ψ_s and ψ_p in eq. (2) should be considered.

B. Derivation of derivative term of ψ_s and ψ_p

A new model for surface potential variation in recessed channel transistor is needed. In the strong inversion region, the surface potential is derived as follows [11]:

$$
\begin{aligned}
V_{GS} - V_{FB} &= \psi_s + \frac{\varepsilon_{si}E_S}{C_{ox,r}} \\
&\approx \psi_s + \frac{\varepsilon_{si}}{C_{ox,r}}\left\{-\frac{2V_t}{R} + \sqrt{\left(\frac{2V_t}{R}\right)^2 + 2V_t^2\delta e^{\frac{\psi_s}{V_t}}}\right\}
\end{aligned} \quad (3)
$$

where ε_{si}(=1.04x10^{-12} F/cm) is the permittivity of silicon, where V_t is the thermal voltage(=k_BT/q) and $C_{ox,r}$ is the oxide capacitance given by [12].

$$C_{ox,r} = -\frac{\varepsilon_{ox}}{R\cdot\ln\left(1 - \dfrac{T_{ox}}{R}\right)} \quad (4)$$

By differentiating the eq. (3) versus V_{gs}, we derived the 1st derivative term of ψ_s as follows:

$$\frac{d\psi_s}{dV_{GS}} \approx \frac{1}{1+\dfrac{\varepsilon_{si}}{C_{ox,r}}\dfrac{V_t\delta e^{\psi_s/V_t}}{\sqrt{\left(\dfrac{2V_t}{R}\right)^2+2V_t^2\delta e^{\frac{\psi_s}{V_t}}}}} \quad (5)$$

The poly depletion effect in the recessed channel structure transistor is reported in [13]. The potential drop in the depletion region in the gate is as follows:

$$\psi_p = \frac{qN_P}{2\varepsilon_{si}}\left\{(R-T_{ox})\cdot x_{dp}-\frac{x_{dp}^2}{2}+(R-T_{ox}-x_{dp})^2\ln\left(1-\frac{x_{dp}}{R-T_{ox}}\right)\right\} \quad (6)$$

where x_{dp} is the depletion width in polysilicon gate.

The potential balance equation including the poly depletion effect is derived as follows.

$$V_G - V_{FB} - \psi_s = \frac{qN_P}{2\varepsilon_{si}}\left\{(R-T_{ox})\cdot x_{dp}-\frac{x_{dp}^2}{2}+(R-T_{ox}-x_{dp})^2\ln\left(1-\frac{x_{dp}}{R-T_{ox}}\right)\right\}$$
$$+\frac{qN_p x_{dp}}{C_{ox,r}}\left(1-\frac{x_{dp}}{2(R-T_{ox})}\right)\left(\frac{R-T_{ox}}{R}\right) \quad (7)$$

The derivative term of ψ_p with respect to the V_{gs} can be derived using following equation.

$$\frac{d\psi_p}{dV_{gs}} = \frac{d\psi_p}{dx_{dp}}\cdot\frac{dx_{dp}}{dV_{gs}} \quad (8)$$

Each term in eq. (8) can be obtained from eq. (6) and eq. (7). Finally, the 1st derivative term of ψ_p can be derived as:

$$\frac{d\psi_p}{dV_{gs}} = \frac{\ln\left(1-\dfrac{x_{dp}}{R}\right)\cdot\left(1-\dfrac{\partial\psi_s}{\partial V_{GS}}\right)}{\ln\left(1-\dfrac{x_{dp}}{R-T_{ox}}\right)-\dfrac{\varepsilon_{si}}{C_{ox,r}}\left(\dfrac{1}{R}\right)} \quad (9)$$

The derivative term of ψ_s and ψ_p at applied gate bias voltage is presented in Fig 3.

III. RESULT AND DISCUSSION

We extracted time constant value from Fig. 1 (b). As we expected τ_c is decreased as increasing the V_{gs} as illustrated in Fig. 4 (a) and time constant ratio with respect to V_{gs} has negative slope as shown in Fig. 4 (b). The vertical position (x_T) of oxide traps was obtained using eq. (2) [8].

The measurement result is applied to both the conventional and the new equations and x_T is found to be 0.87 nm and 0.74 nm from the conventional and the new equation, respectively. The depth of the trap from the conventional method is overestimated by about 19 %. Also, x_T without $d\psi_s/dV_{gs}$ and $d\psi_p/dV_{gs}$ terms is found to be 0.9 nm. It is large difference from x_T with consideration of $d\psi_s/dV_{gs}$ and $d\psi_p/dV_{gs}$. Therefore, the $d\psi_s/dV_{gs}$ and $d\psi_p/dV_{gs}$ have to be included when the accurate trap position is extracted. The extracted trap position is near the substrate rather than the gate electrode.

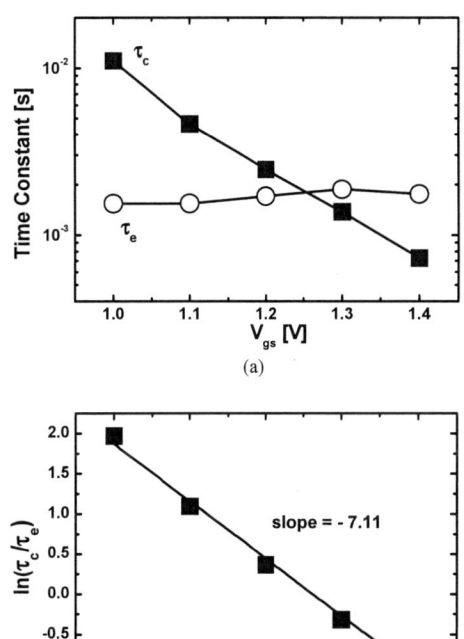

Figure 4. (a) The capture time (τ_c) and the emission time (τ_e) with respect to V_{gs}. (b) The ratio of τ_c and τ_e ($\ln(\tau_c/\tau_e)$) with respect to V_{gs} and its linear fitting.

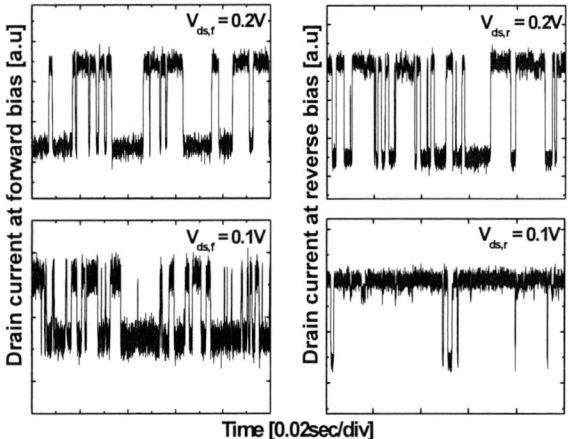

Figure 5. Random telegraph noise in the SRCAT at various $V_{ds,f}$ and $V_{ds,r}$ was observed for extracting the lateral trap position (y_T).

We measured the RTN by varying the drain to source voltage (V_{ds}) to extract the lateral position (y_T). y_T of oxide traps can be obtained by the difference between time constant at forward and reverse drain biases as shown in Fig. 5 and extract the time constant for each case. From eq. (1), the

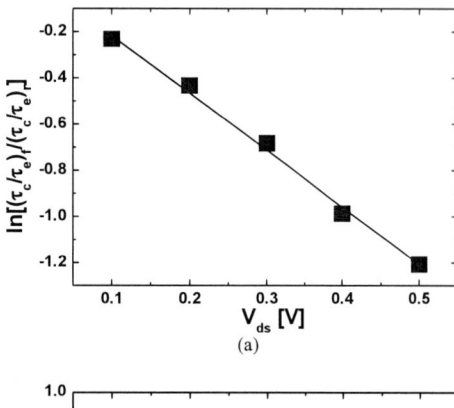

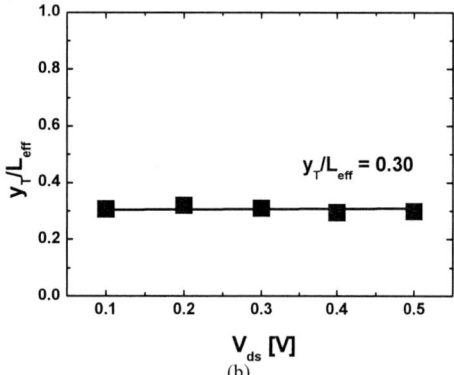

Figure 6. (a) The difference of $\ln(\tau_c/\tau_e)$ with respect to $V_{ds,f}$ and $V_{ds,r}$. (b) The normal-ized distance from source (y_T/L_{eff}) is extracted using data in Fig. 6 (a) and eq. (5).

equation for lateral trap position can be derived. Using those time constant ratio in Fig. 6 (a) and the newly derived equation, we can extract y_T as shown in Fig. 6 (b) [8].

$$\frac{y_T}{L_{eff}} = \frac{\dfrac{k_B T}{q}\dfrac{\ln(1-\dfrac{T_{ox}}{R})}{\ln(1-\dfrac{x_T}{R})}\ln[\dfrac{(\tau_c/\tau_e)_f}{(\tau_c/\tau_e)_r}]+V_{ds,r}}{V_{ds,r}+V_{ds,f}} \quad (10)$$

where L_{eff} is the effective channel length, $V_{ds,f}$ and $V_{ds,r}$ are drain voltage for forward and reverse modes, respectively.

The normalized distance (y_T/L_{eff}) is found to be 0.3 from the source as shown in Fig. 6 (b). Also, we extracted the energy level ($E_{Cox}-E_T$) by using eq. (1) and trap position. $E_{Cox}-E_T$ is found to be 2.88 eV. The x_T and $E_{Cox}-E_T$ are presented in Fig. 2.

IV. CONCLUSION

For the first time, we have measured and characterized RTS noise in SRCAT. The trap interacting with the substrate caused the RTN that has negative slope in time constant ratio with respect to V_{gs}. The new x_T and y_T equations were derived to extract oxide trap location. Also, the variation of surface potential and potential drop in poly gate was also considered for accurate extraction. In addition, $E_{Cox}-E_T$ was extracted by using the newly derived equations.

ACKNOWLEDGMENT

This work was supported by Samsung Electronics co., Ltd.

REFERENCES

[1] K. Bjorkqvist and T. Arnborg "Short Channel Effects in MOS-Transistors," *Phys. Scr.* vol. 24, pp. 418-421, 1981.

[2] Z. H. Liu, C. Hu, J. H. Huang, T.Y. Chan,M. C. Jeng, P. K. KO, and Y. C. Cheng "Threshold voltage model for deep-submicrometer MOSFETs," *IEEE Trans. Electron Device*, vol. 40, pp.86-95, 1993.

[3] J.Y. Kim, D.S. Woo, H.J. Oh, H.J. Kim, S.E. Kim, B.J. Park, J.M. Kwon, M.S. Shim, G.W. Ha, J.W. Song, N.J. Kang, J.M. Park, H.K. Hwang, S.S. Song, Y.S. Hwang, D.I. Kim, D.H. Kim, M. Huh, D.H. Han, C.S. Lee, S.J. Park, Y.R. Kim, Y.S. Lee, M.Y. Jung, Y.I. Kim, B.H. Lee, M.H. Cho, W.T. Choi, H.S. Kim, G.Y. Jin, Y.J. Park, and K. Kim , "The excellent scalability of the RCAT (recess-channel-array-transistor) technology for sub-70nm DRAM feature size and beyond," *VLSI Technical Dig.*, pp. 33-34, 2005.

[4] J.Y. Kim, H.J. Oh, D.S. Woo, Y.S. Lee, H.H. Kim, S.E. Kim, G.W. Ha, H.J. Kim, N.J. Kang, J.M. Park, Y.S. Hwang, D.I. Kim, B.J. Park, M. Huh, B.H. Lee, S.B. Kim, M.H. Cho, M.Y. Jung, Y.I. Kim, C. Jin, D.W. Shin, M.S. Shim, C.S. Lee, W.S. Lee, J.C. Park, G.Y. Jin, Y.J. Park, and K. Kim, "SRCAT (Sphere-shaped-Recess-Channel-Array Transistor) Technology) for 70nm DRAM feature size and beyond," *VLSI Technical Dig.*, pp. 34-35, 2005.

[5] S. Sim, K. Kim, H. K. Lee, J. I. Han, W. H. Kwon, J. H. Han, B. Y.Lee, C. Jung, J. H. Park, D. J. Kim, D. H. Jang, Lee, W.H., C. Park, and K. Kim, "Fully 3-Dimensional NOR Flash Cell with Recessed Channel and Cylindrical Floating Gate - A Scaling Direction for 65nm and Beyond, " in VLSI Symp. Tech. Dig., pp. 17-18, 2006.

[6] H. Lee, D. Kim, B. Choi, G. Cho, S. Chung, W. Kim, M. Chang, Y. Kim, J. Kim, T. Kim, H. Kim, H. Lee, H. Song, S. Park, J. Kim, S. Hong, and S. Park, " Fully integrated and functioned 44nm DRAM technology for 1GB DRAM, " in *VLSI Symp. Tech. Dig.*, pp. 86-87 2008.

[7] C. Surya and T. Y. Hsiang, "Surface mobility fluctuationd in metal-oxide-semiconductor field-effect-transistor." Phys. Rev. B., vol35, pp.6342-6347,1984.

[8] Z. Celik-Butler, P. Vasina, and N. Vibhavie Amarasinghe, "A method for locating the position of oxide traps responsible for random telegraph signals in submicron MOSFETs," *IEEE Trans. Electron Device*, vol. 47, no. 3, pp.646-648, 2000.

[9] H. Lee, Y. Yoon, S. Cho, and H. Shin, "Accurate Extraction of the Trap Depth from RTS Noise Data by Including Poly Depletion Effect and Surface Potential Variation in MOSFETs," *IEICE TRANSACTIONS on Electronics*, vol. E90-C, no. 5, pp. 968-972, May. 2007.

[10] S. Yang, K.H. Yeo, D. Kim, K. Seo, D. Park, G. Jin, K. Oh, and H. Shin, "Random Telegraph Noise in N-type and P-type Silicon Nanowire Transistors," in IEDM Tech. Dig., pp. 765-768, 2008.

[11] Y. Kang, Y. Son, J. Lee, H. Kim, B. Park, J. Lee, and H. Shin, "A surface potential model for recessed channel MOSFETs in strong inversion," *Korean Conference on Semiconductors*, pp.576-577, 2010.

[12] K. Natori, I. Sasaki, and F. Masuoka, "An Analysis of the Concave MOSFET," *IEEE Trans. Electron Devices*, vol. ED-25, no. 4. pp. 448–456, 1978.

[13] Y. Kang, H. Kim, J. H. Lee, Y. Son, B. Park, J. D. Lee, and H. Shin, "Modeling of polysilicon depletion effect in recessed-channel MOSFETs", *IEEE Electron Device Letters*, vol. 30 no. 12 pp.1371-1373, Dec. 2009.

[14] M. J. Kirton and M. J. Uren, "Noise in solid-state microstructures: A new perspective on individual defects, interface states and low-frequency (1/f) noise," *Adv. Phys.*, vol. 38, no. 4, pp. 367-468, 1989.

[15] J.H., Lee, S.Y. Kim, I. Cho, S. Hwang, and J.H. Lee, "1/f Noise Characteristics of Sub-100 nm MOS Transistors," *J. Semicond. Technol.Sci.*, vol. 6, No.1, pp.38-42, 2006.

SOI TFETs: Suppression of Ambipolar Leakage and Low-Frequency Noise Behavior

Jing Wan,[1] Cyrille Le Royer,[2] Alexander Zaslavsky,[3] and Sorin Cristoloveanu[1]

[1]IMEP-INPG/Minatec, 3 Parvis Louis Néel, 38016 Grenoble Cedex 1, France
[2]CEA-LETI, Minatec, 17 avenue des Martyrs 38054 Grenoble Cedex 9, France
[3]Division of Engineering, Brown University, Providence, Rhode Island 02912, USA

Abstract—We report on the thin-body tunneling field-effect transistors (TFETs) built on SOI substrates with both SiO_2 and HfO_2 gate dielectrics. The source-drain leakage current is suppressed by the introduction of intrinsic regions adjacent to the drain side, reducing the electric field at the tunnel junction. We also investigate the temperature dependence of the TFET characteristics, as well as the low frequency noise (LFN) behavior. Unlike conventional MOSFETs, the TFET LFN behaves as $1/f^2$ even for large gate areas, indicating less trapping due to its much smaller effective gate length.

I. INTRODUCTION

As the scaling of conventional CMOS devices is suffering from short-channel effects (SCEs), silicon-compatible devices based on different principles are being studied for their unique properties. In particular, the tunneling field effect transistor (TFET) [1] is of interest due to its complete semiconductor process compatibility and similarity of device layout with the Si MOSFET. The current in TFETs is induced by band-to-band tunneling (BTBT), which makes it theoretically possible to reach extremely low OFF (I_{OFF}) currents as well as an ultra-low subthreshold slope (S) below the ideal MOSFET value of 60 mV/dec at room temperature. A number of studies have focused on power consumption in TFETs, reporting a theoretical advantage over standard MOSFETs [2-3]. In order to enhance the ON current (I_{ON}), multigate [4] and lower bandgap semiconductors, such as Ge and SiGe, have been reported [5-7].

At the same time, experimentally realized TFETs have typically suffered from difficulties in simultaneously achieving low S and high I_{ON}, as well as from high leakage current (I_{LEAK}) due to ambipolar conduction, especially for low bandgap semiconductors.

In this work, we focus on several aspects of Si TFETs. First, compare TFETs with SiO_2 and HfO_2 gate dielectrics to demonstrate lower S and higher I_{ON} values in devices with thinner equivalent oxide thickness (EOT). Second, an asymmetrical TFET layout with an intrinsic region (L_{IN}) is demonstrated to effectively suppress I_{LEAK} while retaining the same I_{ON} level. These experimental results are confirmed using device and process simulations based on Silvaco TCAD. Further, temperature variation tests are performed to examine the validity of Kane's model [8] of the BTBT process. Finally, we present the first low-frequency noise (LFN) measurements on TFETs and qualitatively explain the different LFN properties of TFETs compared to standard MOSFETs (where LFN measurements are widely used to extract trap properties [9-10]).

II. FABRICATION AND DEVICE STRUCTURE

A. Device fabrication

The fabrication of TFETs started from an SOI substrate with 140 nm BOX and 20 nm active Si layer. The MESA process was employed to define isolated device active areas followed by the definition of gate stack which is composed of three layers, as illustrated in Fig. 1. Two different gate oxides were formed for comparison: either a 6 nm SiO_2 grown by dry oxidation or a 3 nm HfO_2 deposited by ALD. After the deposition of a metal gate (10 nm TiN), 50 nm thick poly silicon is deposited. The 1st spacer was formed by the deposition of 10 nm Si_3N_4, then a 10 nm Si layer was epitaxially grown, followed by NLDD and PLDD implantation. The 2nd spacer and Si layer are formed at the same way. Before the implantation of NHDD and PHDD, a Si_3N_4 layer was formed to protect the intrinsic region L_{IN}. Rapid thermal annealing (RTA) was used to activate the dopants, followed by the metallization.

B. Device structure and bias polarity

The structure of the fabricated TFETs is quite similar to that of MOSFET with two exceptions. Firstly, the dopant types in source and drain are different. Secondly, the structure of our TFETs was rendered asymmetrical by the intrinsic regions L_{IN} separating the drain contact from the channel.

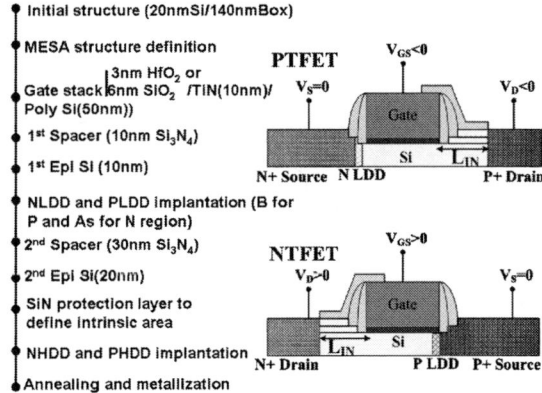

Fig. 1. Fabrication process flow and bias polarity of TFETs. In PTFET, the gate is negatively biased and the BTBT occurs at the n+ doped source. Conversely, in NTFET, the gate is positively biased and the BTBT occurs at the p+ doped source.

978-1-4244-6658-0/10 $26.00 © 2010 IEEE 261

For simplicity, the N+ region in a PTFET is defined as source while the P+ region is defined as drain. This definition is opposite in an NTFET, as can be observed in Fig. 1. The source of both PTFETs and NTFETs is always grounded, while the drain is negatively biased in PTFETs and positively biased in NTFETs. For I_{ON} to flow by tunneling at the source-channel junction, $V_{GS} < 0$ for PTFETs and $V_{GS} > 0$ for NTFETs.

III. ELECTRICAL CHARACTERIZATION AND ANALYSIS

The fabricated devices were systematically characterized. From C-V measurements, the EOT values of TFETs with 6 nm SiO_2 and 3 nm HfO_2 gate oxides were 6 nm and 2.2 nm, respectively. The gate width of all TFETs was 10 μm, the gate length L_G varied from 100 nm to 400 nm, and L_{IN} accounting for the spacer varied from 20 nm to 100 nm.

A. Impact of gate oxide on characteristics

A comparison of I_D-V_{GS} curves of NTFETs and PTFETs with different gate oxides and L_G = 400 nm and $|V_{DS}|$ = 1 V is shown in Fig. 2. Since there exists no unambiguous definition of threshold voltage V_T and S is not a constant value in TFETs, for comparison purposes, we define the V_T as the voltage when drain current starts to surpass 2×10^{-10} A/μm and extract the S value when drain current reaches 2×10^{-12} A/μm, ranging over two decades in drain current. As can be easily observed in Fig. 2, TFETs based on HfO_2 have smaller V_T and higher I_{ON} than those based on SiO_2. The S values are also largely reduced.

The results on devices with different L_G = 100–400 nm show that V_T and S are independent of L_G, as expected in TFETs [11]. This due to the fact the tunneling current is determined by the maximum electric field at the tunneling junction and unaffected by the carrier transport in the channel. The results show that V_T is 4.5V for both NTFETs and PTFETs with SiO_2, whereas V_T is 2.3V for those with HfO_2 gate oxide. As for the S value, both NTFETs and PTFETs with SiO_2 have S value of 1.1 V/dec, which is reduced to 0.33 V/dec in devices with HfO_2. We can note that a formal study

has been performed [5]. Here, we discuss only TFETs with comparable characteristics for both types of oxides.

The comparison reveals that thinner EOT can lead to lower V_T and S values due to the better electrostatic controllability from the gate. The theoretically achievable S < 60 mV/dec value in TFETs requires excellent electrostatic control of the maximum junction field, which requires minimizing gate EOT and sharpening the S/D lateral doping profile. As a result, few experimental reports of low S have been published thus far [5, 7, 12].

B. Impact of device architecture on characteristics

A potential problem of symmetrical TFETs is the large I_{LEAK} under opposite gate bias, because interband tunneling possible occurs at either the source-channel or the drain-channel junction depending on the sign of V_{GS}. This ambipolar I_{LEAK} can be much more severe for TFETs based on low band-gap semiconductors such as Ge. The solution is to introduce an asymmetrical architecture, such as unequal source/drain doping [7, 13], intrinsic regions and even lateral heterojunctions [7].

In our case, we introduce an intrinsic area to make the device asymmetrical, hence to suppress the ambipolar tunneling current. This method has been proposed from simulation work [7, 14] and experimental data have been presented in [5]. Here, we make a systematic study by the combination of simulation and experiment and also show the possibility to completely suppress the ambipolar I_{LEAK}. Figure 3 shows a NTFET with intrinsic area under different V_{GS} and its corresponding simulated maximum electric field (E_{max}) at source and drain sides. Under positive gate bias, the TFET is in ON state and the tunneling occurs at source side. The E_{max} at the source side is almost constant as L_{IN} increases, shown as the rectangular curve in Fig. 3, which thanks to the negligible potential drop on L_{IN} region.

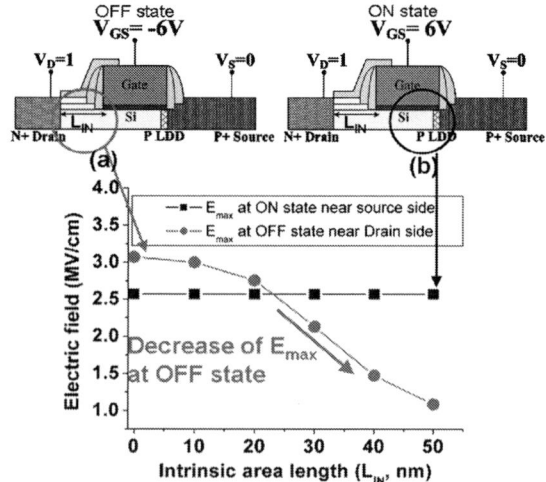

Fig. 3. The maximum electric field (E_{max}) at the tunneling junction of a NTFET with different L_{IN} under positive gate bias and negative gate bias.

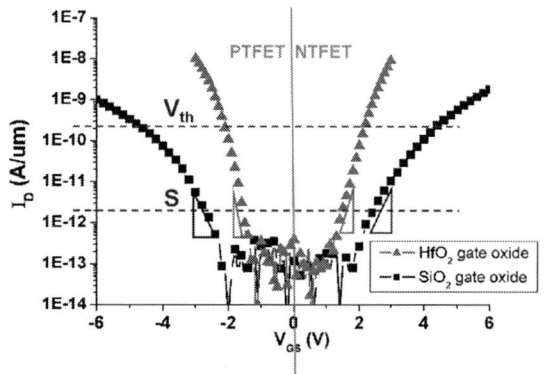

Fig. 2. Measured I_D-V_{GS} curves of both NTFETs and PTFETs based on different gate oxides (L_G =400 nm, $|V_{DS}|$ = 1 V). The inset symbols show the definitions of V_T and S.

978-1-4244-6658-0/10 $26.00 © 2010 IEEE

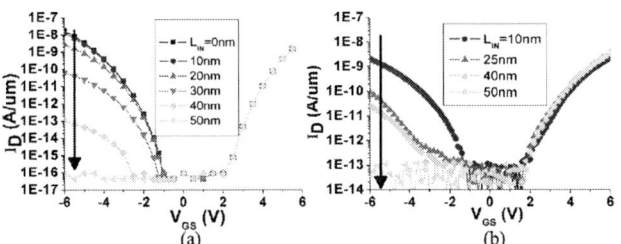

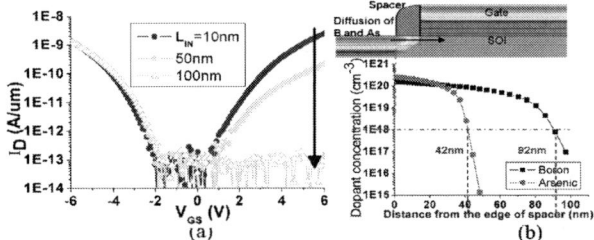

Fig. 4 I_D-V_{GS} curve of NTFETs with different L_{IN} from (a) simulation and (b) experimental results (L_G = 400 nm, V_{DS} = 1 V).

Fig. 5. (a) I_D-V_{GS} curve of PTFETs with different L_{IN} and (b) the simulation about the diffusion length of boron and arsenic.

In contrast, as the gate is negatively biased, the TFET is in OFF state and the tunneling occurs at drain side where the intrinsic area is located. For the TFET with L_{IN} smaller than 20 nm, the E_{max} at OFF state is even slightly larger than that at ON state, see Fig. 3. This due to the fact that the V_{GD} = 7 V at OFF state which is 1 V higher than that of V_{GS} in the ON state. However, as L_{IN} increases larger than 20 nm, the E_{max} at the drain side drops apparently. Since the BTBT rate is directly determined by E_{max}, it can be largely reduced at the drain side due to the increase of L_{IN}. However, at the source side, the E_{max} does not change which means a constant tunneling rate at ON state.

A comparison of the simulated and experimental results for an NTFET with different values of L_{IN} is shown in Fig. 4, where the simulated tunneling current in Fig. 4(a) is obtained from Kane's model as discussed below. The simulated I_{LEAK} of the NTFET with L_{IN} smaller than 20 nm is even slightly higher than the I_{ON} due to the larger E_{max}. As L_{IN} increases from 20 nm to 50 nm, the I_{LEAK} is largely suppressed, whereas I_{ON} is unaffected. Figure 4(b) shows the experimental results for NTFETs with four different L_{IN} from 0 to 50 nm. In agreement with the simulation, the I_{LEAK} of the TFETs without L_{IN} is comparable to I_{ON}, but can be effectively suppressed by increasing L_{IN} to 50 nm.

In PTFETs, the I_{ON} is also unaffected by L_{IN}, as can be seen in Fig. 5(a). However, the full suppression of I_{LEAK} in PTFETs requires a larger L_{IN} = 100 nm. We attribute this difference to the different diffusion coefficients of boron (B) and arsenic (As) in Si. Figure 5(b) shows the simulated doping profiles of implanted B and As after the activation anneal. All parameters in simulation were adjusted according to the fabrication process. As previous work indicates [8], a doping concentration which is lower than 1×10^{18} cm^{-3} can be used to effectively suppress the tunneling. In our case, the I_{LEAK} results from the tunneling at drain side which is doped by As and B in NTFETs and PTFETs, respectively. The characteristic diffusion distance from the edge of spacer to the 10^{18} cm^{-3} value point is 42 nm for As and 92 nm for B, as shown in Fig. 5(b), qualitatively explaining our experimental observations.

C. Impact of temperature on characteristics

Fig. 6(a) shows the I_D-V_{GS} of a NTFET with HfO$_2$ gate oxide, L_G = 400 nm and L_{IN} = 10 nm at temperatures ranging from 77 K to 300 K. The temperature dependence of the TFET I_D can be qualitatively explained by Kane's model in device level as [8]:

$$I_D = A \cdot V_{GS}^2 \cdot \exp(-\frac{B}{V_{GS}}); \quad A \propto E_G^{-0.5}; \quad B \propto E_G^{\frac{3}{2}} \qquad (1)$$

The dominant temperature effect on the TFETs performance comes from the temperature variation of bandgap E_G, which enters in the exponential of (1). The E_G has weakly negative temperature dependence [15]:

$$E_G(T) = E(0) - \frac{\alpha \cdot T^2}{T + \beta} \qquad (2)$$

As temperature increases, the E_G decreases, leading to a corresponding decrease in parameter B and hence an increasing tunneling current.

The validity of Kane's model can be examined by rewriting (1) as:

$$\log(\frac{I_D}{V_{GS}^2}) = \log(A) - \frac{B}{V_{GS}} \qquad (3)$$

The relation between $\log(I_D/V_{GS}^2)$ and $1/V_{GS}$ is quite linear over the entire temperature range, as shown in Fig. 6(b).

Since the behavior of the TFET at various temperature in our work is quite similar to that reported previously [5] and agree with (1), we can conclude that the weakly positive temperature dependence of the TFET I_D can be used as confirmation of BTBT in the device and that the Kane's model is at least qualitatively effective in describing the BTBT process in Si TFETs.

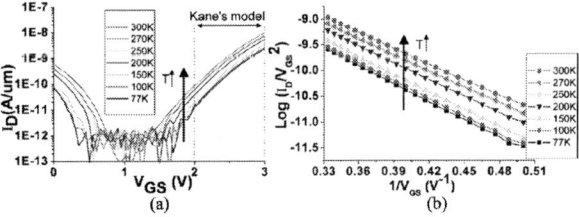

Fig. 6. Relation between (a) I_D-V_{GS} and (b) $\log(I_D/V_{GS}^2)$-$1/V_{GS}$ of an NTFET at various temperatures (L_G = 400nm, V_{DS} = 1 V).

D. Low frequency noise (LFN) characteristics

The LFN caused by the trapping-detrapping process at the channel-dielectric interface has a $1/f$ signature in MOSFETs. As the area of MOSFET decreases below 1 μm^2, only a few interface traps exist in the entire device, leading to a random telegraph signal (RTS) current noise that produces a Lorentzian spectrum whose slope is almost $1/f^2$ [9].

Figure 7 compares the LFN of NMOS and NTFETs with the same 6 nm SiO_2 gate oxides showing totally different spectral behavior. In the NMOS with a gate area of 3.5 μm^2 (L_G = 350 nm) the LFN is $1/f$, whereas in the NTFET, the spectrum of the noise is Lorentzian even though it is much larger (area =25μm^2, L_G = 5 μm).

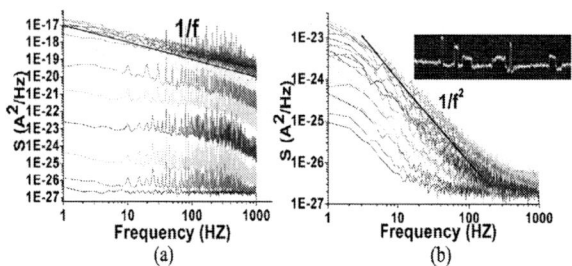

Fig. 7. Comparison of LFN spectrum between (a) NMOS (L_G = 350 nm, V_{DS} = 50 mV, V_{GS} = 0.1-1.5V) and (b) NTFET (L_G = 5 μm, V_{DS} = 1V, V_{GS} = 3-5V). The inset image shows the RTS signal in NTFET.

For TFET, the current is only determined by the tunneling rate, while the channel provides a way for carrier transport. The trapping of carriers at the Si/SiO_2 interface above the channel can only cause the fluctuation of the channel conductance, whereas the tunneling rate at the tunneling junction stays stable. The tunneling rate can only be affected by the trapping process at the Si/SiO_2 interface above the tunneling junction which is very narrow (around 10 nm). Hence, the effective LFN-generating area of the TFET is very small (0.05 μm^2), including only a discrete numbers of traps, just as in a very small MOSFET. This is why the RTS noise is observed in TFET even though its gate area is nominally large.

IV. CONCLUSION

In this paper, we report on the various aspects of Si TFET performance. We demonstrate that the use of HfO_2 gate oxide with smaller EOT leads to a lower threshold V_T and subthreshold slope S than identically processed SiO_2 oxide TFETs, with S decreasing by a factor of ~3. We show that the introduction of an intrinsic region between the channel and the drain in both types of TFETs can largely suppress the I_{LEAK} without impacting I_{ON}: I_{LEAK} decreases from 2×10^{-9} A/μm to sub 1×10^{-13} A/μm as L_{IN} is increased to 50 nm (NTFET) and 100 nm (PTFETs). The difference in the L_{IN} required to suppress the ambipolar I_{LEAK} is explained by the different diffusion profiles of B and As, as confirmed by TCAD simulations. Finally, LFN characterizations are performed on MOSFET and TFETs. The results reveal that, unlike in MOSFETs, the RTS noise is dominant in TFETs even though

the nominal gate area is large, because the relevant area where the interband tunneling takes place is much smaller than the physical L_G.

ACKNOWLEDGMENTS

The work at Minatec is funded by the RTRA program of the Grenoble Nanosciences Foundation. One of the authors (AZ) also acknowledges support by the U.S. National Science Foundation (award ECCS-0701635).

REFERENCES

[1] W. M. Reddick and G. A. J. Amaratunga, "Silicon surface tunnel transistor," Appl. Phys. Lett., vol. 67, no. 4, p. 494, 1995.

[2] Q. Zhang and A. Seabaugh, "Can the interband tunnel FET outperform Si CMOS?," DRC, PP. 73-74, 2008.

[3] Y. Khatami and K. Banerjee, "Steep subthreshold slope n- and p-type tunnel-FET devices for low-power and energy-efficient digital circuits," IEEE Trans. Electron Devices, vol. 56, pp. 2752-2761, 2009.

[4] D. Leonelli, A. Vandooren, R. Rooyackers, A. S. Verhulst, S. De Gendt, M. M. Heyns and G. Groeseneken, "Multiple-gate tunneling field effect transistors with sub-60mV/dec subthreshold slope," SSDM, Sendai, pp. 767-768, 2009.

[5] F. Mayer, C. Le Royer, J.-F. Damlencourt, K. Romanjek, F. Andrieu, C. Tabone, B. Previtali, and S. Deleonibus, "Impact of SOI, $Si_{1-x}Ge_xOI$ and GeOI substrates on CMOS compatible tunnel FET performance" IEDM Tech. Dig., pp. 163-166, 2008.

[6] D. Kazazis, P. Jannaty, A. Zaslavsky, C. Le Royer, C. Tabone, L. Clavelier and S. Cristoloveanu, "Tunneling field-effect transistor with epitaxial junction in thin germanium-on-insulator," Appl. Phys. Lett., vol. 94, p. 263508, 2009.

[7] T. Krishnamohan, D. Kim, S. Raghunathan and K. Saraswat, "Double-gate strained-Ge heterostructure tunneling FET (TFET) with record high drive currents and $\ll$60mV/dec subthreshold slope," IEDM Tech. Dig., pp. 947-949, 2008.

[8] K. K Bhuwalka, J. Schulze and I. Eisele, " A Simulation Approach to Optimize the Electrical Parameters of a Vertical Tunnel FET," IEEE Trans. Electron Devices, vol. 52, pp. 1541-1547, 2005.

[9] K. Akarvardar, B. M. Dufrene, S. Cristoloveanu, P. Gentil, B. J. Blalock and M. M. Mojarradi, "Low-Frequency Noise in SOI Four-Gate Transistors," IEEE Trans. Electron Devices, vol. 53, pp. 829-835, 2006.

[10] G. Ghibaudo and T. Boutchacha, "Electrical noise and RTS fluctuations in advanced CMOS devices," Microelectron Reliab, vol. 42, pp. 573-582, 2002.

[11] C. Aydin, A. Zaslavsky, S. Luryi, S. Cristoloveanu, D. Mariolle, D. Fraboulet and S. Deleonibus, "Lateral interband tunneling transistor in silicon-on-insulator," Appl. Phys. Lett., vol. 84, p. 1780, 2004.

[12] W. Y. Choi, B. G. Park, J. D. Lee, and T. K. Liu, "Tunneling field-effect transistors (TFETs) with subthreshold swing (SS) less than 60 mV/dec," IEEE Electron Device Lett., vol. 28, pp. 743–745, 2007.

[13] V. Nagavarapu, R. Jhaveri, and J. C. S. Woo, "The tunnel source (PNPN) n-MOSFET: a novel high performance transistor," IEEE Trans. Electron Dev., vol. 55, pp. 1013–1019, 2008.

[14] A. S. Verhulst, W. G. Vandenberghe, K. Maex, and G. Groeseneken, "Tunnel field-effect transistor without gate-drain overlap," Appl. Phys. Lett., vol. 91, p. 05312, 2007.

[15] S. M. Sze: Physics of Semiconductor Devices (Wiley, New York, 1981) 2nd ed.

A Simulation-based Study of Sensitivity to Parameter Fluctuations of Silicon Tunnel FETs

Kathy Boucart, Adrian M. Ionescu
Nanolab, EPFL, Lausanne, Switzerland
kathy.boucart@epfl.ch, adrian.ionescu@epfl.ch

Walter Riess
IBM Research - Zurich, Switzerland
wri@zurich.ibm.com

Abstract— **In this paper we study the sensitivity to parameter fluctuations for an optimized double-gate silicon Tunnel FET with a high-k gate dielectric. The impacts of the variability of the dielectric thickness, doping profile at the tunnel junction, silicon body thickness, alignment of the gate dielectric to the tunnel junction, device length, and band gap at the tunnel junction, on the device performance are systematically studied. One parameter is varied at a time to show the resulting fluctuations of the device characteristics. Gate dielectric thickness and doping junction width are pinpointed as the parameters requiring the tightest control during Tunnel FET fabrication in order to limit characteristic fluctuations. Body thickness and gate dielectric alignment with the tunnel junction may also need tight control depending on whether the target values are within a range where the characteristics are highly sensitive.**

I. INTRODUCTION

Silicon Tunnel FETs are promising devices whose simulated characteristics show low power consumption due to a very low off-current and a small subthreshold swing. Many groups are currently working on fabricating Tunnel FETs [1-6] that will live up to the high hopes offered by models and simulations [7-13]. For now, the fabrication technology for these devices is still a new area of exploration, with challenges and design issues that are not necessarily the same as those for conventional MOSFET fabrication. While the basic parameters to optimize in order to have good Tunnel FET characteristics are known – an abrupt doping profile at the tunnel junction, high capacitive coupling from the gate to the tunnel junction, etc. – it is crucial to also understand the sensitivity of device characteristics to parameter fluctuations, now a key ITRS criterion to assess emerging post-CMOS devices [14]. This subject is well-understood for conventional MOSFETs, and the principal sources of fluctuation, including random discrete dopants, line edge roughness, polysilicon granularity, and oxide thickness fluctuations, have been thoroughly studied [15,16]. The goal of this paper is to investigate the influence of the parameter fluctuations which are unavoidable in device fabrication and their impact on Tunnel FET characteristics. The critical parameters that will need to be the most tightly controlled when fabricating these devices are identified, and the parameter variation is quantified. All simulations were carried out in Silvaco Atlas version 5.13.16.C with a non-local band-to-band tunneling model and bandgap narrowing.

II. OPTIMIZED TUNNEL FET STRUCTURE

The starting point of the parameter variation study was an optimized double-gate Tunnel FET, whose device schematic and transfer characteristics are presented in Fig. 1. Its default parameter values are presented in Table I. This all-silicon Tunnel FET has an I_{on}/I_{off} ratio of about 10^{11} at $V_{DS} = V_{GS} = 1$ V, and an average subthreshold swing of 57 mV/decade from turn-on up to threshold (at 10^{-7} A/µm).

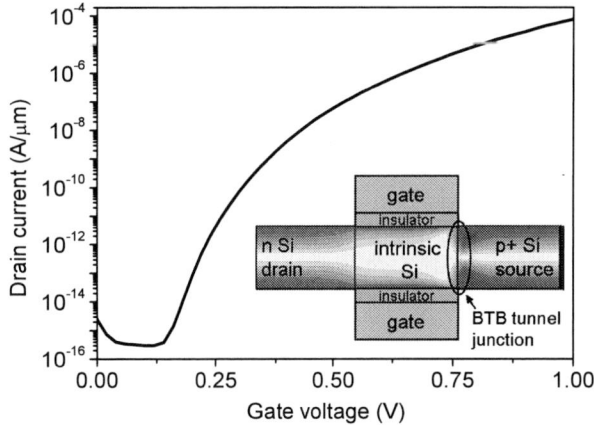

Figure 1. I_{DS}-V_{GS} and device schematic for the optimized silicon Tunnel FET whose parameters are specified in Table I. Current flowlines for $V_{DS} = V_{GS} = 1$V.

TABLE I. Default values for the Tunnel FET used in the parameter variation study.

Parameter	Optimized value
Gate dielectric ε	25
Junction width	12 nm / 5 decades
Body thickness	10 nm
Source doping	1.5×10^{20} atoms/cm³
Gates	Double
Oxide alignment	Over intrinsic
Device/gate length	30 nm

978-1-4244-6658-0/10 $26.00 © 2010 IEEE

III. PARAMETER FLUCTUATION

Since the physics of Tunnel FETs is different from that of conventional MOSFETs, governed by the band-to-band tunneling energy barrier width, the sensitivity of device characteristics to parameter fluctuations is expected to be different as well, and to potentially raise new technology challenges. In the following sections, these parameters are varied: dielectric thickness, tunnel junction doping profile width, body thickness, gate contact and dielectric alignment, device length, and band gap at the tunnel junction.

A. Dielectric thickness

Capacitive coupling between the gate and channel is a critical parameter for highly-scaled conventional MOSFETs, so oxide thickness variations have been a major focus of studies on parameter fluctuations for this type of device [15]. Tunnel FETs are even more sensitive to changes in gate capacitance than conventional MOSFETs [8], thus it is no surprise that changes in dielectric thickness can cause major fluctuations in Tunnel FET characteristics.

In Fig. 2, $t_{ox} = 3$ nm is taken as the target, and the percent change in V_{TG} (gate threshold voltage), S_{point} (the lowest subthreshold swing value on the I_{DS}-V_{GS} curve), and S_{avg} (the average swing from turn-on to threshold at 10^{-7} A/μm) are shown on the left axis. The change in I_{on} (at $V_{DS}=V_{GS}=1$V) is easier to see on an exponential scale, so it is plotted as log $(I_{on}/I_{on,tox=3})$ on the right axis of Fig. 2. The variation of the oxide thickness has a drastic influence on all characteristics.

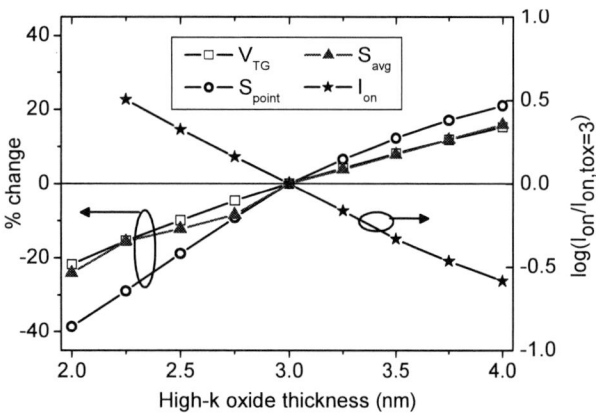

Figure 2. (Left) % change in V_{TG}, S_{point}, and S_{avg} and (right) orders of magnitude by which I_{on} changes when gate dielectric thickness changes.

B. Junction width

Junction width (w_j), or the distance across which the doping falls from its high level in the source to its low level in the intrinsic region, is an important factor that determines the steepness of the energy bands at the tunnel junction, and therefore the minimum possible tunnel barrier width in a Tunnel FET in the on-state.

The dashed curve in Fig. 3 shows the source-side doping profile in the Tunnel FET used as the target in Fig. 4, whose junction width at the source side is about 21 nm for five decades of doping change. Fig. 4 presents the effects of

junction width variation, and as with t_{ox}, all four characteristics show sensitivity to changes in w_j. On-current is especially sensitive, changing by about an order of magnitude for each 10 nm difference in junction width.

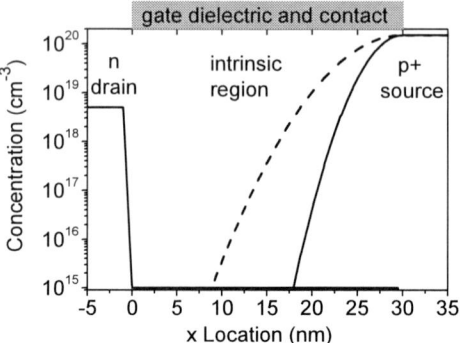

Figure 3. Doping profiles for the Tunnel FET used as the target device in Fig. 4 (dashed line) and elsewhere in this paper (solid line at source side). The source-side doping has a non-abrupt falloff, and is situated to imitate the profile for a self-aligned implantation with diffusion under the gate.

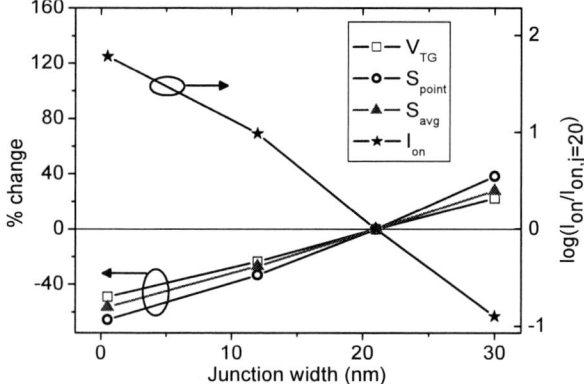

Figure 4. (Left) % change in V_{TG}, S_{point}, and S_{avg} and (right) orders of magnitude by which I_{on} changes when junction width at the tunnel junction changes.

C. Silicon body thickness

Device thickness t_{Si} is a parameter that is becoming increasingly important as more devices are built on thin films and incorporate double or multiple gates, leading to better gate control and the reduction of bulk capacitive effects. However, the choice of Tunnel FET body thickness necessitates a trade-off, as shown in Fig. 5. Here, 10 nm is chosen as the target t_{Si}, but the characteristics are very sensitive around this target point, especially toward thinner layers. Although swing, V_{TG}, and I_{on} can be greatly improved by decreasing body thickness (or nanowire diameter), the fluctuation of characteristics increases at some small t_{Si}, here around 20 nm. In order to fully take advantage of a double-gate design, the body thickness should be under this limit, so tight control of this parameter will be necessary in order to minimize the fluctuations in device behavior.

978-1-4244-6658-0/10 $26.00 © 2010 IEEE   266

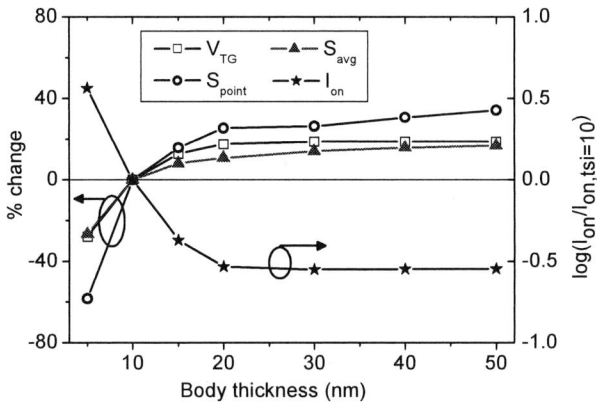

Figure 5. (Left) % change in V_{TG}, S_{point}, and S_{avg} and (right) orders of magnitude by which I_{on} changes when silicon body thickness changes.

D. Gate oxide and gate contact alignment

Gate alignment is another important Tunnel FET design consideration, and there have been recommendations to align the gate dielectric with the tunnel junction in order to take advantage of fringing [10], and to shorten the gate on the drain end in order to increase device speed [17]. It is important to understand how such design choices would influence the magnitude of characteristic fluctuations.

Fig. 6(a) shows a device schematic to clarify how gate alignment is being changed here – the gate contact and the oxide alignment at the source-side junction change together, with negative alignment values indicating underlap, and positive alignment values indicating overlap. Similar to what other groups have shown [5,10], alignment = 0 is defined as that which results from a self-aligned process, in which there is dopant diffusion under the gate. The actual electrical junction (where the n-doping curve and the p-doping curve cross) is 12 nm into the intrinsic region for the default device, as shown by the solid curve at the source side in Fig. 3.

Fig. 6(b) shows that the alignment of the gate oxide has a large influence on characteristics, and one that is not necessarily intuitive on first glance. For a gate overlapped by several nanometers or more, characteristics are stable. This is the case even beyond what is shown in Fig. 6(b), with only a slight performance degradation when the gate covers the entire 100-nm long source region. As the gate alignment moves in the underlapped direction toward the electrical junction marked by the dashed line, the characteristics hit a peak in which the on-current at $V_{DS}=V_{GS}=1V$ goes up by a factor of 4.5, while V_T and subthreshold swing hit a low point. This surprising improvement is further illustrated by the I_{DS}-V_{GS} curves in Fig. 7. While the curve for a gate underlap of -10 nm shows that the gate no longer has good control of the tunnel junction, the curve for a gate underlap of -5 nm shows improvements in on-current, threshold voltage, and swing. The aligned gate (alignment = 0) has the highest current at gate voltages higher than those shown in Fig. 7, but the 5-nm-underlapped gate has better characteristics in the low-voltage region, important for low-power devices.

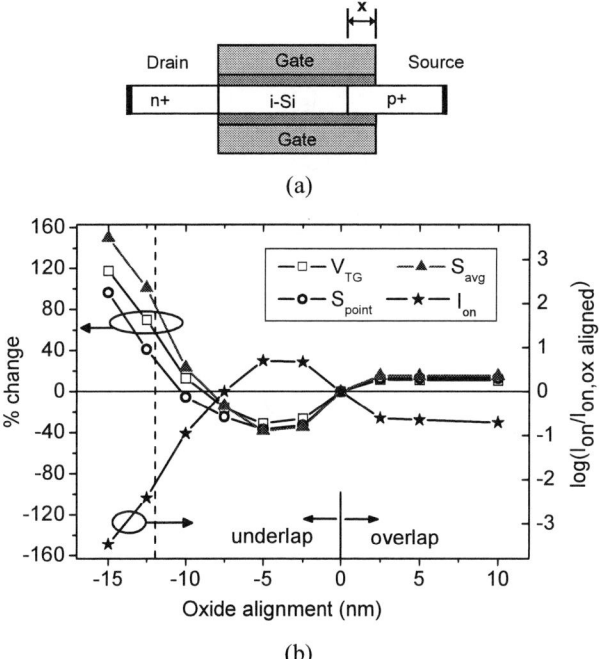

Figure 6. (a) Tunnel FET schematic showing the gate alignment, marked "x." (b) (Left) % change in V_{TG}, S_{point}, and S_{avg} and (right) orders of magnitude by which I_{on} changes when gate alignment changes.

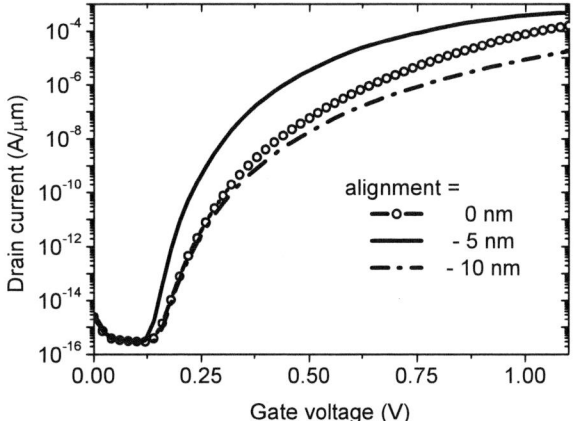

Figure 7. Transfer characteristics for changing gate alignments at the source-side junction, showing the improvement for a slightly underlapped gate.

E. Device length

Device length scaling is a key factor driving conventional MOSFET progress, where scaling happens following a set of predictable guidelines. For Tunnel FETs, due to the different physics within the device, intrinsic region and gate scaling can be carried out without accompanying changes in other dimensions [18]. In terms of characteristic fluctuations, the choice of Tunnel FET i-region (gate) length requires another trade-off decision. Tunnel FET scaling is limited by p-i-n leakage to a critical length (L_{crit}) of about 25 nm for the optimized devices described by Table I.

978-1-4244-6658-0/10 $26.00 © 2010 IEEE   267

TABLE II.   Summary of the results of parameter variations.

Parameter and target value	Parameter variation		V_{TG} %		I_{on} %		S_{point} %		S_{avg} %	
Gate dielectric thickness = 3 nm	-0.75 nm	+0.75 nm	-15	12	220	-66	-29	17	-16	12
Junction width = 21 nm / 5 decades doping	-5.5 nm/dec	+5.5 nm/dec	-13	13	275	-72	-19	19	-18	15
Silicon body thickness = 10 nm	-5 nm	+5 nm	-28	13	266	-57	-58	16	-27	8
Gate alignment at source junction = 0 nm	-2.5 nm	+2.5 nm	-27	12	371	-75	-33	13	-34	16
Intrinsic region/gate length = 30 nm	-5 nm	+5 nm	-0.8	1.1	1.7	-2.2	45	-17	0.7	0
Band gap at tunnel junction = 0.9 eV	-0.05 eV	+0.05 eV	-6	7	30	-24	-4	2	-5	6

Although shorter devices are attractive for reasons of cost and speed, Fig. 8 reveals the downside of designing Tunnel FETs too close to the leakage limit: unacceptably high fluctuation of V_{TG}, S_{point}, and S_{avg}. The standard deviation of the line width, line edge roughness, and other fabrication process-specific sources of variation need to be kept in mind when choosing how close to place the target Tunnel FET length to L_{crit}. Better gate control could reduce L_{crit} slightly, for example by increasing gate capacitance, decreasing body thickness, or using nanowires.

F. Band gap reduction at the tunnel junction

Band gap reduction at the tunnel junction has been proposed to improve on-currents in Tunnel FETs. In principle, there are two primary possibilities: a heterojunction at the source side of the device [1,4,11], or a lateral strain profile with a strain maximum at the tunnel junction [13]. This last performance improvement can be added to the optimized Tunnel FET described in Table I: a smaller band gap of 0.9 eV at the tunnel junction, while keeping the relatively large silicon band gap at the drain-side junction to suppress leakage and maintain low off-current. Fig. 9 shows the impact of the variation of this parameter.

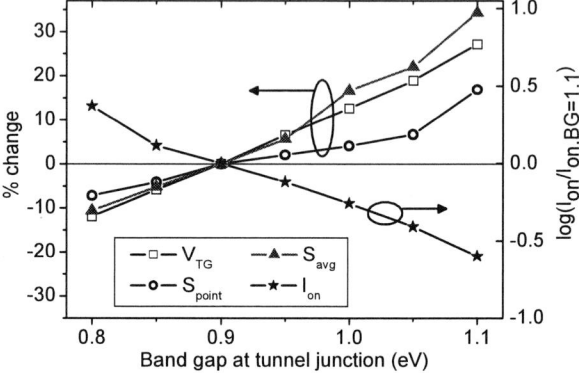

Figure 9. (Left) % change in V_{TG}, S_{point}, and S_{avg} and (right) orders of magnitude by which I_{on} changes when band gap at the tunnel junc. changes.

We have reported the first systematic simulation-based study of the sensitivity of optimized silicon Tunnel FETs to parameter fluctuations. We predict a much reduced sensitivity to gate length scaling (when devices are designed with $L>L_{crit}$) compared to conventional CMOS. A trade-off will need to be made when choosing the desired alignment of the gate dielectric at the source end, between improved characteristics with a design requiring tight control of the gate edge, and non-optimal characteristics with an overlapped gate whose edge variations would not lead to fluctuations in characteristics. Finally, our study suggests that the control of the high-k dielectric thickness, doping profile at the tunnel junction, and film thickness in UTB SOI, with less parameter fluctuation than that required by CMOS, is crucial for future high-performance Tunnel FETs with reproducible characteristics.

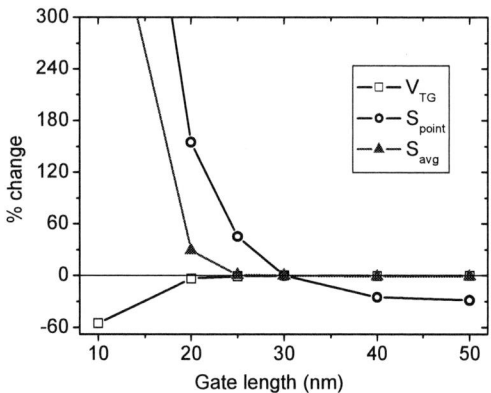

Figure 8. (Left) % change in V_{TG}, S_{point}, and S_{avg} and (right) orders of magnitude by which I_{on} changes when gate / intrinsic region length changes.

IV. DISCUSSION AND CONCLUSION

Table II presents the quantified variation in percentage in key characteristics when the parameters of section III decrease and increase, in an independent way (the possible correlations between varying parameters and the impact on the parameter sensitivity are not taken into account here and should certainly be studied in the future).

REFERENCES

[1] O. Nayfeh, et al., *IEEE TED*, vol. 56, no. 10, 2009, pp.2264-2269.
[2] K. Moselund, et al., in *Proc. ESSDERC*, 2009, pp. 448-451.
[3] C. Le Royer et al., in *Proc. ULIS*, 2009, pp. 53-56.
[4] T. Krishnamohan, et al., in *IEDM Tech. Dig.*, 2008, pp. 947–949.
[5] C. Sandow, et al., *Solid State Electron.*, vol. 53, 2009, pp. 1126-1129.
[6] M. Fulde, et al., in *Proc. INEC*, 2008, pp. 579-584.
[7] K. Bhuwalka, et al., *IEEE TED*, vol. 52, no. 5, 2005, pp. 909–917.
[8] K.Boucart et al., *IEEE TED*, vol. 54, no. 7, 2007, pp. 1725-1733.
[9] J.Knoch, in *Proc. VLSI-TSA*, 2009, pp.45-46.
[10] M.Schlosser, et al., *IEEE TED*, vol. 56, no. 1, 2009, pp.100-108.
[11] A. Verhulst, et al., *J. Appl. Phys.*, vol. 104, 2008, pp. 064514-1-10.
[12] Q. Zhang, et al., in *Proc. DRC*, 2008, pp. 73-74.
[13] K.Boucart, et al., *IEEE EDL*, vol. 30, no. 6, 2009, pp.656-658.
[14] ITRS Roadmap, 2009, available online at http://www.itrs.net.
[15] G. Roy, et al., *IEEE TED*, vol. 53, no. 12, 2006, pp.3063-3070.
[16] H. P. Tuinhout, in *Proc. ESSDERC*, 2002, pp. 95–101.
[17] A. Verhulst, et al., *APL*, vol. 91, 2007, pp. 053102-1-3.
[18] K. Boucart, et al., *SSE*, vol. 51, iss. 11-12, 2007, pp. 1500-1507.

Impact of electron velocity on the I_{ON} of n-TFETs

Hasanali Virani[1], David Esseni[2] and Anil Kottantharayil[1]

[1]Indian Institute of Technology Bombay, Mumbai, India, 400076, e-mail: hgvirani@ee.iitb.ac.in

[2]DIEGM and IU.NET - University of Udine, Udine, Italy, 33100, e-mail: esseni@uniud.it

Abstract—**This paper presents a simulation study concerning the impact of the channel transport on the ON current, I_{ON}, of n-channel tunnel FETs. We show that, when the tunneling generation rate is enhanced to reach I_{ON} values approximately few $\mu A/\mu m$, then the I_{ON} becomes sensitive to the electron velocity in the channel. This is because, at large tunneling rates, the charge due to the generated electrons tends to modify the potential in the channel in such a way that limits the generation rate at the source. Such an electrostatic feedback is smaller the larger is the electron velocity, so that higher tunneling rates and I_{ON} values are observed when increasing the electron velocity.**

I. INTRODUCTION

Tunnel FETs (TFETs) are being explored as an attractive alternative to MOSFETs for low power applications [1–10]. Since the carrier injection from source to channel in a TFET is by tunneling, the subthreshold swing can be lower than 60mV/decade at room temperature. This enables low standby leakage currents and further scaling of supply voltage (V_{DD}).

One of the major hindrance to the use of Tunnel FETs is the poor I_{ON} at low V_{DD}. A lot of work has been reported to improve the I_{ON} by use of low bandgap materials [5, 8], hetero structures [2, 9] and strain [6, 10] to reduce the tunneling barrier. Tunneling currents have also been improved by use of high-κ gate dielectric with scaled EOT to enhance the band bending close to the tunneling junction and thus reduce the tunneling distance [1, 5].

All these technology boosters focus on the improvement of the generation rate at the tunneling junction and little attention has been devoted to the possible role of the carrier transport in the channel. In this respect, however, it is important to note that tunneling is a non local phenomenon that depends on the entire potential profile along the channel, which in turn is affected by the generation rate at the tunneling junction and also by how effectively the generated charge is swept along the channel to the drain contact. With the increase in tunneling generation rate, the transport in the channel may eventually influence the overall potential profile and hence the I_{ON}.

In this paper we use a TCAD modeling approach to address the possible role of transport in silicon n-channel TFET (n-TFET). A toy model is first used, wherein the generation rate is artificially increased by changing the prefactors in the tunneling model. By doing so, we observe that, for I_{ON} values roughly larger than few $\mu A/\mu m$, the I_{ON} does not increase proportionally to the tunneling prefactors. This is essentially because the potential profile is modified by the generated electrons in such a way to partly counteract the increase of the tunneling generation rate. Such an electrostatic feedback is stronger the smaller is the average velocity in the channel, so that in this regime the I_{ON} improves significantly by increasing the carrier velocity.

In second part of the paper we show that a similar electrostatic feedback is also observed when the tunneling generation rate is increased by using the above mentioned technology

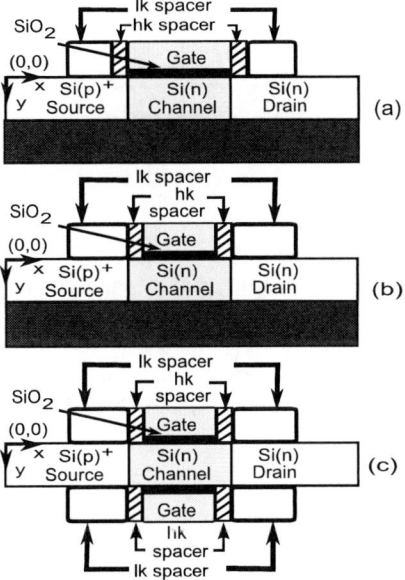

Fig. 1. The n-channel TFET structures with dual-κ spacer {high-κ (hk) with $\epsilon_{hk}=25$ and low-κ (lk) with $\epsilon_{lk}=3.9$} used in this paper. (a) Non underlap SOI structure ($L_g=50$nm). (b) Underlap SOI structure ($L_g=46$nm) (c) Double gate underlap structure ($L_g=46$nm). The SOI thickness is 20nm and body thickness for DG n-TFET is 10nm.

boosters. Also in these practically relevant cases the electron velocity in the channel affects the I_{ON}.

II. DEVICE STRUCTURE AND MODELING APPROACH

The structures of the silicon n-TFET with dual-k spacer used in this work are shown in Fig 1. Gaussian doping profiles are used with a peak density of 10^{20} cm^{-3} for source region and 10^{18} cm^{-3} for drain region, followed by doping gradients of 2 nm/dec. The channel is n-type doped with a concentration of 10^{17} cm^{-3}. The gate workfunction is fixed to 4.1 eV and EOT of gate dielectric is 1.1nm. The 2-D device simulations were performed using Synopsis TCAD tool SENTAURUS [11]. Due to the high doping in the source, Fermi-Dirac statistics and bandgap narrowing are used.

A. Tunneling model

The nonlocal tunneling model is used, which takes into account the potential profile along the entire tunneling path. The tunneling masses m$_c$ and m$_v$ are tuned to calibrate the tunneling model by comparing to experimental data for tunneling diode reported in [12]. The values of the prefactors g_c and g_v, which set the effective Richardson constants, are kept to their default values of 2.1 and 0.66 respectively. Fig. 2 shows fairly good agreement of the simulated with the experimental IV characteristics [12] for a reverse biased tunneling diode obtained with m$_c$=0.65m$_0$ and m$_v$=0.7m$_0$ (where m$_0$ is the electron rest mass).

978-1-4244-6658-0/10 $26.00 © 2010 IEEE

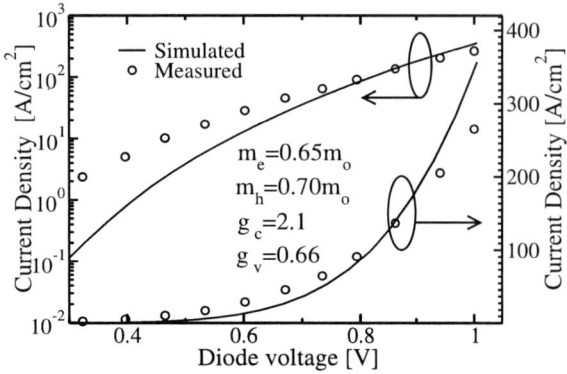

Fig. 2. The simulated (solid line) and experimental (symbols) [12] characteristic of reverse biased silicon tunnel diode showing a fairly good agreement for the parameters indicated in the figure.

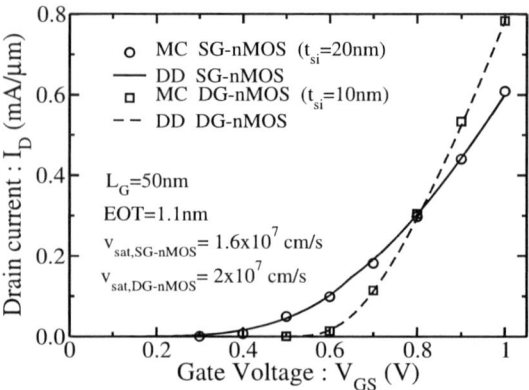

Fig. 3. The simulated characteristic of SG n-MOS and DG n-MOS with v_{sat}=1.6x10^7 V/cm and v_{sat}=2x10^7 respectively. MSMC model as in [15]. t_{si} is the thickness of silicon.

The values of m_c, m_v, g_c and g_v reported above are used throughout the work unless otherwise specified. In particular, they are used for all the simulations in Sec. IV. In Sec. III, instead, g_c and g_v will be varied to artificially change the tunnelling generation rates over a wide range (with m_c and m_v fixed to 0.65m$_0$ and 0.7m$_0$, respectively).

B. Transport model

The use of the Drift Diffusion (DD) model for the channel transport is certainly simplistic for a channel length of about 50 nm. However, we verified that the use of the energy balance model dramatically deteriorates the numerical convergence on the simulations, in spite of a disputable improvement of the physical accuracy [13].

To improve our confidence in the DD model we followed the approach proposed in [14] and adjusted the saturation velocity v_{sat} in the Caughey-Thomas velocity (v) versus field (F) relation

$$v(F) = \frac{\mu_o}{\left[1 + \left(\frac{F\mu_0}{v_{sat}}\right)^\beta\right]^{\frac{1}{\beta}}} \qquad (1)$$

to reproduce the IV characteristics of MOSFETs simulated with the multi-subband Monte Carlo (MSMC) simulator of [15]. To this purpose, we designed MOS transistors with the same channel length and oxide thickness as the TFETs studied throughout the paper. Fig. 3 shows that the DD can reproduce fairly well the MSMC results by increasing the v_{sat} to 1.6x10^7 V/cm for single-gate (SG n-MOS) and to 2x10^7 V/cm for double-gate n-MOSFET (DG n-MOS). The calibrated v_{sat} values are used in the simulations of Fig. 9 to test the effectiveness of different technology boosters. The v_{sat} is instead varied as a parameter in most of the analysis presented below to study the impact of the electron velocity on I_{ON} of the TFETs.

III. IMPACT OF THE CHANNEL TRANSPORT ON THE I_{ON}

The tunneling currents for Si TFET obtained with the calibrated tunneling model are very low for V_{DD}=1V and are reasonably similar to the experimental values [3–7]. Hence a combination of various technology boosters need to be applied to increase the generation rate as detailed in Sec. IV (Fig. 9). However, to gain an insight of the possible role of the carrier transport in n-TFET as the generation rates are increased, in this section we use a toy model wherein the prefactors g_c

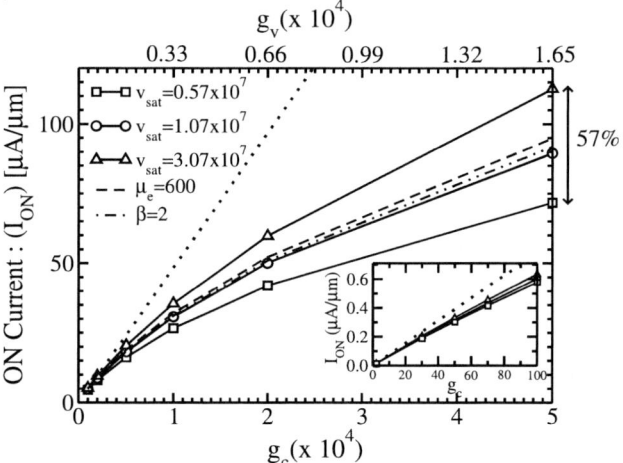

Fig. 4. The effect of increasing the generation rate on I_{ON} by increasing g_c and g_v in the non local tunneling model. I_{ON} for μ_e=600 cm^2/V/s and β=2 are computed for v_{sat}=1.07x10^7 V/cm. The dotted straight line corresponds to an I_{ON} proportional to g_c. Refer to the text for more details.

and g_v in non local tunneling model are increased to enhance the generation rate. The device structure of Fig. 1(a) with ϵ_{hk}=ϵ_{lk}=3.9 is used.

Fig. 4 shows the I_{ON} versus the g_c and g_v, which are always varied by the same factor in this work. For low values of g_c and g_v the I_{ON} improves almost proportionately, as it is seen in the inset of Fig. 4. However for I_{ON} approximately larger than a few μA/μm, the improvement in I_{ON} with increase in g_c and g_v reduces (Fig. 4). The reason behind this behaviour can be understood by examining the inversion density N_{inv} (Fig. 5), the electron velocity (Fig. 6) and the potential energy (Fig. 7) along the channel. As the g_c value is increased, the generation rate and N_{inv} in the channel increase as well (solid lines in Fig. 5). Due to large field at the source-channel tunneling junction (see Fig. 7 where electric field is the x derivative of the band edges), the carriers near the source have a large velocity close to the saturation value. However as the carriers move towards the drain, the velocity tends to decrease (Fig. 6) which is compensated for by the increase of N_{inv} (Fig. 5). The increase of N_{inv} with g_c eventually tends to push up the conduction band edge in the channel, which in turn increases the tunneling distance (solid lines in

978-1-4244-6658-0/10 $26.00 © 2010 IEEE

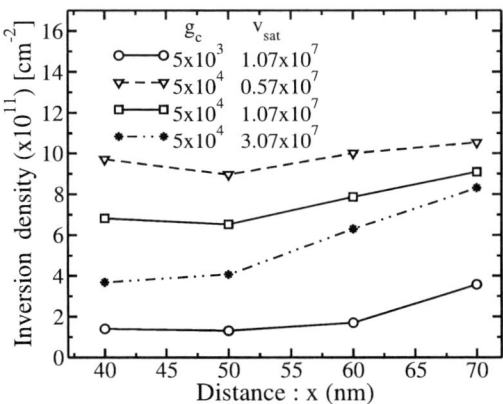

Fig. 5. The inversion density along the channel. Solid line: v_{sat}=1.07x10^7 cm/s (default), dashed line: v_{sat}=0.57x10^7 cm/s and dot-dash line: v_{sat}=3.07x10^7 cm/s.

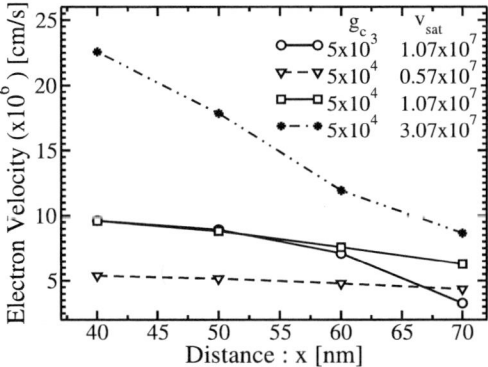

Fig. 6. The electron velocity along the channel. Solid line: v_{sat}=1.07x10^7 cm/s (default), dashed line: v_{sat}=0.57x10^7 cm/s and dot-dash line: v_{sat}=3.07x10^7 cm/s.

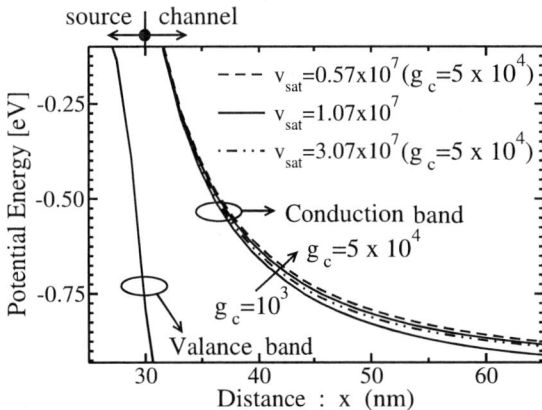

Fig. 7. The Effect of increasing the generation rate on band diagram by increasing g_c and g_v in non local tunneling model. Solid line: v_{sat}=1.07x10^7 cm/s (default), dashed line: v_{sat}=0.57x10^7 cm/s and dot-dash line: v_{sat}=3.07x10^7 cm/s.

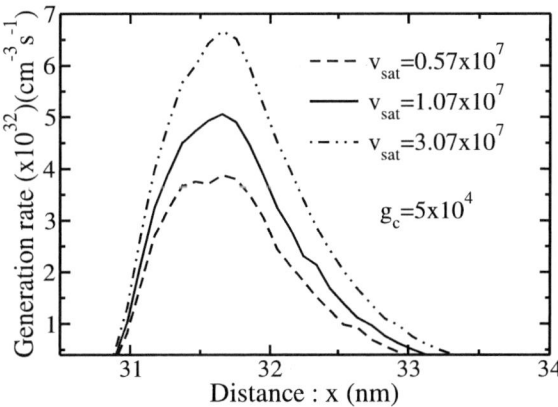

Fig. 8. The electron generation rate along the channel. Solid line: v_{sat}=1.07x10^7 cm/s (default), dashed line: v_{sat}=0.57x10^7 cm/s and dot-dash line: v_{sat}=3.07x10^7 cm/s.

Fig. 7). Therefore the increase in tunneling current, which is exponentially dependent on the tunneling distance, gets limited by the electrostatic feedback.

In Fig. 4 we also observe that, as the velocity saturation is increased, the I_{ON} for a given g_c value tends to increase. This is clearly explained by Fig. 5, 6 and 7. In fact, by increasing v_{sat}, the average carrier velocity is enhanced throughout the channel (Fig. 6), so that a smaller N_{inv} is needed to sustain I_{ON} (see Fig. 5 for different v_{sat}) which in turn reduces the electrostatic feedback on the conduction band edge (see Fig. 7 for different v_{sat}). Hence, for a given g_c, the electrostatic feedback is weaker and the I_{ON} is larger for higher value of v_{sat}. This is clearly observed in Fig. 8, which shows that, for a given g_c, the tunneling generation rate increases with v_{sat}.

In order to complete the analysis concerning the impact of the electron velocity on the I_{ON}, Fig. 4 also reports the I_{ON} obtained for v_{sat}=1.07x10^7 cm/s by changing the low field mobility μ_o and the parameter β of Eq. 1. The curve labelled with μ_o=600 cm^2/V/s corresponds to simulations with a constant mobility value. The curve labelled β=2, instead corresponds to a doubling of the default value β=1. In both cases velocity saturation is included in the simulations. As it can be seen in Fig. 4, the effect of both μ_o and β is quite modest. This is because the electron velocity is very close to v_{sat} near the source-channel junction (Fig. 6), so that the μ_o

and β hardly affect the electron velocity and band profile in the tunneling region.

IV. IMPACT OF CHANNEL TRANSPORT USING TECHNOLGY BOOSTERS

Fig. 9 shows the effect of various technology boosters on silicon n-TFET. The use of high-κ spacer with silicon TFET improves the I_{ON} with negligible impact on the I_{OFF} whereas the use of high-κ gate dielectric improves the I_{ON} along with increase in I_{OFF}. This is because the use of dual κ spacer improves the electric field only at the source-channel tunneling junction through increased fringe field coupling via the high-κ spacer. The underlap structure with dual κ spacer further improves the fringe field coupling thereby improving both the I_{ON} and the subthreshold slope [16]. The I_{ON} improves for the double gate n-TFET (DG n-TFET) (Fig. 1(c)) due to use of multi gate and ultra thin body.

A uniaxial tensile stress in the $\langle 110 \rangle$ source-drain direction of a (100) TFET can enhance the tunneling generation rate by lowering some of the silicon conduction band minima and by shifting upwards the valence band edge. We calculated the strain induced energy shifts of the conduction band Δ_z valleys according to [17] and the shifts of the valence band

978-1-4244-6658-0/10 $26.00 © 2010 IEEE

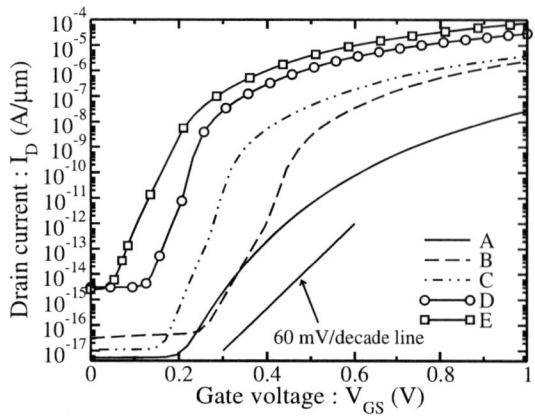

Fig. 9. The I_{DS}-V_{GS} characteristics for different technology boosters. (A) non-underlap n-TFET with SiO2 spacer (Fig. 1(a) with $\epsilon_{hk}=\epsilon_{lk}=3.9$); (B) non-underlap n-TFET with dual κ spacer (Fig. 1(a)); (C) same as B with underlap structure (Fig. 1(b)); (D) same as C with 4GPa uniaxial tensile stress in the source/drain direction; (E) underlap DG n-TFET with dual κ spacer and 4GPa uniaxial tensile stress in the source/drain direction (Fig. 1(c)).

TABLE I

CHANGES IN THE SILICON BAND EDGES AND CORRESPONDING VALUE OF THE BAND-GAP PRODUCED BY A UNIAXIAL STRESS IN THE $\langle 110 \rangle$ SOURCE-DRAIN DIRECTION OF A (100) TFET. THE BAND EDGES ARE CALCULATED ACCORDING TO [17–19].

Stress (GPa)	Conduction Band (eV)	Valence Band (eV)	Bandgap (eV)
0	1.170	0	1.170
1	1.149	0.027	1.122
2	1.121	0.054	1.067
3	1.085	0.081	1.004
4	1.043	0.108	0.935

maxima by using the six bands $\mathbf{k}\cdot\mathbf{p}$ model [18, 19]; the results is reported in Table I and it has then been used in the TCAD simulations. The uniaxial tensile stress is also known to improve the electron mobility [20], however, lacking a reliable model for the stress dependence of the mobility in the TCAD framework and considering the negligible effect of the mobility on the I_{ON} shown in Fig. 4, the strain induced mobility improvements were not accounted for in the simulations of Figs. 9 to 11.

Fig. 10 shows the effect of saturation velocity on the I_{ON} for highest tunneling currents obtained for strained DG n-TFET with dual κ spacer. We observe trends qualitatively similar to those observed in Fig. 4 for the toy model. We verified that the I_{ON} dependence on v_{sat} has the same origin as illustrated in Fig. 5, 6 and 7, namely an increase of the carrier velocity reduces the inversion density in the channel and favours the band bending at the source junction that produces the tunneling (see Fig. 7). Hence the tunneling generation rate increases with v_{sat} as shown in Fig. 11.

V. CONCLUSIONS

By using a calibrated non local tunneling model, this paper has shown that for I_{ON} values larger than few μA/μm, the tunneling generated electrons in TFETs tend to accumulate in the device channel and produce an electrostatic feedback which limits the I_{ON}. In this regime the I_{ON} becomes sensitive to the electron velocity and increases with it.

These results point out that the tunneling barrier is maybe the most important but certainly not the only element to be

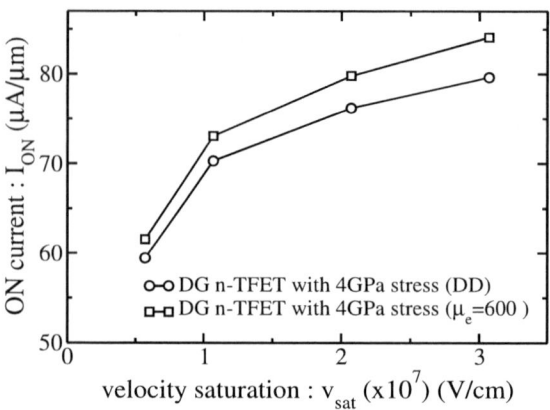

Fig. 10. The ON current versus velocity saturation in DG n-TFET.

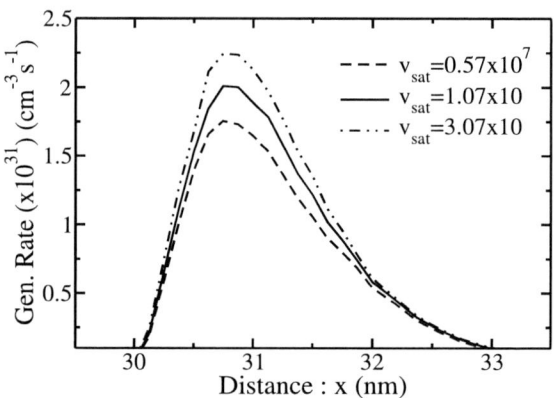

Fig. 11. The generation rate along the channel in DG n-TFET.

considered to achieve large I_{ON} in TFETs.

ACKNOWLEDGMENTS

The authors would like to thank Prof. Luca Selmi and Prof. P. Palestri for many helpful discussions. The work was partially funded by the italiani MIUR through the "Bando per borse a favore di giovani ricercatpri indiani".

REFERENCES

[1] K. Boucart et al., IEEE Trans. Elec. Devices, vol. 54, p. 1725, 2007.
[2] V. Nikam et al., Proceedings of DRC, p. 77, 2008.
[3] K. K. Bhuwalka et al., Jap. J. Appl. Phy., p. 3106, 2006.
[4] O. M. Nayfeh et al., IEEE Electron Device Lett., vol. 29, p. 468, 2008.
[5] F. Mayer et al., IEDM Technical Digest, p. 163, 2008.
[6] T. Krishnamohan et al., IEDM Technical Digest, p. 947, 2008.
[7] M. Fulde et al., Proc. Nanoelectr. Conf., p. 579, 2008.
[8] N. Jain et al., Proceedings of DRC, p. 99, 2008.
[9] E. H. Toh et al., Jap. J. Appl. Phy., p. 2593, 2008.
[10] K. Boucart et al., Proceedings of ESSDERC, p. 452, 2009.
[11] Synopsys sentaurus design suite.
[12] P. M. Solomon et al., J. Appl. Phy., vol. 95, p. 5800, 2004.
[13] C. Jungemann et al., IEEE Trans. Elec. Devices, vol. 52, p. 2404, 2005.
[14] J. D. Bude, Proc. Sim. Semicond. Process and Dev., p. 23, 2000.
[15] L. Lucci et al., IEEE Trans. Elec. Devices, vol. 54, p. 1156, 2007.
[16] H. Virani et al., submitted to IEEE Trans. Elec. Devices.
[17] E. Ungersboeck et al., IEEE Trans. Elec. Devices, vol. 54, p. 2183, 2007.
[18] M. V. Fischetti et al., J. Appl. Phy., vol. 94, p. 1079, 2003.
[19] M. De Michielis et al., IEEE Trans. Elec. Devices, vol. 54, p. 2164, 2007.
[20] S. Thompson et al., IEDM Technical Digest, 2006.

Abrupt Switch based on Internally Combined Band-To-Band and Barrier Tunneling Mechanisms

Livio Lattanzio[1], Luca De Michielis[1], Arnab Biswas[2,1] and Adrian M. Ionescu[1]

[1]Ecole Polytechnique Fédérale de Lausanne, Switzerland
[2]Vellore Institute of Technology, India

Abstract— We report a novel device which exploits the internally combined quantum mechanical Band-To-Band and Barrier Tunneling mechanisms to achieve improved performances and overcome the intrinsic low current drive limitations of conventional Tunnel FETs and the 60 mV/decade limitation of MOSFETs at room temperature. The new structure, including an ultra-thin dielectric between metal source and silicon channel, allows for sub-60 mV/dec average subthreshold slope (SS ~43 mV/dec) and a uniquely high I_{ON}/I_{OFF} ratio (~10^{11}). The device principle and the potential performances are investigated by numerical simulation. We evaluate the impact of the tunneling layer thickness on device performances and compare single and double gate architectures. Finally, we evaluate the impact of device gate length scaling on its performances, which is different from Tunnel FET: we observe an improvement of SS and I_{ON} values at smaller gate lengths.

I. INTRODUCTION

Tunnel FETs (TFETs) are today under intense exploration [1,2] as some of the most promising enablers of future logic circuits operating with a supply voltage smaller than 0.5 V and saving many decades of I_{off}. This is essentially due to the fact that the subthreshold swing of TFETs can be significantly lower than the 60 mV/dec MOSFET limit at room temperature due to band-to-band tunneling (BTBT) in gated reverse-biased p-i-n junctions. Recent reports suggest that TFETs could also be considered as promising candidates for the high performance switch, by using appropriate heterostructure architectures [3,4] that can probably address the I_{on} limitation of all-silicon TFETs. However, the issue with these heterostructures is that also the transistor leakage current is increased, therefore affecting their effective I_{ON}/I_{OFF} ratio.

In parallel, Schottky barrier MOSFETs (SB-MOSFETs), normally fabricated using silicided source/drain, have been proposed to replace conventional doped architectures, as their series resistance is considerably lower [5]. The use of silicides could be interesting for non-conventional TFET implementations, considering that Schottky junctions could provide higher drive currents than p-n junctions.

Fig. 1 shows the comparison between the I_d-V_g of a (a) *p-i-n* TFET and a (b) *metal-i-n* Schottky barrier TFET on silicon-on-insulator (SOI). The common parameters for the two devices are t_{Si} = 40 nm, t_{ox} = 3 nm, ε_{ox} = 25, N_{drain} = 1×10^{18} cm^{-3} (Phosphorous), $N_{channel}$ = 1×10^{16} cm^{-3} (Boron). Source is specific for each device: for (a) we considered doped silicon with N_{source} = 3×10^{20} cm^{-3} (Boron), for (b) a metal with workfunction Φ_M = 5.3 eV. From literature [6] we know that

the subthreshold slope for device (a) can be approximated to:

$$SS \approx \frac{\ln(10)}{|e|}\Delta\Phi = \frac{\ln(10)}{|e|}\left(E_v^{ch} - E_c^s\right)$$

with e the elementary charge, E_v^{ch} the valence band in the channel and E_c^s the conduction band in the source. A band-to-band tunneling device can thus have vanishing values of SS for $\Delta\Phi \to 0$ (low V_g), due to its particular band situation. However, for the same reason, this mechanism can generate only a limited current level compared to thermionic mechanism of a MOSFET. The SS for a TFET with a Schottky barrier at source (b) can be expressed as:

$$SS = \frac{kT}{q}\ln(10)\frac{1}{1-\exp(-d_{tunnel}/\lambda)}$$

with k the Boltzmann constant, T the temperature, q the electron charge magnitude, d_{tunnel} the tunneling distance and λ the screening length [7]. As for a MOSFET, the SS is limited to 60 mV/dec at room temperature, but the generated current at high V_g is much higher than the one in device (a) (see bottom plot in Fig.1).

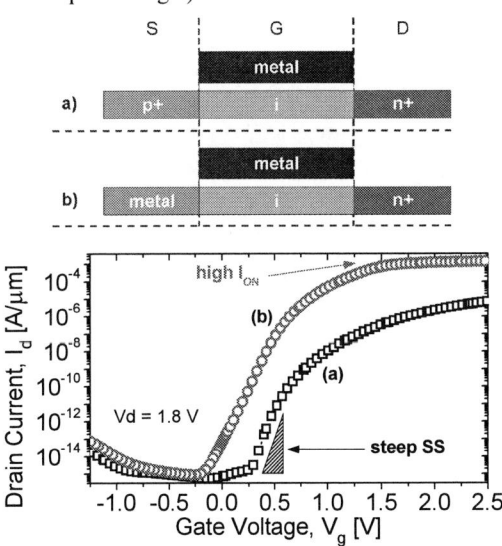

Fig. 1: Structure and transfer characteristics for a (a) *p-i-n* TFET and a (b) *metal-i-n* Schottky barrier TFET. The metal source allows for high ON currents but it shows limitations on subthreshold slope (SS) values due to the kT/q factor given by barrier tunneling. The *p-i-n* TFET, instead, allows for sub-60 mV/dec SS values but it is limited on the drive current by band-to-band tunneling.

In this work we propose a new abrupt switch concept which combines two conduction mechanisms in a single smart device architecture: the abrupt *p-i-n* tunnel FET at low gate voltage and the high I_{ON} Schottky tunnel FET at high gate voltage. This is basically achieved by inserting an ultra-thin dielectric layer between the metal source and the silicon channel of a Schottky barrier TFET.

II. NOVEL TUNNELING DEVICE PRINCIPLE

The new tunneling device architecture is shown in Fig. 2 (top): this device is a modified Schottky barrier TFET (as in Fig. 1(b)), where an ultra-thin (5 Å) SiO$_2$ tunneling layer is inserted between source and channel. Note that a similar device has been recently reported in [8] and its characteristics, different from the ones reported here, are very poorly explained by a different physical principle.

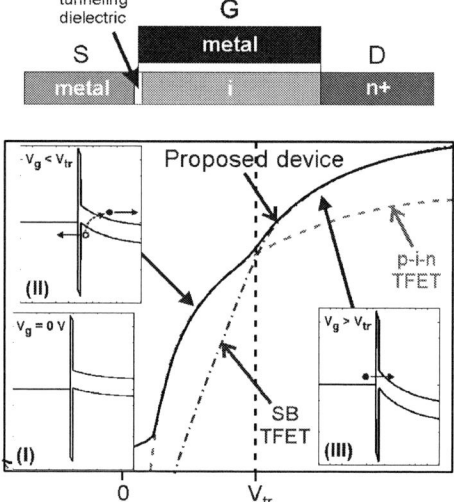

Fig. 2: Schottky Tunnel FET with thin interfacial dielectric at source: schematic structure (top) and source-channel band diagrams under different gate bias conditions (bottom). In (I) the device is off and a very low current flows from source to drain. When V_g is increased (II) the main conduction mechanism is given by band-to-band tunneling. After a certain transition voltage (V_{tr}) barrier tunneling overcomes BTBT and a higher I_{ON} is obtained.

The device is studied by simulation in three different regimes: (I) the "**OFF**" state, when $V_g = 0$ V; (II) the "**BTBT**" state, when the current flow is mainly driven by band-to-band tunneling (low V_g); (III) the "**Barrier**" state, when the barrier tunneling is the dominant current flow mechanism (high V_g). Between (II) and (III) we can define a "transition voltage" (V_{tr}), which determines the threshold between the two regimes.

The fundamental difference between our device and a Schottky TFET without tunneling dielectric is clearly depicted in Fig. 3, by analyzing the energy band diagrams for the different regimes. The two devices have the following common parameters: $t_{Si} = 20$ nm, $t_{ox} = 3$ nm, $\varepsilon_{ox} = 25$, $N_{drain} = 1 \times 10^{18}$ cm^{-3} (Phosphorous), $N_{channel} = 5 \times 10^{16}$ cm^{-3} (Boron), $\Phi_M = 5.65$ eV. The thin SiO$_2$ of device in Fig. 3(b) is 5 Å thick. This thin dielectric layer gives some remarkable advantages: (i) it inhibits thermionic emission from source due

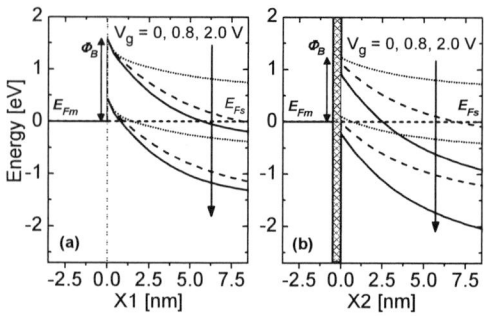

Fig. 3: Simulated band structures for different gate bias conditions at the source/channel interface of a Schottky-source Tunnel FET without (a) and with (b) thin interfacial dielectric layer. The metal work function is $\Phi_M = 5.65$ eV and drain bias is $V_d = 1.8$ V. The Fermi level pinning [10] in structure (a) shows a large barrier height (Φ_B) for electrons and a small band bending with V_g. The pinning is released in structure (b): under same bias conditions, Φ_B decreases with V_g and band bending at the interface occurs faster.

to the high potential barrier; (ii) it makes possible a higher band bending at low V_g; finally, by "de-pinning" the Fermi level in the semiconductor (E_{Fs}) [9]; (iii) it lowers the effective electron energy barrier (Φ_B) at high V_g.

For the Schottky TFET with thin interfacial dielectric at source, the resulting generation rates integrated over the silicon surface and the total drain current as a function of the gate voltage are plotted in Fig. 4.

Both band-to-band and barrier tunneling rates are given but the effective contribution for the I_{ds} is barrier tunneling, as precisely indicate by the carriers that can go from source to channel/drain. Hole barrier tunneling is however simply showing that all the holes generated in the silicon by BTBT or Schockley-Read-Hall (SRH) can directly pass into the source, as there is a negligible energy difference (barrier) between the silicon valence band and the Fermi level in the source metal.

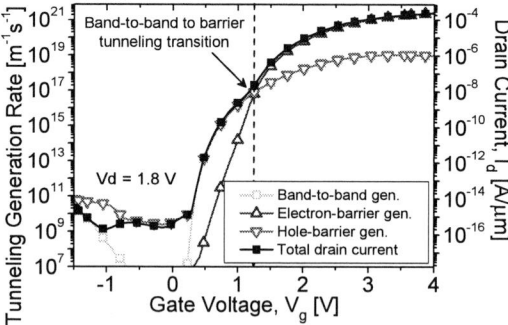

Fig. 4: Simulated drain current (solid squares) and tunneling contributions (open symbols) as a function of the gate bias which validates the mechanism described in Fig. 2.

Fig. 5 illustrates the 2D cross-sections at the source-channel interface with the extracted current density (in A/cm^2). We can distinguish the three regimes: the "OFF" state ($V_g = 0$ V), when a small leakage current is flowing from source to drain; the "BTBT" phase (0 V $< V_g < V_{tr}$), when e/h pairs are generated by BTB tunneling into the channel and are swept towards source and drain; the "Barrier" regime ($V_g > V_{tr}$), when, in correspondence of the surface, electrons tunnel into the channel from the metal source through the thin SiO$_2$.

978-1-4244-6658-0/10 $26.00 © 2010 IEEE

III. RESULTS AND DISCUSSION

Simulations reported in this work have been performed using the commercial tool Sentaurus Device 2009.06c [10]. The band-to-band tunneling process has been modeled according to a dynamic non-local path tunneling model: in this way we have taken into account the non-local generation of electrons and holes caused by direct and phonon-assisted processes [11]. The non-local tunneling at the source-side interfaces is accounted following the approach reported in [12], where the computation of the tunneling probability is based on the WKB approximation.

Due to the small dimensions of our devices, the electric

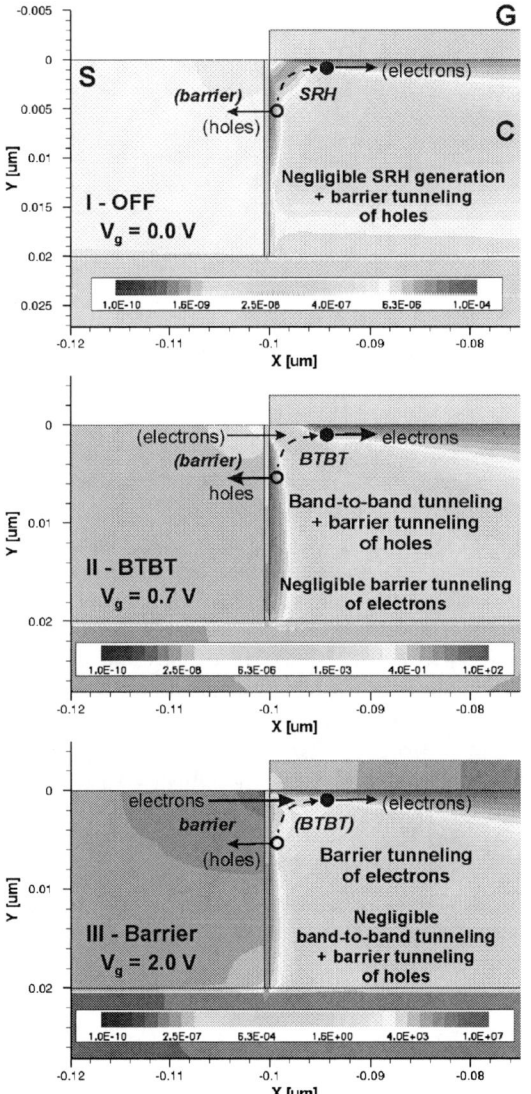

Fig. 5: Total current density 2D distribution (in A/cm²) at the source junction for V_g = 0, 0.7 and 2 V. Source (S), gate (G) and channel (C) regions are shown. In the OFF state (I) a slight leakage current due to Shockley-Read-Hall (SRH) generation, band-to-band and barrier tunneling flows from source to drain. When the transistor switches on (II), BTBT becomes the dominant mechanism, giving a steep subthreshold swing. Further on (III), electron barrier tunneling becomes the dominant mechanism, boosting the drive current.

field at the source/channel junction is to the order of 1 MV/cm: for this reason this junction becomes extremely sensitive to defect-assisted tunneling, which causes electron-hole pair generation before band-to-band tunneling to set in. Thus, in order to model properly the OFF current level of those regions we have used the trap-assisted-tunneling model [13] that properly reduces the SRH recombination lifetimes.

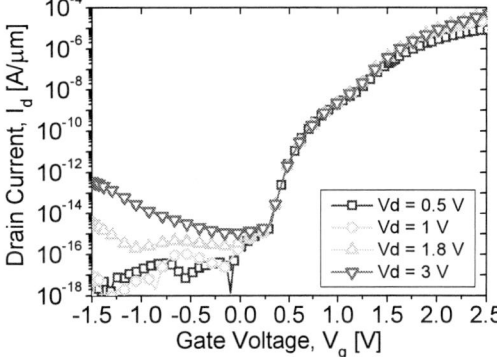

Fig. 6: Simulated I_d-V_g for different V_d. It is clear that the drain affects mainly the OFF region of the curve, by increasing the ambipolarity of the device. The ON current experiences a slight increase, while the region around threshold remains the same with V_d. It has to be noticed that the device can be operated also at relatively low drain voltage (V_d = 0.5 V).

Simulated I_d-V_g are given as a function of V_d in Fig. 6. The relative independence of I_d on V_d at low V_g is just apparent because the V_d effect on the BTB barrier narrowing saturates below 0.5 V. However, barrier tunneling (high V_g) experiences a modest modulation. The leakage region is the one that has the highest variation: the drain bias has in fact a high impact on the device ambipolarity. It can be remarked that this device is able to operate also at relatively low drain voltages (V_d = 0.5 V).

In Fig. 7 are shown the output characteristics of the device. These curves can be categorized in two groups: first, for V_g = 0.5 and 1 V ($V_g < V_{tr}$), and, second, for $V_g > V_{tr}$. The first two curves, in fact, show for some hundreds mV after V_d = 0 V an exponential region (see log scale), typical signature of BTBT.

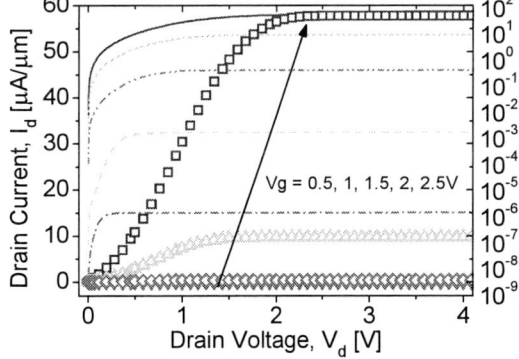

Fig. 7: Simulated I_d-V_d for different V_g. Symbols are the curves in linear scale, lines are in log scale. It can be noted that for V_g = 0.5 and 1 V ($V_g < V_{tr}$) the exponential region close to V_d = 0 V holds for some hundreds of mV (see log scale): this is due to the drain modulation of the band-to-band tunneling barrier. For $V_g > V_{tr}$, instead, due to the nature of barrier tunneling, we have a lower series resistance and a quasi-linear region close to V_d.

978-1-4244-6658-0/10 $26.00 © 2010 IEEE   275

For $V_g > V_{tr}$, this region is quasi-linear, but as we have barrier tunneling it shows however some series resistance at low V_d.

In order to evaluate the role of the tunneling layer at source we have performed simulations for different layer thicknesses (Fig. 8). The plot shows that only by having an ultra-thin (3 to 5 Å) dielectric it is possible to exploit the two mechanisms. With a 1 nm thick SiO_2 the onset of the BTBT occurs after barrier tunneling and no steep slope is achieved. The 3 Å curve represent the extreme case of an oxide monolayer, and we see that a sub-60 mV/dec SS value is obtained for three points, meaning more than 2 decades of drain current.

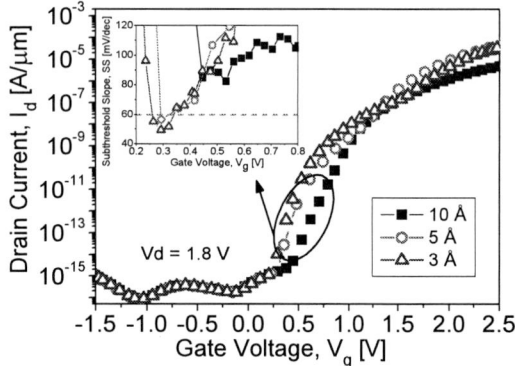

Fig. 8: Simulated transfer characteristics of the device under study for different source tunneling dielectric thickness values. This layer has a big impact on the device performances. A dramatic improvement is obtained for angstrom-scale dielectrics, and a point subthreshold slope lower than 60 mV/dec is obtained for a 0.5 or 0.3 nm thick silicon dioxide (inset).

The importance of the gate electrostatic control is analyzed by comparing a single and a double gate structure (Fig. 9). In these simulations the thin dielectric layer is 5 Å thick. We extracted the subthreshold slope value for 2.3 decades of I_d and the ON current at $V_g = 2$ V. For a double gate structure the slope goes down from 64 to 54 mV/dec and the I_{ON} is more than 10 times higher than in the single gate architecture, with an I_{ON}/I_{OFF} ratio of about 10^{10}.

Fig. 10 shows a simulation-based scaling study of the device. The gate length (L_g) is varied from 300 nm to 100 nm, and the gate dielectric thickness (t_{ox}) is varied from 4.5 nm to

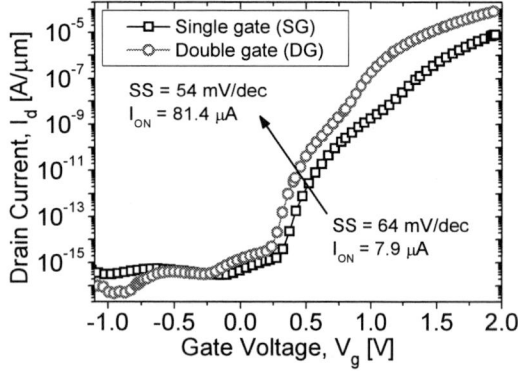

Fig. 9: I_d-V_g of the device in the case of a Single Gate (SG) and a Double Gate structure. A big improvement in the subthreshold slope and in the drive current is observed. The SS goes below 60 mV/dec in the case of a DG structure, and the I_{ON} at $V_g = 2$ V experiences and improvement of more than 1 decade.

1.5 nm. From the plots we observe that the device can be downscaled in terms of subthreshold slope and on current values if the gate length and oxide are simultaneously reduced. An average subthreshold slope value as low as 43 mV/dec over 2.5 decades of drain current and a I_{ON} level of ~204 μA/μm at $V_g = 2.7$ V ($I_{ON}/I_{OFF} = 10^{11}$) are found for $L_g = 100$ nm. Considering just the BTBT contribution, the I_{ON} would have been just 6 μA/μm, meaning a current increase of 198 μA/μm with the proposed structure.

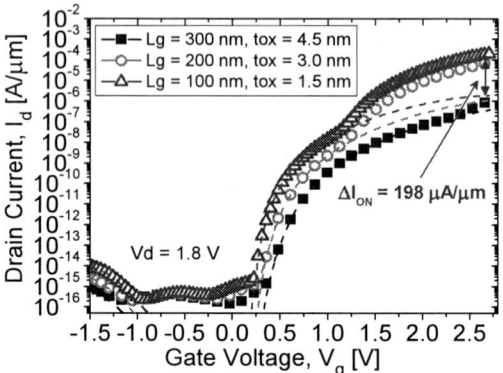

Fig. 10: Simulated scaling study for varying gate lengths and gate oxide thickness. The device seems to benefit from scaling both in terms of subthreshold slope SS and drive current I_{ON}. BTBT benefits just from the scaled gate oxide, while the barrier regime is boosted with smaller L_g.

IV. CONCLUSION

We have presented a new Tunnel FET device concept, implementing a metal-silicon source junction with an ultra-thin tunneling dielectric. The proposed abrupt switch exploits two combined phenomena, band-to-band and barrier tunneling of electrons, to achieve in an all-silicon device steep subthreshold slope values (~43 mV/dec) and high drive currents, keeping the leakage at very low levels ($I_{ON}/I_{OFF} = 10^{11}$). We performed TCAD simulations and explained the new device principle and its characteristics. We show I_d-V_g curves for different source dielectric thicknesses and for single vs double gate architectures. Finally, we performed a scaling study and observe that device performances can improve in terms of SS and I_{ON} when L_g and t_{ox} are scaled together.

ACKNOWLEDGMENT

This work was supported by NOE NANOSIL (FP7-216171). We thank Prof. P.S. Mallick, VIT, India, for useful discussions.

REFERENCES

[1] Q. Zhang, et al., *IEEE Electron Device Lett.*, 2006, 27, (4), 297-300.
[2] W. Y. Choi et al., *IEEE Electron Device Lett.*, 2007, 28, (8), 743-745.
[3] T. Krishnamohan et al., *IEEE IEDM, Technical Digest*, 2008, 947-949.
[4] O. M. Nayfeh et al., *IEEE Electron Device Lett.*, 2008, 29, (9), 1074-1077.
[5] G. Larrieu et al., *IEEE Electron Device Lett.*, 2004, 25, (12), 801–803.
[6] J. Knoch et al., *Solid-State Electronics*, 2007, 51, (4), 572-578.
[7] J. Knoch et al., *Physica Status Solidi (A)*, 2008, 205, (4), 679-694.
[8] P. K. Baghbani, *Solid-State Electronics*, 2010, 54, (1), 48-51.
[9] D. Connelly, *IEEE Trans. on Nanotechnology*, 2004, 3, (1), 98-104.
[10] Sentaurus User Guide, 2009.06C.
[11] E. O. Kane, *J. App. Phys.*, 1961, 32, (1), 83–91.
[12] M. Ieong et al., *IEEE IEDM, Technical Digest*, 1998, 733–736.
[13] A. Schenk, *Solid-State Electronics*, 1992, 35, (11), 1585–1596.

Junctionless Nanowire Transistor (JNT): Properties and Design Guidelines

A. Kranti, R. Yan, C.-W. Lee, I. Ferain, R. Yu, N. Dehdashti Akhavan, P. Razavi, JP Colinge

Tyndall National Institute, University College Cork, Cork, Ireland
E-mail : jean-pierre.colinge@tyndall.ie

Abstract— Conduction mechanisms in junctionless nanowire transistors (gated resistors) are compared to inversion-mode and accumulation-mode MOS devices. The junctionless device uses bulk conduction instead of surface channel. The current drive is controlled by doping concentration and not by gate capacitance. The variation of threshold voltage with physical parameters and intrinsic device performance is analyzed. A scheme is proposed for the fabrication of the devices on bulk silicon.

I. INTRODUCTION

The junctionless nanowire transistor (JNT) is a heavily-doped SOI nanowire resistor with an MOS gate that controls current flow. Doping concentration is constant and uniform throughout the device and typically ranges from 10^{19} and 10^{20} cm^{-3}. The device features bulk conduction instead of surface channel conduction. Junctionless fabrication process is greatly simplified, compared to standard CMOS since there are no doping concentration gradients in the device.[1-3]

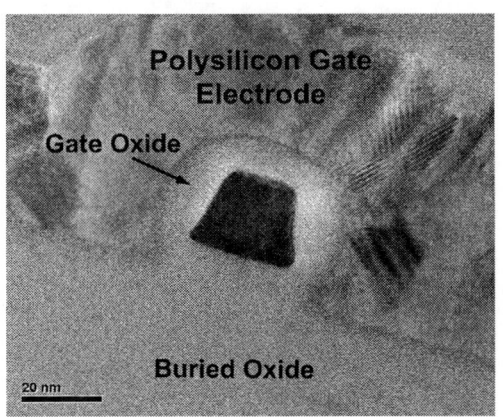

Figure 1: TEM cross section of a junctionless nanowire transistor.

II. DEVICE PHYSICS

The electrical characteristics of the JNT are remarkably identical to those of regular trigate MOSFETs. Figure 2 shows $I_D(V_G)$ characteristics. The device has an effective width of 25nm and L_g=1um. Extrapolating using V_{DS}=1V,

$V_{Goff}=V_{TH}$-0.3V and $V_{Gon}=V_{TH}$+0.7V, L=20nm and a pitch of 50nm one finds that the device is capable of I_{OFF} and I_{ON} of 1nA/µm and 1000µA/µm, respectively, without using any mobility-enhancing technique such as strain.

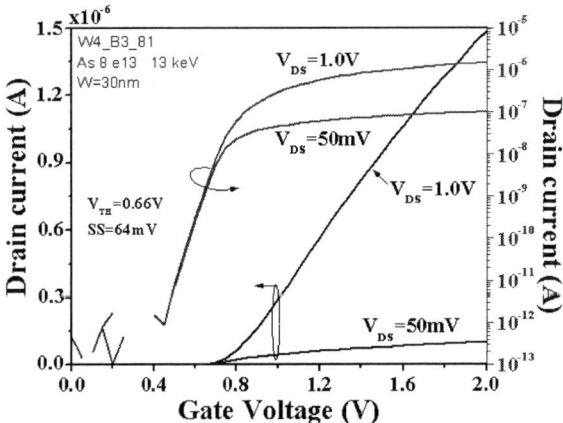

Figure 2: Measured $I_D(V_G)$ characteristics of an n-channel device with W_{eff}=25nm and L=1µm.

The physics of the JNT is quite different from that of standard multigate FETs. Depletion of the heavily doped nanowire creates a large electric field perpendicular to current flow below threshold, but above threshold the field drops to zero. This is the opposite of inversion-mode (IM) or even accumulation-mode (AM) devices where the field is highest when the device is turned on (Table I). The electron concentration profiles in cross sections of IM, AM and JNT devices are shown in Figures 3 and 4.

Table I: Conduction mechanisms and E field perpendicular to current flow in inversion-mode (IM), accumulation-mode (AM) and junctionless (JNT) nanowire MuGFETs

Type	Above Threshold	Subthreshold
IM	Surface conduction High E field	Surface conduction Low E field
AM	Surface conduction High E field	Bulk conduction Low E field
JNT	Bulk conduction Low E field	Bulk conduction High E field

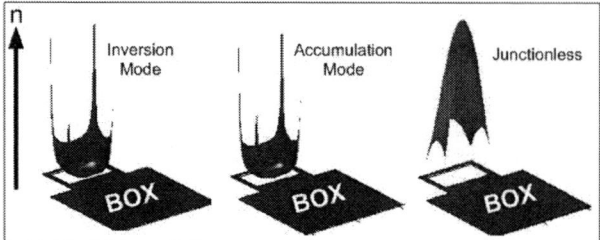

Figure 3: Electron concentration profile above threshold in IM, AM and JNT devices. Surface channels are formed in IM and AM device, while conduction takes place in the bulk of the nanowire in the JNT.

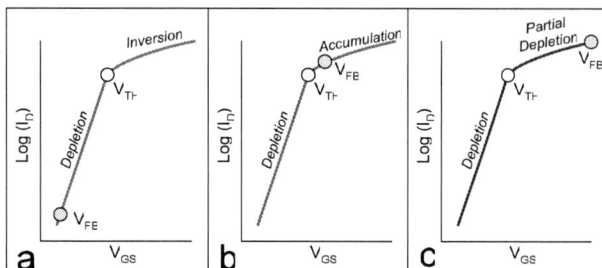

Figure 4: Current in inversion-mode (a), accumulation-mode (b) and junctionless (c) nanowire MuGFETs. Note the very different positions of the flatband voltage, V_{FB}.

The threshold voltage of the JNT depends on doping concentration N_D, EOT, nanowire thickness t_{si} and width W_{si}.(Figs. 5-6)

III. DEVICE DESIGN

The V_{TH} variation with W_{si} offers high flexibility for achieving devices with different threshold voltages. For instance, for $N_D=2\times10^{19}$cm^{-3}, $t_{si}=8$nm and EOT=1nm, A device with $V_{TH}=0.6$V is obtained if $W_{si}=11$nm and $V_{TH}=0.3$V if $W_{si}=17$nm (Fig. 5).

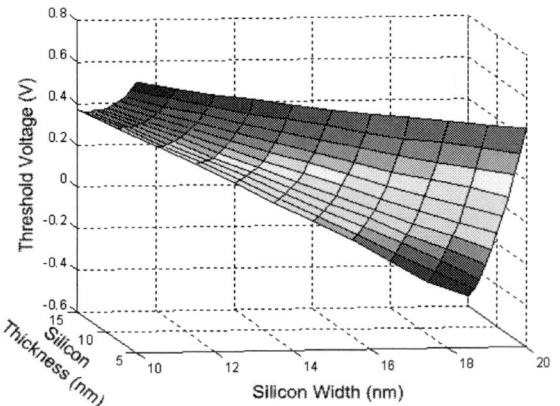

Figure 5: Long-channel threshold voltage *vs.* nanowire width and thickness for $N_D=2\times10^{19}$cm^{-3} and EOT=1nm.

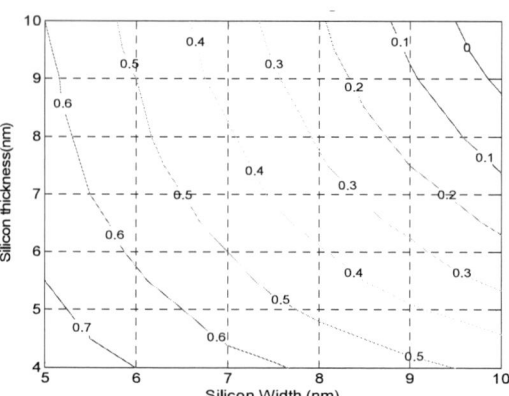

Figure 6: Long-channel threshold voltage *vs.* nanowire width and thickness for $N_D=5\times10^{19}$cm^{-3} and EOT=0.5nm.

When the drain voltage is increased, the JNT enters saturation like a regular MOSFET (Figure 7) and, therefore, presents output characteristics that are identical to those of standard MOS devices.

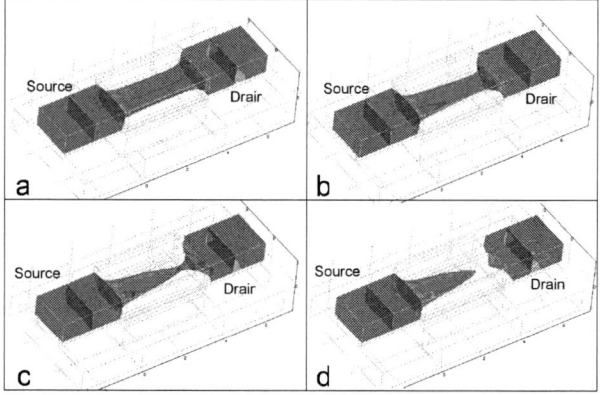

Figure 7: Electron concentration contour plots in an n-type junctionless transistor. A: $V_D = 50$mV; B: $V_D = 200$mV; C: $V_D = 400$mV; D: $V_D = 600$mV. $V_G>V_{TH}$.

IV. BULK VERSION OF THE DEVICE

The feasibility of a bulk silicon version of the JNT was analyzed using 3D simulations. As shown in Fig. 8, the cross-section ($W_{si} \times T_{si}$) of the N$^+$–N$^+$–N$^+$ device is 5 × 5 nm^2. The extension, d, of side gates into the moderately doped (10^{17} cm^{-3}) p–type region was optimized to control SCEs and leakage current. Due to the N$^+$–N$^+$–N$^+$ design of MOSFET no lateral S/D junction (along the current flow path) is formed, but a vertical PN junction is required for device isolation. Although the N$^+$ region is heavily doped (8×10^{19} cm^{-3}) to allow for high current flow in the on–state, the small cross-section of bulk–JLMOS ensures full depletion resulting in low leakage current down to 10 nm devices as demonstrated in Fig. 9. The gate is P$^+$ poly. It should be noted that the leakage current of ~ 10 pA can be achieved in 10 nm bulk JLMOSFET and full device functionality is observed even in

978-1-4244-6658-0/10 $26.00 © 2010 IEEE   278

the absence of reversed biased lateral (source and drain) PN–junctions.

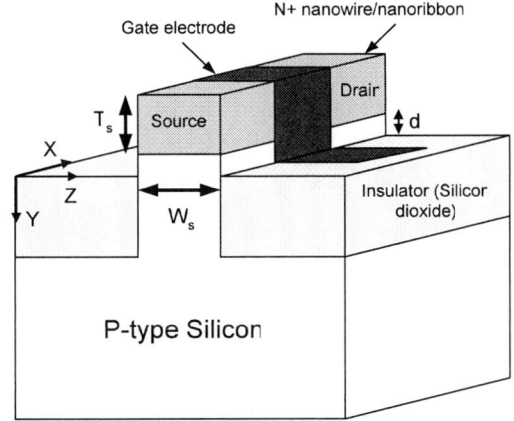

Figure 8: Schematic diagram of a bulk JNT.

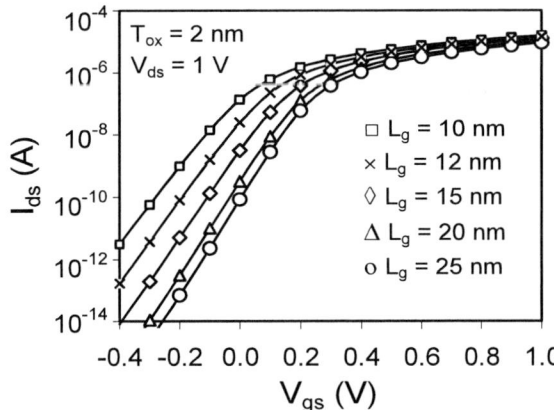

Figure 9: Subthreshold characteristics of bulk multi-gate JMOSFET with gate workfunction of 5.5 eV at drain bias (V_{DS}) of 1 V for different gate lengths (L_g). The extension of side gates (d) into the p–type region is 5 nm.

Fig. 10 shows the dependence of subthreshold slope (S-slope) and drain induced barrier lowering (DIBL) parameters for the bulk–JNT device. Subthreshold slope and DIBL can be limited to less than 80 mV/decade and 100 mV/V, respectively, in bulk JNT devices even with relatively thick gate oxide thickness of 2 nm at drain bias (V_{DS}) of 1 V for gate lengths down to 12 nm. DIBL was extracted as difference in threshold voltages for drain bias of 50 mV and 1V. An inversion mode intrinsic bulk MOSFET designed with buried ground plane ($N_a \sim 10^{19}$ cm^{-3}) and 'idealized' abrupt S/D junction with $L_g = 15$ nm achieves a degraded S-slope of 78 mV/dec and DIBL of 95 mV/V when compared to 70 mV/dec and 74 mV/V respectively, for JLMOSFET with same gate length. The S–slope and *DIBL* parameter for inversion mode devices can be reduced by adopting an

underlap channel architecture.[4] However, such an approach will require precise control of S/D doping gradient in the S/D extension regions and additional process complexity which will be extremely difficult to optimize and be an added source of variability in the nanoscale regime. The bulk–JNT device offers greater flexibility in selecting device parameters to limit SCEs along with simpler fabrication process.

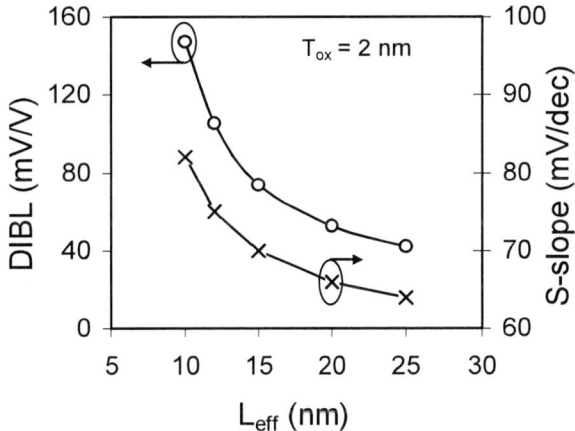

Figure 10: DIBL and Subthreshold slope of bulk multi-gate JNTs for various gate lengths (L_g). S-slope was extracted at $V_{DS} = 1$ V.

Fig. 11 shows the cut-plane of the total current density through the middle of silicon channel at gate bias of 0.4 V. The dominant current flow is through the N$^+$ region where the current density is the highest and not through the moderately doped p–type region or substrate. As the current flow is through the centre of silicon film and not at the Si–SiO$_2$ interface, carriers observe a reduced electric field in the direction perpendicular to flow and carriers travel through the film with higher mobility which is much less influenced by surface roughness scattering as experienced by bulk inversion mode transistors. This gives JNT an advantage in terms of current drive for nanoscale applications.

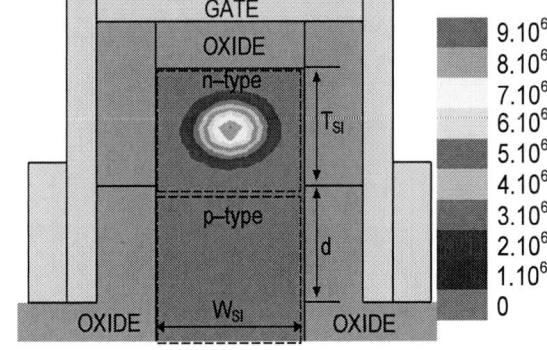

Figure 10: 2D cut-plane showing total current density (A/cm^2) at $V_{GS} = 0.4$ V and $V_{DS} = 50$ mV for 20 nm JNT.

978-1-4244-6658-0/10 $26.00 © 2010 IEEE   279

V. PERFORMANCE

The drain current of the JNT is proportional to the channel doping concentration and the cross section of the nanowire, and not to the gate oxide capacitance. The device meets ITRS constraints at the 25-nm gate length node (Fig. 12).

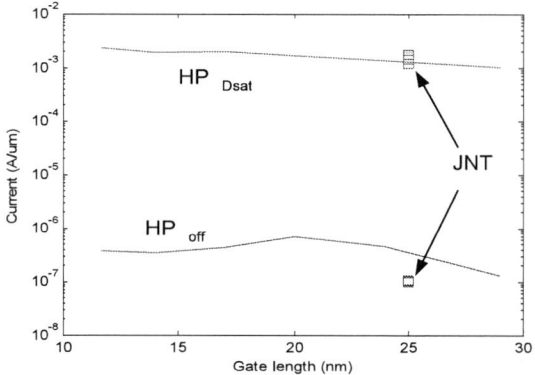

Figure 12: I_{Dsat} and I_{off} for high-performance (HP) CMOS (from ITRS roadmap) and I_{Dsat} and I_{off} for 25-nm JNTs with different cross sections and doping concentrations at $V_{DD}=1V$ and $I_{off}=100nA/um$.

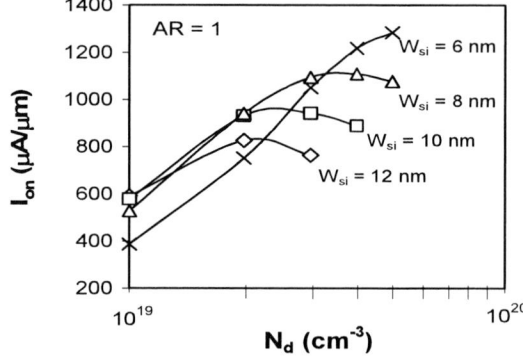

Figure 13: I_{Dsat} in JNTs with $W_{si}=T_{si}$ and L=25nm. $T_{ox}=1nm$ and a P^+-poly gate is used. $V_{DD}=1V$ and $I_{off}=100nA/um$.

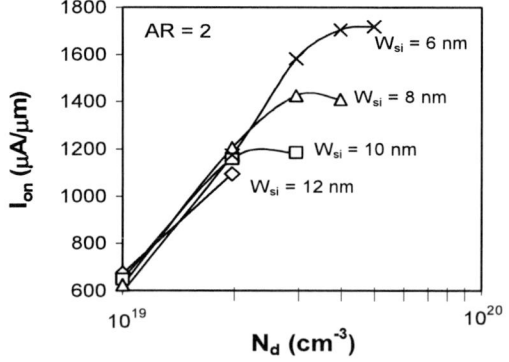

Figure 14: I_{Dsat} in JNTs with $T_{si}=2 \times W_{si}$ and L=25nm. $T_{ox}=1nm$ and a P^+-poly gate is used. $V_{DD}=1V$ and $I_{off}=100nA/um$.

Devices with an aspect ratio ($AR=T_{si}/W_{si}$) equal to 1 and 2 are presented in Figures 13 and 14, respectively. In these simulations, $V_{DD}=1V$ and $I_{off}=100nA/um$; a P^+-poly is used as gate material and the EOT is 1nm. One can see that current drive is increased when doping concentration is increased, even if the cross section of the device needs to be reduced to avoid short-channel effects. One can also see that a 30%-40% increase in drain current is obtained by using an aspect ratio of 2 instead of 1.

VI. CONCLUSION

This paper describes the conduction mechanisms in junctionless nanowire transistors. These devices do not operate in inversion or accumulation, but only in full or partial depletion. The threshold voltage depends on doping, EOT as well as on the width and thickness of the nanowires. In addition, the concept of a bulk multi–gate MOSFET without any lateral source/drain junctions is proposed. It has been demonstrated that JNT can exhibit low leakage currents and excellent short channel behaviour at shorter gate lengths. Simulations show that the JNT is a strong contender for future CMOS as it satisfies the ITRS requirements.

Acknowledgements

This work was supported by the Science Foundation Ireland grant 05/IN/1888: Advanced Scalable Silicon–on–Insulator Devices for Beyond–End–of–Roadmap Semiconductors. This work has also been enabled by the Programme for Research in Third–Level Institutions. This work was supported in part by the European Community (EC) Seventh Framework Program through the Networks of Excellence NANOSIL and EUROSOI+ under contracts 216171 and 216373.

References

[1] CW. Lee, A. Afzalian, N. Dehdashti Akhavan, R. Yan, I. Ferain, JP. Colinge, "Junctionless multigate field-effect transistor", *Applied Physics Letters*, Vol. 94, pp. 053511:1-2, 2009

[2] C.W. Lee, I. Ferain, A. Afzalian, R. Yan, N. Dehdashti Akhavan, P. Razavi, J.P. Colinge, "Performance estimation of junctionless multigate transistors", Solid-State Electronics, Vol. 54, pp. 97-103, 2010

[3] JP Colinge, CW Lee, A. Afzalian, N. Dehdashti Akhavan, R. Yan, I. Ferain, P. Razavi, B. O'Neill, A. Blake, M. White, AM Kelleher, B. McCarthy, R. Murphy, "Nanowire transistors without junctions", Nature Nanotechnology, Vol. 5, No. 3, pp. 225-229, 2010

[4] A. Kranti, G.A. Armstrong, "Design and optimization of FinFETs for ultra-low-voltage analog applications", IEEE Trans. Elect. Dev., Vol. 54, pp. 3308-3316, (2007)

Gate Semi-Around Si Nanowire FET Fabricated by Conventional CMOS Process with Very High Drivability

Soshi Sato[1], Yeonghun Lee[1], Kuniyuki Kakushima[2], Parhat Ahmet[1],
Kenji Ohmori[3], Kenji Natori[1], Keisaku Yamada[3] and Hiroshi Iwai[1]

[1]Frontier Research Center, [2]Interdisciplinary Graduate School of Science and Engineering,
Tokyo Institute of Technology, 4259, Nagatsuta, Midori-ku, Yokohama 226-8502, Japan
[3]Nanotechnology Laboratory, Waseda University, 513, Waseda-Tsurumaki, Shinjuku-ku, Tokyo 162-0041, Japan

Abstract— Gate semi-around silicon nanowire (SiNW) FETs have been fabricated and their electrical characteristics, especially on the drivability, have been assessed for future high performance devices. Among different wire size, a SiNW FET with a cross-section of 12×19 nm^2 has shown an improvement in the on-current (I_{ON}) when normalized by the channel peripheral length. A high I_{ON} over 1600 $\mu A/\mu m$ at an overdrive voltage of 1 V has been achieved with a gate length and an oxide thickness of 65 and 3 nm, respectively. The origin of the high drivability has been speculated by higher carrier density, improved carrier mobility and the reduction in the series resistance.

I. INTRODUCTION

Performance improvements by downsizing the CMOS technology have required new structures, including ultra-thin SOI and double-gate devices, to overcome the severe short channel effects (SCE), which induces an increase in off-state leakage current (I_{OFF}) to degrade the on-off ratio. Recently, silicon nanowire (SiNW) FETs have attracted much attention as they enable strong channel potential controllability by the gate electrode [1]. Beside, improvements in on-current (I_{ON}) with SiNW FETs have been reported, which are advantageous for high-speed with low power consumption application [2]. In this paper, gate semi-around SiNW FETs have been fabricated using conventional CMOS processes and their electrical characteristics, especially on the drivability, have been assessed for future high performance devices.

II. DEVICE FABRICATION PROCESS

A (100)-oriented SOI wafer was used as a starting material with an SOI layer and a buried oxide thickness of 75 and 50 nm, respectively. SiNWs were formed by oxidation of fin patterns with a Si-nitride mask atop. Rectangular-like shape SiNWs with a wire height (h_{NW}) of 12 nm and wire widths (w_{NW}) of 19, 28 and 39 nm were fabricated. The Si-nitride mask suppresses the excess thinning of the embedded source and drain (S/D) pad regions to avoid any unexpected increase in the series resistance (R_{SD}). After striping the nitride and the formed oxides, a conventional self-aligned gate stack formation including gate oxidation and poly-Si deposition was conducted. The gate oxide thickness was set to 3 nm.

Phosphorus ion implantations (boron for pFET) were chosen for extension formation with doses of 1×15 cm^{-2} at 15keV and 5×15 cm^{-2} at 5 keV for S/D region. After S/D activation annealing, a nickel self-aligned silicidation process was applied to reduce the R_{SD}. The key process is summarized in figure 1 and the detailed processes to fabricate semi gate-around SiNW FETs are shown in ref [3]. The SEM and TEM images of the fabricated SiNW FET are shown in figure 2.

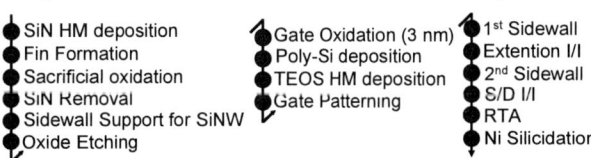

Figure 1. The key process flow for SiNW FETs using CMOS compatible process.

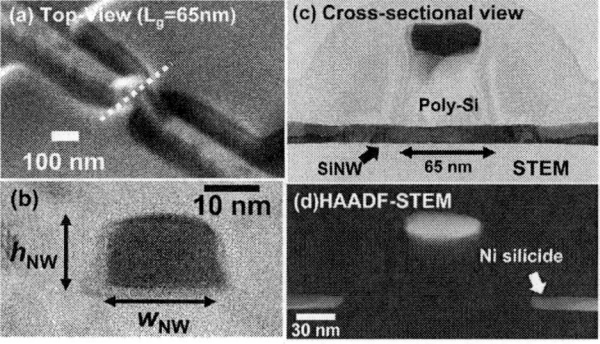

Figure 2. Typical SEM and TEM images of the fabricated SiNW FET.

III. ELECTRICAL CHARACTERISTICS

The dc-characteristics (I_d-V_d and I_d-V_g curves) of the smallest (h_{NW} of 12 nm and w_{NW} of 19 nm) SiNW FETs with a gate length and gate oxide of 65 and 3 nm, respectively, are shown in figure 3. A well behaved transistor operation was confirmed for both *n*- and *p*-FET. The on-current of *n*-FET under an overdrive and a drain voltage of 1 V showed a large

---

This work is supported by NEDO.

value of 60 µA, whereas that of p-FET was as low as 22 µA. The low on-current of pFET is mainly due to large R_{SD} and still needs process optimization. A fairly nice on-off current ratio of $>10^6$ with a DIBL and a favorable SS of 62 mV/V and 70 mV/dec., respectively have been obtained, owing to the large channel potential control of the gate electrode. The smaller w_{NW} was, the better SCE immunity represented by V_{th} roll-off and SS was obtained, which indicate the advantages of small cross section against the SCE, which is in good agreement with previous reports [4]. Moreover, SS of 62-63 mV/dec. with L_g over 150 nm was obtained which strongly suggests that the density of surface states (D_{it}) is negligible with a rectangular-like shape SiNW.

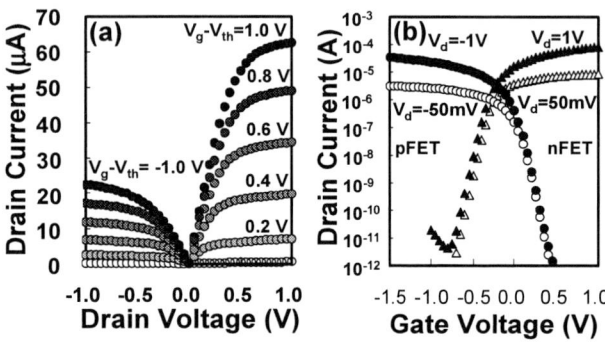

Figure 3.   Dc characteristics of the fabricated gate semi-around SiNW FET (h_{NW}=12nm, w_{NW}=19nm) with L_g=65nm and T_{ox}=3nm.

The on-current per wire and on-current normalized by the channel peripheral width (I_{ON}) of the SiNW FETs with three different w_{NW} were plotted against the effective gate length (L_{eff}) in figure 4. A large on-current per wire can be obtained by wider SiNW mostly owing to the larger channel peripheral width. However, when normalized with channel peripheral length, the I_{ON} showed large improvements with smaller size. As the only difference among the three SiNW is the w_{NW}, one can speculate the effect of the rounded corner of the channel for I_{ON} improvement. Moreover, the I_{ON} showed an increasing trend with L_{eff} scaling, indicating further improvement in the drivability by L_g scaling.

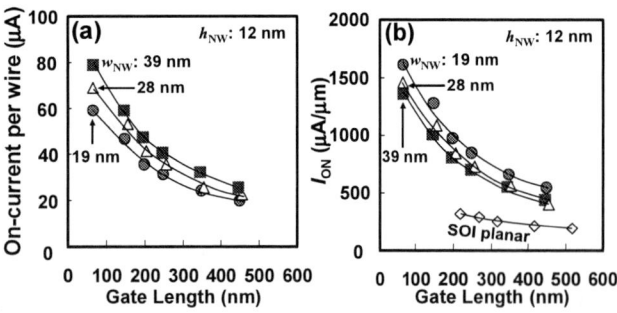

Figure 4.   (a) On-current per wire and (b) I_{ON} normalized by peripheral channel width of three SiNW FET. The I_{ON} of planer SOI FETs fabricated on the same wafer are plotted as references.

To examine the drivability of SiNW FETs, planer SOI FETs with T_{Si} of 28 nm were fabricated simultaneously on the same wafer and are shown in the figure as references. One can observe improved I_{ON} by more than 2 times with SiNW FET, especially with the smallest cross-section.

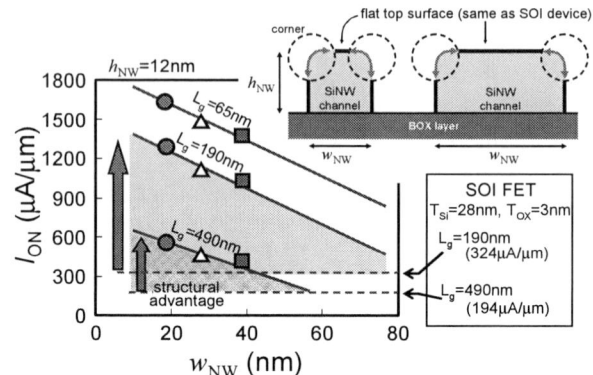

Figure 5.   Structural advantage of SiNW FETs over SOI FETs with smaller w_{NW} for high drivability.

The effect of w_{NW} for SiNW FETs on the I_{ON} drivability is shown in figure 5. Compared to the planar SOI FETs with the same L_{eff}, a structural advantage of rectangular-like shape SiNW is more pronounced at smaller w_{NW}. Note that the existence of multiple V_{th} may be the concern with the rounded corners, however, no kink effect in the subthreshold characteristics was observed with our devices due to low-doped channel [5].

The I_{ON}-I_{OFF} characteristics measured from L_{eff} of 500 to 65 nm are shown in figure 6. Here the I_{OFF} is defined as the drain current normalized by the peripheral channel width at an overdrive voltage of -0.3 V and a drain voltage of 1 V. Although little difference is observed at large L_{eff} region, a distinct improvement on the on-off ratio is obtained with smaller cross-section when L_{eff} scaling is conducted. A large I_{ON} over 1600 µA/µm at an I_{OFF} of 1 nA/µm was achieved with a SiNW FET of 12×19 nm², L_{eff}=65 nm, and T_{ox}=3 nm.

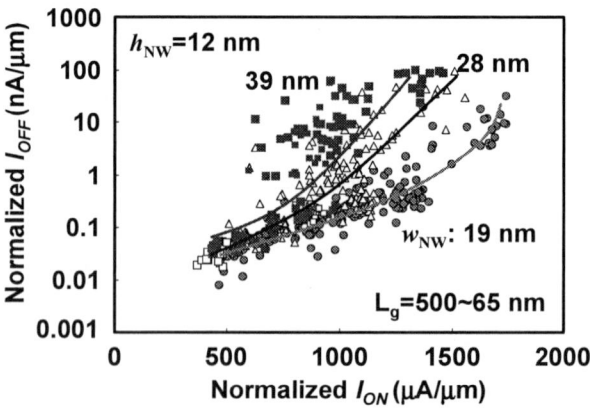

Figure 6. I_{ON}-I_{OFF} characteristics of SiNW with different w_{NW}.

To analyze the origin of the large drivability of I_{ON} with SiNW FETs, the effective electron mobility (μ_{eff}) were extracted by SiNW FETs with multi-wire channel (N=64) using advanced split-CV method, which are shown in figure 7. To avoid any unexpected parasitic effects for gate-to-channel capacitance measurements, an advanced split-CV method was adopted using two different FETs; gate mask length of 550 and 250 nm [6]. The inversion carrier density (N_s) was calculated using the sum of peripheral channel width of the SiNW obtained from TEM images. A large μ_{eff} peaking at 452 cm^2/Vs and 405 cm^2/Vs at N_s of 10^{13} cm^{-2} was extracted with a SiNW of 12×19 nm^2. Compared with the μ_{eff} of the planer SOI FET, a large improvement especially at high N_s can be obtained, presumably due to lower vertical electric field to the nanowire channel. The μ_{eff} showed little dependency on the w_{NW}, indicating small surface roughness at the top of SiNW. However, a large degradation in μ_{eff} was observed when a large h_{NW} was designed, especially at high N_s region where roughness scattering dominates. The reason might be the increase in the area of the etched side-surface to degrade the surface morphology so that process optimization including damage free plasma may improve the performance [7].

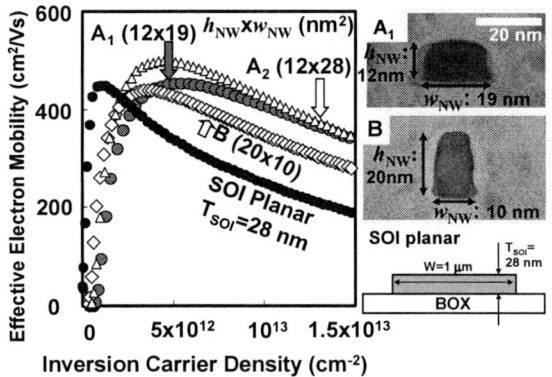

Figure 7. μ_{eff} of SiNW FETs obtained from advanced split-CV method.

The R_{SD} was extracted using the above devices based on Chern's method. A small R_{SD} of 1.5 kΩ was obtained irrespective of the w_{NW}. (fig. 8) This R_{SD} value corresponds to only 10% of the total channel resistance (R_{tot}) for L_{eff} of 65 nm, owing to the process optimization for S/D formation. Note that As implantation instead of P resulted in 10 times higher R_{SD}, presumable due to the damages in the S/D regions as well as the difference in the Ni silicide formation [8].

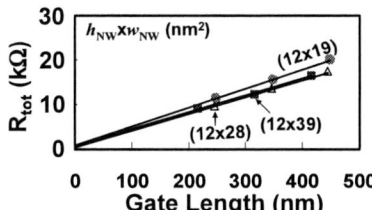

Figure 8. Extracted R_{SD} of the SiNW nFETs usning Chern's method.

Another reason to enhance the drivability of the SiNW FETs might be the distribution in the carrier concentration. Here, a two-dimensional simulation to extract the inversion carrier concentration distribution profile in the SiNW channel were carried out using a Taurus device simulator under modified local density approximation (MLDA) method [9]. The simulated device was designed to have the same cross sectional shape as was obtained experimentally. The work function of the metal electrode was set to 4.1 eV and a non-doped channel was assumed. Figure 9 shows the carrier concentration profile under a gate overdrive voltage of 1 V. High density regions can be observed near the corners of the channels, owing to the electrostatics at the rounded surface and one can see that large amount of carriers located within the SiNW away from the surface. As the w_{NW} decreases, the high-density regions at the corners approach, so that the portion of the high-density region within the cross section increases. Line profiles at the corner and at the top planer regions showed an increase in the carrier concentration by 2.5 times at the corner for both structures, indicating advantage for higher drivability at corners.

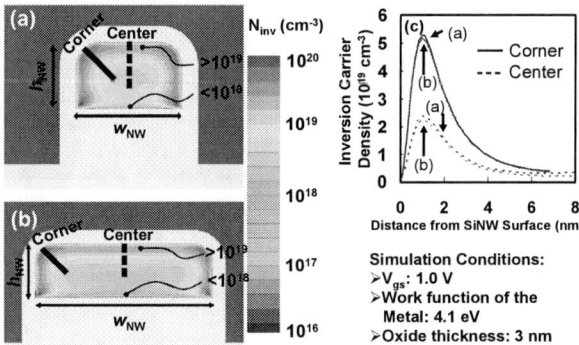

Figure 9. Two-dimensional simulation of the inversion carrier concentration within the SiNW chanels of (a) 12×19 nm^2 and (b) $12\times38nm^2$. (c) Line profile of the carrier concentration at the top planar and at the coner.

Based on the above characterizations, the improvement in the drivability from SOI to SiNW FET and further enhancement with smaller SiNW size could be speculated as attributions of the effect of corners in the cross section of the SiNW channel. As the NWs were fabricated by high temperature thermal oxidation, the corners of the NW have rounded shapes with a curvature radius of 4 nm. Given that the SiNW FETs were fabricated on the same wafer, the peripheral length of the corners can be considered as constant among three devices; only the proportion of the top flat and corner surface within the channel peripheral width changes. Assuming that the top flat surface region performs the same I_{ON} as the planer SOI device with an I_{ON} of 324 μA/μm for L_g=190 nm, the drivability of each corner can be extracted as shown in figure 5. The vertical etched surfaces of the channels at both sides are basically (110) surfaces, so that the μ_{eff} of the carriers at these surfaces may be degraded compared to the flat surface at the top [10]. However, the effect can be considered as a minor factor in this work as the h_{NW} is small compared to the w_{NW}.

TABLE I.   BENCHMARK OF MULTI-GATE FETS REPORTED

	This work	Ref[11]	Ref[12]	Ref[13]	Ref[14]	Ref[15]	Ref[4]	Ref[16]	Ref[17]
NW Cross-sectional shape	Rectangular-like	Rectangular-like	Rectangular-like	Circular	Circular	Elliptical	Elliptical	Fin	Tri-Gate
NW Size (nm)	10×20	10×20	14	10	10	12	13×20		
L_g (nm)	65	25	100	30	8	65	35	25	40
EOT or T_{ox} (nm)	3	1.8	1.8	2	4	3	1.5		
V_d (V)	1.0	1.1	1.2	1.0	1.2	1.2	1.0	1	1.1
I_{ON}(uA) per wire	60.1	102	30.3	26.4	37.4	48.4	43.8		
I_{ON}(uA/um) by circumference	1609	2054	430	841	1191	1283	825	1296	1400
SS (mV/dec.)	70	79	68	71	75	~75	85	83	76
DIBL (mV/V)	62	56	15	13	22	40-82	65	83	89
I_{ON}/I_{OFF}	~1E6	>1E6	>1E5	~1E6	>1E7	>1E7	~2E5	~3E4	~1E4

By solving linear equations of the I_{ON} for the top-flat and corner parts with different SiNW-FETs, a corner with a curvature radius of 4 nm can exhibit a high I_{ON} of 3000 μA/μm can be calculated. On this account, one can calculate that the 80 % of I_{ON} is dominated at the rounded corner with the smallest device of h_{NW} and w_{NW} of 12 and 19 nm, respectively.

IV. DISCUSSION

Table 1 summarizes the benchmarking of the reported SiNW FETs. Although our device has relatively large L_g and T_{ox} among the SiNW FETs listed in the table, the 2nd-highest I_{ON} over 1600 μA/μm was demonstrated. One could expect further higher performances with the dimension scaling. For example, with decreasing both the gate length (65 to 32.5 nm) and gate oxide thickness (3 to 1.5 nm) I_{ON} over 3000 μA/μm by circumference could be expected. Among various cross sectional shape of SiNW shown in the table, including rectangle-like, elliptical and circular shapes, one can see a trend that a large I_{ON} normalized by channel peripheral width is achieved with rectangle-like shape, possibly the effect of rounded corners in the channel.

V. CONCLUSION

Very high on-current of gate semi-around SiNW FETs fabricated with conventional CMOS process have been demonstrated. A rectangular-like shape SiNW with a height and a width of 12 and 19 nm, respectively, has revealed a high on-current over 1600 μA/μm with a gate length and an oxide thickness of 65 and 3 nm, respectively. The origin of high drivability has been explained by fairly nice effective mobility, and low S/D parasitic resistance as well as the enhancement in the carrier concentration at the rounded surface of SiNW. There is a good possibility that the drivability over 3000 μA/μm by circumference could be obtained by future scaling of L_g and EOT even using convention CMOS process.

ACKNOWLEDGEMENT

The authors thank to Front End Process Program in research dept. 1 and ASKA II line staffs in semiconductor leading edge technologies for kind advice and device fabrication, especially H. Watanabe, Y. Ohji, K. Ikeda, T. Aoyama, Y. Sugita, M. Hayashi, T. Morooka and T. Matsuki. S. S. thanks to Dr. Y. Kobayashi and K. Tsutsui for simulations.

REFERENCES

[1] Y-K Choi, et al., "Ultrathin-Body SOI MOSFET for Deep-Sub-Ten Micron Era", IEEE Elec. Dev. Lett., vol. 21, no. 5, pp. 254-255, 2000.

[2] J-P. Colinge, "Mutiple-Gate SOI MOSFETs", Solid-State Elec., vol. 48, no. 6, pp. 897-905, 2004.

[3] S. Sato, et al., "High-Performance Si Nanowire FET with a Semi Gate-Around Structure Suitable for Integration", Proc. of 39th ESSDERC, pp. 249-254, 2009.

[4] S. Bangsaruntip, et al., "High Performance and Highly Uniform Gate-All-Around Silicon Nanowire MOSFETs with Wire Size Dependent Scaling", Tech. Dig. of IEDM, pp. 297-300, 2009.

[5] J. G. Fossum, J. –W. Yang, and V. P. Trivedi, "Suppression of Corner Effects in Triple-Gate MOSFETs", IEEE Elec. Dev. Lett., vol. 24, no. 12, pp. 745-747.

[6] H. Irie and A. Toriumi, "Advanced split-CV technique for accurate extraction of inversion layer mobility in short channel MOSFETs", Extended Abstracts of International Conference of Sol. Stat. Dev. Mat., pp. 864-865, 2005.

[7] K. Endo, et al., "Damage-Free Neutral Beam Etching Technology for High Mobility FinFETs", Tech. Dig. of IEDM, pp 840-843, 2005.

[8] A. Kikuchi, "Phosphorus redistribution during nickel silicide formation", J. Appl. Phys., vol. 64, no. 2, pp. 938-940, 1988.

[9] M. G. Ancona, and H.F. Tiersten, "Macroscopic physics of the silicon inversion layer", Physical Review B, vol. 4, no. 6, pp. 7959-7965, 1987.

[10] T. Sato et al., "Mobility Anisotropy of Electrons in Inversion Layers on Oxidized Silicon Surfaces", Phys. Rev. B, vol. 4, no. 6, pp. 1950-1960, 1971.

[11] G. Bidal, et al., "High velocity Si-nanodot : A candidate for SRAM applications at 16nm node and below", Tech. Dig. of Symp. VLSI Tech., pp. 240-241, 2009.

[12] C. Dupré, et al., "15nm-diameter 3D Stacked Nanowires with Independent Gates Operation: ΦFET", Tech. Dig, of IEDM, pp. 749-752, 2008.

[13] S. D. Suk, et al., "High Performance 5nm radius Twin Silicon Nanowire MOSFET (TSNWFET): fabrication on bulk Si wafer, characteristics, and reliability", Tech. Dig. of IEDM, pp.717-720, 2005.

[14] Y. Jiang, et al., "Performance Breakthrough in 8 nm Gate Length Gate-All-Around Nanowire Transistors using Metallic Nanowire Contacts", Tech. Dig. of Symp. VLSI Tech., pp. 34-35, 2008.

[15] H. S. Wong, et al., "Gate-all-around Quantum-Wire Field-Effect Transistor with Dopant Segregation at Metal-Semiconductor-Metal heterostructure", Tech. Dig. of Symp. VLSI Tech., pp. 92-93, 2009.

[16] C. Y. Chang, et al., "A 25-nm Gate-Length FinFET Transistor Module for 32nm Node", Tech. Dig. of IEDM, pp.293-296, 2009.

[17] J. Kavalieros, et al., "Tri-Gate Transistor Architecture with High-k Gate Dielectrics, Metal Gates and Strain Engineering", Tech. Dig. of Symp. VLSI Tech., pp.50-51, 2006.

Dopant-Independent and Voltage-Selectable Silicon-Nanowire-CMOS Technology for Reconfigurable Logic Applications

Frank Wessely, Tillmann Krauss, Udo Schwalke
Institute for Semiconductor Technology and Nanoelectronics
Darmstadt University of Technology
Schlossgartenstrasse 8
64289 Darmstadt, Germany
e-mail: wessely@iht.tu-darmstadt.de

Abstract— In this paper, we report on the fabrication and characterization of a novel voltage-selectable (VS) nanowire (NW) CMOS technology suitable to extend the flexibility in circuit design and reconfigurable logic applications. Silicon NW-structures with Schottky-S/D-junctions on silicon-on-insulator (SOI) substrate are used to realize dopant-independent unipolar CMOS-like transistors. A selection of the device type (PMOS or NMOS) is performed by application of an appropriate back-gate bias. The versatile programming capability of this approach is demonstrated in a VS-NW-CMOS inverter set-up.

I. INTRODUCTION

Silicon nanowires (Si-NW) are intensively investigated by many research groups and considered as promising replacement for standard MOSFET based transistor technology, since classic geometric downscaling of planar MOSFET devices is reported to come to an end [1]. However, the ambipolar [2] nature of the nanowires turns out to be a roadblock, as p-type and n-type transistors are basic building blocks for today's complementary MOS logic, i.e. its simplest device, the inverter [3]. Concerning bottom-up NW-fabrication, a vapour-liquid-solid growth approach is often not compatible with standard CMOS technology, in view of the used catalyst materials, as well as the need of high growth temperatures during the fabrication process. Other unsolved problems occur during doping [4,5] (e.g. dopant segregation), thus the use of grown nanowires in large-scale integration of integrated circuits is not very likely. As we will show, most of these issues could be circumvented by the top-down fabrication of unipolar Si-NW devices with Schottky contacts for source and drain. Furthermore, our approach is based on controlling the transistor-type (i.e. NMOS or PMOS) of the device via the back-gate voltage, leading the way for switchable transistor characteristic changeable on the fly. Furthermore, when the transistor type, i.e. NMOS or PMOS, could be simply defined by programming via applying a bias,

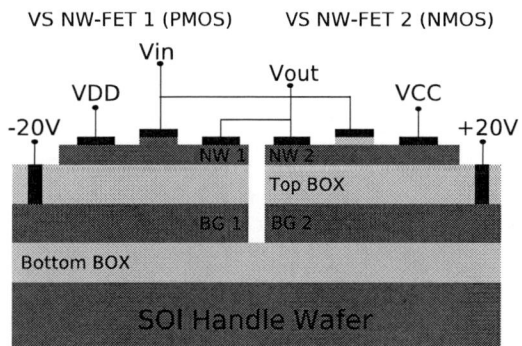

Figure 1. Schematic of realized inverter on multi-SOI substrate. The transistor type (NMOS or PMOS) is selectable by the application of an appropriate back gate bias to the corresponding back-gates, i.e. BG1 and BG2, respectively.

additional flexibility of reconfigurable logic circuit design is expected [6] for programmable logic (FPGA, CPLD) and system-on-chip (SoC) applications. For means of fabrication, a standard top-down technology was used, forming the nanowire by well-known lithography and subsequent reactive ion etching. Fig. 1. gives a schematic view of the device set-up on a MultiSOI substrate [7].

II. FABRICATION

The devices are fabricated on ultrathin-body SOI substrates from SOITEC with a top-silicon thickness of 70 nm, a buried oxide thickness of 145 nm and a boron doping level of 10^{15} cm^{-3}. The substrates are prepared with alignment marks for the subsequent electron beam lithography (EBL). By means of EBL, a 90 nm wide line is defined into the negative resist, and transferred onto the top-Si layer via reactive ion etching in hydrogen bromide (HBr) plasma. In the following the substrate is oxidized in a horizontal furnace at

This work was partially funded by the german federal ministry of education and research (bmbf) under the MEGAEPOS project (13N9259)

978-1-4244-6658-0/10 $26.00 © 2010 IEEE   285

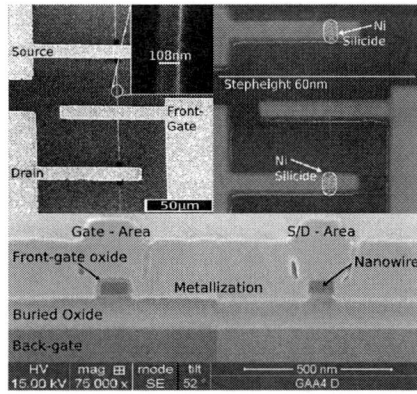

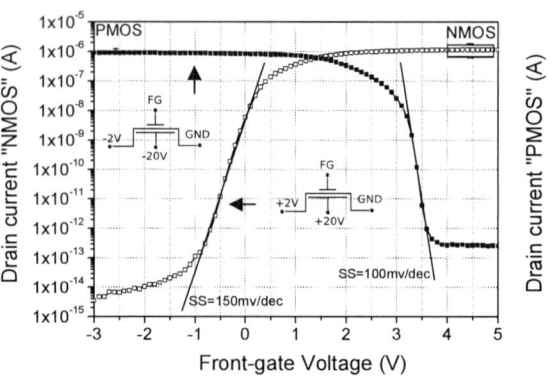

Figure 2 top left: SEM image of the fabricated devices. The nanowire width is measured as 108nm, gate length is 5μm. Top right: AFM scan of the device, the channel height is 60nm, only areas directly beneath source and drain contacts are salicided. Bottom left: Cross-sectional SEM (XSEM) image of the gate area. Note the top tri-gate structure (i.e. front-gate). The gate oxide thickness amounts to 9nm. Bottom right: XSEM of the source/drain region with Schottky S/D strucure.

1000°C for 7 minutes, forming the gate oxide of 9 nm for the tri-gate nanowire device. Subsequently contact holes for S/D contacts are formed via EBL and wet chemical etching in hydro-fluoric acid (HF). Contact pad lithography is also performed by EBL using a dual-layer resist system based on poly methyl methacrylate (PMMA) and novoresist [8] especially developed for EBL lift-off purposes. Metallization of S/D contacts and gate electrode is realized by electron beam evaporation of 70 nm nickel with a 180 nm aluminum capping on top. Subsequently the S/D contacts are silicided at 500°C for 10 minutes in a tube furnace. During this forming gas treatment a mixture of 90% nitrogen and 10% hydrogen is used, where the nickel reacts with the silicon surface forming a mid-gap NiSi-Schottky-barrier contact [9] to the low doped silicon nanowire. Fig. 2 (top) shows several results of process characterization techniques that were used to determine the devices vertical and lateral dimensions. Cross section SEM (XSEM) is used to illustrate the gate area with its tri-gate structure and the S/D area where the salicidation occurrs. From Fig. 2 a channel width of 90nm is determined by subtraction of the gate oxide thickness. The nanowire height is measured to 60 nm with atomic force microscopy (AFM), and the gate to S/D distance corresponds to 20μm Finally, single dies of the SOI-wafer were used to build a voltage-programmable CMOS inverter in a stacked hybrid SOI technology related to the schematic illustrated in Fig. 1.

III. RESULTS AND DISCUSSION

All the properties of the NW-MOSFET structure are determined through the field-effect via front-gate (FG) and back-gate (BG) voltages as shown in Fig. 3. Similar to the conventional MOSFET, the current transport and hence the switching characteristics, is well controlled through front-gate due to the thin oxide and the tri-gate electrode geometry. However, the back-gate bias determines the transistor type, PMOS or NMOS. For example, when applying a sufficiently large negative voltage to the back-gate the thin top-silicon layer will be depleted of electrons and the accumulation of

Figure 3. Subthreshold characteristics of voltage-programmable NW-FETs for fixed positive and negative back-gate bias, respectively. Unipolar transistor characteristics are achieved with high on/off current ratios for both, PMOS-type and NMOS-type NW-FETs. Box-plots of on-currents are included to illustrate the impact of process variations.

holes (here: majority carriers) occurs. The resulting transfer characteristics obtained through a front-gate sweep results in a PMOS-like device characteristic. Otherwise applying a positive BG-voltage, the Si-NW will be depleted of holes and eventually, at a sufficiently large BG-bias an inversion layer of electrons forms. The resulting transfer characteristics obtained through a FG-sweep leads to an NMOS-like device characteristic. Fig. 4. emphasizes on the dependence of the drain current from the back-gate bias, whereas a back-gate swing is performed under floating front-gate conditions. It is clearly found, that the corresponding back-gate bias is able to manipulate the devices charge-carrier type. In the transfer characteristic ON/OFF current ratios up to 10^6 are obtained for both, n- and p-type NW-FETs. However, the PMOS-like NW-FET exhibits the better subthreshold slope of $S_p \approx 100mV/dec$ but a higher off-current. The higher off-current results from the slight p-type doping of the initial substrate, the degraded subthreshold slope of the NMOS-like NW-FET is due to the narrow inversion layer not completely extending the 60nm thick silicon wire and related series resistances at the S/D

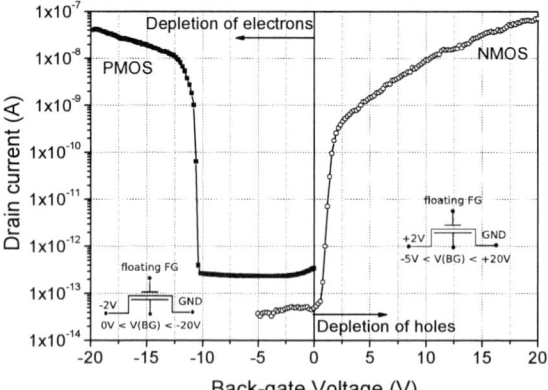

Figure 4. Effect of back-gate bias on NW device properties: Drain current ID measurements as a function of back-gate voltage from -20V < VBG < +20V confirm the change of device types, PMOS vs. NMOS. The front-gate is left floating and a drain bias of -2V (PMOS) and +2V (NMOS) is used, respectively.

978-1-4244-6658-0/10 $26.00 © 2010 IEEE   286

contacts. Nevertheless it is obvious that either n- and p-type behavior of the wire can be selected by the applied back-gate voltage.

Based on these results, a VS-CMOS inverter circuit was set up using two arbitrary nanowire devices. The required PMOS and NMOS transistors were defined by voltage-programming via the back-gates, applying a voltage of -20V (PMOS) and +20V (NMOS), respectively. The transfer characteristic of the realized inverter is presented in Fig. 5. Here, a symmetrical supply voltage of ±2V was provided to the inverter to obtain a HIGH/LOW transition near an input-voltage (V_{in}) of 0V. As can be seen from Fig. 5, a clean inverter characteristic is achieved with a transition from HIGH to LOW state at V_{in} of 0,25V. As expected for a regular CMOS inverter, both NW-devices are momentarily in the ON-state during transition, resulting in a characteristic short current pulse (open symbols in Fig. 5). However, no significant current is drawn from the power supply when the VS-CMOS NW inverter is in its stable HIGH or LOW state.

The versatility of our programmable CMOS-logic is demonstrated in Fig. 6 which shows two traces of the transfer characteristic of a VS-NW-CMOS inverter. The first input-voltage sweep (solid squares in Fig. 6) is performed when using NW-FET 1 as PMOS transistor and NW-FET 2 as NMOS. In this case the HIGH/LOW transition occurs at a V_{IN} of approximately –1.3 V. For the second sweep (open squares in Fig. 6) the supply and back-gate voltages are reversed, so that NW-FET 1 serves as NMOS transistor and NW-FET 2 as PMOS. Again, a CMOS inverter transfer characteristics is observed, however the transition point is shifted by ± 0.5 V. This shift is due to the fact that the drive currents of the devices do not exactly match, since process-variations during fabrication are very likely in this early stage of technology (cf. box-plots in Fig. 3). Nevertheless, a proof of concept of the voltage-programmed CMOS inverter could be demonstrated.

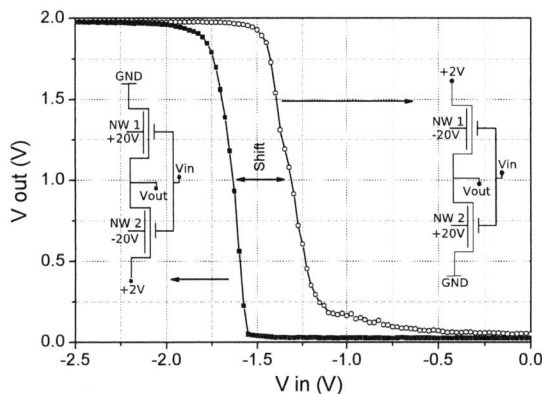

Figure 6. Two traces of the switching characteristics of the VS-NW-CMOS inverter illustrate the versatile programming capability of the VS-NW devices. Filled symbols represent NW-FET1 working as PMOS and NW-FET 2 as NMOS. Interchanging back-gate bias and supply voltage polarity, the transistor types are swapped and again a CMOS-inverter characteristic (open symbols) is observed.

IV. CONCLUSION

In conclusion, we have successfully fabricated unipolar NW-FETs on SOI in which the device type (n- or p-type) is selectable by applying an appropriate back-gate bias. These voltage-programmable nanowire FETs exhibit high on/off ratios and can be used in a universal fashion to extend the flexibility in circuit design of reconfigurable logic. The versatile programming capability of this approach was demonstrated experimentally using a VS-NW-CMOS inverter circuit set-up.

REFERENCES

[1] F. J. Appenzeller, J. Knoch, Mikael T. Björk, H. Riehl, H. Schmid, W. Riess. "Toward Nanowire Electronics", *IEEE Transactions on Electron Devices*, Nov. 2008, pages 2827-2845

[2] S-M. Koo, M D Edelstein, Q. Li, C. A Richter, E.M. Vogel "Silicon nanowires as enhancement mode Schottky barrier field-effect transistors", *Nanotechnology* **16** (2005), pages 1482-1485

[3] M Wanlass, C.T. Sah "Nanowatt Logic Using Field-Effect Metal-Oxide Semiconductor Triodes," *International Solid State Circuits Conference Digest of Technical Papers* (February 20, 1963) pp. 32-33.

[4] S.-D. Kim, C.-M. Park, J. C. S. Woo, "Advanced source/drain engineering for box-shaped ultrashallow junction formation using laser annealing and pre-amorphization implantation in sub-100-nm SOI CMOS*", IEEE Trans. Electron Devices*, Vol. 49, pp. 1748 - 1754, 2002

[5] S.C. Rustagi, N.Sing, W.W. Fang, K.D. Buddharaju, S.R. Omampuliyur, S.H.G. Teo, C.H. Tung, G.Q. Lo, N. Balasubramanian, D.L. Kwong "CMOS Inverter Based on Gate All Around Silicon Nanowire MOSFETs Fabricated Using Top-Down Approach", IEEE Electron Device Letter, Vol.28, Nov. 2007, pages 1021-1024

[6] J. Liu, I. O'Connor, D. Navarro, F. Gaffiot, "Design of a Novel CNTFET-based Reconfigurable Logic Gate," IEEE Computer Society Annual Symposium on VLSI(ISVLSI'07), pp. 285-290, March 2007

[7] SOITEC [Online]. Available: http://www.soitec.com/en/products/pdf/ Tracit_application_specific_substrates.pdf

[8] Allresist GmbH [Online]. Available: http://www.allresist.de/wMedia/ pdf/wEnglish/produkte_ebeamresist/AR_N7500-7520.pdf

[9] E. Bucher, S. Schulz, M. Ch. Lux-Steiner, P. Munz, U.Gubler, F. Greuter, "Work Function and Barrier Heights of Transistion Metal Silicides", *Applied Physics* **A 40**, 71-77 (1986)

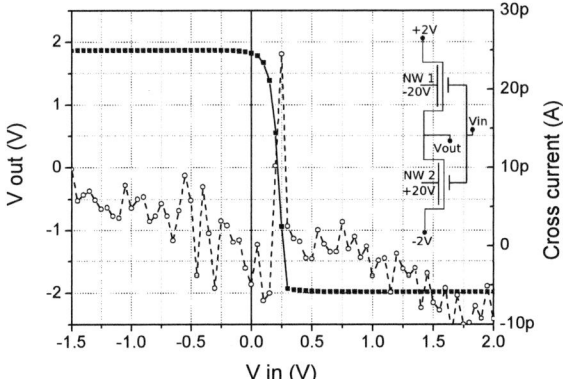

Figure 5. Switching characteristics (solid symbols) of a CMOS inverter circuit formed by CMOS nanowire-FETs via applying the appropriate voltages at the back-gates (see schematic inset). A clear inverter behavior is obtained including the characteristic cross-current peak (open symbols) directly at the HIGH/LOW transition point.

978-1-4244-6658-0/10 $26.00 © 2010 IEEE

3D Source/Drain Doping Optimization in Multi-Channel MOSFET

K. Tachi[1,3,4], N. Vulliet[2], S. Barraud[1], B. Guillaumot[2], V. Maffini-Alvaro[1], C. Vizioz[1], C. Arvet[2], Y. Campidelli[2], P. Gautier[1], J.M. Hartmann[1], T. Skotnicki[2], S. Cristoloveanu[3], H. Iwai[4], O. Faynot[1] and T. Ernst[1]

[1]CEA-LETI, Minatec, 17 rue des Martyrs, 38054 Grenoble, France,
[2]STMicroelectronics, 850 rue J. Monnet, 38926 Crolles, France,
[3]IMEP-LAHC, INPG-MINATEC, 3 Parvis Louis Neel, 38016 Grenoble Cedex 1, France,
[4]Frontier Research Center, Tokyo Institute of Technology, 4259, Nagatsuta, Midori-ku, Yokohama, 226-8502, Japan
Phone: +33 4 38 78 01 39, Fax: +33 4 38 78 30 34, e-mail: kiichi.tachi@cea.fr or thomas.ernst@cea.fr

Abstract—We demonstrate that the integration of in-situ doped Si Source and Drain (S/D) in three-dimensional Multi-Channel Field-Effect Transistors (MCFETs) leads to improved electrical performances. The combination of in-situ doped Selective Epitaxial Growth (SEG) and ion implantation indeed enables to drastically reduce the S/D resistance (down to 72 $\Omega.\mu m$ for nFET and 227 $\Omega.\mu m$ for pFET). Ion implantation induces a small mobility degradation, which becomes negligible in short gate length (L_G) MCFETs. Gate width down-scaling otherwise needed to suppress the overall mobility degradation with L_G and obtain the best electrical properties.

I. INTRODUCTION

The performances of silicon-based large-scale-integrated circuits (LSIs) have improved dramatically in the past 40 years. Those improvements were based on the down-scaling of individual Metal-Oxide-Semiconductor Field-Effect Transistors (MOSFETs). However, the scaling of the planar bulk MOSFETs gate length reduction becomes more and more problematic due to the degraded gate electrostatic control of the channel potential leading to short-channel effects (SCEs) — threshold voltage roll-off, drain-induced barrier lowering (DIBL), subthreshold slope (SS) degradation, etc. Multi-gate devices are nowadays considered as the most promising solutions to minimize SCEs [1]. Vertically-stacked Multi-Channel FETs and 3D-stacked nanowire FETs, because of their reduced footprint, benefit from superior ON current performances [2-6]. Gate-source/drain (S/D) capacitances $C_{GS/GD}$ and S/D resistance R_{SD} have however to be further minimized in order to boost performance and reduce the intrinsic delay in such devices [5,6]. The recent introduction of internal spacers in MCFETs led to 39 % decrease of the intrinsic CV/I delay [4]. In this work, we have used in-situ doped Selective Epitaxial Growth (SEG) in order to fabricate the MCFET S/Ds. A homogenous and efficient S/D doping was thus achieved, resulting in higher electrical performance.

II. MCFET STRUCTURE

A cross-sectional TEM image of the fabricated five-channel MCFET structure on a SOI substrate is shown in Fig.1. This structure is composed of two double-gate (DG) transistors (channels 1&2 and 3&4) and a bottom single-gate transistor on SOI (channel 5). The self-aligned gates completely surround channels 1 to 4, with consequently a better electrostatic control. The SiN internal spacers with "Elevated flat-shaped" Si S/D are located on both sides. HfO$_2$/TiN/Poly-Si was used as the gate stack. The resulting equivalent oxide thickness EOT is ~2.5 nm. The fabrication process is similar to that reported in [4], except for the in-situ doped Si SEG S/D. In this study, we compare three types of samples with different S/D doping schemes in order to investigate their impact on S/D resistance as shown in Table 1. In sample A, S/D were ion-implanted after the intrinsic SEG step. In sample B, the S/D were in-situ doped during the SEG step. In sample C, we cumulated both of them (with the same conditions). Here, the gate length L_G is the average of channels 1 to 5 estimated from cross-sectional TEM images.

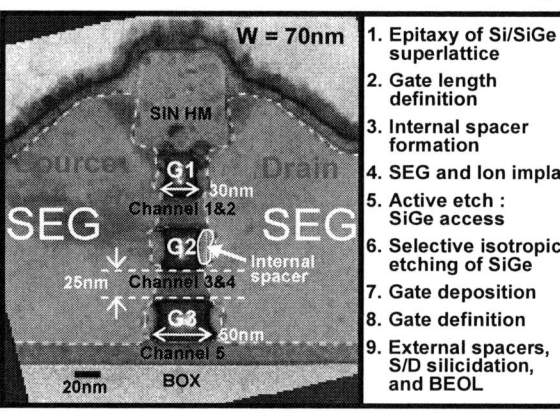

Figure 1. Cross-sectional Transmission Electron Microscopy image of a MCFET with internal spacers and 'flat' shaped S/D, and the process flow.

TABLE I.   SOURCE/DRAIN DOPING CONDITIONS

	Process A	*Process B*	*Process C*
Intrinsic Si SEG	✓		
In-situ doped Si SEG		✓	✓
Ion implantation	✓		✓

III. ELECTRICAL RESULTS

First of all, we compare the impact of the S/D doping conditions on the transistors drive current. Fig. 2 shows the I_{ON}-I_{OFF} characteristics of the MCFETs with 70-nm-L_G. All the currents are normalized by the footprint of the channel width W. Changing the doping process from B to C improves the I_{ON} current by 72 % for nFET and 37 % for pFET at the same I_{OFF} of 10^{-8} A/μm. In order to precisely extract the device parameters and thus understand those improvements, we have used the Y-function-based technique [7-9]. The second order mobility attenuation factor Θ_2 (which is related to the surface roughness effect [10]) is notably taken into account with this method. Fig. 3 shows the I_D and g_m

978-1-4244-6658-0/10 $26.00 © 2010 IEEE

comparisons between the measured data and the model. A good agreement is obtained.

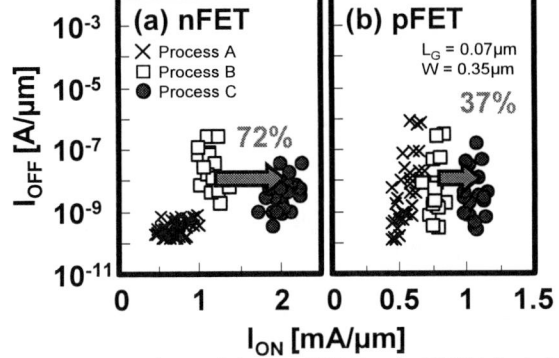

Figure 2. I_{ON}-I_{OFF} characteristics for n-FET (a) and p-FET (b). I_{ON} is the drain current at V_G-V_T = 0.9 V and -0.9 V, I_{OFF} is the off current at V_G-V_T = -0.3 V and 0.3 V for n- and p-FET, respectively. $|V_D|$ = 1.2 V.

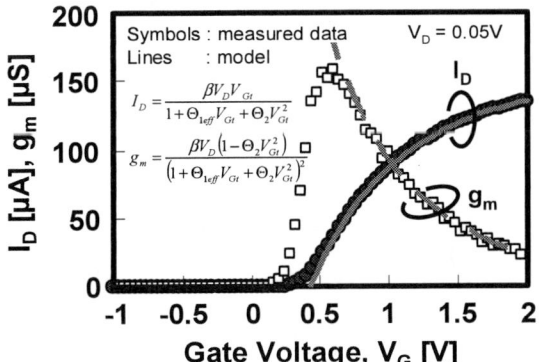

Figure 3. Comparison of I_D-V_G and g_m-V_G curves' fitting with the model shown in the figure. β is the current factor of the transistor ($\beta = \mu_0 W C_{ox}/L_{eff}$), V_{Gt} is the gate drive voltage ($V_{Gt} = V_G - V_T - V_{DS}/2$), and Θ_{1eff} and Θ_2 are the first- and second-order mobility attenuation factors, respectively. Θ_{1eff} represents the mobility limitation caused by phonon scattering and includes the R_{SD}: $\Theta_{1eff} = \Theta_1 + R_{SD}\beta$. The measured device is the C process nMCFET. Gate length and width are 70 nm and 350 nm, respectively.

A. Source/Drain Resistance

R_{SD} was extracted by plotting Θ_{1eff} versus β. Fig.4 shows the good linearity of the relationship between Θ_{1eff} and β for all gate lengths down to 70 nm. The resulting extracted values reveal that R_{SD} reduction was successfully reduced by combining in-situ doped SEG with ion implantation for both n and pFETs.

B. Carriers Mobility

The junction architecture may have a role in the mobility degradation when reducing the gate length due to the possible introduction of neutral defects by the S/D ion implantation [11, 12]. In the following, by comparing process B and process C MCFETs, we will quantify the ion implantation impact on the effective mobility. The latter was extracted versus the inversion charge density for electrons and holes by an improved split C-V method [13]. Indeed, for short-channel devices, extraction errors can be caused by the parasitic capacitances subtraction from the measured gate-to-channel capacitance (C_{GC}). The effective mobility (μ_{eff}) values

extracted by this method were compared with the mobility calculated with the device parameters extracted in the former section, as shown in Fig. 5. It is shown that the μ_{eff} values, both with and without any R_{SD} correction, were well accounted for by the Y-function models.

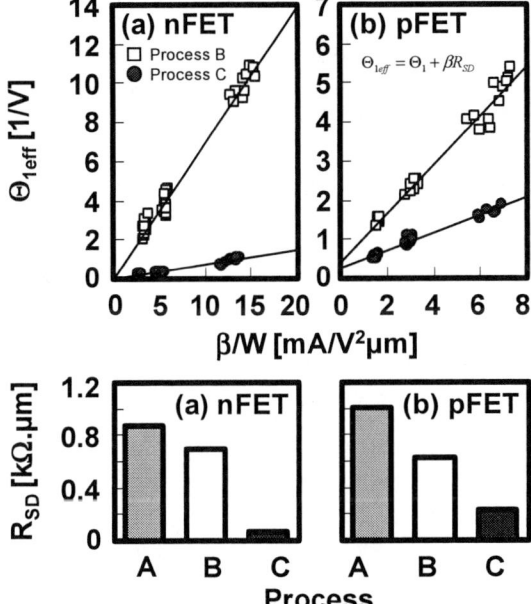

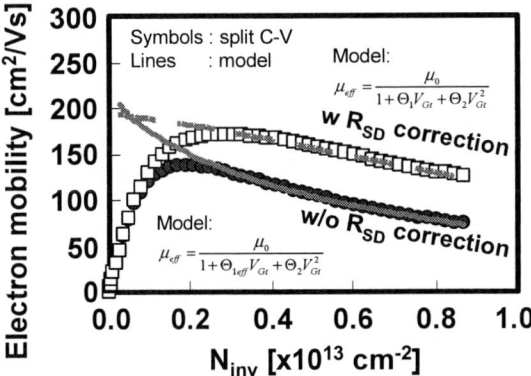

Figure 4. Extraction of R_{SD} from $\Theta_{1eff} - \beta$ plots and comparison of the resulting values for nFET(a) and pFET (b). The MCFETs with process C has a low R_{SD} value of 72Ω.μm and 227 Ω.μm for nFET and pFET, respectively.

Figure 5. Effective mobility extraction with R_{SD} correction. The measured device is a 70-nm-L_G-nMCFET with process C. Gate length and width are 50 nm and 350 nm, respectively. R_{SD} are corrected by $I_{DS0}=I_{DS}/(1-I_{DS}R_{SD}/V_{DS})$

Fig. 6 compares process B with process C for a 70 nm L_G. For process C, μ_{eff} was slightly degraded for both electrons and holes. It may be due to the impact of ionized defects. It can be expected that such small mobility degradation does not significantly impact the I_{ON} current as compared to the benefit of R_{SD} reduction. However, it is important to know for the device scaling whether or not this mobility degradation depends on L_G. Thus, we plotted the low field mobility (μ_0) versus L_G (Fig. 7). For process B and C, μ_0 is degraded for both electrons and holes as L_G decreases. In order to analyze those behaviors, we used the L_G dependent mobility

978-1-4244-6658-0/10 $26.00 © 2010 IEEE

degradation fitting model proposed by G. Bidal *et al.* [12]:

$$\frac{1}{\mu_0(L)} = \frac{1}{\mu_{max}} + \frac{\alpha_\mu}{L} \qquad (1)$$

The two fitting parameters are the maximum mobility μ_{max} [cm^2.V^{-1}.s^{-1}], which is generally equal to the long channel mobility, and a mobility degradation factor α_μ [V.s.cm^{-1}]. First, the C_{GC}-V_G curves were compared for various L_G (Fig. 8). No differences were observed. This means that MCFETs with process C should have electrically the same effective gate length, equivalent gate oxide thickness, parasitic capacitance, and threshold voltage than the ones with process B. The ion implantation has however a large impact on the extracted μ_{max} values. The mobility for sample C is indeed degraded by ion implantation, this even in long channel FETs (as in [12]). On the other hand, α_μ values were similar. Such L_G dependent degradation might partially be attributed to ballistic transport [15].

IV. GATE LENGTH SCALING

In the former section, we have discussed the R_{SD} reduction occurring when combining in-situ doped SEG with ion implantation and the mobility degradation dependence on L_G. In this section, we examine the global MCFETs down-scaling, including SCEs. The I_{OFF} behavior, when normalized by a common threshold voltage V_T, reflects the subthreshold properties such as DIBL and SS (not the V_T roll-off, however). Fig. 9 shows the I_{ON}-I_{OFF} characteristics of n- and p-MCFET with several L_G and W. The currents were normalized by the total channel surface W_{total} (W_{total} = W x 5ch. + T_{Si} x 6side-ch.). When reducing L_G down to 70 nm, I_{OFF} increases progressively due to the enhanced DIBL, while the SS value remains constant at ~70 mV/decade for nFET and ~75 mV/decade for pFET. MCFETs with L_G smaller than 70 nm have degraded subthreshold properties without the expected I_{ON} enhancement. This kind of degradation can be suppressed by adding the lateral gates electrostatic control through W down-scaling [14]. Drive current gains of 14 % for nFET and 20 % for pFET were observed when W was reduced from 350 nm down to 100 nm. We measured in the meantime an improved mobility μ_0 of 11 % and 18 % for nFET and pFET, respectively. This suggests that W down-scaling may reduce the L_G dependent mobility degradation. Additional investigation is needed in order to obtain a physical explanation of this phenomenon which is compatible with volume inversion.

Lastly, we present the scaled MCFETs characteristics with the process C. Fig. 10 shows I_D-V_G and I_D-V_D characteristics associated to 50-nm-L_G 80-nm-W nMCFET and 40-nm-L_G and 70-nm-W pMCFET. We obtained extremely high I_{ON}-currents of 4.1 mA/μm for nFET and a record 2.7 mA/μm for pFET at V_{DD} = 1.2 V. These values are obtained thanks to the 3D configuration of the vertically stacked channels and the enhanced impact of lateral conduction with small gate width. However, I_{OFF} is still high due to the non-optimized threshold voltage on those samples. When normalized at V_{OFF}+V_{DD}, the I_{ON}-currents are 3.3 mA/μm for nFET and 2.0 mA/μm for pFET. When normalized by W_{total}, the I_{ON}-currents at V_{OFF}+V_{DD} for n- and p-FET are 538 μA/μm and 396μA/μm, respectively. These normalized I_{ON} values are comparable to

planar fully depleted – SOIFETs when using the same (unoptimized) gate stack [16].

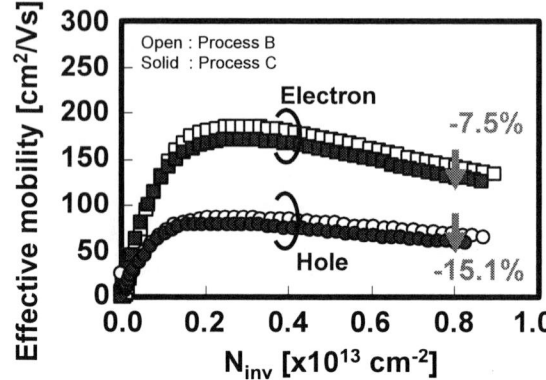

Figure 6. Effective mobility comparison between process B and process C. Gate length and width are 70 nm and 350 nm, respectively.

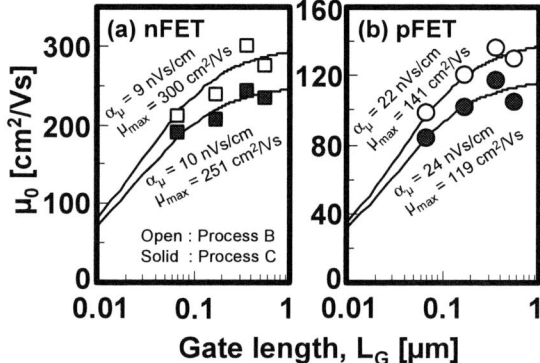

Figure 7. Low-field mobility μ_0 behavior as a function of the gate length. The lines result from L_G dependent mobility degradation modelling.

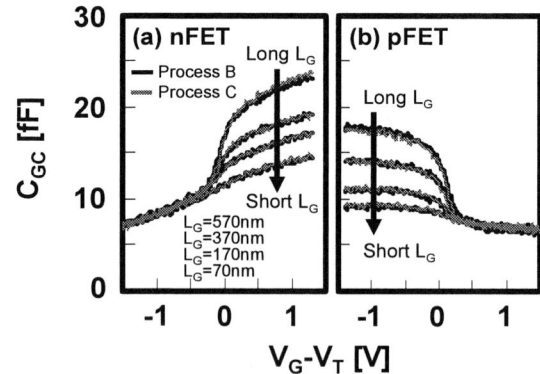

Figure 8. Gate-to-channel Capacitance C_{GC} as a function of V_G for several gate lengths.

V. CONCLUSION

We have successfully reduced the R_{SD} values for both n- and p-MCFETs by combining in-situ doped SEG and ion implantation. These reductions led to dramatic I_{ON} gains. Mobility was however degraded by S/D ion implantation. For short-channels MCFET, this mobility degradation was nevertheless negligible. The gate length scaling maximizes the drive current gains in MCFETs when compared to planar.

Gate width down-scaling enables to obtain the highest drive currents due to both the lateral conduction and the suppression of the L_G dependent mobility degradation.

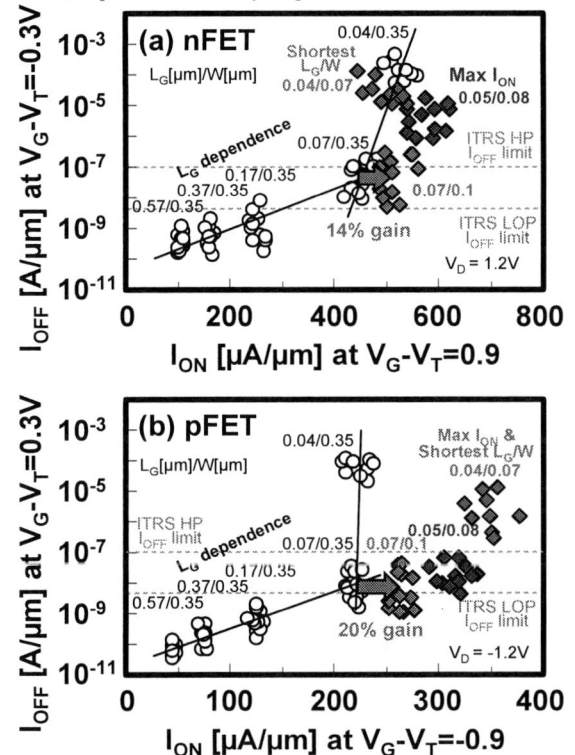

Figure 9. I_{ON}-I_{OFF} characteristics with several channel sizes for nMCFET (a) and pFET (b).

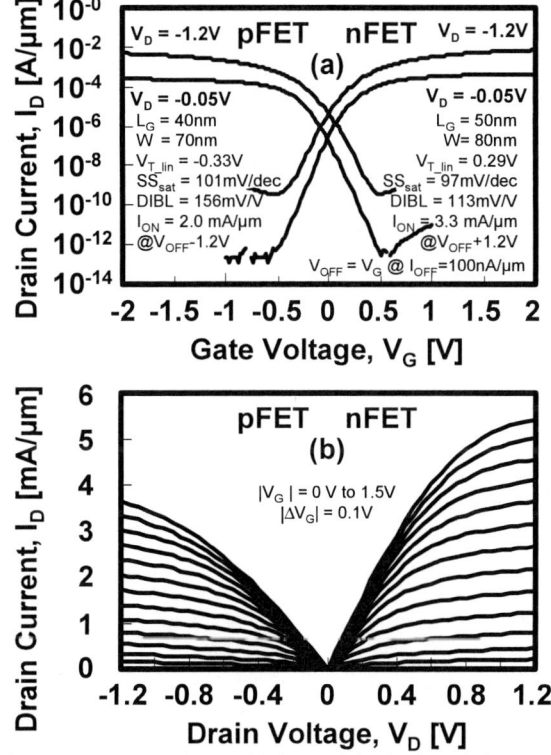

Figure 10. I_D-V_G (a) and I_D-V_D (b) characteristics for the scaled MCFETs.

ACKNOWLEDGMENT

This work was performed as part of the IBM-STMicroelectronics-CEA/LETI-MINATEC Development Alliance. Several authors acknowledge the support from Nanosil, Eurosoi and WCU projects.

REFERENCES

[1] J-T. Park and J-P Colinge, "Multiple-Gate SOI MOSFETs: Device Design Guidelines," *IEEE Trans. Electron Devices*, vol. 49, no. 8, pp. 2222–2229, Dec. 2002.

[2] C. Dupre *et al.*, "15nm-diameter 3D Stacked Nanowires with Independent Gates Operation: ΦFET," in *IEDM Tech Dig.* 2008, pp. 749-752.

[3] K. Tachi *et al.*, "Relationship between mobility and high-k interface properties in advanced Si and SiGe nanowires," in *IEDM Tech Dig.* 2009, pp. 313–316.

[4] E. Bernard *et al.*, "First Internal Spacers' Introduction in Record High I_{ON}/I_{OFF} TiN/HfO$_2$ Gate Multichannel MOSFET Satisfying Both High-Performance and Low Stanby Power Requirements," *IEEE Electron Device Lett.*, vol. 30, no. 2, pp. 148–151, Feb. 2009.

[5] S-Y. Lee *et al.*, "Sub-25nm Single-Metal Gate CMOS Multi-Bridge-Channel MOSFET (MBCFET) for High Performance and Low Power Application," in *VLSI Symp. Tech. Dig.*, 2005, pp. 154–155.

[6] E. Bernard *et al.*, "Multi-Channel Field-Effect Transistor (MCFET) – Part II: Analysis of Gate Stack and Series Resistance Influence on the MCFET Performance," *IEEE Trans. Electron Devices*, vol. 56, no. 6, pp. 1252–1261, June 2009.

[7] G. Ghibaudo, "New method for the extraction of MOSFET parameters," *Electronics Letters*, vol. 24, no. 9, pp. 543–545, 28th Apr 1988.

[8] B. Cretu *et al.*, "New ratio method for effective channel length and threshold voltage extraction in MOS transistors," *Electronics Letters*, vol. 37, no. 11, pp. 717–719, 24th May 2001.

[9] T. Tanaka, "Novel parameter extraction method for low field drain current of nano-scaled MOSFETs," in *Proc. IEEE Int. Conf. Microelectronic Test Structure (ICMTS'07)*, Tokyo, Japan, Mar. 2007, pp. 265–267.

[10] G. Reichert and T. Ouisse, "Relationship between empirical and theoretical mobility models in silicon inversion layers," *IEEE Trans. Electron Devices*, vol. 43, no. 9, pp. 1394–1398, Sep 1996.

[11] A. Cros *et al.*, "Unexpected mobility degradation for very short devices: A new challeng for CMOS scaling," in *IEDM Tech. Dig.*, 2006, pp. 663–666.

[12] G. Bidal *et al.*, "Guidelines for MOSFET Device Optimization acounting for L-dependent Mobility Degradation," in *IEEE 2009 Silicon Nanoelectronics Workshop*, Kyoto, Japan, June 2009, pp. 5–6

[13] K. Romanjek *et al.*, "Improved split C-V method for effective mobility extraction in sub-0.1 µm Si MOSFETs," *IEEE Electron Device Lett.*, vol. 25, no. 8, pp. 583–585, Aug. 2004.

[14] E. Bernard *et al.*, "Novel integration process and performances analysis of Low STandby Power (LSTP) 3D multi-channel CMOSFET (MCFET) on SOI with metal / high-K gate stack," in *VLSI Symp. Tech. Dig.*, 2008, pp. 16-17, 17–19.

[15] K. Huet *et al.*, "Monte Carlo Study of Apparent Mobility Reduction in Nano-MOSFETs," in *ESSDERC proceedings*, 2007, pp. 382-385.

[16] F. Andrieu *et al.*, "Comparative scalability og PVD and CVD TiN on HfO$_2$ as a metal gate stack for FDSOI cMOSFET down to 25 nm gate length and width," in *IEDM Tech. Dig., 2006*, pp. 641–644.

Hole mobilities and electrical characteristics of Ω-gated silicon nanowire array FETs with $\langle 110 \rangle$- and $\langle 100 \rangle$-channel orientation

S. Habicht, S.F. Feste, Q.T. Zhao, and S. Mantl

Institute of Bio and Nanosystems (IBN1) and
JARA-FIT — Fundamentals of Future Information Technology
Forschungszentrum Jülich, 52425 Jülich, Germany.

E-mail: s.habicht@fz-juelich.de

Abstract—**We report on the fabrication and electrical characterization of Ω-gated nanowire (NW) array pFETs on SOI. Devices with gate lengths of $L = 400\,\text{nm}$ and $L = 2\,\mu\text{m}$ and $\langle 110 \rangle$- and $\langle 100 \rangle$-channel orientations were fabricated using a top-down approach. Each device consists of up to 1500 NWs with a crosssection of $20 \times 20\,\text{nm}^2$. The devices feature excellent electrical characteristics with high on-currents, $I_{\text{on}}/I_{\text{off}}$ ratio of 10^8, close to ideal inverse sub-threshold slopes of $64\,\text{mV/dec}$ and low series resistances of $200\,\Omega$. NW-array FETs aligned along the $\langle 110 \rangle$-direction showed $\times 1.4$ larger on-currents and $\times 1.3$ higher transconductances compared to devices aligned along the $\langle 100 \rangle$-direction. Hole mobilities in NW-array pFETs with $\langle 110 \rangle$- and $\langle 100 \rangle$-channel orientation were measured employing a split-CV technique. NW FETs aligned along a $\langle 110 \rangle$-direction display a 40% higher hole mobility at low as well as at high vertical electric field compared to devices along the $\langle 100 \rangle$-direction.**

I. INTRODUCTION

NW-MOSFETs are regarded as candidates for ultimately scale FETs due to the superior electrostatics of the thin body and the multi-gate structure. Alongside scalability and electrostatics, mobility plays a key role for device performance. Current flow on surfaces with different crystal orientations in multi-gate devices allows to take advantage of the mobility anisotropy in silicon by proper choice of the channel direction. For planar devices hole mobilities have been studied intensively [1], [2], [3]. The highest hole mobilities in silicon were measured on $\{110\}$-surfaces along $\langle 110 \rangle$-directions. For NW like structures, such as Fin-FETs, tri-gate, Ω-gate or gate-all around FETs only a small number of experimental studies investigated hole mobility and the dependence of hole mobility on channel orientation [4], [5], [6]. This is in part due to the difficulty of accurately measuring mobilities on NW devices due to the sub fF capacitance of single NWs. Therefore, most authors measured the channel resistance and used calculated values of the channel capacitance to determined the mobility [7], [8]. Only in recent years the first direct measurements of electron and hole mobilities in NW-array devices have been reported [4], [5].

In this study Ω-gated NW-array pMOSFETs were fabricated on $\{100\}$ oriented SOI substrates with $\langle 110 \rangle$- and $\langle 100 \rangle$-channel orientations. Devices with different gate length

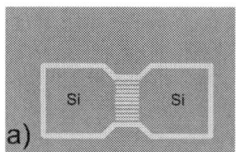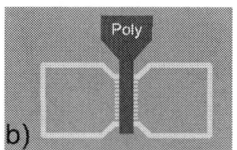
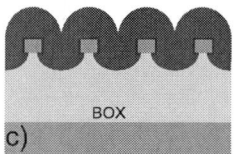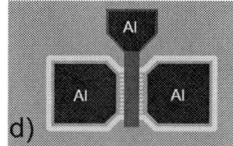

Fig. 1. Schematic of the main fabrication steps. (a) Top view of the structure after mesa etching. Up to 1500 identical NWs were fabricated in parallel; (b) Top view after the selective etching of the poly-Si gate; (c) Cross section of the Ω-gated NW-array with SiO_2/poly-silicon gate stack; (d) Layout after the deposition of Al-contacts on the mesa.

and number of NWs in the channel were electrically characterized and hole mobilities for the different channel orientations were measured using a split-CV technique.

II. DEVICE FABRICATION

Multiple silicon NW FETs were fabricated with a top-down process on 25 nm p-type $\{001\}$ SOI ($N_A = 1 \times 10^{15}\text{cm}^{-3}$) and a buried oxide thickness of 145 nm. Devices with channel lengths of $L = 400\,\text{nm}$ and $L = 2\,\mu\text{m}$ containing from 500 to 1500 parallel NWs aligned along the $\langle 110 \rangle$- and $\langle 100 \rangle$-directions were defined by electron-beam (e-beam) lithography. Reactive ion etching (RIE) was employed to transferred the structures into the SOI layer. To reduce sidewall roughness of the NWs, the samples were subjected to a sacrificial oxidation, followed by diluted HF stripping of the oxide before a 4 nm gate oxide was grown by dry oxidation at $800\,^\circ$C. Subsequently, n$^+$-doped poly-Si was deposited by low pressure chemical vapor deposition (LPCVD) at $580\,^\circ$C. Gate structures were written by e-beam lithography and afterwards the poly-Si was etched with a highly selective HBr/O_2 plasma. Source/drain (S/D) contacts were doped by ion implantation

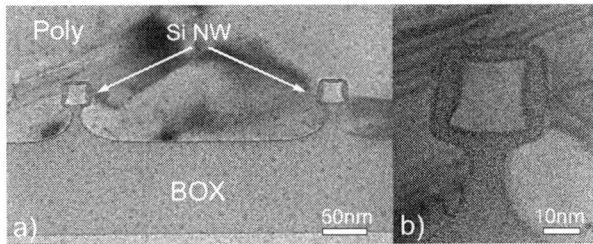

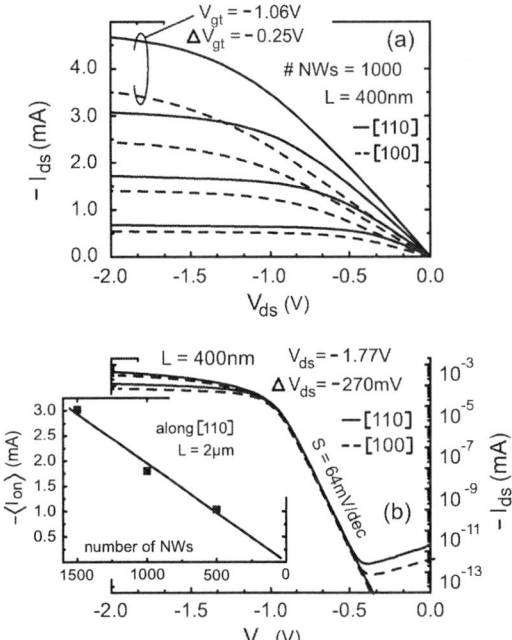

Fig. 2. (a) Cross section-TEM of a Ω-gated NW-array FET with 4 nm SiO$_2$ gate oxide and poly-Si gate. The NWs have a pitch of 300 nm. (b) Close-up of a single NW of 20 nm width and height and Ω-shaped poly-Si gate .

Fig. 3. Output (a) and transfer (b) characteristics for Ω-gated NW-array pMOSFET along $\langle 100\rangle/\{100\}$ - (dashed) and $\langle 110\rangle/\{100\}$ - (solid) channel directions. The transfer characteristics shows the curves for small and large V$_{ds}$ (V$_{ds}$= -0.27V and V$_{ds}$= -1.77V). The inset shows a plot of the on-current of devices with different number of NWs against the number of NWs.

of BF$_2$ at an energy 10 keV of to a dose of 2×10^{15} cm^{-2}. Dopants were activated by rapid thermal annealing for 30sec at 1000°C. To prevent surface leakage currents, a protective SiO$_2$ layer was deposited by plasma enhanced chemical vapor deposition (PECVD). S/D and gate contact windows in the protective oxide layer were defined by optical lithography and opened using RIE. Afterwards a 200 nm Al metallization was deposited with a lift-off process and the samples were annealed for 10 min at 420°C in forming gas atmosphere (H$_2$/N$_2$). Figure 1 schematically shows the main steps of the device fabrication process. Figure 2 shows cross section-TEM images of the Ω-gated NWs. The NWs feature uniform cross sections of 20×20 nm^2 and have a pitch of 300 nm (Fig. 2(a)). Figure 2(b) shows a close-up of a single NW with the 4 nm gate oxide and the poly-Si Ω-gate.

III. ELECTRICAL CHARACTERIZATION

In this paragraph we discuss the electrical properties of the omega-gated p-type NW-array FETs with emphasis on the dependence of on-current, transconductance and mobility on channel direction.

In multi-gate devices current flows on surfaces with different crystal orientation. This allows to improve device performance by proper choice of the channel direction, making use of the anisotropic hole mobility in silicon. The channels of the fabricated NW-array FETs were aligned along the $\langle 110\rangle$- and $\langle 100\rangle$-direction, respectively. For the devices aligned along the $\langle 110\rangle$-direction the vertical sides are $\{110\}$-planes while the top surface is a $\{001\}$-plane, as indicated in Fig. 4. Hole mobility is largest on $\{110\}$-planes along the $\langle 110\rangle$-direction at low as well as at high vertical electrical fields [3]. As in this case two thirds of the inverted surface are $\{110\}$-planes these devices are expected to show superior performance compared to NW-array FETs aligned along a $\langle 100\rangle$-direction. For the devices with a $\langle 100\rangle$-channel direction current flows only on $\{100\}$ planes which feature a significantly lower hole mobility (Fig. 4).

Figure 3 shows measured output and transfer characteristics of NW-array FETs with 1000 parallel NWs and channels aligned a $\langle 110\rangle$- and $\langle 100\rangle$-direction, respectively. The gate length of the devices is $L = 400$ nm and the individual NWs have a width of $W = 20$ nm and a height of $H = 20$ nm with a pitch of 300 nm. The output characteristics of the devices

show nice saturation for both channel orientations while the device aligned along a $\langle 110\rangle$-direction features a larger on-current of $I_{on} = -4.57$ mA compared to the device aligned along a $\langle 100\rangle$-direction with a on-current of $I_{on} = -3.01$ mA at the same gate-overdrive of $V_{gt} = -1.06$ V (Fig. 3(a)). Low off-currents with a high I_{on}/I_{off} ratio of 10^8 and close to ideal inverse sub-threshold swings of $S = 64$ mV/dec at room temperature were measured. The close to ideal sub-threshold behavior indicates the high quality of the gate oxide and a small interface state density. It also indicates a very low variability of the NW dimensions within one array. In the inset of Figure 3(b) the on-current at a gate overdrive of $V_{gt} = |V_g - V_{th}| = 1$ V of NW-array FETs with different number of NWs is plotted against the number of NWs in the channel. The on-current scales linearly with the number of NWs intersecting the point of origin. This indicates that all NWs within an array are functional and that the devices have a low source/drain (S/D) resistance. The S/D resistance was determined from the transfer characteristics to be $R_{S/D} = 200\,\Omega$ by plotting the total resistance as a function of gate length for different gate overdrives (not shown). The statistical significance of the $\times 1.4$ higher on-currents for NW-array FETs aligned along a $\langle 110\rangle$-direction is further shown by plotting the the on-current at a gate overdrive of 1 V averaged over a large number of devices for each gate length ($L = 400$ nm and $L = 2\,\mu$m) and each channel orientation ($\langle 110\rangle$ and $\langle 100\rangle$) in Figure 4. For both channel lengths devices alinged along a

978-1-4244-6658-0/10 $26.00 © 2010 IEEE

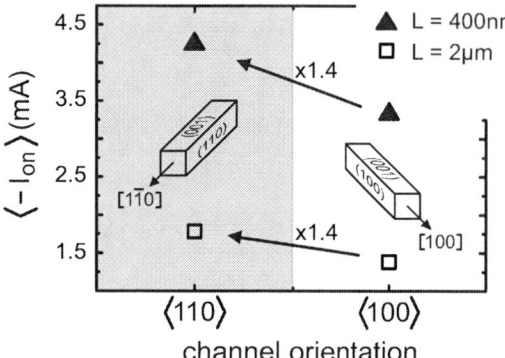

Fig. 4. Average on-current of NW-array FETs with channel length $L = 2\,\mu m$ (square) and $L = 400\,nm$ (triangular) and channels aligned along $\langle 110 \rangle$- and $\langle 100 \rangle$-directions.

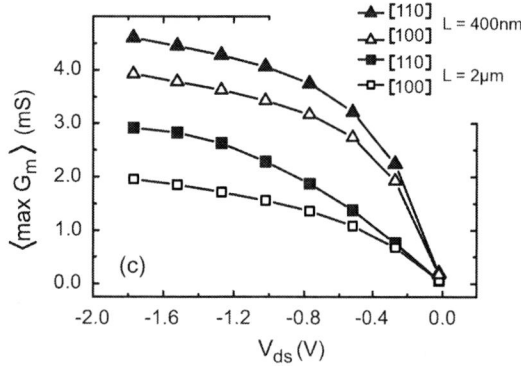

Fig. 5. Maximum transconductance plotted as a function of drain voltage for NW-array pFETs with $L = 400\,nm$ and $L = 2\,\mu m$ for $\langle 100 \rangle/\{100\}$ and $\langle 110 \rangle/\{100\}$ channel orientations.

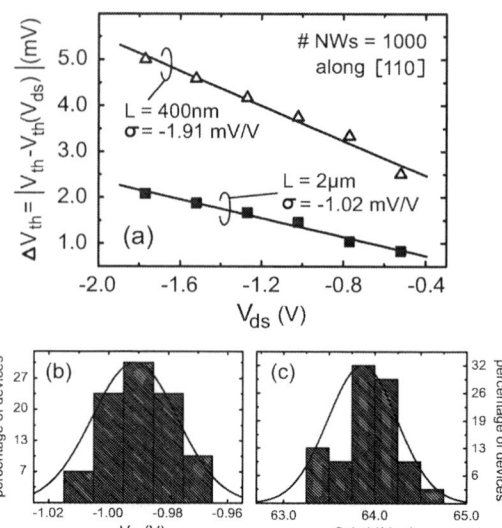

Fig. 6. (a) Extraction of the DIBL parameter σ for NW-array FETs with 1000 NWs from a plot of the threshold voltage shift ΔV_{th} as a function of drain voltage; (b) Histogram of the variation of V_{th} and (c) histogram of the variation of the inverse sub-threshold swing S in NW-array FETs with a 1000 NWs and $L = 400\,nm$.

a frequency of 500kHz was used. After eliminating parasitic capacitances the mobility was extracted from the split C-V curve and the transfer characteristics using the equation:

$$\mu = \frac{L}{W_{el}V_d} \cdot \frac{I_d(V_g)}{Q_i(V_g)} = \frac{L^2}{V_d} \cdot \frac{I_d(V_g)}{\int_{-\infty}^{V_g} C_{gc}(V_g)dV_g} \quad (1)$$

where L/W_{el} is the channel length/width, V_d the drain voltage, I_d the drain current, C_{gc} the total gate capacitance and Q_i the inversion charge density. Figure 7 shows a measured dual sweep of the total gate capacitance C_{gc} as a function of V_g. The curve is nearly hysteresis free. This further verifies the high quality of the gate oxide and the low interface state density. In the inset of Fig. 7 the on-state capacitances at $V_{gt} = -1\,V$ of NW-array FETs with different number of NWs (from 500 to 1500 NWs) and a gate length of $L = 2\,\mu m$ are displayed. $C_{g,on}$ increases linearly with the number of NWs due to the increasing total inverted surface area confirming the homogenous dimensions and functionality of the NWs.

Field effective hole mobilities in Ω-gated NW-array FETs with $\langle 110 \rangle$- and $\langle 100 \rangle$-direction were determined employing Eq. 1. Figure 8 displays the measured hole mobilities for the two channel direction together with the universal hole mobility curve for (100) bulk-Si. Hole mobility in NW-array FETs aligned along the $\langle 110 \rangle$-direction is ≈ 1.4 times larger at low as well as at high vertical electrical fields than in devices aligned along the $\langle 100 \rangle$-direction. This is in good agreement with the observed differences in on-current and transconductance for the two channel orientations. The higher hole mobility on $\{110\}$ surfaces has been attributed to a

$\langle 110 \rangle$-direction show a higher on-current by a factor of 1.4. Similar improvements by a factor of $\times 1.3$ are measured for the transconductances of NW-array FETs with $\langle 110 \rangle$-channel orientation. In Figure 5 the maximum transconductance at $V_{gt} = -1\,V$ averaged over a large number of devices for each channel length and orientation is plotted against drain voltage. The drain induced barrier lowering (DIBL) parameter σ was determined by measuring the threshold voltage V_{th}-shift with increasing drain bias V_{ds}. Figure 6 shows the threshold voltage shift ΔV_{th} as a function of drain voltage V_{ds} for NW-array FETs aligned along the $\langle 110 \rangle$-direction and channel length of $L = 2\,\mu m$ and $L = 400\,nm$. Due to the excellent electrostatic control of the Ω-gate both devices show negligible DIBL ($\sigma < -2\,mV/V$).

In order to investigate the robustness of our process and the variability of fabricated devices, Fig. 6 shows histograms of the threshold voltage (V_{th}) and the sub-threshold swing (S) distribution for devices with 1000 NWs and $L = 400\,nm$. Both distributions show low variations, indicating that all NWs have uniform dimensions across the sample. Note that the threshold voltage of $V_{th} = -1\,V$ is due to the n^+-poly-Si gate.

Mobility measurements were performed using the split C-V technique. For the measurement a 100mV excitation signal at

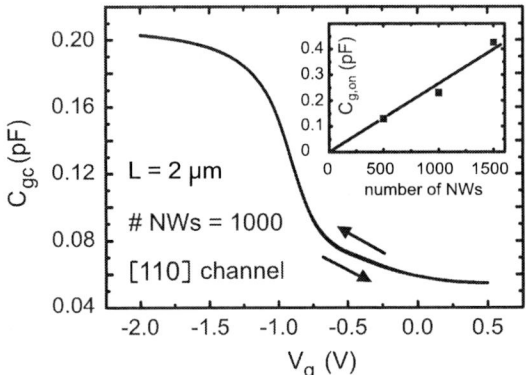

Fig. 7. Dual sweep C-V curve measured by split C-V on a $L = 2\,\mu\text{m}$ device with 1000 NWs. The inset displays the on-state capacitance as function of the number of NWs in the channel. The on-state was defined as $V_{\text{gt}} = V_{\text{th}} - 1\,\text{V}$.

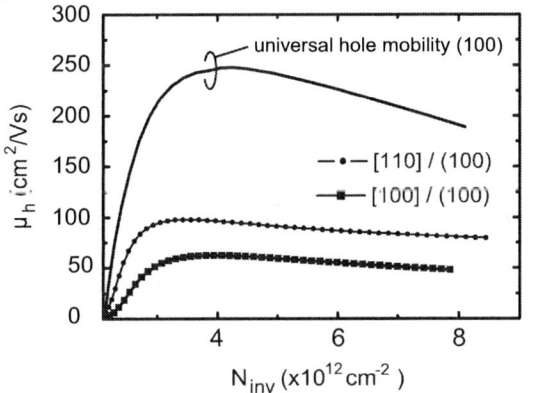

Fig. 8. Hole mobility in Ω-gated NW-array FETs as function of the inversion carrier density N_{inv} for $\langle 110 \rangle$- and $\langle 100 \rangle$-channel orientation. As a reference the universal hole mobility for $\{100\}$ bulk-Si is shown.

smoother interface with the gate dielectric [9]. Thus scattering origination from surface roughness is decreased. Other publications propose that p-MOSFETs on $\{110\}$ surfaces have lower effective electric fields compared to $\{100\}$ surfaces which further reduces surface roughness scattering due to the linear dependence of the scattering rate to the effective electric field [10]. Still the measured hole mobility values for $\langle 110 \rangle$-channel direction are significantly lower than the universal values for bulk Si. This can be due to different reasons: (i) The measured hole mobilities on the NW-array FETs are averaged mobility values over the three inverted surfaces; (ii) for the $\langle 110 \rangle$-aligned devices the sidewalls are not perfect $\{110\}$ planes (Fig. 2) and might therefore feature not the highest possible hole mobilities.

IV. CONCLUSION

Ω-gated NW-array pFETs with up to 1500 parallel NWs aligned along the $\langle 110 \rangle$- and $\langle 100 \rangle$-directions were fabricated on a $\{001\}$ SOI substrate. The devices feature excellent electrical characteristics with high on-currents, $I_{\text{on}}/I_{\text{off}}$ ratio of 10^8, close to ideal inverse sub-threshold slopes of $64\,\text{mV/dec}$

and low series resistances of $200\,\Omega$. NW-array FETs aligned along the $\langle 110 \rangle$-direction showed $\times 1.4$ larger on-currents and $\times 1.3$ higher transconductances compared to devices aligned along the $\langle 100 \rangle$-direction. Hole mobilities in NW-array pFETs with $\langle 110 \rangle$- and $\langle 100 \rangle$-channel orientation were measured employing a split-CV technique. NW FETs aligned along a $\langle 110 \rangle$-direction display a 40% higher hole mobility at low as well as high vertical electric field compared to devices along the $\langle 100 \rangle$-direction. This is explained by the current flow on surfaces with different crystal orientations in multi-gate devices and the hole mobility anisotropy in silicon.

Acknowledgements This work received funding from the EU through the NANOSIL Network of Excellence (Contract NO. ICT-2007-216171) and from the German Federal Ministry of Education and Research via the MEDEA+ project DECISIF (2T104).

REFERENCES

[1] T. Sato, Y. Takeishi and H. Hara, "Effects of Crystallographic Orientation on Mobility, Surface State Density, and Noise in p-Type Inversion Layers on Oxidized Silicon Surfaces ", *Jpn. J. Appl. Phys*, **8**, pp.588, (1969).

[2] F. Gamiz, J.B. Roldan, J.A. Lopez-Villanueva, P. Cartujo-Cassinello and J.E. Carceller, "Surface roughness at the Si-SiO2 interfaces in fully depleted silicon-on-insulator inversion layers", *J. Appl. Phys.*, **86**, pp.6854, (1999).

[3] G. Sun, Y. Sun, T. Nishida and S.E. Thompson, "Hole mobility in silicon inversion layers: Stress and surface orientation ", *J. Appl. Phys.*, **102**, pp.084501, (2007).

[4] Y. Jeong, J. Chen, T. Saraya and Toshiro Hiramoto "Uniaxial Strain Effects on Silicon Nanowire pMOSFET and Single-Hole Transistor at Room Temperature ", *IEDM Tech. Digest.*, pp.761, (2008).

[5] O. Gunawan, L. Sekaric, A. Majumdar, M. Rooks, J. Appenzeller, J.W. Sleight, S. Guha and W. Haensch, "Measurement of Carrier Mobility in Silicon Nanowires ", *Nano Lett.*, **8**, pp.1566, (2008).

[6] J. Ramos, S. Severi, E. Augendre, C. Kerner, T. Chiarella, A. Nackaerts, T. Hoffmann, N. Collaert, M. Jurczak and S. Biesemans, "Effective Mobility Extraction Based on a Split RF C-V Method for Short-Channel FinFETs ", *Solid-State Electron.*, **50**, pp.32, (2006).

[7] N. Singh, A. Agarwal, L.K. Bera, T.Y. Liow, R. Yang, S.C. Rustagi, C.H. Tung, R. Kumar, G.Q. Lo, N. Balasubramanian and D.-L. Kwong, "High-performance fully depleted silicon nanowire (diameter $\leq$ 5nm) gate-all-around CMOS devices ", *IEEE Elec. Dev. Lett.*, **27**, pp.383, (2006).

[8] X. Zhou, J.-Y. Park,S. Huang, J. Lui and P.L. McEuen, "Band Structure, Phonon Scattering, and the Performance Limit of Single-Walled Carbon Nanotube Transistors ", *Phys. Rev. Lett.*, **95**, pp.146805, (2005).

[9] M.M. Chowdhury and J.G. Fossum, "Physical insights on electron mobility in contemporary FinFETs ", *IEEE Elec. Dev. Lett.*, **27**, pp.482, (2006).

[10] K. Lee, J. Choi, S. Sim and C. Kim, "Physical understanding of low-field carrier mobility in silicon MOSFET inversion layer ", *IEEE Trans. Elec. Dev.*, **38**, pp.1905, (1991).

Breaching the kT/q limit with Dopant Segregated Schottky Barrier Resonant Tunneling MOSFETs: A Computationnal Study

Aryan Afzalian, Denis Flandre
Microelectronics Lab, ICTEAM Institute
Université Catholique de Louvain
Louvain-La-Neuve, Belgium
aryan.afzalian@uclouvain.be

Abstract— **We study here, using non-equilibrium Green's function quantum simulations, the impact of dopant segregation (DS) on Schottky barrier (SB) nanoscale transistors for the implementation of ultimate CMOS with low series resistance and steep slope. Owing to their adequate multi-barrier structure, DS-SB transistor can present a gate modulated barrier resonant tunneling (MBRT) effect that allows them to breach the kT/q subthreshold slope limit of classical MOSFET, and therefore pave a way towards steep slope, low S/D resistance electronics. In order to reach their ultimate on-current performances however, new materials with lower SB height and/or means to implant and activate ultra high dopant segregation levels (in the order 10^{21}cm^{-3}) will be needed, especially when considering that Schottky barrier height will be increased through quantum confinement.**

I. INTRODUCTION

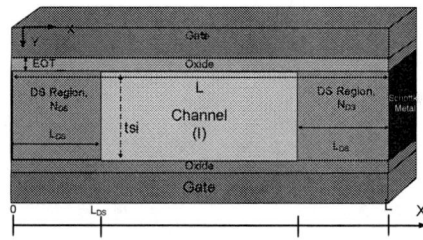

Figure 1. Simulated Double Gate [100] SOI N-type transistor with Schottky contact. tsi=2nm. L=12nm, doping N-=10^{15}cm^{-3}, except in the dopant segregated regions of length L_{DS} and doping $N_+=N_{DS}$. EOT=0.5nm. T=300K.

As transistors are scaled down in the nanoscale regime, scaling alone is not sufficient to achieve performance improvement and new boosters and device concepts are needed. Replacing highly doped source (S) and drain (D) by metallic Schottky contacts, Schottky Barrier (SB) transistors have recently attracted a lot of attention because, doing so, one can potentially solve the increasing problem of S/D resistance

and dopant diffusion [1-4]. This will become crucial for future ultra-scaled transistors, i.e. with channel length below 20nm and ultra-thin body and multigate architectures [1]. Due to the energy difference when compared to vacuum level between the conduction band in the silicon and the Fermi level in the Schottky metal (the Schottky barrier height, SB$_H$) however, a potential shift appears in the conduction band at the interface between Si and Schottky metal after Fermi-level alignment. This creates a barrier in the conduction path that degrades subthreshold slope and current. In order to improve performances one can find a material with lower barrier height such as Er for n-FETs or Pt for p-FETs [3,5]. However midgap materials like Nickel Silicides are presently easier to integrate with Si [4]. Using dopant segregation (DS) of dopants implanted in the Schottky metal, i.e. the natural migration and accumulation in the first Si few nanometers near the interface with the Schottky metal, one can reduce the equivalent barrier height significantly [2,4]. In fact, we will show based on our simulation results that it is more the width of the barrier that decreases as a depletion region is thinned when increasing the doping. Furthermore, as we will show, ultra scaled SB devices, owing to their multi-barrier structure, can present resonant tunneling effects that can strongly influence their characteristics. DS SB-FETs can also present a gate modulated barrier resonant tunneling (MBRT) effect and breach the kT/q limit of a MOSFET, the lower limit for the subthreshold slope (*SS*) of 59.6 mV/dec at 300K, that sets a practical limit to reduction of supply voltage and power consumption of a circuit. We have recently shown the possibility of achieving sub-*kT/q* subthreshold slope near threshold, together with high on-current, by using a similar Si "Multi-barrier boosted" CMOS transistor, the gate modulated resonant tunneling (RT)-FET [6-7]. The tunnel barriers were implemented by local constrictions of the cross-section. Achieving steeper subthreshold slope transistors has become vital for further CMOS downscaling. In order to achieve good performances in Si, these tunnel barriers should however have a length, L_C, of a few nanometers which is quite challenging to

978-1-4244-6658-0/10 $26.00 © 2010 IEEE

process. We study here, using quantum non-equilibrium Green's function (NEGF) simulation, the impact of DS on Schottky barrier nanoscale transistors. Thin barriers and adequate resonant tunneling effect should results from using Schottky contact and dopant segregations in nanoscale device paving a way for steep slope, low S/D resistance electronics.

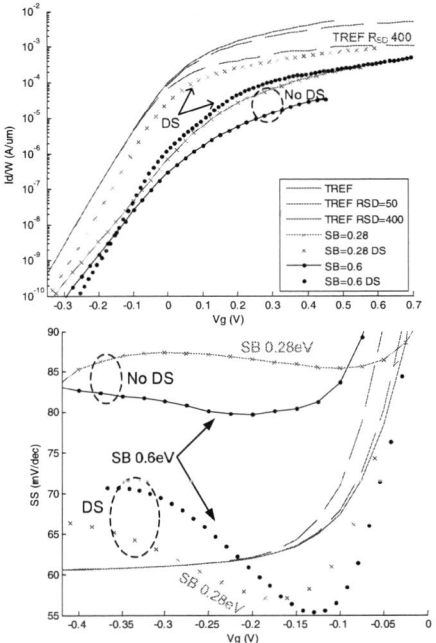

Figure 2. ID(VG) and SS(VG) curves of the DG transistors without (TREF) (1) and with Schottky contacts: with $SB_H=0.28$eV without DS (2), $SB_H=0.28$eV and As dopant DS: $N_{DS} = 2.10^{20}$cm^{-3} and $L_{DS} = 2$nm (3), $SB_H=0.6$eV, without DS (4), $SB_H = 0.6$eV and As DS $N_{DS}=5.10^{20}$cm^{-3} and $L_{DS} = 3$nm (5). Vd=0.5V. L=12nm. EOT=0.5nm. tsi=2nm. Gate workfunction adjusted for V_{th}=50mV for all devices.

II. RESONANT TUNNELING EFFECT IN DG SB-FETs

We have simulated 12nm-long planar double gate (DG) Schottky barrier FETs, with and without DS, using our Fast Coupled Mode Space (FCMS) self-consistent NEGF quantum simulator [8]. The methodology used to simulate the Schottky contact is similar to [6]. The electron concentration is computed self-consistently in the Si using the NEGF. The Schottky barrier is added as boundary condition in the source and drain potential. A potential equal to $-q*(SB_H - (Ec-E_F))$ is added to the source and drain potentials, where E_F is the Fermi potential related to the doping in the Si body and E_C is the bottom of the conduction band in Bulk Silicon. This allows one to take into account the increase of the SB_H due to the increase of E_C through quantum confinement in small cross-sections. The conduction band edge in the Schottky metal is assumed constant and a few 100meV lower than the source and drain Fermi level ($E_{FS}=E_F$ and $E_{FD}=E_{FS}-q*V_D$) in order to ensure sufficient injection (and independent of the exact band edge level of the metal) of carriers in the device.

Fig.1 shows a schematic device representation and denotes the parameters. The body is intrinsic (except for the DS regions near the gate edge of length L_{DS} for the SB-DS transistors) and the channel length is 12nm. For DS, a doped

region with L_{DS} of a few nanometers and doping concentration on the order of a few 10^{20}cm^3 can be achieved [2,4]. The film thickness t_{si} and the equivalent oxide thickness EOT are 2 and 0.5nm respectively. The width, w_{si}, however, is assumed large so that 2D simulations are performed. A classical planar DG-MOSFET, TREF, with N_+ source and drain doping of 10^{20}cm^3 is also simulated for comparison. Note that the impact of the source and drain resistances is not taken into account in these simulations as we are performing ballistic simulations. Therefore, the on-current level achieved for the classical device is its intrinsic on-current, i.e. the maximum current available for an ideal 12nm double gate device with an EOT of 0.5nm without Schottky barrier or contact resistance. A real classical transistor will therefore have a lower on-current depending of the exact geometry of its source and drain extensions and contacts, the presence or not of an overlap, the exact doping profile and the steepness of is transition between source drain value to channel value [9]. Such a modeling is far beyond the scope of this paper. However, in Fig.2 the impact of series resistance on performances of the transistor is estimated for two values of the total series resistance, R_{SD} by including a resistance of $R_{SD}/2$ at S and D electrodes. An optimistic 50 Ω/μm value, which is close to that observed in Bulk transistors with L between 50 and 100nm [9], is equivalent to assuming that the resistance of the whole S or D region can be modeled by that of an equivalent region with same cross-section dimension than the channel, a length of 5nm and uniformly doped to 10^{20}cm^{-3} (we used the universal mobility curve). In an ultra-thin body device as needed for scaling beyond 20nm channel length, this requires however a very steep doping profile. Because of the ultra-thin body, just a few Angstroms of nearly intrinsic region not under the control of the gate could lead to a few kΩ/μm resistance that will strongly degrade on-current performance. An intermediate value of 400 Ω/μm is shown here, which can be obtained for instance assuming a non abrupt junction that could be modeled by an additional equivalent 2 nanometers long region doped to 10^{18}cm^{-3} in series with the highly doped region. For the SB-FETs, the series resistance can be below the 50 Ω/μm and therefore neglected [4]. In the case of TREF with the 400 Ω/μm resistance, the on-current is strongly degraded and close to that of the Erbium silicide SB-FET with DS. Also, the series resistance is always detrimental to the on-current, but has no effect to the off-current, therefore degrading I_{on}/I_{off} ratio. On the contrary, in SB-FETs both on and off current are reduced because of the SBs and I_{on}/I_{off} can be improved compared to TREF in the DS case.

The subthreshold slope and on-current characteristics of the SB without DS are degraded compared to that of the classical transistor. This is because in the SB devices, the current is dominated by the tunneling current through the source barrier. When increasing V_G, this source barrier is thinned, but its height at source cannot be reduced and, therefore, the overall gate coupling is reduced (Fig. 3.A). Also the SB_H is increased by about 0.1eV due to quantum confinement. When using DS, however, the barrier is further thinned (Fig. 3.B) which improve SS and I_{on} considerably as experimentally observed [2]. Also for the devices shown in Fig. 2, the subthreshold slope gets below the kT/q limit. This is due to the presence of modulated barrier resonant tunneling

[6,7]. Compared to the intrinsic channel case, potential wells are created in between SBs and channel barrier (CB) because of the difference of doping in the DS region and the rest of the channel (Fig.3). This creates longitudinal quantum confinement in the device. In these wells, electrons and therefore the current can only propagate through discrete resonant tunneling states (sharp peaks in the electron transmission probability, T, in Fig.4). However due to the faster motion of CB compared to SBs when V_G is increased, the relative shape of the wells with V_G is changed. This modulates the resonant tunneling states as well as their broadening which influences the density of state (DoS), transmission, current, and subthreshold slope in the device. In Fig.4, the depth of the potential well between source SB and CB is reduced when V_G is increased. This increases the energy of levels in this well and transfers them closer or above the top of the channel barrier (TCB) effectively increasing the number of levels that can drive the current. A reduction of *SS* below kT/q is therefore achieved by this means even if the transparency of the SBs decreases in comparison to the channel barrier and gate coupling is degraded. This is because the electron concentration (and therefore I_D) increases faster than the Fermi-Dirac distribution of carriers due to the strong non-linear change of DoS shape with V_G. This change in DoS distribution modifies the transmission (Fig. 4.B). This is not the case in a device without tunnel barrier, where the electrons are free of any confinement effects in the transport direction (x) and where DoS and transmission close to TCB remain constant in subthreshold regime [6,7].

created between source and drain SBs. However in this case, this will usually happen above threshold and be therefore mostly detrimental for device performances. Indeed due to the discrete states, RT reduces the current level in the device so that the on-current is more reduced than the off current. In the DS case on the contrary (Fig.3.B), RT is already present in subthreshold regime owing to the SB-channel barrier wells and slope and I_{on}/I_{off} ratios can be improved through the MBRT effect: at least in a given voltage range, the off-current can be filtered more efficiently than the on-current.

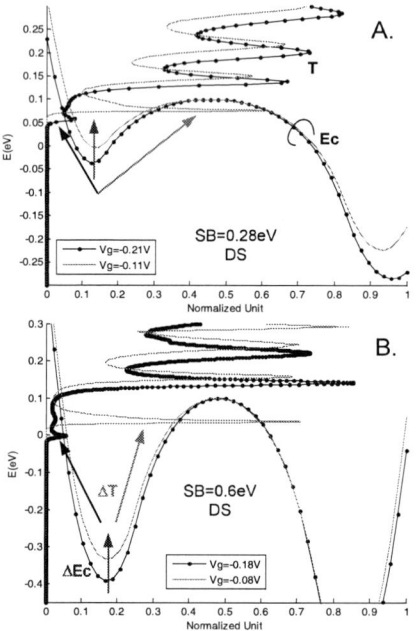

Figure 4. E_C vs. normalized distance (x /L) , and T (rotated by 90° and non normalized) vs. energy for SB DG transistors with DS of Fig.2 in their steep slope region: A) SB_H=0.28eV at V_G=-210mV and 110mV, B) with SB_H=0.6eV, at V_G=-180 and -80mV. All the curves have been shifted down in energy so that E_{TCB}=0.1eV for comparison. The deformation of the quantum well formed between SB_S and channel barrier has increased the relative position of an energy level in the well from below the channel barrier to closer to its top strongly improving its transmission probability. This has allowed for the steep subthreshold slope regions observed in Fig.2.

III.   DG vs. GAA SB RT-FETs

In Fig.5, we compare intrinsic performances of DG and a square (film width, $w_{si} = t_{si}$) Gate-all-around (GAA) nanowires with and without SBs. GAA nanowires are contemplated as a further booster compared to planar architecture device owing to their better electrostatic control. The use of the nanowire architecture boosts the on-current in a similar fashion for all devices. However, the subthreshold characteristics of the DS-SB are differently affected. In the nanowire case, the better gate control allows for a faster reduction of channel and Schottky barriers when increasing V_G. This enables higher on-current but also change the relative motion between channel and SBs. Furthermore, the higher confinement increases SB_H compared to their Bulk values by about 0.25eV (compared to 0.1eV in the DG case). In the case of the Erbium SB nanowire, this strongly decreases the effective transparency of the SB which is favorable for an increase of the thermionic

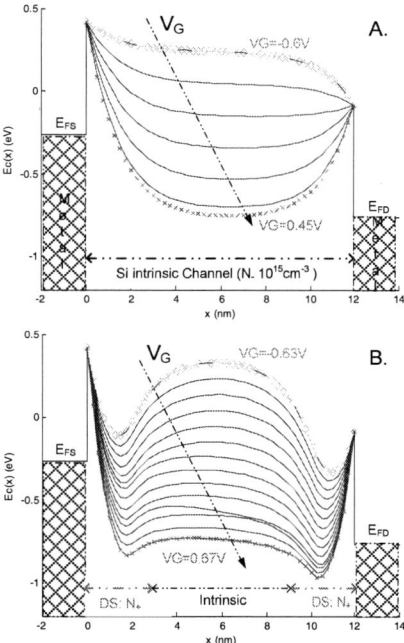

Figure 3. E_C(x) vs. V_G for the SB-DG FET with SB_H=0.6eV of Fig.2: A) without DS and B) with DS. States filled with electrons in the conduction band of the Schottky contacts, i.e. below E_{FS} and E_{FD}, are also shown.

As can be seen in Fig. 3.A, resonant tunnelling can also appear in SB devices without DS due to the potential well

current above the barrier and reduces the effect of MBRT on the subthreshold slope and off-current. This also increases the threshold voltage of this transistor and gives this transistor an intermediate typical regime where both thermionic and RT-currents are flowing and where SS is in between subthreshold and above threshold [7]. Subthreshold characteristics similar to that of the DG Erbium SB-FET could be recovered by increasing the doping level in the DS region however. For the Nickel SB nanowire, the higher barriers and especially the presence of the drain barrier higher above the channel barrier in the steep subthreshold slope region allow for a stronger MBRT effect and very steep SS region of about 35mV/dec. The presence of the higher drain SB-barrier and its further relative increase when increasing V_G also creates, just after the steep slope region, a region of negative resistance (NR): In this region the transmission is suddenly decreased by the relative motion of the drain barrier and the current decreases. The overall impact on the I_{on}/I_{off} ratio is however positive. Also a NR region can be very interesting for creating new and/or added functionalities such as for memory application.

In the case of the GAA devices we also show an optimized constriction tunnel barrier device for comparison [6,7]. As for the DS–SB case, the device achieves low off-current and steep slope owing to the MBRT effect. In this case however the TBs are in good control of the gate so that they move mostly at the channel barrier speed. This allows better and ideal SS out of the steep slope region. It also allows for the constriction RT-FETs to feature a second threshold voltage, V_{th2}, above its first threshold V_{th1}. V_{th1} happens like in a standard transistor when the top of the channel barrier (TCB) passes below the source Fermi level E_{FS}. V_{th2} happens when the TBs passes below E_{FS}, and is related to a change of regime between resonant-tunneling current in the well to thermionic current above the well. This allows for a more efficient filtering of the off than the on-current. For gate voltage above this threshold, an important additional thermionic current will start flowing enabling further improvement of the slope and current ratio and hence very high on-current. The transistor is recovering the on-current level of a transistor without tunnel barrier. In the case of the DS-SB transistors, the tunnel barriers never pass below E_{FS} as the height at the interface is fixed. A similar, but reduced, effect can still be observed, for example for the nanowire or DG with SB=0.6eV of Fig.5 and Fig.2. By increasing V_G the barriers are thinned. The thinner the barrier at a given energy, the higher the tunneling probability of electrons through the barriers, the higher the coupling to the contact and the broader the peaks in the quantum well at this energy. One passes from a close system with very sharp peaks to an open system with continuous longitudinal (x) energy spectra for electrons, transmission and current above this energy. For energies where the barrier is very thin i.e. below 0.5nm, the longitudinal confinement is so weak that the characteristics are very close to that of TREF.

IV. CONCLUSIONS

We have studied here, using NEGF quantum simulation the impact of dopant segregation on Schottky barrier nanoscale transistors. SB transistors together with dopant segregation are promising devices for the implementation of ultra-thin body, ultra scaled transistors with low series resistance. Owing to their multi-barrier structure, they will, however, present resonant tunneling effects that can strongly influence their characteristics and must be taken into account when optimizing the device. DS SB-FETs can also present a modulated barrier resonant tunneling (MBRT) effect, and therefore breach the kT/q limit, have reduced I_{off} and improved I_{on}/I_{off} ratio compared to a classical MOSFET. They could therefore pave a way towards steep slope, low S/D resistance electronics. However their on-current level still lags behind their ultimate performance limit, i.e. the intrinsic on-current of a classical transistor, especially in ultra-thin body architecture where the Schottky barrier height will be increased through quantum confinement. In order to reach this goal, new materials with lower SB height and/or means to implant and activate ultra high dopant segregation levels (in the order $10^{21} cm^{-3}$) will be needed.

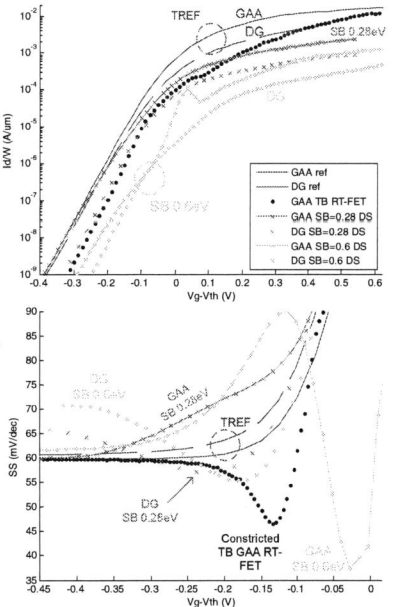

Figure 5. Comparison of ID(VG) and SS(VG) curves for plannar DG and square cross-section nanowire GAA transistors without (TREF) (1) and with Schottky contacts: with SB_H=0.28eV and As DS N_{DS}=2.10²⁰cm⁻³ and L_{DS} = 2nm (2), with SB_H=0.6eV and As DS N_{DS}=5.10²⁰cm⁻³ and L_{DS} = 3nm (3), L=12nm. EOT=0.5nm. tsi=2nm. An optimized GAA constricted TB RT-FET (L_C=2nm and TB_S=0.2eV, TB_D=0.3eV) is also shown for comparison.

ACKNOWLEDGMENT

This material is based upon works supported by FNRS Belgium.

REFERENCES

[1] ITRS Int. Tech. Roadmap for Semiconductors, http://public.itrs.net/.

[2] Q. T. Zhao, et al., Appl. Phys. Lett., 86, 062108 _2005.

[3] G. Larrieu and E. Dubois, IEEE Trans. Elec. Dev., 52, p. 2720, 2005.

[4] S. F. Feste, et al., JAP 107, 044510, 2010.

[5] J. Guo and M. S. Lundstrom, IEEE TED, 49, 11, p.1897, 2002.

[6] A. Afzalian, J.-P. Colinge, and D. Flandre, EUROSOI Conf., 2010.

[7] A. Afzalian, J.-P. Colinge, and D. Flandre, Sol.State E., to be published

[8] A. Afzalian et al., J. of Comp. Electron., 8,3-4, Oct. 2009, pp. 287-306.

[9] S.-D. Kim, C.-M. Park, J. Woo, IEEE TED, 49, 11, p. 457, 2002.

Steep-Slope Nanowire FET with a Superlattice in the Source Extension

E. Gnani, S. Reggiani, A. Gnudi and G. Baccarani

ARCES and DEIS, University of Bologna, Viale Risorgimento 2, 40136 Bologna, Italy
Phone: +39 051 209 3773, Fax: +39 051 209 3779, E-mail: egnani@arces.unibo.it

Abstract—In this work we present an investigation on a novel device concept meant to achieve a steep subthreshold slope by filtering out high-energy electrons entering the device channel. The filtering function is entrusted to a superlattice in the source extension region, which could possibly be fabricated by deposition of a number of appropriate semiconductor layers within a manufacturing process of vertical nanowires. Simulation results indicate that an SS = 26 mV/dec can be achieved using GaAs/AlGaAs as the constituent materials of the superlattice.

I. INTRODUCTION

Power consumption has long been, and still is, the most important limitation for high-performance logic, and one of the main reasons behind its growth across several technology nodes has been the non-scalability of the FET subthreshold slope. From one hand, the leakage current grows exponentially with a decrease of the threshold voltage; on the other hand, supply-voltage scaling at constant leakage is forbidden by performance requirements. Hence, a decrease of dynamic power can only happen at the expense of a growth of static power for an assigned device performance.

The objective of a current turn-on rate much steeper than 60 mV/dec has been pursued by several approaches, which can be classified in two main categories. The first class of devices is based on the introduction of a positive feedback in the turn-on mechanism. Examples of this approach are the impact-ionization MOS (I-MOS) [1], [2], the nano-electromechanical FET (NEMFET) [3], [4], and the negative gate-capacitance ferroelectric FET (Fe-FET) [5], [6]. The second class of devices is based instead on a filtering of the high-energy electrons injected into the channel. The typical example of this device class is the tunnel FET (T-FET) [7], [8], [9], where the filtering function is demanded to the band-to-band tunneling mechanism.

Devices pertaining to the first class suffer severe limitations which make them unpractical for high-performance logic. On the other hand, the T-FET exhibits severe limitations as well. First, the on-current is heavily degraded by the injection mechanism; next, the switching slope is not uniform due to the modulation of the barrier width against gate and drain voltages, and its average value over an extended range of drain currents is fairly disappointing. Finally, the upward curvature of the output characteristics and the related small drain conductance at zero V_{DS} would again prevent rail-to-rail logic swings. Even carefully optimized designs based on strained-SiGe or

Ge heterostructure T-FETs and high-κ gate dielectrics achieve an improved on-current, but can hardly compete with standard CMOS FETs at the same supply voltage [10], [11].

In [12] the filtering function of the energy distribution of the carriers entering the channel can be suitably modified by shaping the density of states in the source extension. Moreover, in a recent IBM patent [13], this shaping of the density of states is entrusted to a superlattice (SL) interposed between the source and the channel of a vertical nanowire FET. However, no simulations are shown to demonstrate the potential of this solution, and no hint is provided on the optimization of the superlattice miniband structure. In this work we investigate the effectiveness of the superlattice to filter out high-energy electrons entering the device channel and propose a simulation-based methodology for device optimization.

II. DEVICE CONCEPT

The MOSFET turn-on characteristic in subthreshold is expressed by the law $I_D = I_{D0} \exp\{q(V_{GS} - V_T)/nk_BT\}$, with k_B the Boltzmann constant, T the lattice temperature and n an ideality factor close to unity. Thus, the subthreshold slope SS $= nk_BT/q \log(10)$ which turns out to be 60 mV/dec at room temperature for $n = 1$. To overcome this limit, the energy distribution of the carriers entering the channel must be suitably modified so as to filter out the high-energy electrons which are responsible for the subthreshold slope referred to above. This energy filter can possibly be implemented by means of a superlattice structure with multiple layers, to be inserted between the source and the FET channel, as suggested in [13], and can be tuned to provide the appropriate energy window both in position and extension, to ensure the desired trade-off between a high on-state current and a low off-state leakage. This goal can be achieved by an accurate selection of the constituent materials and by adjusting the superlattice physical dimensions.

In order to investigate the feasibility of the proposed concept, we first made very simple calculations based on the one-dimensional Krönig-Penney model of the SL assuming $m^* = 0.2 \, m_0$. Fig. 1 shows the first two subband edges vs. several superlattice parameters, namely: the well width a (top left); the barrier width b (top right); the spatial period $a + b$ (bottom left), and the subband offset V_0 (bottom right). The figure shows that the choice of a spatial period $a + b \simeq 2$ nm, with $a = b$ and $V_0 = 0.8$ eV is expected to generate acceptable

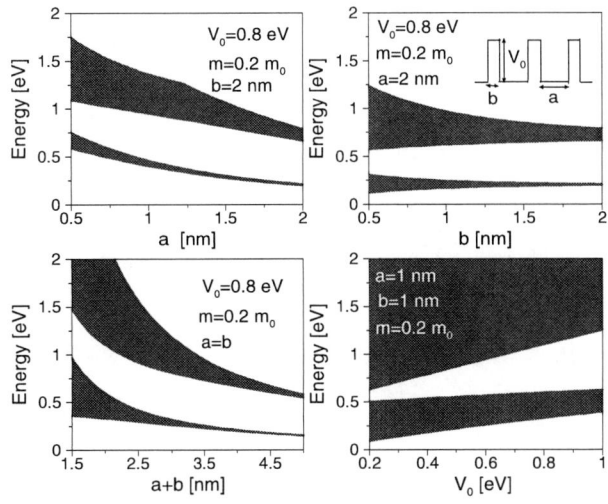

Fig. 1. Energy extension of the first two minibands vs. the well width a (top left), the barrier width b (top right), the spatial period $a + b$ (bottom left) and the barrier height V_0 (bottom right) of the superlattice.

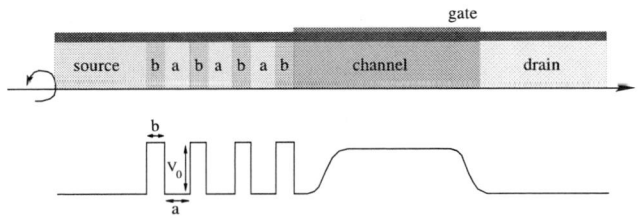

Fig. 2. Top: Pictorial view of the device cross section. Different materials can be used for the source, superlattice, channel and drain regions. Bottom: schematic view of the lowest subband profile within the nanowire.

values of the energy extension of the lowest miniband and of the band gap between the first two minibands. The thickness values are fairly challenging, but do not appear to be beyond the possibilities offered by an advanced MBE machine.

Next, a quantum-based simulation code has been implemented in order to investigate the filtering properties of a superlattice in the source extension of the FET. The code self-consistently solves the open-boundary Schrödinger-Poisson equations in the effective mass approximation with cylindrical coordinates. Figure 2 sketches the cylindrical nanowire geometry of the investigated device as well as a schematic view of its potential profile. The quantum-mechanical treatment of the SL-FET is carried out by decoupling the Schrödinger equation into the transverse and longitudinal transport problems, and the structural confinement in the radial and angular directions gives rise to the splitting of the conduction band into energy subbands along the transport direction. Moreover, every region is characterized by its specific transport mass, dielectric constant and electron affinity. Finally, an energy-adaptive mesh must be used in order to achieve an accurate description of the resonant states generated within the superlattice.

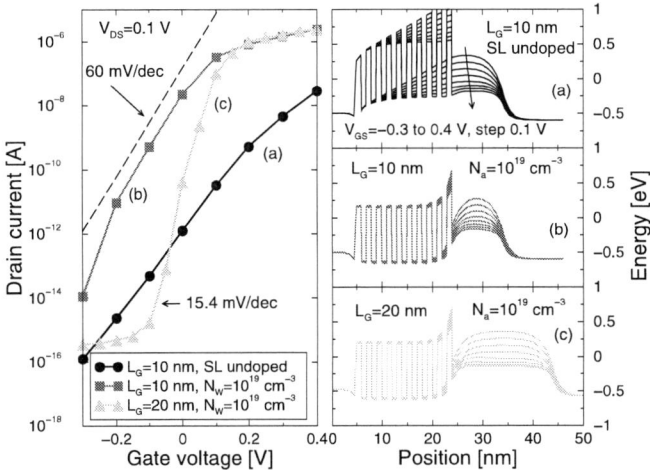

Fig. 3. Left: Turn-on characteristics at $V_{DS} = 0.1$ V for a nanowire FET with undoped SL and gate length $L_G = 10$ nm (device (a)); with a doping density in the well regions $N_W = 2 \times 10^{20}$ cm^{-3} and $L_G = 10$ nm (b), and with a doping density $N_W = 2 \times 10^{20}$ cm^{-3} and $L_G = 20$ nm (c). Device (c) exhibits a subthreshold slope SS $= 15.4$ mV/dec. Right: Energy profiles of the first subband for gate voltages ranging from -0.3 to 0.4 V in steps of 0.1 V for the three above FETs.

III. RESULTS

The first device analysis was carried out with a cylindrical nanowire FET with diameter $D = 5$ nm, oxide thickness $t_{ox} = 1$ nm, a doping density $N_D = 2 \times 10^{20}$ cm^{-3} in the source and drain regions, and a gate contact aligned with the intrinsic channel. The SL parameters were taken from the preliminary results of the Krönig-Penney model, i.e. $V_0 = 0.8$ V, $m^* = 0.2\, m_0$ and $a = b = 1$ nm. A direct band gap is assumed with a spherical conduction band minimum at the Γ point of the Brillouin zone in order to emulate III-V materials. Also, the SL contains ten barriers and nine wells. Fig. 3 (left) compares the simulated turn-on characteristics for three SL-FETs differing for the gate length and the doping density in the wells of the SL. More specifically, device (a) has an undoped superlattice and a gate length $L_G = 10$ nm; device (b) has a doping density $N_W = 2 \times 10^{20}$ cm^{-3} in the well regions of the SL and a gate length $L_G = 10$ nm; device (c) features again a doping density $N_W = 2 \times 10^{20}$ cm^{-3} in the wells, and a gate length $L_G = 20$ nm. The right side of the same figure reports the energy profiles at different gate voltages, with devices (a), (b) and (c) at the top, center and bottom, respectively.

It can be seen that device (a) exhibits a poor subthreshold slope and a degraded on-current (dots) due to the capacitive coupling of the SL with the gate which heavily affects the electrostatic potential in the SL region, as seen by the corresponding energy profile (top right). An improved on-current is exhibited by device (b), but its subthreshold slope hardly departs from the 60 mV/dec limit except in a small region at very low currents. The best characteristic is exhibited by device (c), which features a subthreshold slope SS $\simeq 15$ mV/dec with no degradation of the on-current. As devices (b) and (c) only

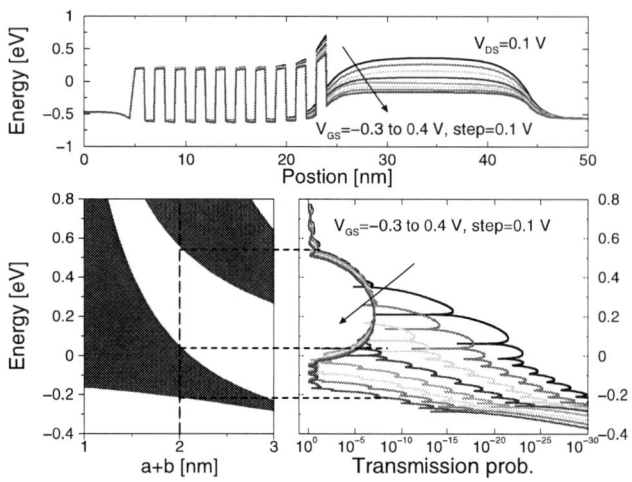

TABLE I

RELEVANT PARAMETERS USED FOR THE SIMULATION OF THE
GaAs/Al$_{0.45}$Ga$_{0.55}$As SUPERLATTICE [14].

	ε	mass	gap	elec. affinity
GaAs	12.9 ε_0	0.067 m_0	direct	4.15 eV
Al$_{0.45}$Ga$_{0.55}$As	14.18 ε_0	0.1 m_0	direct	3.575 eV

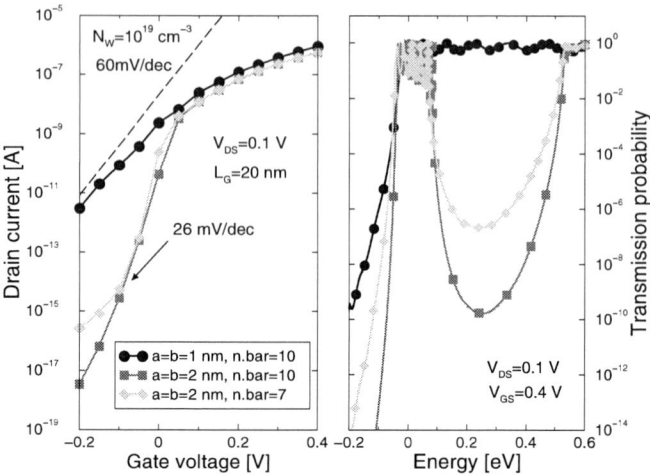

Fig. 4. Top: energy profile of the first subband for gate voltages ranging from -0.3 to 0.4 V in steps of 0.1 V for device (c). Bottom left: lower and upper edges of the first two subbands vs. the spatial period $a+b$ as predicted by the one-dimensional Krönig-Penney model with $V_0 = 0.8$ V and $m* = 0.2\,m_0$. Bottom right: transmission probability of the first subband for gate voltages ranging from -0.3 to 0.4 V in steps of 0.1 V, for device (c).

Fig. 5. Left: turn-on characteristics at $V_{DS} = 0.1$ V for nanowire FETs fabricated with a GaAs/Al$_{0.45}$Ga$_{0.55}$As superlattice; $a = b = 1$ nm, 10 barriers (circles); $a = b = 2$ nm, 10 barriers (squares); $a = b = 2$ nm, 7 barriers (diamonds). Right: transmission probability at $V_{DS} = 0.1$ V and $V_{GS} = 0.4$ V.

differ for the gate length, it turns out that the characteristic of device (b) is degraded by source-to-drain direct tunneling. From this numerical experiment we may conclude that: (i) the SL wells must be heavily doped and, (ii) the gate length cannot be scaled much below 20 nm.

Figure 4 shows the transmission probability in the first subband of device (c) for gate voltages ranging from -0.3 to 0.4 V in steps of 0.1 V, (bottom right). The corresponding subband profiles are shown in the upper part of the figure. With increasing gate voltages, the barrier height in the channel decreases. The figure shows that there are two regions where the transmission probability is close to one, which correspond to the minibands given by the Krönig-Penney model (bottom left). Therefore, this very simple model provides surprisingly-good results, despite the inherent assumption of an infinite number of spatial periods. It may be worth pointing out that the image of the miniband profiles in fig. 4 is a zoom of fig. 1 (bottom left) shifted down in energy by -0.5 eV for consistency with the zero reference of the subband profile in the upper part of the same figure.

Next, an investigation was carried out on the properties of an SL structure made by GaAs/Al$_{0.45}$Ga$_{0.55}$As, i.e. two well-known materials commonly used for the fabrication of heterostructure devices such as HEMTs and lasers. In our simulations, GaAs has been used for the source, SL well, channel and drain regions, while Al$_{0.45}$Ga$_{0.55}$As has been used for the SL barrier regions. The most relevant parameters of these materials are reported in Table I. The doping density in the source and drain regions has been halved with respect to the previous analysis, i.e. $N_D = 10^{20}$ cm^{-3} and that in the wells is assumed to be $N_W = 10^{19}$ cm^{-3}.

Figure 5 (left) shows the turn-on characteristics at $V_{DS} = 0.1$ V for an SL-FET with $L_G = 20$ nm. For $a = b = 1$ nm, an

SS greater than 60 mV/dec is found since no minibands are generated in the SL. This is due to the small effective mass of the SL materials, which makes the transmission probability close to one at every energy, as can be seen on the right of the same figure. Instead, by compensating the small effective mass with larger barriers, i.e. $a = b = 2$ nm, we find a first miniband with a thin energy extension of about 0.2 eV, and an energy gap of about 0.5 eV between the first two minibands. With these parameters, the SS equals 26 mV/dec. We also verified that the subthreshold slope is not heavily degraded if the number of barriers is cut down to 7, despite the higher transmission probability within the miniband gap and the smoother transition from the passband to the stopband. However, due to the sensitivity of the stopband attenuation, the number of barriers cannot be reduced much below 7, as seen in figure 5 (right). The formation of the two minibands when $a = b = 2$ nm is clearly observed by plotting the local density of states at a fixed bias condition (see figure 6). Resonant quasi-bound states form in the superlattice structure for energies below the potential energy of the barriers, while a forbidden region is clearly visible within the SL from about 0.15 eV to 0.55 eV. The upper part of fig. 7 shows the transit time of the carriers across the SL-FET vs. the off-current for two situations, namely: $a = b = 1$ nm (dots) and $a = b = 2$ nm (squares). It is shown that the latter case exhibits a superior performance down to very small off currents, with a transit time ranging between 1 and 2 ps as the off-current varies

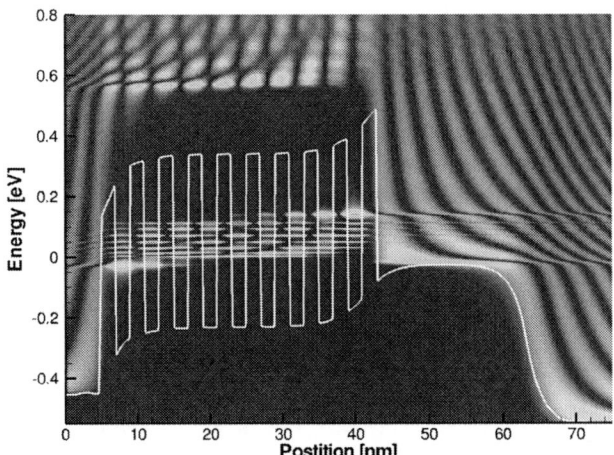

Fig. 6. Local density of states of the first subband at $V_{DS} = 0.1$ V and $V_{GS} = 0.2$ V for the GaAs/Al$_{0.45}$Ga$_{0.55}$As nanowire FET with $a = b = 2$ nm.

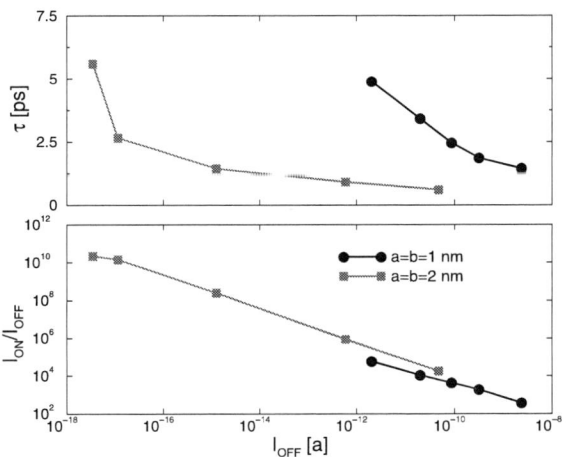

Fig. 7. Transit time τ (top) and I_{ON} over I_{OFF} (bottom) vs. OFF-current for a fixed sliding window of $\Delta V_{GS} = 0.4$ V at $V_{DS} = 0.4$ V. Circles: $a = b = 1$ nm; squares: $a = b = 2$ nm.

between 10^{-10} and 10^{-16} A. The lower part of the same figure shows the ratio $R = I_{ON}/I_{OFF}$ against I_{OFF}, under the assumption of $V_{DD} = 0.4$ V. The ratio $R \propto I_{OFF}^{-0.9}$.

Finally, fig. 8 plots the FET output characteristics, showing that this device can operate with a supply voltage of 0.4 V, at which it would exhibit an on-current $I_{ON} = 0.8$ μA. Assuming that these nanowires can be grown over a square array, with a gross area of 100 nm^2/wire, it would be possible to achieve a global current of 8 mA/μm^2. Also, the drain conductance at zero V_{DS} is $g_d \simeq 6.6$ μA/V. A parallel connection of 100 nanowires, with an area occupancy of about 0.01 μm^2 would achieve a drain conductance of 0.66 mA/V, which largely ensures rail to rail logic switching.

IV. CONCLUSIONS

In this work we investigate the potential of vertical nanowire SL-FETs with the aim to achieve very steep turn-on charac-

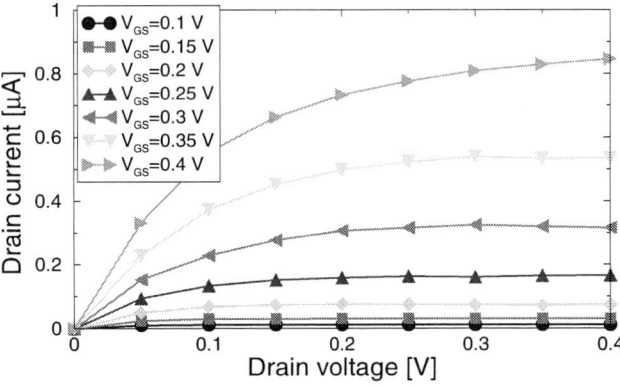

Fig. 8. Output characteristic of the SL-FET with $a = b = 2$ nm.

teristics for low-voltage and low-power logic operation. The investigation was carried out by setting up a Schrödinger-Poisson solver which can accommodate heterostructures with as many layers and materials as appropriate for device optimization. This study leads to the conclusion that an SS = 26 mV/dec could be achieved with a structure featuring seven layers of GaAs/AlGaAs having a thickness of 2 nm per layer. This device can operate at a supply voltage $V_{DD} = 0.4$ V with an estimated $I_{ON} = 8$ mA/μm^2 and is not affected by low drain conductance at zero V_{DS}, which would prevent rail-to-rail logic switching.

ACKNOWLEDGMENTS

This work has been partially supported by the EU Grant No. 257267 (STEEPER) via the IUNET Consortium.

REFERENCES

[1] K. Gopalakrishna et al., *International Electron Devices Meeting (IEDM-2002), Technical Digest*, 2002, pp. 289–292.
[2] ——, *IEEE Trans. on Electron Devices*, vol. 52, pp. 69–76, 2005.
[3] H. Kam et al., *International Electron Devices Meeting (IEDM-2005), Technical Digest*, 2005, pp. 477–480.
[4] N. Abelé et al., *International Electron Devices Meeting (IEDM-2005), Technical Digest*, 2005, pp. 1075–1077.
[5] S. Salahuddin and S. Datta, *Nano Letts.*, vol. 8, no. 2, pp. 405–410, July 2008.
[6] ——, *International Electron Devices Meeting, 2008 – Technical Digest*, 15-17 Dec. 2008, pp. 693–696.
[7] W. M. Reddick and G. A. J. Amaratunga, *Applied Physics Lett.*, vol. 67, no. 4, pp. 494–496, 1995.
[8] K. Boucart and A. M. Ionescu, *Proc. of the 34th European Solid-State Device Research Conference (ESSDERC-2004)*, 2002, pp. 383–386.
[9] W. Y. Choi et al., *IEEE Electron Device Letts.*, vol. 28, no. 8, pp. 743–745, August 2007.
[10] F. Mayer et al., *International Electron Devices Meeting, 2008 – Technical Digest*, 15-17 Dec. 2008, pp. 1–5.
[11] T. Krishnamohan et al., *International Electron Devices Meeting, 2008 – Technical Digest*, 15-17 Dec. 2008, pp. 947–949.
[12] E. Gnani et al., *International Conference on Simulation of Semiconductor Processes and Devices (SISPAD)*, accepted for publication, 2010.
[13] M. Bjoerk et al., *United States Patent Application Publication*, no. US 2009/0200540 A1, 13 August 2009.
[14] S. Adachi et al., *Properties of Group-IV, III-V and II-VI Semiconductors*. Wiley, 1950.

Modeling Impact of Electric Field and Strain on the Leakage of Embedded SiGe Source/Drain Junctions

A. Luque Rodríguez, J.A. Jiménez Tejada, S.Rodríguez-Bolivar

Departamento de Electrónica y Tecnología de los Computadores. Facultad de Ciencias. Universidad de Granada. 18071 Granada, Spain.
abrahamluque@ugr.es

M. Bargallo González[1], G. Eneman[1,2], C. Claeys[1] and E.Simoen

Imec, Kapeldreef 75, 3001 Leuven, Belgium.
[1]Also ESAT-INSYS, K.U. Leuven, Belgium.
[2]AlsoFWO-Vlaanderen, Belgium.

Abstract— **A study of the leakage current in strained p⁺n Si₁₋ₓGeₓ/Si hetero-junctions is presented. The reduction in the band gap, induced by stress forces, and the doping level at the hetero-interface, due to the use of halo implantations, are varied by changing the recess depth. A comparison between simulation results and experimental data is presented to analyze the validity of the models used in this work.**

I. INTRODUCTION

Currently, the study of hetero-junctions is an important topic, since the use of different materials within a single device is commonly considered in order to improve the performance of new devices. The use of different materials is also required in strain engineering which provides a performance booster by means of the stress magnitude. Compressive uniaxial stress in silicon is known to enhance the hole mobility [1]. To induce significant stress in the Si channel, strain engineering using $Si_{1-x}Ge_x$ source/drain (S/D) junctions has been implemented in MOSFET structures [2], (Fig. 1a). The magnitude of the stress can be controlled in the $Si_{1-x}Ge_x$ S/D junctions by the germanium concentration.

Besides the significant enhancement of the transistor's drive current, there are some undesired effects, such as an increased leakage current, resulting from the presence of defects, and from changes in the conduction and valence band structure due to the presence of $Si_{1-x}Ge_x$, and consequently changes in conductive properties of the carriers. This leakage can be severely aggravated by the high electric field at the junction [3], caused by halo implants [4] that are needed to keep short channel effects under control. This field, in turn, enhances the tunnel injection and hence, increases the off-state leakage current. Finding a compromise between a minimum leakage current and a high performance is therefore necessary.

This work presents a study of the dependence of the leakage current on the recess depth from the top of the Si n-substrate where, subsequently, p-type S/D SiGe contacts are grown. The Ge content in the SiGe layers is kept fixed at 20%.

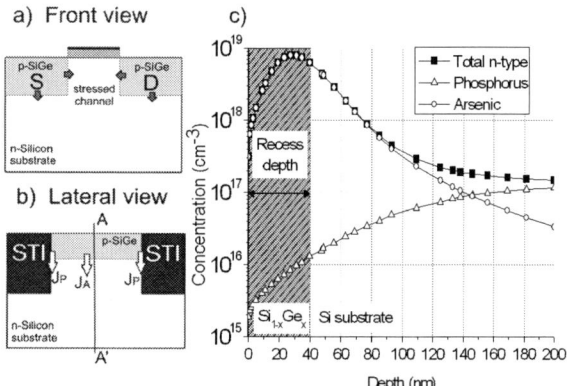

Fig. 1. a) Schematic of a MOSFET with stressed channel (front view). The SiGe S/D contacts compress the channel. b) Detail of one contact (lateral view) and identification of the two leakage current components: areal and peripheral. c) Simulated n-well profile along the line AA' for a deep Phosphorus and a shallow Arsenic implantation, and the total donor concentration in the n-well. The grey zone is the recessed layer latterly refilled with SiGe.

This work shows that a change in the recess depth produces that the stress levels change in both sides of the hetero-junction. The thickness of the SiGe layer has a direct relation with stress level in the Si-substrate region close to the hetero-junction. The thicker the $Si_{1-x}Ge_x$ layer the higher the stress level. The opposite behavior is found inside the SiGe layer. In this region, higher stress levels are achieved with thinner layers.

The recess depth also controls the maximum value of the n-well concentration at the hetero-interface, as highly doped drain (HDD) implantations are not used. This change in the doping level modifies the intrinsic electric field at the hetero-junction. Thereby, this parameter modulates the field-assisted tunneling; either band to band tunneling (BTBT) or trap assisted tunneling (TAT), enhancing the leakage current through the device.

The research was supported financially by Ministerio de Educación y Ciencia and FEDER within the framework of research Project TEC2007-66812/MIC

The leakage current generated by these two tunnel mechanisms is added to the one produced by diffusion or Shockley-Read-Hall (SRH) mechanisms. Experimental ways to detect the dominant mechanism in the leakage current can be found in the literature [5-6]. One of the methods is by determining the activation energies (E_A) from a temperature-dependence study of current-voltage (I-V) curves [5-6]. Around room temperature, the main leakage mechanisms are SRH, TAT, and BTBT, while for devices working at sufficiently high temperatures the effect of the diffusion component [7] is becoming dominant. These temperature-dependence studies can also determine the ranges of reverse-biased voltages where one of the previous mechanisms is dominant [8]. One of these authors' main conclusions is that BTBT is the mechanism which produces a higher level in the leakage currents [8]. Thus, the technological control of this mechanism can be a first step forward in achieving a better design of these structures. However, adequate modeling of BTBT and the other leakage mechanisms is necessary in first place.

A simulation study of the effect of the recess depth in the $Si_{0.8}Ge_{0.2}$ layer on the leakage current in a strained $Si_{0.8}Ge_{0.2}$/Si hetero-junction is presented. Parameters usually employed in homo-junctions are misunderstood or not valid when modeling hetero-junctions. Finding the proper parameters to achieve the best fit between the model and experimental measurements is the main objective of this work.

II. STRUCTURE AND SIMULATION PROCEDURE

This work is focused on the analysis of the leakage current flowing through the S/D p-$Si_{0.8}Ge_{0.2}$ n-Si hetero-junctions (Fig. 1a). In order to isolate this undesired effect from the rest of the transistor and other effects, e.g. like gate induced drain leakage, single hetero-junctions have been fabricated (Fig. 1b). The structures under study are in-situ B-doped ($\sim 10^{20}$cm^{-3}) $Si_{0.8}Ge_{0.2}$ layers, epitaxially grown in the active regions of recessed Shallow-Trench Isolation-patterned (STI) (100) Czochralski (Cz) silicon wafers. The Ge concentration of the SiGe layers is 20%, and the recess depth varies between 20 and 70 nm, as specified in Ref. 8. Figure 1c shows the simulated doping profile in the n-well substrate, along the AA' line in Fig. 1b, resulting from two implantations: the Phosphorus implantation and the Arsenic halo implantation. After these implants, the top silicon is recessed and the trench is refilled by Selective Epitaxial Growth (SEG) of in-situ highly boron-doped $Si_{0.8}Ge_{0.2}$ epitaxial layers, shown in dark in Fig. 1c. The process was concluded with silicidation of the contact and Cu back-end.

TSUPREM4 2D [9] process-simulator is used to simulate the doping profiles and stress components in the structure. These results are then exported to the MEDICI device simulator [10]. The stress components are used as input parameters for the calculation of the stress-induced band gap narrowing. MEDICI computes the conduction band edge shifts using the deformation potential theory [11], and the valence band edge shifts using the 6×6 $k \cdot p$ perturbation theory [12] in the silicon substrate. The $Si_{1-x}Ge_x$ region is treated with a model that relates the shifts in the bands with the Ge content [13]. The leakage current in the hetero-

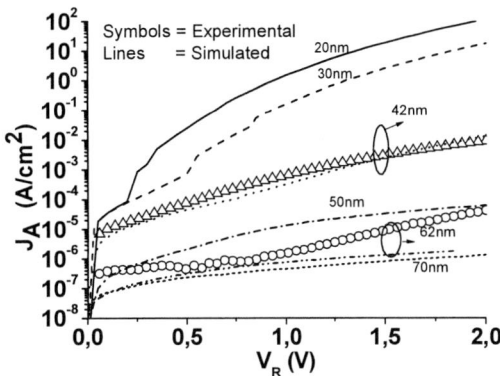

Fig. 2. Comparison between experimental (symbols) and simulated results (lines) for the areal-current density as a function of the applied reverse-voltage. $\tau_n = \tau_p = 6.3 \times 10^{-8}$s, $m_R = 0.25$ (a=1.11×10^{18}, b=28.48×10^6, and c=2.5).

junction is evaluated taking into account different models detailed in the literature for generation-recombination mechanisms: conventional SRH recombination via traps [14,15], TAT [16] and BTBT [17].

To simulate the typical SRH and TAT mechanisms, a bulk trap located at midgap with equal values for the hole and electron lifetimes ($\tau_n = \tau_p = 6.3 \times 10^{-8}$s) is used as possible origin of this leakage current. It is well known that the most effective recombination center is located at midgap, provided the hole and electron lifetimes are the same $\tau_n = \tau_p$ [18]. However, this is not necessarily the case when τ_n and τ_p are different: the most effective recombination center would be displaced from the midgap [19-21]. Thus, similar conclusions can be reached for other centers with different lifetimes $\tau_n \neq \tau_p$. The model used for the TAT mechanism requires one extra physical parameter, the effective mass of the carriers m^*. As a starting point in this work, a generally accepted value of relative mass $m_R = 0.25$ ($m^* = m_R \cdot m_0$) is used [16], where m_0 is the electron rest mass.

In order to simulate the BTBT, the compact expression based on Kane's model can be used as starting point. It can be written as [22]:

$$U_{BTBT} = \frac{q^2 \sqrt{m^*}}{18 \cdot \pi \cdot \hbar^2} \frac{E^2}{E_G^{1/2}} \cdot \exp\left(-\frac{\pi \cdot \sqrt{m^*}}{2 \cdot q \cdot \hbar} \frac{E_G^{3/2}}{E}\right) \qquad (1)$$

with q the charge of the electron, $\hbar$ the reduced constant of Planck, E_G the width of the band gap, and E the value of the local electric field. Frequently, equation (1) is presented in a reduced form that simplifies its use in simulators. It is expressed as a function of three constants:

$$U_{BTBT} = a \frac{E^c}{E_G^{1/2}} \cdot \exp\left(-b \frac{E_G^{3/2}}{E}\right) \qquad (2)$$

Although a similar equation was proposed by Hurkx [16], where the band gap dependence is included in a as well as in b, the choice of (2) is due to its explicit dependence on E_G. In hetero-junctions, the band gap is not necessarily constant but can vary according to the position, since a stress-induced band gap narrowing effect is involved.

Constant c in (2) has different values depending on which transitions are involved in the process [16]. In materials with a direct gap, where the tunneling is through direct transitions, c = 2, whereas in an indirect gap semiconductor $c = 5/2$. In the latter case, an electron-phonon interaction is required for the indirect transitions. For direct tunneling constant a depends only on the effective mass, while in the indirect cases an extra factor is also involved owing to phonon-assisted tunneling.

The following default values are employed in commercial simulators [10]: $a = 3.5 \times 10^{21}$, $b = 22.5 \times 10^6$, and $c = 2$. Comparing (1) and (2), these values can be reproduced by assuming $m_R = 0.156$ and considering a direct semiconductor [17]. However, in order to be consistent with the values of the parameters employed to model the other leakage (i.e. TAT), $m_R = 0.25$ should be considered in both BTBT and TAT models. Looking for a more realistic model for these structures, a value $c = 5/2$ is considered as silicon has an indirect band gap. In that case, the theory predicts that a has to be ~ 4×10^3 lower than the direct transitions [24]. That fact has also been confirmed by experimental measurements [23].

III. RESULTS

In order to validate the simulation study of the preceding section, the leakage current was measured for devices with two different recess depths. The SiGe surfaces are notoriously difficult to passivate. This fact is more pronounced for SiGe-STI interfaces, where surface defects present in the peripheral regions of the S/D result in the so-called peripheral leakage current J_P. This leakage current has to be added to the areal leakage current J_A generated in the depletion layer of the Si and SiGe regions. In order to compare the experimental results with the simulations, the area leakage current was extracted from the total current [8]. This extraction procedure, reported elsewhere [7, 25, 26] consists of measuring the leakage current in junctions with different area to perimeter ratios. Figure 2 shows the areal current density as a function of the applied voltage in samples with different recess depth. It shows the comparison of experimental area-current density (J_A), taken from ref. 8 (open symbols), with the simulation results (solid lines). The impact of the recess depth on the leakage current is remarkably significant. Changing the recess depth from 20 to 70 nm leads to a difference in the leakage current of almost ten decades. This increment in the current can be ascribed to a change in the mechanism (SRH, TAT, or BTBT) dominating the leakage currents. In the etching process, while the doping level of the p-side of the hetero-junction is always kept constant, the maximum value of the n-well changes as a function of the recess depth. Consequently, the intrinsic electric field along the hetero-junctions also varies with the recess depth. In that way, depending on the range of electric field present in the hetero-junction one of the three aforementioned mechanisms plays a prominent role.

For deep recesses a roughly electric-field independent leakage current is observed (70 and 62 nm cases). The low electric fields make that the SRH mechanism controls the leakage currents. As the recess depth is shallower the field-assisted mechanisms become significant. The highest levels of leakages are achieved in the shallowest recess cases, where the BTBT is the dominant mechanism.

The simulated curves seem to be reasonably in agreement with the experimental curves. However, a slight difference is shown at high reverse voltages in the case of 62 nm. In order to give a physical explanation for this discrepancy, the areal-current density as a function of the maximum electric field has been plotted in Fig. 3a. for both experimental (42 nm, circle symbols; and 62 nm, triangle symbols) and simulated cases (lines). The solid grey line depicts the total current which is composed of the three following components: the SRH component (dot-dashed line) which provides a uniform contribution independent of the electric field, the TAT component (dotted line) responsible of the leakage current at medium voltages, and the BTBT component (dashed line) which produces the maximum level of the leakage. However, the experimental data are supposed to lie on one unique line, but this is not the case. Deep and shallow recess cases show a different tendency. Therefore, in order to fit both cases different parameter for TAT/BTBT models have to been considered. The shallow recess depth case, 62 nm, shows a different slope to that of the 42 nm case. Thus, a different value of the m_R should be considered in order to change the slope for TAT and BTBT. This change of the effective mass with the recess depth can be incorporated in (2) giving rise to the following expression for the BTBT mechanism:

$$U_{BTBT} = \frac{3.5 \times 10^{21}}{4 \times 10^3} \sqrt{\frac{m_R(recess)}{0.156}} \frac{E^{2.5}}{E_G^{1/2}}$$
$$\times \exp\left(-22.5 \times 10^6 \sqrt{\frac{m_R(recess)}{0.156}} \frac{E_G^{3/2}}{E}\right) \quad (3)$$

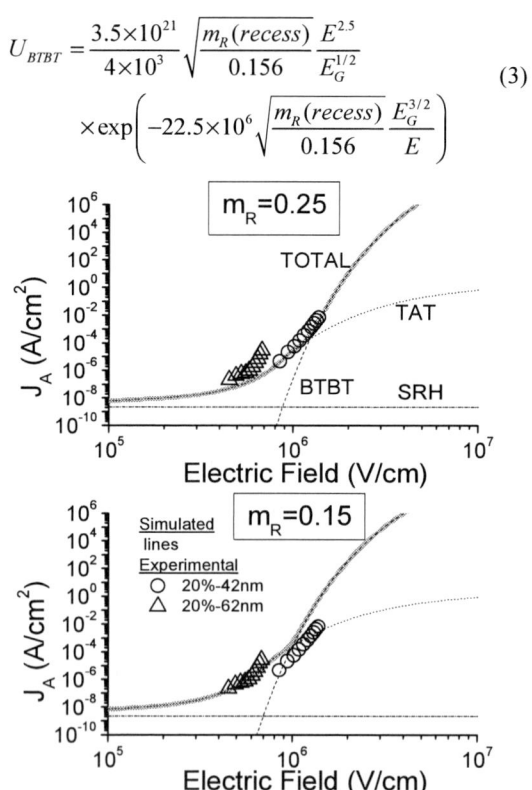

Fig. 3. Areal-current vs electric field obtained from experimental measurements (symbols) and simulated results with its three components (SRH, TAT, and BTBT) for two different effective masses.

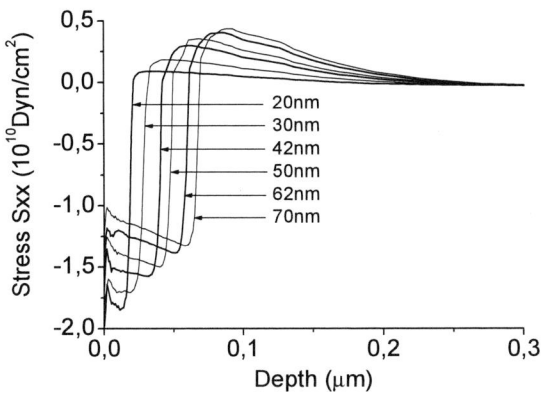

Fig. 4. Stress component for several $Si_{0.8}Ge_{0.2}$/Si samples with different recess depths (line AA' Fig. 1b). Sxx component is parallel to the hetero-interface.

In this model, the m_R and lifetimes are the only degrees of freedom for fitting the curves. While the SRH component only depends on the lifetimes and the BTBT depends on the m_R, the TAT component has a dependence on both. A change in the lifetime only produces a vertical variation in Fig. 3 for the SRH and TAT components. Thus, a value of $m_R = 0.15$ is required in order to obtain a good fitting between experimental and simulated curves in the 62 nm recess depth case, Fig. 3b. This lower value may be ascribed to one of the effects that the stress produces in the hetero-junctions.

Figure 4 shows the S_{xx} stress components along the hetero-junctions for different recess depths. One can see that as the $Si_{1-x}Ge_x$ layer is thicker the stress induced in the Si-substrate is higher. While the aim of using S/D $Si_{1-x}Ge_x$ contacts is to induce compressive uniaxial strain in the channel of the MOS transistors, to lower the hole effective mass, a biaxial tensile strain is produced under the $Si_{1-x}Ge_x$ layer. It is well known that biaxial tensile strain produces a lighter electron effective mass [27]. The combination of both effects may results in the decrease in m_R.

In summary, an expression for BTBT based on Kane's model is presented for hetero-junctions with indirect gap. Explicit band gap dependence is considered in order to consider the band gap narrowing induced by stress. A m_R dependent on the recess depth is considered, achieving a good agreement with experimental results. Further work with different Ge contents is required in order to reinforce the m_R dependence on the stress.

REFERENCES

[1] P. Verheyen, G. Eneman, R. Rooyackers, R. Loo, L. Eeckhout, D. Rondas, et al." Demonstration of recessed SiGe S/D and inserted metal gate on HfO₂ for high performance pFETs", IEDM Tech. Dig., 886 (2005).

[2] C. Claeys, E. Simoen, S. Put, G. Giusi, and F. Crupi," Impact strain engineering on gate stack quality and reliability" Solid-State Electronics 52, 1115 (2008).

[3] G. Eneman, M. Wiot, A. Brugere, O. Sicart I Casain, S. Sonde, D.P. Brunco, et al., "Impact of donor concentration, electric field, and temperature effects on the leakage current in Germanium p⁺/n junctions", IEEE Trans. Electron Dev., 55, 2287 (2008).

[4] Y. Taur, C.H. Wann, and D.J. Frank, "25 nm CMOS design considerations", IEDM Tech. Dig., 789 (1998).

[5] A. Poyai, E. Simoen, C. Claeys, A. Czerwinski, and E. Gaubas, "Improved extraction of the activation energy of the leakage current in silicon p–n junction diodes", Appl. Phys. Lett., 78, 1997 (2001).

[6] A. Czerwinski, E. Simoen, A. Poyai, and C. Claeys, "Activation energy analysis as a tool for extraction and investigation of p–n junction leakage current components", J. Appl. Phys., 94, 1218 (2003).

[7] C. Claeys, M. Bargallo Gonzalez, G. Eneman, P. Verheyen, H. Bender, R. Schreutelkamp, et al., "Leakage current control in recessed SiGe source/drain junctions", J. Electrochem. Soc. 154, H814 (2007).

[8] M. Bargallo Gonzalez, E. Simoen, B. Vissouvanadian, G. Eneman, P. Verheyen, R. Loo, and C. Claeys, "Electric field dependence of trap-assisted-tunneling current in strained SiGe source/drain junctions", Appl. Phys. Lett., 94, 233507-1 (2009).

[9] Taurus TSUPREM, Taurus TSUPREM User Guide, Synopsys Inc., Mountain View, CA, Mar. 2007. Version Z-2007.03.

[10] Taurus MEDICI, Medici User Guide, Synopsys Inc., Mountain View, CA, Mar. 2007. Version Z-2007.03.

[11] G.L. Bir and G. E. Pikus, "Symmetry and strain-induced effects in semiconductors", Wiley, New York (1974).

[12] T. Manku and A. Nathan, "Valence energy-band structure for strained group-IV semiconductors", J. Appl. Phys., 73, 1205 (1993).

[13] SS. Iyer, GL. Patton, JMC. Stork, BS. Meyerson, and DL. Harame, "Heterojunction bipolar transistors using Si-Ge alloys", IEEE Trans. Electron Dev., 36, 2043 (1989).

[14] R.N. Hall, "Electron-hole recombination in Germanium", Phys. Rev., 87, 387 (1952).

[15] W. Shockley and W. T. Read," Statistics of the Recombinations of Holes and Electrons", Phys. Rev., 87, 835 (1952).

[16] G.A.M. Hurkx, D.B.M. Klaassen and M.P.G. Knuvers, "A new recombination model for device simulation including tunneling", IEEE Trans. Electron Dev., 39, 2090 (1992).

[17] E. O. Kane, "Zener tunneling in semiconductor", J. Phys. Chem. Solids, 12, 181 (1959).

[18] S.M. Sze, Physics of Semiconductor Devices Wiley Interscience, New York, (1981).

[19] J.A. Jiménez Tejada, A. Godoy, J.E. Carceller, and J.A. López Villanueva, "Effects of oxygen related defects on the electrical and thermal behavior of a n⁺-p junction", J. Appl. Phys., 95, 561 (2004).

[20] J.A. Jiménez Tejada, P. Lara Bullejos, J.A. López Villanueva, F.M. Gómez-Campos, S. Rodríguez-Bolivar, and M.J. Deen, "Determination of the concentration of recombination centers in thin asymmetrical p-n junctions from capacitance transient spectroscopy", Appl. Phys. Lett., 89, 112107 (2006).

[21] D. K. Schroder, "Carrier lifetimes in silicon", IEEE Trans. Electron Dev. 44, 160 (1997).

[22] M. Takayanagi, and S. Iwabuchi, "Theory of band-to-band tunneling under nonuniform electric fields for subbreakdon leakage currents", IEEE Trans. Electron Dev., 38, 6 (1991).

[23] G.A.M. Hurkx, "On the modelling of tunnelling currents in reverse-biased p-n junctions", Solid-State Electronics, 32, 8, 665 (1989).

[24] E.O. Kane, "Theory of tunneling", J. Appl. Phys., 32, 1, 83 (1961).

[25] A. Czerwinski, E, Simoen, C. Claeys, K. Klima, D. Tomaszewski, J. Dogki, and J. Katcki, "Optimized diode analysis of electrical silicon substrate properties", J. Electrochem. Soc., 145, 2107 (1998).

[26] A. Poyai, E. Simoen, C. Claeys, and A. Czerwinski, "Silicon substrate effects on the current–voltage characteristics of advanced p–n junction diodes", Mater. Sci. Eng., B 73, 191 (2000).

[27] Y. Sun, S. E. Thompson, and T. Nishida, "Physics of strain effects in semiconductors and metal-oxide-semiconductor field-effect transistors", J. Appl. Phys., 101, 104503 (2007).

Modeling Temperature Dependency (6 - 400K) of the Leakage Current Through the SiO₂/High-K Stacks

L. Vandelli, A. Padovani, L. Larcher

DISMI (DIpartimento di Scienze e Metodi dell'Ingegneria)
Università di Modena e Reggio Emilia, Reggio Emilia, Italy
83688@studenti.unimore.it

R.G. Southwick III[1], W.B. Knowlton[1,2]

Dept. of Electrical and Computer Engineering[1], Dept. of
Materials Science and Engineering[2]
Boise State University, Boise, Idaho, USA

G. Bersuker

SEMATECH, 2706 Montopolis Dr.,Austin, Texas, USA

Abstract—**We investigate the mechanism of the gate leakage current in the Si/SiO₂/HfO₂/TiN stacks in a wide temperature range (6 – 400 K) by simulating the electron transport using a multi-phonon trap assisted tunneling model. Good agreement between simulations and measurements allows indentifying the dominant physical processes controlling the temperature dependency of the gate current. In depletion/weak inversion, the current is limited by the supply of carrier. In strong inversion, the electron-phonon interaction is found to be the dominant factor determining the current voltage and temperature dependencies. These simulations allowed to extract important defect parameters, e.g. the trap relaxation energy and phonon effective energy, which defines the defect atomic structure.**

I. INTRODUCTION

Hafnium oxide (HfO₂) has emerged as the high-K gate dielectric solution for ultra-scaled MOSFETs. Despite the significant use of this material, a deep knowledge of the physical mechanisms governing the conduction through the high-k dielectric stacks and of the nature of the contributing defects is still lacking. Several papers published in the literature have investigated the temperature (T) dependence of the gate current in order to gain insight in the carrier transport mechanisms [1],[2]. Only one of these studies [3] examined cryogenic T, demonstrating that the Poole-Frenkel mechanism cannot describe the carrier transport behavior over a broad range of T and voltages. Hence, a complete explanation supported by physical simulations of the observed T dependence is still missing. In order to understand the physical mechanisms controlling the electron transport across the HfO₂, we simulated the gate currents in the HfO₂ stacks measured in a very wide T range (6 - 400K) [2] using the Multi-phonon Trap Assisted Tunneling (MTAT) model [4], [5].

II. EXPERIMENTAL RESULTS

Measurements were performed on large area (30x30μm²) MOSFETs. The gate stack of each device consists of a titanium nitride (TiN) metal gate and a dielectric bi-layer of 3nm or 5nm of ALD HfO₂ on a 1.1nm chemically grown SiO₂ interfacial layer (IL). A control wafer composed of 2nm SiO₂ is used as a base to compare the HfO₂ samples. A Janis custom built variable range probe station (5.6-450K) with actively cooled Kelvin probes combined with a Keithley 4200SCS with remote pre-amps were used in the measurements. The

gate current versus gate voltage ($J_{gate} - V_{gate}$) curves measured over a T range from 6K to 400K on the different stacks are shown in Fig. 1. Experimental results indicate the presence of two regimes [2]: i) in the carrier-limited regime (e.g. in depletion-weak inversion conditions, $V_{gate} < V_{TH}$) the current is limited by the supply of carriers at the injection interface; ii) in the conduction-limited regime (e.g. in strong inversion conditions, $V_{gate} > V_{TH}$) the current dependence is limited by the conduction mechanism. Interestingly, the T dependence of the gate current density J_{gate} is larger in HfO₂/SiO₂ stacks (it increases with the HfO₂ thickness) compared pure SiO₂ stacks. In order to fully understand the observed T dependence, we performed current simulations through our MTAT model.

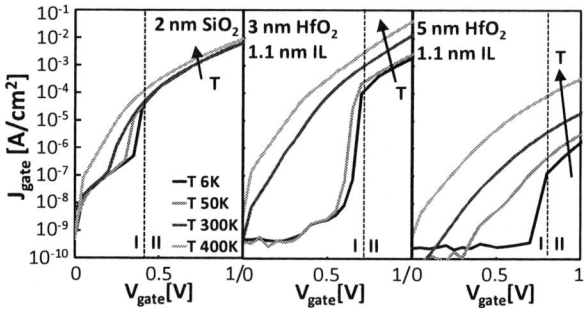

Figure 1. J_{gate} vs V_{gate} measured for T ranging from 6K to 400K on 2nm SiO₂ nMOS and on 3 and 5nm HfO₂/SiO₂ nMOS. Carrier limited regime (I) and conduction limited regime (II) are indicated.

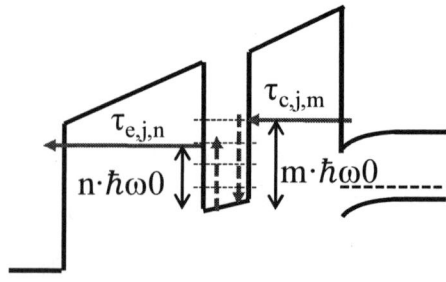

Figure 2. Multiphonon assisted tunneling mechanism on a monolayer stack: capture and emission processes in the case of a single-trap conductive path.

III. MODEL DESCRIPTION

Gate current simulations are performed using the statistical MTAT model presented in [4], extended here to model T effects on a SiO$_2$/high-K dielectric stack. The electric field across the stack is calculated taking into account charge quantization effects at the Si/SiO$_2$ interface and the freeze-out of carriers at very low T [6]. The gate current includes direct tunneling (DT) and trap assisted tunneling (TAT) components. The DT current is calculated through the semi-classical approach [7], calculating the tunnel probability through the WKB method. Defects are randomly placed within SiO$_2$ and HfO$_2$ layers according to their spatial and energetic distributions. The TAT current is calculated by taking into account both single-trap and multi-trap conductive paths contributions [4]. The current driven by a conductive path I_{cp} is determined by the slowest trap of the path through

$$I_{cp} = q/(\tau_{c,\max} + \tau_{e,\max}); \qquad (1)$$

where q is the electron charge; $\tau_{c,\max}$ and $\tau_{e,\max}$ are the time constants associated to the electron capture and emission by and from the slowest trap, respectively. The model considers electron coupling to oxide phonons [8], whose effective energy is $\hbar\omega0$. The calculation of the time constants associated to the j-th trap $\tau_{c,j}$ and $\tau_{e,j}$ is performed by summing over the discretized phonon energy (indexes m and n) the time constants $\tau_{c,j,m}$ and $\tau_{e,j,n}$:

$$\tau_{c,j}^{-1} = \sum_m (\tau_{c,j,m})^{-1}, \quad (2) \qquad \tau_{e,j}^{-1} = \sum_n (\tau_{e,j,n})^{-1} . \quad (3)$$

As sketched in Fig. 2 in the case of a monolayer stack, $\tau_{c,j,m}$ is the time required for the electron tunneling (from the substrate or from the previous trap of the conductive path) into the trap and the release of the electron energy $m\hbar\omega0$. $\tau_{e,j,n}$ is the time required for the absorption of the energy $n\hbar\omega0$ and for the electron tunneling out of the trap (to the gate or to the next trap):

$$(\tau_{c,j,m})^{-1} = N(E_{j-1,m}) f(E_{j-1,m}) C_{j,m} P_T(E_{j-1,m}, E_j), (4)$$

$$(\tau_{e,j,n})^{-1} = N(E_{j+1,n}) Em_{j,n} P_T(E_{j,n}, E_{j+1}), \qquad (5)$$

$$E_{j,l} = E_{c,j} + l\hbar\omega0. \qquad (6)$$

$E_{c,j}$ is the conduction band edge or the j-th trap energy level; N is the density of states; f is the Fermi-Dirac occupation probability; P_T is the WKB tunnel probability; $C_{j,m}$ and $Em_{j,n}$ are the trap capture and emission rates that account for the electron-phonon interaction:

$$C_{j,m} = c_0 L(m), \quad (7) \quad Em_{j,n} = c_0 L(n)\exp\left(\frac{-nh\omega0}{kT}\right). \quad (8)$$

Here c_0 is a constant factor dependent on the electric field and on the capture cross section of the trap [4]. $L(m)$ is the multiphonon transition probability [8]:

$$L(m) = \left(\frac{f_B+1}{f_B}\right)^{\frac{m}{2}} e^{-2(2f_B+1)} I_m(2S\sqrt{f_B(f_B+1)}), \quad (9)$$

$$f_B = 1/\left(\exp\left(\frac{\hbar\omega0}{kT}\right) + 1\right). \qquad (10)$$

I_m is the modified Bessel function of order m; f_B is the Bose function, providing the phonon occupation number; S is the Huang-Rhys factor, which is the most important parameter describing the electron coupling with lattice phonons [9]. The role played by S will be explained in Section IV. Notably, the emission usually occurs from the ground state of the trap (n=0), since the term $\exp\left(\frac{-nh\omega0}{kT}\right)$ in (8) decreases exponentially with n.

IV. SIMULATION RESULTS AND DISCUSSION

The parameters of the traps in the SiO$_2$ interfacial layer (IL) and in HfO$_2$ considered in the simulations are reported in Table I. ρ is the density, σ_T is the capture cross section, E$_T$ is the energy of the defects counted from the dielectric conduction band edge. The spatial defect distribution is uniform for both SiO$_2$ and HfO$_2$ layers, whereas it linearly increases approaching the HfO$_2$ interface in the IL, according to frequency dependent charge pumping measurements [10].

A. 2nm SiO$_2$ stack

Fig. 3 shows the gate currents measured and simulated on the 2nm SiO$_2$ stack. Experimental data are well reproduced by simulations within the entire V$_{gate}$ and T ranges. The TAT contribution (I_{TAT}), shown by dashed lines in Fig. 3, is negligible in the considered voltage range, hence the total gate current (I_{TOT}) is dominated by the DT current (I_{DT}).

In depletion/weak inversion, the DT current is limited by the supply of carriers available for tunneling. Thus, the T and

TABLE I.   TRAP PARAMETERS USED IN THE MTAT SIMULATION

	ρ [cm^{-3}]	σ_T [m^2]	E$_T$ [eV]	hω_0 [eV]	S
SiO$_2$	Uniform (ρ=1·10^{16})	1·10^{-14}	2.4-2.8	0.06	6
IL	Non uniform [10]	1·10^{-14}	2.3-3.2	0.06	6
HfO$_2$	Uniform (ρ=4·10^{19})	2.5·10^{-14}	1.6-2.6	0.07	17

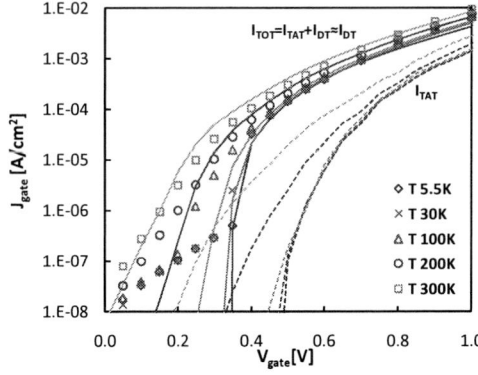

Figure 3.   Experimental (symbols) and simulated total (solid lines) and TAT (dashed lines) gate currents on the 2nm SiO$_2$ stack at different T.

voltage dependences of the current are determined by the occupation of the quantized states at the Si/SiO$_2$ interface, governed by the Fermi-Dirac statistics. As T decreases, the carrier population and the DT current exponentially reduces. Increasing V$_{gate}$ gradually populates quantized states and the device enters the strong inversion regime, as depicted by the shallower slope at higher V$_{gate}$ in Fig. 3. Since the carrier generation is a thermally-assisted process, at lower T the inversion is reached at higher gate voltages, and the I-V dependency is steeper because the Fermi-Dirac distribution approaches a step function. In strong inversion, the T dependence of the gate current is mainly due to the tunneling probability from the quantized states. As T increases the thermal energy distribution broadens allowing electrons in the inversion layer to occupy higher energy levels, with higher tunnel probabilities (due to the lower tunnel barrier), resulting in a larger DT current.

B. IL/3nm HfO$_2$ stack

As shown in Fig. 4, the model accurately reproduces the gate currents measured on the IL/3nm HfO$_2$ stack. Differing from the SiO$_2$ stack, the DT component (dashed lines), is negligible; hence the current conduction is dominated by the TAT current. Interestingly, if HfO$_2$ traps are not considered, current simulations do not change significantly (not shown here for brevity), indicating that the conduction is mainly driven by traps located in the IL. Thus, the analysis of the conduction mechanisms on this stack is suitable to extract the properties of IL defects.

According to (5) and (6), the TAT current includes two major T dependences associated with i) the carrier supply, which is described by the Fermi-Dirac occupation probability $f(E)$ and ii) the electron-phonon interaction, which is described by the multi-phonon transition probability $L(m)$. In weak inversion/depletion, because of the limited population of minority carriers, the carrier supply dependency dominates. Thus, in this regime the TAT current shows the same T and V$_{gate}$ dependences observed for the DT current. Conversely, in strong inversion the T dependency is mainly due to the electron-phonon interaction. This is the reason why the TAT current exhibits a stronger T dependence compared to the DT current, where phonons are not involved. The strength of the electron-phonon interaction is described by the Huang-Rhys factor S, which is the number of the phonons required for the dielectric lattice to re-arrange its atomic configuration around the defect to accommodate the trapped/detrapped charge. $S\hbar\omega0$ is the relaxation energy associated with these atomic displacements [8], [9]. We assumed the effective phonon energy of $\hbar\omega0 = 0.06eV$, in agreement with previous studies of the TAT current [4], [5] and with Inelastic Electron Tunneling Spectroscopy (IETS) data [11] in SiO$_2$. In order to understand the T dependence of the TAT current in strong inversion conditions, we should consider the $L(m)$ plots shown in Fig. 5 for different T and S values. $L(m)$ represents the probability of a transition assisted by m phonons. $L(m)$ reaches its maximum at $m \approx S$. Its T dependence increases with the $(m\text{-}S)$ difference, and therefore a higher S increases the $L(0)$ dependence on T. Since the electric field in HfO$_2$ is much lower than in the IL layer (due to its higher K) the electron emission toward the gate is lower than the

electron capture from the substrate ($\tau_{e,max} \gg \tau_{c,max}$), hence from (1) $I_{cp} \approx q/\tau_{e,max}$. Since, as stated in Section III, the carrier emission is dominated by the emission from the ground state, the T dependence is mainly determined by $L(0)$ and therefore by S. Fig. 6 reports the Arrhenius plot of the gate current densities measured and simulated at different voltages in a strong inversion regime. The T dependence is well reproduced by considering $S = 6$ for IL traps, consistent with

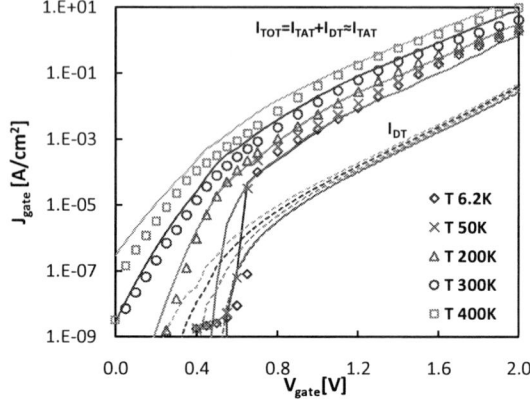

Figure 4. Experimental (symbols) and simulated total (solid lines) and DT (dashed lines) gate currents on the IL/3nm HfO$_2$ stack at different T.

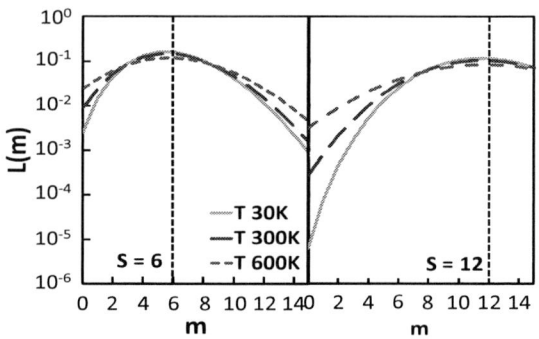

Figure 5. Multiphonon transition probability L(m) at different T for S=6 and S=12.

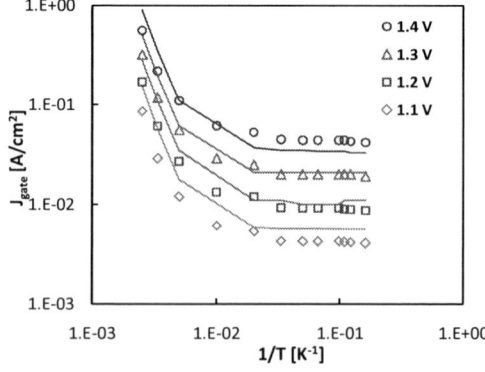

Figure 6. Arrhenius plot of the experimental (symbols) and simulated (lines) gate current densities at different voltages of an IL/3nm HfO$_2$ stack.

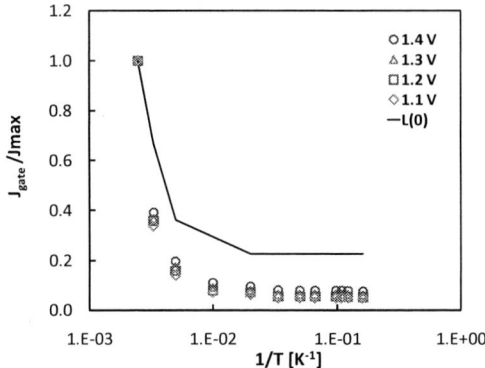

Figure 7. Arrhenius plot of the experimental currents of an IL/3nm HfO₂ stack (symbols) and of $L(0)$ (line), normalized to their value at the highest T.

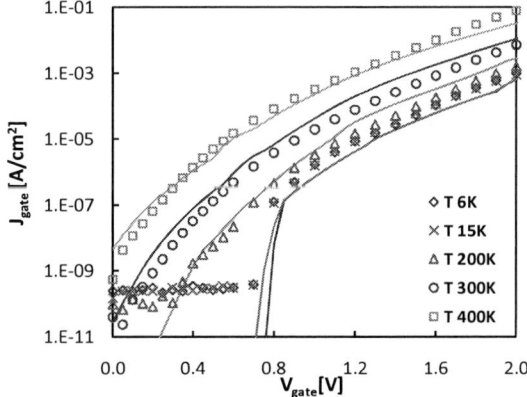

Figure 8. Experimental (symbols) and simulated (lines) gate currents on the IL/5nm HfO₂ stack at different T.

literature reports [4], [5], resulting in a relaxation energy $S\hbar\omega0 = 0.36eV$. At very low T the model correctly predicts the strongly reduced sensitivity to T of the current [2]. Fig. 7 reports the Arrhenius plot of the experimental currents and of $L(0)$ normalized to their corresponding values at the highest T (400K), showing that the experimental currents are strongly correlated to $L(0)$, as discussed above. It is worth noticing that the effect of tunnel barrier lowering at higher T, discussed for DT current, is still present, resulting in a wider variation of the experimental current than the one induced by $L(0)$.

C. IL/5nm HfO₂ stack

Fig. 8 shows the gate currents measured and simulated on the IL/5nm HfO₂ stack. The agreement between measurements and simulations performed using the same set of parameters is very good. Differently from the 3nm HfO₂ stack, these simulations do not reproduce the experimental currents if the HfO₂ traps are not considered (not shown here for brevity), thus indicating that defects in the HfO₂ bulk play a significant role in the current conduction, as can be expected in the case of a thicker HfO₂ layer. Thus, characteristics of bulk defects in the HfO₂ layer can be extracted from the thicker high-K stack. Comparing to the 3nm HfO₂ stacks, a stronger temperature dependence of the gate current is observed in this stack in strong inversion conditions. This

indicates, according to the above discussion, a stronger electron-phonon interaction in the bulk HfO₂ defects compared to the traps in the IL, thus yielding a higher S value. The effective phonon energy $\hbar\omega0$ is assumed 0.07eV for HfO₂, consistent with IETS data [11]. Considering $S = 17$ for bulk HfO₂ traps, resulting in a relaxation energy $S\hbar\omega0 =1.19eV$, reproduces very well the experimental T dependence of J_{gate}, as shown in Fig. 8.

V. Conclusions

We simulated the T dependence of the gate leakage current in SiO₂/HfO₂ nMOSFET devices using a statistical MTAT model, which reproduces accurately the experimental J_{gate} in the entire measured T range (6 – 400K). Simulations showed that the gate current in the 2nm pure SiO₂ stacks is due to the DT current, while in the SiO₂/HfO₂ stacks the TAT current component prevails. The observed strong J_{gate} dependence on T in depletion/weak-inversion is due to the limited supply of carriers. In strong inversion the T dependence is instead dominated by the electron-phonon interaction, which determines the rate of the electron transfer in the TAT process. As the HfO₂ layer is scaled, TAT is mainly driven by defects located in the IL, hence the effect of HfO₂ traps decreases. Modeling the T dependence of the gate current provides a means of estimating important parameters of both HfO₂ and IL traps, e.g. the Huang Rhys factor and the phonon effective energy, which may help to identify the defect structure.

VI. Acknowledgment

We thank J. Reed, C. Buu and R. Butler for their contributions to the low temperature measurements.

VII. References

[1] L. F. Mao, "Modeling of temperature dependence of the leakage current through a hafnium silicate gate dielectric in a MOS device", Semicond. Sci. and Tech., vol. 22, 2007 pp. 1203-1208.

[2] R. Southwick et al., "Temperature (5.6-300K) dependence comparison of carrier transport mechanisms in HfO₂/SiO₂ and SiO₂ MOS Gate Stacks", IIRW 2008, pp.48–54.

[3] R. Southwick et al., "Limitations of Poole-Frenkel Conduction in Bilayer HfO₂/SiO₂ MOS Devices", IEEE Trans. Device Mater. Reliab., in press.

[4] L. Larcher, "Simulation of leakage currents in MOS and Flash memory devices with a new multiphonon trap-assisted-tunneling model", IEEE Trans. on Elec. Dev., vol. 50 (5), 2003, pp. 1246-1253.

[5] M. R. Herrmann, M. Ciappa, A. Schenk, "A new model for long term charge loss in EPROM's", SSDM 1994, pp. 494-496.

[6] S. M. Sze. "Physics of semiconductor devices". Wiley-Interscience, 1981.

[7] N. Yang, W.K. Henson, J. Hauser JR, Wortman JJ "Modeling study of ultrathin gate oxides using direct tunneling current and capacitance–voltage measurements in MOS devices", IEEE Trans. on Elec. Dev., vol. 46(7), 1999, pp.1464-1471.

[8] C.H. Henry, D. V. Lang, "Non radiative capture and recombination by multiphonon emission in GaAs and GaP", Phys. Rev. B vol. 15(2), 1977, pp. 989-1016.

[9] K. Huang, A. Rhys, "Theory of light absorption and non-radiative transition in F-centres" Proc. R. Soc. London, vol. 204A, 1950, pp. 406–423.

[10] G. Bersuker et al., "Breakdown in the metal/high-k gate stack: identifying "weak link" in the multilayer dielectric," IEDM 2008, pp.791-794.

[11] W. He, T.P. Ma, "Inelastic Electron Tunneling Spectroscopy Study of Ultra-thin HfO₂ and HfAlO", Appl. Phy. Lett., vol. 83, 2003, p. 2605.

978-1-4244-6658-0/10 $26.00 © 2010 IEEE

Oxide-Based RRAM: Physical Based Retention Projection

B. Gao, *J.F. Kang, H.W. Zhang, B. Sun, B. Chen, L.F. Liu, X.Y. Liu, R.Q. Han, Y.Y. Wang,
Institute of Microelectronics, Peking University
Beijing 100871, China
*E-mail: kangjf@pku.edu.cn

B. Yu
College of Nanoscale Science and Engineering
State University of New York, Albany, NY 12203, USA

Z. Fang, H.Y. Yu
EEE/Nanyang Technological University, Singapore 639798

D.-L. Kwong
Institute of Microelectronics, A*STAR, Singapore 117685

Abstract—Based on the retention failure mechanism of high resistance state due to reconstruction of oxygen vacancy filament in the rupture region, a physical model is proposed to quantify the retention failure behavior of oxide-based RRAM devices, supported by experiments. A new data retention evaluation methodology is proposed to predict the failure probability and lifetime of the memory devices.

I. INTRODUCTION

Transition metal-oxide based resistive random access memory (RRAM) has demonstrated superior performance (operation voltage and power, P/E speed) and scalability [1-4]. Despite of many rapid progresses made very recently, several key challenges still need to be addressed towards mass manufacturing, including device operation mechanisms, switching parameters variation, and limited endurance and retention. Among them, data retention is of great importance for nonvolatile memory (NVM) operation. In particular, retention of high resistance state (HRS) remains one of the major reliability concerns in RRAM [3].

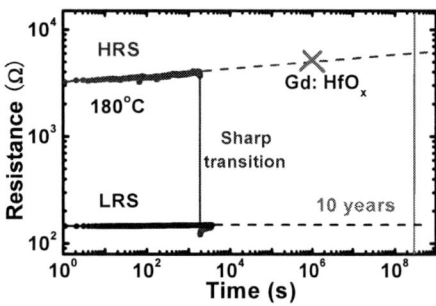

Fig. 1. Typical retention behavior of RRAM. Sharp transition between high and low resistance states is observed, indicating that the traditional evaluation method is invalid for RRAM devices.

This work is partially supported by 973 & 863 Programs (2006CB3027002 & 2008AA031401).

As shown in Fig. 1, the typical retention failure of RRAM is represented by a sharp resistance transition, significantly different from the traditional NVM devices. In this case, traditional data retention evaluation method cannot be used for RRAM. Therefore, understanding the unique retention failure mechanism and developing new retention evaluation methodology is critical. However, there have been only very few studies exploring the retention mechanism and the evaluation methodology [3,5].

In this paper, a comprehensive physical model for the HRS retention failure in bipolar RRAM cells is proposed for the first time, verified by experiments. Furthermore, a new evaluating method is developed to predict the HRS retention failure probability and the lifetime of RRAM devices based on both temperature- and voltage-accelerating techniques.

II. DEVICE FABRICATION & MEASUREMENTS

In this experiment, undoped and Gd-doped thin-film HfO_x based RRAM devices were fabricated. About 20nm HfO_x layer was deposited on $Pt/Ti/SiO_2$-coated silicon substrates by reactive sputtering, followed by a furnace annealing at 600°C

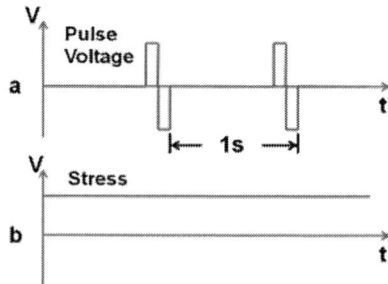

Fig. 2. Schematic of the applied voltage signal for retention measurement: a) read mode; and b) stress mode. Under read mode, positive-negative pulses are alternately applied to eliminate the dielectric polarization effect.

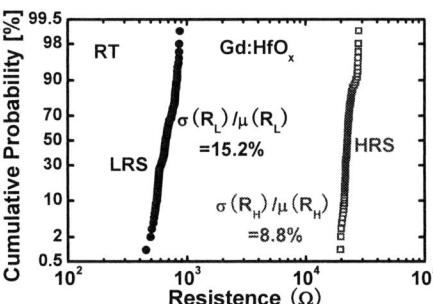

Fig. 3. Distribution of R_{LRS} and R_{HRS} for the Gd-doped HfO$_x$ devices under pulse voltage switching. Excellent uniformity is achieved. The width of pulse is 1μs for SET and 100μs for RESET, respectively.

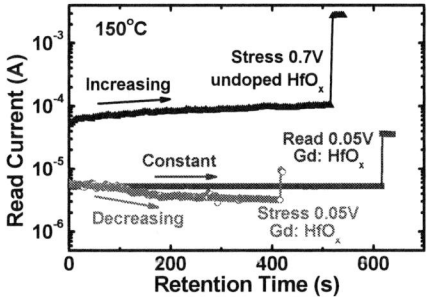

Fig. 4. HRS retention behavior under different stresses: 1) current is almost constant under read mode; 2) current decreases at low voltage stress; 3) current increases at high voltage stress.

in O$_2$. Gd implantation was performed for Gd-doped devices prior to the 800°C annealing process in N$_2$ ambient. Finally, TiN layer was deposited and patterned. Electrical measurements were performed using Agilent 4156C and Keithley 4200 analyzers. Two modes of measurement (read mode and stress mode) were applied to evaluate the data retention behavior. The corresponding testing pulse series (read mode and stress mode) are schematically shown in Fig. 2. Both positive and negative pulses are applied under read mode to eliminate the influence of dielectric polarization effect.

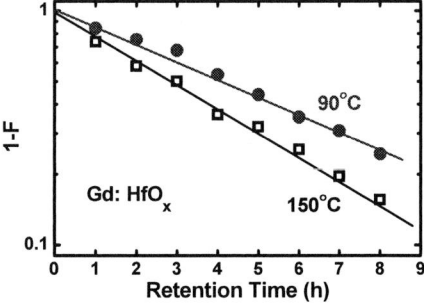

Fig. 5. Measured retention failure probability with time at various temperatures. Amount of devices were measured at the same time. The resistances were read every hour to find the retention failure time. Exponential dependence with time is observed. Retention failure time decreases with temperature.

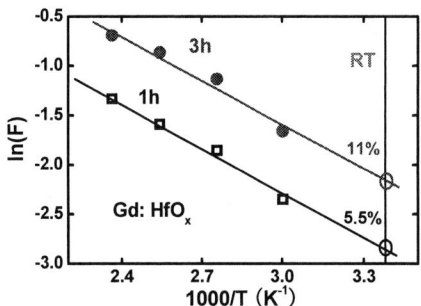

Fig. 6. Measured temperature dependence on failure probability for the Gd-doped HfO$_x$ RRAM devices. The failure probability at a fixed time increases with temperature.

III. CHARACTERISTICS OF RETENTION FAILURE

Despite excellent device uniformity, shorter retention time is observed in Gd-doped HfO$_x$ devices, as shown in Figs. 3&4. The shorter retention time is attributed to the low formation energy of oxygen vacancy (E_a) with Gd doping [6]. The typical retention behavior (shown in Fig. 4) indicates that current decreases with time at low stress due to relaxation effect and current increases until retention failure occurs at large stress, much resembling the TDDB process.

To quantify the failure behavior, the failure probability as a function of time ($F(t)$) is introduced. Fig. 5 shows the measured retention failure characteristics under read mode. In contrast to the excellent uniformity of resistance (as shown in Fig. 3), large variation of retention failure time is observed, which is fitted well using an exponential model. Fig. 6 shows the temperature dependence of failure probability. The failure probability at a fixed time exponentially depends on 1/T with the same slope. Fig. 7 shows the retention failure time of one device under read and stress modes illustrated in Fig. 2. Different voltage stresses were applied on the device under stress mode. The exponential dependence on retention time is also measured. The failure time decreases when voltage increasing. This result indicates that temperature and stress voltage can accelerate RRAM's retention failure.

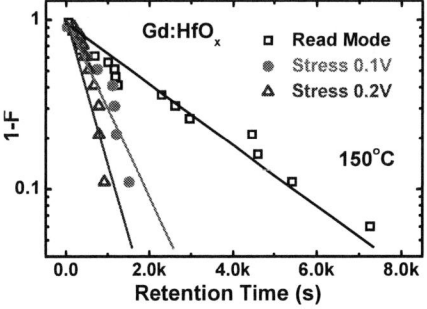

Fig. 7. Measured retention failure probability with time under no stress (read mode) and different stress voltages (stress mode). Only one device is measured to eliminate device variation. The resistance was read every second. Exponential dependence with time is also observed. Failure time decreases with stress voltage.

978-1-4244-6658-0/10 $26.00 © 2010 IEEE

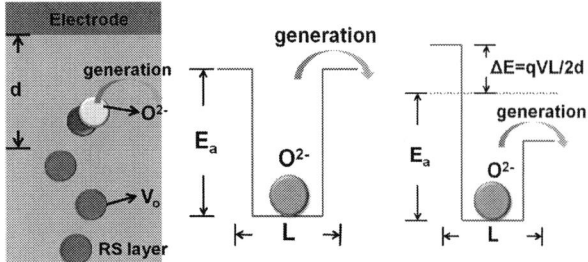

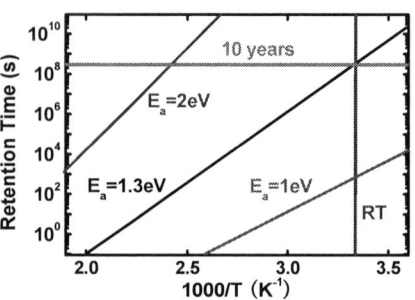

Fig. 8. Schematic view of HRS failure mechanism due to reconstruction of the ruptured oxygen vacancy (Vo) filaments (left) induced by a thermal activation process (middle) or an electric field effect (right). Where Ea is the formation energy of Vo, d, L, and V is the length of rupture region of filament, lattice constant, and applied voltage, respectively.

Fig. 9. Retention time as a function of temperature for different E_a. Once E_a is measured, the retention time can be estimated easily. The expected retention time is >10 yrs if E_a is larger than 1.3eV.

IV. RETENTION FAILURE MODEL

Based on the conductive filament (CF) model of resistive switching in RRAM, it is reasonable to assume that the HRS retention failure is due to generation of oxygen vacancy (V_o) in the rupture region of CF by thermal activation process [7], which causes reconstruction of the ruptured CFs, as illustrated in Fig. 8. The generation of new V_o in the rupture region causes significant reduction of HRS resistance, resulting in retention failure. Based on crystal defect and probability theories, the retention failure model is developed. The failure probability is expressed as

$$F(t) = 1 - (1-p)^{nt/t_0} \qquad (1)$$

where n is the total amount of possible places for V_o generation, and t_0 is the vibration period for lattice oxygen atom. p is the generation probability of V_o, and is expressed as

$$p = \exp(-E_a / k_B T) \qquad (2)$$

where E_a is the formation energy of V_o [6]. As $p \ll 1$, the expected retention time (t_E) is estimated to be

$$t_E = t_0 / (n|\ln(1-p)|) \approx t_0 / np \qquad (3)$$

When a stress voltage is applied on the device, the probability of V_o generation will increase due to the reduction of the formation potential barrier as illustrated in Fig. 8. Therefore, the failure probability as a function of time and voltage is expressed as

$$F(t,V) = 1 - [1 - p\exp(qLV / 2dk_B T)]^{nt/t_0}$$
$$\approx F(t)\exp(qLV / 2dk_B T) \qquad (4)$$

where q, d, L, V and T refer to electric charge quantity of V_o, effective length of rupture region of CF, lattice constant, applied voltage, and temperature, respectively. Excellent agreement between data and model predictions is achieved, as shown in Figs. 5-7. Moreover, the dependence suggests that retention time can be extracted from temperature- and voltage-stress- accelerated retention time measurements.

V. METHODOLOGY FOR EVALUATION & PREDICTION

In the new model, the formation energy E_a is a key factor determining RRAM retention behavior. Increasing E_a is an effective way to enhance the device retention performance. As reported in Ref. [6], doping is an effective method to control data variation, but causes reduced E_a. Therefore, the trade-off between data uniformity and retention needs to be considered in selecting storage material of RRAM devices.

Fig. 9 shows the calculated temperature dependence of retention time at different E_a. Based on such dependence, if E_a is determined for the selected storage layer, HRS retention can be easily extracted. Based on the above discussion, two methods are developed to evaluate the data retention failure behavior, as summarized in Tables 1&2. Based on the model,

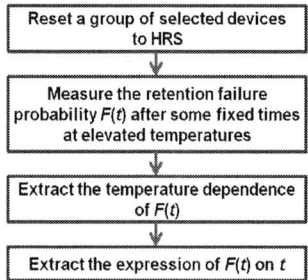

Table 1. The flow of the proposed method I to extract the expression of $F(t)$. From the expression, one can predict $F(t)$ at the targeted temperature and time.

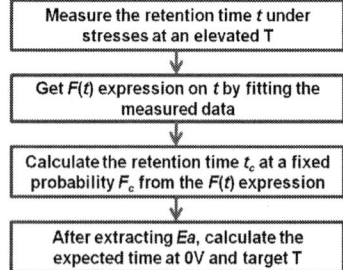

Table 2. The flow of the proposed method II to extract the expected HRS retention time using temperature and voltage acceleration method.

978-1-4244-6658-0/10 $26.00 © 2010 IEEE   314

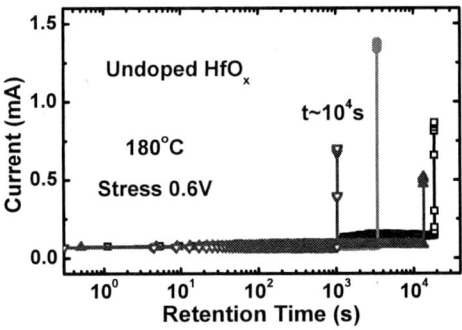

Fig. 10. Retention time of undoped HfO$_x$ devices at fixed stress. About 10^4s retention time is estimated.

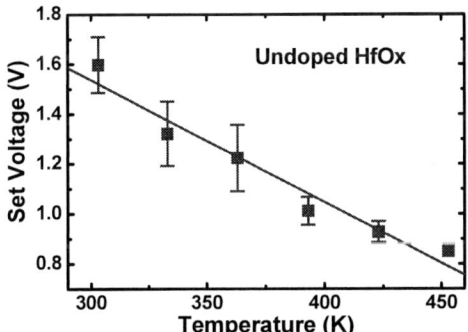

Fig. 11. Measured temperature dependence of SET voltage of the undoped HfO$_x$ devices. SET voltage decreases with temperature. According to the model, relationship between E_a and x is extracted based on data.

the failure probability at the targeted temperature and time (method I) and retention time at the targeted temperature (method II) can be estimated based on the measured data at elevated T- and V-stress conditions.

As shown in Fig. 5, the device retention failure probability at room temperature (RT) can be predicted following the method I. Failure probability 5.5% @ 1 hour and 11% @ 3 hours is estimated for Gd doped HfO$_x$ RRAM devices, respectively. This is a quick with an acceptable accuracy to estimate failure probability at a fixed time. To estimate the expected retention failure time, the expression of $F(t)$ should be extracted first, as following the method II. This is more precise but time-consuming measurement method since parameters extracting for the expression would take a great amount of measuring time for failure, especially when the devices show excellent retention. As shown in Fig. 10, undoped HfO$_x$ device can maintain the high resistance state for about 10^4s, even under 180°C and 0.6V voltage stress. In addition, we also develop the method III to roughly estimate expected retention time.

According to the model, when applying voltage on the device, the effective generation probability can be expressed as

$$p = \exp[(xV - E_a) / k_B T] \qquad (5)$$

where $x=qL/2d$. Based on equation (1) and Fig. 10, it can be estimated that E_a-xV approximately equals to 1.5eV. To estimate E_a, the temperature dependence of SET voltage is measured, as shown in Fig. 11. Because the SET process can be regard as a rapid HRS retention process, it is reasonable to infer that the generation probability of V_o should be the same value under different temperature, which is equivalent to

$$(E_a - xV) / k_B T = const \qquad (6)$$

Based on the fitting data from Fig. 11, it can be estimated that E_a/xV is approximately equal to 5.0. Therefore, E_a is estimated to be about 1.9eV for the undoped HfO$_x$ RRAM device. Although this value is much smaller than the theoretical formation energy [6], the device has already got excellent date retention. By extrapolating stress to zero, the expected average retention time of undoped HfO$_x$ devices is estimated to be about 10^8s at 180°C and 10^{18}s at room temperature.

VI. CONCLUSION

For the first time, a physical based model for RRAM's HRS retention failure is proposed and verified experimentally. The proposed model captures the effects of temperature and voltage stress on the HRS retention behavior. Based on the proposed model, new evaluating methods are developed to successfully predict the failure probability at targeted temperature and time, as well as the projected retention time at targeted temperature in bipolar RRAM devices.

ACKNOWLEDGMENT

The authors would like to thank Dr. M.-J. Tsai at ITRI for their valuable discussions, and the lab staff in IME/PKU for help in device sample fabrication.

REFERENCES

[1] D. Lee, D.-J. Seong, H.J. Choi, I. Jo, R. Dong ,W. Xiang, S. Oh, M. Pyun, S.-O. Seo, S. Heo, M. Jo, D.-K. Hwang, H.K. Park, M. Chang, M. Hasan and H. Hwang, "Excellent uniformity and reproducible resistance switching characteristics of doped binary metal oxides for non-volatile resistance memory applications", *IEDM Tech. Dig.*, 2006, pp. 796-799.

[2] N. Xu, B. Gao, L.F. Liu, X.Y. Liu, R.Q. Han, J.F. Kang and B. Yu, "A Unified Physical Model of Switching Behavior in Oxide-Based RRAM", *VLSI Symp. Tech .Dig.*, 2008, pp. 100-101.

[3] Y.S. Chen, H.Y. Lee, P.S. Chen, P.Y. Gu, C.W. Chen, W.P. Lin, W.H. Liu, Y.Y. Hsu, S.S. Sheu, P.C. Chiang, W.S. Chen, F.T. Chen, C.H. Lien and M.-J. Tsai, "Highly Scalable Hafnium Oxide Memory with Improvements of Resistive Distribution and Read Disturb Immunity", *IEDM Tech. Dig.*, 2009, pp. 105-108.

[4] R. Waser and M. Aono, "Nanoionics-based resistive switching memorics," Nature Materials, vol. 6, pp. 833-840, 2007.

[5] C. Cagli, D. Ielmini, F. Nardi and A.L. Lacaita, "Evidence for threshold switching in the set process of NiO-based RRAM and physical modeling for set, reset, retention and disturb prediction", *IEDM Tech. Dig.*, 2008, pp. 301-304.

[6] B. Gao, H. W. Zhang, S. Yu, B. Sun, L. F. Liu, X. Y. Liu, Y. Wang, R. Q. Han, J. F. Kang, B. Yu, and Y.Y. Wang, "Oxide-Based RRAM: Uniformity Improvement Using A New Material-Oriented Methodology", *VLSI Symp. Tech .Dig.*, 2009, pp. 30-31.

[7] B. Gao, S. Yu, N. Xu, L. F. Liu, B. Sun, X. Y. Liu, R. Q. Han, J. F. Kang, B. Yu, and Y. Y. Wang, "Oxide-Based RRAM Switching Mechanism: A New Ion-Transport-Recombination Model", *IEDM Tech. Dig.*, 2008, pp.563-566.

A Stochastic Model of Bipolar Resistive Switching in Metal-Oxide-Based Memory

Alexander Makarov, Viktor Sverdlov, and Siegfried Selberherr

Institute for Microelectronics, TU Wien

Vienna, Austria

Email: {makarov | sverdlov | selberherr}@iue.tuwien.ac.at

Abstract — **A stochastic model of the resistive switching mechanism in bipolar metal-oxide-based resistive random access memory (RRAM) is presented. The distribution of electron occupation probabilities obtained is in agreement with previous work. In particular, a low occupation region is formed near the cathode. Our simulations of the temperature dependence of the electron occupation probability near the anode and the cathode demonstrate a high robustness of the low occupation region. This result indicates that a decrease of the switching time with increasing temperature cannot be explained only by reduced occupations of the vacancies in the low occupation region, but is related to an increase of the mobility of the oxide ions. A hysteresis cycle of a RRAM simulated with our stochastic model is in good agreement with experimental results.**

I. INTRODUCTION

With flash memories rapidly approaching the physical limits of scalability, research on new nonvolatile memory concepts has significantly accelerated. Several new memory concepts as potential substitutes of the flash memory were invented and developed: a technology of phase change RAM (PCRAM), spin transfer torque RAM (STTRAM), carbon nanotube RAM (NRAM), copper bridge RAM (CBRAM), racetrack memory, and resistive RAM (RRAM). A new type of nonvolatile memory must exhibit low operating voltages, low power consumption, high operation speed, long retention time, high endurance, simple structure, and small size [1].

One of the most promising candidates for future universal memory is the resistive random access memory (RRAM). It is based on new materials, such as metal oxides [2-4] and perovskite oxides [5]. This type of memory is characterized by high density, excellent scalability, low operating voltages (<2V), fast switching times (<10ns), and long retention time.

In the literature a broad spectrum of electronic and/or ionic switching mechanisms for oxide-based memory has been suggested: a model based on trapping of charge carriers [6], electrochemical migration of oxygen vacancies [7, 8], electrochemical migration of oxygen ions [9, 10], a unified

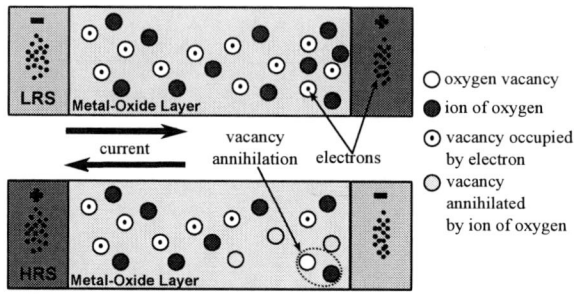

Figure 1. Schematic illustration of the conducting filament in the low resistance state (top) and the high resistance state (bottom).

physical model [11, 12], a domain model [13], a filament anodization model [14], a thermal dissolution model [15], and others. Unfortunately, a proper fundamental understanding of the switching mechanism is still missing.

In this work we present a stochastic model of the bipolar resistive switching mechanism based on electron hopping between the oxygen vacancies along the conductive filament in an oxide layer.

II. MODEL DESCRIPTION

We associate the resistive switching behavior in oxide-based memory with the formation and rupture of a conductive filament (CF). The CF is formed by localized oxygen vacancies (V_o) [11, 12] or domains of V_o (Fig. 1). Formation and rupture of a CF is due to a redox reaction in the oxide layer under a voltage bias. The conduction is due to electron hopping between these V_o.

For modeling the resistive switching in bipolar oxide-based memory by Monte Carlo techniques, we describe the dynamics of oxygen ions (O^{2-}) and electrons in an oxide layer as follows:

This research is supported by the European Research Council through the grant #247056 MOSILSPIN.

978-1-4244-6658-0/10 $26.00 © 2010 IEEE

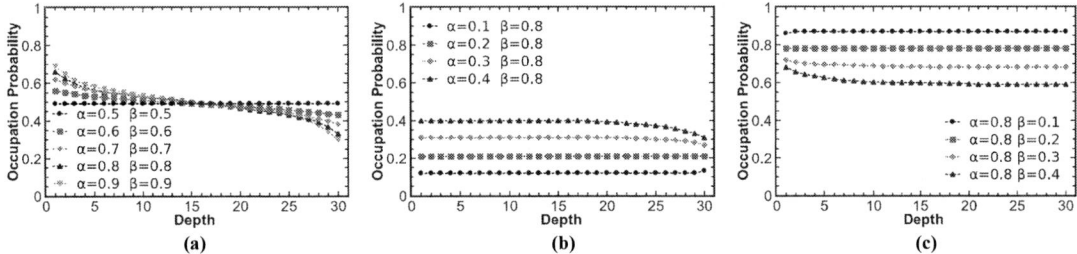

Figure 2. Calculated distribution of electron occupation probabilities for unidirectional next nearest neighbor hopping between the V_o (the 1st V_o is near the cathode, the last V_o is near the anode): (a) $\alpha>0.5$ and $\beta>0.5$, $p_c=0.5$; (b) $\alpha<0.5$ and $\alpha<\beta$, $p_c=\alpha$; (c) $\beta<0.5$ and $\beta<\alpha$, $p_c=1-\beta$.

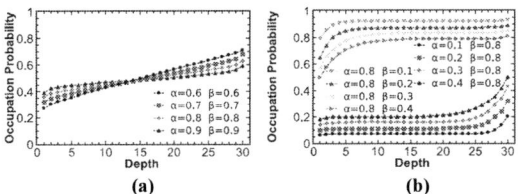

Figure 3. Calculated distribution of electron occupation probabilities, if unidirectional hopping is allowed not only to/from the closest V_o: (a) $\alpha>0.5$ and $\beta>0.5$; (b) $\alpha<0.5$ and $\alpha<\beta$; $\beta<0.5$ and $\beta<\alpha$.

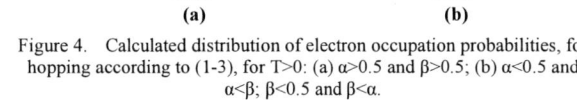

Figure 4. Calculated distribution of electron occupation probabilities, for hopping according to (1-3), for T>0: (a) $\alpha>0.5$ and $\beta>0.5$; (b) $\alpha<0.5$ and $\alpha<\beta$; $\beta<0.5$ and $\beta<\alpha$.

- formation of V_o by O^{2-} moving to an interstitial position;

- annihilation of V_o by moving O^{2-} to V_o;

- an electron hop into V_o from an electrode;

- an electron hop from V_o to an electrode;

- an electron hop between two V_o.

In order to model the dependences of transport on the applied voltage and temperature we choose the hopping rates for electrons as [16]:

$$\Gamma_{nm} = A_e \cdot \frac{dE}{1-\exp(-dE/T)} \cdot \exp(-R_{nm}/a) \quad (1)$$

Here, A_e is a coefficient, $dE=E_n-E_m$ is the difference between the energies of an electron positioned at sites n and m, R_{nm} is the hopping distance, a is the localization radius. The hopping rates between an electrode (0 or $N+1$) and an oxygen vacancy m are described [12]:

$$\Gamma_m^{iC} = \alpha \cdot \Gamma_{0m}, \Gamma_m^{oC} = \alpha \cdot \Gamma_{m0} \quad (2)$$

$$\Gamma_m^{iA} = \beta \cdot \Gamma_{(N+1)m}, \Gamma_m^{oA} = \beta \cdot \Gamma_{m(N+1)} \quad (3)$$

Here, α and β are the coefficients of the boundary conditions on the cathode and anode, respectively, N is the number of sites, A and C stand for cathode and anode, and i and o for hopping on the site and out from the site, respectively.

The current generated by hopping is calculated as:

$$I = q_e \cdot \sum dx / \sum \left(1 / \sum_m \Gamma_m \right) \quad (4)$$

III. MODEL VERIFICATION

All calculations are made on one or/and two-dimensional lattices, the distances between two nearest neighboring V_o in all directions are equal. All V_o are at the same energy level, if no voltage or temperature is applied. Despite the fact that in the binary metal oxides, oxygen vacancies can have three different charge states with charge 0, +1, +2, to simplify the calculations, we assume that the oxygen vacancy is either empty or occupied by one electron.

A. Calculation of electron ocupation probability

To verify the proposed model, we first evaluate the average electron occupations of hopping sites under different conditions. For comparison with previous works all calculations in this subsection are made on a one-dimension lattice consisting of thirty equivalent, equidistantly positioned hopping sites V_o.

Following [17], we first allow hopping in one direction and only to/from the closest V_o. The occupation probability of central oxygen vacancies, p_c, is described, depending on the boundary conditions as follows: 1) for $\alpha>0.5$ and $\beta>0.5$, $p_c=0.5$; 2) for $\alpha<0.5$ and $\alpha<\beta$, $p_c=\alpha$; 3) for $\beta<0.5$ and $\beta<\alpha$, $p_c=1-\beta$. Fig.2 shows simulation results of our stochastic model, which are fully consistent with theoretical predictions [17].

978-1-4244-6658-0/10 $26.00 © 2010 IEEE   317

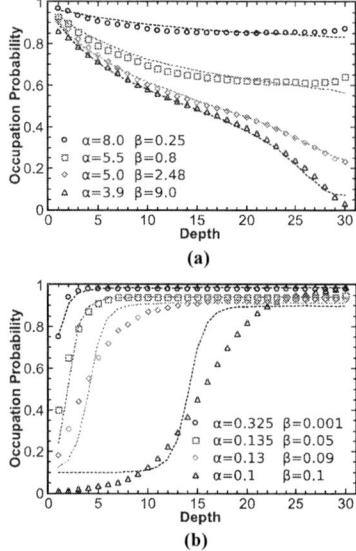

(a)

(b)

Figure 5. Calculated distribution of electron occupation probabilities under different biasing voltages. Lines are from [12], symbols are obtained from our stochastic model.

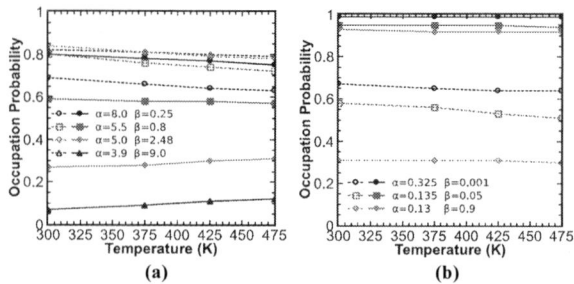

(a)   **(b)**

Figure 6. Temperature dependence of electron occupation probability near the anode (filled symbols) and the cathode (open symbols).

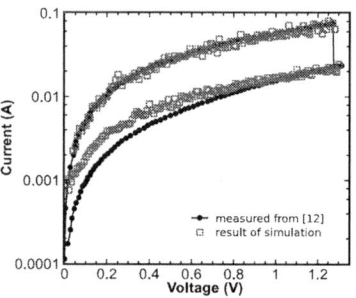

Figure 7. Current-voltage curves during reset process. Lines are measured result from [12], symbols are obtained from our stochastic model.

To move from a model system [17] to a more realistic structure, we calculated the distribution of electron occupations for a chain, where hopping is allowed not only to/from the nearest V_o (Fig. 3), and for systems, where hopping (1-3) is allowed in both directions (Fig. 4). Note that for $\alpha>0.5$ and $\beta>0.5$ (Fig. 3a and Fig. 4a) we still have p_c=0.5 in the center, while for other values α, β we observe a decrease in p_c for $\alpha<\beta$ and an increase in p_c for $\beta<\alpha$.

We have calibrated our model in a manner to reproduce the results reported in [12], for V=0.4 V to V=1.6 V. Fig. 5a shows a case, when the hopping rate between the electrodes and V_o is larger than the rate between two V_o (i.e. α, $\beta > 1$). In this case the low occupation region is formed near the anode (unipolar behavior).

Fig. 5b shows a case, where the hopping rate between two V_o is larger than the rate between the electrodes and V_o (i.e. α, $\beta < 1$). In this case a low occupation region is formed near the cathode (bipolar behavior).

Note that when V=0 V, the probability of occupation of all vacancies becomes equal to 0.5 regardless of the voltage applied before. This result indicates that, after the voltage is turned off, the probability of a CF rupture should increase substantially due to a decrease of the occupation probability of the vacancies from 1 to 0.5. But in practice we do not observe this, so to associate a CF rupture only with the formation of the low occupation region is wrong.

B. Modeling of temperature dependence

With the model calibrated to [12] we simulated the temperature dependence of the site occupations in the low occupation region. The results shown in Fig. 6a and Fig. 6b indicate high robustness of the low occupation region demonstrating changes of less than 10%, when the temperature is elevated from 25° C to 200° C. At the same time this result indicates that the measured decrease of switching time with increasing temperature reported in [12] stems from the increased mobility of oxide ions, rather than from reduced occupations of V_o in the low occupation region.

C. Modeling of the RESET process

Results obtained from simulations of the temperature dependence of the site occupations in the low occupation region demonstrate a necessity to include the dynamics of oxygen ions.

To describe the motion of ions we have chosen the ion rates similar to (1):

$$\Gamma_n' = A_i \cdot \frac{dE}{1-\exp(-dE/T)} \qquad (5)$$

Here we assume that O^{2-} can only move to the nearest interstitial. A distance-dependent term is thus included in A_i. dE includes the formation energy for the m-th V_o /annihilation energy of the m-th V_o, when O^{2-} is moving to an interstitial or back to V_o, respectively.

To verify the model, we simulated the reset I-V characteristics for a single-CF device [12]. For the simulations we have used a one-dimensional lattice consisting of thirty equivalent, equidistantly positioned hopping sites V_o. Near each V_o an oxygen ion is placed. Fig.7. shows the simulation results of the stochastic model, which are in perfect agreement with measurements from [12].

978-1-4244-6658-0/10 $26.00 © 2010 IEEE   318

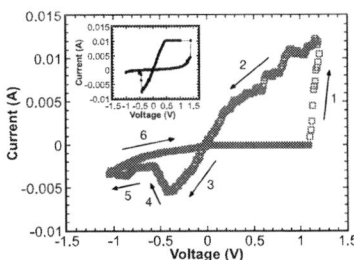

Figure 8. *I-V* characteristics showing the hysteresis cycle obtained from our stochastic model (α=0.1 and β=0.1). The inset shows the hysteresis cycle for *M-ZnO-M* from [4].

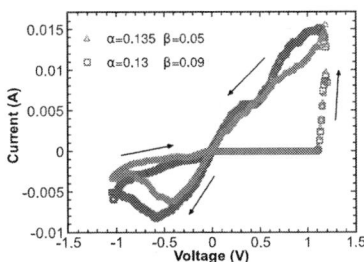

Figure 9. *I-V* characteristics showing hysteresis cycles obtained from the stochastic model for different parameters.

D. Modeling of the hysteresis cycle

All calculations of RRAM *I-V* characteristics are now performed on a two-dimensional lattice (*2×30*). We have investigated the *I-V* hysteresis by applying a saw-tooth like voltage *V*. We have assumed that the coefficients of the boundary conditions are constant and equal to 0.1. The simulated RRAM switching hysteresis cycle is shown in Fig.8. The cycle is in good agreement with the experimental one from [4] shown in the inset of Fig.8.

The interpretation of the RRAM hysteresis cycle obtained from the stochastic model is as follows. If a positive voltage is applied, the formation of a CF begins, when the voltage reaches a critical value sufficient to create V_o by moving O^{2-} to an interstitial position. This leads to a sharp increase in the current (Fig. 8 Segment 1) signifying a transition to a state with low resistance. When a reverse negative voltage is applied, the current increases linearly (Fig. 8 Segment 3), until the applied voltage reaches the value at which annihilation of V_o is triggered by means of moving O^{2-} to V_o. The CF is ruptured and so the current decreases (Fig. 8 Segment 4). This is the transition to a state with high resistance.

Interestingly, for parameters different from those in Fig.8 (α=0.135, β=0.05 and α=0.13, β=0.09) we still see a hysteresis cycle (Fig. 9), although the region of low occupation which, according to [12], is responsible for the rupture of the CF is almost absent. This result supports the observation that the ion dynamics is critical in describing the RRAM switching mechanism.

IV. CONCLUSION

In this work we have presented a stochastic model of the bipolar resistive switching mechanism. The distribution of the electron occupation probabilities calculated with the model is in excellent agreement with previous work. The simulated RRAM switching hysteresis cycle is in good agreement with the experimental result. We have shown that the process of rupture of the CF is determined by the dynamics of oxygen ions and not only by the formation of the low occupation region. The proposed stochastic model can be used for performance optimization of RRAM devices.

REFERENCES

[1] M. H. Kryder, C. S. Kim, "After Hard Drives - What Comes Next?," IEEE Trans. Magn., vol. 45, no. 10, pp. 3406-3413, 2009.

[2] C. Kugeler, C. Nauenheim, M. Meier, A. Rudiger, R. Waser, "Fast Resistance Switching of TiO_2 and MSQ Thin Films for Non-Volatile Memory Applications (RRAM)," NVM Tech. Symp., p. 6, 2008.

[3] Y. S. Chen, T. Y. Wu, P. J. Tzeng, "Forming-free HfO_2 Bipolar RRAM Device with Improved Endurance and High Speed Operation," Symp. on VLSI Tech., pp. 37-38, 2009.

[4] S. Lee, H. Kim, D. J. Yun, S. W. Rhee, K. Yong, "Resistive Switching Characteristics of ZnO Thin Film Grown on Stainless Steel for Flexible Nonvolatile Memory Devices," APL, vol. 95, no. 26, pp. 262113/1-3, 2009.

[5] C. C. Lin, C. Y. Lin, M. H. Lin, "Voltage-Polarity-Independent and High-Speed Resistive Switching Properties of V-Doped SrZrO3 Thin Films," IEEE Trans. Electron Devices, vol. 54, no. 12, pp. 3146-3151, 2007.

[6] T. Fujii, M. Kawasaki, A. Sawa, H. Akoh, Y. Kawazoe, and Y. Tokura, "Hysteretic Current–Voltage Characteristics and Resistance Switching at an Epitaxial Oxide Schottky Junction SrRuO3/SrTi0.99Nb0.01O3," APL, vol. 86, no. 1, art. no. 012107, 2005.

[7] Y. B. Nian, J. Strozier, N. J. Wu, X. Chen, A. Ignatiev, "Evidence for an Oxygen Diffusion Model for the Electric Pulse Induced Resistance Change Effect in Transition-Metal Oxides," PRL, vol. 98, no. 14, pp. 146403/1-4, 2007.

[8] S. X. Wu, L. M. Xu, X. J Xing, "Reverse-Bias-Induced Bipolar Resistance Switching in Pt/TiO2/SrTi0.99Nb0.01O3/Pt Devices," APL, vol. 93, no. 4, pp. 043502/1-3, 2008.

[9] K. Szot, W. Speier, G. Bihlmayer and R. Waser, "Switching the Electrical Resistance of Individual Dislocations in Single-Crystalline SrTiO3," Nature Materials, vol. 5, pp. 312-320, 2006.

[10] Y. Nishi, J. R. Jameson, "Recent Progress in Resistance Change Memory," Dev. Res. Conf. 2008, pp. 271-274, 2008.

[11] N. Xu, B. Gao, L. F. Liu, B. Sun, X. Y. Liu, R. Q. Han, J. F. Kang, and B. Yu, "A Unified Physical Model of Switching Behavior in Oxide-Based RRAM," Symp. on VLSI Tech., pp. 100-101, 2008.

[12] B. Gao, B. Sun, H. Zhang, L. Liu, X. Liu, R. Han, J. Kang, B. Yu, "Unified Physical Model of Bipolar Oxide-Based Resistive Switching Memory," IEEE Electron Device Lett., vol. 30, no. 12, pp. 1326-1328, 2009.

[13] M. J. Rozenberg, I. H. Inoue, and M. J. Sanchez, "Nonvolatile Memory with Multilevel Switching: A Basic Model," PRL, vol. 92, no. 17, pp. 178302-1, 2004.

[14] K. Kinoshita, T. Tamura, H. Aso, H. Noshiro, C. Yoshida, M. Aoki, Y. Sugiyama, H. Tanaka, "New Model Proposed for Switching Mechanism of ReRAM," IEEE Non-Volatile Semicond. Memory Workshop 2006, pp. 84 – 85, 2006.

[15] U. Russo, D. Ielmini, C. Cagli, A.L. Lacaita, S. Spiga, C. Wiemer, M. Perego, and M. Fanciulli, "Conductive-Filament Switching Analysis and Self-Accelerated Thermal Dissolution Model for Reset in NiO-Based RRAM," IEDM Tech. Dig., pp. 775-778, 2007.

[16] V. Sverdlov, A. N. Korotkov, K. K. Likharev, "Shot-Noise Suppression at Two-Dimensional Hopping," PRB, vol.63, 081302, 2001.

[17] B. Derrida, "An Exactly Soluble Non-Equilibrium System: The Asymmetric Simple Exclusion Process," Phys. Rep., vol. 301, no. 1-3, pp. 65-83, 1998.

Dependence of the Switching Characteristics of Resistance Random Access Memory on the Type of Transition Metal Oxide

Wan Gee Kim[1], Min Gyu Sung, Sook Joo Kim, Ja Yong Kim, Ji Won Moon, Sung Joon Yoon, Jung Nam Kim, Byung Gu Gyun, Taeh Wan Kim, Chi Ho Kim, Jun Young Byun, Won Kim, Te One Youn, Jong Hee Yoo, Jang Won Oh, Ho Joung Kim, Moon Sig Joo, Jae Sung Roh, and Sung Ki Park

[1]R&D Division, Hynix Semiconductor Inc.,
San 136-1 Ami-ri, Bubal-eub, Ichon-si, Kyoungki-do, 467-701, Korea
Tel) +82-31-639-0837, Fax) +82-31-645-8139, E-mail) wangee.kim@hynix.com

Abstract

In this paper, a systematic approach using HfO_2, ZrO_2 and TiO_2 with TiN or Ti/TiN electrode has been conducted to research the best material for ReRAM device integration. From the experimental results and proposed model, the proper electrical properties such as the stability of switching variation, low current and voltage operation, long endurance and retention characteristics are obtained with a TiN/Ti/HfO_2/TiN structure.

Introduction

ReRAM (Resistance random access memory) which has many advantages over the conventional non-volatile memories such as floating gate flash memory and phase change memory has gained much interest. Among the candidates of resistive materials and electrodes for the ReRAM, conventional transition metal oxide (TMO) and metal nitride electrode, which are CMOS (complementary metal oxide semiconductor) friendly materials, can be easily implemented into mass production.

Recently, superior performance of HfO_2 switching using Ti/TiN electrode was reported [1]. It showed lower operating voltage and current and satisfactory endurance properties than other transition oxide with Pt electrode [2]. However, systematic studies to understand the different switching behaviors depending on the kind of TMO have not been made.

In this paper, we conducted a systematic approach using the HfO_2, ZrO_2, and TiO_2 with TiN or Ti/TiN electrode. From the comparison of different switching behaviors and physical analysis, we report fundamental explanations of the better switching properties for the HfO_2 based ReRAM.

Experimental procedure

The TiN/Ti/TMO/TiN devices were fabricated using sub-50 nm process technology with 300 mm wafers. Various TMOs such as the TiO_2, HfO_2, and ZrO_2 were deposited by atomic layer deposition (ALD) method on the bottom TiN electrode contact (< 50 nm). After the TMO deposition, formation of the Ti/TiN top electrode was followed. The cross sectional schematic process flow of the fabricated ReRAM structure is shown in Fig. 1. DC I-V electrical measurements as well as physical failure analysis with a focused ion-beam (FIB) transmission electron microscopy (TEM) were conducted.

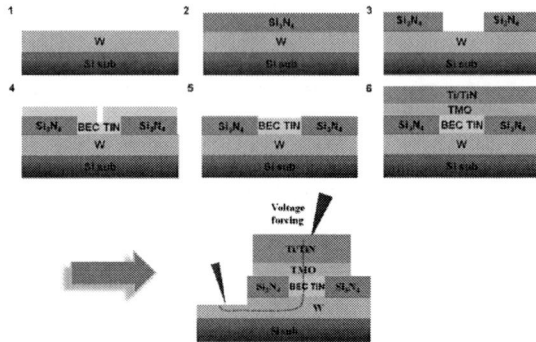

Figure 1. Schematic process flow of fabricated ReRAM structure.

Results and Discussion

A. Resistive switching characteristics of binary TMOs

The typical bipolar resistive switching curves and their switching parameters depending on the resistive switching materials (TiO_2, ZrO_2, and HfO_2) are shown in Fig. 2 and Fig. 3, respectively. For the desirable device integration with a high density of cell arrays, the low current operation is absolutely needed because the current level from sensing device such as transistor or diode is limited. Moreover, the stability of switching parameters such as set voltage (Vset), reset current (Ireset), on/off ratio, the excellent endurance and retention characteristics are essential for scaled downed devices.

With the TiN/TMO/TiN cell structure, there are some problems to solve in order to obtain high density cell arrays in ReRAM device. Vset (including the forming step) and Ireset are very high, as shown in Fig 3. Under the low current compliance (CC) mode, which is essential to obtain low level of operation current, HRS (high resistance state) doesn't completely converted into LRS (low resistance state) leading to the reset fail (Fig. 2(d)). It means conducting filaments are not formed perfectly under the low current of CC. In addition, the variations of switching parameters such as Ireset, Vset, and on/off ratio are large, and these switching properties result in poor endurance and retention performance (not shown here).

To overcome these limitations, increasing the oxygen vacancies in the TMO by adding a reactive metal layer such as Ti on the TMO layer has been proposed recently [1], [3]. Accordingly, different resistive switching characteristics of the TiN/Ti/TMO/TiN structure can be estimated using the TiO_2, ZrO_2, and HfO_2 as resistive switching materials.

978-1-4244-6658-0/10 $26.00 © 2010 IEEE

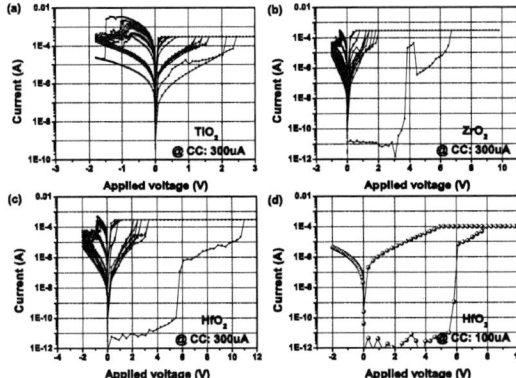

Figure 2. Typical bipolar I-V characteristics depending on the TMO material of the TiN/TMO/TiN structure. The TMO materials are (a) TiO_2, (b) ZrO_2, and (c) and (d) HfO_2, respectively.

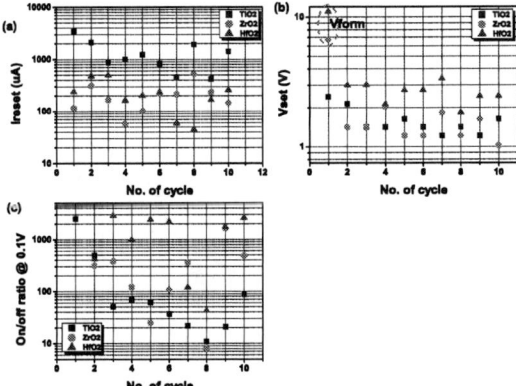

Figure 3. Resistive switching parameters depending on the resistive material; (a) Ireset, (b) Vset, and (c) On/off ratio.

B. Resistive switching characteristics of Ti added-binary TMOs

Figure 4 shows the initial current level of the TiN/Ti/TMO/TiN structure measured at 0.2 V depending on the TMO materials. From these results, it is confirmed that the Ti/TiO_2 is in LRS in the beginning, while the Ti/ZrO_2 and the Ti/HfO_2 are still in HRS. Considering the heat of formation of the TiO_2, ZrO_2, and HfO_2, the TiO_2 has a lower value than those of the ZrO_2 and HfO_2 [4]. It is expected that oxygen gettering of the Ti [5] makes the TiO_2 much more deficient in oxygen than the other materials, which leads to LRS in initial state eventually.

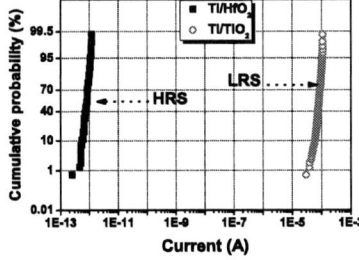

Figure 4. Distribution of initial current at 0.2V depending on the TMO materials of the TiN/Ti/TMO/TiN structure.

The typical bipolar resistive switching curves for the TiN/Ti/TMO/TiN structure and their switching parameters depending on the resistive material are shown in Fig. 5 and Fig. 6, respectively. On the whole, the switching parameters such as Ireset and Vset are decreased and become stable with a decrease in on/off ratio after the Ti layer is added. From these results, it is expected that the increase in oxygen vacancies created by adding the Ti layer makes the switching more stable and reduces the on/off ratio by the reduction of resistance in HRS.

By comparing these three different resistive switching materials, we can select the best material for ReRAM device. In terms of Ireset, Vset, and stable on/off ratio, the HfO_2 is a best candidate as a resistive switching material. On the other hand, there is a serious problem with the TiO_2 and the ZrO_2. They have a first abnormal high Ireset, as shown in Fig. 5(a) and (b). Having a high density of the oxygen vacancy in the Ti/TiO_2, the high Ireset is due to the current through the initial bulk TiO_2, which was mentioned in Fig. 4. Therefore, it is impossible to control the first reset current level. On the other hand, the first Ireset of the Ti/ZrO_2 flow through the filaments created during the forming step. However, the current can not be controlled by the CC level, as shown in Fig. 5(b). The scale downed device can not be operated under the abnormal high Ireset because the sensing device such as the transistor or diode cannot supply the high current level over 1mA. So, to explain this strange phenomenon, we propose one simple model.

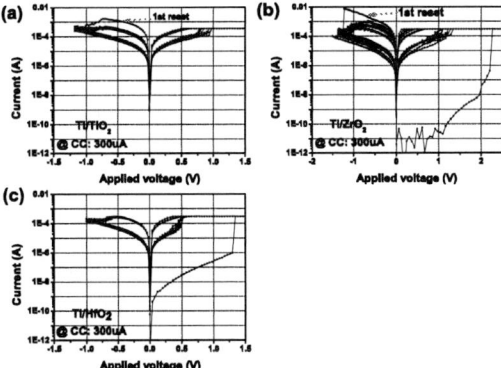

Figure 5. Typical bipolar I-V characteristics depending on the TMO material of the TiN/Ti/TMO/TiN structure. The TMO materials are (a) TiO_2, (b) ZrO_2, and (c), (d) HfO_2, respectively.

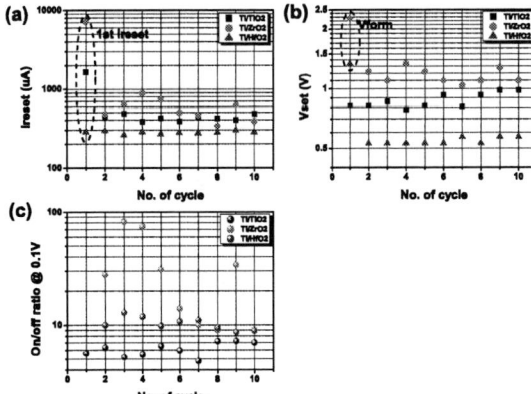

Figure 6. Resistive switching parameters depending on the resistive material; (a) Ireset, (b) Vset, and (c) On/off ratio.

978-1-4244-6658-0/10 $26.00 © 2010 IEEE

Generally, the forming process needs higher voltage than the set process. It produces higher power leading to the crystallization which makes the dielectric constant much higher. Moreover, the crystallization could be also made from thermal budget during the fabrication process. It is commonly known that the ZrO_2 is much easier to be crystallized than the HfO_2 [6]. Eventually, the Ti/ZrO_2 makes more charges accumulate to the TMO than the Ti/HfO_2, as shown in Fig. 7. Therefore, when the conductive filaments are created during the forming process, the capacitive charges stored in the crystallized ZrO_2 layer are added to the resistive charges passing through the filaments resulting in the abnormal first high reset current, as shown in Fig. 5(b) [7].

In the TiN/TMO/TiN structure, more charges are accumulated to the TMO due to the higher forming voltage than that of the TiN/Ti/TMO/TiN structure. Fig. 2 shows more than 8V is needed for the ZrO_2 and HfO_2. However, much smaller density of the oxygen vacancy in the TMO leads to the thinner and smaller number of the conductive filaments. Therefore, it is believed that the increased resistance of filaments limits the flow of the capacitive charge avoiding the abnormal first high reset current.

Figure 8 shows a cross sectional FIB TEM images of the TiN/Ti/TMO/TiN structures after the resistive switching. In the case of Ti/HfO_2 structure (Fig. 8(a)), there is no difference between the fresh sample and the sample which was measured over 10^4 cycles. In the TiN/Ti/ZrO_2/TiN structure, however, the TiN metal in BEC (bottom electrode contact) region is melted due to locally generated heating after just one switching cycle (Fig. 8(b)). Therefore, it is expected that the high power generated by the first abnormal Ireset damages the device structure, which means that scaling downed device with the abnormal high Ireset is difficult to operate properly.

Figure 9 shows the resistive switching curves of the TiN/Ti/HfO_2/TiN structure depending on the CC. From this result, we can say that Ireset can be controlled as low as 15μA by CC level.

The switching endurance characteristics of the TiN/Ti/HfO_2TiN structure were measured in DC sweep mode and achieved more than $3X10^4$ cycles without the degradation of switching properties, as shown in Fig. 10.

During retention test at the temperature of 200°C, the current level of the TiN/Ti/HfO_2/TiN structure is stable for more than 2000 sec., as shown in Fig. 11.

Accordingly, from the series of the experimental results and proposed model, we can conclude that, the HfO_2 is an appropriate candidate as a resistive switching material for the scaled downed ReRAM device in the future.

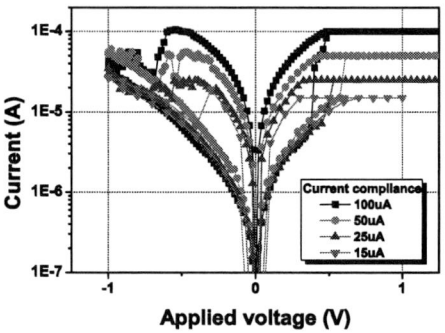

Figure 9. Typical bipolar I-V characteristics of the TiN/Ti/HfO_2/TiN structure depending on the current compliance.

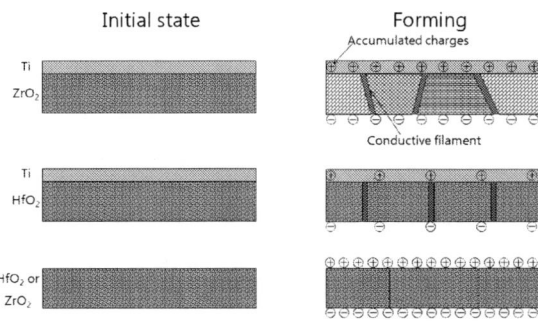

Figure 7. Model of abnormal first high reset current

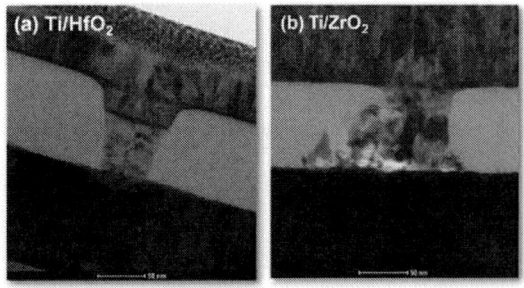

Figure 8. TEM images of the TiN/Ti/TMO/TiN structure after (a) the resistive switching over 10^4 DC cycles, and (b) the first resistive switching cycle.

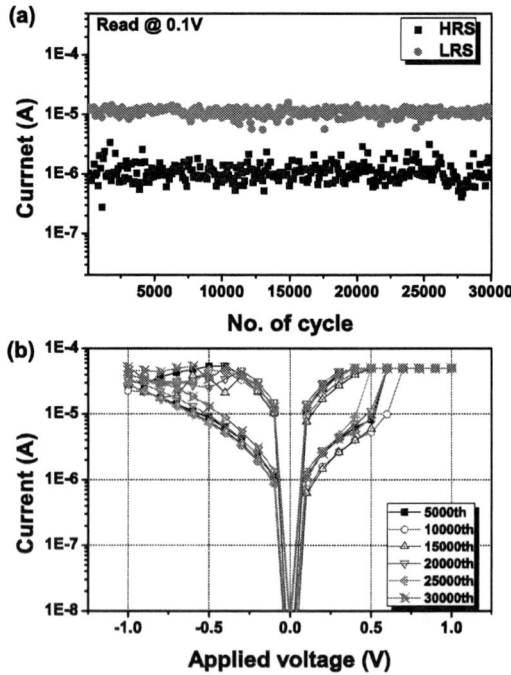

Figure 10. (a) DC switching endurance characteristics of the TiN/Ti/HfO_2/TiN structure during $3x10^4$ and (b) typical bipolar I-V curves every 5000 cycles.

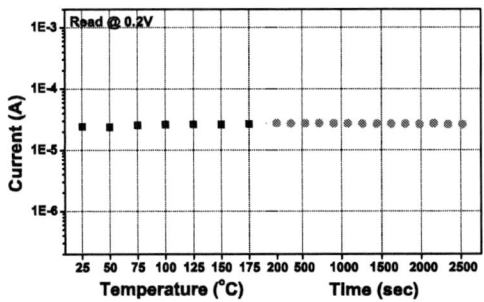

Figure 11. Retention property of the TiN/Ti/HfO$_2$/TiN structure at 200°C.

Conclusions

In this work, satisfactory electrical properties for ReRAM device such as the stability of switching variation, low current, voltage operation, long endurance, and retention characteristics are obtained with the TiN/Ti/HfO$_2$/TiN structure, which shows better switching performances compared to other TMO materials. Theses phenomena can be explained by different material characteristics of the HfO$_2$ in terms of the heat of formation and the capacitive charge.

References

[1] S.S Sheu, et. al., Symp. on VLSI Tech., 82 (2009).

[2] J. Joshua Yang, et. al., Nanotech., 20, 215201 (2009).

[3] C. -Y. Lin, et. al., IEEE Elec. Dev. Let., 28 (5), 366 (2007)

[4] O. Sharia, et. al., Phys. Review B, 79, 125305 (2009).

[5] Sawa, T. Fujii, et. al., Appl. Phys. Lett., 85, 4073 (2004).

[6] J. Muller, et. al., Microelec. Eng., 86, 1818 (2009).

[7] S. J. Song, et. al., Appl. Phys. Lett., 96, 112904 (2010).

A 3D Stackable Carbon Nanotube-based Nonvolatile Memory (NRAM)

Sohrab Kianian, Glen Rosendale, Monte Manning, Darlene Hamilton, X. M. Henry Huang, Karl Robinson, Young Weon Kim, Thomas Rueckes

Nantero, Inc.
25D Olympia Ave.
Woburn, MA 01801
glen@nantero.com

Abstract—A 4Mbit nonvolatile memory with a Carbon Nanotube (CNT) storage element has been manufactured in a 0.25 μm CMOS process at a production fab. The CNT storage element is integrated in BEOL, requires minimal additional processing steps, and only a single additional mask. The memory can be RESET in 50 nanoseconds and SET in 500 nanoseconds. Demonstrated read access time of the development vehicle is 50 nanoseconds. Write endurance is in excess of 10,000 cycles, and robust data retention has been demonstrated. The CNT storage element is scalable to <5 nm, and voltage and current consumption during write operations are low. As intrinsic NRAM SET & RESET times are < 1 nanosecond, improvements in performance are anticipated.

I. INTRODUCTION

The world's first publicly disclosed Nanotube RAM (NRAM) product is described. This memory device is fabricated in 0.25 μm CMOS, and the Carbon Nanotube (CNT) memory element is integrated in the Back End of Line (BEOL) portion of the process[1]. NOR architecture is chosen over NAND to take greater advantage of higher NRAM performance, although NRAM is suitable for NAND as well as NOR. The use of a 0.25 μm foundry was strictly for cost reasons.

The memory test chip is organized as a 4Mb×1, and the target application is nonvolatile storage. The CNT memory element is a 2-terminal variable resistance device, and is selected with a single NMOS transistor. Device performance shows clear advantages compared to Flash memory.

Previous implementations of carbon related semiconductor devices have focused on replacing the silicon substrate (e.g., graphene); or have emphasized novel devices manually assembled in small quantity, in a laboratory environment. These implementations suffer from either the need to overhaul a significant portion of semiconductor manufacturing capability (e.g. elimination or modification of silicon wafers) or do not support manufacturability. The NRAM described here follows a novel approach, emphasizing the use of existing semiconductor manufacturing equipment and processes while achieving the gains of incorporating NRAM as the memory

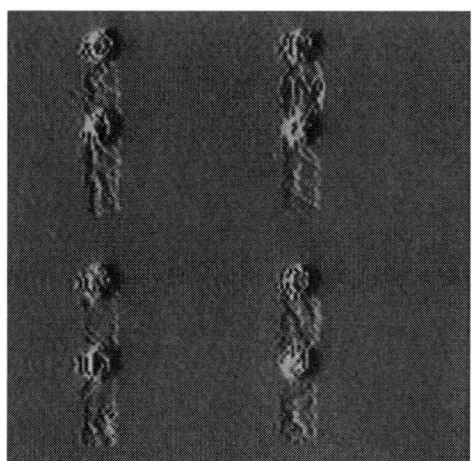

Figure 1.   AFM of patterned CNT on silicon

storage element of an NVM. As a result it is not necessary to significantly change the manufacturing flow. Most significantly NRAM relies on utilizing an IC-grade (i.e. ppb level purity) suspended CNT solution, deposited and patterned via standard fab tools and processes resulting in a consistent film of controlled quality, uniformity, resistance and thickness of CNT. The CNT film formation process is thus key to a controlled manufacturability of NRAM and its integration into base-line process thereby. Fig. 1 is an AFM image of patterned and etched carbon nanotube on a silicon surface overlaying W vias.

II. DEVICE AND OPERATION

Fig. 2 is a diagram of the NRAM bitcell, where the 2–terminal CNT is represented as a switch. Terminals WL (Wordline), BL (Bitline), and SL (Sourceline), respectively, connect the bitcell within the array. The sense of these terms is conventional, with the exception that the Sourceline terminal SL is a common return line for the CNT switch on the opposite side from its connection to the select transistor, rather than the connection to an MOS source junction. The NMOS transistor acts as the selector for the CNT data storage element in this design, although many active elements (diode, BJT, etc) could serve equally as well. Fig. 3 is a chip micro-photograph of the 4Mb characterization vehicle.

---

[1] While numerous implementations are possible, in this implementation (4Mb×1) the CNT switch is integrated between Metal 3 and Metal 4 and the access device consists of an NMOS transistor

978-1-4244-6658-0/10 $26.00 © 2010 IEEE

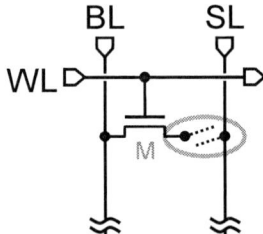

Figure 2. NRAM Bitcell

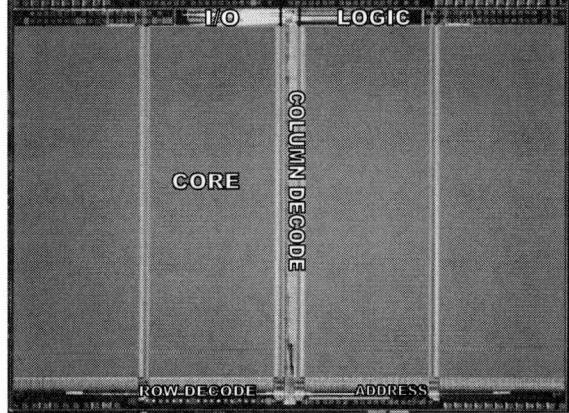

Figure 3. 4Mb NRAM microphotograph

A. Definition of SET and RESET States

The CNT bitcell as described in this paper has 2 states, SET (logical '1') and RESET (logical '0'); although multiple levels are possible (the CNT element can be programmed to a very wide range of resistances). The circuits described in this paper consider a SET cell to have a resistance in the KΩ range, and a RESET cell to have a resistance in the MΩ range.

B. Write Operation

A system of CNT film produces multiple CNT junctions (i.e., one CNT touching another) which provide for a conductive path. These junctions have nano–size surface area and require very small currents to experience Joule heating. A CNT junction (CNT–CNT space <3Å) requires energy of <~1 picojoule to separate (>10Å) and form a non-conducting path (Fig. 4).

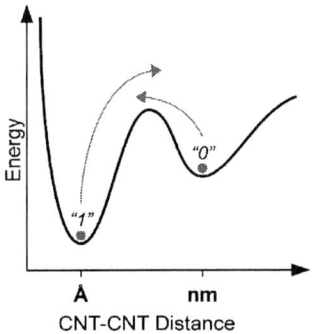

Figure 4. Relative energy states of CNT vs. distance

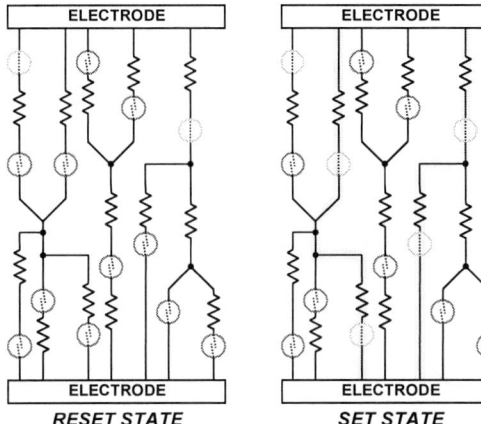

RESET STATE   SET STATE

Figure 5. Electrical equivalent of CNT network in RESET and SET states. Resistors represent intrinsic CNT resistance, and circled switches represent 'junctions' between adjacent touching nanotubes.

The CNT bitcell is written to a RESET state by applying a voltage pulse of <5V to terminal BL of the memory element (Fig. 2) while the other terminal (SL) is grounded. Driver current and bias on transistor M are such that an arbitrarily large current (a few tens of microamps) is allowed to pass through the CNT switch. The pulse is 50 nanoseconds in duration, and results in a high resistance state (RESET, or '0') due to transformation of the CNT network from one of closely touching nanotubes to a network of separated nanotube junctions (Fig. 5).

The bitcell is written to the complementary SET state by applying a similar bias (<5V on BL, 0V on SL) but modifying the driver and bias on device M such that current through the switch is limited to a much smaller current during the 500 nanosecond SET pulse.

Fig. 5 illustrates the resulting electrical equivalent of the CNT network, with a reduced resistance due to the increased number of CNT junctions which have been 'closed' by electrostatic attraction ("closed" switches). Table I summarizes voltage and time requirements for RESET and SET operations.

TABLE I. EXAMPLE PROGRAM AND READ CONDITIONS

RESET	SET	READ TIME
VPP: 4.5V	VPP: 5.0V	"1", 50 nSec
Pulse Width: 50 nSec	Pulse Width: 500 nSec	"0", 30 nSec
* SET is differentiated from RESET by current control. The NRAM CNT is unipolar and bilateral; the same potential is used for both SET & RESET, and can be reversed in either case with no difference in result.		

C. Read Operation

The 4Mb NRAM bitcell is read in NOR fashion with terminal SL at ground potential, and terminal BL at about 1V. Under these bias conditions a cell current (ICELL) of a few µA is typical for a SET bit, and < 10 nA for a RESET bit. The 4Mb NRAM uses a current comparison sense amplifier with an internal reference level of 800nA, adjustable from 100 nA to 10 µA.

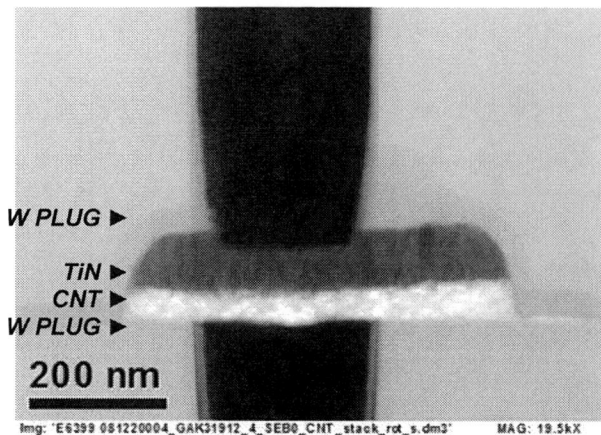

Figure 6.   CNT Bitcell Integration

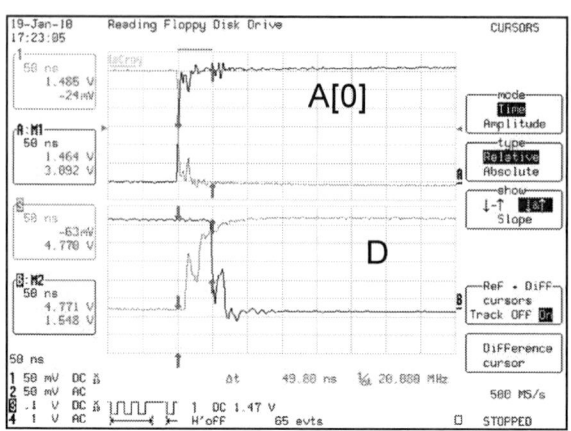

Figure 7. t_{ACC} = 49.8 nSec

III.   INTEGRATION

The NRAM bitcell is comprised of the 2-terminal CNT element and a select device. In this design an NMOS transistor is used as the selector, and the CNT element is integrated between METAL3 and METAL 4 (Fig. 6, transistor not shown).

The CNT element is composed of carbon nanotubes which have been placed in solution and deposited on the wafer surface via spin coat. An electrode is then deposited on top of the CNT layer, then both the CNT and Metal layers are patterned with photoresist to define the desired geometry (all steps are compatible with CMOS fab equipment and the CNT solution can be applied with a standard spin coat track). The resultant structure is a vertical stack of metal (W plugs and TiN metal cladding in this case) interconnects above and below a CNT layer. A standard planarized oxide is deposited as a passivation barrier around the structure. A single extra photomask is required, and only those additional steps necessary to add the CNT layer (depositions, photo patterning, etch & clean). The resulting structure is a vertical stack connecting the metal layers above and below, which are unaffected by this step. In this implementation the metal layers are Al, but there is no barrier to implementing the CNT bitcell in a process with a Cu back end.

The resulting structure has a variable, unipolar resistance from one electrode (W pillar, in this design) to the other. It is electrically identical in either direction, and SET & RESET operations can be identically performed with either polarity. The CNT layer provides the varying resistance (KΩ – GΩ) depending on the stimulus, as described in the "Write Operation" section above.

This integration method makes NRAM particularly suitable for 3-D structures, in which successive layers of bitcells can be fabricated vertically – ideally with an integrated selector. The CNT material is also electrically unaffected by manufacturing temperatures to >700°C.

IV.   ELECTRICAL RESULTS

A.   Access Time

The NRAM development test chip has a single data pin D. Fig. 7 shows a single random read access of the NRAM, illustrating 50 nSec random address access time (tACC) at room temperature. Although this design is organized as 4Mb×1, any organization (e.g. ×8, ×16, ×32) is possible; as well as pipelining techniques to boost sequential throughput to CMOS clock speeds.

B.   Endurance

NRAM has no known intrinsic wear mechanism. Fig. 8 shows ICELL over 10,000 cycles of SET/RESET on a typical array bit from the 4Mb NRAM. Read points are at every cycle for the first 10, then every 50 cycles to completion (10K). The bit passed verify on every cycle, including those whose values are not shown on the graph.

C.   Current Consumption

The NRAM described here has a NOR architecture, and current consumption of the 4Mb NRAM during READ is comparable to high-performance random access NOR memory (e.g. NOR Flash, ROM, etc). As individual bitcell current measurement can be complex, current during SET and

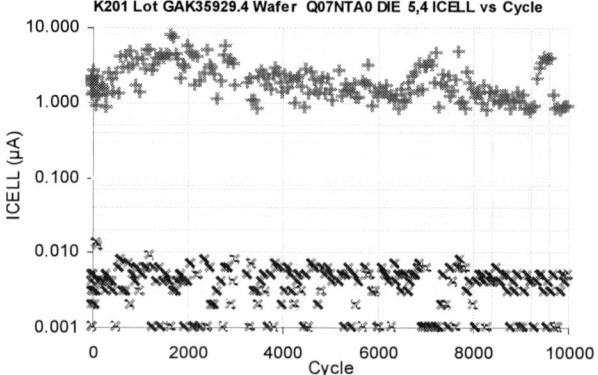

Figure 8.   I$_{CELL}$ over 10,000 SET/RESET cycles

RESET are inferred from bias conditions on a series transistor during write operations as described below.

D. Measuring ISET, IRESET:

The NRAM SET and RESET transitions are intrinsically fast (< 1nSec) and difficult to measure directly, particularly in an array. In order to determine current flow through the CNT element, array selector M (see Fig. 2) can be biased so as to limit the series current to the CNT during SET and RESET. The current required to write data to the CNT can then be determined from the bias voltages on M (WL, BL, SL) and from measured characterization data of device M in a test structure.

Using this configuration, typical current required for RESET of a 4Mb NRAM bit is measured for the operation condition outlined in Table I at ~15 μA and typical current required for SET of a bit is measured at ~1 μA. In addition, the RESET current is intrinsically self-limiting: once the bit goes to the RESET state, current drops to <<1 μA regardless of bias voltages. Table II compares NRAM current consumption during RESET (the higher-current operation) to existing and emerging NVM technologies.

TABLE II.   PROGRAM AND READ CONDITIONS

	Max I during WRITE/PGM
Flash*	~200 μA
MRAM**	~170 μA/~625 μ (1mSec/2nSec)
PCRAM	~200 μA
NRAM	~15 μA
* Flash, ETOX using hot carrier injection	
** MRAM, current is inverse to write time	

E. Retention

NRAM has demonstrated 24 hours of data retention at 125°C. Fig. 9 shows before and after distribution of RESET and SET bits. In neither case is failure seen.

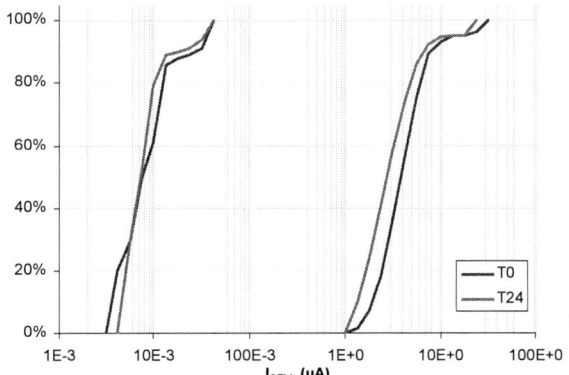

Figure 9. I_{CELL} over 10,000 SET/RESET cycles

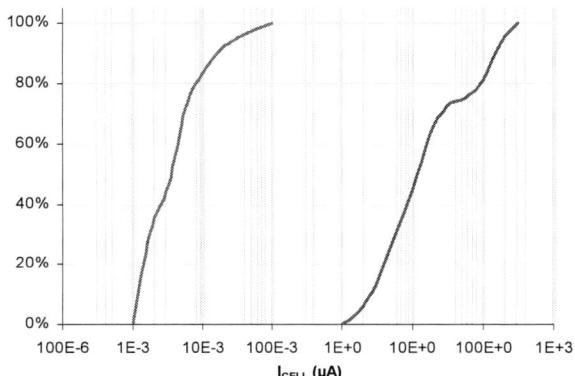

Figure 10. 20nm CNT Bitcell I_{CELL} distribution

V.   SCALING

NRAM has been fabricated in our Woburn laboratory using EBL (E-Beam Lithography) at dimensions smaller than available Fab capability. Fig. 10 illustrates the SET/RESET current distributions of an array of 20nm bitcells.

VI.   CONCLUSION

The first publicly disclosed CNT-based NVM has been demonstrated. It is manufactured in a CMOS production fab, provides write and read performance superior to existing Flash memory, low power consumption during read and write, and is scalable to <5 nm. Endurance and data retention show no wear or failure and are intrinsically expected to surpass that of Flash. The memory storage element (the CNT material) can be paired with any choice of select device (various diode, N/P MOSFET, BJT, etc) and due to its temperature tolerance is suitable for 3-D manufacturing techniques with multiple levels of 'stacked' CNT switches in a monolithic structure as opposed to a multi-die stack approach to 3-D with its associated assembly and interconnect requirements.

REFERENCES

[1]   Yuji Awano, "Graphene for VLSI: FET and Interconnect Applications" *IEDM 2009 Technical Digest*, pp. 10.1.1-10.1.4.

[2]   [Y. Jiang, T.Y. Liow, N. Singh, L.H. Tan, G.Q. Lo, D.S.H. Chan, D.L. Kwong, "Performance Breakthrough in 8nm Gate Length Gate-All-Around Nanowire Transistors using Metallic Nanowire Contacts" *2008 Symposium on VLSI Technology Digest of Technical Papers*, pp 34-35.

[3]   Yiming Huai, "Spin-Transfer Torque MRAM (STT-MRAM): Challenges and Prospects" AAPS Bulletin December 2008, Vol. 18, No. 6, pp 33-40.

[4]   [M. Crowley, A. Al-Shamma, D. Bosch, M. Farmwald, L. Fasoli, A. Ilkbahar, M. Johnson, B. Kleveland, T. Lee, T. Liu, Q. Nguyen, R. Scheuerlein, K. So, and T. Thorp, "512Mb PROM with 8 layers of antifuse/Diode cells," *IEEE International Solid-State Circuits Conference, vol. XLVI*, pp. 284 - 285, February 2003.

[5]   K. M. Brown, "System in package "The Rebirth of SIP"," *2004 IEEE Custom Integrated Circuits Conference*, May 2004.

[6]   G. Servalli, "A 45nm Generation Phase Change Memory Technology", *IEDM 2009 Technical Digest*, pp 5.7.1-5.7.4.

Experimental and simulation study of the program efficiency of HfO$_2$ based charge trapping memories

Sabina Spiga, Gabriele Congedo, Ugo Russo,
Alessio Lamperti, Olivier Salicio
Laboratorio MDM, IMM-CNR
Via C. Olivetti 2,
Agrate Brianza (MB) 20041, Italy
sabina.spiga@mdm.imm.cnr.it

Francesco Driussi and Elisa Vianello
DIEGM and IU.NET
Università di Udine
Via delle Scienze, 208
Udine 33100, Italy
francesco.driussi@uniud.it

Abstract— **This work addresses the use of HfO$_2$ as trapping layer in TaN/Al$_2$O$_3$/HfO$_2$/SiO$_2$/Si (TAHOS) stacks for scaled non-volatile memories, by a complete characterization of the physical properties as a function of thermal budget and film thickness and of the program characteristics, with the aim of a benchmarking with the conventional Si$_3$N$_4$ (TANOS). The TAHOS stack withstands high temperature budget (>1000 °C) and shows similar program speed with respect to TANOS devices, thus revealing similar electron injection conditions. Moreover, the high dielectric constant of HfO$_2$ allows for an efficient EOT scaling and/or a large physical thickness for an improved trapping efficiency. Modeling of program transients contributed to the understanding of the trapping in TANOS/TAHOS devices and to the identification of physical processes possibly limiting the gate stack scaling.**

I. INTRODUCTION

Charge trapping memories based on the TaN/Al$_2$O$_3$/Si$_3$N$_4$/SiO$_2$/Si (TANOS) stack represent the mainstream solution to extend the NAND FLASH scalability to the ultrascaled technology nodes (<2x nm) [1]. The implementation of new materials such as high-dielectric constant (high-*k*) oxides in this stack could further improve the cell performance in term of trade off between program/erase window, program/erase speed and retention. Moreover, the use of high-*k* dielectric as charge trapping layer could help in decreasing the full stack equivalent oxide thickness (EOT) or, alternatively, to implement an oxide with large physical thickness to improve retention [2]. HfO$_2$ could be interesting as charge trapping layer due to its high dielectric constant and thermal stability, and previous studies of Hf-based charge trapping layer in SONOS devices have shown promising results and better program efficiency than Si$_3$N$_4$ [3-5].

In this work, we carried out a comprehensive study, based on physical/electrical analyses and a physics-based modeling, of the properties of HfO$_2$ layers employed as trapping material in the TaN/Al$_2$O$_3$/HfO$_2$/SiO$_2$/Si (TAHOS) stack. Our aim is to perform a thorough analysis of the HfO$_2$ integration capabilities in a charge trapping memory process, of the program/erase characteristics and of the physical processes

involved in the cell operation, in order to allow for an in-depth benchmarking of the HfO$_2$ and the conventional Si$_3$N$_4$ used in non-volatile memory applications.

II. STACK DEPOSITION AND DEVICE FABRICATION

The TAHOS stacks with HfO$_2$ as charge trapping layer were fabricated by in-situ atomic layer deposition (in the Savannah 200 reactor from Cambridge Nanotech) at 300 °C of HfO$_2$/Al$_2$O$_3$ films on 4.5 nm-tunnel-SiO$_2$/Si substrates. The metalorganic compound (MeCp)$_2$Hf(Me)(OMe) and TMA were used, respectively as Hf and Al precursors, while O$_3$ was used as oxygen source. The growth rate is 0.9±0.1 Å/cycle for HfO$_2$ and 0.8±0.1 Å/cycle for Al$_2$O$_3$. After deposition, films were subjected to rapid thermal annealing (RTA) in N$_2$ atmosphere in the 900 °C–1030 °C temperature range for 30 s for Al$_2$O$_3$ crystallization. The selected thicknesses for the HfO$_2$ charge trapping layer were between 10 and 16 nm, while the thickness of crystallized Al$_2$O$_3$ was fixed at 14 nm. A reference TANOS stack with 6 nm thick LPCVD Si$_3$N$_4$, and 14 nm thick Al$_2$O$_3$ blocking layer (O$_3$ based process as for HfO$_2$) was also analysed as benchmarking (same EOT of the 16 nm HfO$_2$-based stack). For the electrical characterization, the stacks were grown on n-Si or p-Si substrates for program and erase curves, respectively. The TaN-W metal gate was deposited after Al$_2$O$_3$ crystallization and annealed at 900 °C in N$_2$ for TaN crystallization. Capacitors with area of 8x10^{-4} cm^2 were defined by optical lithography and TaN etching.

III. PHYSICAL CHARACTERIZATION

As-deposited and RTA films and stacks were characterized by grazing incidence X-ray diffraction (GIXRD), X-ray reflectivity (XRR) and time of flight secondary ion mass spectrometry (ToF-SIMS). Fig. 1 shows the GIRXD patterns of the 10 nm thick HfO$_2$ films incorporated in the TAHOS stacks after RTA at 900 °C and 1030 °C. The as deposited HfO$_2$ is crystallized polymorph (mix of monoclinic and orthorhombic phases). Upon annealing no significant evolution of the HfO$_2$ crystallographic phases is seen, apart from marginal increase of the monoclinic phase. Al$_2$O$_3$ crystallization is achieved for annealing temperatures ≥ 900°C

The work is supported by the European Project GOSSAMER (FP7-research contract 214431) and by Italian MIUR under the FIRB RBIP06YSJJ

978-1-4244-6658-0/10 $26.00 © 2010 IEEE

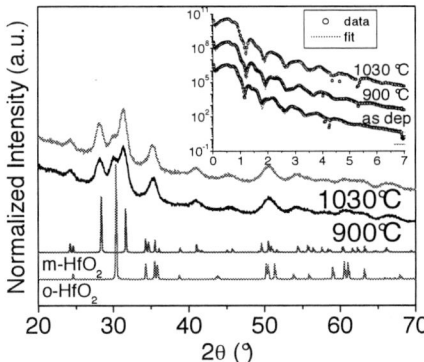

Figure 1. GIRXD patterns of TAHOS stacks annealed at 900 °C and 1030 °C. Reference patterns for monoclinic and orthorombic HfO_2 are also shown. Inset: XRR curves for as deposited and annealed TAHOS stacks (symbol: data, line: fit).

in the cubic gamma phase [6]. XRR data and fitting (inset of Fig. 1) shows that the as deposited stack consists of a tri-layer with thickness of 4.5 nm for SiO_2, 10 nm for HfO_2 and 16.5 nm for Al_2O_3. The same layered structure is preserved after RTA. HfO_2 layer thickness and roughness remain unaltered after annealing, within the error (± 1 nm). On the other hand, after RTA at 900°C the Al_2O_3 electron density increases from 0.95±0.05 e⁻/Å³ (as deposited value) to 1.17±0.05 e⁻/Å³ and film thickness shrinks decreasing to a value around 14 nm. No further evolution is detected for higher annealing temperatures. Similar structural properties were measured for the TAHOS stack incorporating the 16 nm thick HfO_2 layer.

ToF-SIMS depth profiles (not shown) for the as-deposited and after RTA stacks confirm that the layered structure is preserved for all temperatures. Interface stability is fully kept up to 900°C. For RTA of 1030 °C, a limited diffusion of Hf into Al_2O_3 and Al into HfO_2 is observed, while the SiO_2/HfO_2 interface is stable. For temperatures higher than 1030 °C the diffusion phenomena increase significantly.

Therefore, the thermal budget in the 900 °C-1030 °C is suitable for the TAHOS stacks. While RTA at 900 °C guarantees a complete thermal stability of the interfaces, the RTA at 1030 °C could be the best trade-off for a full Al_2O_3 crystallization [6] with limited diffusion phenomena.

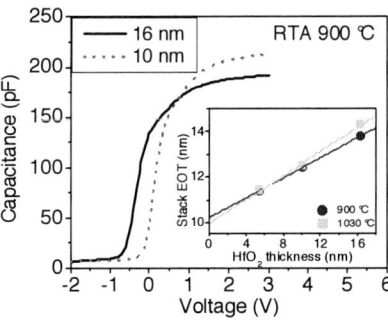

Figure 2. Capacitance-voltage of the TAHOS stacks (capacitor area: 8×10^{-4} cm²). Inset: Plot of the stack EOT versus the HfO_2 physical thickness.

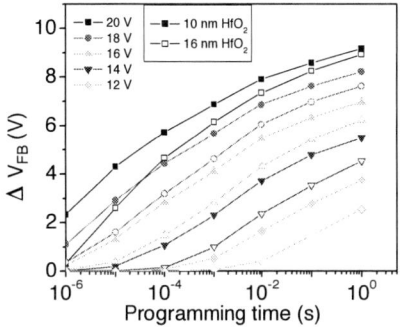

Figure 3. Programming transient for the TAHOS stacks annealed at 900 °C with HfO_2 thickness of 10 nm (full symbols) and 16 nm (open symbols).

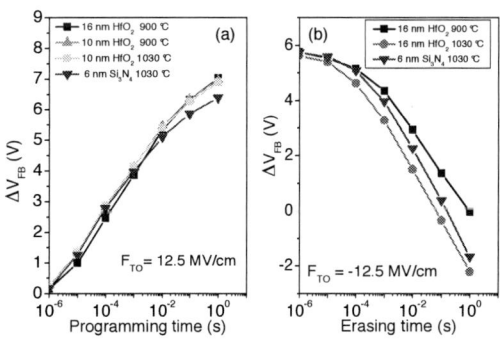

Figure 4. Programming (a) and erasing (b) transients acquired at the same applied field to the tunnel oxide F_{TO}.

I.   ELECTRICAL CHARACTERIZATION

Fig. 2 shows the capacitance-voltage (CV) curves of the TAHOS stacks acquired at 100 kHz, while the inset of Fig. 2 shows the experimental EOT values of the stacks versus the HfO_2 physical thickness. The extracted k value for HfO_2 is 16 for the annealing temperature of 900 °C and slightly decreases to 15 for the annealing at 1030 °C, likely related to the observed diffusion phenomena described in the previous section. The EOT of the intercept (10.0±0.2 nm) is consistent with the sum of a 4.5 nm SiO_2 and a 14 nm thick Al_2O_3 with a k value around 10 (i.e. EOT of Al_2O_3 around 5.5 nm).

Fig. 3 shows the programming transients (the ΔV_{FB} is extracted from the shift of the flat band of CV curves after each programming pulse, with respect to the fresh sample) acquired at various gate voltages (V_G) for the TAHOS stacks with 16 nm and 10 nm thick HfO_2. The memory stack can be efficiently programmed with a large V_{FB} shift of 5-7 V at typical operative condition (18-20 V, 100 μs-1 ms). Moreover, the TAHOS stack with reduced EOT (10 nm HfO_2) can be efficiently programmed with a reduced V_G (Fig. 3). From the comparison of programming transients of TAHOS and TANOS stacks at the same electric field applied to the tunnel oxide F_{TO} (Fig. 4(a)), the HfO_2 charge trapping layer shows a V_{FB} shift comparable to the reference TANOS stack (same EOT of the 16 nm HfO_2-TAHOS stack) for programming time

978-1-4244-6658-0/10 $26.00 © 2010 IEEE   329

$\leq$ 1 ms and a slightly improved V_{FB} shift for larger programming times. This small slowdown of the programming speed of the TANOS could indicate the beginning of the saturation of the number of available traps in the thin Si_3N_4 layer with respect to the HfO_2 films, which are thicker and thus possibly more trap rich. No significant effect of HfO_2 thickness is observed in the 10-16 nm range. Finally, TAHOS erase operation is slightly better than TANOS (Fig.4(b)), for the same stack thermal budget.

II. MODELING

To investigate in detail the program efficiency of TAHOS devices and to better benchmark their operation with the standard TANOS stack, we performed simulations of the program transients of the measured cells. We used the numerical model described in [7], that solves the electrostatics self–consistently with the in and out electron tunneling fluxes, the drift–diffusion transport in the Conduction Band (CB) of the trapping layer, and the SRH generation/recombination inside the trapping film (considering Poole–Frenkel emission).

Fig. 5(a) shows the comparison between the experimental program curves of the TAHOS cell with 16 nm HfO_2 and the modeling results. For these simulations, we used the band alignment among different materials and the effective tunneling masses (m^*) reported in Fig. 6, which have been obtained from the literature [5,8,9] and by comparison

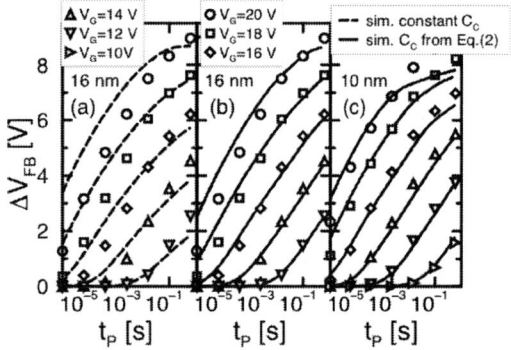

Figure 5. Experimental and simulated program curves of the 16 nm HfO_2 (a), (b) and of the 10 nm HfO_2 (c) TAHOS cells. (a) The dashed lines represent simulations with the model in [7]; (b), (c) the solid lines are simulations with the modified model that accounts for the electron energy dependence of trapping in the HfO_2 layer. Trap energy depth: 1.6 eV.

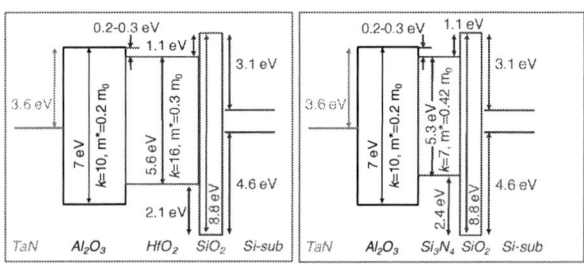

Figure 6. Schematic energy band alignment of the TAHOS and TANOS stacks. m^* is the effective tunneling mass, m_0 is the free electron mass.

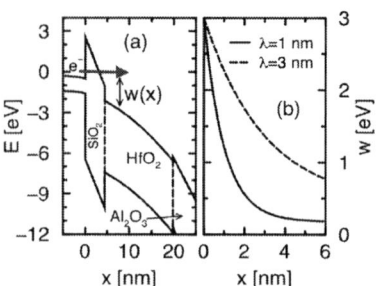

Figure 7. (a) Energy band diagram of the TAHOS cell during programming at V_G=18 V. The injected electrons reach the HfO_2 with high energy with respect the CB minimum. (b) Energy relaxation calculated with Eq.(1), while electrons are travelling in the CB of HfO_2. F=5 MV/cm.

between simulations and experiments of tunneling current through Al_2O_3 layers (not shown). Note the reduced conduction band misalignment between the HfO_2 and Al_2O_3.

The other model parameters are taken from [8] and, in particular, we used a trap concentration N_T=4·10^{19} cm^{-3} and a constant capture coefficient C_C=5·10^8 cm^{3}s^{-1} (in agreement with the modeling of SONOS cells with thin tunnel oxide), that are the main parameters governing the capture into the traps of the electrons traveling in the CB (see Eq.(7) of [7]).

We note a poor agreement between the model and the experiments (see Fig. 5(a)). In particular, the model does not grasp the correct V_G dependence of the curves. The experiments show a reduced program efficiency at large V_G [2], thus suggesting a decreasing trapping of the injected electrons with increasing V_G. In this respect, during the programming of SONOS cells with thin tunnel oxides (about 2 nm), the electrons are essentially injected at the bottom of the CB of the trapping layer, so the trapping is expected to be constant along the Si_3N_4 thickness [7]. However, as observed in Fig. 7(a), the Fowler-Nordheim tunneling take place due to the thicker tunnel oxide in TANOS/TAHOS cells (4.5 nm), and the electrons reach the trapping layer with an high average energy (w) with respect to the CB minimum. The w energy is higher for larger V_G and this may reduce the capability of the HfO_2 traps to capture the high energetic electrons traveling in the CB, as proposed in [10]. Therefore, to simulate the programming of TANOS and TAHOS stacks, we implemented in the model a variable capture coefficient C_C for the traps, which accounts for the energy of the electrons traveling in the CB. To describe the energy relaxation during the electron motion, we consider $w(x)$ dependent on the electric field ($F(x)$) at the position x inside the trapping layer and on a decay length (λ) that summarize the scattering events occurring in the CB of the HfO_2 film [11]

$$\frac{dw}{dx} = \frac{3}{5}qF - \frac{w - w_0}{\lambda} \qquad (1)$$

In Eq.(1), q is the elementary charge and w_0 is the average energy of an equilibrium distribution. Fig. 7(b) shows the w evolution along the HfO_2 thickness for different λ values and for a constant electric field in the layer.

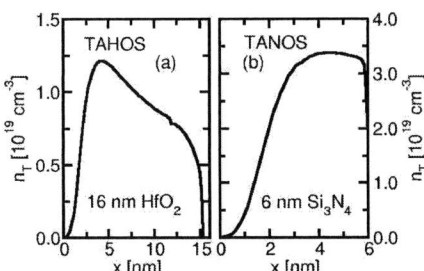

Figure 8. Simulated trapped charge distribution along the trapping layer thickness for the TAHOS (a) and TANOS (b) cells programmed for 1 ms at V_G=18 V.

To account for the possible impact of the average energy of the CB electrons on the trapping probability of the HfO_2 layer, we assumed an empirical exponential dependence of the capture coefficient C_C

$$C_C(x) = C_0 \cdot \exp\left\{-r \cdot \left[w(x) - w_0\right]\right\} \quad (2)$$

where r is an empirical parameter, while C_0 is the capture coefficient related to electrons travelling at the bottom of the CB. We assumed C_0=5·10⁸ cm⁻³s⁻¹, in agreement with [7] for the thin tunnel oxide SONOS devices.

Therefore, Eqs. (1) and (2) make possible to take into account of the average energy of the electrons for each applied V_G, for any programming level (ΔV_{FB}) and at any distance from the tunnel oxide. We validated such approach by well reproducing both the TANOS reference data (not shown) and the TAHOS characteristics. In both cases, we assumed λ=1.1 nm and r=3.5 eV⁻¹ and good agreement has been found between the model and the program curves of the 16 nm and 10 nm HfO_2 device, as shown by Fig. 5(b) and (c).

The high energy electrons injected at the interface between the tunnel oxide and the trapping layer are difficult to be trapped by the HfO_2 film. However, while travelling, these electrons loose energy as shown in Fig. 7(b) and are more prone to be trapped [2]. Therefore, the calculated trapped charge profiles (Fig. 8) present a very low concentration near the tunnel oxide, while they increase at the centre of the trapping layer. This is much more evident in the TANOS with thinner Si_3N_4 (b), while the TAHOS (a) is less affected by the reduced trapping near the tunnel oxide because of the thicker HfO_2. Near the blocking oxide there is again a decrease of the trapped charge because of the few CB electrons reaching the last part of the layer and of the tunnelling towards the gate.

III. CONCLUSIONS

The physical analyses show that polycrystalline HfO_2 can be efficiently integrated as charge trapping layer in the TAHOS stack, as withstands the high temperature annealing requested for Al_2O_3 crystallization, without degrading the layers and interface properties up to 900-1030 °C.

The TANOS and TAHOS samples show similar program characteristics for similar electric fields across the tunnel oxide of the two devices, thus revealing very similar electron injection conditions. This confirms very similar band alignments between the SiO_2/Si_3N_4 and the SiO_2/HfO_2 interfaces, and the high-k value of HfO_2 allows for an efficient scaling of the EOT and of the programming voltages in TAHOS cells. Moreover, at large ΔV_{FB}, the TAHOS devices show improved programming with respect to TANOS, possibly because of the beginning of the saturation of the available traps in the thinner Si_3N_4 layer. This could be relevant for multi-bit applications, in which large threshold voltage shifts are required to the cell.

The simulation analysis revealed that, due to the thick tunnel oxide, the electrons are injected with high energy in the trapping layer in both TANOS and TAHOS devices. A model approach to account, in the trapping process, for the average energy of the electrons traveling in the CB of the HfO_2 has been proposed and good agreement with experiments has been found.

Concerning the modeling of the HfO_2 film, the parameters used for the simulations are in good agreement with those for the standard TANOS stacks (similar trap density and same parameters for the electron energy relaxation), revealing that the trapping processes in the two materials are similar.

The charge distributions after programming show a limited trapping near the interface with the tunnel oxide. This can be an important constraint in the viewpoint of the vertical gate stack scaling, since can impose limits to the minimum thickness of cells with good program efficiency.

ACKNOWLEDGMENT

The authors would like to thank Numonyx for providing the Si/SiO₂ and Si/SiO₂/Si₃N₄ stacks and Prof. L. Selmi for helpful discussions.

REFERENCES

[1] C. H. Lee et al., "A Novel SONOS Structure of SiO₂/SiN/Al₂O₃ with TaN metal gate for multi-giga bit flah memories", in *IEDM Tech Dig*, 2003, pp. 613-616.

[2] E. Vianello et al., "Program efficiency and high temperature retention of SiN/high-k based memories", *Microel. Engin.* vol.86, p.1830, Jul. 2009.

[3] P.-H. Tsai et al., "Charge Trapping-type Flash Memory Devices with stacked high-k charge trapping layer", *IEEE Electron Device Letters*, vol. 30, pp 775-777, July 2009

[4] Y. Q. Wang et al., "Fast erasing and highly reliable MONOS type memory with HfO₂ high-k trapping layer and Si₃N₄/SiO₂ tunneling stack", in *IEDM Tech. Dig. 2006*, pp.1-4, Dec. 2006

[5] Y.-N. Tan et al., "Over-Erase Phenomenon in SONOS-Type lash Memory and its Minimization Using a Hafnium Oxide Charge storage layer", *IEEE Trans. Electron Devices.*, vol. 51, p 1143-1147, July 2004

[6] M. Alessandri et al., "High-k materials in Flash Memories", *ECS Transactions* 1, pp. 91 (2006)

[7] E. Vianello et al., "Experimental and Simulation Analysis of Program/Retention Transients in Silicon Nitride-Based NVM Cells", *IEEE Trans. Electron Devices*, vol.56, pp. 1980-1990, Sept. 2009

[8] V.V. Afanas'ev et al., "Internal photoemission of electrons and holes from (100) Si into HfO₂"*Appl. Phys. Lett.* 81, pp 1053-1055, Aug. 2002

[9] H. Y. Yu et al., "Energy gap and band alignment for (HfO₂)ₓ(Al₂O₃)₁₋ₓ on (100) Si", *Appl. Phys. Lett* 81, pp. 376-378, July 2002

[10] C. M. Compagnoni et al., "Physical modeling for programming of TANOS memories in the Fowler-Nordheim regime," *IEEE Trans. Electron Devices*, vol. 56, pp. 2008-2015, Sept. 2009.

[11] L. Selmi and C. Fiegna, "Physical aspects of cell operation and reliability", in Flash Memories, Kluwer Academic Publishers, 1999

Carrier Transport Characteristics of Strained N-MOSFET Featuring Channel Proximate Silicon-Carbon Source/Drain Stressors for Performance Boost

Shao-Ming Koh,[1] Peng Zhang,[1] Shu-Feng Ren,[1] Chee-Mang Ng,[2] Ganesh S. Samudra,[1] and Yee-Chia Yeo.[1,*]

[1] Dept. of Electrical & Computer Engineering, National University of Singapore (NUS), 117576, Singapore.
[2] GLOBALFOUNDRIES Singapore Pte. Ltd., 60 Woodlands Industrial Park D, St 2, 738406, Singapore.
* Email: eleyeoyc@nus.edu.sg or yeo@ieee.org

Abstract — We report a study of carrier transport in strained n-channel MOSFETs (nFETs) with embedded silicon-carbon (Si:C) source/drain (S/D) stressors formed in close proximity to the channel, taking parasitic resistance into account in the extraction of carrier transport parameters. While bringing the Si:C S/D stressors closer to the channel improves their effectiveness in imparting tensile strain to the channel, a degradation in ballistic efficiency B_{sat} due to increased carrier scattering is observed. This is compensated, however, by an increase in the carrier injection velocity v_{inj}, thereby resulting in an on-state current I_{On} enhancement of ~7 % in nFETs with channel-proximate Si:C S/D over nFETs with conventional *e*-Si:C S/D. In addition, the impact of channel orientation on carrier transport characteristics for the new process integration scheme is also evaluated in this paper.

I. INTRODUCTION

Channel strain engineering for n-channel MOSFETs or nFETs using stress memorization technique [1] and high tensile stress silicon nitride etch-stop layer [2] have been widely adopted in integrated circuit manufacturing. Recently, the embedded silicon-carbon (*e*-Si:C) source/drain (S/D) stressor formed by ion implantation of carbon followed by solid-phase epitaxy (SPE) has gained attention as another viable option for strain engineering of silicon nFETs [3]-[5].

With reduction of gate pitch for higher circuit density, performance degradation associated with effective stress loss in the channel from various strain engineering schemes can be substantial [6]. To enable further pitch reduction with minimal performance compromise, strained nFETs with enhanced strain effects by deploying Si:C S/D stressors in closer proximity to the channel, hereafter referred to as channel-proximate (CP) Si:C S/D stressors, has been proposed [5]. Fig. 1(a) shows a transmission electron microscopy image of an nFET with CP Si:C S/D studied in this work.

In this paper, further improvement in the on-state current I_{On} with CP Si:C S/D as compared to the conventional Si:C S/D is elucidated from the standpoint of the carrier transport characteristics extracted using a characterization approach based on the temperature dependent channel backscattering model [7]. For the first time, the impact of channel orientation

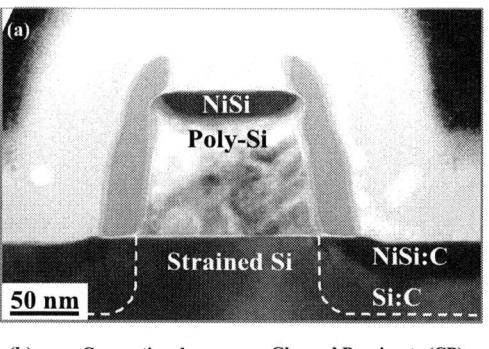

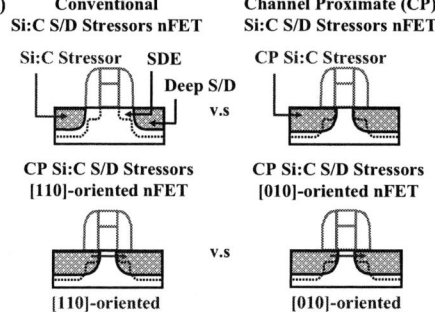

Figure 1. (a) Transmission electron microscopy (TEM) image of an nFET with CP Si:C S/D. The spacer dimension is ~25 nm for all devices. (b) Experimental splits studied in this work. Carrier transport characteristics of Si:C S/D and CP Si:C S/D nFETs are analyzed. The effect of channel orientation on carrier transport is also investigated.

on carrier transport characteristics of CP Si:C S/D nFETs is also evaluated.

II. TEMPERATURE DEPENDENT BACKSCATTERING MODEL AND EXTRACTION METHODOLOGY

Experimental splits investigated in this paper are shown in Fig. 1(b). Fig. 2(a) summarizes the extracted R_{series} of various nFET splits. The R_{series} extraction method is found in Ref. 8. Generally, R_{series} is higher for Si:C S/D and CP Si:C S/D than for the Si S/D. Incorporating carbon (C) in the S/D

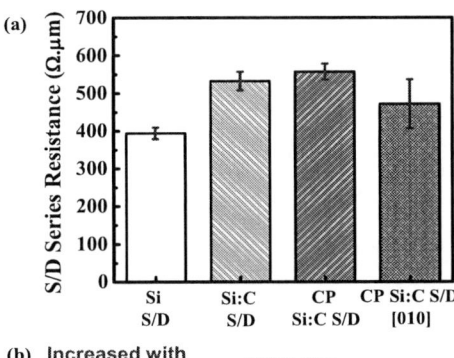

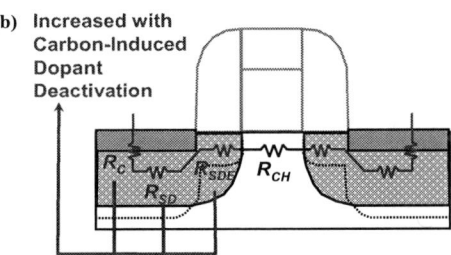

Figure 2. (a) Parasitic resistance R_{series} of various device splits. (b) Schematic illustrating the various components of R_{series}. Reduced dopant activation could occur in regions where carbon is present.

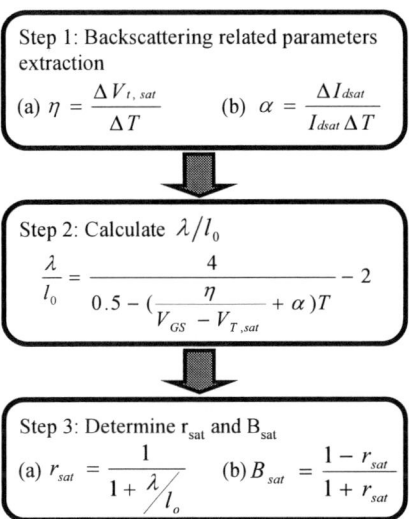

Figure 3. Experimental procedure for the extraction of carrier backscattering coefficient r_{sat} and ballistic efficiency B_{sat} based on a temperature dependent channel backscattering model [7],[9].

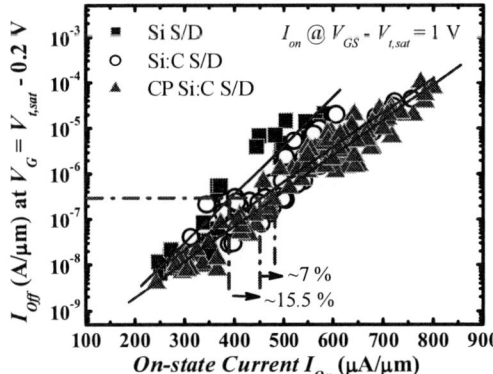

Figure 4. At an I_{Off} of 300 nA/μm, nFET with CP Si:C S/D has 23 % and 7 % I_{On} enhancement over unstrained nFET and nFET with conventional Si:C S/D, respectively.

increases R_{series} due to reduced dopant activation [Fig. 2(b)]. The extracted R_{series} for CP Si:C S/D, where C is present in the As-doped S/D extension (SDE), is slightly higher than that of Si:C S/D, where C is absent in the SDE. This is attributed to reduced activation of As in the presence of C.

In the backscattering theory, the near equilibrium mean free path λ and the critical length l_o for carrier backscattering are considerably affected by the intrinsic voltages V_{GS} and V_{DS} [9]. As R_{series} will affect the V_{GS} and V_{DS} applied for the device splits differently, it is vital to account for the R_{series} effect in the extraction methodology.

Fig. 3 explains the measurement procedure and analytical expressions employed in the extraction of the carrier transport characteristics. Carrier backscattering coefficients are determined from the parameters η and α, which represent the temperature dependence of threshold voltage $V_{t,sat}$ and drive current I_{dsat}, respectively. $V_{t,sat}$ is extracted at constant current of $0.1×(W/L)$ μA, where W and L are the physical gate width and length of the device. Using $V_{t,sat}$ instead of $V_{t,lin}$ in estimating Q_{inv} allows accurate reflection of the Drain-Induced Barrier Lowering (DIBL) effect on threshold voltage reduction [10]. I_{dsat} is determined at a gate overdrive of 1 V, with R_{series} accounted for. The effect of R_{series} is included in the calculation $λ/l_o$ by replacing V_{GS} in the $λ/l_o$ equation with $V_G - 0.5I_{dsat}R_{series}$, where V_G is the applied terminal voltage.

III. RESULTS AND DISCUSSION

A. Carrier Transport Characteristics of CP Si:C S/D

Fig. 4 shows the I_{On}-I_{Off} plot of nFETs with Si S/D, Si:C S/D and CP Si:C S/D after R_{series} correction. At a fixed I_{Off} of 300 nA/μm, Si:C S/D achieves an I_{On} enhancement of ~15.5 %

over unstrained nFETs with Si S/D. This is due to the lateral tensile strain induced in the Si channel by the Si:C S/D stressors, which enhances electron mobility. With CP Si:C S/D, the strain effect is higher, therefore giving a further ~7% improvement in I_{On} improvement over nFETs with Si:C S/D.

Fig. 5 depicts r_{sat} as a function of gate length L_G for nFETs with CP Si:C S/D and conventional Si:C S/D. Standard deviation of r_{sat} at each L_G is plotted as error bars. NFETs with CP Si:C S/D have higher r_{sat} than nFETs with conventional Si:C S/D for all L_G, implying higher carrier scattering due to the CP Si:C S/D. An increase in strain has been reported to reduce r_{sat} due to a lower l_o as a result of conduction band barrier modulation [11]. Here, the higher carrier scattering observed in nFETs with CP Si:C S/D is probably caused by the diffusion of carbon into the strained silicon channel, since the CP Si:C S/D stressors are close to the channel. The increased r_{sat} in nFETs with CP Si:C S/D degrades B_{sat}, as compared with nFETs with Si:C S/D [Fig. 6].

978-1-4244-6658-0/10 $26.00 © 2010 IEEE

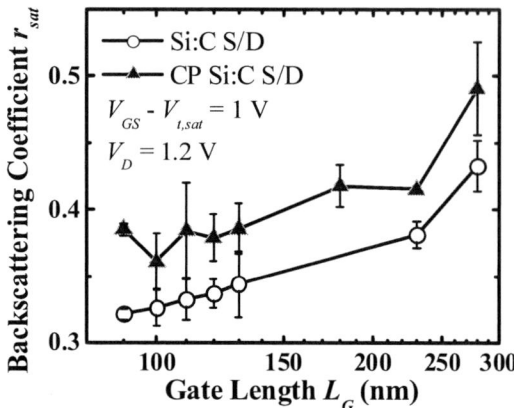

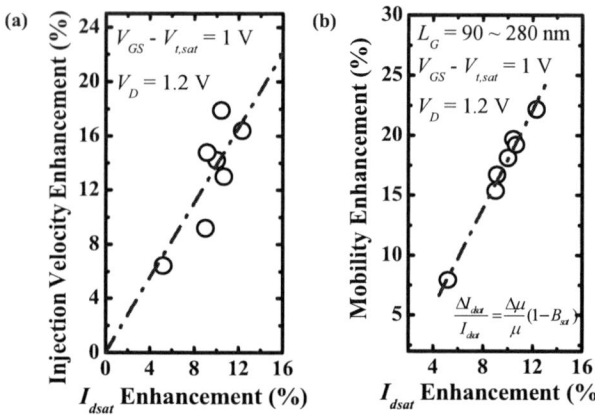

Figure 5. Devices with CP Si:C S/D show higher r_{sat} than devices with conventional Si:C S/D (formed after spacers). The increased carrier scattering could be caused by the diffusion of carbon into the strained silicon channel.

Figure 7. (a) Significant increase in v_{inj} accounts for the I_{dsat} enhancement observed in CP Si:C S/D nFETs over Si:C S/D nFETs. (b) Correlation of μ and I_{dsat} enhancement.

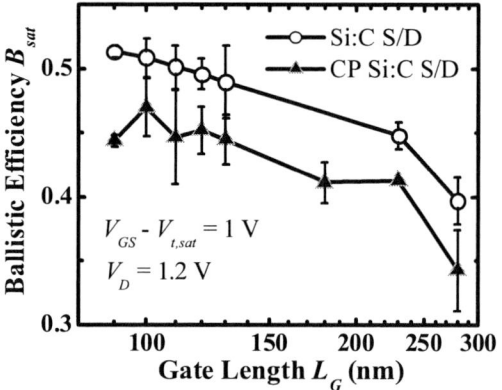

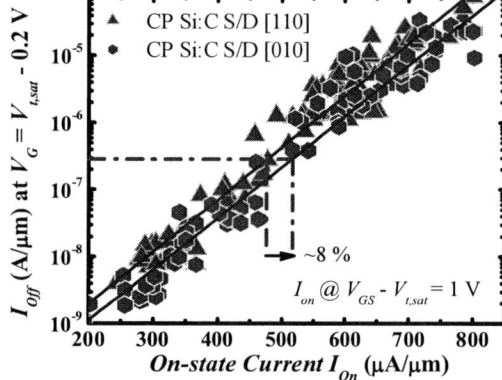

Figure 6. B_{sat} is lower in nFETs with CP Si:C S/D than in nFETs with Si:C S/D. A lower B_{sat} implies increased carrier scattering in the channel near the source.

Figure 8. CP Si:C S/D stressors with [010]-oriented channel give ~8 % higher I_{On} than CP Si:C S/D with [110]-oriented channel.

Next, the carrier injection velocity enhancement Δv_{inj} of strained nFETs with CP Si:C S/D is investigated. Fig. 7(a) plots the percentage increase in v_{inj} against I_{dsat} enhancement of strained nFETs with CP Si:C S/D. CP Si:C S/D nFETs have higher v_{inj} as compared to nFETs with conventional Si:C S/D. The v_{inj} enhancement is attributed to further reduction in electron effective mass m^*_e from the enhanced strain effects. I_{dsat} enhancement is lower than v_{inj} enhancement due to the degraded B_{sat} of CP Si:C S/D nFETs. The increase in v_{inj} is responsible for the I_{dsat} enhancement.

Fig. 7(b) shows the relationship between the enhancement of carrier mobility μ and the enhancement in I_{dsat} of CP Si:C S/D nFETs. Through a best fitting to the μ versus I_{dsat} enhancement data, a linear relationship with a gradient of ~2 is obtained for CP Si:C S/D nFETs with $L_G = 90 - 280$ nm, which is consistent with the value reported in Ref. 12. This implies that μ continues to be an important device parameter affecting the drive current performance of short channel MOSFETs.

B. Impact of Channel Orientation on Carrier Transport Characteristics of CP Si:C S/D

Here, the impact of channel orientation on carrier transport characteristics will be investigated. Fig. 8 plots the I_{On}-I_{Off} plot of CP Si:C S/D nFETs with [110]-oriented and [010]-oriented channels. The [010] channel gives ~8 % higher I_{On} than the [110]-oriented channel. This is qualitatively consistent with the piezoresistance coefficients.

The B_{sat} and v_{inj} of [010]-oriented CP Si:C S/D nFETs were also extracted. Fig. 9 shows that B_{sat} improves slightly with the incorporation of tensile strain induced by CP Si:C S/D stressors along the [010] Si channel direction. The slight improvement in B_{sat} for CP Si:C S/D nFETs with [010] oriented channel is due to a reduced r_{sat}. However, the small improvement in B_{sat} is insufficient to account for the I_{dsat} enhancement observed. Further investigation is carried out by correlating v_{inj} enhancement with I_{dsat} enhancement [Fig. 10]. Increase in v_{inj} contributes predominantly to I_{dsat} enhancement.

978-1-4244-6658-0/10 $26.00 © 2010 IEEE

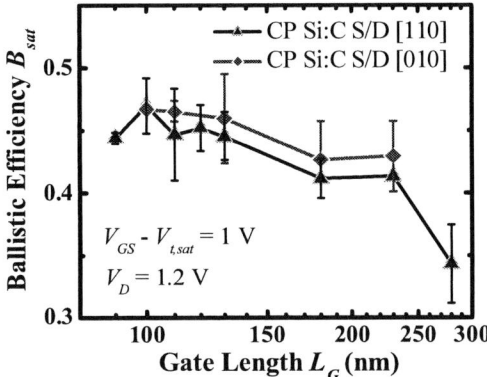

Figure 9. B_{sat} in [010]-oriented nFETs with CP Si:C S/D nFETs is slightly higher than that in [110]-oriented nFETs with CP Si:C S/D.

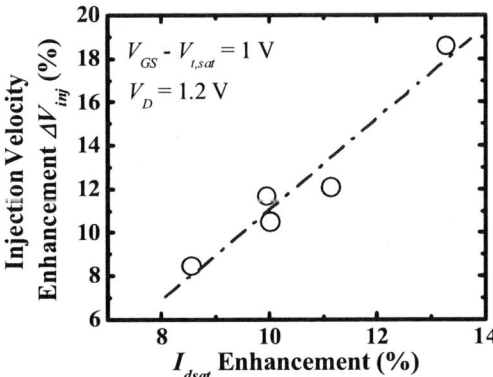

Figure 10. For nFETs with CP Si:C S/D, I_{dsat} enhancement for [010]-oriented devices over [110]-oriented devices is contributed by a higher v_{inj}.

Fig. 11 compares the percentage change in I_{dsat}, B_{sat}, and v_{inj} for nFETs with CP Si:C S/D with that of nFETs with conventional Si:C S/D and [110] channel direction. While CP Si:C S/D nFETs have lower B_{sat} due to increased carrier scattering, the large improvement in v_{inj} results in I_{dsat} enhancement over nFETs with conventional Si:C S/D. The v_{inj} improvement is higher for the [010] channel orientation as compared with the [110] channel orientation. A combination of CP Si:C S/D stressors and the [010] channel orientation may be pursued for performance optimization.

IV. CONCLUSION

Backscattering characterization with R_{series} correction was performed to investigate enhancement in carrier transport in CP Si:C S/D nFETs. As compared with unstrained control or nFET with Si:C S/D, CP Si:C S/D nFETs have a degraded B_{sat} but an enhanced v_{inj}, which contributed to a higher I_{dsat}. Carrier transport characteristics of CP Si:C S/D nFETs with [010] channel direction was also investigated. Greater B_{sat} and v_{inj} improvements observed in CP Si:C S/D [010]-oriented devices make it a good option for possible use in future technology generations.

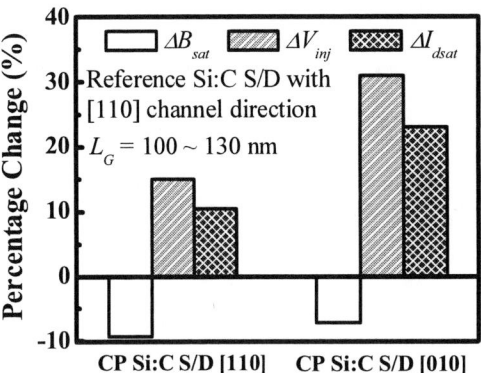

Figure 11. Comparison of carrier transport characteristics of nFETs with CP Si:C S/D with that of [110]-oriented nFETs with Si:C S/D. The I_{dsat} enhancement is primarily due to v_{inj} improvement.

ACKNOWLEDGMENT

Research Grant from National Research Foundation (NRF-RF2008-09) is acknowledged. S.-M. Koh acknowledges GLOBALFOUNDRIES Singapore and the Economic Development Board of Singapore for a Graduate Scholarship.

REFERENCES

[1] C. H. Chen et al., "Stress memorization technique (SMT) by selectively strained-nitride capping for sub-65 nm high performance strained-Si device application," in VLSI Symp. Tech. Dig., 2004, pp. 56-57.

[2] T. Ghani et al., "A 90 nm high volume manufacturing logic technology featuring novel 45 nm gate length strained silicon CMOS transistors," in IEDM Tech. Dig., 2003, pp. 978–980.

[3] Y. C. Liu et al., "Strained Si channel MOSFETs with embedded silicon carbon formed by solid phase epitaxy," in VLSI Symp. Tech. Dig., 2007, pp. 44–45.

[4] S.-M. Koh et al., "N-channel MOSFETs with embedded silicon carbon source/drain stressors formed using Cluster-Carbon implant and excimer laser-induced solid phase epitaxy," IEEE Elect. Dev. Lett., 29, pp. 1315, 2008.

[5] S.-M. Koh et al., "Channel-proximate silicon-carbon source/drain stressors for performance boost in strained n-channel field-effect transistors," in Ext. Abst. of the 2009 Int. Conf. on Solid State Dev. and Mat., Oct. 7-9, 2009, pp. 18-19.

[6] J. W. Sleight et al., "Challenges and opportunities for high performance 32 nm CMOS technology," in IEDM Tech. Dig., 2006, pp. 697–700.

[7] M.-J. Chen et al., "Temperature dependent channel backscattering coefficients in nanoscale MOSFETs," in IEDM Tech. Dig., 2002, pp. 39-42.

[8] A. Dixit et al., "Analysis of the parasitic S/D resistance in multiple-gate FETs," in IEEE Trans. Elect. Dev., 52, pp. 6, 2005.

[9] A. Rahman and M. S. Lundstrom, "A compact scattering model for the nanoscale double-gate MOSFET," IEEE Trans. Elect. Dev., 49, pp. 481, 2002.

[10] W. Lee and P. Su, "On the experimental determination of channel backscattering characteristics–limitation and application for the process monitoring purpose," in IEEE Trans. Elect. Dev., 56, pp. 2285, 2009.

[11] K.-W. Ang et al., "Carrier backscattering characteristics of strained N-MOSFET featuring silicon-carbon source/drain regions," 36th European Solid-State Dev. Res. Conf., Sep. 18-22, 2006, pp. 89-92.

[12] M. S. Lundstrom, "On the mobility versus drain current relation for a nanoscale MOSFET," IEEE Elect. Dev. Lett., 22, pp. 293, 2001.

Fluorinated CMOS HfO₂ for High Performance (HP) and Low Stand-by Power (LSTP) application by pre- and post-CF₄ Plasma Passivation

Woei-Cherng Wu[1], Chao-Sung Lai[2*], Huai-Hsien Chiu[2], Jer-Chyi Wang[2], Pai-Chi Chou[2], and Tien-Sheng Chao[1*]

[1]Department of Electrophysics, National Chiao Tung University, 1001 Ta Hsueh Rd, Hsinchu, Taiwan
[2]Department of Electronic Engineering, Chang Gung University, 259 Wen-Hwa 1st Road, Kwei-Shan, Tao-Yuan, Taiwan
Tel: +886-3-2118800 ext 5786 Fax: +886-3-2118507 cslai@mail.cgu.edu.tw
Tel: +886-3-5712121 ext 31367 Fax: +886-3-5725230 tschao@mail.nctu.edu.tw

Abstract

Fluorine distribution engineering in HfO2 to get higher performance (HP EOT=1.2nm) and low stand-by power (LSTP EOT=1.8nm) CMOS device with 48% driving current enhancement for HP, 73% leakage reduction for LSTP were demonstrated. Better uniformity (50% VtU and 9% IdU reduction) and excellent reliability, P/NBTI, of n-/p-MOSFET were achieved A new physical model of fluorine re-incorporation was proposed to explain the turnaround of NBTI and the much improvement of NBTI for HP and LP device.

Introduction

HfO₂ gate dielectrics are considered to be the most promising high-k dielectrics to meet the 28nm node ULSI application. Recently, fluorine (F) incorporation into the high-k gate dielectric has been shown to improve device performance and reliability (1-7). However, a CMOS compatible F incorporation technology to meet both HP and LSTP application has never clarified yet. In this paper, a CMOS compatible CF₄ plasma treatment technology is demonstrated. Material science, performance enhancement and reliability improvement for the CF₄ plasma treated CMOS HfO₂ are deeply proposed.

Device Fabrication

HfO₂ thin film was deposited on a HF-last Si surface by PVD system. Before and after HfO₂ thin film deposition, CF₄ plasma was used to treat the Si-wafer (denoted as pre) and HfO₂ thin film (denoted as post), respectively. The reactive pressure and the flow rate of the CF₄ gas were 600 mtorr and 500 sccm, respectively with 40 W RF power. For the normal HfO₂ gate dielectrics samples (denoted as as-dep.), there was no CF₄ plasma treatment before and after the hafnium dioxide deposition. In a later phase of the investigation, a 50 nm TaN metal gate was also deposited by the RF sputter method. The source and drain regions in the active device region were implanted with phosphorus (15 keV at 5×10^{15} cm^{-2}) and boron (25 keV at 5×10^{15} cm^{-2}) for nFET and pFET, respectively. The S/D were activated at 900 °C for 30 s annealing in a N₂ ambient. After the patterning of source/drain contact holes, TaN gate was then deposited for use as the gate electrode.

Results and Discussion

A. Material science of F incorporated HfO₂

Figure 1 shows that CF₄ pre-treated HfO₂ has the largest permittivity and thinnest EOT owing to interfacial reaction suppression, as shown in TEM image. Generally, speaking, a PVD HfO₂ thin film, which tended to have interfacial layers (IL) like Hf-silicate at the HfO₂/Si interfaces. However, for the CF₄ treated Si-wafer, the native oxide would disappear and been replaced by a significantly weaker absorption band which is similar to that observed for amorphous SiO₂ films on Si [1]. Therefore, the IL was effectively suppressed for the sample with CF₄ plasma pre-treatment (TEM image). The growth of interfacial layer is inhibited by F passivation of the Si substrate and the blocking of oxygen diffusion into the Si. The permittivity of pre-treat sample can be enhanced 16% for the pre-treated HfO₂ (due to IL elimination), achieving thin EOT(1.2nm), which is suitable to HP application. On the other hand, after HfO₂ deposition, CF₄ plasma was used to treat the HfO₂ thin film to form the fluorinated HfO₂, which is almost the same as as-deposited one, as approved in inset Fig. 1 TEM image. The gate leakage current is much decreased (73%) for the post-treat HfO₂ with same EOT(1.8nm), which is suitable to LSTP application. This gate leakage reduction also indicates that fluorinated HfO₂ has better gate oxide quality.

Compared to the Hf–O bonds in Hf 4*f* spectra of the HfO₂ thin film (Fig. 2(a)), the Hf 4*f* spectra of the fluorinated HfO₂ thin film is shifted roughly 0.7 eV, indicating that Hf–F bonding formation after CF₄ plasma treatment (Fig. 2(b)). Hf–Si bondings are also observed, meaning that Hf silicide was easily formed during the HfO₂ film deposition. Fortunately, this formation of Hf–Si bonding was effectively suppressed for the CF₄ plasma treated samples. On the other hand, the pre-CF₄ plasma treatment can effectively improve Si surface roughness due to native oxide disappearing (Fig.3)[6]. For the CF₄ treated Si-wafer, the native oxide band at 1221 cm^{-1} has disappeared and been replaced by a significantly weaker absorption band centered near 1180 cm^{-1} which is similar to that observed for amorphous SiO₂ films on Si. The growth of interfacial layer is inhibited by F passivation of the Si substrate and the blocking of oxygen diffusion into the Si. This hypothesis is supported by both the TEM imaging and FTIR spectroscopy. The RMS of CF₄ treated Si is only 0.75nm while the as-deposited one is 1.64nm, resulting in better device uniformity(discussed in next paragraph). In addition, TEM images(Fig.1) indicated that the thicknesses of as-deposited and fluorinated HfO₂ thin films are the same, implying that the CF₄ plasma etching effect during the treatment of HfO₂ thin films is negligible. The root-mean-square (RMS) variations of the surfaces of the as-deposited and post-treated HfO_2 thin film, extracted from AFM images, are 1.05 and 1.74 Å, respectively (Fig. 4).

B. CMOS TaN/HfO_2 gate engineering with F incorporation

Figure 5 shows I_D-V_G transfer characteristics of the as-deposited and CF_4 treated HfO_2 nFETs and pFETs. Decreased V_{TH} and I_{OFF} reduction can be observed for F-HfO_2 nFETs. Besides, Driving current has 48% and 45% increase for pre-CF_4 treated HfO_2 nFETs and pFETs, respectively, Fig. 6. The mobility enhancement for the fluorinated sample was more pronounced in the high electric field. A roughly 65% (E_{eff} = 1 MV/cm) and 91% (E_{eff} = 2 MV/cm) increase can be observed for electron and hole mobility (Fig. 7), respectively. This can be speculated to the improved surface roughness. Fluorine incorporation into HfO_2 film can effectively passivate HfO_2/SiO_2 and the SiO_2/Si interfaces [1], resulting in the improved surface roughness of HfO_2/SiO_2 and the SiO_2/Si interfaces. Figure 8 shows that better uniformity can be observed in F-HfO_2 nFETs and pFETs. Both V_{TH} and subthreshold slope (S.S.) 1-sigma is largely decreased, especially for pre-treat one owing to IL elimination and good HfO_2/Si interface. Besides, 10% S.S. reduction for the CF_4 treated samples, indicating that fluorinated CMOS HfO_2 has better interface characterization. The performance enhancement, leakage reduction and uniformity improvement for the CMOS HP and LSTP device are summarized in Table I.

To understand the reliability of fluorinated CMOS HfO_2, NBTI and PBTI characterization(stress V_G=±1.5V) are discussed in this paper. Figure 8 shows NBTI characteristics of all samples. The post-CF_4 treated sample shows obvious NBTI improvement, including decreased V_{TH} shift, negligible I_{OFF} increase, less I_D and g_m degradation, is suitable in LP application. On the other hand, the pre-CF_4 treated sample shows quite different NBTI characterization, such as decreased V_{TH} and I_{OFF}, increased I_D and g_m during NBTI stress, as shown in Fig. 9 and 10. As discussed before, the EOT of pre-CF_4 treated sample is only 1.26nm, which is 0.6nm smaller than the other samples. However, the stress voltage is the same for all samples. Therefore, the stress electric field of pre-CF_4 treated sample is 1.5 times than that of the other samples(EOT=1.8nm). As a result, the hole energy is high enough to break the strong Si-F bond(5.73eV) and create an interface trap by releasing fluorine species at the IL/Si interface(Fig. 11). In addition, the released F would re-incorporate into HfO_2 film, resulting in trapping level increase, as shown in Fig. 12. Therefore, the NBTI characterization of CF_4 pre-treated sample is quite different from conventional NBTI phenomenon. Figure 13 shows that PBTI power-law exponent value changes from n ≈ 0.208 (as-dep.) to 0.184 (pre), 0.134 (post) due to F incorporation into HfO_2 film.

Furthermore, the Id_{OFF} increase during PBTI stress can be largely released for post-treat sample (Fig. 14), mainly resulted from Ib_{OFF} reduction [1]. This obvious improvement of V_{TH} shift and Id_{OFF} increase in post-treat device suggested that the F-incorporated device has deep electron traps [1], which is extremely suitable for LP application, as summarized in Table II. Fig. 15 shows that the gate induce drain leakage (GIDL) can be largely decreased for the CF_4 treated one. In addition, the GIDL current will increase a lot after PBTI stress, resulting in worse device reliability. However, the GIDL increase can be much suppressed for the CMOS HfO_2 with fluorine incorporation. The physical band diagrams show that direct tunneling and trap-assist-tunneling (TAT) contribute to GIDL current under large band bending. For the CMOS HfO_2 with F incorporation, the TAT can be effectively eliminated, resulting in less GIDL increase after PBTI stress.

Conclusion

For the first time, a CMOS compatible F incorporation into high-k technology in both HP and LSTP application has been successfully demonstrated. F tends to segregate to the HfO_2/SiO_2 and the SiO_2/Si interfaces passivating oxygen vacancy and interface traps by forming stronger Hf-F and Si-F bonds compared to Hf-H and Si-H bonds, effectively improving HfO_2 quality and its reliability. This advanced CMOS compatible technology is very useful to future HP and LSTP metal-gate/high-k gate engineering.

Acknowledgement

This work was supported by the National Science Council under the contract of NSC 98-2221-E-182-057-MY3

Reference

(1) W. C. Wu, et al., "Fluorinated HfO_2 Gate Dielectrics Engineering for CMOS by pre- and post-CF_4 Plasma Passivation," in *IEDM Tech. Dig.* P.405, 2008

(2) H. H. Tseng, et al., "Defect Passivation with Fluorine in a TaxCy/High-K Gate Stack for Enhanced Device Threshold Voltage Stability and Performance," in *IEDM Tech. Dig.* pp.29.4.1-29.4.4, 2005

(2) M. Inoue, et al., "Fluorine Incorporation into HfSiON Dielectric for V_{th} Control and Its Impact on Reliability for Poly-Si Gate pFET," in *IEDM Tech. Dig.*, pp. 17.1.1–17.1.4. 2005.

(3) K. Seo, et al., "Improvement in High-*k* (HfO_2/SiO_2) Reliability by Incorporation of Fluorine," in *IEDM Tech. Dig.*, pp. 17.2.1–17.2.4. 2005.

(4) Y. Yasuda, et al., "Effect of Fluorine Incorporation on 1/f Noise of HfSiON FETs for Future Mixed-Signal CMOS," in *IEDM Tech. Dig.*, pp. 10.5.1–10.5.4. 2006.

(5) W. C. Wu, et al., "Carrier Transportation Mechanism of TaN/HfO_2/IL/Si Structure with Silicon Surface Fluorine Implantation (SSFI)," *IEEE Trans. Electron Devices*, vol. 55, pp. 1639-1646, 2008.

(6) C. S. Lai, et al., "Suppression of Interfacial Reaction for HfO_2 on Silicon by Pre-CF_4 Plasma Treatment," *Appl. Phys. Lett.*, vol. 89, pp. 072904-072906, 2006.

(7) C. S. Lai, et al., "The Characterization of CF_4 Plasma Fluorinated HfO_2 Gate Dielectrics with TaN Metal Gate," *Appl. Phys. Lett.* vol. 86, pp. 22905-22907, 2005.

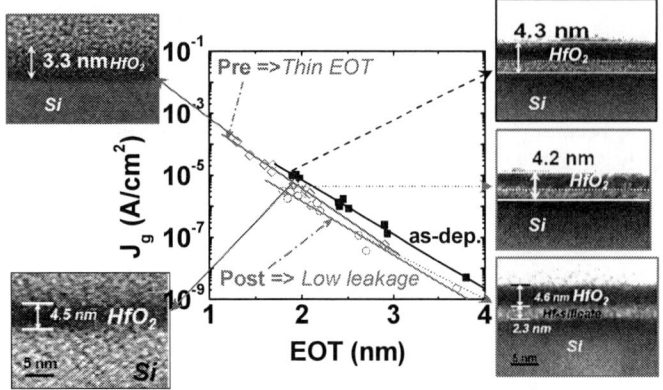

Fig. 1. Jg vs. CET for NMOS compared with CF_4 plasma treatment. Pre-treat samples(HP) has large leakage due to ZIL(thinner CET).

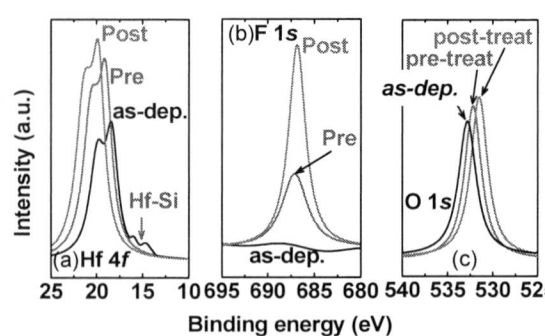

Fig.2. (a)Hf $4f$, (b)F $1s$, where the F $1s$ peak is at 687 eV. (c)O $1s$ XPS spectra of all samples. Energy peak shift indicates F incorporated into HfO_2.

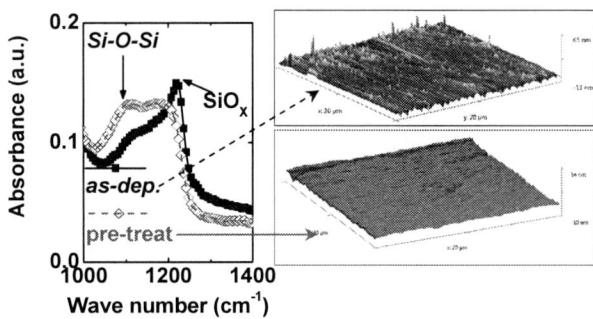

Fig.3. FTIR analysis for as-dep. and CF_4 pre-treat Si-wafers. The RMS of CF_4 treated Si (0.75nm) is smaller than that of as-deposited one (1.64nm) due to native oxide elimination.

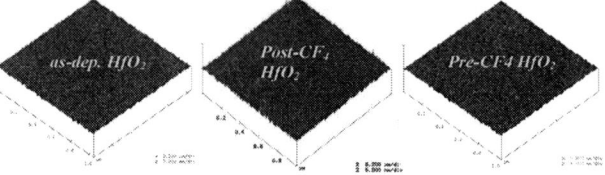

Fig.4. AFM images show that CF_4 plasma did not damage the HfO_2 film roughness.

S.S. 1-sigma (mV/dec.)					
as-dep.(N)	as-dep.(P)	Pre(N)	Pre(P)	Post(N)	Post(P)
1.61	2.23	1.14	1.23	1.15	1.43

Vth 1-sigma (mV)					
as-dep.(N)	as-dep.(P)	Pre(N)	Pre(P)	Post(N)	Post(P)
15.7	18.7	6.6	8.7	10.2	13.2

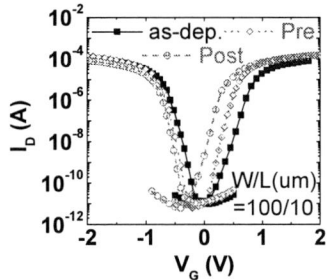

Fig.5. nFET and pFet I_D-V_G curves at linear region for transistors with and without CF_4 plasma treatment.

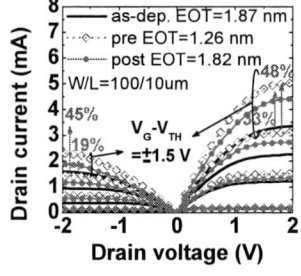

Fig.6. 48% and 45% drain current enhancement can be observed for pre-treated(HP) nFET and pFet.

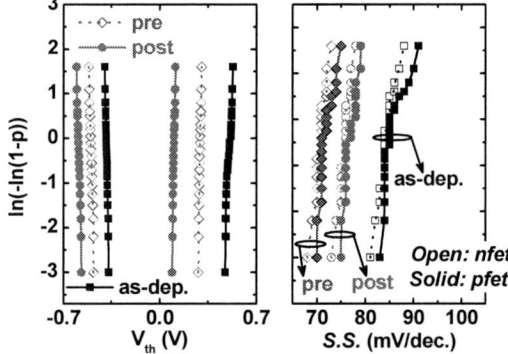

Fig.7. Uniform Vt and *S.S.* can be observed for nFET and pFet with pre-CF_4 plasma treatment(HP).

Table I Device performance summary of as-deposited, CF_4 pre-treat(HP), CF_4 pose-treat(LSTP) CMOS HfO_2. Device channel width and length are 100um and 10um, respectively.

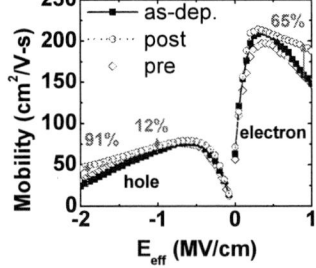

Fig.8. Electron and hole mobility increase a lot for post-CF_4 plasma treated samples at high electric field.

Performance		EOT(nm)	Jg(A/cm²)@Vfb-1V	Vth(V)	VtU(1-sigma)	Ids(uA)	IdU(1-sigma)	S.S.(mV/dec.)	Idoff(pA)@V_G-V_th=0V
as-dep.	nfet	1.87	1.03E-05	0.48	15.7mV	335	12.2%	83	73
	pfet			0.38	18.7mV	160	15.6%	85	106
Pre (HP)	nfet	1.26	1.28E-04	0.29	6.6mV	504	3.5%	76	16
	pfet			0.46	8.7mV	290	6.3%	76	23
Post (LSTP)	nfet	1.82	1.77E-06	0.11	10.2mV	443	8.0%	72	0.3
	pfet			0.55	13.2mV	229	10.1%	73	1.2

978-1-4244-6658-0/10 $26.00 © 2010 IEEE

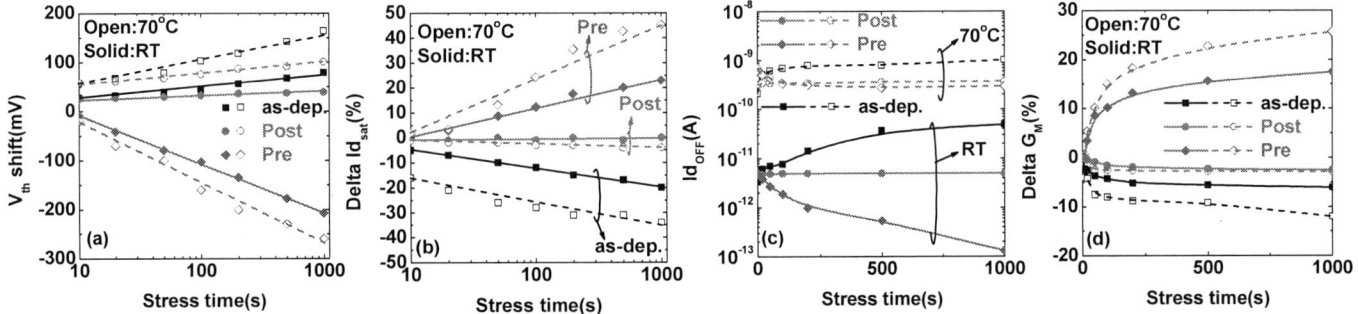

Fig.9. NBTI characteristics(@V_G-V_{th}=-1.5V) of all samples: (a)V_{th} shift, (b)delta I_{dsat}, (c)Id_{off}, (d)delta g_m, respectively. Pre-treat sample shows quietly different NBTI characteristics from the others.

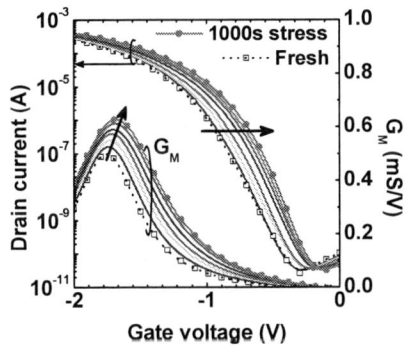

Fig.10. I_D-V_G and g_m curves (@70°C) of pre-CF$_4$ plasma treated sample during NBTI stress(V_G-V_{th}=-1.5V).

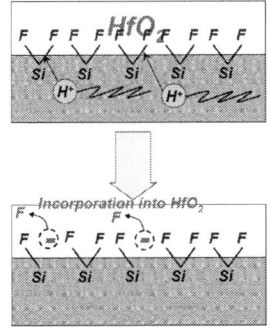

Fig.11. Physical NBTI model of CF$_4$ pre-treat HfO$_2$ dielectric. Si-F broken results in Vth decrease during NBTI stress.

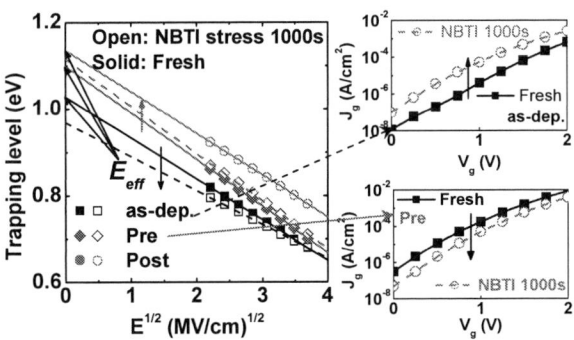

Fig.12. Extracted F-P trapping level of pret-treat HfO$_2$ increased after NBTI stress 1000s (1.09→1.11 eV) due to F re-incorporation.

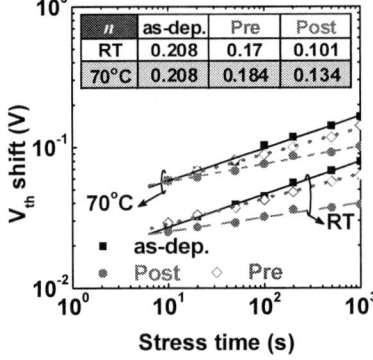

Fig.13. PBTI (V_G-V_{th}=1.5V) reliability is much improved for fluorinated nMOSFETs. n value is significantly reduced due to F incorporation.

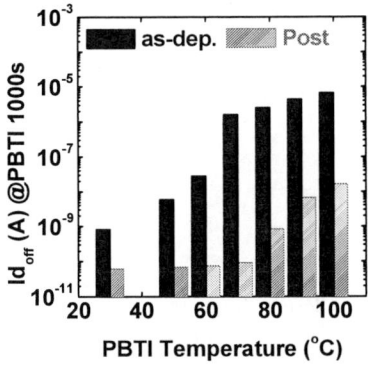

Fig.14. Id_{off} increasing in high temperature PBTI stress(@V_G-V_{th}=1.5V) can be reduced for post-CF$_4$ plasma treated sample(LSTP).

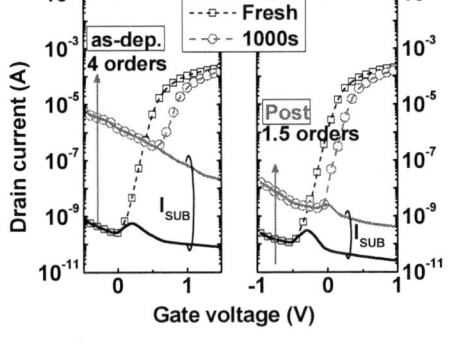

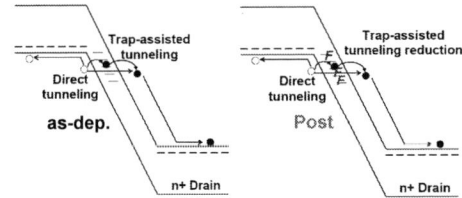

Fig.15. I_D-V_G and g_m curves (@100°C) of as-deposited and post-CF$_4$ plasma treated samples after PBTI stress(V_G-V_{th} =1.5V) 1000s. GIDL current reduction due to TAT suppression for the F incorporated device.

Table II Reliability summary of as-deposited, CF$_4$ pre-treat(HP), CF$_4$ pose-treat(LSTP) CMOS HfO$_2$.

Reliability	NBTI@1000s V_G-V_{th}=-1.5V(70°C)			PBTI@1000s V_G-V_{th}=1.5V(70°C)		
	V_{th} shift	Ids degradation	Id_{off} increase	V_{th} shift	Ids degradation	Id_{off} increase
as-dep.	172 mV	-35%	381%	168 mV	-38%	~2 orders
Pre(HP)	-253 mV	45%	-53%	136 mV	-22%	77%
Post(LSTP)	96 mV	-4%	16%	93 mV	-3%	13%

Origins of Universal Mobility Violation in SOI MOSFETs

N. Rodriguez[1], S. Cristoloveanu[2], F. Gamiz[1]

[1]Departamento de Electrónica. University of Granada. 18071 Granada, Spain
[2]Institute of Microelectronics, Electromagnetism and Photonics (UMR 5130), Grenoble INP Minatec, 38016 Grenoble, France

e-mail: noel@ugr.es

Abstract—The relation between the effective mobility and the transversal field is systematically investigated for SOI-MOSFETs operated in single-gate and double-gate modes. We point out several practical situations where the Universal Mobility (UM) concept is not applicable. In particular, different carrier distributions can lead to the same value of effective field and to distinct mobility values, breaking the foundations of the UM curve. The presence of two interfaces and channels with different quality or the carrier redistribution within the transistor body cannot be accounted in the UM. In particular cases, very unusual mobility-field curves are obtained.

I. INTRODUCTION

The Universal Mobility (UM) concept is an important milestone for electronic device characterization. It states that the inversion channel mobility in Si MOSFETs satisfies a universal relationship when is represented against the transverse effective field, E_{eff}. This universality is believed to hold regardless the substrate impurity concentration, oxide thickness or substrate biasing [1], [2], [3]. The UM law has been introduced based on systematic measurements on high quality bulk MOSFETs. Numerical simulations under diverse complexity approximations have corroborated the experimental data [4].

Nevertheless most of the results recently shown in the literature do not lay on the UM curve. There are two types of restrictions :

Extrinsic aspects – The material quality may not be perfect, in particular in thin Si layers. The interface defect density varies according to the gate dielectric. The scattering mechanisms are often inhomogeneous along the channel of short devices. Finally, mobility boosters are being included in the CMOS process. In all these cases there is no room for a universal mobility. If the UM curve needs to be adapted to each particular case, this merely means that there is no universality at all. The UM curve is actually used as a reference to measure the benefits or drawbacks of a particular process and optimization.

Intrinsic aspects – The definition of effective mobility and field are debatable, especially in ultrathin and multiple-gate transistors. Note that in FinFETs the field distribution is 2D or 3D which makes the definition of an effective field rather complex. In ultrathin layers subband splitting occurs, leading to mobile carriers featuring several effective masses. In addition, volume inversion causes the spreading of carriers from the interface into the transistor body.

In this paper, we focus on the intrinsic reasons which may invalidate the UM curve. Numerical simulations, based on the Poisson and Schroedinger equations, were performed in order to determine the parameter range where the related approximations are reasonably satisfied. We have recently pointed out several pragmatic examples of UM violation in single-gate and double-gate SOI MOSFETs [5]. The aim of the present work is to provide additional results confirming the inadequacy of the UM law.

II. EFFECTIVE FIELD

The effective field is defined as the pondered average of the magnitude of the local transverse electric field with the local electron concentration [6]:

$$E_{eff} = \frac{\langle n(z)|E(z)|\rangle}{\langle n(z)\rangle} \qquad (1)$$

where the brackets indicate integrals along the transverse direction, from the interface to the border of the depletion region (bulk and partially depleted MOSFETs) or to the bottom interface (FD SOI).

Unfortunately Eq. 1 is unpractical from an experimental point of view, since the in-depth electric field distribution and minority carrier concentration cannot be accurately measured. Assuming that the inversion layer remains closer to the interface, the effective field can be related to the depletion charge and a portion of the inversion charge [1]:

$$E_{eff} = \frac{1}{\gamma \varepsilon_{Si}} (\eta Q_{inv} + Q_{dep}) \qquad (2)$$

where normally $\gamma = 1$.

Despite basic Gauss law based calculations would give $\eta = 0.5$, $\gamma = 1$, the actual value of η to produce UM curves depends on the particular device structure. It is commonly

978-1-4244-6658-0/10 $26.00 © 2010 IEEE

accepted that η equals 1/2 and 1/3 for electron and holes respectively on Si (100) surfaces [1], [7]. The reason for that is that $n(z)$ in Eq. (1) should be evaluated accounting the each particular band structure (i.e. $n(z)=\sum_i N_i|\psi_i(z)|^2$, where N_i and ψ_i are respectively the density of electrons per unit area and the wave function in the ith subband). Therefore, η has become more like a fitting parameter to adjust Eq.(2) to (1) rather than a physical meaningful parameter [8].

Figure 1 illustrates the actual unintuitive location of the effective field point in the inversion layer of a bulk MOS transistor. The position corresponding to the effective field value calculated with Equation 1 (1.4nm) does not match the maximum of the electron distribution (0.9nm) neither the inversion layer centroid (2.3nm). The conclusion is that E_{eff} is an average field, but not necessarily the one seen by the *average carriers*.

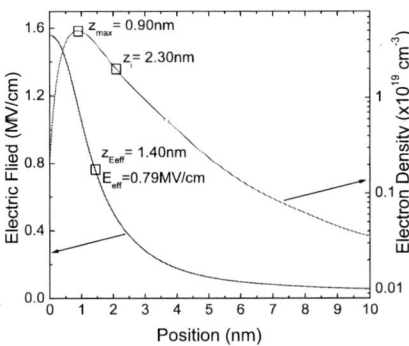

Fig 1. Transversal electric field and electron density in a bulk-MOSFET as a function of the distance from the SiO$_2$ interface. The effective field point location does not coincide with neither the maximum of the electron concentration distribution nor the centroid of the inversion charge. $T_{ox}=2nm$, $V_G=1.5V$, $N_A=10^{15}cm^{-3}$, Metal gate $\chi=4.4eV$.

The situation becomes far more complex in SOI MOSFETs. Equation 2 has been adapted to Fully Depleted FD-SOI devices (the case of partially depleted (PD) SOI transistors is reduced to Eq. 2 since the electric field becomes zero in the neutral region above the back oxide interface). In FD-SOI the BOX interface field is rarely zero and is currently accounted by a modified effective field equation [9]:

$$F_{eff}=\frac{1}{\varepsilon_{Si}}(\eta Q_{inv}+Q_{dep})+E_{sb} \qquad (3)$$

where E_{sb} is the field at the back Si/SiO$_2$ interface. In Figure 2, Eq. (3) is compared with Eq. (2) and definition (1) for SOI transistors with two values -5V and 5V of the substrate bias (back gate). When the back interface is depleted (Fig.2a) the electric field monotonically decreases; Eq.2 fails to reproduce

the actual values of E_{eff}, whereas corrected Eq. (3) match the definition. By contrast as far as the back interface is weakly inverted (Fig.2b) the electric field is no-longer monotonic, having positive and negative values depending on the in-depth position. Eq. (2) overestimates the actual value of E_{eff}, whereas Eq. (3) is only valid when the front interface in strongly inverted and the contribution of the back interface becomes negligible.

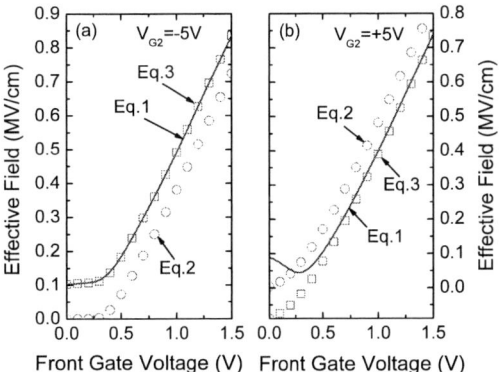

Fig 2. Effective field as a function of the front-gate voltage in a SOI MOSFET for two values of the back-gate bias. Results given by equation (1), (2) and (3) are compared. $V_{G2} = -5V$, depleted back interface (a), $V_{G2} = +5V$ weakly inverted back interface (b). $T_{ox}=2nm$, $T_{BOX}=145nm$, $T_{Si}=20nm$, $N_A=10^{15}cm^{-3}$, $\chi=4.4eV$.

Further analysis on Figure 2 brings unexpected conclusions. When the back interface is positively biased, increasing the voltage at the front interface first decreases the effective field to some minimum value. Then, the field increases again as a consequence of the further minority carriers enrichemend at the front interface and the increased potential difference between the two interfaces. The consequence is stunning: one must move back and forth on the mobility curve! Equations 2 and 3 are unable to integrate the non-monotonic variation of the effective field (Fig. 2b).

Another important case is that of the Double-Gate (DG) transistors. If the device is symmetrical the field is zero in the middle of the film. Extrapolating this to Equation 2 leads to a modification of coefficient γ from 1 to 2 [10].

The effective field given by Eqs. (1) and (2) is compared in Figure 3. For a perfect symmetrical device (3.a) Eq. 2 with $\gamma=2$, fits the exact values given by Eq. (1). By contrast Eq. 2 with $\gamma=1$, becomes totally inappropriate overestimating the value of E_{eff} by a factor of two. However technological variations can break the symmetry of the DG-transistor; Figure 3b, shows how a difference of *0.2eV* between the work-functions of front and back gates can make Eq. 2 deviate from the actual value of E_{eff}.

978-1-4244-6658-0/10 $26.00 © 2010 IEEE

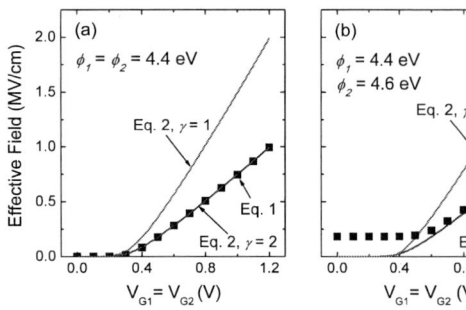
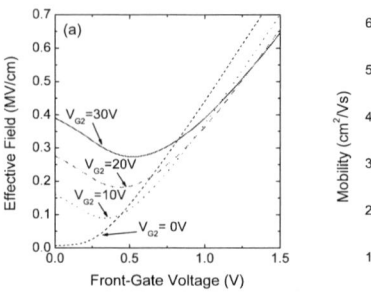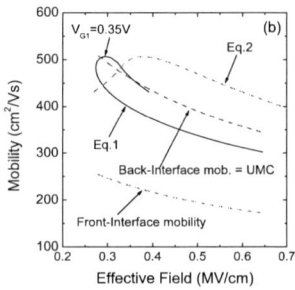

Fig 3. Effective field versus gate voltage for an ultrathin DG-SOI transistor: $T_{si} = 5nm$, $T_{OX} = T_{BOX} = 1nm$, $N_A = 10^{14}cm^{-3}$. Metal gates: symmetrical work-functions (a), asymmetrical work-functions (b).

III. EFFECTIVE MOBILITY

The key point of Eq. (1) is the multivalent nature of the effective field magnitude: there are different carrier distributions conducting to the same value of E_{eff} (Fig. 2b).

We have considered an SOI transistor with the back interface biased in inversion and the front interface swept from depletion to inversion. The effective mobility has been calculated by weighting the local mobility with the number of carriers:

$$\mu = \frac{\langle n(z)\,\mu(z, E(z))\rangle}{\langle n(z)\rangle} \qquad (4)$$

The local mobility is depending on the local electric field, E_f, and the in-depth position. In Eq. (5), we assumed

$$\mu = 1500 \frac{0.5\left(1+\dfrac{z}{T_{Si}}\right)}{1+\left(\dfrac{|E_f|}{100000}\right)^{0.65}} \qquad (5)$$

According to Eq. (5) the mobility at the back interface ($z=T_{Si}$) follows the UM curve whereas the front interface mobility is half of it, as usually in MOSFETs with high-k gate dielectric [11].

From this simulation, increasing the front gate voltage initially decreases the effective field (Fig. 4a, $V_{G2} = 30V$), and subsequently the mobility *increases* (mostly influenced by the back interface, Fig. 4b). It is worthnoting that the effective mobility initially exceeds that of the UM curve, this is due to the fact that the local field at the back-interface is very low contributing with high mobility values in Eq. 1. However the increasing role of the low-mobility front interface starts to prevail (before the effective field reaches the minimum value) and degrades the mobility. The overall mobility–field dependence is qualitatively and quantitatively different from the UM curve.

Fig 4. Effective field as a function of front gate voltage for different values of the back-gate bias (a). Electron mobility as a function of effective field calculated with Eqs. 1 and 2. The mobility model corresponds to Eq. 5. Back-interface mobility matches the UM ($z=T_{Si}$) whereas back-interface mobility ($z=0$) is half of it. Same structure as in Fig. 2 with $V_{G2} = 30$ V.

Another case of UM violation is illustrated in Fig. 5, where the mobility profile is:

$$\mu = 1500 \frac{1-0.95\dfrac{z}{T_{Si}}}{1+\left(\dfrac{|E_f|}{100000}\right)^{0.65}} \qquad (6)$$

For $z = 0$ (front interface) the UMC is recovered; by contrast, the back interface ($z = T_{Si}$) is assumed to have a very poor mobility (20 times lower, as in SOS films). When the back gate is biased in inversion, the result again is unconventional. At $V_{G1} = 0V$, the back interface conduction dominates, leading to a low value of $\mu = 81cm^2/Vs$. Increasing the front-gate voltage reduces the effective field according to Eq. (1). The mobility increases, actually much more than the corresponding effect of the effective field reduction, due to the increasing role of the high mobility interface. Further increase in V_{G1}, continues enriching the front interface at the expense of the poor back-interface mobility.

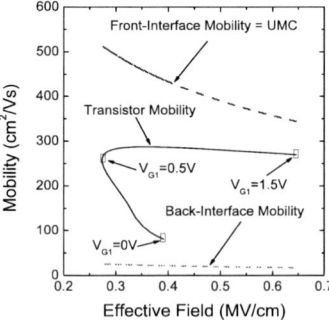

Fig 5. Electron mobility as a function of the effective field calculated with Eq. 1. The mobility model corresponds to Eq. 6. Top interface mobility matches the UMC ($z=0$) whereas back-interface mobility ($z = T_{Si}$) is severely degraded. Same structure as in Fig. 2 with $V_{G2} = 30$ V.

978-1-4244-6658-0/10 $26.00 © 2010 IEEE   342

There are three striking features:

i) The mobility curve has two branches in the range of 0.29-0.4MV/cm: different carrier distributions (leading to different effective mobilities) can correspond to the same value of E_{eff}.

ii) The mobility increase as V_{G1} increases is due to the enrichment of the high mobility interface and not to a decrease in Coulomb scattering.

iii) The mobility degradation rate ($d\mu_{eff}/dE_{eff}$), for fields higher than 0.4MV/cm, is almost null. This is again due to the redistribution of carriers between the two channels. But such a flat curve may be misinterpreted as a improvement in surface roughness scattering.

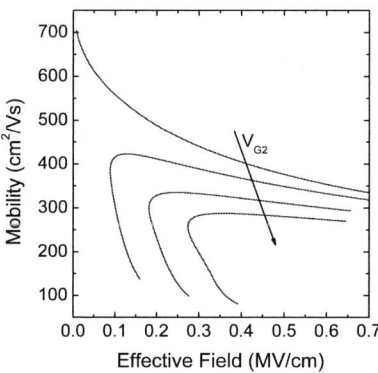

Fig 6. Electron mobility as a function of the effective field calculated with Eq. 1. The mobility model corresponds to Eq. 6. Different V_{G2} were applied (0, 10, 20, 30 V). Same structure as in Fig. 2.

We have studied the impact of the back-gate bias on the UM representation. As observed in Figure 6, the universality of the mobility curves is broken. Several curves are obtained according to the back-gate bias. More remarkably, each curve features a special non-monotonic variation. This is due to the fact pointed above: the mobility obtained for a given effective field depends on the particular carrier distribution. Further examples will be documented showing the impact of film and BOX thickness.

Conclusions

The effective field and the universal mobility concepts have been revisited by focusing on SOI MOSFETs. Analytical models fail to reproduce the actual behavior of the effective field: Back gate biasing can lead to a decrease in the average field as the inversion charge increases. Selected examples have shown that the universality can be contradicted. The quality of interfaces and the local values of the carrier concentration and mobility can lead to complex mobility-field curves, definitely not accounted for by the UM law.

Acknowledgments

Authors are partially supported by the Spanish Government under contract TEC2008-06758-C02-01, by Junta de Analucía under project TIC-1899, and by EU within FP7-216171-NoE (NANOSIL) and FP7-216373-CA (EUROSOI+).

References

[1] A.G. Sabins and J. t. Clemens, "Characterization of electron mobility in inverted Si-<100> interface," International Electr. Dev. Meeting, 1979, pp. 18-21.

[2] S. Takagi, A. Toriumi, M. Iwase and H. Tango, "On the universality of inversion layer mobility in Si MOSFETs: part I. Effects of substrate impurity concentration," IEEE Trans. Electr. Dev., vol. 41, no. 12, pp. 2357-2362, Dec. 1994.

[3] ---, "On the universality of inversion layer mobility in Si MOSFETs: part II. Effects of surface orientation," IEEE Trans. Electr. Dev., vol. 41, no. 12, pp. 2363-2368, Dec. 1994.

[4] F. Gámiz, J.A. López-Villanueva, J. Banqueri, J.E. Carceller and P. Cartujo, "Universality of electron mobility curves in MOSFETs: a Monte Carlo study," IEEE Trans. Electr. Dev., vol. 42, no. 2, pp-258-265, Feb. 1995.

[5] S. Cristoloveanu, N. Rodriguez, F. Gámiz, "Why the Universal Mobility is not," IEEE Trans. Electr. Dev., vol. 57, no 6. pp. 1327-1333, 2010.

[6] M. Shoji and S. Horiuchi, "Electronic structures and phonon limited electron mobility of double-gate silicon-on-insulator Si inversion layers,", J. Appl. Phys., vol. 85, no. 5, pp. 2722-2731, 1999.

[7] N. Arora and G. Gildenblat, "A semi-empirical model of the MOSFET inversion layer mobility for low-temperatura operation", IEEE Trans. Eletr. Dev., vol. 34, no 1. pp. 89-93, 1987.

[8] M. Saitoh, S. Kobayashi and K. Uchida, "Physical understanding of fundamental properties of Si (110) pMOSFETs inversion-layer, capacitance, mobility universallity, and uniaxial stress effects", IEDM, 2007, pp. 711-714

[9] M.J. Sherony, L.T. Su, J.E. Chung and D.A. Antoniadis, "SOI MOSFETs effective channle mobility", IEEE Trans. Eletr. Dev. Vol.41 No. 2, pp. 276-278, Feb. 1994

[10] D. Esseni et al., "An experimental of low mobility enhacement in ultrathin SOI transistors operated in double-gate mode", IEEE Trans. Eletr. Dev. Vol. 50, pp. 802-808, 2003.

[11] L. Pham-Nguyen, "In situ comparison of Si/High-k and Si/SiO2 channel properties in SOI MOSFETs", Oct. 2009, vol 30. Issue 10. pp. 1075-1077.

Optimization of III-V FET architectures for high frequency and low consumption applications

Ming Shi*, Jérôme Saint-Martin, Arnaud Bournel, Philippe Dollfus

IEF, CNRS UMR 8622 / Université Paris Sud 11, Bât 220

F 91405 Orsay cedex, France
* ming.shi@u-psud.fr

Abstract— **To fulfill high-speed and low-power specifications for intelligent applications, III-V FETs (Field Effect Transistor) with high-κ gate dielectric stack are very appealing. Indeed, combining weak gate leakage of standard MOSFETs and good RF performance of HEMTs, they could enhance device scalability. Using full 2D Poisson-Schrödinger solver and then semi-classical Ensemble Monte Carlo device simulator, MOSFET (Metal Oxide Semiconductor Field Effect Transistor) and HEMT (High Electron Mobility Transistor) structures are investigated in terms of gate charge control and both static and dynamic I-V performance. In particular, Y parameters are carefully extracted from time-varying currents. This comparative study allows us to propose optimized nanoscale III-V FET with high-frequency performance under low power supply.**

I. INTRODUCTION

Next generations of Si-CMOS (Silicon based Complementary Metal Oxide Semiconductor) circuits are expected not to meet future high-speed and low power specifications for ambient intelligent functions [1]. Indeed, their optimal frequency-performance/power-consumption trade-off is limited by low Si carrier mobility and relatively large supply voltage required for circuit operation. Among promising emerging high-mobility materials, III-V heterostructures appear to be the most convenient in terms of patterning and do not raise fundamental placement problem such as carbon nanotubes or semiconducting nanowires [2]. Standard industrial HEMTs exhibit performance-frequency to on-power-consumption ratios which outperform those of silicon channel FETs [3]. However, due to high off-state leakage I_{OFF} inducing high DC power consumption, they tend to reach their scaling limits in terms of gate length and layer structure. High-κ dielectric gate stacks compatible with III-V channel [4] can improve the electrostatic gate control and subthreshold slope, and therefore reduce their DC consumption [5,6]. Nevertheless, high mobility III-V MOSFET performance might be affected by rough channel/oxide interface, by their low Density Of States (DOS) and by band to band tunneling current leakage [7]. The originality of this work is to asses the potentiality of

III-V MOSFETs operating at low supply voltage to minimize their intrinsic limitations.

Both electrostatic gate charge control via capacitance-voltage characteristics and static/dynamic I-V characteristics are theoretically investigated. Four InGaAs MOSFET and one HEMT architectures are compared using 1D-2D Poisson-Schrödinger simulations and a semi-classical Ensemble Monte Carlo simulator.

II. SIMULATED DEVICE

The first considered structure is a 50 nm-long-T-shaped gate HEMT whose layer structure and device parameters were described in Ref [8]. It consists of a 100 nm-thick $In_{0.52}Al_{0.48}As$ buffer layer on (001) InP substrate, a 7.5 nm $In_{0.53}Ga_{0.47}As$ sub-channel, a 7.5 nm $In_{0.8}Ga_{0.2}As$ main channel, a 3 nm undoped $In_{0.52}Al_{0.48}As$ spacer, a 3 nm planar Si-doped layer ($1\times10^{13}/cm^2$), a 10 nm $In_{0.52}Al_{0.48}As$ Schottky barrier layer and a 26 nm Si-doped $In_{0.53}Ga_{0.47}As$ cap layer. Other devices are 50 nm-long-MOSFET structures (cf. Fig. 1) with $In_{0.53}Ga_{0.47}As$ channel and self-aligned Al_2O_3/TaN gate stack. Since the performance of conventional bulk-MOS structures is known to be affected by scattering on rough dielectric interface and doping impurities, three other alternative structures likely to limit these detrimental effects are proposed in Fig. 1. The so called Thin Body TB-MOS structure is inspired by the Silicon On Insulator (SOI) technology. In the HEMT-MOS structure, a wide bandgap III-V spacer is inserted between the channel and the dielectric. The COMB-MOS transistor is a combination of the two previous ones.

III. CHARGE CONTROL- POISSON SCHRÖDINGER ANALYSIS

Thanks to a self-consistent Poisson-Schrödinger solver [8], the charge control in bulk MOS structures has been investigated in terms of gate capacitance-voltage characteristics. Both quantum and degeneracy effects were carefully considered. We first examine the 1D-behavior of the long-channel gate capacitance C_G which results from a series connection of the inversion charge capacitance C_{INV} and the oxide capacitance C_{OX}. The surface confinement orientation

978-1-4244-6658-0/10 $26.00 © 2010 IEEE

is (001). In Fig. 2, we plot C_{INV} and C_G as a function of the gate voltage for an Al_2O_3 gate oxide thickness T_{OX} of 6 nm. The inset shows the electron distribution over Γ and L valleys as a function of gate voltage and also of the total electron density. At low gate voltage, only the low-DOS Γ valley is occupied and the inversion charge capacitance C_{INV} is small. The limitation in overall gate capacitance C_G comes from this finite inversion-layer capacitance in this gate voltage range [9].

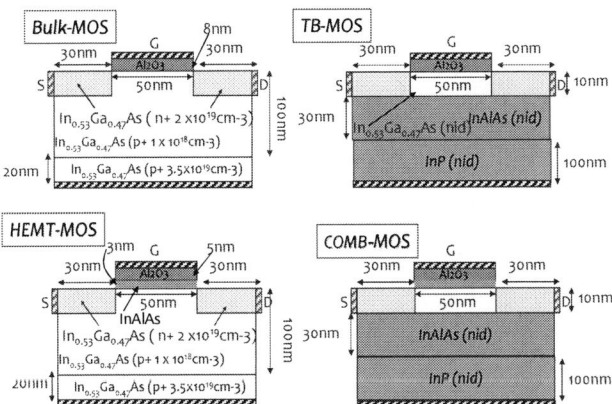

Figure 1 Cross section of the four MOSFET structures under investigation (nid: non intentionally doped).

When increasing the gate voltage, the electron density in L valleys is enhanced, which leads to higher C_{INV} and C_G. The transfer into L valleys - with much higher conduction mass - is favorable to get an efficient charge control with a high C_G capacitance close to the limit value $C_{OX} = 2\ \mu F/cm^2$, but detrimental to the carrier transport properties. Thus, a V_G swing from 500 mV to 1 V is appropriate to provide enough 'fast' Γ electrons for good RF performance.

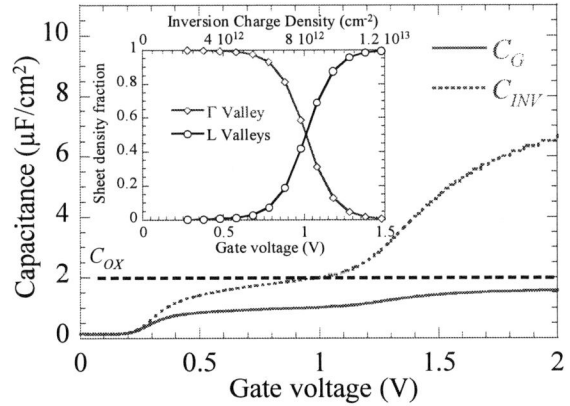

Figure 2 Capacitances C_G and C_{inv} as a function of V_G for a 1D Bulk MOS capacitance with T_{ox} = 6 nm. Inset: fraction of electron density in L and Γ valleys vs. V_G and electron density.

This gate capacitance degradation becomes larger with decreasing T_{OX}. In Fig. 3, the gate capacitance C_G, the oxide

capacitance C_{OX} and the ratio C_G/C_{OX} are plotted as a function of the equivalent silicon oxide thickness (EOT). In the considered regime, electrons occupy only the Γ valley. Due to C_{INV}, the gate capacitance ratio drastically decreases with reducing EOT below 5 nm-6 nm. It corresponds to a physical aluminum oxide thickness of 8 nm-10 nm. Such an oxide thickness may be a good compromise to achieve quasi optimal charge control, sheet electron density higher than $10^{12}\ cm^{-2}$ (cf. inset of Fig. 2) and weak gate leakage current.

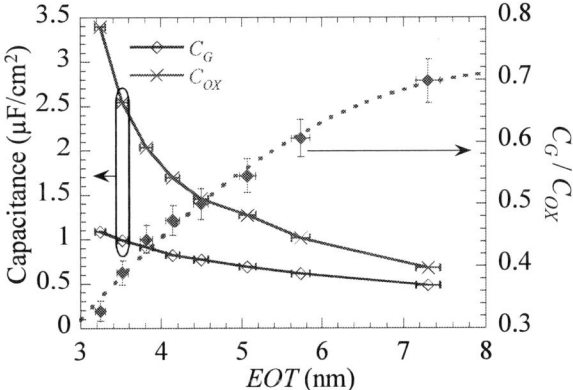

Figure 3 Oxide and gate capacitance and ratio (C_G/C_{ox}) in Γ valley as a function of EOT for Bulk-MOS structure.

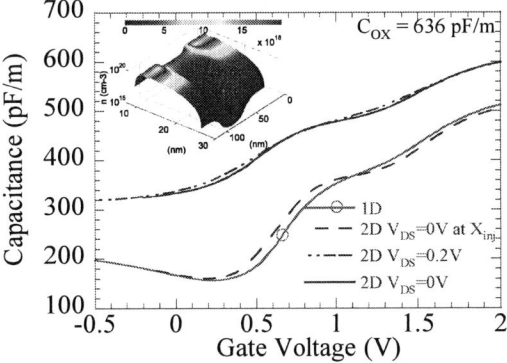

Figure 4 Gate capacitance obtained from the 2D charge density profile and that of 1D profile in the 50 nm Bulk-MOS. Inset: 2D cartography of electron density at V_{GS} = 1 V and V_{DS} = 0 V.

The above 1D approach is questionable to evaluate the capacitances for deca-nanometer III-V FETs. Indeed III-V semiconductors have a higher permittivity than Si, which could enhance short-channel effects (SCE) and fringing effects. Hence, 2D Poisson-Schrödinger analysis is used to refine the previous 1D estimations. Fig. 4 shows the results for Bulk-MOS structure (cf. Fig. 1). In the subthreshold regime, a high value of C_G is obtained in 2D simulation due to influence of fringing capacitance C_F (about 120 pF/m). Fringing capacitances are independent of bias and gate length (not shown). We also calculated the charge in a small slice at the top of the source-side injection barrier X_{inj} in 2D

978-1-4244-6658-0/10 $26.00 © 2010 IEEE   345

simulation for $V_{DS} = 0$ V in order to eliminate C_F. SCE induces a shift between the capacitance-voltage curves deduced from the integration at X_{inj} compared to C_G obtained from simple 1D calculation. Furthermore, 2D simulation can estimate the influence of source/drain bias, i.e. the lateral field along the channel and the drain-induced barrier lowering (DIBL). The drain effect is implemented by considering pseudo Fermi levels in recess regions. DIBL induces a shift in the C-V curves toward negative values. Due to high doping density in the channel, the Bulk-MOS structure (presented here) is less degraded by SCE and DIBL effects than the HEMT and TB-MOS structures with undoped channel.

IV. MONTE CARLO SIMULATION

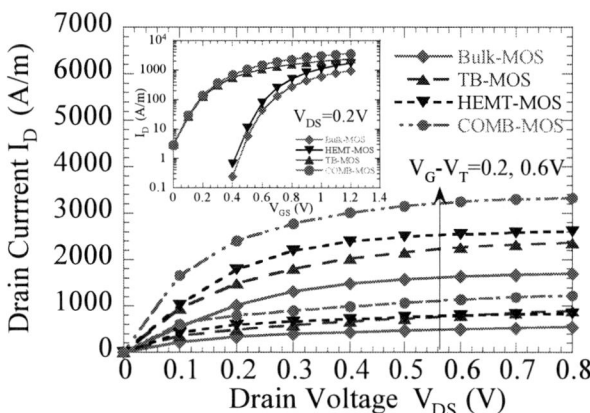

Figure 5 I_D-V_{DS} **intrinsic characteristics for four 50 nm MOSFET structures. Inset:** I_D-V_{GS} **curves.**

Besides, using a semi-classical (no quantum correction) Monte Carlo (MC) device simulator with convenient scattering mechanisms (phonon, roughness, alloy and piezo defects) and energy band model (analytical non parabolic Γ and L valleys) described in Refs [8,10], we carried out I-V studies. The simulator has been previously calibrated in static regime against experimental data for HEMTs [8]. Fig. 5 shows intrinsic I_D-V_{DS} characteristics for the four MOSFET structures at V_G - $V_T = 0.2$ V and 0.6 V. The transfer characteristics at low V_{DS} (0.2 V i.e. in the Ohmic regime) are presented in the inset of Fig. 5. For low power specifications, the leakage current and the subthreshold slope (SS) should be as low as possible. According to these criteria, structures with undoped channels appear disappointing. Indeed, even with a 10 nm-thick body, the subthreshold control in TB-MOS and COMB-MOS is poor. In contrast, Bulk-MOS and HEMT-MOS exhibit a good normally-off behaviour. To ensure good frequency performance, high transconductance g_m is mandatory. Thanks to the effective mobility improvement in the channel, I_D and g_m are enhanced for the three alternative structures, in particular for devices with undoped channel (i.e. TB-MOS and COMB-MOS) in comparison with Bulk-MOS. Besides, the percentage of fully ballistic electrons in these devices (no scattering mechanisms undergone) at the end of

channel was calculated: it is 8% for Bulk-MOS, 22% for HEMT-MOS, 37% for TB-MOS, and 55% for COMB-MOS. Thanks to strongly out of equilibrium or even quasi ballistic transport, these alternative III-V structures present high performance even at low V_{DS} (0.2 V), which is mandatory to achieve low dynamic power consumption.

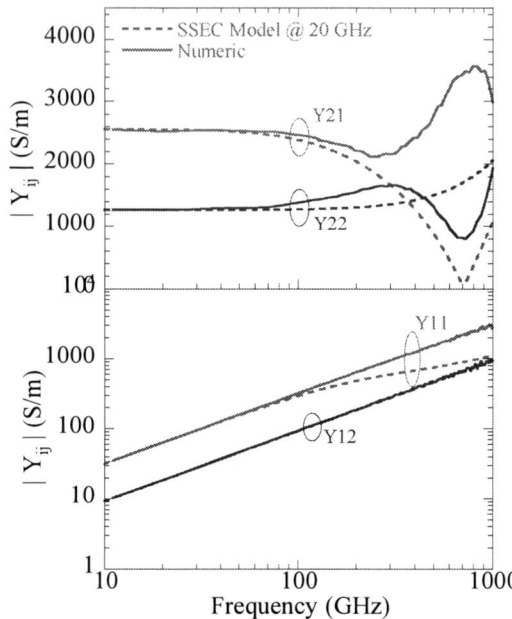

Figure 6 Calculated admittance parameters Y_{ij} **as a function of frequency for TB-MOS at** $V_{GS} = 0.5$ V **and** $V_{DS} = 0.2$ V.

Among the numerous advantages of semi-classical Monte Carlo simulation, the transient analysis of the device allows advanced investigation of the frequency response. Accurate complex frequency-dependent two-port admittance parameters Y_{ij} can be directly extracted over a wide frequency range from transient currents provided by MC simulation. The instantaneous terminal currents were calculated on the basis of the Ramo-Shockley theorem for time-varying electrode potentials [11-14]. In RF analysis, the DC bias point is tuned to reach maximum transconductance. Thus V_{GS} is chosen equal to 1.0 V for Bulk-MOS, 0.5 V for TB-MOS, 0.9 V and 0.6 V for HEMT-MOS and COMB-MOS, all biased at $V_{DS} = 0.2$ V. Fig. 6 shows the magnitude of the Y parameter extraction for MOSTB structure as a function of frequency up to 1 THz. The small-signal equivalent circuit (SSEC) is also frequently used to provide a more functional characterization of the device. We considered the typical intrinsic SSEC for the studied MOSFETs in common-source configuration, whose different elements were calculated from the Y parameters [13]. The Y parameters extracted using these SSEC elements at 20 GHz are also plotted in dotted lines in Fig. 6. Y_{11}, Y_{21} and Y_{22} from SSEC show significant differences above 200 GHz, but the proposed equivalent circuit is valid only for lower frequencies. Thanks to Y parameter extraction, we can derive different RF figures of merit such as the intrinsic cut-off frequency f_{T_exact} via the absolute value of current gain $h_{21} = |Y_{21}|/|Y_{11}|$ directly calculated from MC simulation. Fig. 7 shows the current gains with drain short-circuited as a function of frequency for the four MOSFET structures. Above 200 GHz their behavior cannot be described within the quasi-static approximation (-20dB/dec extrapolation to obtain f_{T_extrap}). These MOS structures and in particular HEMT-MOS and COMB-MOS exhibit performance comparable to or even higher than that of HEMT, as summarized in Table I. It should be noted that external parasitic impedances were not included here. Though their effects should strongly reduce the actual performance, these results are fully relevant to achieve a benchmark between investigated devices.

Our simulator allows us to evaluate other RF figures of merit such as the maximum oscillation frequency f_{MAX} [15]. The inset of Fig. 7 shows the behavior of the unilateral gain G_u as a function of frequency for COMB-MOS. The decay of 20dB per decade corresponds to the 'low frequency' behavior. At about 500 GHz, G_u shows a resonance peak. At the resonance frequency, another pole is added to the frequency behavior of G_u [16], which can lead to a substantial difference between the extrapolated f_{max_extrap} (200 GHz in the inset of Fig. 7) within quasi-static approximation and the exact value of f_{MAX} (600 GHz in the inset of Fig. 7).

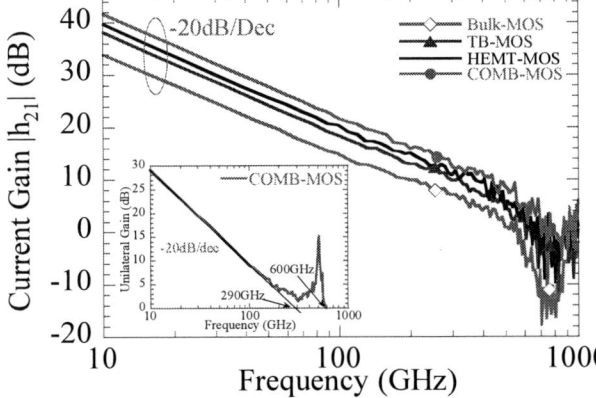

Figure 7 Current gain against frequency with drain short-circuited. Inset: unilateral gain versus frequency.

V. CONCLUSION

The charge control in several bulk and thin body structures of InGaAs-based capacitors has been studied via simulation of coupled Schrödinger/Poisson equations. A high gate voltage can generate an electron transfer into the heavier L valleys that enhances the gate capacitance but is detrimental to the transport properties. These calculations were able to predict a voltage range of relevant control for electrons in Γ valley and to provide an optimization of the oxide thickness.

Besides, the 2D solutions show that geometry effects such as fringing capacitances and DIBL are important in undoped 50 nm long FET structures. 2D charge and potential profiles must be considered to obtain realistic capacitance-voltage characteristics.

Besides, thanks to Monte Carlo device simulation, accurate intrinsic electric performance of four different III-V FETs has been assessed in both static and dynamic regimes. The validity of the small-signal equivalent circuit model appears questionable for frequency beyond 200 GHz. MC time-varying simulations are mandatory to investigate device operation at higher frequency.

Finally, III-V Bulk-MOS and even better HEMT-MOS devices with reasonably thin oxide (8 nm) present very promising performance in the off-state and in the RF domain as well, even at low V_{DS} (0.2 V). High-speed and low-power specifications using optimized structures such as HEMT-MOS and COMB-MOS should be reachable.

L_{ch} 50nm	f_{T_extrap} (GHz)	f_T (GHz)	f_{MAX} (GHz)	SS (mV/dec)
Bulk-MOS	530	520	200	75
TB-MOS	800	600	300	102
HEMT-MOS	990	690	580	80
COMB-MOS	1230	750	600	102
HEMT	900	630	540	110

Table 1 Summary of intrinsic RF performance at low V_{DS}

ACKNOWLEDGMENT

This work was supported by the Agence Nationale de la Recherche through Project MOS35 (#ANR-08-NANO-022).

REFERENCES

[1] ITRS 2007 update RF and Anal/Mixed-Signal technology

[2] R. Chau et al., in Proc. Devices Res. Conf., pp. 3-4. 2006.

[3] D. H. Kim et al. IEEE Trans on Elec Dev, vol. 54, pp. 2606-2613, 2007.

[4] D. Lin, et al. IEDM Tech Dig, pp. 337, 2009

[5] S. Datta et al., Microelectronic Engineering, vol. 84, pp. 2133–2137 2007.

[6] S. Takagi et al., IEEE Trans. Electron Dev., vol. 55, pp. 21-39, 2008.

[7] M. V. Fischetti et al., J. Appl. Phys, vol. 90, pp. 4587-4609, 2001

[8] M. Shi et al., J. Nanosci. Nanotechnol., vol. 10, to be published, 2010.

[9] D. Jin et al., IEDM Tech. Dig., pp. 495-498, 2009.

[10] P. Dollfus et al., J. Appl. Phys., vol .73, pp. 804-812, 1993.

[11] H. Kim et al., Solid-State Electron., vol. 34, pp. 1251-1253, 1991.

[12] S. Babiker et al., IEEE Trans. Electron Dev., vol. 45, pp. 1644-1652, 1998.

[13] T. Gonzalez et al., IEEE Trans. Electron Dev., vol 42, pp. 605-611, 1995.

[14] J. Mateos Lopez et al., IEEE Trans. Electron Dev., vol. 51, pp. 521-528, 2004

[15] S. J. Mason, IRE Trans. Circ. Theor. vol 1, pp. 20-25, 1954.

[16] M. B. Steer, IEEE Elec Device Lett, vol. 7, pp. 640-642, 1986.

Vertical Design of InN Field Effect Transistors

Ralf Granzner, Mario Kittler, Frank Schwierz
Fachgebiet Festkörperelektronik, TU Ilmenau
Postfach 100565
D-98684 Ilmenau, Germany
ralf.granzner@tu-ilmenau.de

Vladimir M. Polyakov
Fraunhofer Institut für Angewandte Festkörperphysik
Tullastrasse 72
D-79108 Freiburg, Germany

Abstract—The vertical design of indium nitride field effect transistors is investigated by numerical simulation. To this end, the Schrödinger equations for electrons and holes and Poisson's equation are solved self-consistently. It is shown that in several layer sequences simultaneously two-dimensional electron and hole gases are formed in the InN channel. It is demonstrated that because of the high unintentional n-type doping only thin InN layers are useful for proper transistor operation. Strain in the InN layer leads to the formation of parasitic hole channels which can dramatically deteriorate transistor characteristics. Finally it is shown that thin relaxed InN channels on GaN or AlInN buffers are a viable option for InN transistors.

I. Introduction

Due to its excellent electron transport properties (mobility up to 14000cm²/Vs, peak velocity up to 5×10⁷cm/s), InN is a promising material for high-speed transistors [1]-[4]. While the fabrication of high quality InN layers has already been demonstrated [5]-[8], so far only one experimental InN-based FET showing transistor characteristics has been reported [9]. Moreover, only few theoretical studies on the layer design for InN-based FETs can be found in the literature [10], [11]. A drawback of these studies is that holes are not considered which, as we will show, can play a significant role in InN channels.

Some specific properties of InN, such as the narrow band gap, strong polarization effects, or the high unintentional n-type doping, have a strong impact on device characteristics and complicate the design and fabrication of InN transistors. Particularly, the formation of a two-dimensional electron gas (2DEG), and for certain designs even a hole gas, is very sensitive to the detailed layer structure used for the device.

The intention of our study is to investigate possible layer sequences for c-plane InN channel FETs using self-consistent solutions of the Schrödinger and Poisson equations. Starting with the structure proposed by Kong et al. [11], i.e. a thin strained InGaN barrier on bulk InN, we show that the high unintentional n-type doping (~10¹⁸cm⁻³) usually observed in InN layers [7] is problematic for transistor operation. We demonstrate that thin, but relaxed InN layers are preferable for FET applications. We further show, that strained InN layers can be problematic, because of the formation of hole gases with very high density.

II. Models and Structures

Our study is based on self-consistent solutions of the one-dimensional Schrödinger and Poisson equations [12].

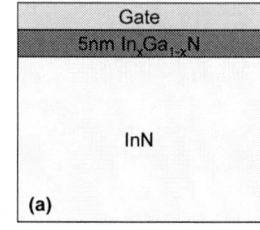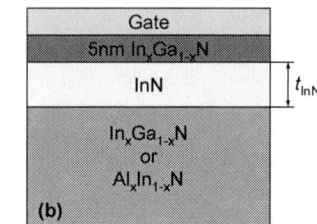

Fig. 1. Layer structures investigated in this work: (a) A 5nm thin strained $In_xGa_{1-x}N$ barrier on bulk InN (In-face) and (b) structures using thin InN layers on $In_xGa_{1-x}N$ or $Al_xIn_{1-x}N$ buffers. The gate is modeled by Dirichlet boundary conditions.

Although this study is focused on n-channel FETs, the formation of a 2D hole gas (2DHG) has to be taken into account because of the small band gap of InN. Thus, in our calculations the Schrödinger equation is solved for both electrons and holes.

The layer sequences considered in this work are sketched in Fig. 1. The material parameters of InN, GaN, and AlN used in our simulations are listed in Table 1. All parameters for $In_xGa_{1-x}N$ and $Al_xIn_{1-x}N$ alloys were calculated by linear interpolation, except the band gaps, for which we used the quadratic model

$$E_G(A_xB_{1-x}N) = xE_G(AN) + (1-x)E_G(BN) - cx(1-x), \quad (1)$$

where the bowing parameter c is 1.4 for $In_xGa_{1-x}N$ and 2.5 for $Al_xIn_{1-x}N$ [17]. The conduction band offset was assumed to be $0.8\Delta E_G$ at each interface. We used equations (13), (44), (46), and (50) of [15] to calculate interface bound charges due to spontaneous and piezoelectric polarization. Figure 2 shows these interface charges as function of the composition. As can be seen, the interface charge strongly depends on the strain in the layer structure. While spontaneous polarization in relaxed structures can cause interface charges up to ~10¹³cm⁻², the effect of piezoelectric polarization in strained layers is about one order of magnitude stronger. Note that the sign of the interface charge can be opposite for structures with strained layers compared to fully relaxed structures, particularly at InN/InGaN interfaces.

III. Results and Discussion

Let us start with the basic structure from Fig. 1(a). Figure 3 shows the calculated electron and hole sheet densities, n_s and

978-1-4244-6658-0/10 $26.00 © 2010 IEEE

TABLE I.   MATERIAL PARAMETERS USED FOR THE SELF-CONSISTENT SOLUTION OF THE SCHRÖDINGER AND POISSON EQUATIONS.

	InN	GaN	AlN
$m_{e,\parallel}$ (m_0) [13]	0.065	0.186	0.322
$m_{e,\perp}$ (m_0) [13]	0.068	0.209	0.329
$m_{h,d}$ (m_0) [14]	1.65	1.50	7.26
E_G (eV) [13]	0.69	3.24	6.47
dielectric constant [15]	14.61	10.28	10.31
nonparabolicity [3], [16]	1.43	0.363	0.237

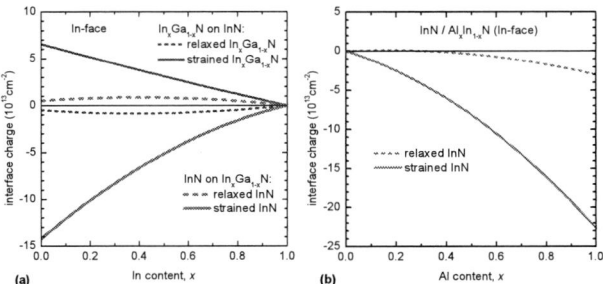

Fig. 2. Polarization charge as function of composition fraction at (a) InGaN/InN and InN/InGaN interfaces, and (b) InN/AlInN interfaces [15].

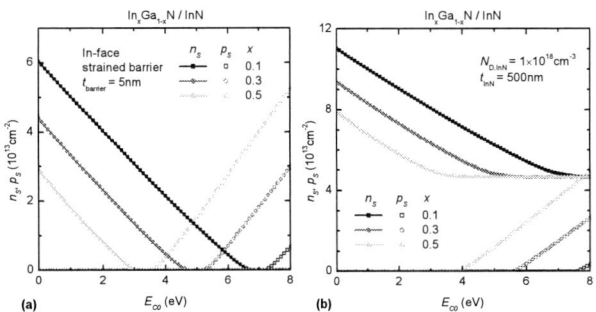

Fig. 3. Calculated electron and hole sheet densities as function of the applied surface potential for the structure of Fig. 1(a) with different In-contents in the barrier assuming (a) undoped and (b) n-type doped InN.

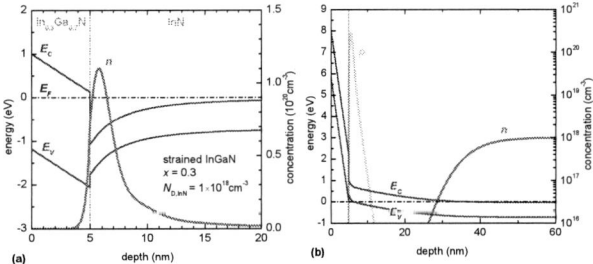

Fig. 4. Band diagram and electron and hole distributions in a structure with $x = 0.3$ biased in (a) the on-state, and (b) in the off-state.

p_s respectively, as function of the applied surface potential E_{C0}, i.e. the conduction band edge at the gate contact. The gate voltage is related to the surface potential via $V_G = -(E_{C0} - \Phi_B)/q$, where Φ_B is the Schottky barrier height and q is the elementary charge. The results in Fig. 3(a) obtained assuming a completely undoped InN body suggest that this simple layer structure is perfectly suited for transistor applications: the maximum n_s-values and the linear slopes are very high, thus suggesting a large transconductance, and the electron channel can be switched off. The threshold voltage and the n_s-value in the on state can easily be adjusted by choosing an appropriate In content in the barrier. Note that because of the small band gap of InN, a hole channel is formed approximately 0.7eV after the electron channel is switched off. These hole concentrations, however, should not be a problem for a HEMT, as long as they only occur directly underneath the gate and no complete hole channel is formed between source and drain.

Assuming the InN to be n-type doped with $N_D = 1 \times 10^{18}$ cm^{-3}, a concentration typical for epitaxial InN layers, the picture changes dramatically: the electron concentration cannot fully be controlled by the gate anymore (Fig. 3(b)). For the considered structures with 500nm thick InN, n_S saturates at a very high value around 5×10^{13}cm^{-2}. A transistor with a similar transfer characteristics would not be suited for high-speed RF applications, because it does not switch off and shows a large drain conductance. Figure 4 shows the electron and hole distributions in a structure with $x = 0.3$ biased in the on-state, Fig. 4(a), and in the off-state, Fig. 4(b). In the on-state, a 2DEG is formed with electron concentrations in excess

of 1×10^{20}cm^{-3}, which is 2 orders of magnitude larger than N_D. In the off-state on the other hand, only the top 30…40nm of InN are fully depleted from electrons, while in the largest part of the InN body the electron concentration remains at 1×10^{18}cm^{-3}. The depletion layer does not expand further with negative bias, since a 2DHG is formed at the InGaN/InN interface, when the valence band edge crosses the Fermi level. This explains the off-state saturation value of $n_S \approx N_D \times t_{InN}$.

There are two options to reduce the off-state n_S: either reducing N_D or reducing the InN layer thickness t_{InN}. While the first option is related to the technology, the latter is a design issue, and thus corresponds to the focus of this work.

For a structure with a thin InN layer as shown in Fig. 1(b), the choice of an appropriate buffer material is very important. Let us start with an InGaN buffer. Figure 5 shows $n_S(E_{C0})$ curves for structures with different t_{InN}. The In content was 0.3 in all InGaN layers considered here. The curves in Fig. 5(a) were calculated assuming fully relaxed InN layers. It can be seen that t_{InN} has to be reduced to ~10nm to guarantee a good switch-off behavior. This is much less than the maximum depletion layer width in bulk InN. The reason for that is the positive bound charge at the back interface, which creates a second electron channel. The second channel can only be controlled by the gate if the InN layer is thin enough. Figure 5(b) shows the band diagram and the electron distribution of a structure with $t_{InN} = 10$nm biased in the on-state. Although the InN-layer is very thin, one can clearly see two maxima in the electron density related to each interface.

978-1-4244-6658-0/10 $26.00 © 2010 IEEE   349

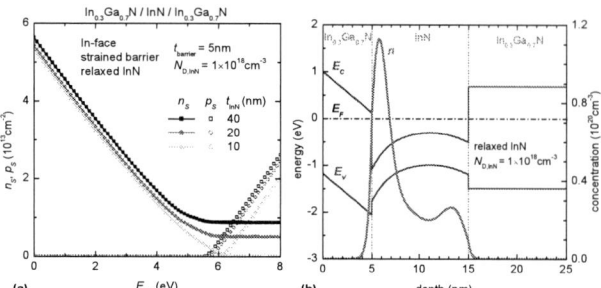

Fig. 5. InGaN/InN/InGaN structures with relaxed InN layers. (a) Electron and hole sheet densities as function of the surface potential. (b) Band diagram and electron distribution in a structure with t_{InN} = 10nm biased in the on-state.

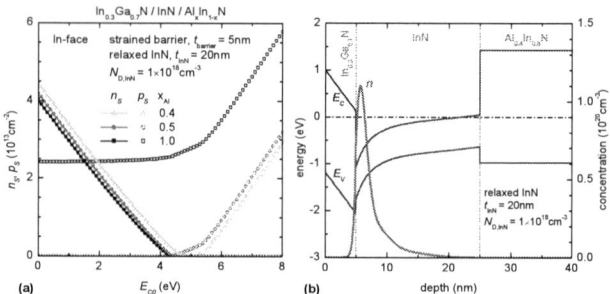

Fig. 7. InGaN/InN/AlInN structures with relaxed InN layers. (a) Electron and hole sheet densities as function of the surface potential. (b) Band diagram and electron distribution in a structure with an Al-content of 0.4 biased in the on-state.

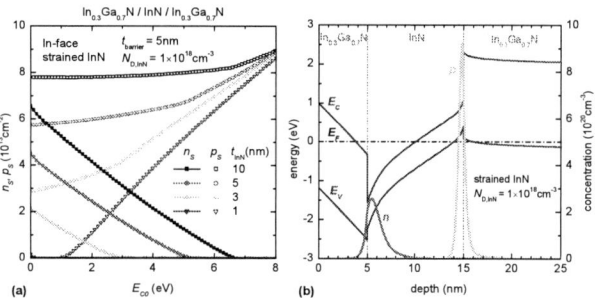

Fig. 6. InGaN/InN/InGaN structures with strained InN layers. (a) Electron and hole sheet densities as function of the surface potential. (b) Band diagram, electron and hole distributions in a structure with t_{InN} = 10nm biased in the on-state.

The thinner the InN-layer is made, the higher is the probability that it keeps the lattice constant of the buffer and strain becomes an issue. In this case the bound charge at the back interface is dominated by piezoelectric polarization which is much stronger than spontaneous polarization and it has the opposite sign. Figure 6 shows results for thin strained InN layers. Obviously the piezoelectric polarization leads to extremely large p_S, which can be larger than n_S even in the on-state. In Fig. 6(b) it can be seen, that at the top interface a 2DEG is formed while a 2DHG is formed at the back interface. Such a layer structure is not suited for transistors since a persistent hole channel always exists between source and drain. Therefore it will be impossible to switch off the drain current. As can be seen from Fig. 6(a), the hole density becomes more and more dependent on the gate potential as t_{InN} is reduced. However, the InN layer has to be extremely thin, i.e. $t_{InN} \sim$ 1nm, to switch off the hole channel. Unfortunately, the on-state n_S decreases by almost the same amount as p_S when t_{InN} is reduced, and for t_{InN} = 1nm the maximum n_S is just in the order of 10^{12}cm^{-2}. In other words, the InGaN/InN/InGaN structure with a thin strained InN-layer is not a good choice for transistor applications. However, in case a $t_{InN} \sim$ 1nm is feasible, one could improve the structure by using a different material for the barrier layer. One option could be a strained AlInN-barrier, since it would increase the positive bound charge at the top interface due to piezoelectric polarization. Another possibility could be the use of a "real"

insulator like HfO. This would allow normally-off operation and switch on for large positive gate voltages.

The shift of the $n_S(E_{C0})$ curves in Fig. 6(a) with decreasing t_{InN} can be explained with the shift of the electron subband energies due to stronger quantum confinement. Moreover, the energy separation between the subbands becomes larger, which decreases the quantum capacitance of the structure. Thus the linear slope of the $n_S(E_{C0})$ curves decreases with t_{InN}.

To avoid strain in the InN layer, t_{InN} should be larger than 10nm. In this case the positive bound charge at the back interface has to be reduced. This can be achieved by using a GaN buffer instead of In$_{0.3}$Ga$_{0.7}$N, since for relaxed layers the interface charge has a maximum near x = 0.4 as can be seen from Fig. 2(a).

Another option is the use of an Al$_x$In$_{1-x}$N buffer. The advantage of Al$_x$In$_{1-x}$N is that the bound charge at the back interface can be adjusted in a wide range by the Al-content (Fig. 2), e.g. it becomes negative for x > 0.31. Figure 7(a) shows n_S and p_S as function of the surface potential for structures with t_{InN} = 20nm and different x. It turns out that x should not be much larger than 0.4, otherwise the hole concentration would be too large in the entire gate voltage range. Figure 7(b) shows the band diagram and electron distribution for a structure with x = 0.4 biased in the on-state. Such a carrier distribution is desirable for transistor operation since neither a second electron channel nor a 2DHG is created at the back interface, and the 2DEG is closely confined at the InGaN/InN interface.

IV. CONCLUSION

Our simulations show that the choice of a proper layer sequence is crucial for the design of InN channel FETs. The specific properties of InN layers make the vertical design a challenging task. Due to the high unintentional n-type doping only thin InN layers can be used for transistor applications. On the other hand, strain in the InN layer leads to the formation of parasitic hole channels which will dramatically deteriorate transistor characteristics. We therefore suggest designs with relaxed 10 to 20nm thick InN layers, depending on the doping level and the buffer material. For the latter we favor either GaN or AlInN with an Al-content of about 0.4.

978-1-4244-6658-0/10 $26.00 © 2010 IEEE

ACKNOWLEDGMENT

This work has been supported by DFG under contract number SCHW 729/9-1. The authors thank V. Lebedev of Fraunhofer IAF Freiburg for helpful discussions.

REFERENCES

[1] S. K. O'Leary, B. E. Foutz, M. S. Shur, and L. F. Eastman, Potential performance of indium-nitride-based devices, Appl. Phys. Lett. 88, 152113, 2006.

[2] V. M. Polyakov and F. Schwierz, Low-field electron mobility in wurtzite InN, Appl. Phys. Lett. 88, 032101, 2006.

[3] V. M. Polyakov and F. Schwierz, Nonparabolicity effect on bulk transport properties in wurtzite InN, J. Appl. Phys. 99, 113705, 2006.

[4] S. K. O'Leary, B. E. Foutz, M. S. Shur, and L. F. Eastman, The sensitivity of the electron transport within bulk wurtzite indium nitride to variations in the crystal temperature, the doping concentration, and the non-parabolicity coefficient: an updated Monte Carlo analysis, J Mater Sci: Mater Electron 21, pp. 218–230, 2010.

[5] E. Dimakis, E. Iliopoulos, M. Kayambaki, K. Tsagaraki, A. Kostopoulos, G. Konstantinidis, and A. Georgakilas, Growth Optimization of an Electron Confining InN/GaN Quantum Well Heterostructure, J. Electron. Mater. 36, pp. 373-378, 2007.

[6] V. Yu. Davydov, A. A. Klochikhin, R. P. Seisyan, V. V. Emtsev, S. V. Ivanov, F. Bechstedt, J. Furthmüller, H. Harima, A. V. Mudryi, J. Aderhold, O. Semchinova, and J. Graul, Absorption and Emission of Hexagonal InN. Evidence of Narrow Fundamental Band Gap, Phys. phys. stat. sol. (b) 229, No. 3, pp. R1–R3, 2002.

[7] D. C. Look, H. Lu, W. J. Schaff, J. Jasinski, and Z. Liliental Weber, Donor and acceptor concentrations in degenerate InN, Appl. Phys. Lett. 80, 258, 2002.

[8] T. Richter, H. Lüth, Th. Schäpers, R. Meijers, K. Jeganathan, S. Estévez Hernández, R. Calarco, and M. Marso, Electrical transport properties of single undoped and n-type doped InN nanowires, Nanotechnology 20, 405206, 2009.

[9] Y.-S. Lin, S.-H. Koa, C.-Y. Chan, S. S. H. Hsu, H.-M. Lee, and S. Gwo, High current density InN/AlN heterojunction field-effect transistor with a SiN_x gate dielectric layer, Appl. Phys. Lett. 90, 142111, 2007.

[10] M. Singh and J. Singh, Design of high electron mobility devices with composite nitride channels, J. Appl. Phys. 94, 2498, 2003.

[11] Y.C. Kong, Y.D. Zheng, C.H. Zhou, Y.Z. Deng, B. Shen, S.L. Gu, R. Zhang, P. Han, R.L. Jiang, Y. Shi, A novel $In_xGa_{1-x}N/InN$ heterostructure field-effect transistor with extremely high two-dimensional electron-gas sheet density, Solid-State Electronics 49, pp. 199–203, 2005.

[12] V. M. Polyakov and F. Schwierz, Formation of two-dimensional electron gases in polytypic SiC heterostructures, J. Appl. Phys. 98, 023709, 2005.

[13] P. Rinke, M. Winkelnkemper, A. Qteish, D. Bimberg, J. Neugebauer, and M. Scheffler, Consistent set of band parameters for the group-III nitrides AlN, GaN, and InN, Phys. Rev. B 77, 075202, 2008.

[14] Electronic archive: New Semiconductor Materials. Characteristics and Properties, Ioffe Institute, St. Petersburg, www.ioffe.rssi.ru/SVA/NSM/, 2010.

[15] O. Ambacher, J. Majewski, C. Miskys, A. Link, M. Hermann, M. Eickhoff, M. Stutzmann, F. Bernardini, V. Fiorentini, V. Tilak, B. Schaff, and L. F. Eastman, Pyroelectric properties of Al(In)GaN/GaN hetero- and quantum well structures, J. Phys.: Condens. Matter 14, pp. 3399–3434, 2002.

[16] V. M. Polyakov, F. Schwierz, I. Cimalla, M. Kittler, B. Lübbers, and A. Schober, Intrinsically limited mobility of the two-dimensional electron gas in gated AlGaN/GaN and AlGaN/AlN/GaN heterostructures, J. Appl. Phys. 106, 023715, 2009.

[17] I. Vurgaftman and J. R. Meyer, Band parameters for nitrogen-containing semiconductors, J. Appl. Phys. 94, 3675, 2003.

A Continuous Physics-Based Electrothermal Compact Model for the Study of Non-Linearities in III-V HEMTs

Toufik Sadi and Frank Schwierz

Fachgebiet Festkörperelektronik, Technische Universität Ilmenau, PF 100565,
D-98684 Ilmenau, Germany
Email: toufik.sadi@tu-ilmenau.de

Abstract— **We present a newly developed continuous physics-based electrothermal I-V compact model suitable for the study intermodulation distortion in GaAs HEMTs and MESFETs. The model, which is an improvement of the standard Chalmers model, accurately includes self-heating while significantly minimizing the need for parameter fitting. The model, which is carefully calibrated using experimental data for submicrometer arsenide pHEMTs, is employed to calculate and analyze intermodulation products using the Volterra series method.**

I. Introduction

In the last three decades, there have been intense activities focused on the development of reliable non-linear compact models for the simulation of III-V MESFETs and HEMTs [1]−[5]. These models are necessary for the design of modern microwave circuits and systems. Accurate non-linear models are also important for the development of HEMT technology and for the minimization of intermodulation distortion in communication systems. Such models have been continuously improved to study several effects, including temperature effects, frequency dispersion, charge conservation and soft-breakdown [1], [6], [7]. Perhaps one of the most efficient compact models for the study of non-linearities in HEMTs is the Chalmers model introduced by Angelov *et al* [8]. This empirical model is capable of accurately modeling the current-voltage characteristics and the associated higher order derivatives, which is necessary for intermodulation distortion studies [9], while simplifying the procedure of model parameter extraction.

For medium and high power applications, power dissipation can reach levels causing the heating of the FETs to a given temperature T [10]. In the case of III-V HEMTs, such temperature rise may be significant due to the extremely low thermal conductivities of the III-V alloy-based thin layers forming such structures. As a result, there is an increasing demand for reliable electrothermal modeling techniques that account for self-heating in these devices. Several publications have discussed the possibility of considering temperature effects in the framework of the Chalmers model (see for example [1] and [7]), but most of them focused on the effect of ambient device temperature change rather than temperature rise due to self-heating. This work presents a continuous compact DC model for HEMTs (and MESFETs) self-consistently including device self-heating. This involves the extension of the

Chalmers model to consider such effects, with a reduced number of fitting parameters while maintaining one of the main advantages of this model: simple model parameter extraction. To demonstrate the continuity of the model, we study intermodulation distortion in submicrometer GaAs pHEMTs using the Volterra series method.

II. Device Model

The model presented here is an extention of the Chalmers model for the simulation of MESFETs and HEMTs. Assuming that self-heating modifies the isothermal current I_{iso} (usually obtained from pulsed measurements) in a multiplicative manner, the DC drain current I_{eth} (at electrothermal conditions) is calculated as follows:

$$I_{eth} = I_{iso} \, f(T), \tag{1}$$

where f is a function depending on the device temperature T at a given bias. The isothermal currents are calculated using the improved Chalmers model [1], at quiescent biasing conditions and pulse widths where trapping and thermal effects are negligible. As is well known, there exists a linear relationship between the logarithm of mobility and the logarithm of temperature [11]. Assuming that the channel current varies in a similar fashion with temperature, $f(T)$ can be written as:

$$f(T) = \exp(A) \left[\frac{T_0 + \Delta T}{T_0} \right]^C. \tag{2}$$

T_0 is the ambient temperature and ΔT is the temperature rise at a given bias ($T = T_0 + \Delta T$). A ($|A| << 1$) and C ($C = -1 + \alpha$ where $|\alpha| << 1$) are constant parameters physically associated with the temperature dependence of electron mobility in the device channel. The reliable study of non-linearities in HEMTs necessitates a continuous model allowing the calculation of the current-voltage characteristics and the associated high-order derivatives in the whole simulation biasing range. Therefore, a closed form analytical expression of f, solely dependent on the applied bias (drain-to-source bias V_{DS} and gate-to-source bias V_{GS}) instead of the device temperatures, must be developed to guarantee such continuity. In structures exhibiting strong self-heating, the

978-1-4244-6658-0/10 $26.00 © 2010 IEEE

device temperature at any bias can be calculated as follows [12]:

$$T = \frac{T_0}{\left[1 - \left(\frac{\theta(T_0)P_d}{4T_0}\right)\right]^4}, \tag{3}$$

where $\theta(T_0)$ is the device thermal impedance at T_0, and P_d is the power dissipation in the device:

$$P_d = V_{DS}\, I_{eth}. \tag{4}$$

In this case, a negligible gate current is assumed. These equations lead to an obvious complication: the power dissipation dependends on the DC current I_{eth}, which is the quantity to be evaluated from the electrothermal model. A simple analytical solution to this problem exists, involving a rough estimate of I_{eth} to be used in equation 4 (referred to as $I_{eth-est}$) using the isothermal model (see equation 5), but with the main model parameters corrected to account (as accurately as possible) for self-heating. This is explained in the next paragraph.

According to the Chalmers model, the isothermal drain currents are calculated as follows:

$$I_{iso} = I_{pk}(1 + \tanh(\Psi))\tanh(\alpha V_{DS})(1 + \lambda V_{DS}), \tag{5}$$

where

$$\Psi = P_1\,(V_{GS} - Vpk) + P_2\,(V_{GS} - Vpk)^2 + P_3\,(V_{GS} - Vpk)^3. \tag{6}$$

The parameters in these equations are defined in [8]. $I_{eth-est}$ can be estimated using equation 5, but with the main parameters P_1, Vpk and Ipk slightly modified to (approximately) consider the temperature rise ΔT due to self-heating. It is generally accepted that such parameters (we refer to all the parameters P_1, Vpk and Ipk generally as P_{rm}) vary with temperature as follows [1]:

$$P_{rm} = P_{rm0} + P_{rmT}\,\Delta T, \tag{7}$$

where P_{rm0} is the corresponding parameter value at T_0 and P_{rmT} is an additional fitting parameter to consider temperature rise. As published theoretical and experimental work show that ΔT varies in a near-linear fashion with V_{DS} (see for example [10]), equation 7 can be rewritten as follows to describe the dependence of P_{rm} on V_{DS}:

$$P_{rm} = P_{rm0}(1 + \lambda_{Prm}\,V_{DS}). \tag{8}$$

Note that one could always use a higher order polynomial to improve the description of P_{rm} (representing any of the parameters P_1, Vpk and Ipk) as a function of V_{DS}. As discussed earlier, the approximated electrothermal model employing equations 5 and 8 is only used to roughly estimate the DC current ($I_{eth-est}$). In this case, visible deviations are expected to be observed when comparing the measured electrothermal DC curves with the calculated $I_{eth-est} - V_{DS}$ curves. However, when using the estimated $I_{eth-est}$ values to calculate the power dissipation P_d to be used to determine device current reduction due to self-heating (using equation 2), much more accurate electrothermal current values (and hence $I_{eth} - V_{DS}$ characteristics) are obtained from equation 1. This is because equation 2 is not very sensitive to relatively small variations in the macroscopic power dissipation P_d.

The model developed here presents several advantages to the user. This physics-based model provides us with a simple tool to correctly account for self-heating using a limited number of input parameters. These include the channel material parameters (A and C) and the room temperature device thermal resistance ($\theta(T_0)$), all of which can be directly extracted from experimental measurements. The model also uses a minimized number of fitting parameters, including λ_{P1}, λ_{Vpk} and λ_{Ipk} for parameters P_1, Vpk and Ipk, respectively. These model parameters can be directly extracted from the experimental DC and pulsed data. The presented model is continuous, since one closed form expression is applied to accurately calculate DC currents in the linear and saturation operating regimes without using conditional functions. Since the drain current is the main source of non-linearities in HEMTs, continuity and accuracy are important issues when studying intermodulation distortion.

III. RESULTS, ANALYSIS AND DISCUSSIONS

We apply our model to generate the pulsed (isothermal) and DC characteristics of the 0.25μm gate GaAs pHEMT simulated in [13]. This device exhibits significant self-heating, and therefore is ideal to demonstrate the efficiency of our model. Figs. 1(a) and 1(b) show the experimental and simulated pulsed and DC $I_d - V_{DS}$ characteristics, respectively. The pulsed measurements are performed at quiescent biasing conditions and pulse durations chosen such that thermal and trapping effects are negligible. Fig. 1(c) shows the electrothermal $I_d - V_{DS}$ (or $I_{eth-est} - V_{DS}$) characteristics calculated from the simple correction model described by equations 5 and 8. Figs. 1(a) and 1(b) show excellent agreement between the measured data from [13] and the simulation results obtained from our model, which accurately reconstructs the thermal droop effect (negative output conductance) observed at the highest applied V_{GS}. Fig. 1(c) shows visible discrepancies between the measured and simulation results obtained from the simple correction electrothermal model. Clearly, this model would also lead to significant discrepancies in the output conductance and transconductance values, and hence it may not be suitable for intermodulation distortion modeling.

The continuous electrothermal model, using a closed form expression to calculate device currents, can be used within the Volterra series framework to calculate the linear (P_{lin}), second- (P_{IM2}) and third-order (P_{IM3}) intermodulation output power, as well as the second- (I_{P2}) and third-order (I_{P3}) intercept points, as a function of, for example, bias or input power. Such quantities are calculated using the rigorous formula given in [14], which are derived by modeling the small-signal drain current as a third-degree two-dimensional Taylor series. The Taylor series coefficients, including the output conductance and the transconductance and their higher order derivatives, as well as the cross-terms defined in [14], are extracted from both the simulated pulsed (isothermal) and DC (electrothermal) characteristics. It should be noted that the calculations performed here assume a common-source

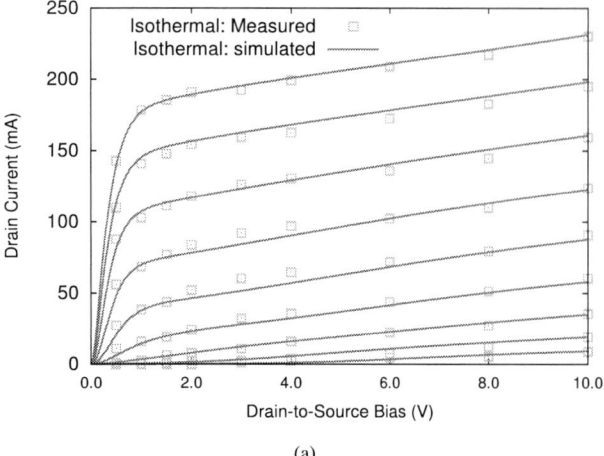

(a)

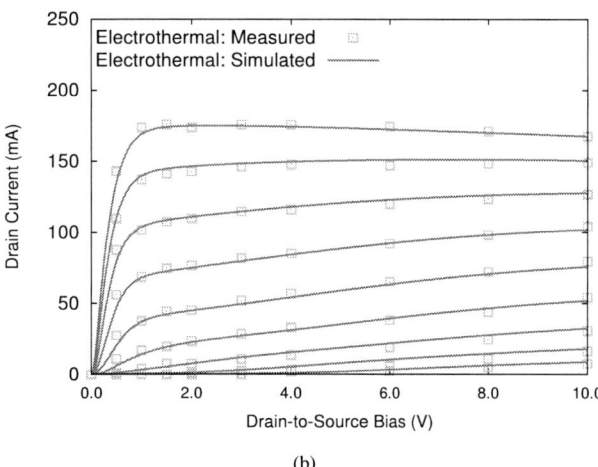

(b)

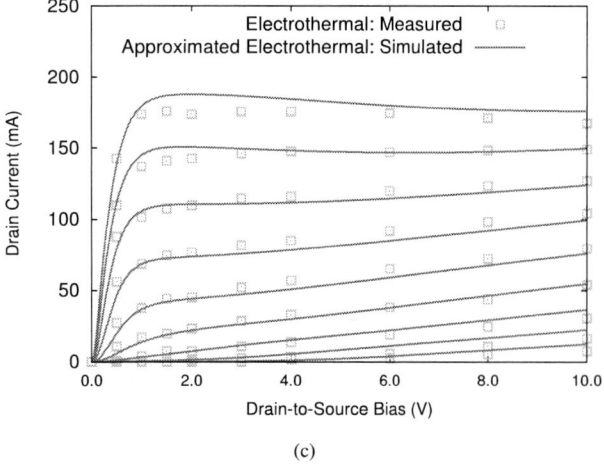

(c)

Fig. 1. Measured (from [13]) and simulated Id-V_{DS} characteristics: (a) the isothermal (pulsed) characteristics, (b) the electrothermal Id-V_{DS} characteristics obtained from the self-consistent physics-based electrothermal model developed here, (c) the electrothermal Id-V_{DS} characteristics obtained from the approximated electrothermal model using equations 5 and 8. The gate biases are varied over the range of -1.2V to -0.4V with a step of 0.1V.

configuration, with the small-signal input v_{in} being taken to be a two-tone excitation of the following form [14]:

$$v_{in} = V_s \left[cos(\omega_1 t) + cos(\omega_2 t) \right]. \qquad (9)$$

The load resistance is taken to be 50Ω, the signal source resistance is fixed to the same value, and the available input power is set to -30dBm.

Fig. 2 shows the variation of P_{lin}, P_{IM2} and P_{IM3} with V_{GS} for different V_{DS} values. Clearly, the effect of self-heating is more visible on the fundamental linear power P_{lin}, which is reduced at higher drain biases. While isothermal data show higher P_{lin} at higher drain biases, electrothermal data show lower P_{lin} values at higher drain biases, at gate biases (> -0.7V) where the influence of self-heating is significant. Fig. 3 shows the variation of second- (I_{P2}) and third-order (I_{P3}) intercept points as a function of V_{GS} for several V_{DS} values. I_{P2} and I_{P3} are important parameters for the characterization of distortion performance in devices; Fig. 3 can give us a very useful guide for the selection of gate biases at which lower intermodulation can be obtained. In this example, and for the three drain biases used, one would expect minimized second order intermodulation products by selecting a gate bias value in the range [-0.7V : -0.5V] where the maximum of I_{P2} occurs. Minimized third order intermodulation products can be obtained by applying gate biases at which the peak of I_{P3} occurs and the I_{P2} values are at a reasonably high level. While the general shapes of the I_{P2} and I_{P3} curves remain unchanged, the inclusion of self-heating clearly affects I_{P2} and I_{P3} levels: self-heating may modify both their peak values as well as the location of such peaks, affecting the device range at which the effect of intermodulation is minimal.

IV. CONCLUSIONS

In this paper, we propose a new continuous electrothermal compact HEMT model for calculating the current-voltage characteristics. The model, which is based on the standard Chalmers model, uses simple analytical expressions to accurately account for self-heating, while minimizing the need for parameter fitting and allowing simple model parameter extraction. The continuous nature of this model makes it suitable for intermodulation distortion, as demonstrated using the Volterra series method.

ACKNOWLEDGMENT

This work is funded by the European Union, in the framework of the COMON project.

REFERENCES

[1] I. Angelov, L. Bengtsson, and M. Garcia, "Extensions of the Chalmers nonlinear HEMT and MESFET model," *IEEE Trans. Microwave Theory and Techniques*, vol. 44, pp. 1664–1674, Oct. 1996.

[2] G. Qu and A. Parker, "Continuous HEMT model for SPICE," *IEE Electronics Letters*, vol. 32, pp. 1321–1323, July 1996.

[3] W. R. Curtice and M. Ettenberg, "A nonlinear GaAs FET model for use in the design of output circuits for power amplifiers," *IEEE Trans. Microwave Theory and Techniques*, vol. 33, pp. 1383–1394, Dec. 1985.

[4] A. Materka and T. Kacprzak, "Computer calculation of large-signal GaAs FET amplifier characteristics," *IEEE Trans. Microwave Theory and Techniques*, vol. 33, pp. 129–135, Feb. 1985.

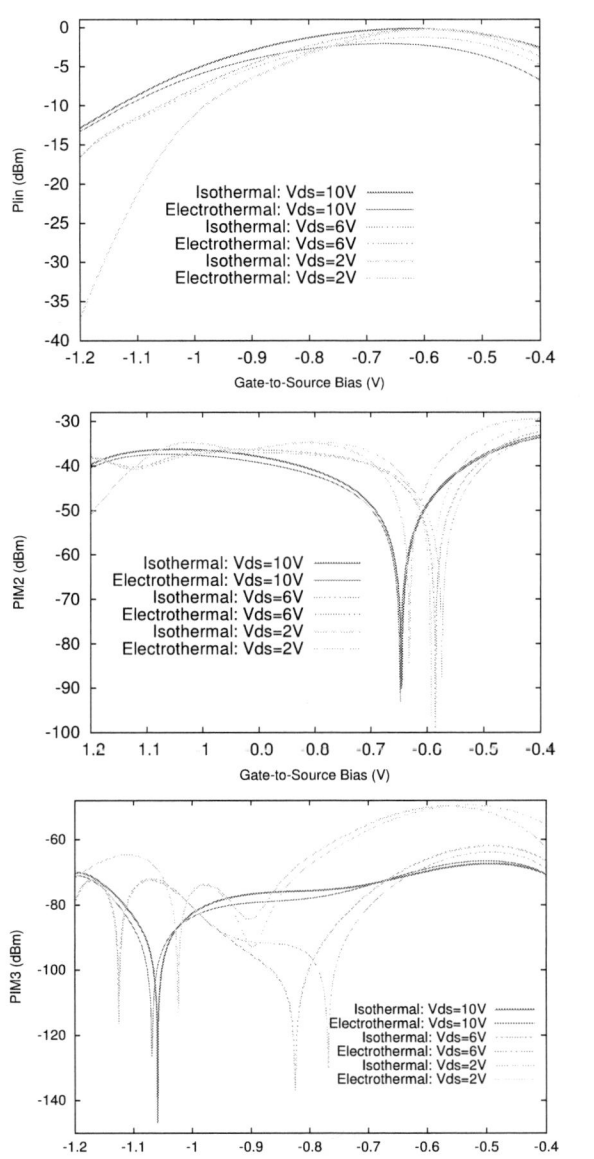

Fig. 2. Variation of the linear (P_{lin}), the second- (P_{IM2}) and the third-order (P_{IM3}) intermodulation output power, with V_{GS} for three V_{DS} values.

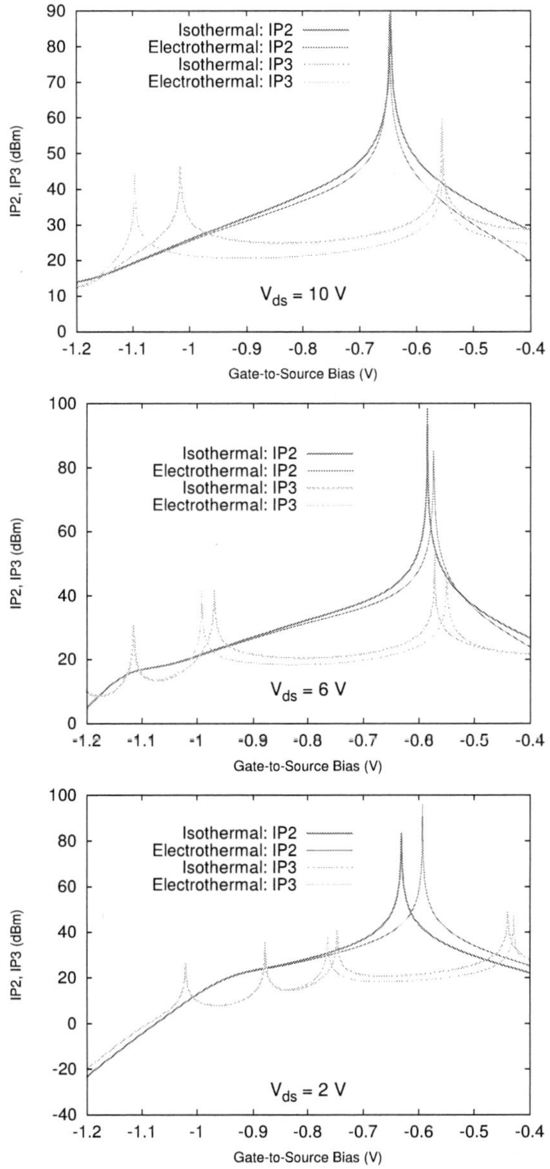

Fig. 3. Variation of second- (I_{P2}) and third-order (I_{P3}) intercept points as a function of V_{GS} for three V_{DS} values.

[5] Y. Tajima and P. D. Miller, "Design of broad-band power GaAs FET amplifiers," *IEEE Trans. Microwave Theory and Techniques*, vol. 32, pp. 261–267, Mar. 1984.

[6] P. C. Canfield, S. C. F. Lam, and D. J. Allstot, "Modeling of frequency and temperature effects in GaAs MESFETs," *IEEE Journal of Solid-State Circuits*, vol. 25, pp. 299–306, Feb. 1990.

[7] L.-S. Liu, J.-G. Ma, and G.-I. Ng, "Electrothermal large-signal model of III-V FETs accounting for frequency dispersion and charge conservation," in *Proc. IEEE International Microwave Symposium Digest (IEEE MTT-S) '09*, June 2009, pp. 749–752.

[8] I. Angelov, H. Zirath, and N. Rorsman, "A new empirical nonlinear model for HEMT and MESFET devices," *IEEE Trans. Microwave Theory and Techniques*, vol. 40, pp. 2258–2266, Dec. 1992.

[9] S. A. Maas, "How to model intermodulation distortion," in *Proc. IEEE International Microwave Symposium Digest (IEEE MTT-S) '91*, July 1991, pp. 149–151.

[10] T. Sadi, R. Kelsall, and N. Pilgrim, "Simulation of electron transport in InGaAs/AlGaAs HEMTs using an electrothermal Monte Carlo method," *IEEE Trans. Electron Devices*, vol. 53, no. 8, pp. 1768–1774, Aug. 2006.

[11] O. Madelung, *Data in Science and Technology: Semiconductors - Group IV Elements and III-V Compounds*. New York: Springer-Verlag, 1991.

[12] R. Anholt, *Electrical and Thermal Characterization of MESFETs, HEMTs and HBTs*. Norwood, MA: Artech House, 1995.

[13] K. Koh, H.-M. Park, and S. Hong, "A large signal FET model including thermal and trap effects with pulsed I-V measurements," in *Proc. IEEE International Microwave Symposium Digest (IEEE MTT-S) '03*, July 2003, pp. 467–470.

[14] G. Qu, "Characterizing intermodulation in high electron mobility transistors," Ph.D. dissertation, Macquarie University, Australia, Feb. 1999.

Investigation of rare-earth aluminates as alternative trapping materials in Flash memories

A. Cacciato, A. Suhane, O. Richard, A. Arreghini, C. Adelmann, J. Swerts, A. Rothschild, G. Van den bosch, L. Breuil, H. Bender, M. Jurczak, I. Debusschere, J.A. Kittl, J. Van Houdt

Imec vzw

Kapeldreef 75, B-3001, Leuven, Belgium

Abstract- The integration of La, Gd, and Lu aluminates in a Charge-Trapping Flash (CTF) memory flow as alternative trapping materials is evaluated. It is found that, in order to control the mixing of the aluminates with the tunnel oxide, nitride (for Gd) or nitride + oxide (for La and Lu) buffer layers have to be used. It is also found that during the post-deposition annealing treatments, the nitride buffer layer mixes with the tunnel oxide. This results in very good erase and endurance performance, which is attributed to enhanced hole tunneling from the Si substrate.

I. INTRODUCTION

The TANOS type ($TaN/Al_2O_3/Si_3N_4/SiO_2$) charge trapping cell is a technology that could replace conventional floating gate technology for multilevel non-volatile memories (NAND applications) [1]. However, to become a feasible alternative to floating gate memories, the erase and retention performances of TANOS devices have to improve. A way to improve both could be the use of high-k dielectric materials such as La_2O_3 or HfO_2 as alternative trapping layers. In fact, these materials are expected to have deeper traps than Si_3N_4, which could improve retention [2, 3]. Moreover, having a k value higher than Si_3N_4, they allow a greater voltage drop at the tunnel oxide which would also improve the program and erase operations.

Rare-earth aluminates have high k values (> 12). Moreover, deep traps (> 2 eV) have been detected in $GdALO_3$ by trap spectroscopy by charge injection and sensing (TSCIS) [4]. In this paper we therefore investigate the integration of Gd, Lu and La aluminates as alternative trapping materials into a charge-trapping Flash memory flow. In particular, their stability, with respect to the other components of the stack, upon the thermal treatments necessary to cure the stack and activate the dopants is investigated and ways to improve it are proposed.

II. EXPERIMENTAL

TANOS capacitors were fabricated on 300 mm, p-type Si wafers with n$^+$ junctions, the ANO stack consisting of 4 nm thermally grown SiO_2, 6 nm LPCVD Si_3N_4 deposited at 650 °C, 12 nm Al_2O_3 deposited by Atomic-Layer deposition (ALD). Capacitors with alternative trapping materials have

been fabricated by substituting the 6 nm Si_3N_4 layer with rare-earth aluminates, while keeping the recipes for tunnel and blocking oxides identical to those of the standard TANOS stack. $La_{0.5}Al_{0.5}O_3$, $Gd_{0.5}Al_{0.5}O_3$ and $Lu_{0.5}Al_{0.5}O_3$ rare-earth aluminates have been deposited in an ALD reactor using $La(thd)_3$, $Lu(thd)_3$, $Gd(i\text{-}PrCp)_3$ and Trimethyl Aluminum (TMA) as precursors, and O_3 (for $LaAlO_3$ and $LuAlO_3$) or H_2O (for $GdAlO_3$) as oxidants. After each ALD deposition, layers received a 60 s Post Deposition Annealing (PDA) in O_2 or N_2 ambient, at temperatures varying from 600 °C to 1100 °C. Finally, 10 nm physical vapor deposition (PVD) of the metal gate (TaN or TiN) was performed. The metal gate was capped with 50 nm in-situ-doped poly-Si.

High Resolution Transmission Electron Microscopy (HRTEM) was carried out in a Tecnai F30 microscope operating at 200 kV. High Angular Annular Dark Field Scanning Transmission Electron Microscopy (HAADF-STEM) was performed on the same instrument operating at 300kV. HAADF-STEM contrast is proportional to the TEM specimen thickness and to $<Z>^2$, where Z is the atomic number. Crystallization of the alternative trapping layer was analyzed by x-ray diffraction (XRD).

Memory characteristics were assessed by monitoring the shift of the capacitor flat band voltage (ΔV_{fb}) upon the program and erase operations.

III. RESULTS AND DISCUSSION

Post deposition annealing (PDA) at high temperature (up to 1100 °C) is required to cure the Al_2O_3 blocking dielectric and ensure good program and erase (P/E) performance and improved data retention of the TANOS stack [5, 6]. For a successful integration, rare-earth aluminates need therefore to be stable upon such PDA treatments. Unfortunately, $LaAlO_3$, $GdAlO_3$ and $LuAlO_3$ layers easily intermix both with the tunnel and the blocking oxides. This is shown in Fig. 1 in the case of a 6 nm $LaAlO_3$ trapping layer. In the figure, cross section TEM (XTEM) pictures of the TANOS and the stack with $LaAlO_3$ are compared after the Al_2O_3 PDA (60 s anneal in O_2 at 1100 °C). The physical (as obtained by XTEM) and the electrical equivalent oxide thicknesses (EOT), as obtained by CV measurements, are reported in Table 1. The La- based trapping layer is amorphous and much thicker (14 nm) than

978-1-4244-6658-0/10 $26.00 © 2010 IEEE

the as-deposited one (6 nm). From the data in Table 1, its dielectric constant is estimated to be ≈ 6.4. This value is significantly lower than that expected for polycrystalline $LaAlO_3$ (≈ 17) suggesting that the as-deposited layer has transformed into an amorphous silicate layer.

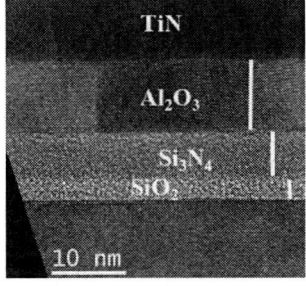 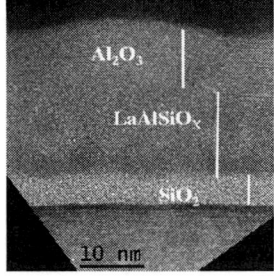

Figure 1.   XTEM for a standard TANOS stack (left) and for a stack with 6 nm (as deposited) $LaAlO_3$ as trapping layer (right). $LaAlO_3$ received a PDA in N_2 at 600 °C for 60 s before the Al_2O_3 deposition and the final PDA PDA (60 s anneal in O_2 at 1100 °C).

Most probably, during the high temperature anneal in O_2 ambient, a dynamic process sets in, in which, on one hand, the $LaAlO_3$ layer intermixes with the underlying tunnel oxide and, on the other hand, new oxide is continuously grown at the Si surface [7]. The $LaAlO_3$ trapping layer also interacts with the Al_2O_3 blocking dielectric, as suggested by the rough interface with the metal gate.

TABLE 1 PHYSICAL AND ELECTRICAL THICKNESSES (IN NM) OF THE STACKS IN FIG.1

	Tunnel oxide	Trapping layer	Al_2O_3	EOT
TANOS stack	3.2	5.6	9.4	11.4
$LaAlO_3$ stack	3.8	11.5-14.5	10.9-12.1	16.8

Uncontrolled interaction with the tunnel and the blocking oxides could be minimized by inserting a thin Si_3N_4 buffer layer at the SiO_2/trapping layer interface and by crystallizing the aluminates before the Al_2O_3 deposition. This would require temperatures in the 900 °C-1050°C range, depending on the layer stoichiometry.

Figure 2 shows the XRD spectra of 6nm $GdAlO_3$ films deposited on nitride buffer layer of different thickness and annealed at 1050 °C in N_2 for 60 s to crystallize the layer. Indeed, for nitride thicknesses down to 2 nm, two peaks corresponding to the (002) and (004) diffraction peaks of the $GdAlO_3$ hexagonal crystalline phase are clearly detected. No silicate is therefore formed in this case.

Intermixing of $LaAlO_3$ and $LuAlO_3$ films with the tunnel oxide is more difficult to control. This is illustrated in Fig. 3 where the XRD spectra of $LaAlO_3$, $GdAlO_3$ and $LuAlO_3$ layers deposited on a 2 nm-thick nitride buffer layer are compared after PDA at 1050 °C in N_2 for 60 s. Unlike for $GdAlO_3$, no diffraction peaks are detected in the $LaAlO_3$ and $LuAlO_3$ case. Since the PDA temperature is high enough to crystallize the layers, this suggests intermixing and silicate formation.

This is confirmed by the XTEM pictures Fig. 4. In the case of La and Lu aluminates (Fig. 4a and 4b, respectively), the aluminate, the nitride buffer and the tunnel oxide are completely mixed in an amorphous layer, about 6.8 nm and 9.5 nm thick. From the values reported in Table 2, k values of

≈ 10 and 7.5 are calculated for the La and Lu case, respectively. These values are much lower than those expected for polycrystalline $LaAlO_3$ (≈ 17) or $LuAlO_3$ (≈ 16) further confirming that silicates are formed.

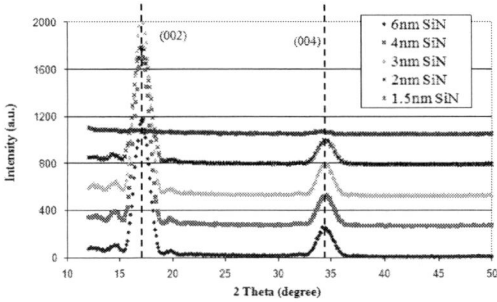

Figure 2.   XRD spectra of 6 nm $GdAlO_3$ films deposited on a 4 nm SiO_2/ Si_3N_4 stack (nitride thicknesses varied from 1.5 nm to 6 nm) after PDA at 1050 °C in N_2 for 60 s

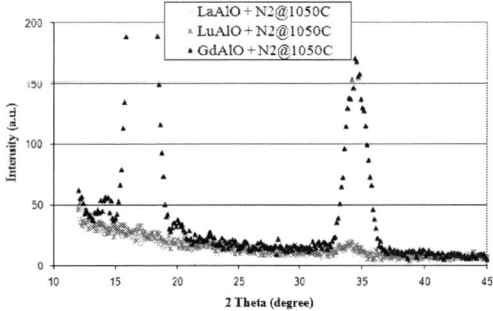

Figure 3.   XRD spectra of 6nm $LaAlO_3$, 6 nm $GdAlO_3$ and 6 nm $LuAlO_3$ films deposited on a 4 nm SiO_2/ 2 nm Si_3N_4 stack after PDA at 1050 °C in N_2 for 60 s

In the case of $GdAlO_3$ instead (Fig. 4c), the Moiré fringes visible in the TEM picture prove that a strongly-textured polycrystalline $GdAlO_3$ layer is present on top of an amorphous layer, resulting by the intermixing of the SiO_2 and Si_3N_4 layers. The HAADF-STEM picture in Fig.4d shows that this amorphous layer is not uniform and exhibits a darker contrast than the polycrystalline $GdAlO_3$ layer confirming the presence of lower Z elements (like Si, N, O). Moreover, a thin interfacial layer (most likely oxide) is clearly observed at the SiO_2+Si_3N_4/Si interface on the HAADF-STEM image. From the values reported in Table 2 and assuming $k = 16$ for polycrystalline $GdAlO_3$, the k values of the amorphous layer can be estimated to be about 6.

Data in Figs. 3 and 4 demonstrate that simply inserting a nitride layer is not enough to prevent La or Lu aluminates to intermix with the tunnel oxide. The reactivity of rare-earth aluminates could be reduced by inducing a "controlled" silicate formation via a thin layer of High Temperature Oxide (HTO) deposited at the Si_3N_4/Aluminate interface. The idea is that, during the first instants of the PDA anneal, the La, Gd and Lu aluminates would mix with the HTO layer and this would decrease the driving force for further mixing with the underlying SiO_2 and Si_3N_4 layers. XTEM in Fig. 5 demonstrates that this idea works.

978-1-4244-6658-0/10 $26.00 © 2010 IEEE   357

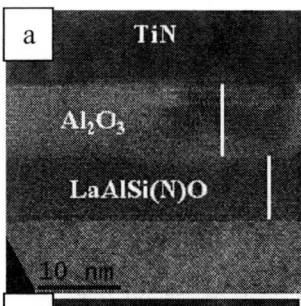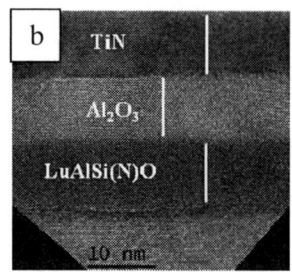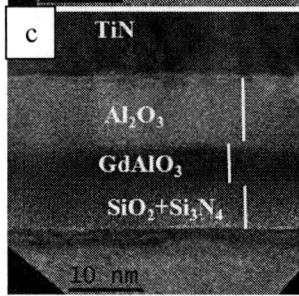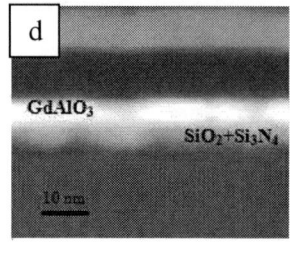

Figure 4.   XTEM pictures for 4 nm SiO$_2$/2 nm Si$_3$N$_4$/6 nm XAlO$_3$ + PDA1/12 nm Al$_2$O$_3$ + PDA2 stacks, with X= La in (4a), X= Lu in (4b), X= Gd in (4c). PDA1= N2@1050 °C for 60 s; PDA2= N2@1000 °C for 60 s. Fig. 4d is the HAADF-STEM picture of the stack in Fig. 4c. Thickness values refer to the as-deposited/grown layers

TABLE 2   PHYSICAL AND ELECTRICAL THICKNESSES (IN NM) OF THE STACKS IN FIG. 4

	Tunnel dielectric	Trapping layer	Al$_2$O$_3$	EOT
Si$_3$N$_4$ /LaAlO$_3$	6.2-7.3		8.6	6.4
Si$_3$N$_4$ /LuAlO$_3$	8.7-10.2		9.3	8.7
Si$_3$N$_4$ /GdAlO$_3$	6.1	4.7	10	9.3

Indeed, a three-layer stack is clearly observed for the LaAlO$_3$ case when the HTO buffer layer is added: a textured polycrystalline Al$_2$O$_3$, at the top, an amorphous silicate layer in the middle and an intermixed (SiO$_2$ + Si$_3$N$_4$) amorphous dielectric at the bottom. The HAADF-STEM picture shows that the intermediate layer is uniform and much brighter than the non-uniform bottom dielectric, confirming the presence of Lanthanum (high Z element). Similar images are obtained for the Lu- and Gd-based stacks (not shown).

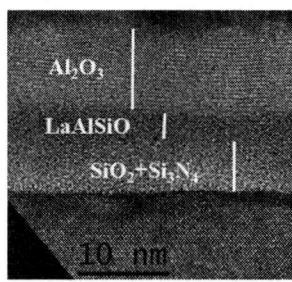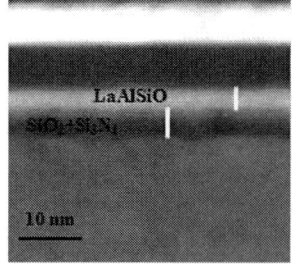

Figure 5.   XTEM (left) and HAADF STEM (right) pictures for the 4nm SiO$_2$/2 nm Si$_3$N$_4$/2 nm HTO/3 nm LaAlO$_3$ + PDA1/12 nm Al$_2$O$_3$ + PDA2 stack (PDA1= N2@1050 °C for 60 s; PDA2= N2@1000 °C for 60 s). Thickness values refer to the as-deposited/grown layers

TABLE 3 PHYSICAL AND ELECTRICAL THICKNESSES (IN NM) OF THE STACKS WITH Si$_3$N$_4$ AND HTO BUFFER LAYERS

	Tunnel dielectric	Trapping layer	Al$_2$O$_3$	EOT
Si$_3$N$_4$ /HTO/LaAlO$_3$	4.8	3.3-4.1	9.3	9.2
Si$_3$N$_4$ /HTO /LuAlO$_3$	4.8	5	9.3	10.6
Si$_3$N$_4$ /HTO /GdAlO$_3$	4.9	3.7-4.5	9.6	9.5

From the electrical and physical thicknesses in Table 3, assuming $k = 6$ for the mixed bottom dielectric (i.e. the same value as for the stack in Fig. 4c), the dielectric constant for the La, Gd and Lu silicate trapping layers is calculated to be ≈ 7 for the La- and Gd- and 5.6 for the Lu-based stack. These values are consistent with silicate formation.

The electrical performance of the stacks in Tables 2 and 3 is evaluated in Figs. 6 to 10. Figures 6 and 7 show the flat band voltage shift (ΔV_{fb}) versus the electric field applied to the stack for the program (Fig. 6) and erase (Fig. 7) operation, respectively. The electric field is calculated dividing (V_G - V_{fb}) by the stack EOT (V_G is the applied voltage, V_{fb} the initial flat band voltage). The program performance deteriorates with respect to the standard TANOS stack, being the worst for the fully mixed stacks (stacks in Figs. 4a and 4b). Worse program performance could be attributed to a lower band offset with respect to Al$_2$O$_3$, which deteriorate the trapping efficiency [8]. On the contrary, the erase performance is much improved compared to the TANOS reference (≈ 2 V difference for the same erase conditions) for all of the stacks in which the original 4 nm tunnel oxide is mixed with the 2 nm Si$_3$N$_4$ buffer layer, i.e. all of the stacks with "controlled" silicate formation (Table 3) and with polycrystalline GdAlO$_3$ (Fig. 4c).

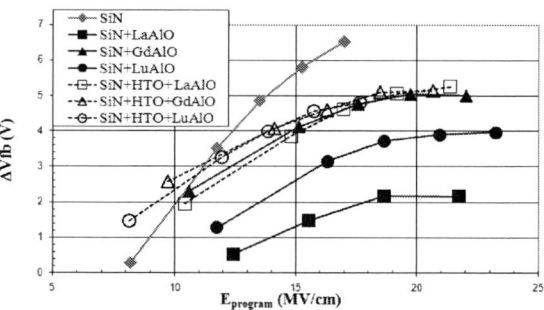

Figure 6.   Flat band voltage shift (ΔV_{fb}) versus the applied electric field during programming (pulse duration 100 µs). Full symbols refer to the stacks in Table 2, open symbols to those in Table 3

A possible explanation of the improved erase performance could be enhanced hole tunneling from the substrate due to a decreased valence band offset (with respect to the Si substrate) caused by the SiO$_2$/Si$_3$N$_4$ mixing. Simulations of the erase transient carried out assuming a decrease (compared to that of stoichiometric SiO$_2$) of the valence band offset of ≈ 2 eV show indeed good agreement with the experimental results (Fig. 8). In the figure simulations performed taking into account (continuous line), or neglecting (dashed-dotted line), trapping in the mixed tunnel dielectric are compared. Clearly trapping in the tunnel dielectric must be considered to properly explain the experimental data. The high-temperature retention performance is clearly worse for the alternative trapping materials than for Si$_3$N$_4$ (Fig. 9). In the case of La, Lu and Gd

978-1-4244-6658-0/10 $26.00 © 2010 IEEE   358

silicates this could be due to shallow traps present in the amorphous trapping layer. However, this cannot be the case for the polycrystalline GdAlO$_3$ layer, where traps deeper than those in Si$_3$N$_4$ have been observed [4]. Most probably then, the worse retention is caused by the poor quality of and/or the presence of trapped charges in the mixed tunnel dielectric. Further optimization of this oxide is therefore required to achieve better high-temperature retention performance.

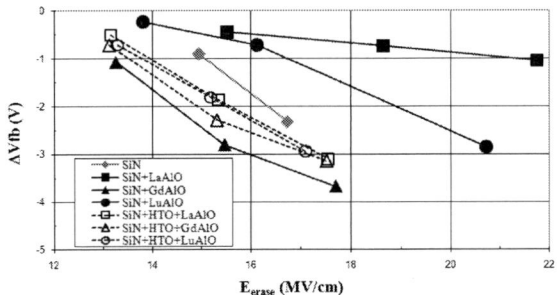

Figure 7. ΔV_{fb} versus the applied electric field during erase (pulse duration 1 ms). Full symbols refer to the stacks in Table 2, open symbols to those in Table 3

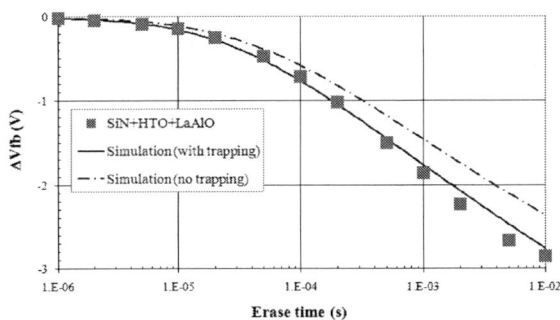

Figure 8. Simulations of the erase transient at -14V for the stack in Fig. 5. Simulation parameters for the tunnel dielectric are: k = 6, bandgap = 7 eV, Electron affinity= 1.1 eV, thickness = 4.8 nm.

The poor quality of the tunnel dielectric, however, does not impact the endurance performance, which as shown in Fig. 10 is very good in all cases. This is because the fast erase operation allows good program/erase (P/E) window to be achieved using low erase voltages [9].

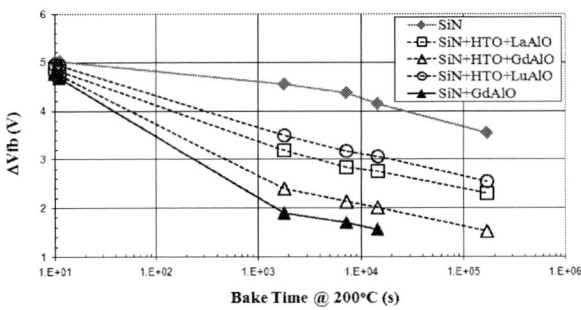

Figure 9. ΔV_{fb} shift from the program state upon retention bakes at 200 °C for the stack in Fig. 4c and in Table 3.

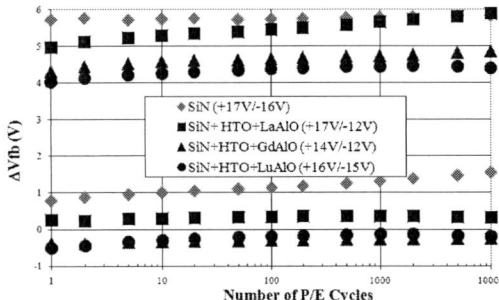

Figure 10. Endurance for the stacks in Table 3 and a P/E window of about 5 V. Program and Erase voltages are reported in brackets

IV. CONCLUSIONS

In this paper we have studied the integration of La, Gd, and Lu aluminates in a charge-trapping Flash memory flow. It is found that, in order to control the mixing of the rare-earth aluminates with the tunnel oxide, nitride (for Gd) or nitride + HTO (for La and Lu) buffer layers have to be used. It is also found that during the post-deposition annealing treatments necessary to cure the layers, the nitride buffer layer mixes with the underlying tunnel oxide. This results in very good erase and endurance performance. The improved erase performance is attributed to a decreased valence band offset of the intermixed SiO$_2$/Si$_3$N$_4$ tunnel dielectric with respect to the Si substrate, which causes enhanced hole tunneling.

ACKNOWLEDGMENT

This work has been carried out within the Imec's Industrial Affiliation Program on Advanced Flash memory and within the EC project GOSSAMER "Gigascale oriented solid-state-Flash memory for Europe".

References

[1] Y. Park, *et al.* "Highly Manufacturable 32 Gb Multi-Level NAND Flash Memory with 0.0098 µm^2 Cell Size using TANOS cell technology", IEDM Tech. Dig, p. 1-4, 2006

[2] Y.N. Tan, W.K. Chim, W.K. Choi, M.S. Joo, and B.J. Cho, "Hafnium Aluminum Oxide as charge storage and bloacking layers in SONOS-type nonvolatile memory for high-speed operation", IEEE Trans. Electron Devices, vol. 53, p. 654-662, 2006

[3] Y. Lin, C. Chien, T. Yang, and T. Lei "Two-bit Lanthanum oxide trapping layer nonvolatile flash memory" J. Electrochemical Society, vol. 154, p.H619-H622, 2007

[4] M.B. Zahid, R. Degraeve, unpublished

[5] A. Rothschild, *et al.* " O2 post deposition anneal of Al$_2$O$_3$ blocking dielectric for higher performance and reliability of TANOS flash memory" Proc. of ESSDERC, p. 272-275, 2009

[6] A. Cacciato, A. Furnémont, L. Breuil, J. De Vos, L. Haspeslagh, J. Van Houdt. "Effect of Al$_2$O$_3$ morphology on the erase saturation performance in SANOS-type memory cells" Proc. of ICMTD, p.217-220, 2007

[7] S. Van Elshocht, *et al.* "Silicate formation and thermal stability of ternary rare earth oxides as high-k dielectrics", J. Vac. Sci. Technol., vol A26, p.724-730, 2008

[8] A. Suhane *et al.*, "Experimental evaluation of trapping efficiency in silicon nitride based charge trapping memories" Proc. of of ESSDERC, p. 276-279, 2009

[9] G. Van den bosch, L. Breuil, A. Cacciato, A. Rothschild, M. Jurczak, J. Van Houdt, "Investigation of window instability in program/erase cycling of TANOS NAND Flash memories" Proc. of IMW, p. 84-85, 2009

Optimization of the Crystallization Phase of Rare-Earth Aluminates For Blocking Dielectric Application In TANOS Type Flash Memories.

L. Breuil*, C. Adelmann, G. Van den bosch, A. Cacciato, M.B. Zahid, M. Toledano-Luque, A Suhane, A. Arreghini, R. Degraeve, S. Van Elshocht, I. Debusschere, J. Kittl, M. Jurczak, J. Van Houdt

imec, kapeldreef 75, 3010 Leuven. *breuil@imec.be

Abstract

Rare-Earth aluminates GdAlO and LuAlO are investigated as blocking dielectric for Al_2O_3 replacement in TANOS flash memory devices. Since the energy bandgap of aluminates strongly depends on their crystallization phase and it is the highest for orthorombic phase, both materials were engineered using templates to assure the highest Eg and k-value after crystallization. As a consequence, the memory stack performances are significantly improved. Compared to Al_2O_3 reference top dielectric, retention can be improved.

Introduction

Charge-trapping-based TANOS devices are promising candidates to overcome the scaling limitation of Floating Gate based NAND flash memories. In particular, the discrete nature of the storage element makes it more resistant to SILC and coupling interferences issues [1]. The key element of this device derived from the SONOS concept [2] is the use of the higher K value Al_2O_3 material for top dielectric, which allows erase operation even with sufficiently thick bottom oxide for retention. Indeed the high dielectric constant reduces the field in the top dielectric and prevents electron injection from the gate during erase. However, Al_2O_3 still has a moderate K value (~9), and further improvement requires a material with even higher dielectric constant. Previous papers have introduced GdAlO [3,4], LaAlO [5], or HfAlO [6] high-k top dielectrics. However, in general, the program / erase window is still too small for multi-level applications. In particular, the erase saturation level is not deep enough, probably because of a too small energy barrier of the top dielectric to the metal gate, which promotes parasitic injection of electrons from the gate. In this paper, we investigate LuAlO and GdAlO for replacement of Al_2O_3 top dielectric in TANOS stacks, focusing on the crystallization phases. After showing the limitations of GdAlO and LuAlO naturally crystallized into hexagonal phase, we engineered these materials to target their orthorombic phase, featuring a high K value together with a larger energy bandgap (Eg) [7], as required for improved performance.

Part I : GdAlO / LuAlO in amorphous or hexagonal phase

Device fabrication

TANOS capacitors were fabricated in order to focus on the intrinsic properties of the gate stack materials, without any process integration related issues. Memory characteristics are thus investigated by measuring the flatband voltage shift of the capacitors. n+ junctions are implanted around the gate area in the p-type substrate in order to provide minority carriers for the programming operation. As for the gate stack, 4nm ISSG SiO_2 tunnel oxide is grown, followed by a 6nm LPCVD stoechiometric nitride trapping layer. Water based ALD Al_2O_3 of 12nm, annealed for 60s at 1100C in diluted O_2 atmosphere, is used as a reference top dielectric [8]. 10nm PVD-TaN is used for metal gate. In the different splits of this experiment, the Al_2O_3 reference layer is replaced by GdAlO and LuAlO deposited by ALD, using $Gd(i-PrCp)_3$ / water, and $Lu(thd)_3$ / ozone precursors, respectively. The composition is tuned to 50% Al by changing the precursor ratios, as verified by XPS measurements. Annealing after top dielectric deposition was performed in N_2 atmosphere at 700C and 1000C for 60s, targeting respectively amorphous and hexagonal phases, as evidenced by PV-TEM with Selective Area Electron Diffraction (SAED). In some cases, samples with Al_2O_3 / GdAlO, and Al_2O_3 / LuAlO bi-layers top dielectric stacks were also prepared. Table 1 summarizes the different stacks presented in this part. The details of physical analysis of the different layers will be the subject of another publication [7].

top oxide	PDA after high-k (60s)	Th. (nm)	CET of Entire TANOS stack (nm)
GdAlO	700C-N2	15	10.8
GdAlO	1000C-N2	15	10.8
LuAlO	1000C-N2	18.5	13
Al_2O_3 / GdAlO	1100C-O2 / 700C-N2	5 / 9	10.7
Al_2O_3 / GdAlO	1100C-O2 / 1000C-N2	5 / 9	11
Al_2O_3 / LuAO	1100C-O2 /1000C-N2	5 / 12	12. 4
Al_2O_3	1100C-O2	12	11.5

Table 1: Summary of different top dielectrics investigated in part I.

Electrical results :

Table 2 lists the dielectric constants of the different materials extracted from the accumulation capacitance of stacks with different top dielectric thicknesses. It also lists the energy bandgap (Eg) and the conduction band offsets to silicon (ΔEc), extracted from Internal Photo Emission (IPE) measurements on different wafers with the layers of interest. As expected, the K value is higher for GdAlO and LuAlO, but Eg and ΔEc are typically smaller.

978-1-4244-6658-0/10 $26.00 © 2010 IEEE

Top oxide	K	Eg (eV)	ΔEc(eV)
GdAlO (hex)	15	5.7	2.1
LuAlO (hex)	10	6.1	2
Al$_2$O$_3$	9	8.8 [9]	2.8 [9]

Table 2

Fig. 1 shows the Program / Erase behavior of the different TANOS stacks. Devices with GdAlO top dielectrics show poor performance as compared to the Al$_2$O$_3$ reference, whether GdAlO is in amorphous or hexagonal phase. Devices with bi-layers Al$_2$O$_3$ / GdAlO show some improvement, in particular for program operation, but still insufficient for erase operation. The same observations are made in case of LuAlO and Al$_2$O$_3$ / LuAlO top dielectrics (not shown here for brevity). This is most probably due to the lower ΔEc of GdAlO and LuAlO compared to Al$_2$O$_3$ (table 2). This is detrimental for programming operation because it can limit program efficiency [10]. Fig. 2 shows the measured program efficiency of the different top dielectrics, defined as the amount of charge being trapped in the device divided by the amount of injected charge during application of the pulse [10]. The charge is assumed to be uniformly stored in the nitride layer. A low program efficiency is clearly seen for GdAlO and LuAlO top dielectrics, while bi-layers using Al$_2$O$_3$ underneath show similar efficiency as pure Al$_2$O$_3$ top dielectric, consistently with programming characteristics in Fig. 1a. For erase operation, a lower barrier to the gate material will promote parasitic electron injection from the gate during erase, which therefore saturates at a higher level, as seen in Fig. 1b. Therefore, adding the Al$_2$O$_3$ pre-layer helps in blocking parasitic carriers flows.

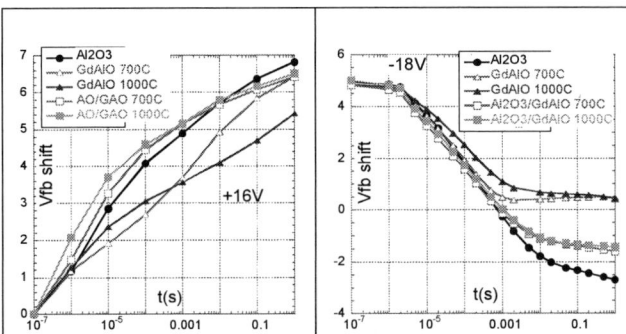

Fig. 1 : Program (a) , Erase (b) characteristics of TANOS capacitors with various top dielectric stacks.

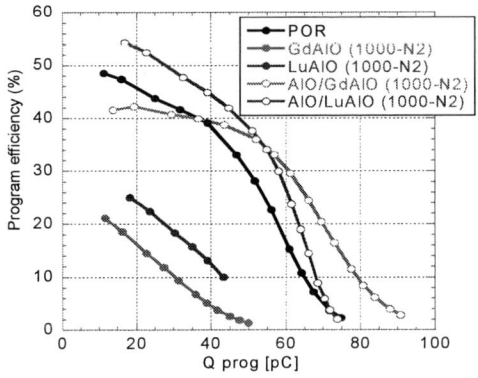

Fig. 2 : Program efficiency of TANOS capacitors with various top dielectric stacks, as a function of trapped charge during programming.

Also retention is typically worse than with Al$_2$O$_3$ (Fig. 3), especially in case of GdAlO. At high temperature, the lower ΔEc is critical : thermally emitted electrons from the nitride, can easily escape above the top oxide barrier. [3]. Additionally, it was found by trap spectroscopy (TSCIS [11]) that GdAlO and LuAlO contain a significant amount of shallow traps, (Fig. 4). Those can also compromise the retention by detrapping of charges from the top dielectric, even at lower temperature. Fast detrapping measurements, referred to as "Post Program Discharge" (PPD) [12] confirm that charge trapping occurs in GdAlO or LuAlO during programming, but charges get easily detrapped, as seen in Fig. 5. An Al$_2$O$_3$ pre-layer can thus improve retention by providing a higher ΔEc, and by limiting parasitic trapping in the high-k shallow traps. Fig. 4 shows indeed that traps are deeper in Al$_2$O$_3$ so that charges trapped here will not easily escape.

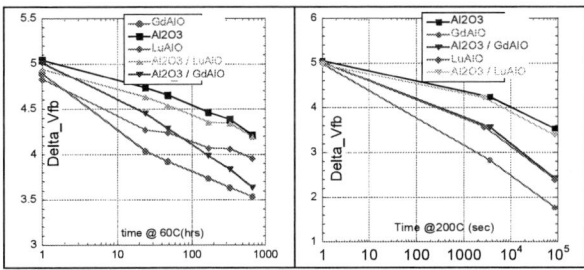

Fig. 3 : Retention at 60C (a) and 200C (b) of TANOS capacitors with various top dielectric stacks.

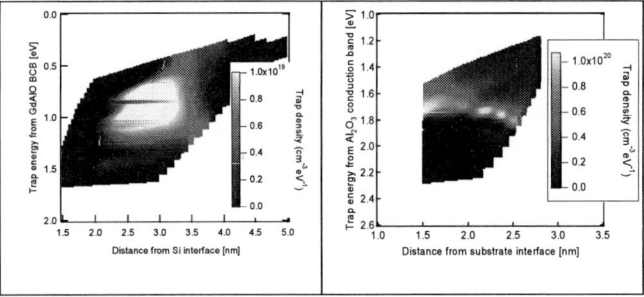

Fig. 4: Trap distributions in GdAlO (left) compared to Al$_2$O$_3$ (right) as measured by TSCIS [11]. Despite the reduced trap density in the GdAlO, the significantly shallower energy level of the traps (0.9 vs. 1.8eV) makes them very suited for fast discharging. LuAlO shows a similar picture as GdAlO.

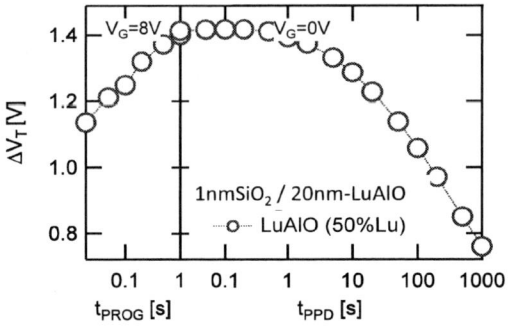

Fig. 5: Post Program Discharge measurement on SiO$_2$ / LuAlO samples. Charge is stored in LuAlO after programming. Discharge starting only after 1sec indicates that trapping occurs in the bulk of the material [12]. Similar behavior for GdAlO.

Part II : Orthorombic phase using template

In order to achieve a higher Eg for GdAlO or LuAlO high-k materials, we need to crystallize them in the right phase. Indeed, it has been found from ab-initio calculations [13] that the orthorombic phase of Rare-Earth aluminates presents a significantly higher Eg as compared to amorphous and hexagonal phases. Unfortunately, the orthorombic phase could not be obtained directly. In particular, Orth-LuAlO is reported to be not stable at temperatures compatible with device processing [14]. On the other hand, it is possible to obtain it when deposited on an appropriate substrate, which can induce the crystal structure thanks to similar lattice parameters (template approach [7]). It was shown indeed that LaAlO can crystallize into an orthorombic phase when deposited on g-Al_2O_3 ; but unfortunately, this material does not present a large Eg [13]. On the other hand, GdAlO or LuAlO cannot be obtained in orthorombic phase directly on g-Al_2O_3. However, when GdAlO or LuAlO are deposited on top of an orthorombic LaAlO, they will form an orthorombic phase as well upon annealing, as shown by GI-XRD. g-Al_2O_3 / orth-LaAlO can thus serve the role of a template for growing a large Eg orthorombic phase of GdAlO and LuAlO. It was confirmed experimentally that orthorombic LuAlO and GdAlO obtained by this template approach have a larger Eg, compared to amorphous or hexagonal phase (table 3). It is therefore promising for use as top dielectric stack in TANOS.

Fig. 6 shows the different sequences for the orthorombic top dielectric deposition in a TANOS stack. The full stack is implemented in a capacitor, as in part I. After a 4nm ISSG oxidation and 6nm SiN deposition, a thin (4nm) ALD-Al_2O_3 layer is deposited on top of the trapping nitride and annealed at 1100C in diluted O_2, similarly to part I experiments. Then, a thin (5nm) ALD-LaAlO (50%La) is deposited and annealed at 1000C to reach an orthorombic LaAlO layer that will serve as a template. Finally, the dielectric of interest (GdAlO or LuAlO) with 50% Al is deposited with a thickness of around 10nm. Then, annealing is done at 1050C in N2 to obtain the orthorombic phase, induced by the LaAlO template, as shown by GI-XRD (Fig. 7). In some cases, annealing of the high-k dielectric was performed also in diluted O_2 ambient. Finally, in these experiments, PVD-TiN was used as metal gate. Comparison samples were prepared with GdAlO or LuAlO deposited directly on top of thin g-Al_2O_3, and thus, in hexagonal phase (same cases as in part I). CET target of the full stack is kept quasi constant for all samples, around 11.5nm.

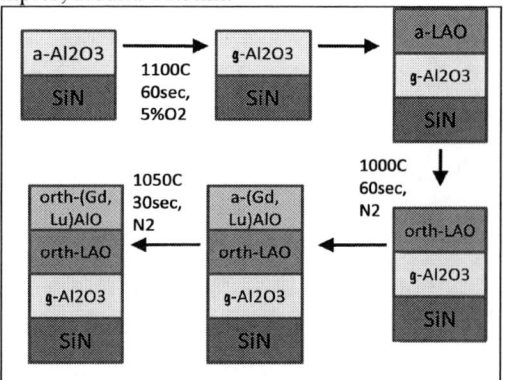

Fig. 6: Crystallization of Gd and Lu aluminates into orthorombic phase using template method : process flow .

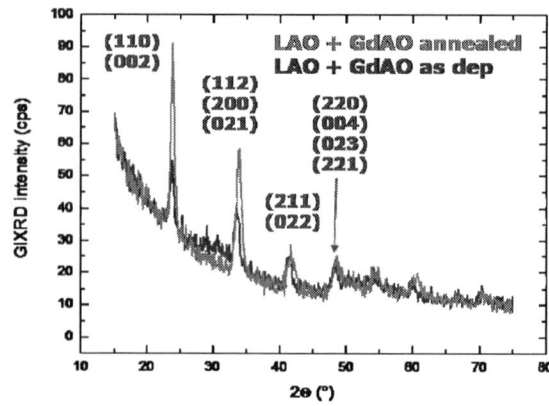

Fig. 7: GI-XRD measurement after GdAlO deposition and after GdAlO annealing. The pattern can be indexed by orthorombic LaAlO and GdAlO with similar lattice parameters. The intensity of the peaks increases after GdAlO annealing due to its crystallization into orthorombic phase. Similar observation with LuAlO.

Additionally to the gain in Eg value, the dielectric constant was found to increase with orthorombic phases as compared to hexagonal phases. Table 3 summarizes the K value of the different materials, obtained from accumulation capacitance of stacks with different top oxide thicknesses. Also are indicated the values of Eg for the corresponding materials, obtained from IPE and XPS measurements.

Material / phase	K value	Eg (eV)
g-Al2O3	9	8.8 [9]
Hex LaAlO	15	5.9
Orth-LaAlO	25	5.9
Hex GdAlO	16	5.7
Orth- GdAlO	24	**6.7**
Hex LuAlO	10	6.1
Orth-LuAlO	12	**7.4**

Table 3

Hexagonal GdAlO and LuAlO deposited on top of a thin Al_2O_3 pre-layer lead to slightly faster programming than Al_2O_3 reference, as seen in part I (Fig. 1). An orthorombic phase only slightly improves this further (not shown for brevity). More important are the erase characteristics, shown in Fig. 8. The characteristics are plotted as a function of the electric field in the tunnel oxide at the beginning of the erase operation (Eerase), defined as (Vg-Vfbo) / EOT. Erase is improved from hexagonal to orthorombic phase, especially at high field in case of GdAlO. However, it is still slightly inferior to Al_2O_3 reference in case of LuAlO. Annealing the GdAlO or LuAlO layer in diluted O_2 instead of pure N2 does not further improves the Program / Erase behavior.

978-1-4244-6658-0/10 $26.00 © 2010 IEEE

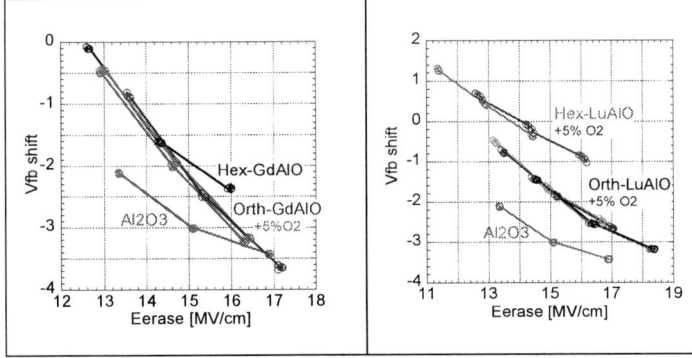

Fig. 8 : Erase Vfb shift (after program to ΔVfb = +5V), as a function of erase electric field for orthorombic and hexagonal phases of GdAlO or LuAlO, compared to Al_2O_3 reference. Erase time 100ms

Fig. 9 shows the retention characteristics. As seen also in part I, GdAlO samples typically give worse retention than LuAlO. In the latter case, the orthorombic phase shows an improved retention w.r.t. the hexagonal phase. This can be explained by the remarkably large Eg of the orthorombic LuAlO. Furthermore, PPD measurements show that the programmed state features less instability in the orthorombic phase, suggesting a lower trap density in the layer (Fig. 10). Retention is further improved in case of annealing in a diluted O_2 ambient for both GdAlO and LuAlO orthorombic phases. This might be due to a reduction of defect density and will be further investigated using TSCIS. As compared to the reference Al_2O_3, slightly higher initial charge loss is still observed, but the charge loss rate is smaller in case of the orthorombic phase. Retention is therefore expected to be significantly improved.

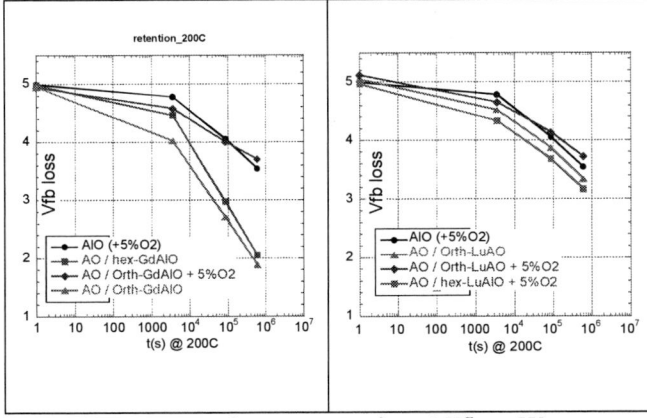

Fig.9 : Retention characteristics from ΔVfb=+5V program state, with GdAlO and LuAlO based top dielectrics, comparing orthorombic and hexagonal phases

Fig. 11 summarizes erase, and program state retention relative performances of the different stacks. Similar erase performance (as well as program, not shown here) to Al_2O_3 is obtained, with slightly better retention.

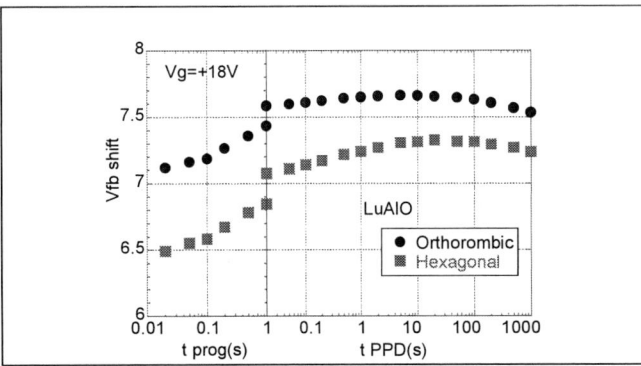

Fig. 10 : Post Program Discharge measurement on hexagonal and orthorombic phase LuAlO. After 1 sec programming, less Vfb instability is observed in case of the orthorombic phase.

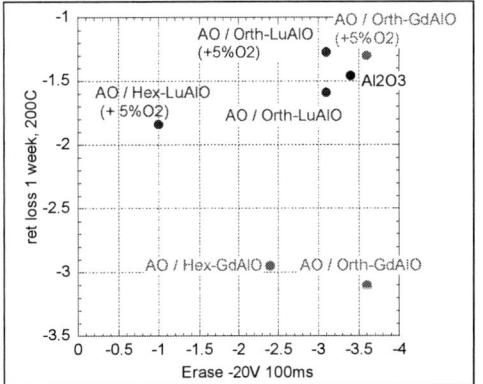

Fig. 11 : Erase / retention performance of the different stacks. (AO = Al_2O_3)

Conclusion

GdAlO and LuAlO high-k dielectrics used as blocking oxide in TANOS memory suffer from a too small Eg, and high density of shallow traps. This limits the program / erase window and compromises the retention of the high programmed level, as required for multi-level application. An important step forward is realized by engineering these materials for the orthorombic phase with the template approach. Higher K value, larger Eg and a reduced trap density have been obtained, leading to significantly improved program / erase window and retention.

References

[1] Y. Shin et al. Tech. Dig. IEDM, 2005
[2] M. H. White et al., Sol. State. Elec., Vol. 33, No. 1, p.105, 1990
[3] J. Pu et al., Trans. Electron. Dev., Vol. 56, No 11, 2009
[4] Y. Park et al., Appl. Phys. Lett. 96, 052907, 2010
[5] W. He et al., Trans. Electon. Dev., Vol. 56, No 11, 2009
[6] Y. Ny Tan et al. Trans. Elec. Dev. Vol. 53, No4, p.654, 2006
[7] C. Adelmann et al., to be submitted
[8] A. Rothschild et al., Proc. ESSDERC, 2009
[9] J. Robertson et al.,J. Vac. Sci. Tech.B,Vol.18, No3,p1785, 2000
[10] A. Suhane et al., Proc. ESSDERC, 2009
[11] R. Degraeve et al., Tech. Dig. IEDM, 2008
[12] M. Toledano Luque et al., Tech. Dig. IEDM, 2009
[13] J. Pourtois et al., Proc. ECS, p. 2216, 2009
[14] P. Wu et A.D. Pelton, J. Alloys Comp. **179**, 259 (1992).

Acknowledgements

This work was partially funded by the FP7 research contract 214431 "GOSSAMER". M. Toledano stay at imec was possible thanks to the Spanish MEC projects:TEC2007-63318, and JC2009/00052.

Investigation of the ISPP dynamics and of the programming efficiency of charge-trap memories

Alessandro Maconi, Christian Monzio Compagnoni,
Salvatore M. Amoroso, Evelyne Mascellino,
Michele Ghidotti, Giorgio Padovini,
Alessandro S. Spinelli, Andrea L. Lacaita

Dipartimento di Elettronica e Informazione,
Politecnico di Milano–IU.NET, 20133 Milano, Italy
Email: maconi@elet.polimi.it

Aurelio Mauri, Gabriella Ghidini,
Nadia Galbiati, Alessandro Sebastiani,
Claudia Scozzari, Eugenio Greco,
Elisa Camozzi, Paolo Tessariol

Numonyx, R&D–Technology Development,
20041 Agrate Brianza (MI), Italy

Abstract—This paper presents a detailed investigation of the ISPP dynamics of charge-trap memory capacitors, considering not only the flat-band voltage but also the bottom oxide electric field and tunneling current evolution during programming. Differently from the floating-gate case, results on nitride-based memories show that the flat-band increase per step does not equal the step amplitude of the gate staircase, decreasing, moreover, as programming proceeds. As a consequence, the electric field and tunneling current through the bottom oxide are shown to largely increase. Using results at different temperatures and on samples with different stack compositions, this dynamics is explained in terms of a drop of the programming efficiency as more and more charge is stored in the nitride layer, due to the reduction of the number of free traps available for capturing the injected electrons.

Index Terms—Charge trap memories, SONOS memories, TANOS memories, incremental step pulse programming, programming efficiency.

I. INTRODUCTION

The incremental step pulse programming (ISPP) algorithm represents a mandatory programming scheme for multi-level deca-nanometer NAND Flash memories, allowing very narrow threshold voltage (V_T) distributions to be obtained [1], [2]. With this algorithm, accurate placement of cell V_T by Fowler-Nordheim tunneling is achieved by applying short pulses of equal duration τ_s and increasing amplitude to the gate, keeping the channel grounded. In the case of floating-gate devices, a constant increase V_s of the pulse amplitude gives the possibility to reach a stationary condition where the V_T variation per step ($\Delta V_{T,s}$) equals V_s [1], [3], which can be exploited to tighten the V_T distribution by means of verify operations after each programming pulse [2], [4]. Different results are, instead, reported for nitride-based memories, displaying a V_T variation per step that is usually lower than V_s [5], [6]. A comprehensive understanding of this behavior and of its impact on device performance represents a mandatory issue for the development of next generation multi-level SONOS and TANOS memories.

In this paper we present a detailed analysis of the ISPP dynamics of charge-trap memory capacitors, investigating the bottom oxide electric field (F_b) and the tunneling current

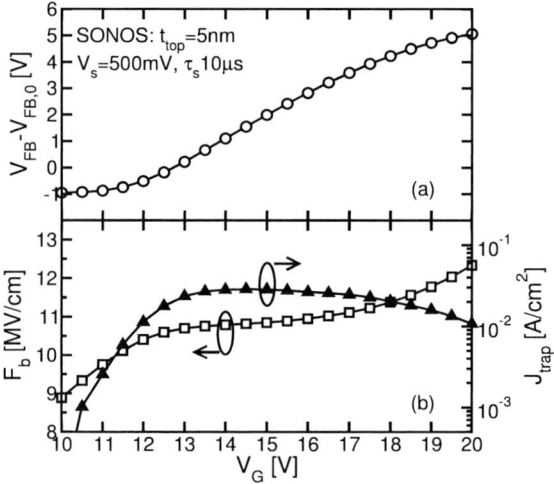

Fig. 1. V_{FB} transient during ISPP ($V_s = 500$ mV, $\tau_s = 10$ μs) of a SONOS capacitor (a) and calculated F_b and J_{trap} (b). $V_{FB,0}$ is the neutral (no charge in the nitride) flat-band voltage of the device.

(J_{tun}) evolution during programming. As a result of the decrease of the flat-band voltage shift per step ($\Delta V_{FB,s}$) as programming proceeds, both F_b and J_{tun} are shown to largely increase when high flat-band voltages (V_{FB}) are reached, potentially compromising device reliability. Results from ISPP experiments with different τ_s, at different temperatures and on samples with different stack compositions, show that this dynamics is due to a drop of the programming efficiency as more and more charge is stored in the nitride layer, critically reducing the number of free traps available for capturing the injected electrons.

II. ISPP DYNAMICS ON SONOS DEVICES

Fig. 1a shows the V_{FB} transient as a function of the gate bias V_G during ISPP with $V_s = 500$ mV and $\tau_s = 10$ μs, for a SONOS capacitor featuring p-type substrate and poly-gate, n^+-ring, thickness of the bottom oxide $t_{bot} = 4$ nm,

978-1-4244-6658-0/10 $26.00 © 2010 IEEE

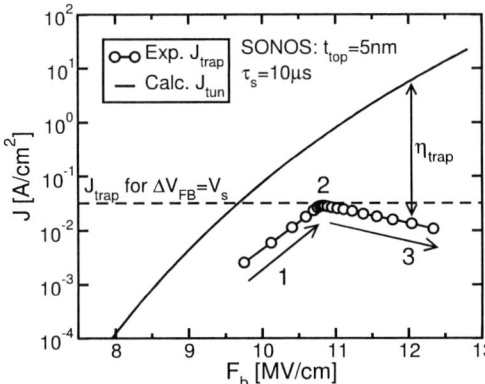

Fig. 2. Experimentally extracted J_{trap} as a function of F_b (symbols) from the V_{FB} transient of Fig. 1. The calculated J_{tun} through the bottom oxide is also shown (solid line).

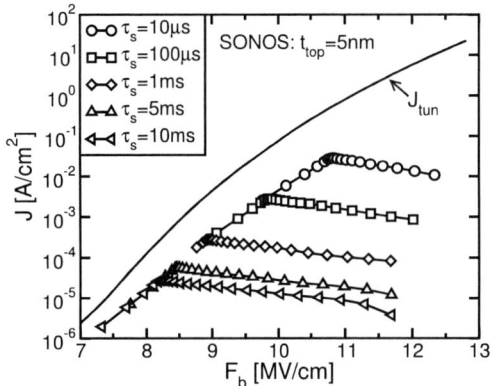

Fig. 3. Same as in Fig. 2, but for different τ_s.

of the stoichiometric nitride $t_N = 6$ nm and of the top oxide $t_{top} = 5$ nm. When V_G rises above 12 V, a significant increase of V_{FB} starts taking place as a consequence of electron storage in the nitride, with $\Delta V_{FB,s}$ remaining nearly equal to the 80% of V_s up to $V_G = 17$ V. However, for higher V_G a strong reduction of $\Delta V_{FB,s}$ appears, resulting in a clear sub-linear V_{FB} growth.

In order to study in more detail the programming dynamics, we calculated F_b and the average trapped current density (J_{trap}) during each programming step of the ISPP staircase from the V_{FB} transient according to:

$$F_b = \frac{V_G - \phi_i - (V_{FB} - V_{FB,0})}{EOT} \quad (1)$$

$$J_{trap} = \frac{C_{NG} \cdot \Delta V_{FB,s}}{\tau_s} \quad (2)$$

where $\phi_i \simeq 1.3$ V takes into account the voltage drop on the substrate and on the gate during programming (assumed constant for simplicity), $V_{FB,0}$ is the neutral device flat-band voltage and $EOT = t_{bot} + t_N \epsilon_{ox}/\epsilon_N + t_{top}$ is the equivalent oxide thickness of the gate stack (ϵ_{ox} and ϵ_N are the oxide and nitride dielectric constants, respectively). C_{NG} represents the capacitance (per unit area) from the position of the stored-charge centroid in the nitride (assumed in the middle of the nitride) to the gate. Results are reported in Fig. 1b and in Fig. 2, where J_{trap} is directly shown as a function of F_b. Three programming phases clearly appear during the ISPP transient, as highlighted in Fig. 2. In phase 1, corresponding to the V_G increase from 10 to 12 V in Fig. 1, F_b and J_{trap} rapidly increase, allowing J_{trap} to become nearly equal to $\overline{J_{trap}} = C_{NG}\Delta V_s/\tau_s = 32$ mA/cm² , representing the stationary trapping current allowing the condition $\Delta V_{FB,s} = V_s$ to hold during ISPP. When this condition is reached, F_b from (1) remains nearly constant and a nearly stationary working point for ISPP is reached in the F_b–J_{trap} diagram (phase 2, V_G from 12 to 17 V), similarly to what is usually reported for floating-gate devices [2], [3]. However, as programming proceeds J_{trap}

starts decreasing (phase 3, V_G higher than 17 V), resulting into a growth of F_b due to the V_G increase from one step to the next. Note that the electric field increase given by the V_G growth in this phase is not enough to increase J_{trap} to maintain a stationary condition during ISPP.

An additional strong difference between the ISPP dynamics of SONOS and floating-gate devices can be derived from the comparison of the J_{trap} characteristics in Fig. 2 with the calculated [7] J_{tun} through the bottom oxide (solid line). In all the ISPP phases of Fig. 2, J_{trap} is always lower than J_{tun}, revealing a trapping efficiency $\eta_{trap} = J_{trap}/J_{tun}$ lower than 1, as commonly reported for charge-trap devices [8].

III. INVESTIGATION OF THE TRAPPING EFFICIENCY

The decrease of J_{trap} and the consequent increase of F_b during phase 3 of Fig. 2 represents an important feature of the ISPP dynamics of charge-trap memories. In fact, besides being a drawback for the programming speed and efficiency, the increase of F_b and of the tunneling current J_{tun} flowing through the bottom oxide is a potential drawback also for device reliability. In order to understand the physical reasons behind this effect, we investigated η_{trap} for different τ_s on the previously considered SONOS capacitors. Fig. 3 shows the experimentally extracted J_{trap} vs. F_b characteristics for τ_s ranging from 10 μs to 10 ms. The overlap of the growing part of the curves (phase 1) associated to different τ_s confirms the carefullness of our analysis, revealing that J_{trap} depends in this phase only on F_b and J_{tun}. The curves reach, then, a different $\overline{J_{trap}}$ during phase 2 as a result of the different τ_s and, finally, decrease during phase 3 displaying, at the same F_b, a different distance from the J_{tun} curve. This reveals that during phase 3, η_{trap} is not driven by F_b, as instead observed during phase 1.

To better understand the previous results, Fig. 4 shows η_{trap} extracted from Fig. 3 as a function of the calculated nitride stored charge $Q_T = -(V_{FB} - V_{FB,0}) \cdot C_{NG}$. At the beginning of the programming transient ($|Q_T| < 2$ μC/cm²), η_{trap} remains nearly constant and equal to 10%, in agreement with typical trapping efficiencies reported for nitride memories [8].

978-1-4244-6658-0/10 $26.00 © 2010 IEEE

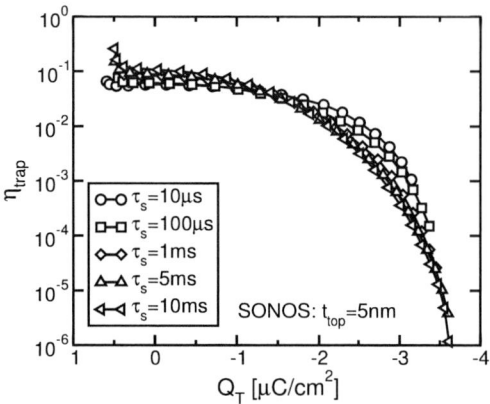

Fig. 4. Trapping efficiency for the SONOS devices with $t_{top} = 5$ nm from ISPP transients with different τ_s, as a function of the nitride stored charge.

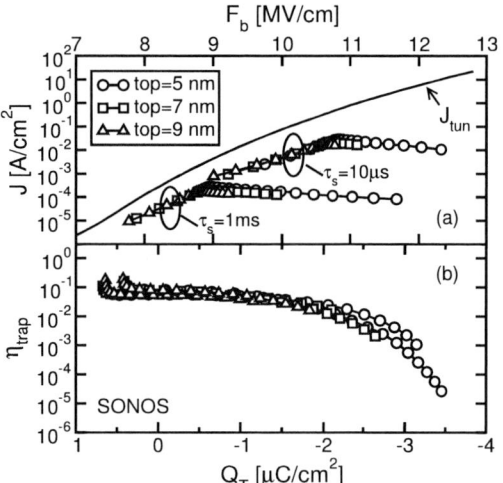

Fig. 5. J_{trap} vs. F_b characteristics for SONOS capacitors of different t_{top} (a) and corresponding η_{trap} (b).

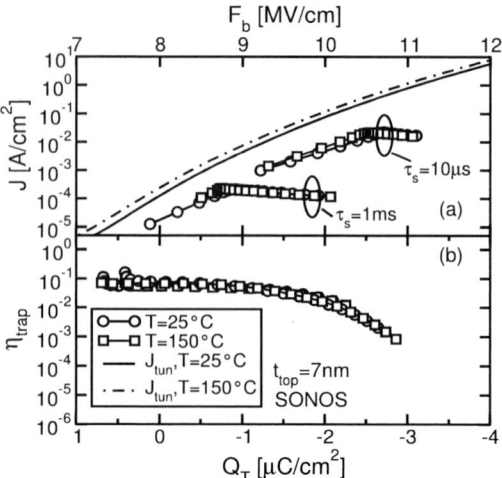

Fig. 6. Same as in Fig. 5, but on the same SONOS stack having $t_{top} = 7$ nm and at different temperatures T.

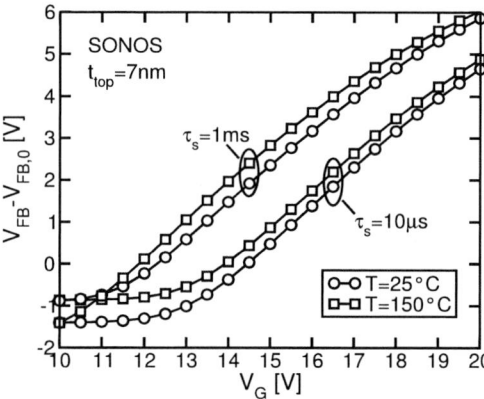

Fig. 7. V_{FB} transients during ISPP on SONOS capacitors with $t_{top} = 7$ nm. Results at different temperatures are reported.

This corresponds to phase 1 and 2 of the ISPP transients, where J_{trap} is ruled only by F_b and J_{tun}. However, when $|Q_T|$ increases above 2 μC/cm^2, a rapid reduction of η_{trap} on the logarithmic axis clearly appears, corresponding to phase 3 of the ISPP dynamics of Figs. 2-3. The good matching of the η_{trap} results for different τ_s even in this regime reveals that the J_{trap} reduction is driven by the amount of charge stored in the nitride and, in turn, by the number of free electron traps available for trapping in the nitride.

To further check the previous results, Fig. 5 shows the J_{trap} vs. F_b characteristics (a) and η_{trap} (b) for SONOS capacitors with different t_{top} and the same t_{bot} and t_N of the previous devices. Results confirm that during phase 1 and 2, a constant efficiency is obtained, whose value is independent from the stack composition. During this phases, therefore, J_{trap} depends only on F_b and J_{tun}, as highlighted

by the overlap of the J_{trap} curves associated to different t_{top}. Moreover, Fig. 5b shows that the η_{trap} curves overlap also in the high Q_T regime (phase 3 of the ISPP transients), revealing that the efficiency drop does not change when modifying the top oxide layer. This confirms that the decrease of J_{trap} in this regime is mainly a result of the finite number of traps in the nitride and not, for instance, of a significant electron emission from the filled traps when large fields in the top oxide layer are reached. This is further confirmed by the results of Fig. 6, referring to ISPP transients at different temperatures T on SONOS capacitors with $t_{top} = 7$ nm. In this case, the J_{trap} curves do not perfectly overlap in their rising regime, revealing a larger electron storage as T is increased. Note, however, that this is due to the increase of J_{tun} at higher T, as shown in Fig. 6a, and that the resulting η_{trap} is almost constant with T (see Fig. 6b), confirming that the trapping dynamics are ruled by a temperature-independent capture cross-section [9]. As a

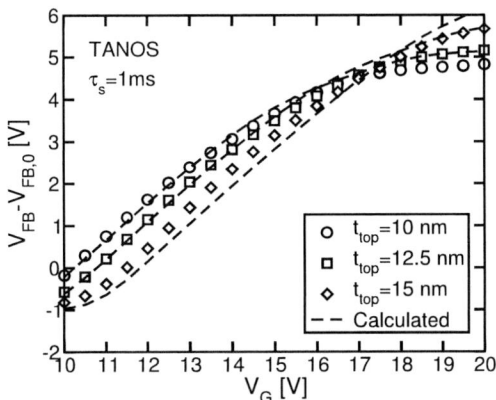

Fig. 8. Experimental V_{FB} transients during ISPP on TANOS capacitors with different alumina thickness. Calculated results obtained assuming the same trapping efficiency of SONOS capacitors are also shown.

consequence, the V_{FB} transients during ISPP are a little bit faster at higher temperatures, as shown in Fig. 7. Finally, note that similar results are found for η_{trap} in Fig. 6b even in the high Q_T regime, confirming that electron emission from the traps is not responsible for the J_{trap} reduction during phase 3 of the ISPP transient.

IV. INVESTIGATION OF TANOS DEVICES

We investigated the ISPP dynamics on TANOS capacitors having bottom oxide thickness $t_{bot} = 4.5$ nm, nitride thickness $t_N = 6$ nm and top alumina thickness ranging from 10 to 15 nm. In order to highlight any possible impact of alumina non-idealities (mainly, trapping and leakage) on the programming transients, we used the η_{trap} results obtained from SONOS capacitors to describe the nitride trapping dynamics in the TANOS devices, solving the following equation [9] for the subsequent steps of the ISPP algorithm:

$$\frac{dV_{FB}}{dt} = \frac{J_{tun}}{C_{NG}} \eta_{trap} \qquad (3)$$

where η_{trap} is adjusted at the beginning of each step to follow the dependence on Q_T, and in turn V_{FB}, obtained from Fig. 4. Calculated results are shown in Fig. 8, where they are compared with experimental results. A good agreement between data and calculations appears for $t_{top} = 10$ and 12.5 nm in the initial part of the transient, with differences appearing only when very large V_{FB} shifts nearly equal to 6 V are reached, where electron emission through the alumina may become significant. This reveals that thin alumina layers behave as ideal dielectrics on a large part of the ISPP transient, which is still ruled by the electron trapping dynamics in the nitride. However, when a thick alumina of 15 nm is used, differences between calculations and experimental results appear since the beginning of the ISPP transient, revealing that alumina trapping is not negligible and impacts the ISPP dynamics.

V. CONCLUSIONS

We presented a detailed analysis of the ISPP dynamics on charge-trap memories, investigating the F_b, J_{trap} and J_{tun} evolution during programming. Differently from the floating-gate case, after the stationary working point of the ISPP algorithm is reached on the J_{trap}–F_b diagram, a reduction of J_{trap} determines an increase of F_b and of J_{tun}. Results from ISPP experiments with different τ_s, at different temperatures and on samples with different stack compositions show that this is due to a drop of η_{trap} as more and more charge is stored in the nitride layer, as a result of the reduction of the number of free traps available for capturing the electrons injected from the substrate to the nitride.

VI. ACKNOWLEDGMENTS

Authors would like to thank P. Cappelletti, E. Camerlenghi, R. Bez and L. Baldi from Numonyx for discussions and support. This work has been partially supported by the European Commission under the FP7 research contract 214431 "GOSSAMER" and by MIUR under the FIRB Project No. RBIP06YSJJ.

REFERENCES

[1] G. J. Hemink, T. Tanaka, T. Endoh, S. Aritome, and R. Shirota, "Fast and accurate programming method for multi-level NAND EEPROMs," in *1995 Symp. VLSI Tech. Dig.*, pp. 129–130, 1995.

[2] C. Monzio Compagnoni, A. S. Spinelli, R. Gusmeroli, S. Beltrami, A. Ghetti, and A. Visconti, "Ultimate accuracy for the NAND Flash program algorithm due to the electron injection statistics," *IEEE Trans. Electron Devices*, vol. 55, pp. 2695–2702, Oct. 2008.

[3] C. Monzio Compagnoni, R. Gusmeroli, A. S. Spinelli, and A. Visconti, "Analytical model for the electron-injection statistics during programming of nanoscale NAND Flash memories," *IEEE Trans. Electron Devices*, vol. 55, pp. 3192–3199, Nov. 2008.

[4] C. Friederich, J. Hayek, A. Kux, T. Muller, N. Chan, G. Kobernik, M. Specht, D. Richter, and D. Schmitt-Landsiedel, "Novel model for cell-system interaction MCSI in NAND Flash," in *IEDM Tech. Dig.*, pp. 831–834, 2008.

[5] H.-T. Lue, T.-H. Hsu, S.-Y. Wang, E.-K. Lai, K.-Y. Hsieh, R. Liu, and C.-Y. Lu, "Study of incremental step pulse programming ISPP and STI edge effect of BE-SONOS NAND Flash," in *Proc. IRPS*, pp. 693–694, 2008.

[6] H.-T. Lue, T.-H. Hsu, Y.-H. Hsiao, S.-C. Lai, E.-K. Lai, S.-P. Hong, M.-T. Wu, F. H. Hsu, N. Z. Lien, C.-P. Lu, S.-Y. Wang, J.-Y. Hsieh, L.-W. Yang, T. Yang, K.-C. Chen, K.-Y. Hsieh, R. Liu, and C.-Y. Lu, "Understanding STI edge fringing field effect on the scaling of charge-trapping (CT) NAND Flash and modeling of incremental step pulse programming (ISPP)," in *IEDM Tech. Dig.*, pp. 839–842, 2009.

[7] P. Palestri, N. Barin, D. Brunel, C. Busseret, A. Campera, P. A. Childs, F. Driussi, C. Fiegna, G. Fiori, R. Gusmeroli, G. Iannaccone, M. Karner, H. Kosina, A. L. Lacaita, E. Langer, B. Majkusiak, C. Monzio Compagnoni, A. Poncet, E. Sangiorgi, L. Selmi, A. S. Spinelli, and J. Walczak, "Comparison of modeling approaches for the capacitance-voltage and current-voltage characteristics of advanced gate stacks," *IEEE Trans. Electron Devices*, vol. 54, pp. 106–114, January 2007.

[8] A. Suhane, A. Arreghini, G. Van den bosch, L. Breuil, A. Cacciato, A. Rothschild, M. Jurczak, J. Van Houdt, and K. De Meyer, "Experimental evaluation of trapping efficiency in silicon nitride based charge trapping memories," in *Proc. ESSDERC*, pp. 276–279, 2009.

[9] C. Monzio Compagnoni, A. Mauri, S. M. Amoroso, A. Maconi, and A. S. Spinelli, "Physical modeling for programming of TANOS memories in the Fowler-Nordheim regime," *IEEE Trans. Electron Devices*, vol. 56, pp. 2008–2015, Sept. 2009.

3D Analytical Modelling of Subthreshold Characteristics in Pi-gate FinFET Transistors

R. Ritzenthaler, F. Lime, B. Iñiguez
Department of Electrical Engineering
Rovira i Virgili University
Tarragona, Spain
romain.ritzenthaler@urv.cat

O. Faynot
CEA-LETI Minatec,
17 rue des Martyrs,
38054 Grenoble, France

S. Cristoloveanu
IMEP-LAHC (UMR 5130)
BP 257, Grenoble INP Minatec
Grenoble, France

Abstract— **A core model is developed in order to obtain, for the first time, analytical expressions of the subthreshold characteristics of advanced Pi-gate Multiple-gate FET transistors. Based on the resolution of the 3D Laplace's equation, the interface coupling in the structure is accurately described. The short-channel characteristics (Subthreshold Slope and DIBL) are calculated and compared to experimental data with an excellent agreement. Additionally, it is shown that the proposed analytical equation for the 3D potential distribution can be used to determine the scalability of a wide range of Multiple-gate FET transistors.**

I. INTRODUCTION

As the downscaling race continues to improve the performance of integrated circuits, the 'bulk transistor' architecture is facing serious physical problems. SOI fully depleted (FDSOI) structures are known to improve the scalability of CMOS transistors; among these FDSOI structures, Multiple-gate transistors (MuGFETs) allow to relax the tight scaling rules and therefore to improve the scalability of transistors [1]. FinFET-type non-planar transistors are very attractive, with demonstrated gate lengths down to 5 nm and intrinsic self-alignment of the lateral gates (contrary to planar double-gate FETs (DGFETs)). In the absence of a hard mask on top of the body, the front-gate will control the three faces (one top channel and two lateral channels) of the silicon body; this type of transistor is called Triple-gate FETs (or TGFETs, Fig. 1.a and 1.b). If the Buried Oxide (BOX) is recessed due to the Fin overetch and if the gate penetrates into the BOX, the transistors are named Pi-gateFETs (Fig. 1.c, [2]). It has been shown previously using numerical simulations [1][2] that Pi-gateFETs are improving the electrostatic control of the gate on the channel while keeping a pragmatic process flow [3].

Several analytical approaches have been developed for Triple-gate FETs [4][5][6], but so far no model has been proposed for Pi-gateFETs. In this work, an electrostatic potential expression is developed for short-channels Pi-gate FETs. Based on the electrostatic potential analytical model,

the short-channel characteristics are derived and validated with measurements.

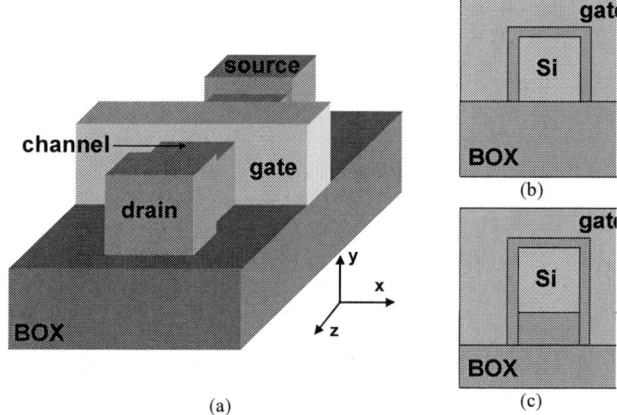

Fig. 1. (a) 3D scheme of a Triple-gate FET transistor (TGFET). Transversal cross-section in the channel of Triple-gate (b) and Pi-gateFET (c) transistor.

II. DERIVATION OF THE 3D POTENTIAL

Considering undoped channels (a necessary approach in TGFETs and Pi-gate FETs due to the impossiblity to implant uniformly the Fin), and negligible mobile charge in the subthreshold regime, the Poisson's equation can be approximated by the Laplace's equation:

$$\frac{\partial^2 \psi(x,y,z)}{\partial x^2} + \frac{\partial^2 \psi(x,y,z)}{\partial y^2} + \frac{\partial^2 \psi(x,y,z)}{\partial z^2} \approx 0 \qquad (1)$$

with $\psi(x,y,z)$ the electrostatic potential in the channel.

In this work, the source and drain junctions have been assumed abrupt; the influence of field penetration from the source, drain and overlapped regions of the gate through the BOX (i.e., the so called 'DIVSB' effect [7]) have been neglected, as well as quantum confinement effects (which means that the model will be valid down to gate length L_G, width W, and heigth H around 10 nm [8]). As for the

---

This work was supported by the European Union IAPP FP7 'Marie Curie' COMON (COmpact MOdeling Network) project (ref. pro. 218255). Support from NANOSIL (contract 216171) and EUROSOI+ (contract 216373) contracts of the European Commission, the Spanish Ministry of Science under TEC2008-06758-C02-02/TEC, PGIR/15 grant from URV and ICREA Academia Prize are also greatly acknowledged.

978-1-4244-6658-0/10 $26.00 © 2010 IEEE

boundary conditions, the potential in the lateral direction has been assumed parabolic at the interfaces between channel and overetch region, and between overetch region and BOX, respectively (Fig. 2).

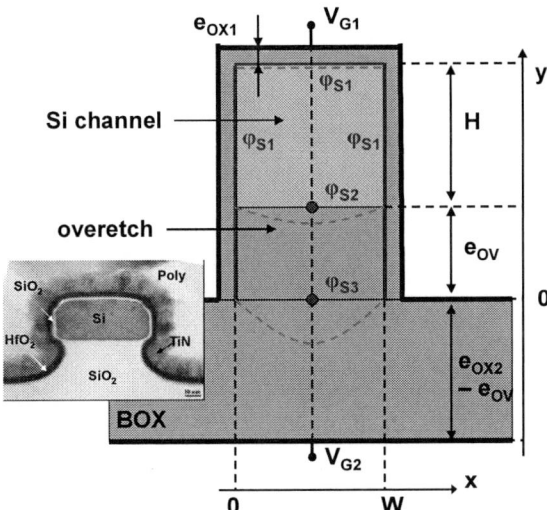

Fig. 2. Scheme of the transversal cross-section of a Pi-gateFET with the notations used in this work. A TGFET corresponds to $e_{OV} = 0$ nm, and a Pi-gateFET to $e_{OV} \neq 0$. Inset shows TEM image of the devices [3].

In order to calculate the potential in the structure, the influences of the six terminals (source, drain, the three sides of the front gate, and the back gate) are considered separately ('superposition theorem'). For each terminal, the potential is developed in Fourier series (a similar approach as in [4]) while setting the other terminals to zero. This approach has the advantage of creating symmetries and simplifying the calculation of the Fourier series coefficients; mathematical details can be found in [9], and the values of the series coefficient are given in the Appendix. Finally, the Gauss's theorem is applied at the back-interface in order to take into account the effect of the back-gate and the Pi-shape of the transistor.

The obtained potential formula is given in Eqs. (2) to (7). The comparison with numerical simulations (using COMSOL Femlab [10], Figs. 3 and 4, show an excellent agreement. It can be seen that the parabolic approximation in the transversal direction yields very good results (Fig. 3). Along the longitudinal axis (Source/Drain axis), the potential is influenced by the source and drain biases, as well as by the lateral gates. Their influence extends in the channel and in the overetched region (Fig. 4). This enhanced electrostatic control of the gate on the channel limits the influence of the source and drain, and accounts for improved Short-Channel Effects (SCEs) of the TGFETs and Pi-FETs architectures. The potential value in Pi-gate FETs is higher than for TGFETs (Fig. 4); this highlights the better electrostatic control from the front-gate in Pi-gate FETs architectures and accounts for their better subthreshold characteristics compared to TGFETs (see section IV).

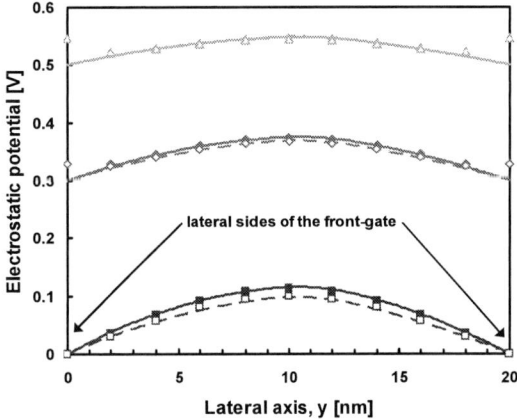

Figure 3. Transversal cross-section at mid-channel ($z = L_G/2$) and at the body/BOX interface ($y = e_{OV}$) of the electrostatic potential obtained with a numerical simulator (solid lines, using COMSOL Femlab) and with the analytical model for TGFETs (closed symbols) and Pi-gateFETs (open symbols). Front gate bias $V_{G1} = 0$ V (squares), 0.3 V (diamonds) and 0.5 V (triangles). $L_G = 40$ nm, $W = 20$ nm, $H = 20$ nm, $e_{OV} = 20$ nm, $e_{OX1} = 2$ nm, $e_{OX2} = 100$ nm, $V_{DS} = 0.8$ V, $V_{G2} = 0$ V.

$$\psi(x,y,z) = \psi_{Top-gate(TG)}(x,y,z) + \psi_{Back-gate(BG)}(x,y,z) + \psi_{Lateral-gates(LG)}(x,y,z) + \psi_{Source/Drain(SD)}(x,y,z) \tag{2}$$

$$\psi_{TG}(x,y,z) = (V_{G1} - V_{FB1}) \sum_{m=1}^{+\infty} \sum_{n=1}^{+\infty} F_P(m) F_P(n) \sin\left(\frac{m\pi x}{W}\right) \sin\left(\frac{n\pi z}{L_G}\right) \left[\cosh\left(\sqrt{\left(\frac{m}{W}\right)^2 + \left(\frac{n}{L_G}\right)^2}\,\pi y\right) \Big/ \cosh\left(\sqrt{\left(\frac{m}{W}\right)^2 + \left(\frac{n}{L_G}\right)^2}\,\pi(H + e_{OV})\right) \right] \tag{3}$$

$$\psi_{LG}(x,y,z) = (V_{G1} - V_{FB1}) \sum_{m=1}^{+\infty} \sum_{n=0}^{+\infty} F_P(m) F_n(n) \sin\left(\frac{m\pi z}{L_G}\right) \sin\left(\frac{(2n+1)\pi(H + e_{OV} - y)}{2(H + e_{OV})}\right) \left[\frac{\sinh\left(\sqrt{\left(\frac{m}{L_G}\right)^2 + \left(\frac{2n+1}{2(H + e_{OV})}\right)^2}\,\pi(W-x)\right) + \sinh\left(\sqrt{\left(\frac{m}{L_G}\right)^2 + \left(\frac{2n+1}{2(H + e_{OV})}\right)^2}\,\pi x\right)}{\sinh\left(\sqrt{\left(\frac{m}{L_G}\right)^2 + \left(\frac{2n+1}{2(H + e_{OV})}\right)^2}\,\pi W\right)} \right] \tag{4}$$

$$\psi_{SD}(x,y,z) = \sum_{m=1}^{+\infty} \sum_{n=0}^{+\infty} F_P(m) F_n(n) \sin\left(\frac{m\pi x}{W}\right) \sin\left(\frac{(2n+1)\pi(H + e_{OV} - y)}{2(H + e_{OV})}\right) \left[\frac{V_S \sinh\left(\sqrt{\left(\frac{m}{W}\right)^2 + \left(\frac{2n+1}{2(H + e_{OV})}\right)^2}\,\pi(L_G - z)\right) + V_D \sinh\left(\sqrt{\left(\frac{m}{W}\right)^2 + \left(\frac{2n+1}{2(H + e_{OV})}\right)^2}\,\pi z\right)}{\sinh\left(\sqrt{\left(\frac{m}{W}\right)^2 + \left(\frac{2n+1}{2(H + e_{OV})}\right)^2}\,\pi L_G\right)} \right] \tag{5}$$

$$\psi_{BG}(x,y,z) = \varphi_{S3} \sum_{m=1}^{+\infty} \sum_{n=1}^{+\infty} F_P(m) F_P(n) \sin\left(\frac{m\pi x}{W}\right) \sin\left(\frac{n\pi z}{L_G}\right) \left[\sinh\left(\sqrt{\left(\frac{m}{W}\right)^2 + \left(\frac{n}{L_G}\right)^2}\,\pi(H + e_{OV} - y)\right) \Big/ \sinh\left(\sqrt{\left(\frac{m}{W}\right)^2 + \left(\frac{n}{L_G}\right)^2}\,\pi(H + e_{OV})\right) \right] \tag{6}$$

with:

$$\varphi_{S3} = (V_{G2} - V_{FB2}) \Big/ \left(1 + \frac{\varepsilon_{Si}}{\varepsilon_{BOX}/(e_{OX2} - e_{OV})} \sum_{m=1}^{+\infty} \sum_{n=1}^{+\infty} F_P(m) F_P(n) \sin\left(\frac{m\pi}{2}\right) \sin\left(\frac{n\pi}{2}\right) \left[\sqrt{\left(\frac{m}{W}\right)^2 + \left(\frac{n}{L_G}\right)^2}\,\pi \Big/ \tanh\left(\sqrt{\left(\frac{m}{W}\right)^2 + \left(\frac{n}{L_G}\right)^2}\,\pi(H + e_{OV})\right) \right]\right) \tag{7}$$

W being the fin width, H the fin height, L_G the gate length, e_{OV} the overetch depth, ε_{BOX} the BOX permittivity, ε_{Si} the silicon permittivity, V_{FB1} (resp. V_{FB2}) the front-gate (resp. back-gate) flat band voltage. The series coefficient F_p, F_n, and F_c are defined in the Appendix.

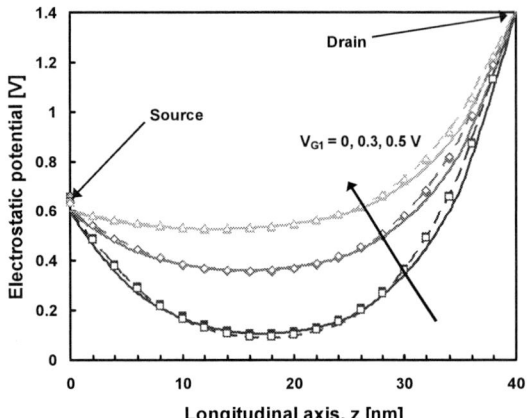

Figure 4. Longitudinal cross-section at mid-channel (x = W/2) and at the body/BOX interface (y = e_{OV}) of the electrostatic potential obtained with a numerical simulator (solid lines, using COMSOL Femlab) and with the analytical model for TGFETs (closed symbols) and Pi-gateFETs (open symbols). Front gate bias $V_{G1} = 0$ V (squares), 0.3 V (diamonds) and 0.5 V (triangles). $L_G = 40$ nm, W = 20 nm, H = 20 nm, $e_{OV} = 20$ nm, $e_{OX1} = 2$ nm, $e_{OX2} = 100$ nm, $V_{DS} = 0.8$ V, $V_{G2} = 0$ V.

III. SHORT-CHANNEL DEVICE CHARACTERISTICS

For short channels, it is often considered that the subthreshold current is flowing preferentially where the electrostatic control of the front-gate is the weakest (the so called 'most leaky path' aproach [6]). Therefore, calculating the minimum of the potential barrier and its location, the subthreshold characteristics of the transistor can be derived (a detailed description of the procedure can be found in [6] and in the BSIM-CMG model for planar DGFETs [1]). In TGFETs and Pi-gate FETs devices, the minimum of potential is located at mid-channel (x = W/2, fig. 2) and at the body/overetched region interface (y = e_{OV}, fig. 2).

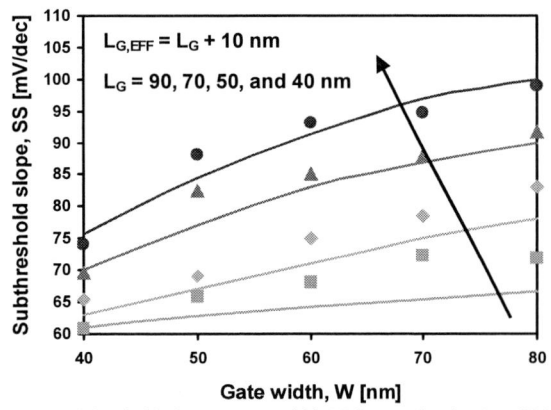

Figure 5. Subthreshold slope vs. gate width W for gate lenghts L_G of 90, 70, 50, and 40 nm. Comparison between experimental extractions (symbols) and our analytical model (solid lines). $e_{OX1} = 1.95$ nm, $e_{OX2} = 100$ nm, H = 26 nm.

The model is compared with experimental data for ΩFET devices [1,3], where the encroachment under the channel is small enough (≈ 5 nm, Fig. 2) to consider them as Pi-gate FETs. A correct agreement is found in Fig. 5. It is shown that the scaling of

the channel width W together with the gate length L_G allows controlling the SCEs. Additionally, for a given gate length L_G, reducing the gate width W leads to a better electrostatic control of the front-gate on the channel, and therefore to an improved subthreshold slope. Similarly, the variation of the minimum potential between low and high values of the drain voltage yields straightforwardly the DIBL values. Similar trends as for the subthreshold slope are observed, and a reasonable agreement is obtained (Fig. 6).

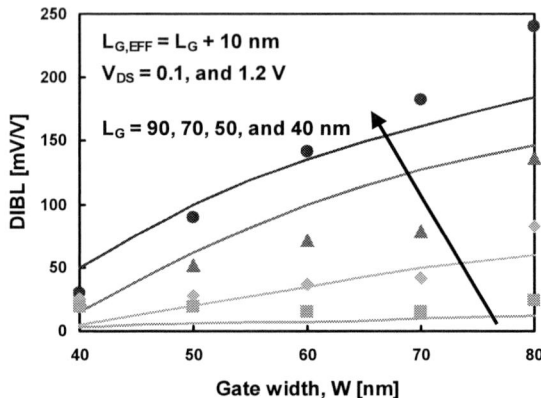

Figure 6. DIBL vs. gate width W for gate lenghts L_G of 90, 70, 50, and 40 nm. Comparison between experimental extractions (symbols) and analytical model (solid lines). $e_{OX1} = 1.95$ nm, $e_{OX2} = 100$ nm, H = 26 nm.

IV. DEVICE SCALABILITY

The Pi-gate FET structure is very informative since it contains in itself all the other multi-gate transistors by simply changing the device parameters (see Tab. 1). Therefore, the analytical formulas derived for Pi-gateFETs can be used as a core model to depict the subthreshold slope SS vs. gate length L_G for a wide range of devices (Fig. 7). An excellent agreement is found with numerical simulations (data taken from [2], except for planar FDSOI where experimental measurements were used). Increasing the 'number of gates' makes remarkably easier the scalability of the SOI devices. Pi-gate FETs, acting like a device with a number of gates between 3 and 4 (depending of the overetch depth and gate width), allow increasing the front-gate control almost as in a four gate FET. The advantage of Pi-gate FETs comes from a very pragmatic process flow, as opposed to 4-gates Gate-All-Around devices.

TABLE I.   VARIATIONS OF THE CORE STRUCTURE

Structure	Features
Pi-gateFET (core structure)	$e_{OV} \neq 0$
TGFET	$e_{OV} \approx 0$
Planar FDSOI	$e_{OV} \approx 0$, W>>H
DGFET/FinFET	$e_{OV} \approx 0$, W<<H
Gate All Around	$e_{OV} \approx 0$, $\varphi_{S3} = V_{G1} - V_{FB1}$

978-1-4244-6658-0/10 $26.00 © 2010 IEEE

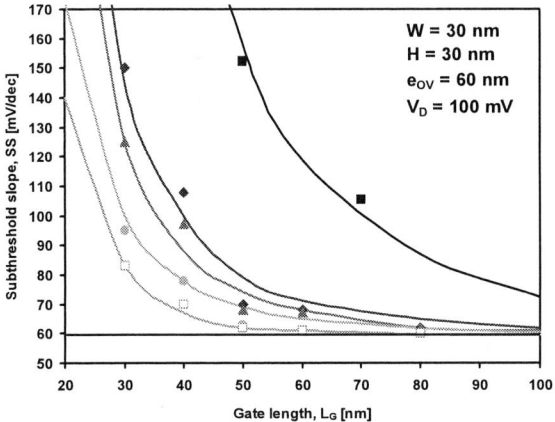

Figure 7. Subthreshold slope SS vs. gate length L_G obtained with numerical simulations (symbols) and the analytical model (solid lines) for planar FDSOI FETs (squares), DGFET (diamonds), TGFET (triangles), Pi-gateFET (circles), and GAA (open squares) transistors. All numerical simulations results are obtained from [2], except for planar FDSOI FETs (planar FDSOI: experimental measurements, H = 26 nm, W = 10 μm).

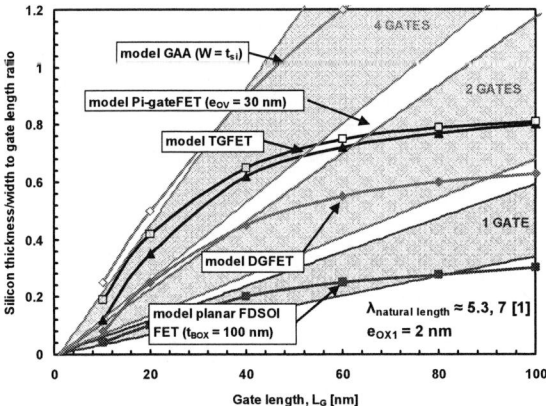

Figure 8. Silicon width/thickness to gate length ratio vs. gate length Lg for FDSOI, DGFET, TGFET, Pi-gateFET, and GAA transistors (solid lines) obtained for a subthreshold slope criterion of 75 mV/dec. Coloured areas show the results obtained with the 'natural length' approaches [1]. e_{OX1} = 2 nm, e_{OX2} = 100 nm, V_{DS} = 100 mV.

The silicon gate width/heigth necessary to keep the SCEs under control (using a criterion of 75 mV/dec) can be calculated as a function of the gate length for MuGFETs (Fig. 8). The better scalability achieved by increasing the number of gates is hightlighted. The 3D potential formula can be used for the precise calibration of MuGFET devices. It should be noted that the scaling rules presented in [1] based on the 'natural length' concepts are also represented (coloured areas, Fig. 8), and are remarquably coherents with the formula developed in this work.

V. CONCLUSIONS

Analytical models for the short-channel subthreshold performance of Pi-gateFET transistors were developed. It is shown that the effect of the gate penetration in the BOX is of great importance since it improves considerably the subthreshold performance of the device. Additionally, it is demonstrated that without the use of any fitting parameter, the model can be extended to nearly all the MuGFETs devices, and therefore can be used as a core model for the scaling and calibration of a wide range of MuGFETs. The 3D potential formula still needs to be simplified into a compact form, with the ultimate goal of its inclusion into compact models for MuGFETs transistors (SPICE simulators).

APPENDIX – FOURIER SERIES COEFFICIENTS

- Constant potential boundary condition :

$$F_c(n) = \frac{2(1 - \cos(n\pi))}{n\pi}$$

- Parabolic potential boundary condition

$$F_P(n) = \frac{2(1 - \cos(n\pi)) - n\pi \sin(n\pi)}{(\frac{n\pi}{2})^3}$$

- Neumann boundary condition :

$$F_n(n) = \frac{4}{(2n+1)\pi}(1 - \cos(\frac{(2n+1)\pi}{2}))$$

REFERENCES

[1] J.-P. Colinge et al., "FinFETs and Other Multi-Gate Transistors", *Springer*, ISBN 978-0-387-71751-7, 2007.

[2] J.-T. Park et al., "Pi-Gate SOI MOSFET," *IEEE Electron Device Letters*, vol. 22, no. 8, pp. 405-406, Aug. 2001.

[3] C. Jahan et al., "10nm ΩFETs transistors with TiN metal gate and HfO₂," *Digest of Technical Papers, 2005 Symposium on VLSI Technology*, pp. 112-113, 2005.

[4] G. Pei et al., "FinFET Design Considerations Based on 3D Simulation and Analytical Modeling," *Electron Devices, IEEE Transactions on*, vol. 49, no. 8, pp. 1411-1419, Aug. 2002.

[5] D. Havaldar et al., "Subthreshold Current Model of FinFETs based on Analytical Solution of 3D Poisson's Equation," *Electron Devices, IEEE Transactions on*, vol. 53, no. 4, pp. 737-742, Apr. 2006.

[6] H. Abd El Hamid et al., "A 3D Analytical Physically Based Model for the Subthreshold Swing in Undoped Trigate FinFETs," *Electron Devices, IEEE Transactions on*, vol. 54, no. 9, pp. 2487-2496, Sept. 2007.

[7] T. Ernst et al., "A Model of Fringing Fields in Short-Channel Planar and Triple-Gate SOI MOSFETs," *Electron Devices, IEEE Transactions on*, vol. 54, no. 6, pp. 1366-1375, 2007.

[8] J.-P. Colinge et al., "Quantum-Mechanical Effects in Trigate SOI MOSFETs", *Electron Devices, IEEE Transactions on*, vol. 53, no. 5, pp. 1131-1136, 2006.

[9] K.-T. Tang, "Mathematical method for engineers and scientists 3," *Springer*, ISBN 978-3540446958, 2007.

[10] COMSOL Multiphysics [online]. Available: http://www.comsol.com.

A compact model for double gate carbon nanotube FET

Sebastien Fregonese, Cristell Maneux, Thomas Zimmer

Laboratoire IMS, UMR5218
CNRS, Université Bordeaux
Bordeaux, France
Sebastien.fregonese@ims-bordeaux.fr

Abstract—A compact model for Double gate carbon nanotube FET is presented. This compact model includes the most significant mechanism present in DGCNTFET such as Schottky barrier at the metallic-nanotube interface, charge, electrostatic modelling and quasi-ballistic transport through the Landauer equation. Then, this compact model is compared to measurement. Finally, a simple ring oscillator circuit has been simulated using ten identical devices highlighting new technology concepts.

I. INTRODUCTION

The pursuit of Moore's Law, as predicted by the International Technology Roadmap for Semiconductors (ITRS) has pointed out future intrinsic device difficulties (such as leakage, interconnect, power, quantum effects) to the capability of realizing system architectures using CMOS transistors with the performance levels required by future applications. These fundamental and economic limitations enforce the semiconductor industry to explore the use of novel materials and devices able to complement or even replace the CMOS transistor or its channel in systems on chip within the next decade and before silicon based technology will reach its limits. Following the latest update of ITRS, the most promising devices are carbon based nanodevices such as the carbon nanotube field-effect transistor (CNTFET), the graphene nano-ribbon field-effect transistor (GNRFET) or GFET.

Specific properties of the CNTFET can be used, allowing the creation of completely new logic functions, inaccessible to MOSFET-based circuits. In particular, the DG-CNTFET device design is capable of producing pure n- and/or p-type CNFETs behaviour with improved OFF-state performance and abrupt switching behaviour close to theoretical limits [1]. The use of a specific property in double-gate CNTFETs (DG-CNTFETs) allows building logic cells that offer fine-grain reconfigurability [2]; this opportunity is not available with MOSFET technology. To develop, optimize and evaluate these logic cells in term of electrical performances, the compact model is an efficient tool [9], widely used by circuit designers. For this, the compact DG-CNTFET models must be physic based and must accurately model specific CNTFET characteristics such as ambipolarity, source exhaustion [3], Schottky barriers (SB) at metallic contact, and electrostatic

control of the back gate and front gate through accurate charge calculation in the channel. Hence, this work focuses on compact modelling development of DGCNTFETs and its evaluation of their potential in a circuit context. A first version of DGCNTFET model has been presented in [3]: this model is limited to thermionic transport (SB modelling was not included in [3]). Also, the model presented in this paper uses a more computationally efficient charge model and an improved equivalent circuit.

First part of this paper will describe the full compact model including Schottky barriers modelling, charge modelling, electrostatic modelling, and thermionic current modelling. Then, a validation of the compact model is performed through comparisons with measurements from an IBM technology [1]. Last part shows a design application example of the DGCNTFET in a basic circuit cell context.

II. COMPACT MODELLING OF DGCNT-FET

The DGCNTFET structure proposed in [1] is described in fig. 1 and is made of three different regions: source access, inner part and drain access. The inner part is the active part of the device.

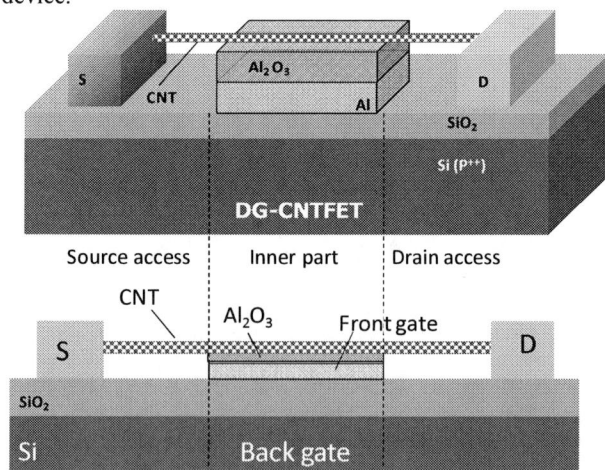

Figure 1. Schematic cross-sectional view of a DG-CNTFET

CNRS, French National Research Agency ANR

978-1-4244-6658-0/10 $26.00 © 2010 IEEE

This device is built with two independent gates: a back gate which alter the electrostatic doping of the source access and the drain access region; the front gate modulates the potential in the inner part of the device and controls the carrier flow through the device. Depending on the back gate bias, the carrier flow can be dominated by electrons or holes to make a N or P-type device. In this structure, four energy barriers appear in the device: at the metal to source access (or drain access) junction, two SB like barriers appear, while between the source access (or drain access) to the inner part junction, the barrier is more conventional and is of PN junction shape.

A. Schottky barrier modelling at source and drain contact

In DGCNTFET, depending on the work function difference between the metal contact and the CNT, carriers at the metal-CNT interface encounter different barrier heights: Carriers with energies above the Schottky barrier height reach the channel by thermionic emission. On the other hand, carriers with energies below the SB height have a probability to reach the channel according to a transmission function describing the tunnel effect which can be calculated from WKB approximation.

To overcome the complexity of WKB expression for compact modelling, an approximation based on recent works from [4] is applied. This effective SB height is expressed as [4]:

$$\varphi_{Sn} = \left(\varphi_{SB} - \left(sbbd[1] - V_{CNTS,Si}\right)\exp\left(-d_{tunnel}/\lambda\right)\right) + \left(sbbd[1] - V_{CNTS,Si}\right) \tag{1}$$

The effective Schottky barrier φ_{Sn} at the source side for electrons is defined by the barrier height φ_{SB} between the Fermi level of the metal and the conduction band of the nanotube, by the screening length λ at the metal/nanotube junction and by a fitting parameter d_{tunnel}. V(vcnts,si) is the potential difference between the source access and the source metal contact; and sbbd[1] is the minimum of the first sub-band. Hence, a unitary transmission coefficient is considered for carriers having energy higher than the effective barrier height, while carriers having energy lower than the effective barrier height are assumed to be reflected.

The same procedure is applied to calculate the drain effective SB φ_{Dn} and also Schottky barriers in the valence band (φ_{Sp} and φ_{Dp}). Then, these barriers (source and drain side) are used when computing the current and the charge.

B. Current source including Schottky barrier modelling

The electron (hole) current is calculated through the Landauer equation, by integrating over energy from the dominating barrier to infinite. The dominating barrier position depends on the applied bias. In fact, the electron current can be limited by three barriers: (i) the SB from source φ_{Sn}, (ii) the SB from drain φ_{Dn} and (iii) the conduction (valence) band of the inner part $CBi = sbbd[p] - V_{CNTi,S}$. The dominating or limiting potential is calculated through the function: $\max\left(\varphi_{Sn}, sbbd[p] - V_{CNTi,Si}, \varphi_{Dn}\right)$.

$$I_{DS-e^-} = \frac{4ek_BT}{h}\sum_{p=1}^{nb\_sbbd}\left[\begin{array}{l}\ln\left(1+e^{\frac{-\max\left(\varphi_{Sn},sbbd[p]-V_{CNTi,Si},\varphi_{Dn}\right)}{k_BT}}\right)\\ -\ln\left(1+e^{\frac{-\max\left(\varphi_{Sn},sbbd[p]-V_{CNTi,Si},\varphi_{Dn}\right)-V_{Di,Si}}{k_BT}}\right)\end{array}\right] \tag{2}$$

The same procedure is applied for the holes.

C. Charge modelling including Schottky barrier modelling

In each region, the charge is computed considering that the carriers do not change their energy (ballistic or elastic scattering hypothesis) - even if they are back-scattered.

In each region four contributions may appear depending on bias conditions: carrier can be transmitted from source and drain and/or reflected on the both sides of the zone. The fig. 2 highlights how the charge is calculated in the inner part of the device. In the conduction band, three contributions appear: i) charge coming from the source and filling +k states controlled by φ_{Sn}; ii) charge coming from the drain filling –k states controlled by the conduction band of the inner part CBi; iii) charge coming from the drain, reflected on source SB, and therefore filling –k states.

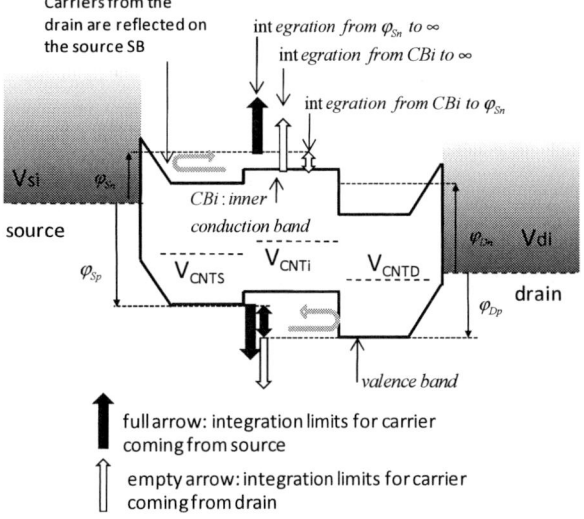

Figure 2. Scketch showing conduction and valence band of the DGCNTFET: arrows highlight integration limits for charge calculation concerning the inner part.

The charge in each region is the integral over energy of the density of states multiplied by the corresponding Fermi distribution source or drain. Electron charge in the inner part of the device can be calculated through the equation eq. 3.

$$Q_I = -e \left[\begin{array}{l} \int_{MAX(\varphi_{Sn}, CB_I)}^{+\infty} \dfrac{D(E)}{2} f_S(E) dE \\[2mm] + \int_{MAX(CB_I, \varphi_{Dn})}^{+\infty} \dfrac{D(E)}{2} f_D(E) dE \\[2mm] + \int_{MAX(CB_I, \varphi_{Dn})}^{MAX(\varphi_{Sn}, CB_I, \varphi_{Dn})} \dfrac{D(E)}{2} f_D(E) dE \\[2mm] + \int_{MAX(CB_I, \varphi_{Sn})}^{MAX(\varphi_{Sn}, CB_I, CB_D)} \dfrac{D(E)}{2} f_S(E) dE \end{array} \right] \qquad (3)$$

Where, $D(E)$ is the density of states (DOS), f_S and f_D are the Fermi distribution of source and drain respectively and CB_S, CB_I, CB_D are respectively the conduction band in the source access, inner part and drain access. The same approach can be used for the source and drain access regions.

Each contribution can be calculated analytically using an approach close to [5]. This charge model is computed from 1D DOS based on the non-parabolic energy dispersion relation [6].

D. Potential calculation and equivalent circuit

The BG (back gate) and the FG (front gate) control the energy bands of the access regions and of the inner part respectively: the BG voltage V_{BGS} permits electrons or holes to enter in the source access and drain access of the device and then, carriers can flow through the device depending on the FG voltage V_{FGS}. The energy bands of each region are shifted according to the local channel potential V_{CNT} (V_{CNTs-S}, V_{CNTi-S}, and V_{CNTd-S} for source access, inner part, and drain access): a positive (negative) V_{CNT} decreases (increases) the associated energy bands. These potentials are function of the local gate potential and of the charge accumulated in the associated region (Q_S, Q_I, and Q_D for the source access, inner part and drain access respectively): the charge in a region lowers the effect of the local gate bias according to the following relations:

$$\begin{cases} V_{CNTS,S} = \dfrac{C_{SEi}V_{CNTi,S} + C_{SEx}V_{Si,S} + V_{BGiS,S}C_{BGS}L_S + Q_S}{C_{BGS}L_S + C_{SEi} + C_{SEx}} \\[3mm] V_{CNTi,S} = \dfrac{C_{DEi}V_{CNT,S} + C_{SEi}V_{CNT,S} + V_{Gi,S}C_{FGS}L_I + Q_i}{C_{FGS}L_I + C_{SEi} + C_{DEi}} \\[3mm] V_{CNTD,S} = \dfrac{C_{DEi}V_{CNTi,S} + C_{DEx}V_{Di,S} + V_{BGiD,S}C_{BGD}L_D + Q_D}{C_{BGD}L_D + C_{DEi} + C_{DEx}} \end{cases} \quad (4)$$

where C_{BGS}, C_{FG}, and C_{BGD} are the source access BG capacitance, the FG capacitance, and the drain access BG capacitance respectively. L_S, L_I and L_D are the length of the source access, inner part and drain access respectively. C_{SEi} and C_{DEi} are the electrostatic capacitances correlated to the source access (drain)/inner part junction while C_{SEx} and C_{DEx} are the electrostatic capacitances correlated to the Schottky barriers.

The equivalent circuit of the compact model is described on Fig. 3: lumped serial resistance, flat band voltages, V_{FB}, gate insulator capacitances, junction capacitances, charges and current source are shown. Moreover, the front gate and back gate coupling is also taken

into account. This effect is of importance to analyse dynamic performances of DGCNTFET since these metallic contact are in vis-à-vis.

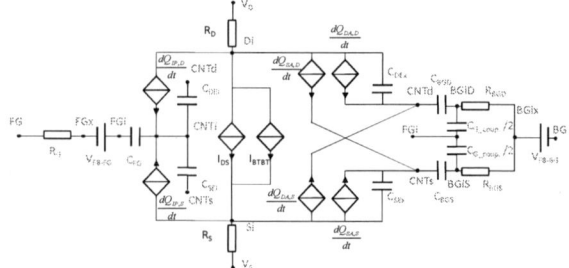

Figure 3. Equivalent circuit for the DGCNTFET

III. COMPARISON TO MEASUREMENT

In order to validate our modelling approach, we have performed a comparison of the simulation results obtained from our compact model to measurements of the literature [1]. With the aim of highlighting the physical basement of the compact model, the model parameters have been determined as close as possible from technological information: i) energy sub-bands, effective mass, non-parabolicity parameters have been calculated from nanotube diameter; ii) SB height is calculated from metal material and nanotube diameter using work from [7]; iii) finally, insulator capacitances of back gate and front gate have been calculated from gate insulator thickness and permittivity and nanotube diameter, and then have been optimized using the theoretical value as a starting point.

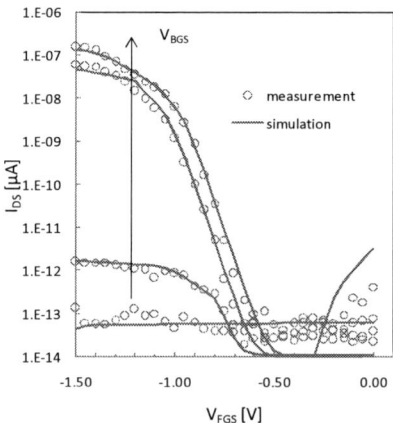

Figure 4. I_{DS} in log scale versus V_{FGS} and for different V_{BGS} (=-0.5, -1, -2, -4 V) for V_{DS}=0.5 V. Comparison between measurement of the IBM [1] and compact model.

Remaining parameters are locally optimised on current-voltage characteristics. Fig. 4 shows drain current versus front gate voltage and the effect of the back-gate control on the I_{ON} current: A strong decrease of the I_{ON} current with small back gate voltage can be observed. This effect can be attributed to the source exhaustion effect [8]. In fact, at small negative back gate voltage, the "electrostatic doping" is small and source exhaustion effect appears sooner and limits the I_{ON} current.

978-1-4244-6658-0/10 $26.00 © 2010 IEEE

Finally, the drain current is plotted versus front gate voltage for different V_{DS} and at a constant back gate voltage (see Fig. 5). This measurement shows a shift in threshold voltage compared to fig. 4, which may be due to trapped charge at the insulator interface. This can be taken into account by the compact model through an additive constant charge included in equation 4.

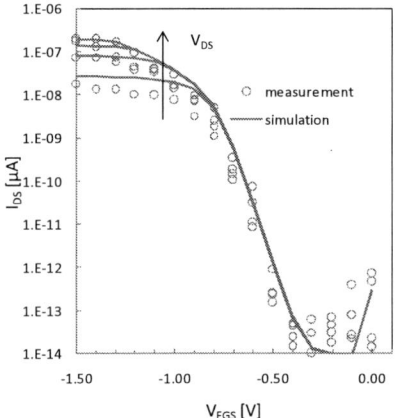

Figure 5. I_{DS} in log scale versus V_{FGS} and for different V_{DS} (=-0.1, -0.3, -0.5; -0.7 V) for V_{BGS}=-4 V. Comparison between measurement of the IBM [1] and compact model.

IV. APPLICATION

Finally, we use our model to simulate a 5 stages ring oscillator using 10 identical transistors. The P like transistors are biased with a negative back gate bias and the N like devices are biased with a positive front gate bias to build the elementary inverter. The parameters used in the simulation are the one extracted from the IBM technology except for the flat band voltage of the front gate. We shift this flat band voltage in order to obtain a symmetrical $V_{out}(V_{in})$ characteristic. We assume a 1fF load capacitance on each inverter to take into account parasitic effects. Fig 6 shows simulation results of the ring oscillator for different back gate voltages. Changing the back gate voltage allows controlling the I_{ON} current of each transistor and hence changing the oscillation frequency of the ring oscillator on a large frequency range since the I_{ON} current can be modulated from pA to nearly μA. This design example highlights a new technology concept where n and p type devices are realised with the same physical device.

V. CONCLUSION

We have developed an electrical compact model for the Double Gate Carbon NanoTube FET. This compact model is physics-based and includes the most significant mechanism such as Schottky barrier at the metallic-nanotube interface, charge and electrostatic modelling, and quasi-ballistic transport through the Landauer equation. This compact model is compared to measurements of a DGCNTFET device from an IBM technology. This comparison shows the ability of the compact model to represent accurately real structures. Finally, a simple ring oscillator circuit has been simulated using ten identical devices proposing a new technology concept

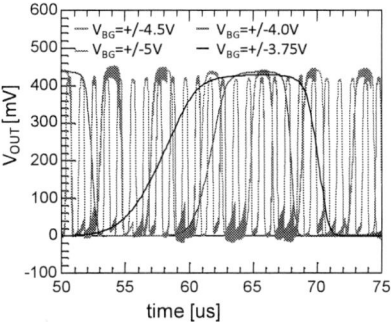

Figure 6. Simulation of a five stages ring oscillator for different V_{BG} values (+/-3.75V, +/- 4V, +/- 4.5 V, +/-5V): parasitic capacitance for each stage= 1fF. Parameters used in the simulation are extracted from IBM measurement except the flat band voltage of the front gate.

ACKNOWLEDGMENT

This work was supported by the french National Research Agency ANR through ARPEGE "NANOGRAIN" project. The authors would also like to thank all partners of these projects for the fruitful discussions.

REFERENCES

[1] Y.-M. Lin, J. Appenzeller, J. Knoch, and Ph. Avouris, "High performance carbon nanotube field-effect transistor with tunable polarities," IEEE Trans. On Nanotech., vol. 4, N° 5, pp. 481–489, September 2005.

[2] I. O'Connor, J. Liu, F. Gaffiot, F. Prégaldiny, C. Lallement, C. Maneux, J. Goguet, S. Frégonèse, T. Zimmer, L. Anghel, TT Dang, R. Leveugle, "CNTFET Modeling and Reconfigurable Logic-Circuit Design", IEEE Trans. On Circuits And Systems I: Regular Papers, Vol. 54, No. 11, November 2007 pp 2365-2379.

[3] Goguet, J.; Fregonese, S.; Maneux, C.; Zimmer, T.; "Compact Model of a Dual Gate CNTFET: Description and Circuit Application", IEEE Conference on Nanotechnology, 2008. IEEE NANO '08., Publication Year: 2008, Page(s): 388 – 389.

[4] Knoch, Joachim; Appenzeller, Joerg "Tunneling phenomena in carbon nanotube field- effect transistors", physica status solidi (a), vol. 205, issue 4, pp. 679-694 (2008).

[5] S. Frégonèse, H. Cazin d'Honincthun, J. Goguet, C. Maneux, T. Zimmer, JP. Bourgoin, P. Dollfus, S. Galdin-Retailleau, "Computationally Efficient Physics-Based Compact CNTFET Model for Circuit Design" IEEE TED, Vol. 55, No. 6, pp1317-1327, 2008.

[6] S. Fregonese, C. Maneux, T. Zimmer, "Implementation of Tunnelling Phenomena in a CNTFET Compact Model », IEEE TED, October 2009.

[7] Zhihong Chen, Joerg Appenzeller, Joachim Knoch, Yu-ming Lin, and Phaedon Avouris, "The Role of Metal-Nanotube Contact in the Performance of Carbon Nanotube Field-Effect Transistors", Nano Letters, 2005, Vol. 5, No. 7, 1497-1502

[8] Neophytou, N., Rakshit, T., Lundstrom, M.S., "Performance analysis of 60-nm gate-length III-V InGaAs HEMTs: Simulations versus experiments", IEEE TED 56 (7), pp. 1377-1387

[9] Deng J., Wong H.-S.P. "A compact SPICE model for carbon-nanotube field-effect transistors including nonidealities and its application - Part I: Model of the intrinsic channel region (2007)", IEEE TED, 54 (12), pp. 3186-3194.

[10] D. Akinwande,a_ J. Liang, S. Chong, Y. Nishi, and H.-S. Philip Wong "Analytical ballistic theory of carbon nanotube transistors: Experimental validation, device physics, parameter extraction, and performance projection", JAP 104, 124514 _2008

978-1-4244-6658-0/10 $26.00 © 2010 IEEE   375

Modeling of partial-RESET dynamics in Phase Change Memories

S. Braga, A. Sanasi, A. Cabrini, and G. Torelli
University of Pavia, Department of Electronics, Pavia, Italy
Email: stefania.braga@unipv.it

Abstract—In this work, a physics-based analytical model for the partial-RESET operation in Phase Change Memories is proposed. The model describes the electro-thermal behavior of the memory cell and gives an insight into the dynamical phenomena involved in the amorphization process inside the chalcogenide layer. Simulations are compared to measurements carried out on a 180-nm PCM experimental chip based on the μ-trench cell architecture, showing good accuracy of the proposed model both in the case of single pulse programming and in the case of staircase-up partial-RESET programming.

Index Terms—Phase Change Memories, partial-RESET programming, multilevel storage.

I. INTRODUCTION

In Phase Change Memories (PCMs), digital information is stored as the value of the electrical resistance of a thin layer of a chalcogenide alloy (typically, Ge$_2$Sb$_2$Te$_5$, GST) [2], [3]. The electrical resistance of a PCM cell depends on the structural phase of a small portion (hereinafter referred to as the active volume) of the GST film, which can be reversibly switched between a polycrystalline and an amorphous state by means of suitable electrical current or voltage pulses. During the SET operation, the GST is generally heated to a temperature between a critical temperature for the crystallization process and its melting point for a predetermined time interval. This way, the activated nucleation and the microcrystal growth inside the GST [4] lead to a (poly)crystalline active volume (full-SET state, minimum cell resistance). During the RESET operation, a high-amplitude electrical pulse with fast falling edge is applied to the cell, in order to heat the active GST

volume to above its melting temperature ($T_{melt} = 600°$ C) and then rapidly cool it down to the normal operating temperature. The cooling rate must be faster than the crystallization rate so as to freeze the atoms in a disordered structure, thus producing an amorphous volume (full-RESET state, maximum cell resistance). Intermediate resistance states between the full-SET and the full-RESET levels can be achieved by applying to the cell in the full-SET state a partial-RESET pulse so as to partially amorphize the GST layer (partial-RESET programming approach). This way, multilevel storage can be implemented to increase the bit density and reduce the cost. In this work, the dynamics of the RESET process is experimentally investigated and an analytical model of the partial-RESET operation that takes its dynamic behavior into account is proposed. Simulations are compared to measurements carried out on a 180-nm 4M-cell MOSFET-selected PCM experimental chip based on the μ-trench cell architecture [5].

In order to model the RESET operation of the PCM cell, we first considered its static behavior by means of a linear approximation of the temperature obtained inside the GST layer [1], as shown in Fig. 1. Then, we can write the GST resistance as

$$ R_{GST} \simeq \frac{\rho_A d}{A} \left(1 - \frac{T_{melt} - T_0}{T_{max} - T_0} \right), \tag{1} $$

where ρ_A is the electrical resistivity of amorphous GST, T_{max} is the maximum temperature inside the GST layer (which is assumed to be achieved at the heater-GST interface), T_0 is room temperature (20° C), $d \simeq 80$ nm is the thickness of the GST layer, and $A = 1600$ nm^2 is the contact area between the heater and the GST.

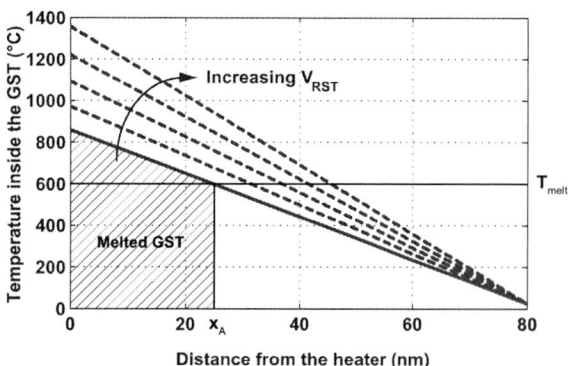

Fig. 1. Linearized temperature profile inside the GST layer during the RESET operation [1]. The amorphous volume thickness is equal to x_A.

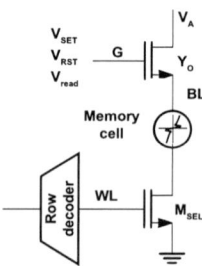

Fig. 2. Schematic of the circuit used to program and read the memory cell. Transistors M_{SEL} and Y_O are the word-line (WL) select transistor and the bit-line (BL) bias device, respectively.

978-1-4244-6658-0/10 $26.00 © 2010 IEEE   376

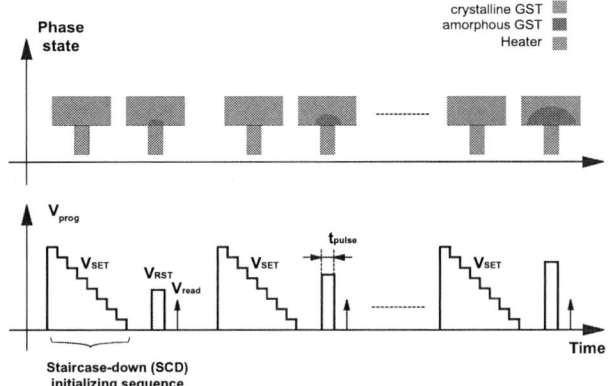

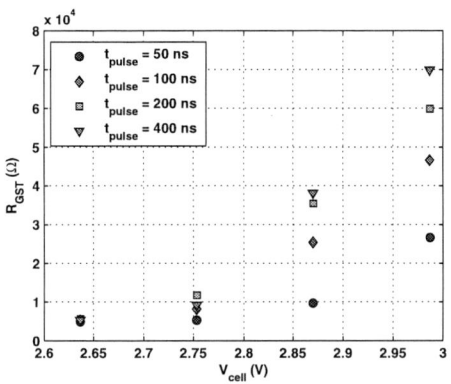

Fig. 3. Sequence of program and read pulses used in single-pulse (SP) programming of PCM cells.

Fig. 5. Detail of the SP programming curves showing the dependence of $V_{RST,min}$ upon the pulse duration.

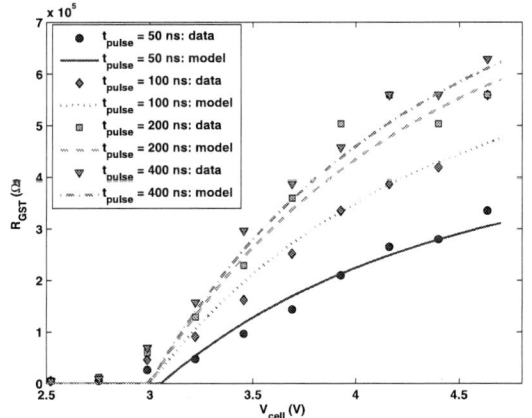

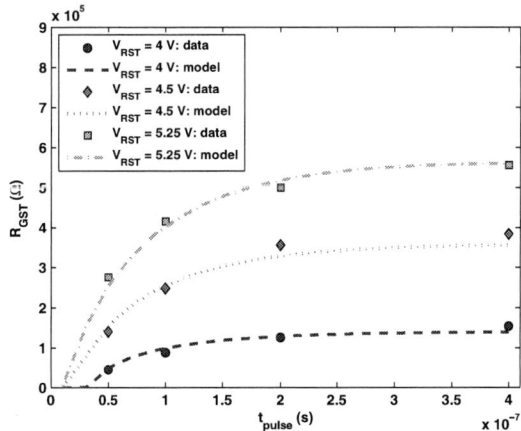

Fig. 4. Single-pulse programming curve obtained by means of the procedure shown in Fig. 3: experimental data (dots) and model in Eq. (6) (dashed lines).

Fig. 6. Dynamics of the RESET process in single-pulse partial-RESET programming: experimental data (dots) and model in Eq. (6) (dashed lines).

II. MEASUREMENTS AND PROPOSED MODEL

The memory cell is biased by applying adequate voltage levels to the selected bit-line (BL) through a high-voltage natural NMOS transistor, Y_O, which operates as a source follower (Fig. 2) [6]. In particular, voltage pulses having an amplitude V_{RST} are applied to the gate terminal of Y_O for partial-RESET operations. The stored information is read out by sensing the current, I_{cell}, flowing through the cell (hereinafter referred to as cell current) when a suitable read voltage V_{read} is applied to the gate terminal of Y_O. In both program and read operation, device Y_O is kept in saturation by applying a sufficient high voltage V_A to its drain terminal. Word-line (WL) select transistor M_{SEL} operates in triode and is turned on by applying a high voltage level to its gate. A high-voltage device is used to implement transistor Y_O so as to safely sustain the high-voltage levels required in memory operations. In order to measure the read cell current I_{cell} with high accuracy, the PCM array can also be operated in direct memory access (DMA) mode. The GST resistance, R_{GST}, is evaluated by first calculating the ratio between the read voltage

across the cell, I_{cell}, and then subtracting the heater resistance, R_h.

First, we measured the single-pulse (SP) programming curve of a memory cell. To this end, we used the sequence of program and read pulse depicted in Fig. 3. We first programmed the cell to the SET state by means of a staircase-down (SCD) initializing sequence and, then, applied a single partial-RESET voltage pulse having predetermined values of amplitude V_{RST} and duration (t_{pulse}), followed by a readout operation. In Fig. 4, V_{cell} is the programming voltage across the PCM cell ($V_{cell} = V_{RST} - V_{GS,Y_0}$, where V_{GS,Y_0} is the gate-to-source voltage of Y_0).

The above sequence was repeated with different values of V_{RST} and t_{pulse}, thus obtaining the programming curves in Fig. 4. A significant dependence of the obtained cell resistance over the pulse duration is observed.

As highlighted in Fig. 5, also the minimum programming voltage ($V_{RST,min}$) required to increase the cell resistance decreases with increasing pulse duration as a consequence of the temperature dynamics inside the GST layer. This difference

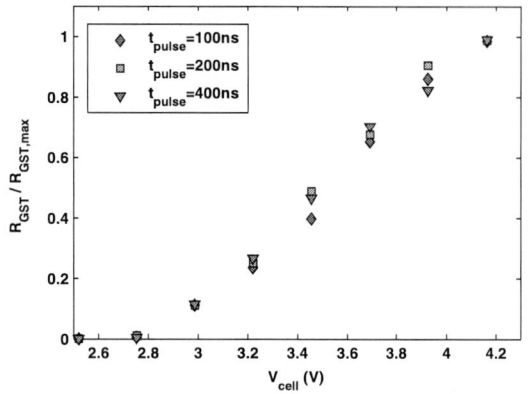

Fig. 7. Single-pulse programming curves obtained by means of the programming procedure shown in Fig. 3 normalized with respect to the resistance value obtained with $V_{cell} \simeq 4.2$ V.

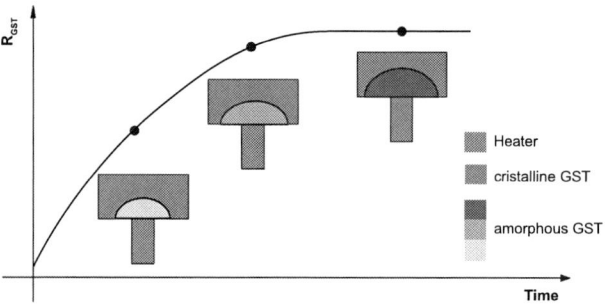

Fig. 8. Modeling of the effect of the pulse duration on the amorphous GST resistivity.

is more evident when considering low values of t_{pulse} (50 ns and 100 ns).

These phenomena are responsible for the overall RESET dynamics, which is apparent in Fig. 6, where the GST resistance obtained with the SP programming procedure for different values of V_{RST} is plotted against the pulse duration.

In order to model the RESET dynamics, we assumed, to a first approximation, that the temperature dynamics inside the GST layer is independent of the distance from the heater-GST contact (this approximation is supported by means of finite-element simulations of the memory cell [7]). Then, Eq. (1) can still be used to calculate the GST resistance as a function of time, provided that T_{max} is substituted by $T(t)$, that is the transient temperature at the heater-GST interface:

$$T(t) = (T_{max} - T_0) \cdot \left(1 - e^{-\frac{t}{\tau_T}}\right) + T_0, \qquad (2)$$

where τ_T is the time constant for the temperature dynamics inside the GST.

The time constant τ_T takes the effects of both the purely electrical and the purely thermal dynamics into account. Since

$$T_{max} = \frac{R_{th} \cdot V_{cell}^2}{R_h} + T_0, \qquad (3)$$

where R_{th} is the equivalent thermal resistance of the PCM cell, Eq. (2) can be written as

$$T(t) = \frac{R_{th} \cdot V_{cell}^2}{R_h} \cdot \left(1 - e^{-\frac{t}{\tau_T}}\right) + T_0. \qquad (4)$$

The highest sensitivity to the temperature dynamics is observed in the case of low t_{pulse}.

To better analyze the impact of the pulse duration over the obtained resistance, we normalized SP programming curve with respect to the resistance value obtained with $V_{cell} \simeq 4.2$ V. As shown in Fig. 7, the obtained curves are substantially superimposed (only the curves with negligible difference in $V_{RST,min}$ are shown). Since the GST resistance is proportional to the amorphous GST resistivity, this behavior suggests that the pulse duration affects the value of this parameter. We model this dependence by means of the following equation:

$$\rho_{A,eff} = (\rho_A - \rho_C) \cdot \left(1 - e^{-\frac{t}{\tau_A}}\right) + \rho_C, \qquad (5)$$

where $\rho_{A,eff}$ is the effective resistivity of the programmed amorphous GST, ρ_A is the resistivity of the fully-amorphous GST, ρ_C is the resistivity of GST in the crystalline form, and τ_A is the time constant for the amorphization process. Fig. 8 schematically depicts the increase of the effective amorphous resistivity $\rho_{A,eff}$ with time. It can be noticed that also the thickness of the amorphous GST cap increases due to the temperature dynamics.

Then, by extending Eq. (1) to the dynamic case by means of Eqs. (2) and (5), we model the GST resistance as

$$R_{GST} = \frac{\rho_{A,eff} \cdot d}{A} \cdot \left(1 - \frac{T_{melt} - T_0}{T(t)}\right). \qquad (6)$$

The proposed model is able to reproduce the experimental SP programming curves and the RESET dynamics in Figs. 4 and 6. The values of the used parameters are summarized in Table I.

III. MODEL VALIDATION

The model proposed in previous Section was validated by means of additional measurements on the same PCM cell. In particular, we considered the partial-RESET staircase-up (SCU) programming sequence shown in Fig. 9. In SCU programming, the cell is first brought to the SET state by means of a staircase-down (SCD) initializing sequence and, then, a staircase-up sequence of voltage pulses, each followed by a read pulse, is applied.

First, we considered the SCU curves obtained with different values of t_{pulse} (the voltage increment ΔV between two adjacent programming pulses was set to 25 mV), shown in

TABLE I
SIMULATION PARAMETERS

Parameter	Value	Parameter	Value
ρ_A	0.02 Ωm	ρ_C	10^{-4} Ωm
τ_A	70 ns	τ_T	15 ns
R_h	5 kΩ	R_{th}	320 k $^\circ$C/W

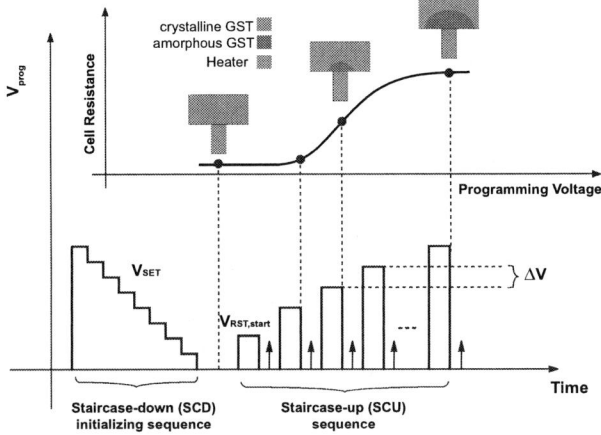

Fig. 9. Sequence of program and read pulses used in staircase-up (SCU) programming of PCM cells.

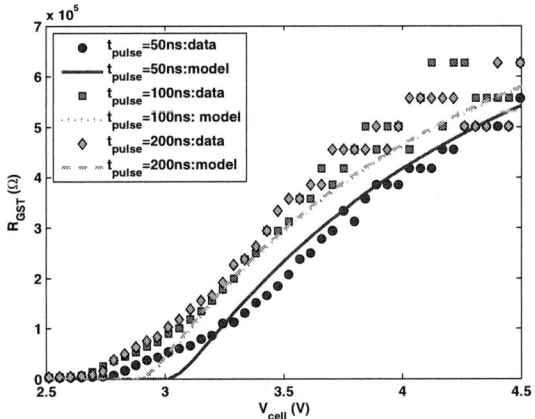

Fig. 10. SCU programming curve obtained by means of the procedure shown in Fig. 9: experimental data (dots) and model in Eq. (6) (dashed lines). $V_{RST,start}$ was set to 2.5 V.

Fig. 10. It can be noticed that the SCU programming curves are steeper than the corresponding (i.e., obtained with the same pulse duration) SP curves, due to the fact that in SCU programming each partial-RESET pulse is applied to a cell which is already partially amorphized. The model is able to reproduce this phenomenon due to $\rho_{A,eff}$ and the dependence of the minimum RESET voltage on the pulse duration. The significant disagreement between model and data that can be noticed for low-voltage RESET pulses is ascribed to the effects of self-heating inside the GST layer. Although self-heating is not implemented by our model, it may be taken into account by adding its contribution to the temperature T_{max}.

Then, we considered the SCU programming curves obtained with different values of ΔV, ranging from 150 mV to 1 V (Fig. 11). A good agreement is observed also in this case. In particular, our model adequately reproduces the dependence of the slope of the SCU programming curves on the value of ΔV.

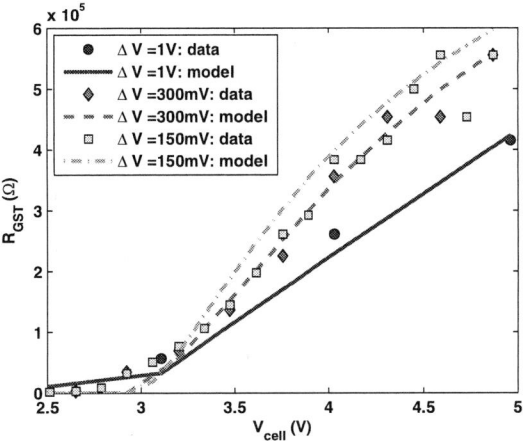

Fig. 11. SCU programming curves obtained with different values of ΔV and $t_{pulse} = 50$ ns: experimental data (dots) and model in Eq.(6) (lines). $V_{RST,start}$ was set to 2.5 V.

IV. CONCLUSIONS

In this work, we proposed a physics-based analytical model for partial-RESET programming in Phase Change Memories. Our model implements the temperature dynamics inside the PCM cell and the observed dependence of the amorphous GST resistivity on the pulse duration. Simulations are compared to measurements carried out on a 180-nm PCM experimental chip based on the μ-trench cell architecture, showing good accuracy both in the case of single pulse programming and staircase-up partial-RESET programming.

ACKNOWLEDGMENT

This work has been supported by Italian MIUR in the frame of its National FIRB Project RBAP06L4S5.

REFERENCES

[1] S. Braga, A. Cabrini, and G. Torelli, "Theoretical analysis of the RE-SET operation in phase-change memories," *Semiconductor Science and Technology*, vol. 24, no. 11, p. 115008 (6pp), 2009.

[2] A. Lacaita, "Phase change memories: State-of-the-art, challenges and perspectives," *Solid-State Electronics*, vol. 50, no. 1, pp. 24 – 31, 2006, papers selected from the 2005 ULIS Conference.

[3] G. W. Burr, M. J. Breitwisch, M. Franceschini, D. Garetto, K. Gopalakrishnan, B. Jackson, B. Kurdi, C. Lam, L. A. Lastras, A. Padilla, B. Rajendran, S. Raoux, and R. S. Shenoy, "Phase change memory technology," *Journal of Vacuum Science & Technology B: Microelectronics and Nanometer Structures*, vol. 28, no. 2, pp. 223–262, 2010.

[4] S. Senkader and C. D. Wright, "Models for phase-change of $Ge_2Sb_2Te_5$ in optical and electrical memory devices," *Journal of Applied Physics*, vol. 95, pp. 504–511, Jan. 2004.

[5] F. Pellizzer *et al.*, "Novel μtrench phase-change memory cell for embedded and stand-alone non-volatile memory applications," *Symposium on VLSI Technology. Digest of Technical Papers.*, pp. 18 – 19, June 2004.

[6] F. Bedeschi *et al.*, "Bit-line biasing technique for phase-change memories," *Proc. of Int. Conference on Signals and Electronic Systems*, pp. 229–232, Sept. 2004.

[7] S. Braga, A. Cabrini, and G. Torelli, "An integrated multi-physics approach to the modeling of a phase change memory device," *Proc. of European Solid-State Device Research Conference*, pp. 154–157, Sept. 2008.

On the Modelling and Optimisation of a novel Schottky based Silicon Rectifier

T. van Hemert, R.J.E. Hueting, B. Rajasekharan, C. Salm and J. Schmitz

MESA$^+$ Institute for Nanotechnology, University of Twente, Enschede, The Netherlands

email: t.vanhemert@utwente.nl

Abstract—The charge plasma (CP) diode is a novel silicon rectifier using Schottky barriers, to circumvent the requirement for doping and related problems when small device dimensions are used. We present a model for the DC current voltage characteristics and verify this using device simulations. The model revealed an exponential dependence of the current on the metal work functions. And approximate linear dependence on the device geometry. The model is used to optimise the device performance. We show a factor 30 improvement in on/off current ratio (and hence rectification) toward 10E7 by appropriate sizing of the lateral device dimensions at given specific metal work functions.

I. INTRODUCTION

Rectifying pn-junctions are essential in nearly every electronic component. In recent years, these junctions have been downscaled to such dimensions that doping control has become a major issue. In particular, doping fluctuation [1] and [2], doping activation [3] and steep doping profiles that yield a low temperature budget have become difficult demands.

Alternatively, Schottky based devices in SOI are being investigated [4] in which abrupt source and drain contacts are formed between the metallic contact and the silicon body. An alternative to a Schottky based rectifier is the charged plasma (CP) diode [5] shown in Fig. 1 (a). Here two separate gates having different work functions are placed on top of a thin and lowly doped silicon body. The gates are isolated from the top of the body by a dielectric. Each of the metals forms a contact at the side of the silicon body. First realizations of this diode are presented in [6].

Recently this diode has been investigated using device simulations [7]. They concluded that the CP-diode shows good rectifying behavior depending on the dimensions and metal work functions. In this work we present a model and compare this with device simulations. Furthermore we optimise the rectifying behavior quantified by the on/off current ratio.

II. THEORY

1) Thermal Equilibrium: For a well chosen cathode gate work function ϕ_{mc} an elevated electron concentration (n) is induced in the underlying silicon body, here referred to as electron plasma. The hole barrier height at the cathode silicon interface is given by $\phi_{bc} = \chi_{Si} + E_g - \phi_{mc}$, where χ_{Si} is the silicon electron affinity and E_g the silicon band gap. The anode work function ϕ_{ma} on the other hand induces a hole plasma. The electron barrier height at the anode silicon interface is given by $\phi_{ba} = \phi_{ma} - \chi_{Si}$. Fig. 1 (b) shows a schematic band

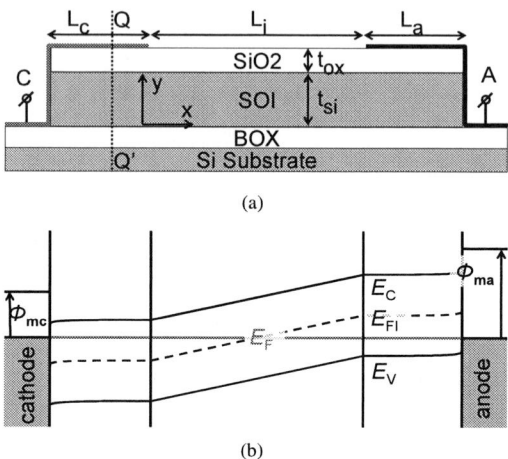

Fig. 1. (a) schematic cross section of the CP-diode, the cathode gate with length L_c induces an electron plasma due to $\phi_{mc} < \chi_{Si} + E_g/2$. The anode gate with length L_a induces a hole plasma because $\phi_{ma} > \chi_{Si} + E_g/2$. L_i is the length of the intrinsic region. (b) schematic band diagram of the CP-diode in horizontal direction just underneath the oxide.

diagram along a horizontal axis through the silicon. On the cathode side the Fermi level is close too the conduction band, indicating a high n. A high hole concentration (p) is formed at the anode side. The center region is not influenced by a gate and is lowly doped. Thus, there is hardly any charge present and the electric field is almost constant ($\frac{dE}{dx} = \rho/\epsilon$).

Fig. 2 shows a band diagram along the vertical dimension through the cathode of the CP-diode. The electron Schottky barrier between the SOI layer and cathode is relatively low because of the electrostatic effect. Let us neglect the influence of the buried oxide, silicon substrate (valid when $t_{box} > t_{ox}$), interface states and oxide charge. Also we neglect image force barrier lowering [8](important for high electric fields) and tunneling (important for very short regions). In the cathode region a positive charge is formed, the concentration of which varies along y.

To derive $n(y)$ it is necessary to have solution for the potential, for intrinsic silicon this is given by $q\Psi(y) = E_F - E_{Fi}(y)$. A similar problem shows up for the inversion charge in the subthreshold regime of a double gate MOSFET. This was presented earlier [9], [10] being,

$$\Psi(y) = \Psi(0) - 2u_T \ln\cos(\beta y), \tag{1}$$

978-1-4244-6658-0/10 $26.00 © 2010 IEEE

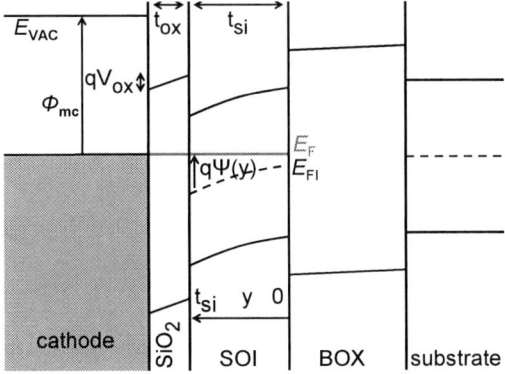

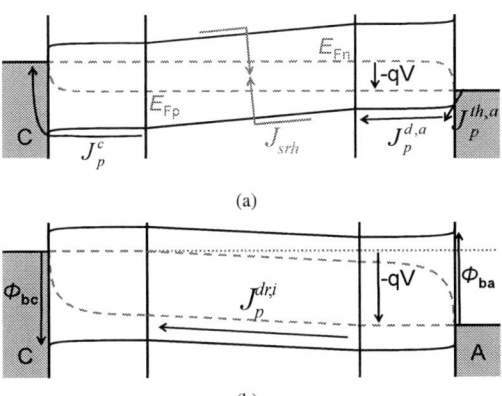

Fig. 2. Schematic band diagram taken along QQ' in Fig. 1 (a). The cathode work function ϕ_{mc} pulls the silicon bands downward inducing an electron plasma. A low metal work function, thin oxide and silicon layers clearly result in an elevated electron concentration in the SOI.

Fig. 3. Schematic band diagram of the CP-diode (Fig. 1) in horizontal direction of Fig. 1 just underneath the gates. (a) For small V the quasi Fermi levels E_{Fn} and E_{Fp} are splitted by a distance V. This results in thermionic emission and diffusion currents. Furthermore SRH generation/recombination could be taken into account. (b) For forward bias $V > V_{FB}$ drift through the intrinsic regions becomes important.

here u_{T} is the thermal voltage, $\beta = \sqrt{exp(\Psi(0)/u_{\mathrm{T}})\delta/2}$, $\delta = n_i/\epsilon_{Si}u_{\mathrm{T}}$ is a measure for influence of carriers on $\Psi(y)$ and ϵ_{Si} is the dielectric constant of silicon. The band diagram illustrates that the work function difference is equal to the potential drop across the oxide V_{ox} and the potential at the oxide silicon interface $\Psi(t_{\mathrm{si}})$. Furthermore the dielectric displacement at this interface is constant. Hence

$$\Psi(0) = \phi_{ms} - t_{\mathrm{ox}}\frac{\epsilon_{Si}}{\epsilon_{ox}}\frac{\delta\Psi}{\delta y} - 2u_t \ln\cos(\beta y)|_{y=t_{\mathrm{si}}}. \quad (2)$$

This equation can be solved numerically to find $\Psi(0)$. Note that we neglect the influence of the substrate. The highest minority concentration mainly determines the diffusion current density J, therefore, we use $y = 0$. The electron concentration under the cathode gate is given by $n_{\mathrm{c}} = n_i \exp\frac{\Psi(y)}{u_{\mathrm{T}}}$ and the hole concentration by $p_{\mathrm{c}} = n_i^2/n_{\mathrm{c}}$. Analogously the carrier concentrations under the anode gate (p_{a} and n_{a}) can be found.

2) Reverse and small forward bias: Fig. 3 (a) shows the schematic band diagram along the silicon body for a small bias on the anode. Following the conventional p-n junction theory, [11] and [12], the majority quasi Fermi levels E_{Fn} and E_{Fp} are constant because the current is limited by minority carrier diffusion (constant J). Under the gates the majority quasi-Fermi levels are fixed by the gate-semiconductor work function difference. Hence an anode bias V results in a shift $qV = E_{\mathrm{Fn}} - E_{\mathrm{Fp}}$. This raises the minority carrier concentration under the gates by a factor $\exp V/u_{\mathrm{T}}$. Which results in a diffusion current of both carriers in the gate regions.

At the silicon cathode interface the hole concentration is governed by the effective surface recombination rate ($S_{\mathrm{p,eff}} = A_p^* T^2/qN_V$) [13], here A_p^* is the Richardson constant and T the temperature. Near the intrinsic region the hole concentration is increased by the applied bias. For the diffusion and thermionic emission J in the cathode region we find

$$J_p^c = q\frac{p_{\mathrm{c}}\left(\exp\frac{V}{u_{\mathrm{T}}} - 1\right)}{L_{\mathrm{c}}/D_{\mathrm{p}} + 1/S_{\mathrm{p,eff}}}, \quad (3)$$

where D_{p} is the hole diffusion constant and k Boltzmann's constant. The hole concentration at the right hand side of the intrinsic region is high compared to the left hand side, see Fig. 3 (a). Hence we observe that holes can easily diffuse through this region and that this region doesn't affect the hole J.

3) Far Forward: When $V > V_{FB} = \phi_{\mathrm{ma}} - \phi_{\mathrm{mc}}$, see Fig. 3 (b), the holes have to drift through the intrinsic region. The J is found by multiplying the hole concentration with the electric field,

$$J_p^{dr,\mathrm{i}} = \frac{n_i\mu_{\mathrm{p}}q(V - V_{FB})}{L_{\mathrm{i}}}e^{\frac{\phi_{\mathrm{ba}} - E_{\mathrm{g}}}{kT}}, \quad (4)$$

where μ_{p} is the hole mobility and L_{i} the length of the intrinsic region. The potential is constant under the gate, hence holes diffuse through the anode region. The p at the anode metal interface is high compared to the anode intrinsic region interface hole concentration p_{a}. As a result the diffusion component is given by,

$$J_p^{d,\mathrm{a}} = \frac{N_V qD_{\mathrm{p}}}{L_{\mathrm{a}}}e^{\frac{\phi_{\mathrm{ba}} - E_G}{kT}}, \quad (5)$$

where $N_V \exp(\phi_{\mathrm{ba}} - E_G)/kT$ gives the hole concentration (p) at the anode silicon interface. Holes travel by means of thermionic emission [14] from the metal anode into the silicon. The hole barrier at this interface is given by $E_{\mathrm{g}} - \phi_{\mathrm{ba}}$ resulting in a thermionic emission J,

$$J_p^{th,\mathrm{a}} = A_p^* T^2 e^{\frac{\phi_{\mathrm{ba}} - E_{\mathrm{g}}}{kT}}. \quad (6)$$

4) Current Model: The current components in Eqs. (3 - 6) are connected in series. Hence the smallest of them determines the total hole current density J_p^t. This can be modeled by

$$\frac{1}{J_p^t} \approx \frac{1}{J_p^{th,\mathrm{a}}} + \frac{1}{J_p^{d,\mathrm{a}}} + \frac{1}{J_p^{dr,\mathrm{i}}} + \frac{1}{J_p^c}. \quad (7)$$

978-1-4244-6658-0/10 $26.00 © 2010 IEEE

Analogously for the electrons J_n^t can be found. For negative or small positive biases the J may be far in excess from what is predicted by the diffusion and thermionic emission theory [15], [16]. This results from Shockley-Read-Hall generation/recombination which is modeled using theory from [17], [18]. We obtain

$$J_{\text{SRH}} = \frac{q n_{\text{i}} L_{\text{i}}}{\tau_n + \tau_p} \left(e^{\frac{V}{2u_{\text{T}}}} - 1 \right), \qquad (8)$$

where τ_n and τ_p are the electron and hole life time respectively. The total J can be found by summation of the electron, hole and SRH components.

III. SIMULATION

To evaluate our model we use the Synsopsys Sentaurus Device simulator [13]. The following models were used in the simulations: Schottky contact [19], electron effective density [20], hole effective density [21], carrier lifetimes [22], [23], enhanced Lombardi model for surface scattering and temperature dependent mobility [24] and Philips unified mobility model [25] for carrier concentration dependent mobilities.

Two different metal work function combinations were used in this work. Combination A ($\phi_{\text{mc}} = 4.47$ eV, $\phi_{\text{ma}} = 4.90$ eV) results from characterization of Schottky junctions fabricated in our cleanroom. Combination B ($\phi_{\text{mc}} = 4.20$ eV, $\phi_{\text{ma}} = 5.10$ eV) is chosen such that the metal work functions are very close to the silicon band edges, giving high carrier concentrations.

Fig. 4 shows the results. The theory predicts that for combination A the total off current is limited by the hole diffusion through the cathode region (J_p^c). The on current by diffusion through the anode region ($J_p^{d,\text{a}}$). For A we also show a measurement result [6]. In general the characteristics are as expected, the differences can be attributed to differences in the metal work functions and the presence of interface states.

For combination B the currents are limited by both electron and hole diffusion. Due to the workfunctions being close to the band edges the diffusion current becomes rather small and the SRH current shows up for small or negative V.

In combination A the total current is determined by J_p^t, see Eq. (7). Current dependence on L_{a} and L_{c} is predicted by Eqs. (3) and (5). This dependence is shown in Fig. 5. Again the model is in good agreement with the simulation results. For long gate lengths the hole current is reduced so much that the electron current becomes important. This explains the reduced dependence on the gate lengths of the on and off currents for long gate lengths.

IV. OPTIMISATION

The model can be used to optimise the rectifying performance of the CP-diode. For a good rectifier we need at least: (1) a large maximum current through the device, (2) a high on/off current ratio. As shown in Fig. 5 scaling L_{c} and L_{a} helps to meet these requirements. In Figs. 6 and 7 the gate lengths have been scaled for combination A. The on current scales with L_{a}. When $L_{\text{a}} < 10$ nm the on current

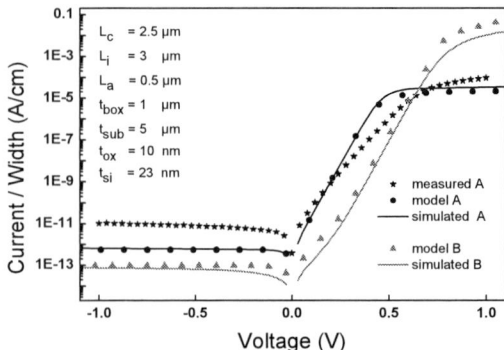

Fig. 4. Modeled and simulated IV characteristics for CP-diodes for work function combinations A and B. The dimensions are $L_{\text{c}} = 2.5$ μm, $L_{\text{i}} = 3$ μm, $L_{\text{a}} = 0.5$ μm, $t_{\text{box}} = 1$ μm, $t_{\text{sub}} = 5$ μm, $t_{\text{ox}} = 10$ nm and $t_{\text{si}} = 23$ nm.

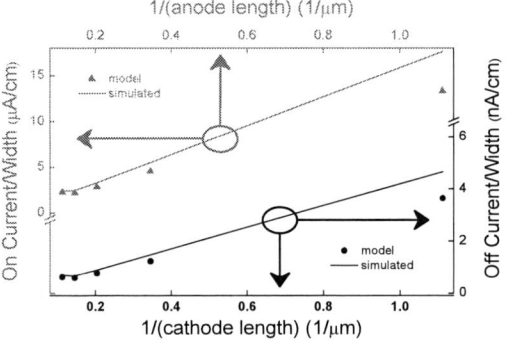

Fig. 5. The on and off current scaling with both gate lengths for combination A. The on current is extracted at $V = 1.5V_{\text{FB}}$ and scales with $1/L_{\text{a}}$. The off current is extracted at with $V_{\text{FB}}/2$ and scales with $1/L_{\text{c}}$. The device dimensions are $L_{\text{i}} = 0.2$ μm, $t_{\text{box}} = 1$ μm, $t_{\text{sub}} = 5$ μm, $t_{\text{ox}} = 15$ nm and $t_{\text{si}} = 20$ nm.

is maximized and limited only by thermionic emission of holes from the anode, see Eq. (6). For very long L_{a} the electron current contribution becomes important and the L_{a} dependence becomes less.

The off current is determined by Eq. (3). Hence increasing L_{c} decreases the off current and results in an better on/off current ratio. For large L_{c} the electron off current becomes dominant and for small L_{a} the hole on current becomes limited by thermionic emission.

A short L_{i} is preferable because (1) the SRH current, Eq. (8), could enhance the off current, (2) the drift current, Eq. (4), could limit the on current. Regarding t_{si} and t_{ox}, good rectifying behavior is reached when both the cathode and anode region have a high carrier concentration, requiring a small t_{ox}. By tuning t_{si} a trade-off can be made between current level and reduced on/off current ratio.

For combination B the metal work functions are almost symmetric with respect to the silicon work function. Therefore, both the electron and hole current are equally important. Thus L_{c} and L_{a} cannot be used to optimise for either the hole or electron current. Here short gates results in a high on (and

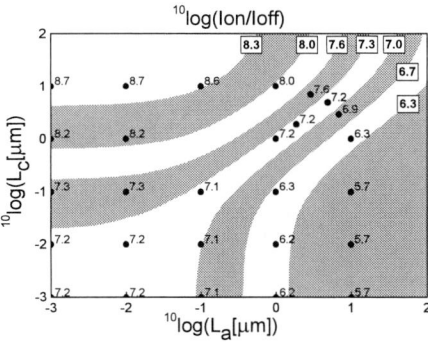

Fig. 6. Simulated (dots) and modeled (contours) on/off current ratio for combination A. Fixed parameters are $L_i = 0.1, t_{box} = 1, t_{sub} = 5$ μm, $t_{ox} = 15, t_{si} = 20$ nm. Maximum on/off current ratio is obtained for short L_a and long L_c.

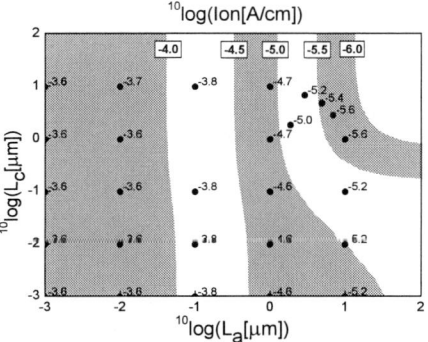

Fig. 7. Simulated (dots) and modeled (contours) on current for the same devices as in Fig. 6. Maximal on current is obtained for minimal L_a irrespective of L_c.

off) current, but good on/off current ratio. Increasing both gate lengths result in a lower on and off current yielding a worse device performance.

However in a real CP-diode the metal work functions are unlikely to be symmetric with respect to the silicon work function and thus appropriate L_c and L_a sizing can improve the performance.

V. CONCLUSION

This work shows that the device dimensions of the CP-diode can be used to optimise the rectifying performance given the metal work functions. If the work functions are approximately equally distant from midgap, then a CP-diode without gates would give the best possible rectifier. However when the work functions are not equally distant from midgap finetuning the gate length gives an improved performance.

Our model can be used for optimizing the rectifying performance by adjusting the metal work functions, oxide and silicon thickness and lateral dimensions. For combination A (shown in Fig. 6) a maximum on/off current ratio of 5×10^8 was found, the on current is 1.2×10^{-4} A/cm and off current is 4×10^{-13} A/cm. Compared to the on/off current ratio of 1.5×10^7 for a device without gates this does yield an important improvement.

REFERENCES

[1] M.-H. Chiang *et al.*, "Random dopant fluctuation in limited-width finfet technologies," *IEEE Trans. El. Dev.*, vol. 54, no. 8, pp. 2055–2060, Aug. 2007.

[2] A. Martinez *et al.*, "The impact of random dopant aggregation in source and drain on the performance of ballistic dg nano-mosfets: A negf study," *IEEE Trans. Nanotech.*, vol. 6, no. 4, pp. 438–445, July 2007.

[3] J. C. Ho *et al.*, "Controlled nanoscale doping of semiconductors via molecular monolayers," *Nature Materials*, vol. 6, no. 1, pp. 62–67, Jan. 2008.

[4] J. Larson and J. Snyder, "Overview and status of metal s/d schottky-barrier mosfet technology," *IEEE Trans. El. Dev.*, vol. 53, no. 5, pp. 1048–1058, 2006.

[5] B. Rajasekharan *et al.*, "Dimensional scaling effects on transport properties of ultrathin body p-i-n diodes," *ULIS*, pp. 195–198, 2008.

[6] ——, "Fabrication and characterization of the charge plasma diode," accepted for publication in IEEE El. Dev. Lett..

[7] R. Hueting *et al.*, "The charge-plasma pn-diode," *IEEE El. Dev. Lett.*, vol. 29, no. 12, pp. 1367–1369, 2008.

[8] E. Rhoderick and R. Williams, *Metal-Semiconductor Contacts*, 2nd ed. Oxford, U.K.: Clarendon, 1988.

[9] Y. Taur, "An analytical solution to a double-gate mosfet with undoped body," *IEEE El. Dev. Lett.*, vol. 21, no. 2, p. 254, 2000.

[10] ——, "Analytical solutions of charge and capacitance in symmetric and asymmetric double-gate mosfets," *IEEE Trans. El. Dev.*, vol. 48, no. 12, p. 2861, 2001.

[11] W. Shockley, "The theory of $p - n$ junctions in semiconductors and $p - n$ junction transistors," *Bell. Syst. Tech. J.*, vol. 28, p. 435, 1949.

[12] S. Sze and K. K. Ng, *Physics of Semiconductor Devices*. Wiley & Sons Inc., 2007.

[13] Synopsys Inc, "Sentaurus device user guide," *Version A-2008.09, 1.3, ia32*.

[14] H. Bethe, "Theory of the boundary layer of crystal rectifiers," *MIT Radiation Lab Rep.*, vol. 43-12, 1948.

[15] W. Shockley and W. T. Read, "Statistics of the recombinations of holes and electrons," *Phys. Rev.*, vol. 87, no. 5, pp. 835–842, September 1952.

[16] R. N. Hall, "Electron-hole recombination in germanium," *Phys. Rev.*, vol. 87, no. 2, p. 387, July 1952.

[17] C.-T. Sah *et al.*, "Carrier generation and recombination in p-n junctions and p-n junction characteristics," *Proc. of the IRE*, vol. 45, no. 9, pp. 1228–1243, Sept. 1957.

[18] R. Pierret, *Semiconductor Devices Explained*. Pearson Education Inc., 1996.

[19] A. Schenk, *Advanced Physical Models for Silicon Device Simulation*. Wien: Springer, 1998.

[20] M. A. Green, "Intrinsic concentration, effective densities of states and effective mass in silicon," *J. Appl. Phys.*, vol. 67, no. 6, pp. 2944–2954, 1990.

[21] J. Lang *et al.*, "Temperature dependent density of states effective mass in nonparabolic p-type silicon," *J. Appl. Phys.*, vol. 54, no. 6, pp. 3612–3612, 1983.

[22] M. Tyagi and R. van Overstraeten, "Minority Carrier Recombination in Heavily-Doped Silicon," *Solid State El.*, vol. 26, no. 6, pp. 577–597, 1983.

[23] H. Goebels and K. Hoffman, "Full dynamic power diode model including temperature behavior for use in circuit simulators," *ISPSD*, pp. 130–135, 1992.

[24] C. Lombardi *et al.*, "A physically based mobility model for numerical simulation of nonplanar devices," *IEEE Trans. Comput.-Aided Design Integr. Circuits Syst.*, vol. 7, no. 11, pp. 1164–1171, Nov 1988.

[25] D. Klaassen, "A unified mobility model for device simulation–i. model equations and concentration dependence," *Solid State El.*, vol. 35, no. 7, pp. 953 – 959, 1992.

CMOS-MEMS free-free beam resonators

JL.López, E.Marigó, J.Giner, JL.Muñoz-Gamarra, F.Torres, A.Uranga, N.Barniol

Dept. Electronics Engineering.
Universitat Autónoma de Barcelona
08193-Bellaterra-Barcelona. Spain
Nuria.barniol@uab.cat

Abstract—**In this paper a 25 MHz free-free beam flexural resonator monolithically integrated in a 0.35 um CMOS technology is presented. A comparison between the frequency response and electrical characteristics between free-free beam and clamped-clamped beams shows higher qualities factor for free-free beams which will allow better oscillators for frequency references in terms of phase noise.**

I. INTRODUCTION

It is widely accepted that MEMS resonators would be one of the most promising candidates to substitute crystal quartz in frequency references for communications systems [1,2]. One of the reasons for that interest is their capability to be scaled and integrated with a CMOS technology. One of the problems which must be overcome in MEMS monolithically integrated using CMOS technology, is their low quality factor. In this paper a free-free beam resonator with improved quality factor compared with a clamped-clamped beam, due to a decrease on the losses over the anchors is presented. The free-free beam resonator is monolithically integrated in the CMOS 0.35 two polysilicon and 4 metals layers from Austria Microsystems.

II. FREE-FREE BEAM RESONATORS DESIGN

A. Free-free beam design

Following the Euler-Bernouilli theory [3], it is shown that the deflection of a general beam (moving in the x direction and with dimensions: L for the length, h for its thickness and w for its width) as a function of the position along the beam axis (y) can be written as:

$$\frac{\partial^4 x(y)}{\partial y^4} = \kappa_n^4 \cdot x(y)$$

(1)

with

$$\kappa_n^4 = \omega_n^2 \frac{\rho h W}{EI}$$

(2)

Where ω_n represents the n-th mode resonance frequency, ρ is the mass density, E is the Young Modulus, I is the inertia momentum, and h and W are the thickness and width of the beam, respectively.

General solution of (1) can be written as:

$$x(y) = A_n \cdot \sin(\kappa_n \cdot y) + B_n \cdot \cos(\kappa_n \cdot y) + C_n \cdot \sinh(\kappa_n \cdot y) + D_n \cdot \cosh(\kappa_n \cdot y)$$

(3)

Where An, Bn, Cn and Dn are integration constants which value is determined by the boundary conditions. Once found these values, equation (3) will determine the profile of the deflection of the beam.

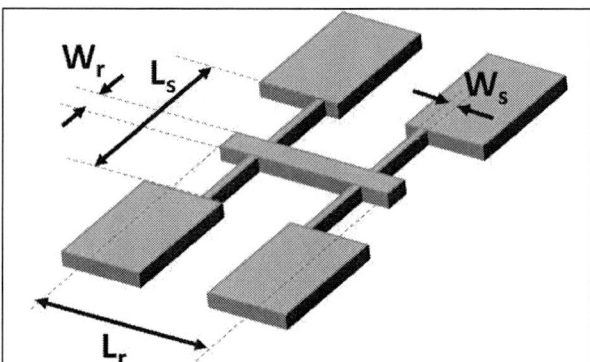

Figure 1. Design variables of the lateral free-free beam. W_r: resonator width; L_r=resonator length; W_s: support width and L_s: support length.

In free-free beams, both ends are not anchored (see figure 1), and they can move freely. Therefore the boundary conditions for the ends of this kind of resonators are related to the second and third derivative of the displacement:

$$\left.\frac{\partial^2 x}{\partial y^2}\right|_{y=0}=0; \quad \left.\frac{\partial^3 x}{\partial y^3}\right|_{y=0}=0; \quad \left.\frac{\partial^2 x}{\partial y^2}\right|_{y=L}=0; \quad \left.\frac{\partial^3 x}{\partial y^3}\right|_{y=L}=0$$

(4)

When the first and second boundary conditions are applied to (3), the following relationships are found:

978-1-4244-6658-0/10 $26.00 © 2010 IEEE

$$A_n = C_n$$
$$B_n = D_n \qquad (5)$$

Whereas the third and fourth boundary conditions lead to:

$$C_n \cdot \left[\sinh(\kappa_n \cdot L) - \sin(\kappa_n \cdot L) \right] + D_n \cdot \left[\cosh(\kappa_n \cdot L) - \cos(\kappa_n \cdot L) \right] = 0$$

$$C_n \cdot \left[\cosh(\kappa_n \cdot L) - \cos(\kappa_n \cdot L) \right] + D_n \cdot \left[\sinh(\kappa_n \cdot L) + \sin(\kappa_n \cdot L) \right] = 0$$

And the conditions between the coefficients are:

$$\frac{D_n}{C_n} = \frac{\sin(\kappa_n \cdot L) - \sinh(\kappa_n \cdot L)}{\cosh(\kappa_n \cdot L) - \cos(\kappa_n \cdot L)} \qquad (6)$$

$$\frac{D_n}{C_n} = \frac{\cos(\kappa_n \cdot L) - \cosh(\kappa_n \cdot L)}{\sinh(\kappa_n \cdot L) + \sin(\kappa_n \cdot L)}$$

Combining these equations, $\kappa_n L$ parameter for a free-free beam is obtained:

$$\cosh(\kappa_n \cdot L) \cdot \cos(\kappa_n \cdot L) = 1 \qquad (7)$$

Note that equation (7) formula free-free beams, is the same expression than for for clamped-clamped beams, and therefore the values of $\kappa_n L$ are: 0.73, 7.853, 11 and 14.14 for the first four vibration modes [3].

The bending profile can be obtained by combining equation (3) with equations (5,6,7):

$$x(y) = \cosh(\kappa_n \cdot L) + \cos(\kappa_n \cdot L) - \xi \cdot \left[\sinh(\kappa_n \cdot L) + \sin(\kappa_n \cdot L) \right] \qquad (8)$$

Where:

$$\xi \equiv \frac{\cosh(\kappa_n \cdot L) - \cos(\kappa_n \cdot L)}{\sinh(\kappa_n \cdot L) - \sin(\kappa_n \cdot L)} \qquad (9)$$

Whereas for clamped-clamped beams this profile is not at all important, it has to be calculated for free-free beams to determine the zero displacement points (nodal points) that will be used to anchor the structure (figure 2).

The anchoring of these structures in lateral mode is done by using two clamped-clamped beams orthogonal to the free-free beam connected through its nodal points. The key point of the design is the use of second-mode clamped-clamped beams so that the nodal point of the CC-beams (located in the middle of the structure) corresponds to the nodal points of the free-free beam, and therefore the support beam movement does not affect the free-free beam resonance.

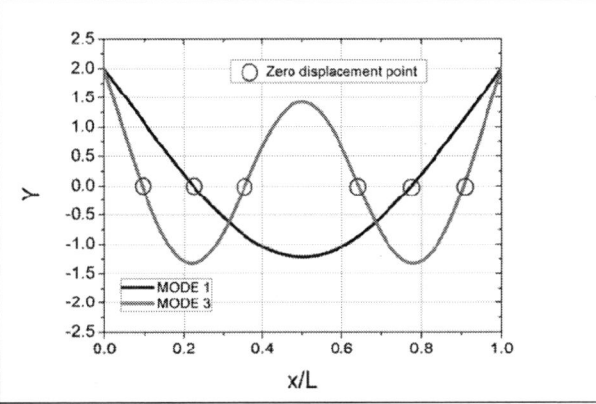

Figure 2. Diagram of the bending profile of a free-free beam for the first and third lateral mode

Following these design equations, the free free beam dimensions are: W_r=800nm, L_r=17um, W_s=400nm, L_s=20 um. The thickness of the structure is 282 nm which corresponds to the polysilicon 1 layer of the 0.35 um CMOS technology. For the calculations a Young modulus of 160GPa and mass density of 2230kg/m³ have been considered (polysilicon material).

Figure 3 shows the Coventor FEM simulations a free-free beam resonator. In this simulation it is shown how the joints of the supports and the structures present zero lateral displacement (areas in blue). These mechanical simulations show a resonance frequency of 23.98MHz for the first mode free-free beam resonator.

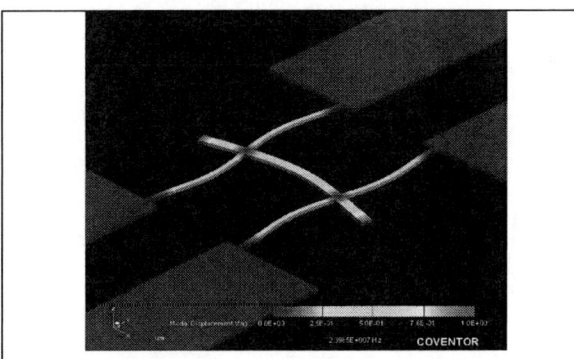

Figure 3. FEM mechanical simulation of the designed free-free resonators (using Coventor software) in its first lateral mode at 24 MHz.

B. Free-free beam fabrication on CMOS technology

The free-free beam resonator is actuated electrostatically using a driver electrode situated in parallel to the beam. Additionally we are using capacitive transduction with a parallel electrode, similar to our previous reported works with clamped-clamped beams [4-6].

The CMOS technology used is the 0.35 um CMOS from Austria Microsystems. Following the spacer technique [4] and using the 2 polysilicon layers of the technology as structural material for the driver electrodes (poly2) and the resonator (poly1), a gap of only 40nm between them is defined. The resonator and electrodes are fabricated simultaneously to the CMOS and only a mask-less wet etching post-process is made after receiving the chip from the foundry to release the resonator from the sacrificial oxide which surrounds the beam. In this free-free beam resonator, the driver dimensions are reduced to 5 um due to the location of the supports. Figure 4 shows a SEM image of the released free-free beam resonator.

III. EXPERIMENTAL RESULTS

Electrical characterization is made using a probe station and network analyzer to acquire the frequency response of the beam. S21 magnitude and phase measurements were performed in air conditions under different applied DC voltages (*Figure* 5).

From S21 measurements it can be observed that a quality factor Q=197 is measured in air from the 3dB peak for V_{DC}=12V, and a similar value is obtained using the phase curves. This value is twice the best obtained with the 24MHz clamped-clamped beam, demonstrating the higher Q of the free-free beams, as expected..

The natural resonance frequency of the free-free beam is obtained using the linear fit of the f_{RES}vs.V_{DC}^2 relationship. This value is of 25.3MHz, a 6% higher than the obtained using Coventor simulations. The frequency tuning of this resonator is of 113ppm/V^2, lower than the obtained for the CC-beam with s=40nm, making this resonator more robust against DC voltage drifts. Because of this robustness, the required V_{DC} to obtain the target 24MHz resonance frequency with the free-free beam resonator is 23V, higher than the required for s=40nm clamped-clamped beam.

A $RLC//C_p$ model is used to fit the frequency response of the s=40nm 24MHz FF-beam. *Figure* 6 shows the fitting curve for P_{in}=0dBm and V_{DC}=12V. From this figure, a motional resistance of 4.77MΩ is obtained.. In general and due to a decrease on the coupling area between the resonator and the drivers, motional resistance for free-free beams resonators would be higher than the one for the clamped-clamped beams for the same bias voltage and gap (s=40nm). We can evaluate the ratio between the motional resistance of the two resonators (equation 10).

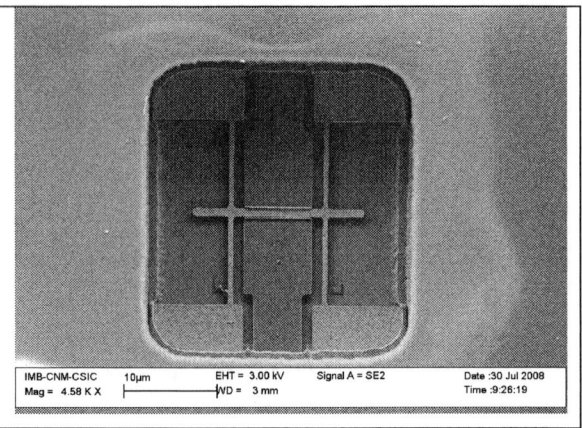

Figure 4. SEM image of the 24MHz free-free beam, s= 40 nm

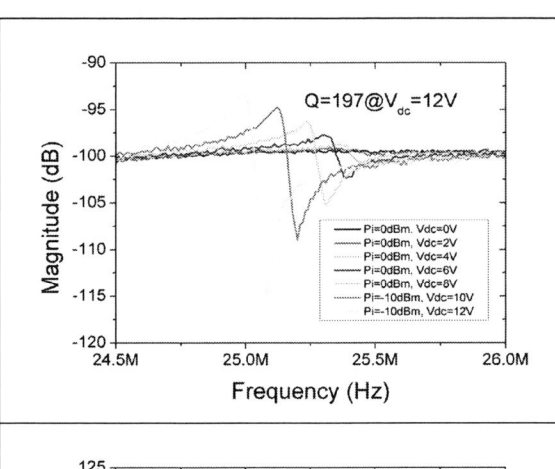

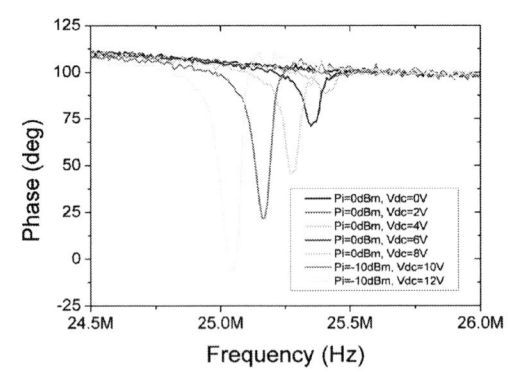

Figure 5: 24 MHz free-free beam frequency response. (a) Magnitude and (b) Phase

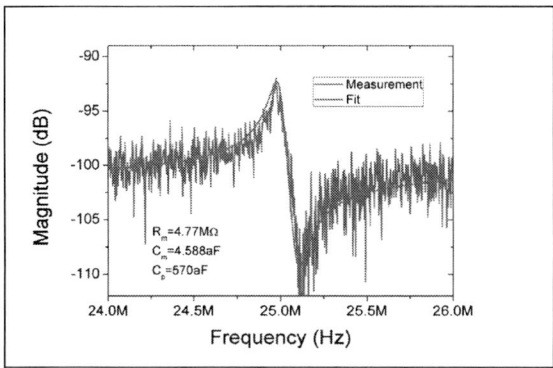

Figure 6: Curve fitting of the 24MHz FF-beam. Applied input power is P_i=0dBm and V_{DC}=12V.

$$\frac{R_m^{FF-Beam}}{R_m^{CC-Beam}} = \frac{Q_{CC-Beam} \cdot A_{CC-BEAM}^2 \cdot k_{FF-Beam}}{Q_{FF-Beam} \cdot A_{FF-BEAM}^2 \cdot k_{CC-Beam}} \quad (10)$$

To evaluate expression (10), Q values are obtained from previous measurements ($Q_{CC-Beam}$=80 and $Q_{FF-Beam}$=197, the coupling areas are determined by the coupling length ($L_{cCC-Beam}$=9µm and $L_{cFF-Beam}$=7µm) and the mechanical elastic constants are calculated from Coventor FEM simulations ($k_{CC-Beam}$=38.7N/m and $k_{FF-Beam}$=90.6N/m), using these values the motional resistance for a free-free resonator is 1.6 times higher than the clamped-clamped beam, even though the higher Q of the free-free beams.

In these experimental results, as we are using a capacitive detection, the parasitic capacitance between the drivers and the resonator provides an antiresonance or parallel resonance which masks the quantification of the quality factor. In order to overcome this intrinsic measurement problem, we have used the mixing technique [4]. In this technique the resonator is actuated with a different frequency that its resonance mode, and thus the parasitic current is avoided.

Figure 7 shows the mixing measurement of the 24MHz free-free beam with applied powers of P_i=10dBm, P_{LO}=17dBm and V_{DC}=12V, the LO frequency is fixed to 5MHz and the input signal is swept from 29MHz to 31MHz. Comparing this curve with the previously obtained for a CC-beam, it can be observed that this response shows a less noisy signal and a higher measured Q of 334. The maximum achieved with a clamped-clamped beam in air and with the same technology, was around 200 [6], so a 1.7x higher Q factor is obtained for the free-free beam.

IV. CONCLUSIONS

We have presented a free-free resonator MEMS integrated in a CMOS technology which presents high Q factor. Particularly, we have measured a Q of 334 in a 25MHz

CMOS-MEMS resonator. These numbers represents a figure of merit, fxQ=8.35x10^9 which is higher (a factor of 3) than other reported in a CMOS integration process operating in air [7] and only a factor of 2 lower than [8] which only reports results on vacuum environment. We expect further improvement on the Q factor in vacuum operation. In this way this new device can be the core of a new class of oscillators for fully integrable frequency references.

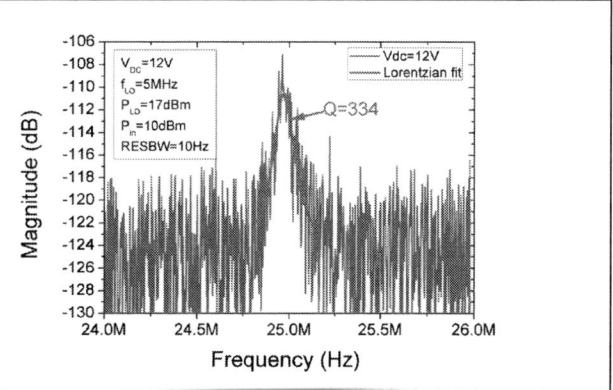

Figure 7: Mixing measurement for a 24MHz free-free beam with P_i=0dBm, P_{LO}=10dmb, and V_{DC}=12V

ACKNOWLEDGMENT

This work has been supported by project NEMESYS TEC2009-9008 funded by the spanish Ministerio de Ciencia y Tecnología .

REFERENCES

[1] C. T. C. Nguyen, "MEMS technology for timing and frequency control," *Ultrasonics, Ferroelectrics and Frequency Control, IEEE Transactions on,* vol. 54, pp. 251-270, 2007

[2] ITRS 2009

[3] M.Bao, Micro mechanical Transducers: pressure, sensors, accelerometers ang gyroscopes. Amsterdam-New York, Elsevier 2000.

[4] J. L. Lopez, J. Verd, J. Teva, G. Murillo, J. Giner, F. Torres, A. Uranga, G. Abadal, and N. Barniol, "Integration of RF-MEMS resonators on submicrometric commercial CMOS technologies," *Journal of Micromechanics and Microengineering,* vol. 19, p. 015002, 2009.

[5] J. Verd, A. Uranga, G. Abadal, J. L. Teva, F. Torres, J. L. Lopez, E. Perez-Murano, J. Esteve, and N. Barniol, "Monolithic CMOS MEMS Oscillator Circuit for Sensing in the Attogram Range," *Electron Device Letters, IEEE,* vol. 29, pp. 146-148, 2008

[6] J. L. Lopez, J. Verd, A. Uranga J. Giner, G. Murillo, F. Torres, G. Abadal, and N. Barniol. A CMOS MEMS RF tunable bandpass filter based on two high Q polysilicon clamped clamped beam resonators. *Electron Device Letters, IEEE,* vol.30, pp. 718-20, 2009

[7] C. C. Lo and G. K. Fedder, "On-Chip High Quality Factor CMOS-MEMS Silicon-Fin Resonators," in *Solid-State Sensors, Actuators and Microsystems Conference, 2007. TRANSDUCERS 2007. International,* 2007, pp. 2449-2452

[8] W.-L. Huang, Z. Ren, Y.-W. Lin, H.-Y. Chen, J. Lahann, and C. T. C. Nguyen, "Fully monolithic CMOS nickel micromechanical resonator oscillator," in *Micro Electro Mechanical Systems, 2008. MEMS 2008. IEEE 21st International Conference on,* 2008, pp. 10-13

Electro-Thermal Analysis of RF MEM Capacitive Switches for High-Power Applications

Francesco Solazzi, Cristiano Palego, Subrata Halder,
and James C. M. Hwang
Lehigh University
Bethlehem, PA, USA

Alessandro Faes, Viviana Mulloni, and Benno Margesin
Fondazione Bruno Kessler-IRST
Trento, Italy

Paola Farinelli, and Roberto Sorrentino
RF Microtech
Perugia, Italy

Abstract—Self heating in electrostatically actuated RF MEM capacitive shunt switches is analyzed by coupled electrical and thermal simulations using three-dimensional finite-element analysis. The result shows that despite highly nonuniform current and temperature distributions, the self-heating effect can be approximated by lumped thermal resistances of the switch membrane and the substrate. Additionally, since the thermal resistance of thermally insulating substrates such as quartz is significant compared to that of the membrane, it is important to consider the heat transfer across both the membrane and the substrate.

I. INTRODUCTION

Microelectromechanical (MEM) capacitive shunt switches exhibit low loss and high linearity and, hence, are attractive for applications in reconfigurable RF front ends [1]. With significant improvement in packaging and reliability of the switches over the past decade, increasing attention has been paid to improving their power-handling capacity [2]. To date, the attention has been mostly focused on improving the material [3] or geometry [4] of the movable membrane in the switches. This is because, under a large RF input signal, small yet significant amount of power can be dissipated on the membrane even when it is suspended between the anchors at both ends. The dissipated power can raise the temperature, relax the stress, and soften the membrane, so that with increasing power the pull-in voltage of the membrane changes significantly until the switch fails to operate. Recently, we found [5] that in addition to the membrane, the substrate on which the switch is fabricated plays an important role in limiting its power-handling capacity. Specifically, although

---

[1]This work was supported in part by a grant from the Commonwealth of Pennsylvania, Department of Community and Economy Development, through the Pennsylvania Infrastructure Technology Alliance (PITA) and by the European Space Agency under Contract No. ITT AO/1-5288/06/NL/GLC. Additional support was received from the OPTEL InP Consortium.

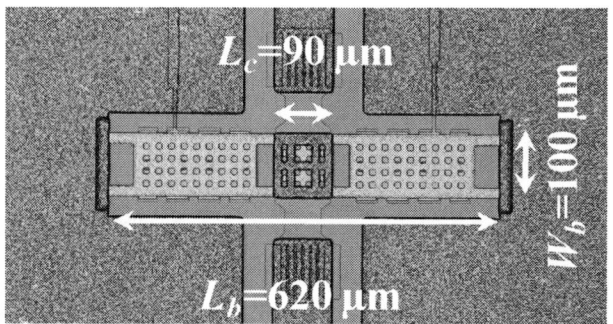

(a)

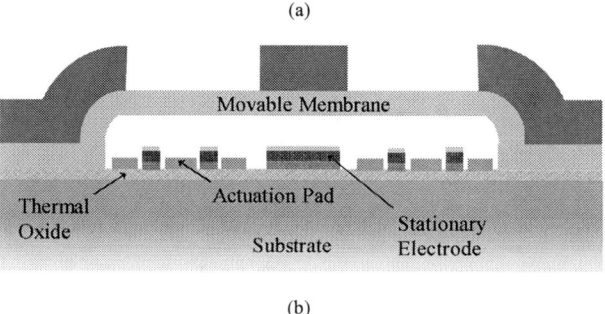

(b)

Fig. 1. (a) Top view and (b) cross-sectional schematic of the electrostatically actuated RF MEM capacitive shunt switch with separate and non-contacting actuation pads.

the switch fabricated on a silicon substrate is capable of hot switching up to at least 5.6 W, the same switch fabricated on a quartz substrate has lower power capacity and its pull-in voltage has stronger power dependence. This was attributed to greater electrical and thermal mismatches in the switch fabricated on quartz than in that fabricated on silicon. This

TABLE I
SUBSTRATE PROPERTIES AND EQUIVALENT-CIRCUIT PARAMETERS

Parameter	Silicon	Quartz
Substrate dielectric constant ε	11	3.8
Substrate resistivity ρ ($\Omega \cdot$cm)	5000	10^{19}
Substrate loss tangent δ	0.006	0.0001
Substrate thermal conductivity K_{SUB} (W/cm°C)	1.56	0.02
Substrate thermal expansion coeff. α (ppm/°C)	2.6	0.4
Substrate shunt resistance R_{SUB} (Ω)	2000	6000
Substrate shunt capacitance C_{SUB} (pF)	1.89	0.0015
Substrate thermal resistance R^{TH}_{SUB} (°C)/W)	48	680
Transmission line impedance Z_{TL} (Ω)	50	75
Transmission line length θ_{TL} (°)	25	16
Transmission line loss R_{TL} (Ω)	0.4	
On-state membrane capacitance C_U (fF)	40	
Off-state membrane capacitance C_D (fF)	300	
Membrane inductance L_{MEM} (pH)	2	
Membrane resistance R_{MEM} (Ω)	0.2	
Membrane stress σ_0 (MPa)	60	80
Membrane pull-in voltage V_P (V)	38	45
Membrane thermal resistance R^{TH}_{MEM} (°C/W)	4900	6100

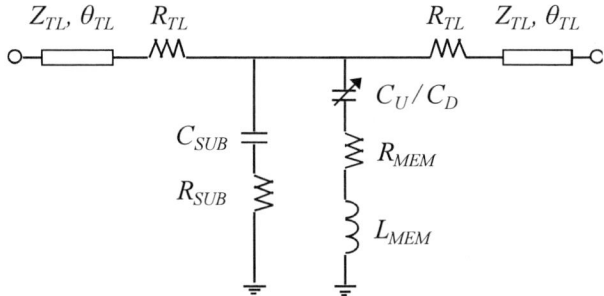

Fig. 2.   Equivalent circuit of the switch.

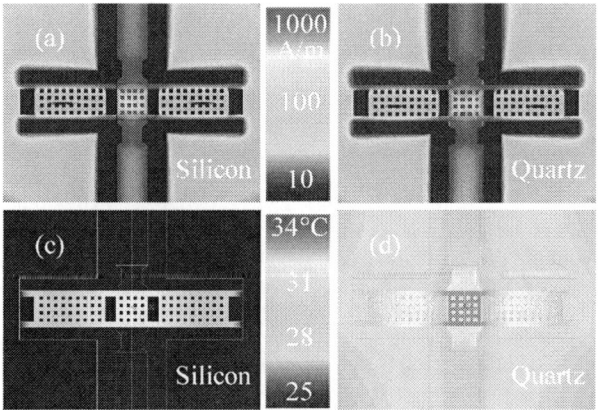

Fig. 3.   Simulated surface (a, b) current and (c, d) temperature distributions for the switches on (a, c) silicon and (b, d) quartz under 1 W input signal at 15 GHz.

paper expands on [5] by showing that the relatively poor thermal conductivity of quartz is yet another important factor, so that the dissipated powers and thermal resistances of both the membrane and the substrate need to be considered.

The coupled electrical, thermal and mechanical effects described above can be very involved. Fortunately, they involve rather different time constants so that they can be analyzed separately and iteratively. Following [6]-[8], we first use a 3D finite-element electromagnetic simulator *HFSS* [9] to simulate the current distribution over the entire switch die of $1944 \times 1330 \times 530$ μm³ under different RF input powers. The losses associated with the simulated current distribution are then ported to a 3D finite-element multiphysics simulator *ANSYS* [9] to simulate the temperature distribution. Using *HFSS*, Maxwell equations are solved on the surface of conductors and in the bulk of insulators, resulting in surface losses and volume losses, respectively. Using ANSYS, surface losses are mapped into heat fluxes while volume losses are mapped into heat generators. A scaling factor is used to account for the different mesh sizes of *HFSS* and *ANSYS*. The heat transfer is assumed to be through conduction with negligible convection and radiation [10]. Finally, the resulted temperature rise is used to derive [3] the power dependence of the pull-in voltage for comparison with that measured experimentally as described below.

II. EXPERIMENTAL

Fig. 1 illustrates the switch fabricated on either silicon or quartz by using a combination of CMOS processes and micromachining techniques [11]. The switch consists of a rectangle-shaped membrane suspended over the central

conductor of a coplanar transmission line and two actuation pads beside the center conductor. The membrane is anchored on both ends to the ground conductors of the coplanar transmission line. The membrane is 620 μm long and 100 μm wide and is mostly made of 2 μm-thick electroplated gold, except the 100 μm × 90 μm portion directly above the centre conductor is stiffened by additional 3.5 μm of electroplated gold. The overall stiffness of the membrane is reduced by two thin springs at either anchor of the membrane. The portion of the center conductor directly under the membrane serves as the stationary electrode of the switch and is made of 0.63 μm-thick Ti/TiN/Al/Ti/TiN and 0.1 μm-thick SiO₂. Beside the stationary electrode are two actuation pads made of 0.63 μm-thick polysilicon, which are electrically isolated from either the membrane or the stationary electrode. Standoffs made of the same material as the stationary electrode are embedded in the actuation pads to prevent them from contacting the membrane when it is pulled in. Such separate and non-contacting actuation pads help minimize dielectric charging making the switch not only more robust, but also more reliable [12], [13].

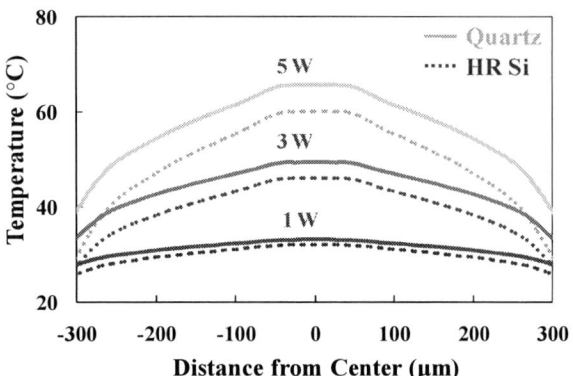

Fig. 4. Simulated anchor-to-anchor temperature profile along the center of the membrane under different RF powers at 15 GHz.

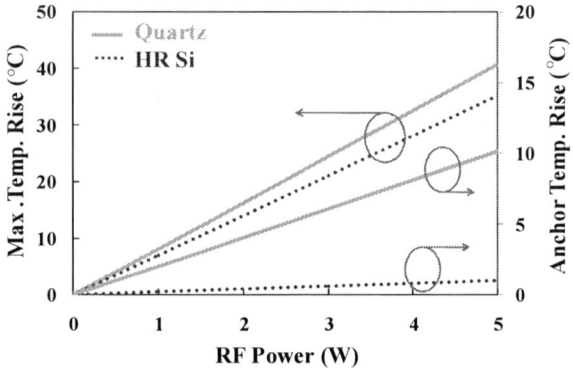

Fig. 5. Simulated temperature rise at the (left axis) center and (right axis) anchor of the membrane on (---) silicon and (—) quartz under different RF powers at 15 GHz.

To evaluate the substrate effect, the same switch was fabricated on either high-resistivity (HR) silicon or quartz, with the silicon substrate insulated by 1-μm-thick SiO_2. Table I compares the properties of the silicon and quartz substrates. Table I shows also that the switches on silicon and quartz have slightly different residual stresses and pull-in voltages due to uncontrolled deviations in fabrication. Table I lists also the equivalent-circuit (Fig. 2) parameters extracted from small-signal RF measurements.

III. RESULTS AND DISCUSSION

Figs. 3(a) and 3(b) compare the simulated surface current distributions on the silicon and quartz switches, both under a 1-W input power at 15 GHz. It can be seen that in both cases the current concentrates around the edges of the center and ground conductors of the coplanar transmission line. Additional currents concentrate along the leading and trailing edges of the membrane. From the simulation the dissipated powers on the membrane and transmission line, P_{MEM} and P_{TL}, are estimated to be 2 mW and 18 mW on silicon, and 1.7 mW and 14 mW on quartz. The total dissipated power $P_{MEM} + P_{TL}$ scales approximately linearly with the input power and is in general agreement with that extracted from small-signal RF measurements. The finding that the total dissipated power is dominated by the transmission line instead of the membrane is consistent with [7] but opposite of [6]. Subtle differences in the switch design may cause such discrepancy. For example, in the present switch the transmission line extends far beyond what is shown in Fig. 1(a) for a total length of 580 μm on either side of the MEM section. The relatively large loss in the transmission line, when dissipated through a thermally insulating substrate, contributes significantly to the temperature rise in the membrane as described below.

Figs. 3(c) and 3(d) show vividly the difference between the simulated surface temperatures on silicon and quartz. Despite the higher dissipated power on silicon than that on quartz, the silicon is mostly deep blue while the quartz is mostly light blue. According to the color scale of the figure, on silicon there is only appreciable temperature rise on the membrane and it peaks near the middle of the membrane. In comparison, the peak temperature on quartz is higher and the entire switch including the center and ground conductors of the coplanar transmission line is heated up.

Fig. 4 shows the anchor-to-anchor temperature profile along the middle of the membrane under different RF powers at 15 GHz. It can be seen that in general the membrane on quartz is always hotter than that on silicon and the difference increases with increasing power. In addition, the anchor temperature also increases with increasing power, especially on quartz. This implies that for thermally insulating substrates such as quartz, the anchor should not be assumed to be at the ambient temperature and heat transfer needs to be considered across both the membrane and the substrate.

The above observation is further illustrated in Fig. 5. It can be seen that the temperatures at the center and anchor of the membrane both increase linearly with increasing power. Based on such linear relationships, "average" thermal resistances R^{TH}_{MEM} and R^{TH}_{SUB} can be extracted for the membrane and the substrate, respectively, in spite of the highly nonuniform current and temperature distributions.

$$\Delta T_{MAX} = \Delta T_{MEM} + \Delta T_{ANC} = P_{MEM} R^{TH}_{MEM} + (P_{MEM} + P_{TL}) R^{TH}_{SUB} \quad (1)$$

where ΔT_{MAX} is the maximum temperature rise at the center of the membrane, ΔT_{MEM} is the temperature difference between the center and the anchor of the membrane, and ΔT_{ANC} is the temperature rise at the anchor. Thus, the complicated 3D heat transfer can be approximated by a lumped circuit model as shown in Fig. 6 with the extracted thermal resistances listed in Table I.

Using the above-simulated maximum temperature rise ΔT_{MAX}, power dependence of the pull-in voltage can be derived [3]. Fig. 7 shows that while the power dependence of

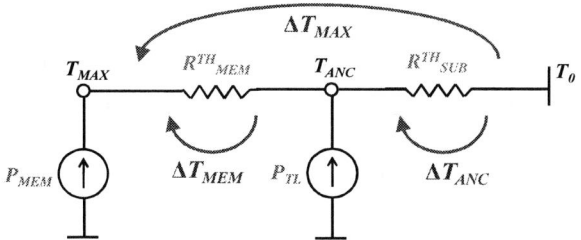

Fig. 6. Lumped circuit model of heat transfer across the membrane and substrate.

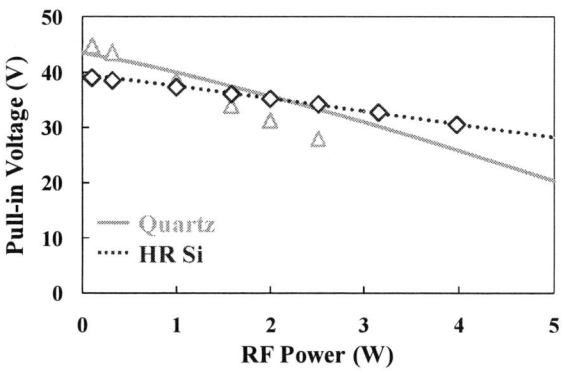

Fig. 7. (curve) Modeled vs. (symbol) measured pull-in voltages of silicon and quartz switches under different RF powers at 15 GHz.

the silicon switch is correctly modeled, the measured power dependence of the quartz switch is even stronger than that modeled. This discrepancy is being investigated with multiphysics simulation by iteratively coupling the electrothermal simulation with thermomechanical and electromechanical simulations to account for 3D effects in more detail.

The above analysis focuses on the power-handling capacity of a switch with its membrane in the resting (suspended) state. However, [8] found it to be limited by the state when the membrane is pulled-in to contact the stationary electrode so that the heat transfer is mainly limited by the substrate instead of the membrane. Since the switch of [8] was built on a polymer, a poor thermal conductor, the heat transfer was probably limited by the substrate whether the membrane was suspended or pulled in. In general, whether the power-handling capacity of a switch is limited by its suspended or pulled-in state is probably dependent on the ratios of P_{MEM}/P_{TL} and $R^{TH}_{MEM}/R^{TH}_{SUB}$ in both states.

IV. CONCLUSION

According to Table I, $R^{TH}_{MEM} / R^{TH}_{SUB} \approx 100$ on silicon while $R^{TH}_{MEM} / R^{TH}_{SUB} \approx 10$ on quartz. Since $P_{MEM}/P_{TL} \approx 0.1$ in both cases, from (1) ΔT_{MEM} and ΔT_{ANC} can be of the same order of magnitude on quartz. Therefore, in improving the power-handling capacity of MEM switches, it is important to consider the electrical, thermal and mechanical properties of not only the membrane, but also the substrate.

REFERENCES

[1] G. M. Rebeiz, *RF MEMS, Theory, Design and Technology.* Hoboken, NJ: Wiley, 2003.

[2] J. C. M. Hwang, and C. L. Goldsmith, "Robust RF MEMS switches and phase shifters for aerospace applications," in *Proc. IEEE Radio-Frequency Integration Technology Symp.*, Dec. 2009, pp. 245-248.

[3] C. Palego, J. Deng, Z. Peng, S. Halder, J. C. M. Hwang, D. Forehand, D. Scarbrough, C. L. Goldsmith, I. Johnston, S. K. Sampath, and A. Datta, "Robustness of RF MEMS capacitive switches with molybdenum membranes," *IEEE Trans. Microwave Theory Techniques,* vol. 57, pp. 3262-3269, Dec. 2009.

[4] I. Reines, B. Pillans, and G. M. Rebeiz, "A stress-tolerant temperature-stable RF-MEMS switched capacitor," in *Proc. IEEE Int. Conf. Microelectromechanical Systems,* Jan. 2009, pp. 880-883.

[5] C. Palego, F. Solazzi, S. Halder, J. C. M. Hwang, P. Farinelli, R. Sorentino, A. Faes, V. Mulloni, and B. Margesin, "Effec of substrate on temperature range and power capacity of RF MEMS capacitive switches" *European Microwave Conf.,* Sept. 2010, submitted for publication.

[6] W. Thiel, K. Tornquist, R. Reano, and L. P. B. Katehi, "A study of thermal effects in RF-MEM-swtiches using a time domain approach," in *IEEE MTT-S Int. Microwave Symp. Dig.,* June 2002, pp. 235-238.

[7] J. B. Rizk, E. Chaiban, and G. M. Rebeiz, "Steady state thermal analysis and high-power reliability considerations of RF MEMS capacitive switches," in *IEEE MTT-S Int. Microwave Symp. Dig.,* June 2002, pp. 239-242.

[8] F. Coccetti, B. Ducarouge, E. Scheid, D. Dubuc, K. Grenier, and R. Plana, "Thermal analysis of RF-MEMS switches for power handling front-end," in *Proc. 13th GAAS Symp.,* Sept. 2005, pp. 513-516.

[9] www.ansys.com.

[10] J. R. Reid, L. A. Starman, and R. T. Webster, "RF actuation of capacitive MEMS switches", in *IEEE MTT-S Int. Microwave Symp. Dig,* June 2003, pp. 1919-1922.

[11] P. Farinelli, B. Margesin, F. Giacomozzi, G. Mannocchi, S. Catoni, R. Marcelli, L. Vietzorreck, F. Vitulli, R. Sorentino, and F. Deborgies, "A low contact-resistance winged-bridge RF-MEMS series switch for wide-band applications," *Proc. European Microwave Assoc.,* vol. 10, pp. 1-8, Dec. 2009.

[12] P. Blondy, A. Crunteanu, C. Champeaux, A. Catherinot, P. Tristant, O. Vendier, J. L. Cazaux, and L. Marchand, "Dielectric less capacitive switches," in *IEEE MTT-S Int. Microwave Symp. Dig.,* June 2004, pp. 573-576.

[13] A. Tazzoli, E. Autizi, M. Barbato, G. Meneghesso, F. Solazzi, P. Farinelli, F. Giacomozzi, J. Iannacci, B. Margesin, R. Sorrentino., "Evolution of electrical parameters of dielectric-less ohmic RF-MEMS switches during continuous actuation stress," in *Dig. European Solid-State Device Resaerch Conf.,* Sep. 2009, pp 343-346.

Active NEM Filters for Communications Applications Based on Vibrating Body Transistors

A. Lovera, S. Bartsch, D. Grogg, S. Ayöz, R. Kaunisto[#], A. M. Ionescu

Nanoelectronic Devices Laboratory, Ecole Polytechnique Fédérale de Lausanne (EPFL), CH-1015, Lausanne, Switzerland.
Email: andrea.lovera@epfl.ch, adrian.ionescu@epfl.ch
#Nokia Research, Finland

Abstract — **In this work we propose and demonstrate the first active Nano-Electro-Mechanical (NEM) filters based on scaled vibrating body field effect transistor (VB-FET) with mechanically coupled flexural-mode beam resonators working at a fundamental resonant frequency of *115MHz*. The VB-FET filters are fabricated on a *200 nm* thin SOI substrate using E-Beam lithography and sacrificial layer etching. Numerical simulations prove the validity of the design and allow a fine control of the center frequency and bandwidth via applied DC voltages. The measured DC characteristics show a working FET with threshold voltage at *-3V* and short channel effects. Despite the low signal-to-background ratio, direct S-parameter mesurements demonstrate the functionality and the tuneable gain of VB-FET based NEM-filters.**

I. INTRODUCTION

The co-integration of electronic and MEMS components leads to an important advantage in size and power consumption for many telecommunication applications. Current wireless architectures, like super heterodyne transceiver, require discrete components for signal processing. Off-chip devices like quartz resonators and surface acoustic wave filters (SAW) represent the main part and are the building blocks of the high-bandpass filters commonly used in the radio frequency (RF) and intermediate frequency (IF) stages. These off-chip components limit the scaling and the final dimension for applications where monolithic integration is beneficial.

On chip components allow not only to reduce the size of components but also to reduce power consumption and production prices. In particular MEMS /NEMS considered strategical devices for integrated applications due to their compatibility with CMOS technology [1] and their potential for low power operation.

Many examples of MEMS filters can be find in the literature [2,3,4], most of them exploiting capacitive detection. The main bottleneck of these architectures is that output signal level decreases along side with the dimension until noise and parasitic effects tart to limit the usability. For this reason other transduction mechanisms like piezoelectricity [5] piezoresistance [6] or the field effect transistors [7] are under investigation. We recently presented a novel structure exploiting the built-in amplification of transistors, the so called vibrating body field effect transistor (VB-FET) [7] to improve the performances of MEMS resonators.

In this work is presented for the first time an active NEMS filter composed by mechanically coupled flexural mode beams combined with the VB-FET detection. Design and fabrication of these devices is reported together with experimental results showing static and RF characteristics of these devices.

II. DESIGN OF ACTIVE NEM FILTER BASED ON VB-FET

A. Working Principle

An SEM image of the MEMS filter presented in this work is given in Fig. 1(a). The structure is composed of two beams coupled together at their fixed ends by a third small beam anchored to the substrate and forming double-ended tuning-fork geometry. While the center frequency of the device is mainly given by the dimensions of the tuning tines, the filter bandwidth can be designed by the coupling beam. A detailed study is reported below in the mechanical design section.

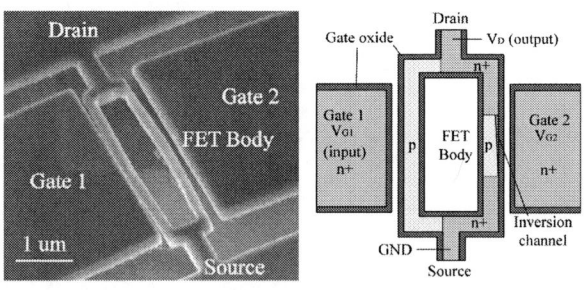

Figure 1: (a) SEM image of a fabricated double-ended tuning-fork filter. The two beams are normally bended probably due the stress induced by oxide gate that cover all the structure. (b) Schematic representation of the device.

Fig. 1a shows a SEM image of a VB-FET filter next to a schematic view of the filter (Fig.1b). Electrodes are placed at the two side of the structure. The filter and the electrode are separated by 105nm wide air gaps, with an additional 27nm layer of gate oxide (t_{ox}) on either side.

978-1-4244-6658-0/10 $26.00 © 2010 IEEE

Fig. 2 shows a simplified electrical equivalent circuit for the two-resonator filter. Here, the $L_m C_m R_m$ and the $L'_m C'_m R'_m$ circuits represent the single beam resonators on the left and on the right, respectively. A capacitive T-network represents the two coupling beams.

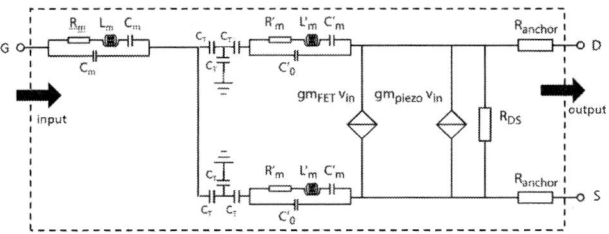

Figure 2: A simplified, lumped electrical equivalent circuit with filter input and output indicated.

The mechanical beam is put into vibration electrostatically by an AC signal applied on the left electrode (input v_{in}). Via mechanical coupling, the entire structure is driven into resonance. The resulting modulation of the right-sided gap causes a modulation of the drain current (output i_{out}) in the channel of the FET. The total output current is constituted by the field-effect current; a possible contribution by a piezoresistive current is also indicated [8]. The right-sided gate electrode sets the point of operation of the transistor, while the overall frequency response of the mechanical filter depends on the voltage applied on both electrodes (V_{G1} and V_{G2}). In contrast to passive MEMS filters the output signal amplitude can be amplified by the transistor effect ($g_{m,FET}$ and $g_{m,\,piezo}$), which is extremely useful especially at high frequency where the control of parasitic effects and the design of HF amplifiers are difficult. Note that the current flow on the left side of the tuning fork is prevented by the p-doped region.

B. Mechanical Designs

The filters are designed for in-plane fundamental flexural mode with a maximum displacement in the center of the beam. The number of modes of the filter is equal to the number of resonators. In the case of double tuning-fork there are two in-plane modes, one oscillating in-phase (Fig. 3a), the other moving out-of-phase (Fig. 3b).

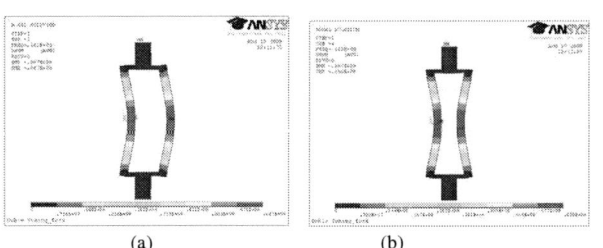

Figure 3: Modal simulations done by finite element analysis of the double tuning-fork. In-phase (a) and out-of-phase (b) flexural modes of the NEM filter are presented.

The center frequency of the mechanical filter is well approximated as the resonant frequency of the beam resonators and can be designed accordingly. The targeted center frequency is 150MHz and can be achieved for flexural mode resonators that are 200nm wide and 3.1μm

long. Numerical simulations are used to design the coupling element and verify the center frequency and the bandwidth of the filter. Both the coupling elements and the anchors have to be taken into account since they modify the position of the nodal points of the system. Furthermore these elements are linked to the bandwidth.

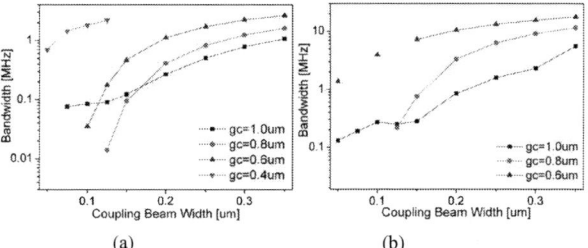

Figure 4: Finite element analysis of the bandwidth of the filter for different coupling beam widths. In the legend **gc** represent the distance between the middle points of the two beams. Anchor width is (a) 0.4 μm (b) 0.6 μm.

Fig. 4 shows numerical simulations of the bandwidth of the doubly ended tuning-fork filters for various coupling beam and anchor geometries. From the results we see that the bandwidth is proportional to the stiffness of these elements [9]; the softer the coupling elements the smaller the bandwidth.

One interesting advantage of this filter is the fine control of the center frequency and of the bandwidth with the applied DC voltages. Fig. 4 presents simulation results of the frequency response for a double-ended tuning fork *2.7μm* long and 200nm wide for different applied gate voltages. The center frequency can be shifted without affecting the characteristic shape by varying the two gate voltages symmetrically (Fig. 5b). With increased asymmetrically applied gate voltage the bandwidth is widened, however, trading off with the signal amplitude (Fig. 5a).

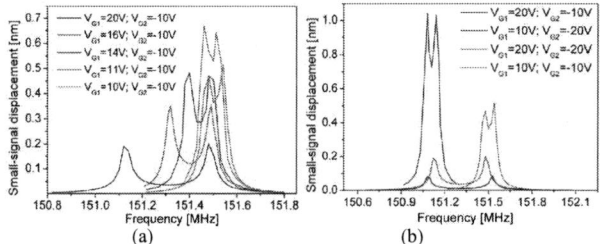

Figure 5: Harmonic simulations of the NEM filter for different applied DC voltages on both gates. On both cases the coupling beam length and width are fixed to *1.2μm* and *100nm* while the anchor width is *0.6μm*. Symmetric potentials shift the center frequency while unsymmetrical tunes the bandwidth.

The theoretical pull-in voltage [10] is calculated to be ~ *50V* for a *2.7μm* long and *200nm* wide and thick beam. This allows applying high voltages on the electrodes which is necessary to drive the device into its optimal point of operation.

C. FET Design

Designing the FET is the second important aspect of the VB-FET filter. The transistor should provide a high transconductance at the point of operation. Further, the threshold voltage (V_{th}) should be moderately low with respect to future integration. Fig. 6 shows schematic cross-sections of the VB-FET. The inversion channel is formed on the lateral side of the right-sided beam. The FET width is given by the thickness of the silicon layer (*200nm*) and its length is defined lithographically. To keep V_{th} low, a channel doping of $5x10^{16}$ cm^{-2} is selected which sets the V_{th} to approximately *+10V* (assuming an air-gap of 100 nm and t_{ox}=*30nm*).

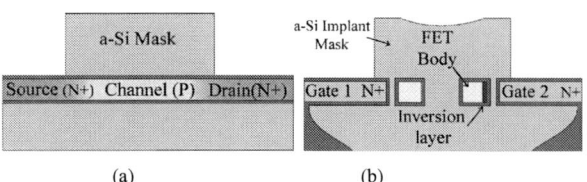

(a)   (b)

Figure 6: (a) Cross-section of the filter along the beam length. Atoms diffuse laterally under the mask for around *250nm*. (b) Cross-section of the filter perpendicular to the beam length. The inversion layer is formed only on the beam on the right.

The drawback of the low doping and the thick equivalent oxide thickness is a limited control on the channel that leads to short channel effects. The VB-FETs described here have a channel length of 1.3μm. The source, drain and gate regions are highly phosphorus doped, with a concentration close to $1x10^{20}$atoms/cm^{-3}.

III. FABRICATION

The fabrication process for suspended filters with VB-FET detection is presented in Fig. 7.

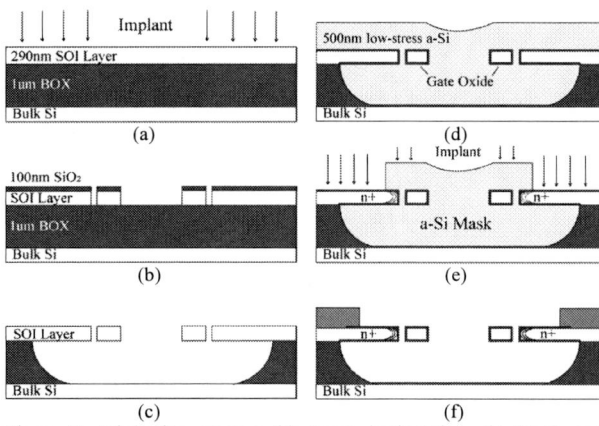

Figure 7: Fabrication process: (a) Boron implantation. (b) Hard mask growing and device layer etching. (c) Structure releasing in BHF + supercritical point drying. (d) Gate oxide growth and a-Si deposition. (e)a-Si RIE+ Phosphorous implantation and RTA annealing. (f) Gate oxide etching in the contact region, Al deposition by Lift-off and a-Si isotropic dry etching.

It includes two E-beam lithographic steps and two optical masks. The filters are fabricated on SOI wafers with *290nm* thick device layer, initially p-type doped with N_A=$9x10^{14}cm^{-3}$ and a 1μm thick buried oxide. Boron is first implanted using 40keV (Fig. 7a), then activated and diffused to achieve a smooth profile along the depth by a thermal oxidation process. 100nm wet oxide are grown (Fig. 7b) as a hard mask for the RIE process and to thin down the substrate to *240nm*. Filters and gaps are created inside the silicon layer with e-beam lithography at *5nm* resolution. The hard mask is then transferred into the silicon exploiting a RIE process based on chlorine chemistry. A controlled notching effect is useful to obtain vertical sidewalls that otherwise would be slightly

inclined causing uncontrolled variation of the threshold voltage of the MOSFET along the depth. After the etching the devices are released in BHF (Fig. 7c) and dried in a CO$_2$ super-critical point dryer. To protect MOSFET interfaces and filters a *40nm* gate oxide is grown (the silicon is thin down to *200nm*) and 500nm low-stress LPCVD a-Si is deposited (Fig. 7d). This layer is used as implantation mask for protect the channel region during gate and electrodes implantation. To obtain *50nm* precision alignment for high energy implantation E-beam lithography is required. A-Si etch with bromine hydrate (Fig. 7e) is particularly well suited for vertical etching on high aspect ratio structures. Phosphorous is then implanted using *120keV* and activated by a Rapid Thermal Annealing (RTA) to minimize any further diffusion. During the final metallization step two optical masks are used, one for opening the gate oxide, etched in BHF in the contact region and a second to pattern with a two-layer lift-off process the 400nm PVD aluminum layer. Finally, the a-Si is removed by a fluorine based isotropic dry silicon etching (Fig. 7f) and the metal annealed at 400°C.

IV. ELECTRICAL DEVICE CHARACTERISTICS

A. Measurement set-up

The electrical characterization of the MEM filters is done with a Süss Microtec on-wafer probe station (PMC 150) (Fig. 8). The DC characteristics are measured using an Agilent 4156 Parametric Analyzer, the RF performance using a HP 8753D vector network analyzer (VNA) in 2-port configuration, respectively. Figure 5 shows the basic RF set-up. Bias-T's are used to isolate the VNA from the DC signal. All measurements are done in a vacuum (< 10^{-5} mbar) and at room temperature (300 K).

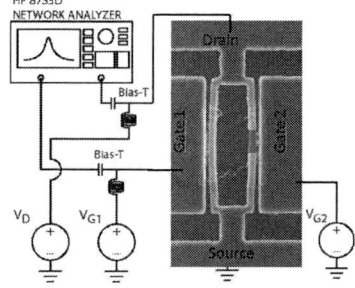

Fig. 8: The RF-measurement set-up is indicated around the SEM image of the fabricated device. The left tine does not contribute to the drain current.

B. Static characteristics

Measurements of the $I_D - V_D$ and $I_D - V_G$ characteristics of the VB-FET based filter are shown in Fig. 9 and 10. Note that the gate voltage is applied on one electrode ($V_{G1} = 0$) for these measurements since only the laterally vibrating beam with the transistor contributes to the measured drain modulation. The FET is designed to work as *fully depleted normally-off* enhancement mode device with a target threshold voltage of $V_{th} = +10V$. A considerable shift of V_{th} to a negative value is observed making the FET to a normally-on device. This might be caused by charges on top of the gate oxide, i.e. at the SiO$_2$-air interface. The slight dependence of the drain current on the drain voltage in the saturation region may indicate that the FET is operated at the onset of short-channel effects (i.e.

drain-induced barrier lowering, DIBL). The positive slope of the I_D-V_D curve in saturation (Fig. 10) also insinuates the presence of DIBL effect.

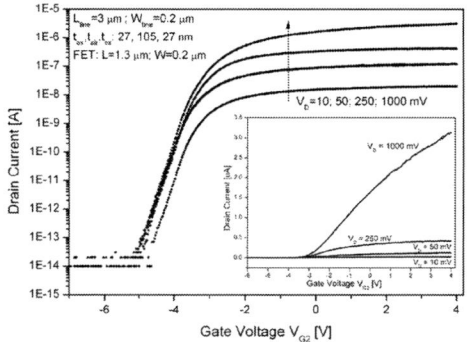

Fig. 9: Experimental I_D-V_G characteristic of the doubly-ended tuning fork with a lateral suspended body nMOSFET and one-sided bias. The inset shows the I_D-V_G characteristic in linear scale.

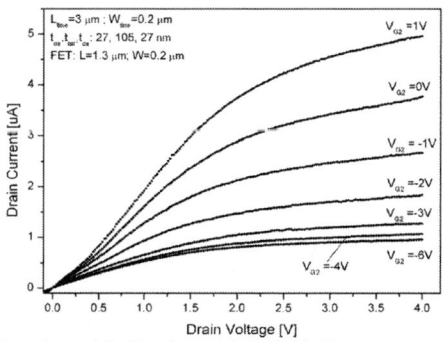

Fig. 10: Experimental I_D-V_D characteristic (with V_{G2} as parameter) of the doubly-ended tuning fork with a laterally suspended body nMOSFET on one tine.

C. Exprimental Investigation of RF performance

For the scattering (S-) parameter characterization of the tuning fork VBFET filter, the small-signal excitation is applied on one gate ($v_{G1} = V_{G1} + V_0 \cos(2\pi f_0 t)$), while V_0 is the voltage amplitude and f_0 the input RF frequency. The other gate electrode is biased with a DC voltage V_{G2}. In these measurements, the small-signal input power for linear operation is set to P_{ac}=-25dBm.

The centre resonant frequency is found to be 115 MHz (Fig. 11) and differs from the initial design, which targeted a 150MHz center frequency. This is probably caused by the relatively thick gate oxide that induces mechanical stress and lowers the effective Young's modulus.

The influence of the gate voltage on the transmission characteristics of the device can be seen in Fig. 11. The gate voltage $V_{G1} = V_{G2}$ is set between *10V* and *25V*, while the drain voltage is kept constant at $V_D = 1V$. The MOSFET is operated in the linear region. The strong change of signal amplitude at resonance is attributed to the increase in the effective g_m of the FET at higher V_{G2}, therefore to the increase of the active NEM filter gain. The impact of V_D on the transmission properties is shown in Fig. 12. For high V_D the measured filter characteristics correspond well with the theory and two distinct peaks are visible, corresponding to the two modes of vibration. However, at lower V_D only one peak can be distinguished. It is not yet fully understood why the second

peak is not measurable, but this may be related to the change of the effective gate-to-beam voltage (V_{GD}). This may influence the energy transferred into the second mode (insufficient amplitude for direct detection) and the separation of the two peaks (peaks too close together). The change in V_{GD} also probably accounts for the frequency shift due to the electrical spring softening, as previously reported [7].

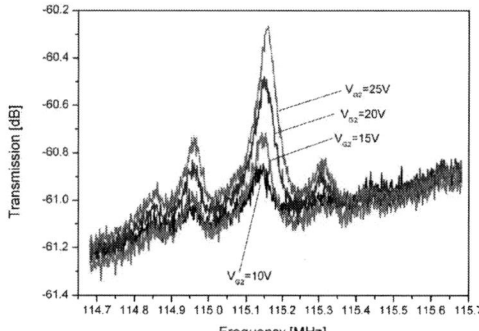

Fig.11: Magnitude of the transmission coefficient versus frequency of the tuning fork filter, with V_{G2} as parameter and $V_D = 1V$.

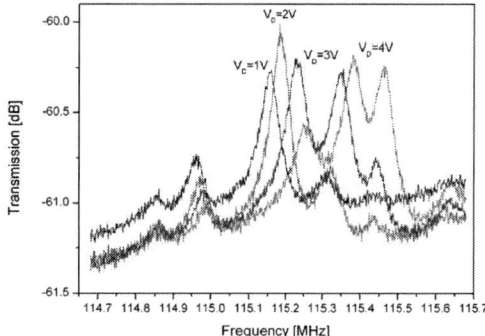

Fig.12: Magnitude of the transmission coefficient versus frequency of the tuning fork filter, with V_D as parameter. The gate voltage is constant with $V_{G1} = V_{G2} = +25V$.

V. CONCLUSIONS

In this work the applicability of VB-FET for building high-frequency active filters is demonstrated. At *115MHz* direct RF measurements are still possible, but the parasitic feed through in the structure limits the peak to background separation. We have reported the effect of the gate and drain voltages on the filter transmission characteristics. The use of a lock-in amplifier directly connected to the output of the filters (currently under investigation) may improve the signal-to-background ratio and enable a more appropriate characterization of the NEM filter.

REFERENCES

[1]   J. Teva et al., Int. Conf. on Micro Electro Mechanical System, 2007.
[2]   J. Yan et al., Int. Conf. on Micro Electro Mechanical System, 2008.
[3]   Sheng-Shian Li et al., Int. Ultrasonics, Ferroelectrics, and Frequency Control Conference, 2004.
[4]   F. D. Bannon III, Journal of Solid-State Circuits, Vol. 35, No. 4, 2000
[5]   J. Baborowski et al., Intl. Frequency Control Symposium, 2007.
[6]   J. T. M. van Beek et al., Intl. Electron Devices Meeting, 2008.
[7]   D. Grogg et al., Intl. Electron Devices Meeting, 2008.
[8]   D. Grogg et al., Intl. Electron Devices Meeting, 2009.
[9]   Mohammed M. Shalaby et al., Transactions on industrial electronics, Vol. 56, No. 4, April, 2009.
[10]   Nicolas Abele, Ph. D. Thesis, EPFL, 2007.

Piezoresistivity and Electrical Properties of Poly-SiGe Deposited at CMOS-Compatible Temperatures

Pilar Gonzalez[a], Luc Haspeslagh, Simone Severi, Kristin De Meyer[a] and Ann Witvrouw

IMEC, Kapeldreef 75, 3001 Leuven (Belgium)

[a] also with K.U. Leuven, Kasteelpark Arenberg 10, 3001 Leuven (Belgium)

Phone:+32 16-287-847. Email: Pilar.Gonzalez @imec.be

Abstract—**In this work the effect of doping concentration on the piezoresistive and electrical properties of poly-SiGe deposited at temperatures compatible with MEMS integration on top of CMOS are evaluated for the first time. With proper tuning of the boron content, a gauge factor around 14 and a TCR close to 0 are achievable. These results prove the potential of using poly-SiGe as a sensing layer for MEMS-above-CMOS applications.**

I. INTRODUCTION

The monolithic integration of micro-electromechanical systems (MEMS) with the driving, controlling and signal processing electronics on the same CMOS substrate can improve performance and reliability as well as lower the manufacturing, packaging and instrumentation costs [1]. The post-processing route (fabricating MEMS directly on top of CMOS) is the most promising approach for CMOS-MEMS monolithic integration as it enables integrating MEMS without introducing any changes in standard foundry CMOS processes. However post-processing limits the maximum fabrication temperature of MEMS to 450°C [2], to avoid introducing any damage or degradation in the performance of the existing electronics or interconnects. This temperature constraint is quite restrictive for post-processing surface micromachined MEMS, as it affects relevant physical properties such as crystallinity, growth rate, mechanical properties, doping activation, etc. Polycrystalline silicon germanium (poly-SiGe) has emerged as an attractive alternative to polycrystalline silicon (poly-Si) for MEMS above-CMOS applications thanks to its lower electrical resistivity [3] and lower amorphous to polycrystalline transition temperature [4,5], which can be as low as 400°C (with the appropriate germanium content) [4].

In earlier work [6] we already studied the piezoresistive and electrical properties of poly-$Si_{51}Ge_{49}$ and poly-$Si_{36}Ge_{64}$ for different doping concentrations (from $5\cdot10^{17}$ to $1\cdot10^{20}$ cm^{-3}), obtaining very promising results. However, the processing temperatures used during the deposition and annealing of these layers were too high (above 500°C) to allow for the monolithic integration of MEMS above CMOS. In this work we characterize a new poly-SiGe layer processed at CMOS compatible temperatures. 400nm-thick poly-$Si_{38}Ge_{62}$ layers were deposited at a temperature around 450 °C using chemical vapor deposition. The films were later boron implanted with a dose varying from $4\cdot10^{13}$ to $4\cdot10^{15}$cm^{-2} and annealed at 450 °C.

The resistivity is measured in the temperature range of 25 to 150°C using a four point probe. Hall measurements are carried out to evaluate the hole mobility at room temperature. The piezoresistivity is estimated by measuring the resistance variation when a uniform and uniaxial stress provided by a four point bending fixture is applied to the films. The obtained results are compared to those of our previous poly-$Si_{36}Ge_{64}$ layer, deposited at 500°C and annealed at 570°C, reported in [6]. Finally, finite element simulations of a possible application to a pressure sensor are performed to illustrate the potential of using this new layer for MEMS piezoresistive sensor applications.

II. EXPERIMENTAL

Silicon wafers (100) covered with 500nm-thick SiO_2 were used as starting substrates. The poly-SiGe layers have been deposited on top of the SiO_2 using chemical vapor deposition (CVD) at ~450°C and 275 Torr in an Applied Materials (AMAT) Centura platform. The silicon gas source is pure silane, whereas 1% germane in hydrogen has been used as the germanium gas source. The germanium content of the layers was determined by Rutherford Back Scattering (RBS) to be 62%. The films were doped through ion implantation of boron at 65 keV with dosages between $4\cdot10^{13}$ to $4\cdot10^{15}$cm^{-2}. After implantation, the films were annealed in a conventional furnace for 30 min at 450°C. Secondary Ion Mass Spectrometry (SIMS) was used to determine the chemical boron concentrations (Fig. 1).

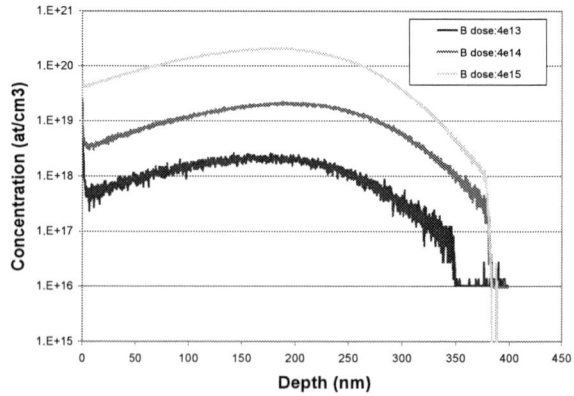

Figure 1. SIMS results for layers with three different doping doses:$4\cdot10^{13}$, $4\cdot10^{14}$ and $4\cdot10^{15}$ cm^{-2}

The resistors are created by etching the poly-SiGe in the central region of each sample using a dry etch process. Each resistor is contacted by 1μm-thick aluminum copper (AlCu 0.1%) traces to four bonding pads in order to measure the resistance by the four-point probe method, eliminating any spurious contribution of contacts and metal traces. Resistors of different lengths (from 0.5mm to 18mm) and widths (from 25 μm to 150 μm) have been patterned. Fig. 2 shows microscope pictures of typical samples used for piezoresistive and Hall measurements, respectively.

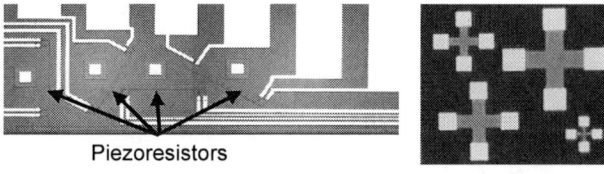

Piezoresistors

Figure 2. Microscope picture of a) typical sample used for piezoresistive measurements with 4 differently orientated poly-SiGe resistors and b) four cross-shaped resistors of different sizes used for Hall measurements

III. ELECTRICAL PROPERTIES

The resistivity has been measured on blanket wafers using a four-point probe. From Fig. 3 we can observe that the resistivity of poly-$Si_{38}Ge_{62}$ reduces with increasing boron concentration, as expected. The layer exhibits slightly higher resistivity than our previous layer [6], which is probably caused by the smaller grain size and lower number of active carriers (due to the lower deposition and annealing temperatures).

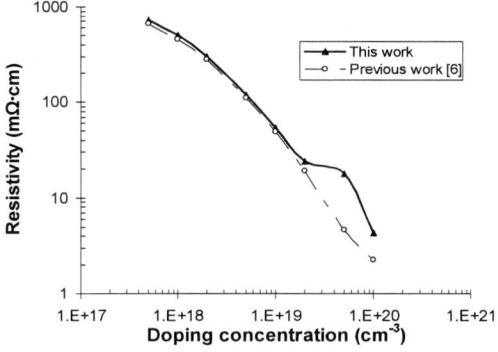

Figure 3. Dependence of resistivity on doping dose. The continuous line represents the values obtained for poly-$Si_{38}Ge_{62}$ while the dashed line corresponds to values reported in [6] for poly-$Si_{36}Ge_{64}$

To study the temperature coefficient of resistance (TCR) of the layers, four point resistance measurements of patterned structures have been performed in the temperature range of 25 to 150°C. The relative resistance variation with temperature is shown in Fig. 4 with the doping concentration as the varying parameter. As expected for a polycrystalline material, at moderate doping levels (up to $3 \cdot 10^{19}$ cm^{-3}) the thermionic emission mechanism is dominant and the temperature dependence is non-linear with negative TCR [7]. At higher doping levels the carrier mobility mechanism is dominant and the temperature dependence is nearly linear with positive TCR [8].

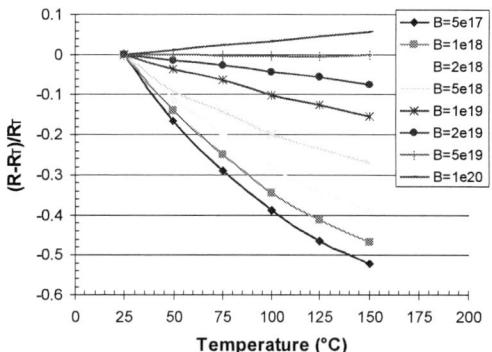

Figure 4. Dependence of resistivity on temperature for different boron concentrations (cm^{-3}). RT represents resistance at room temperature

In Fig. 5 the temperature coefficient of resistance (TCR) of poly-$Si_{38}Ge_{62}$ vs. doping dose is plotted together with the TCR of poly-$Si_{36}Ge_{64}$ (dashed line) from our previous work [6]. The TCR of this new layer exhibits similar behaviour as the film from [6], varying from strongly negative values for low doping to slightly positive values for high doping, with a cross-over point with TCR~0 for doping doses around $2 \cdot 10^{15}$ cm^{-2}. From the figure we can observe that the TCR of the new layer is slightly lower than that reported in [6], especially at higher doping levels, which might be due to smaller grain size [10]. In conclusion, similar as for our previous poly-SiGe layers and also poly-Si, the TCR of this new poly-SiGe layer deposited at CMOS compatible temperatures can also be modulated with doping concentration. This effect makes it possible to obtain a positive, negative or even zero TCR depending on the specifications of the considered application.

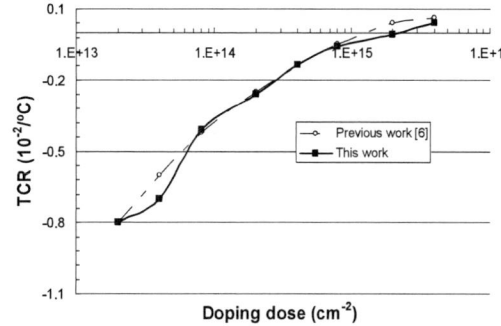

Figure 5. Dependence of TCR (calculated from the slope at room temperature) on doping dose. The continuous line represents the values obtained for poly-$Si_{38}Ge_{62}$ while the dashed line corresponds to values reported in [6] for poly-$Si_{36}Ge_{64}$

Hall measurements varying the magnetic field from -1 to 1T and using cross-shaped resistors (Fig. 2) were performed on layers with three different doping doses: $4 \cdot 10^{13}$, $4 \cdot 10^{14}$ and $4 \cdot 10^{15}$ cm^{-2}. From Fig. 6 we can see that poly-$Si_{38}Ge_{62}$ exhibits a dip in mobility for a doping concentration of around $2 \cdot 10^{19}$ cm^{-3}. The same behaviour was observed in poly-Si [10] and attributed to a modulation of barrier potential with doping density. Table I contains the measured Hall concentrations and mobilities together with the chemical concentration of boron (SIMS) for the three different layers. The obtained Hall concentrations are higher than the chemical concentrations

978-1-4244-6658-0/10 $26.00 © 2010 IEEE

measured by SIMS, what suggests a hall scattering factor less than 1, in agreement with previously reported data for poly-SiGe [11].

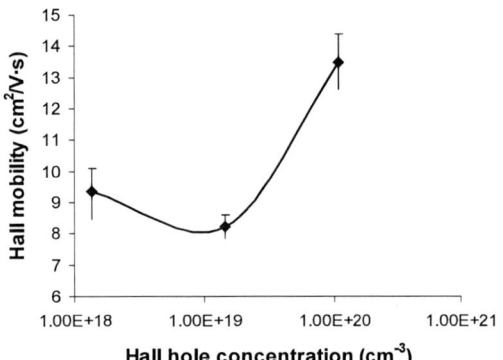

Figure 6. Hall mobilities of holes in poly-Si$_{38}$Ge$_{62}$ plotted against the carrier concentration at RT

TABLE I. Measured boron concentrations and hall mobilities

B dose (cm^{-2})	SIMS B concentration (cm^{-3})	Hall concentration (cm^{-3})	Hall mobility (cm^2/V·s)
4·10^{13}	1.05 ·10^{18}	1.36 ·10^{18}	9.34
4·10^{14}	1.01 ·10^{19}	1.42 ·10^{19}	8.23
4·10^{15}	1 ·10^{20}	1.085 ·10^{20}	13.45

IV. PIEZORESISTIVITY

The piezoresistive effect describes the changing electrical resistance of a material due to an applied mechanical stress and it is extensively used as sensing principle in many mechanical sensors. Monocrystalline Silicon is the dominant material for the fabrication of piezoresistive MEMS sensors thanks to its high gauge factor and excellent mechanical properties. In the past years polysilicon has gained importance in the MEMS piezoresistive sensors market despite its lower gauge factor (compared to monocrystalline silicon). However, neither silicon nor poly-Si allows monolithic integration of MEMS directly on top of the CMOS [1]. The lower deposition temperature of poly-SiGe compared to poly-Si makes it a very interesting material for CMOS-MEMS monolithic integration. In [6] we already demonstrated the piezoresistive properties of poly-SiGe. In this work we study the piezoresistivity of poly-SiGe processed at a lower temperature than the layers in [6], allowing the post-processing on top of CMOS.

To evaluate the piezoresistive properties of poly-Si$_{38}$Ge$_{62}$, the samples were stressed in a four point bending fixture (figure 7), which creates a pure bending condition. Four blades exert a force F resulting in a uniform and unidirectional stress on top of the region in between the inner blades, where the piezoresistors are placed. To determine the value of the stress σ induced in the piezoresistors, finite element simulations (COMSOL3.4) were performed [12], assuming a Young's

modulus of E=147GPa, based on a weighted average of the values for poly-Si and poly-Ge [13].

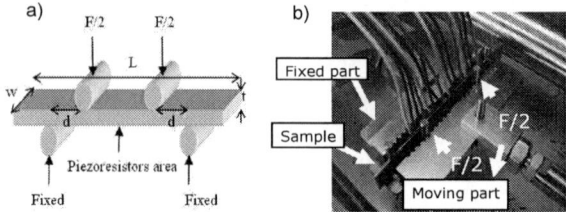

Figure 7. a) Sample stressed in a 4 point-bending fixture, where F is the applied force; Scheme (a) and measurement setup (b). w=8mm; L=80mm; t=725.5µm; d=14mm

With the measured relative resistance variation and the simulated value of the stress, the piezoresistive coefficients were found. Table II contains the obtained longitudinal and transversal piezoresistive coefficients together with those reported in [6] for poly-Si$_{36}$Ge$_{64}$ deposited at 500°C and annealed at 570°C. Figure 8 plots the obtained longitudinal gauge factor (defined as the relative resistance variation per unit strain across the length of the resistor) as a function of the doping concentration. We found a similar behavior to that of our previous poly-Si$_{36}$Ge$_{64}$ layer [6], with the gauge factor tailing off for high and low doping levels and reaching a maximum for a doping concentration of around $2 \cdot 10^{15}$ cm^{-3}. A maximum gauge factor of 14 was found, which is somewhat smaller than the maximum value of 20 reported in [6]. The fact that this new layer shows slightly lower piezoresistive sensitivity than our previous layer can be explained by considering the smaller grain size due to the lower deposition and annealing temperatures. In this work the layer was deposited (and annealed) at a temperature 50°C (120 °C) lower than the film in [6]. According to [9], the gauge factor of a polycrystalline material is expected to increase with grain size.

In conclusion, the maximum gauge factor found for this layer is smaller than maximum values reported for poly-Si (22-29) [14] and poly-SiGe [6] processed at higher temperatures.However the possibility to post-process on top of CMOS still makes it a very interesting material for MEMS piezoresistive sensors as monolithic integration leads to a higher signal/noise ratio which might offset the smaller gauge factor.

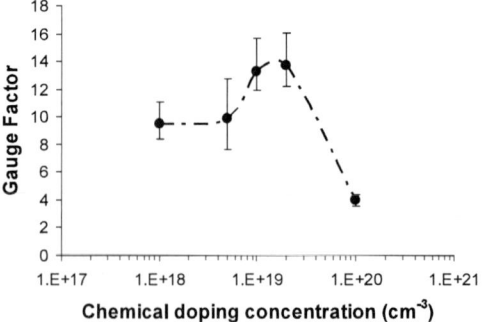

Figure 8. Longitudinal Gauge factor as a function of boron concentration

Table II. Obtained longitudinal and transversal piezoresistive coefficients(in Pa^{-1}) Poly-$Si_{38}Ge_{62}$. The two right columns include values from [6]

B (cm^{-2})	Poly-$Si_{38}Ge_{62}$		Poly$Si_{36}Ge_{64}$ [6]	
	Πl (10^{-11})	$\Pi t(10^{-11})$	$\Pi l(10^{-11})$	$\Pi t(10^{-11})$
$4 \cdot 10^{13}$	6.46	2.3	8.8	0.7
$2 \cdot 10^{14}$	6.713	1.55	12.4	1.5
$4 \cdot 10^{14}$	9	1.23	13.7	0.94
$8 \cdot 10^{14}$	9.3	1.34	8.6	-0.43
$4 \cdot 10^{15}$	2.7	-0.85	4	-0.5

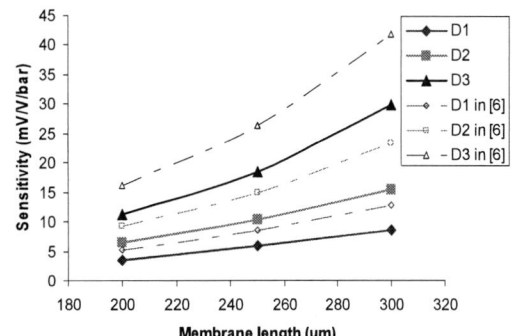

Figure 10. Obtained sensitivities (mV/V/bar) for the different designs studied. The dashed lines correspond to the values reported in [6].

V. APPLICATION TO A PRESSURE SENSOR

To illustrate the potential of this material as a sensing layer for MEMS-above-CMOS applications, the performance of a possible piezoresisistive pressure sensor using poly-$Si_{38}Ge_{62}$ was studied . The sensitivity of such a sensor was evaluated by means of finite element simulations along with the experimentally obtained piezoresistive coefficients for the optimum layer (with $B=2 \cdot 10^{19}$ cm^{-3}). To allow for a direct comparison with our previous layer (poly-$Si_{36}Ge_{64}$ deposited at 500°C and annealed at 570 °C), the same sensor designs as in [6] were considered (Fig. 9).

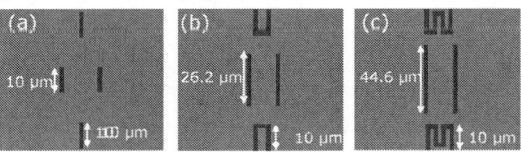

Figure 9. Designs considered. Membrane and piezoresistors thicknesses are 0.4 and 4 µm, respectively.

The program COMSOL is used to perform finite-element simulations of the deformation and stress induced in the membrane and the piezoresistors when a uniform pressure is applied. Due to the positive transversal piezoresistive coefficient (πt), it is best to place the transversal piezoresistors in the centre of the membrane where the stress is negative, instead of on the edge, as it is usually done. Similar as in [6], thanks to this placement of the transversal piezoresistors the sensitivity could be improved by a factor of around 2.

Fig. 10 shows the obtained sensitivities vs. membrane length for the different designs considered. The values reported in our previous work [6] are also included for comparison. We observe the same behaviour of the sensor sensitivity with respect to membrane area and piezoresistor shape. However as our new poly-$Si_{38}Ge_{62}$ layer exhibits a lower piezoresistive effect than our previous layer, the new sensitivities for the same sensor design are smaller than those reported in [6]. In this work a maximum sensitivity of 30 mV/V/bar was found, which, even though slightly lower than the 42mV/V/bar reported in [6] for the same sensor, is still higher than previously reported values for poly-Si and poly-SiGe sensors [8,12]. Moreover, unlike the poly-$Si_{36}Ge_{64}$ layer presented in [6], this new poly-$Si_{38}Ge_{62}$ layer allows the monolithic integration of MEMS on top of CMOS, which leads to better signal to noise ratio that might offset the lower sensitivity .

VI. CONCLUSION

The piezoresistive and electrical properties of poly-$Si_{38}Ge_{62}$ deposited at CMOS compatible temperatures were studied as a function of doping concentration and compared to those of poly-$Si_{36}Ge_{64}$ deposited and annealed at temperatures 50 °C and 120°C higher. Results show that with proper tuning of the boron concentration, a gauge factor of 14 and a TCR around 0, which is ideal for piezoresistive sensor applications, are achievable. Sensitivites over 30 mV/V/bar are predicted for a possible piezoresistive pressure sensor, proving the potential of poly-$Si_{38}Ge_{62}$ for MEMS sensor applications.

REFERENCES

[1] A. Witvrouw et al, "Poly SiGe: a Superb Material for MEMS", proc. MRS, Vol. 782, pp. 25-36, 2004.

[2] S. Sedky et al,"Experimetal Determination of the maximum post-processing temperature of MEMS on top of standard CMOS wafers" IEEE Trans. Electron Devices, Vol. 48 (2), pp.1-9,2001

[3] D. Bang et al, "Resistivity of boron and phosphorus doped polycrystalline $Si_{1-x}Ge_x$ films", Applied Physics Letters,66 (2), pp. 195-197 (1995)

[4] T.J. King et al, "Deposition and properties of low-pressure chemical vapor deposited polycrystalline silicon-germanium films", Journal of The Electrochemical Society, 141, 2235-2241 (1994).

[5] B. Guo et al, "Improvement of PECVD Silicon-Germanium Crystallization for CMOS compatible MEMS applications", Journal of The Electrochemical Society, 157 (2) D103-D110 (2010).

[6] P. Gonzalez et al, "Evaluation of the piezoresistive and electrical properties of polycrystalline silicon germanium for MEMS sensor applications", Proc. MEMS 2010

[7] L. Jiang et al, "Micromachined polycrystalline thin film temperature sensors", Meas. Sci. Technol., Vol. 10, pp. 653-664 (1999).

[8] K. N. Bhat, "Silicon Micromachined Pressure Sensors", Journal of the Indian Institute of Science, Vol. 87:1, 2007

[9] S.V. Spoutai, "Practical model for electrical properties of highly doped p-type polysilicon", APEIE-98

[10] J. Y. W. Seto, "The electrical properties of polycrystalline silicon films", J. Appl. Physics 46 , 5247 (1975).

[11] C. Salm et al, J. Electrochem. Soc., Vol. 144, No. 10 (1997).

[12] P. Gonzalez et al, "Evaluation of the Piezoresistivity and 1/f Noise of Polycrystalline Silicon Germanium for MEMS sensors applications", Proc. Eurosensors ´08.

[13] A. Franke et al, "Polycrystalline Silicon-Germanium Films for Integrated Microsystems", J. of MEMS, Vol. 12, No. 2 (2003)

[14] P.J. French, A.G.R Evans, "Piezoresistance of Polysilicon", Electronics Letters,Vol 20 (24),1984

AUTHOR INDEX

Adelmann, Christoph 356, 360
Afzalian, Aryan 296
Ahmet, Perhat 221, 281
Akhavan, Nima Dehdashti 277
Alatise, Olayiwola 193
Allain, Fabienne 126
Altimime, Laith 74
Amara, Amara 130
Amoroso, Salvatore M. 364
Andricciola, Pietro 46
Andrieu, François 126, 130, 245
Annunziata, R. 233
Aoulaiche, Marc 74
Arreghini, Antonio 356, 360
Arvet, Christian 288
Asenov, Asen 50
Asenov, Plamen 50
Aulnette, C. 126
Aumont, Christophe 26
Averty, Dominique 38
Aymerich, Xavier 58
Ayoz, Suat 392
Baccarani, Giorgio 189, 300
Bajolet, Aurelie 42
Barnett, Joel 82
Barniol, Nuria 384
Barraud, Sylvain 288
Bartsch, Sebastian 392
Baudot, Sophie 126
Bauer, Guenther 217
Baumgartner, Peter 213
Bawedin, Maryline 70
Becherer, Markus 134
Bender, Hugo 356
Beneventi, Giovanni Betti 233
Bensahel, D. 233
Bersuker, Gennadi 253, 308
Besland, Marie-Paule 38
Billon, T. 233
Biswas, Arnab 273
Blachier, Denis 233
Boeuf, Frédéric 130
Bolivar, Salvador Rodríguez 304
Bonaiuti, Matteo 205
Boucart, Kathy 265
Boulanger, F. 233, 245
Bourdelle, K. K. 126
Bournel, Arnaud 344
Bouvet, Didier 138
Bouzid, Samira 201
Braga, Stefania 376
Breitkreutz, Stephan 134
Breuil, Laurent 360
Brevard, Laurent 126
Burenkov, Alexander 209

Byun, Jun Young 320
Cabrini, Alessandro 376
Cacciato, Antonio 356, 360
Calderoni, Alessandro 237
Camozzi, Elisa 364
Carbone, Beatrice 185
Carboni, Francesca 241
Carrere, Jean-Pierre 26
Cassé, Mikaël 126, 245
Cathignol, Augustin 42
Cervenka, Johann 217
Challali, Fatiha 38
Chang, Chung Fu 62
Chao, Tien-Sheng 336
Charpentier, Alain 38
Chen, Bing 312
Chiarella, Thomas 225
Chini, Alessandro 205
Chiu, Huai-Hsien 336
Chiu, Kai Ling 62
Chou, Pai-Chi 336
Chuang, Ching-Te 118
Chung, J. W. 1
Claeys, Cor 304
Colinge, Jean-Pierre 277
Collaert, Nadine 74
Compagnoni, Christian Monzio 364
Condorelli, Giovanni 185
Congedo, Gabriele 328
Conte, Fabrizio Lo 114
Cremer, Sébastien 78
Crespo-Yepes, Albert 58
Cristoloveanu, Sorin 70, 261, 288, 340, 368
Csaba, Gyorgy 134
Damlencourt, Jean-François 126
Daval, N. 126
De Gendt, Stefan 90
De Jaeger, Jean-Claude 201
De Keersgieter, An 74
De Meyer, Kristin 166, 396
De Michielis, Luca 273
De Salvo, B. 233
De Vittorio, Massimo 102
De Wachter, Bart 74
Debusschere, Ingrid 356, 360
Decoutere, Stefaan 166
Defrance, Nicolas 201
Degraeve, Robin 13, 360
Denison, Marie 189, 197
Di Lecce, Valerio 205
Dollfus, Philippe 344
Douvry, Yannick 201
Drapatz, Stefan 66
Driussi, Francesco 328
Endo, K. 122

AUTHOR INDEX

Eneman, Geert 304
Ernst, Thomas 70, 288
Esposto, Michele 205
Esseni, David 154, 269
Estl, Hannes 197
Faes, Alessandro 388
Fallica, Piero Giorgio 185
Fan, Ming-Long 118
Fang, Zheng 312
Fantini, Andrea 233
Fantini, Fausto 205
Fantini, Paolo 237
Faraone, Lorenzo 102
Farinelli, Paola 388
Faynot, Olivier 70, 130, 288, 368
Feldis, H. 233
Fenouillet-Beranger, Claire 130
Ferain, Isabelle 277
Feste, Sebastian 292
Figuet, C. 126
Flandre, Denis 296
Fregonese, Sebastien 372
Frey, Lothar 249
Fu, Chuan-Shian 62
Gaberl, Wolfgang 170
Gaillard, Sébastien 78
Galbiati, Nadia 364
Galy, Philippe 110
Gámiz, Francisco 158, 340
Gao, Bin 312
Garcia-Loureiro, Antonio 158
Garros, Xavier 245
Gatefait, Maxime 26
Georgakos, Georg 66
Ghibaudo, Gérard 42, 225, 229
Ghidini, Gabriella 364
Ghidotti, Michele 364
Giner, Joan 384
Gnani, Elena 189, 300
Gnudi, Antonio 189, 300
Godoy, Andrés 158
Goducheau, Olivier 78
Goldbach, Matthias 249
Gonzalez, Mireia Bargallo 304
Gonzalez, Pilar 396
Goullet, Antoine 38
Gourvest, Emmanuel 233
Grant, Lindsay 177
Granzner, Ralf 348
Greco, Eugenio 364
Groeseneken, Guido 13, 90
Grogg, Daniel 392
Guarin, Fernando 98
Guegan, Georges 70
Guillaumot, Bernard 288

Gutierrez, Edmundo 98
Gyun, Byung Gu 320
Habicht, Stefan 292
Hahn, Berthold 6
Halder, Subrata 388
Hamilton, Darlene 324
Han, Ruqi 312
Hartmann, Jean-Michel 288
Haspeslagh, Luc 396
Hattori, Takeo 221
Hayashida, T. 122
Heerkens, Carel 22
Henderson, Robert 177
Heppenstall, Keith 193
Heyns, Marc 90
Hilleringmann, Ulrich 146
Hochschulz, Frank 174
Hoel, Virginie 201
Hoffmann, Thomas 225
Hofmann, Karl 66
Hrauda, Nina 217
Hsieh, Chien-Yu 118
Hu, Chenming 82
Hu, Vita Pi-Ho 118
Huang, Henry 324
Hubert, Alexandre 70
Hueting, Ray 154, 380
Hung, Pui-Yee 86
Hutin, Louis 126
Hwang, James C. M. 388
Iglesias, Vanessa 253
Iniguez, Benjamin 368
Ionescu, Adrian 94, 138, 265, 273, 392
Ishihara, Takamitsu 54
Ishikawa, Yuki 122
Iwai, Hiroshi 221, 281, 288
Jahan, Carine 233
Jakschik, Stefan 249
Jammy, Raj 82, 86, 253
Jang, Doyoung 225
Jaud, Marie-Anne 130
Jenny, Cecile 26, 78
Jeon, Kanghoon 82
Jimenez, Jean 110
Jin, Gyoyoung 257
Joo, Moon Sig 320
Jovanovic, Vladimir 162
Ju, Xueming 134
Jungemann, Christoph 150
Jurczak, Malgorzata 74, 356, 360
Kaczer, Ben 13
Kakushima, Kuniyuki 221, 281
Kamei, T. 122
Kampen, Christian 209
Kamsani, Noor Ain 50

AUTHOR INDEX

Kang, Chang Yong 82
Kang, Jinfeng 312
Kang, Yeonsung 257
Kao, Chih Yang 62
Kaunisto, Risto 392
Kawanago, Takamasa 221
Kayal, Maher 114
Kennedy, Ian 193
Khalilyulin, Ruslan 106
Khan, Khalid 193
Kianian, Sohrab 324
Kiermaier, Josef 134
Kikkawa, Takamaro 34
Kim, Chi Ho 320
Kim, Gyu Tae 225
Kim, Ho Joung 320
Kim, Ja Yong 320
Kim, Jung Nam 320
Kim, Sook Joo 320
Kim, Taeh Wan 320
Kim, Wan Gee 320
Kim, Won 320
Kim, Young Weon 324
Kimoto, K. 34
Kirsch, Paul 253
Kittl, Jorge 356, 360
Kittler, Mario 348
Klein, Peter 213
Knowlton, Bill 308
Koh, Adrian 193
Koh, Shao-Ming 332
Kooijman, Kees 22
Kosina, Hans 217
Kostov, Plamen 170
Kottantharayil, Anil 269
Kranti, Abhinav 277
Krauss, Tillmann 285
Kubota, S. 34
Kwong, Dim-Lee 312
Lacaita, Andrea 364
Lai, Chao-Sung 336
Lalanne, Frédéric 78
Lamperti, Alessio 328
Landesman, Jean-Pierre 38
Larcher, L. 233, 241, 308
Lattanzio, Livio 138, 273
Le Royer, Cyrille 126, 261
Lecourt, François 201
Lee, Chi-Woo 277
Lee, Jae Woo 225
Lee, Jong-Ho 257
Lee, Jooyoung 257
Lee, Sanghoon 257
Lee, Yeonghun 221, 281
Leonelli, Daniele 90

Lhostis, S. 233
Liang, Chia Wen 62
Libertino, Sebania 185
Lime, Francois 368
Lin, Guan Shyan 62
Liu, Lifeng 312
Liu, Tsu-Jae King 62, 82
Liu, Xiaoyan 312
Liu, Yongxun 122
Liu, You Ren 62
Loh, Wei Yip 82
Loi, Alberto 189
Lombardo, Salvatore 185
Lopez, Joan Lluis 384
Lorenz, Jürgen 209
Loubriat, Sebastien 233
Lovera, Andrea 392
Lu, B. 1
Maconi, Alessandro 364
Maffini-Alvaro, Virginie 288
Maher, Hassan 201
Mai, Andreas 30
Maitrejean, Sylvain 233
Majhi, Prashant 82, 86
Makarov, Alexander 316
Maneux, Cristell 372
Manning, Monte 324
Mantl, Siegfried 292
Marca, Vincenzo Della 241
Margesin, Benno 388
Marigo, Eloi 384
Martens, Koen 13
Martin-Martinez, Javier 58
Marzegalli, Anna 217
Masahara, Meishoku 122
Mascellino, Evelyne 364
Matsukawa, Takashi 122
Mauri, Aurelio 364
Mazoyer, P. 233
McKenna, Keith 253
Meinerzhagen, Bernd 150
Meuris, Peter 110
Mezzomo, Cecilia 42
Miglio, Leo 217
Mikolajick, Thomas 249
Millar, Campbell 50
Ming, Xue 324
Moon, Ji Won 320
Mouis, Mireille 225, 229
Mulloni, Viviana 388
Muminovic, Hajro 213
Munoz-Gamarra, Jose Luis 384
Nafria, Montse 58, 253
Nakabayashi, Yukio 54
Nanver, Lis 22

AUTHOR INDEX

Narasimhamoorthy, Subramanian 229
Natori, Kenji 221, 281
Nemecek, Alexander 181
Ng, Chee-Mang 332
Nguyen, Bich-Yen 126
Nikolic, Borivoje 62
Nishiyama, Akira 221
Nodin, Jean-François 233
Noel, Jean-Philippe 130
Numata, Toshinori 54
Oddou, Jean-Pierre 26
Ogura, A. 122
Oh, Jang Won 320
Oh, Jungwoo 82, 86
Ohmori, Kenji 281
Ok, Injo 86
Ota, Kensuke 54
O'Uchi, Shin-Ichi 122
Padovani, Andrea 241, 308
Padovini, Giorgio 364
Pagano, Roberto 185
Palacios, Tomás 1
Palego, Cristiano 388
Palestri, Pierpaolo 154
Park, Byung-Gook 257
Park, Chanro 82
Park, Sung Ki 320
Park, Sunyoung 257
Paschen, Uwe 174
Passaseo, Adriana 102
Patel, Pratik 82
Pavan, Paolo 233, 241
Pendharkar, Sameer 189
Perniola, Luca 233
Perreau, Pierre 126
Persico, Alain 233
Petiton, Hervé 78
Petkos, George 193
Pham, Anh-Tuan 150
Place, Sebastien 26
Poiroux, Thierry 130
Poli, Stefano 189
Poljak, Mirko 162
Polyakov, Vladimir 348
Porti, Marc 253
Rajasekharan, Bijoy 380
Rauer, Caroline 126
Razavi, Pedram 277
Reggiani, Susanna 189, 300
Reid, David 50
Reimbold, Gilles 233, 245
Ren, Shu-Feng 332
Renvoise, Michel 201
Rhallabi, Ahmed 38
Richard, Claire 26

Richard, Olivier 356
Richardson, Justin 177
Riess, Philipp 213
Riess, Walter 265
Ritzenthaler, Romain 368
Rivallin, Pierrette 130
Robinson, Karl 324
Rodríguez, Abraham Luque 304
Rodriguez, Noel 158, 340
Rodriguez, Rosana 58
Roh, Jae Sung 320
Roll, Guntrade 249
Rooyackers, Rita 90
Rosendale, Glen 324
Rothschild, Aude 58, 356
Roule, Anne 233
Roy, Francois 26
Roy, Scott 50
Rücker, Holger 30
Rueckes, Thomas 324
Russo, Ugo 328
Rusu, Alexandru 94
Rutter, Phil 193
Ryu, K. 1
Sadi, Toufik 352
Saint-Martin, Jérôme 344
Saitoh, Masumi 54
Sakamoto, Kunihiro 122
Sakic, Agata 22
Salicio, Olivier 328
Salimy, Siamak 38
Sallese, Jean-Michel 114
Salm, Cora 380
Salvatore, Giovanni Antonio 94, 138
Sampedro, Carlos 158
Samudra, Ganesh S. 332
Sanasi, Alessandro 376
Sanfilippo, Delfo Nunzio 185
Sato, Soshi 281
Saubat, Jean-Claude 38
Scheiblin, Pascal 126, 130
Schmitt-Landsiedel, Doris 66, 134
Schmitz, Jurriaan 380
Schoenmaker, Wim 110
Scholtes, Tom 22
Schrag, Gabriele 106
Schwalke, Udo 285
Schwierz, Frank 348, 352
Scozzari, Claudia 364
Sebastiani, Alessandro 364
Seetharaman, Sridhar 189
Selberherr, Siegfried 217, 316
Severi, Simone 396
Shannon, John 142
Shi, Ming 344

AUTHOR INDEX

Shin, Changhwan 62
Shin, Hyungcheol 257
Shluger, Alex 253
Sibaja-Hernandez, Arturo.....................166
Silva, Ravi 142
Simoen, Eddy 304
Skotnicki, Thomas 288
Smith, Casey 82
Smith, Derek 201
Solazzi, Francesco 388
Sorrentino, Roberto 388
Sousa, Veronique 233
Southwick, Richard..............................308
Spiga, Sabina 328
Spinelli, Alessandro S. 364
Sporea, Radu 142
Stangl, Julian 217
Steinhuber, Thomas 106
Stomeo, Tiziana 102
Su, Pin 118
Sugii, Nobuyuki 221
Suhane, Amit 356, 360
Suligoj, Tomislav 162
Sun, Bing 312
Sung, Min Gyu 320
Sverdlov, Viktor 316
Swerts, Johan 356
Tabone, Claude 126
Tachi, Kiichi 288
Tasco, Vittorianna 102
Tejada, Juan Antonio Jimenez.................304
Tessariol, Paolo 364
Thomas, Olivier 130
Toffoli, Alain 233
Toledano-Luque, Maria..........................360
Torelli, Guido 376
Torres, Francesc..................................384
Tournier, Arnaud..................................26
Toutain, Serge 38
Trainor, Mike 142
Tsai, Chen Hua 62
Tsai, Cheng-Tzung 62
Tseng, Hsing-Huang 82
Tsukada, Junichi 122
Tsutsui, Kazuo 221
Tuinhout, Hans.....................................46
Uchida, Ken 54
Umana-Membreno, Gilberto.....................102
Uranga, Arantxa 384
Valín, Raul 158
Valvo, Giuseppina 185
Van Den Bosch, Geert............................356, 360
Van Der Steen, Jan-Laurens 154
Van Hemert, Tom 380
Van Houdt, Jan 356, 360

Van Huylenbroeck, Stefaan 166
Van Veen, Gerard 22
Vandelli, Luca 308
Vandooren, Anne...................................90
Vastola, Guglielmo 217
Veksler, Dmitry 86
Venegas, Rafael 166
Ventrice, Domenico 237
Vernet, Marc..78
Vianello, Elisa 328
Vidal, Gabriel 384
Virani, Hasanali....................................269
Vizioz, Christian 288
Vogelsang, Patrick.................................22
Vogt, Holger 174
Vulliet, Nathalie 288
Wachutka, Gerhard 106
Wan, Jing 261
Wang, Jer-Chyi 336
Wang, Yangyuan 312
Weber, Olivier 126, 130, 245
Webster, Eric 177
Wessely, Frank 285
Wise, Rick 189
Witvrouw, Ann 396
Wolff, Karsten 146
Wu, Mei Hsuan 62
Wu, Woei-Cherng...................................336
Xie, Yizhong 197
Yamada, Keisaku 281
Yamauchi, Hiromi 122
Yan, Ran 277
Yeo, Yee-Chia 332
Yesilada, Emek 78
Yoo, Jong Hee 320
Yoon, Sung Joon 320
You, Shuzhen 166
Youn, Te One 320
Young, Nigel 142
Yu, Bin 312
Yu, Hongyu 312
Yu, Ran 277
Yum, Jung 253
Zahid, Mohammed B. 360
Zaslavsky, Alexander 261
Zhang, Haowei 312
Zhang, Jianjun 217
Zhang, Peng 332
Zhao, Qing Tai 292
Zimmer, Thomas 372
Zimmermann, Horst 170, 181

9781424466580